Geophysical Monograph Series
Including
Maurice Ewing Volumes
Mineral Physics Volumes

GEOPHYSICAL MONOGRAPH SERIES

Geophysical Monograph Volumes

1 **Antarctica in the International Geophysical Year** *A. P. Crary, L. M. Gould, E. O. Hulburt, Hugh Odishaw, and Waldo E. Smith (Eds.)*
2 **Geophysics and the IGY** *Hugh Odishaw and Stanley Ruttenberg (Eds.)*
3 **Atmospheric Chemistry of Chlorine and Sulfur Compounds** *James P. Lodge, Jr. (Ed.)*
4 **Contemporary Geodesy** *Charles A. Whitten and Kenneth H. Drummond (Eds.)*
5 **Physics of Precipitation** *Helmut Weickmann (Ed.)*
6 **The Crust of the Pacific Basin** *Gordon A. Macdonald and Hisashi Kuno (Eds.)*
7 **Antarctic Research: The Matthew Fontaine Maury Memorial Symposium** *H. Wexler, M. J. Rubin, and J. E. Caskey, Jr. (Eds.)*
8 **Terrestrial Heat Flow** *William H. K. Lee (Ed.)*
9 **Gravity Anomalies: Unsurveyed Areas** *Hyman Orlin (Ed.)*
10 **The Earth Beneath the Continents: A Volume of Geophysical Studies in Honor of Merle A. Tuve** *John S. Steinhart and T. Jefferson Smith (Eds.)*
11 **Isotope Techniques in the Hydrologic Cycle** *Glenn E. Stout (Ed.)*
12 **The Crust and Upper Mantle of the Pacific Area** *Leon Knopoff, Charles L. Drake, and Pembroke J. Hart (Eds.)*
13 **The Earth's Crust and Upper Mantle** *Pembroke J. Hart (Ed.)*
14 **The Structure and Physical Properties of the Earth's Crust** *John G. Heacock (Ed.)*
15 **The Use of Artificial Satellites for Geodesy** *Soren W. Henriksen, Armando Mancini, and Bernard H. Chovitz (Eds.)*
16 **Flow and Fracture of Rocks** *H. C. Heard, I. Y. Borg, N. L. Carter, and C. B. Raleigh (Eds.)*
17 **Man-Made Lakes: Their Problems and Environmental Effects** *William C. Ackermann, Gilbert F. White, and E. B. Worthington (Eds.)*
18 **The Upper Atmosphere in Motion: A Selection of Papers With Annotation** *C. O. Hines and Colleagues*
19 **The Geophysics of the Pacific Ocean Basin and Its Margin: A Volume in Honor of George P. Woollard** *George H. Sutton, Murli H. Manghnani, and Ralph Moberly (Eds.)*
20 **The Earth's Crust: Its Nature and Physical Properties** *John G. Heacock (Ed.)*
21 **Quantitative Modeling of Magnetospheric Processes** *W. P. Olson (Ed.)*
22 **Derivation, Meaning, and Use of Geomagnetic Indices** *P. N. Mayaud*
23 **The Tectonic and Geologic Evolution of Southeast Asian Seas and Islands** *Dennis E. Hayes (Ed.)*
24 **Mechanical Behavior of Crustal Rocks: The Handin Volume** *N. L. Carter, M. Friedman, J. M. Logan, and D. W. Stearns (Eds.)*
25 **Physics of Auroral Arc Formation** *S.-I. Akasofu and J. R. Kan (Eds.)*
26 **Heterogeneous Atmospheric Chemistry** *David R. Schryer (Ed.)*
27 **The Tectonic and Geologic Evolution of Southeast Asian Seas and Islands: Part 2** *Dennis E. Hayes (Ed.)*
28 **Magnetospheric Currents** *Thomas A. Potemra (Ed.)*
29 **Climate Processes and Climate Sensitivity (Maurice Ewing Volume 5)** *James E. Hansen and Taro Takahashi (Eds.)*
30 **Magnetic Reconnection in Space and Laboratory Plasmas** *Edward W. Hones, Jr. (Ed.)*
31 **Point Defects in Minerals (Mineral Physics Volume 1)** *Robert N. Schock (Ed.)*
32 **The Carbon Cycle and Atmospheric CO_2: Natural Variations Archean to Present** *E. T. Sundquist and W. S. Broecker (Eds.)*
33 **Greenland Ice Core: Geophysics, Geochemistry, and the Environment** *C. C. Langway, Jr., H. Oeschger, and W. Dansgaard (Eds.)*
34 **Collisionless Shocks in the Heliosphere: A Tutorial Review** *Robert G. Stone and Bruce T. Tsurutani (Eds.)*
35 **Collisionless Shocks in the Heliosphere: Reviews of Current Research** *Bruce T. Tsurutani and Robert G. Stone (Eds.)*
36 **Mineral and Rock Deformation: Laboratory Studies—The Paterson Volume** *B. E. Hobbs and H. C. Heard (Eds.)*
37 **Earthquake Source Mechanics (Maurice Ewing Volume 6)** *Shamita Das, John Boatwright, and Christopher H. Scholz (Eds.)*

Maurice Ewing Volumes

1 **Island Arcs, Deep Sea Trenches, and Back-Arc Basins** *Manik Talwani and Walter C. Pitman III (Eds.)*
2 **Deep Drilling Results in the Atlantic Ocean: Ocean Crust** *Manik Talwani, Christopher G. Harrison, and Dennis E. Hayes (Eds.)*
3 **Deep Drilling Results in the Atlantic Ocean: Continental Margins and Paleoenvironment** *Manik Talwani, William Hay, and William B. F. Ryan (Eds.)*
4 **Earthquake Prediction—An International Review** *David W. Simpson and Paul G. Richards (Eds.)*
5 **Climate Processes and Climate Sensitivity** *James E. Hansen and Taro Takahashi (Eds.)*
6 **Earthquake Source Mechanics** *Shamita Das, John Boatwright, and Christopher H. Scholz (Eds.)*

Mineral Physics Volumes

1 **Point Defects in Minerals** *Robert N. Schock (Ed.)*

Geophysical Monograph 38

Ion Acceleration in the Magnetosphere and Ionosphere

Tom Chang
Editor-in-Chief

M. K. Hudson
J. R. Jasperse
R. G. Johnson
P. M. Kintner
M. Schulz
Co-Editors

American Geophysical Union
Washington, D.C.
1986

Published under the aegis of AGU Geophysical Monograph Board: Patrick Muffler, Chairman; Wolfgang Berger, Donald Forsyth, and Janet Luhmann, members.

G. B. Crew, *Assistant Editor*
D. M. Klumpar, *Editorial Advisor*
J. Buchholz, *Editorial Assistant*

Library of Congress Cataloging in Publication Data

Main entry under title:

Ion acceleration in the magnetosphere and ionosphere.

(Geophysical monograph, ISSN 0065-8448 ; 38)
Includes bibliographies and index.
1. Magnetosphere-Congresses.   2. Ionosphere—Congresses.   3. Ion flow dynamics—Congresses.
4. Space plasmas—Congresses.   I. Chang, Thomas.
II. Series.
QC809.M35I587   1986   538'.766   86-10800
ISBN 0-87590-063-1
ISSN 0065-8448

Copyright 1986 by the American Geophysical Union, 2000 Florida Avenue, NW, Washington, DC 20009

Figures, tables, and short excerpts may be reprinted in scientific books and journals if the source is properly cited.

Authorization to photocopy items for internal or personal use, or the internal or personal use of specific clients, is granted by the American Geophysical Union for libraries and other users registered with the Copyright Clearance Center (CCC) Transactional Reporting Service, provided that the base fee of $1.00 per copy, plus $0.10 per page is paid directly to CCC, 21 Congress Street, Salem, MA 01970. 0065-8448/86/$01. + .10.
This consent does not extend to other kinds of copying, such as copying for creating new collective works or for resale. The reproduction of multiple copies and the use of full articles or the use of extracts, including figures and tables, for commercial purposes requires permission from AGU.

Printed in the United States of America.

CHAPMAN CONFERENCE ON ION ACCELERATION IN THE MAGNETOSPHERE
Wellesley, Massachusetts, June 3-7, 1985
List of Participants

Andre, Mats
Kiruna Geophysical Institute
University of Umea
S-901 87 Umea, Sweden

Arnoldy, Roger L.
Space Science Center, Demeritt Hall
University of New Hampshire
Durham, New Hampshire 03824

Ashour-Abdalla, Maha
Institute of Geophysics and Planetary Physics
University of California
Los Angeles, California 90024

Baker, Daniel N.
Space Plasma Physics Group, MS D438
Los Alamos National Laboratory
Los Alamos, New Mexico 87545

Bakshi, P.
Physics Department
Boston College
Chestnut Hill, Massachusetts 02167

Basinska-Lewin, E. W.
Max-Planck-Institut fur Extraterr. Physik
8046 Garching, Munich
Federal Republic of Germany

Belcher, John
Center for Space Research, Bldg. 37-695
Massachusetts Institute of Technology
Cambridge, Massachusetts 02139

Bergmann, Rachelle
Physics Department
Dartmouth College
Hanover, New Hampshire 03755

Boardsen, Scott A.
Department of Physics and Astronomy
University of Iowa
Iowa City, Iowa 52242

Borovsky, Joseph E.
Space Plasma Physics Group, MS D348
Los Alamos National Laboratory
Los Alamos, New Mexico 87545

Bosqued, Jean-Michel
Centre d'Etude Spatiale de Rayonnement
B.P. 4346
31029 Toulouse, France

Brown, Dale Calvin
Department of Physics and Astronomy
University of Maryland
College Park, Maryland 20742

Burke, William J.
Air Force Geophysics Laboratory
Hanscom Air Force Base
Bedford, Massachusetts 01731

Cattell, Cynthia
Space Sciences Laboratory
University of California
Berkeley, California 94720

Chan, Chung
Center for Electromagnetics Research
Northeastern University
Boston, Massachusetts 02115

Chang, Tom
Massachusetts Institute of Technology
77 Massachusetts Avenue, Rm. 37-261
Cambridge, Massachusetts 02139

Chao, Yu-Faye
2911 8th Street
Tucson, Arizona 87516

Chapman, Sandra C.
School of Mathematical Sciences
Queen Mary College
London E1 4NS, United Kingdom

Cladis, John B.
Lockheed Palo Alto Research Laboratory
Department 91-20, Building 255
3251 Hanover Street
Palo Alto, California 94304

Collin, H. L.
Lockheed Palo Alto Research Laboratory
Department 91-20, Building 255
3251 Hanover Street
Palo Alto, California 94304

Cornwall, John M.
Department of Physics
University of California, Los Angeles
Los Angeles, California 90024

Crew, Geoffrey B.
Massachusetts Institute of Technology
Center for Space Research, Rm. 37-271
Cambridge, Massachusetts 02139

Del Pozo, Carlos
Massachusetts Institute of Technology
Haystack Observatory
Westford, Massachusetts 01886

Delcourt, Dominique
Centre d'Etude Spatiale de Rayonnement
B.P. 4346
31029 Toulouse, France

Eastman, Tim E.
Department of Physics and Astronomy
University of Iowa
Iowa City, Iowa 52242

Erlandson, Robert E.
University of Minnesota
Space Science Center

100 Union Street S.E.
Minneapolis, Minnesota 55455

Feynman, Joan
Jet Propulsion Laboratory
4800 Oak Grove Drive
Pasadena, California 91109

Frahm, Rudy A.
Department of Space Physics and Astronomy
Rice University, P.O. Box 1892
Houston, Texas 77251

Frank, L. A.
Department of Physics and Astronomy
The University of Iowa
Iowa City, Iowa 52242

Fritz, Theodore A.
Los Alamos National Laboratory
Mail Stop D438
Los Alamos, New Mexico 87545

Ganguli, Gurudas I.
Science Applications International Corporation
8200 Greensboro Drive
McLean, Virginia 22102

Ghielmetti, Arthur G.
Lockheed Palo Alto Research Laboratory
Department 91-20, Building 255
3251 Hanover Street
Palo Alto, California 94304

Gloeckler, George
Department of Physics and Astronomy
University of Maryland
College Park, Maryland 20742

Gombosi, Tamas I.
Space Physics Research Laboratory
The University of Michigan
Space Research Bldg., 2455 Hayward
Ann Arbor, Michigan 48109

Gorney, David J.
The Aerospace Corporation
M2-260, P.O. Box 92957
Los Angeles, California 90009

Haerendel, Gerhard
Max-Planck-Institut fur Extraterr. Physik
8046 Garching, Munich
Federal Republic of Germany

Hamilton, Douglas C.
Department of Physics and Astronomy
University of Maryland
College Park, Maryland 20742

Hasegawa, Akira
AT&T Bell Laboratories
Room 1E-351, 600 Mountain Avenue
Murray Hill, New Jersey 07974

Hayes, Dallas
Rome Air Development Center
Hanscom Air Force Base
Bedford, Massachusetts 01731

Hershkowitz, Noah
Nuclear Engineering Department
University of Wisconsin-Madison
Madison, Wisconsin 53706

Hones, Edward W., Jr.
Los Alamos National Laboratory
Los Alamos, New Mexico 87545

Horwitz, J. L.
Department of Physics
The University of Alabama in Huntsville
Huntsville, Alabama 35899

Hu, Xiwei
IPST, University of Maryland
College Park, Maryland 20742

Hudson, Mary K.
Dartmouth College
Department of Physics & Astronomy
Hanover, New Hampshire 03755

Hultqvist, Bengt
Kiruna Geophysical Institute
P.O. Box 704
S-981 27 Kiruna, Sweden

Ipavich, Fred M.
Department of Physics and Astronomy
University of Maryland
College Park, Maryland 20742

Jasperse, John R.
Air Force Geophysics Laboratory
Hanscom Air Force Base
Bedford, Massachusetts 01731

Johnson, Richard G.
Executive Office of the President
OSTP, NEOB, Room 5026
17th & Pennsylvania Avenue, N.W.
Washington, D.C. 20506

Johnstone, Alan D.
Mullard Space Science Laboratory
University College London
Holmbury St. Mary, Dorking
Surrey, RH5 6NT, United Kingdom

Joiner, R. G.
Office of Naval Research, Code 414
800 N. Quincy Street
Arlington, Virginia 22217

Kaufmann, Dick
Department of Physics
University of new Hampshire
Durham, New Hampshire 03824

Kintner, Paul M.
School of Electrical Engineering
Cornell University, 115 Phillips Hall
Ithaca, New York 14853

Klecker, Berndt
Max-Planck-Institut fur Extraterr. Physik
8046 Garching, Munich
Federal Republic of Germany

Klumpar, David M.
Lockheed Palo Alto Research Laboratory
Department 91-20, Building 255
3251 Hanover Street
Palo Alto, California 94304

Koskinen, Hannu E. J.
Uppsala Ionospheric Observatory
S-755 90 Uppsala, Sweden

LaBelle, James
School of Electrical Engineering
Cornell University, 109 Phillips Hall
Ithaca, New York 14853

Lai, Shu
Air Force Geophysics Laboratory
Hanscom Air Force Base
Bedford, Massachusetts 01731

Lennartsson, Walter
Lockheed Palo Alto Research Laboratory
Department 91-20, Building 255
3251 Hanover Street
Palo Alto, California 94304

Litwin, Christof

1500 Johnson Drive
Madison, Wisconsin 53705

Longtin, Mary
Physics Department
Dartmouth College
Hanover, New Hampshire 03755

Loranc, Mark
University of Texas, Dallas
P.O. Box 830688, MS F023
Richardson, Texas 75081

Lotko, William
Thayer School of Engineering
Dartmouth College
Hanover, New Hampshire 03755

Ludlow, G. R.
Department of Physics
University of New Hampshire
Durham, New Hampshire 03824

Lui, A. T. Y.
The Johns Hopkins University
Applied Physics Laboratory
Laurel, Maryland 20707

Lysak, Robert L.
School of Physics and Astronomy
University of Minnesota
116 Church St., S.E.
Minneapolis, Minnesota 55455

Malcolm, Perry
Air Force Geophysics Laboratory/PHA
Hanscom Air Force Base
Bedford, Massachusetts 01731

Martin, Richard F., Jr.
NOAA Space Environment Laboratory, R/E/SE
Boulder, Colorado 80303

Mauk, Barry H.
The Johns Hopkins University
Applied Physics Laboratory
Laurel, Maryland 20707

Maynard, Nelson
Air Force Geophysics Laboratory/PHG
Hanscom Air Force Base
Bedford, Massachusetts 01731

McEntire, Richard W.
The Johns Hopkins University
Applied Physics Laboratory
Laurel, Maryland 20707

McWilliams, Roger D.
University of California, Irvine
Department of Physics 420, Physical Sciences
Irvine, California 92717

Menietti, J. Douglas
Department of Space Sciences
Southwest Research Institute
P.O. Drawer 28510
San Antonio, Texas 78284

Migliuolo, Stefano
Massachusetts Institute of Technology
Room 26-205
Cambridge, Massachusetts 02139

Mobius, Eberhard
Max-Planck-Institut fur Extraterr. Physik
8046 Garching, Munich
Federal Republic of Germany

Moffett, R. J.
Dept. of Applied and Computational Mathematics
University of Sheffield
Sheffield S10 2TN, United Kingdom

Moore, Thomas E.
Space Science Laboratory
NASA Marshall Space Flight Center
Huntsville, Alabama 35812

Nishikawa, K.-I.
Department of Physics and Astronomy
University of Iowa
Iowa City, Iowa 52242

Okuda, H.
Plasma Physics Laboratory
Princeton University
Princeton, New Jersey 08544

Olsen, Richard C.
Physics Department
University of Alabama
Huntsville, Alabama 35899

Olson, Lynn
Physics Department, 405 Hilgard
University of California
Los Angeles, California 90024

Orsini, Stefano
Inst. di Fisica dello Spazio Interplanetario
Consiglio Nazionale delle Ricerche, C.P. 27
00044 Frascati (Roma), Italy

Ossakow, Sidney
Naval Research Laboratory, Code 4700
Washington, D.C. 20375

Palmadesso, Peter J.
Naval Research Laboratory, Code 4700.IP
Washington, D.C. 20375

Peterson, William K.
Lockheed Palo Alto Research Laboratory
Department 91-20, Building 255
3251 Hanover Street
Palo Alto, California 94304

Pollock, Craig J.
Space Science Center, Demeritt Hall
University of New Hampshire
Durham, New Hampshire 03824

Pu, Yikang
Massachusetts Institute of Technology
Room 26-209
Cambridge, Massachusetts 02139

Reiff, Patricia H.
Department of Space Physics and Astronomy
Rice University
Houston, Texas 77251

Retterer, John M.
Space Data Analysis Laboratory
Boston College
Chestnut Hill, Massachusetts 02167

Rhodes, Barry c/o J. R. Jasperse
Air Force Geophysics Laboratory
Hanscom Air Force Base
Bedford, Massachusetts 01731

Rich, Frederick
Air Force Geophysics Laboratory
Hanscom Air Force Base
Bedford, Massachusetts 01731

Roth, Ilan
Space Sciences Laboratory
University of California at Berkeley
Berkeley, California 94720

Rothwell, Paul
Air Force Geophysics Laboratory
Hanscom Air Force Base
Bedford, Massachusetts 01731

Rynn, Nathan
  Department of Physics
  420 Physical Sciences
  University of California, Irvine
  Irvine, California 92717
Sachs, Walter
  Max-Planck-Institut fur Extraterr. Physik
  8046 Garching, Munich
  Federal Republic of Germany
Sagawa, Eiichi
  National Research Council
  100 Sussex Drive
  Ottawa, Ontario, Canada K1A 0R6
Samir, Uri
  Space Physics Research Laboratory
  University of Michigan
  Ann Arbor, Michigan 48109
Scales, Wayne A.
  School of Electrical Engineering
  Cornell University, 112 Phillips Hall
  Ithaca, New York 14853
Scholer, Manfred
  Max-Planck-Institut fur Extraterr. Physik
  8046 Garching, Munich
  Federal Republic of Germany
Schriver, David
  13929 Marquesas Way, #103
  Marina Del Rey, California 90292
Schulz, Michael
  Space Sciences Laboratory
  The Aerospace Corporation
  M2-259, P.O. Box 92957
  Los Angeles, California 90009
Schunk, Robert W.
  Center for Atmospheric and Space Sciences
  Utah State University, UMC 3400
  Logan, Utah 84322
Sharber, James R.
  Southwest Research Institute
  Division 15, P.O. Drawer 28510
  San Antonio, Texas 78284
Shelley, Edward G.
  Lockheed Palo Alto Research Laboratory
  Department 91-20, Building 255
  3251 Hanover Street
  Palo Alto, California 94304
Shi, Bingren
  IPST, University of Maryland
  College Park, Maryland 20742
Silevitch, Michael B.
  Northeastern University
  235 Forsyth Building
  Boston, Massachusetts 02115
Singh, Nagendra
  Center for Atmospheric and Space Sciences
  Utah State University, UMC 3400
  Logan, Utah 8432
Stenzel, Reiner L.
  University of California, Los Angeles
  Department of Physics, 405 Hilgard Avenue
  Los Angeles, California 90024
Studemann, Wolfgang
  Max-Planck-Institute fur Aeronomie
  Postfach 20, 3411 Katlenburg-Lindau
  Federal Republic of Germany
Tanskanen, Pekka
  Department of Physics
  University of Oulu
  SF-90570 Oulu, Finland
Temerin, Michael A.
  Space Science Laboratory
  University of California
  Berkeley, California 94720
Trefall, Harald
  University of Bergen
  5000 Bergen, Norway
Tsunoda, Roland T.
  SRI International
  33 Ravenswood Avenue
  Menlo Park, California 94025
Ungstrup, Eigil
  Danish Space Research Institute
  Lundtoftevej 7
  DK-2800 Lyngby, Denmark
Waite, J. Hunter, Jr.
  Space Science Laboratory
  NASA Marshall Space Flight Center
  Huntsville, Alabama 35812
Walker, David N.
  Plasma Physics Division
  Naval Research Laboratory
  4445 Overlook Avenue
  Washington, D.C. 20375
Weimer, Daniel R.
  Regis College Research Center
  235 Wellesley Street
  Weston, Massachusetts 02193
Winckler, John R.
  School of Physics and Astronomy
  University of Minnesota
  116 Church Street, S.E.
  Minneapolis, Minnesota 55455
Witt, Earl F.
  Mission Research Corporation
  735 State Street, P. O. Drawer 719
  Santa Barbara, California 93102
Wong, Alfred Y.
  Department of Physics, 405 Hilgard Avenue
  University of California
  Los Angeles, California 90024
Yau, Andrew W.
  Herzberg Institute of Astrophysics
  National Research Council of Canada
  100 Sussex Drive
  Ottawa, Ontario K1A 0R6, CANADA
Yeh, Huey-Ching
  Physics Department
  Boston College
  Chestnut Hill, Massachusetts 02167
Young, David T.
  Space Plasma Physics Group
  Los Alamos National Laboratory
  Los Alamos, New Mexico 87545
Zhu, Meimei
  Dartmouth College
  Hanover, New Hampshire 03755

CONTENTS

List of Conference Participants v

Preface *Tom Chang* xiii

I. INTRODUCTION

Magnetospheric Ion Acceleration Processes *John M. Cornwall* 3

Experimental Aspects of Ion Acceleration in the Earth's Magnetosphere *David T. Young* 17

II. HIGH LATITUDE PROCESSES

Low-Altitude Transverse Ionospheric Ion Acceleration *A. W. Yau, B. A. Whalen, and P. M. Kintner* 39

Transverse Auroral Ion Energization Observed on DE-1 With Simultaneous Plasma Wave and Ion Composition Measurements *W. K. Peterson, E. G. Shelley, S. A. Boardsen, and D. A. Gurnett* 43

Observations of Coherent Transverse Ion Acceleration *T. E. Moore, J. H. Waite, Jr., M. Lockwood, and C. R. Chappell* 50

Transport of Accelerated Low-Energy Ions in the Polar Magnetosphere *J. L. Horwitz, M. Lockwood, J. H. Waite, Jr., T. E. Moore, C. R. Chappell, and M. O. Chandler* 56

Ion Energization in Upwelling Ion Events *J. H. Waite, Jr., T. E. Moore, M. O. Chandler, M. Lockwood, A. Persoon, and M. Suguira* 61

Observations of Transverse and Parallel Acceleration of Terrestrial Ions at High Latitudes *H. L. Collin, E. G. Shelley, A. G. Ghielmetti, and R. D. Sharp* 67

Accelerated Auroral and Polar-Cap Ions: Outflow at DE-1 Altitudes *A. W. Yau, E. G. Shelley, and W. K. Peterson* 72

Ion Specific Differences in Energetic Field Aligned Upflowing Ions at 1 R_E *A. G. Ghielmetti, E. G. Shelley, H. L. Collin, and R. D. Sharp* 77

Heating of Upflowing Ionospheric Ions on Auroral Field Lines *P. H. Reiff, H. L. Collin, E. G. Shelley, J. L. Burch, and J. D. Winningham* 83

Interaction of H^+ and O^+ Beams: Observations at 2 and 3 R_E *R. L. Kaufmann and G. R. Ludlow* 92

Banded Ion Morphology: Main and Recovery Storm Phases   *R. A. Frahm, P. H. Reiff, J. D. Winningham, and J. L. Burch*   98

The Conductance of Auroral Magnetic Field Lines   *D. R. Weimer, D. A. Gurnett, and C. K. Goertz*   108

III. PLASMA SHEET AND BOUNDARY LAYER PROCESSES

Velocity Distributions of Ion Beams in the Plasma Sheet Boundary Layer   *T. E. Eastman, R. J. DeCoster, and L. A. Frank*   117

Ion Interactions in the Magnetospheric Boundary Layer   *B. Hultqvist, R. Lundin, and K. Stasiewicz*   127

Ion Acceleration During Steady-State Reconnection at the Dayside Magnetopause   *A. D. Johnstone, D. J. Rodgers, A. J. Coates, M. F. Smith, and D. J. Southwood*   136

The Effect of Plasma Sheet Thickness on Ion Acceleration Near a Magnetic Neutral Line   *R. F. Martin, Jr.*   141

IV. EQUATORIAL REGION PROCESSES

Acceleration of Energetic Oxygen ($E > 137$ keV) in the Storm-Time Ring Current   *A. T. Y. Lui, R. W. McEntire, S. M. Krimigis, and E. P. Keath*   149

On the Loss of O^+ Ions (<17 keV/e) in the Ring Current During the Recovery Phase of a Storm   *J. B. Cladis and O. W. Lennartsson*   153

Eigenfunction Methods in Magnetospheric Radial-Diffusion Theory   *Michael Schulz*   158

Ionospheric Ion Streams at Altitudes Below 14 R_E *S. Orsini and M. Candidi*   164

Statistical Study of Enhanced Ion Fluxes in the Outer Plasmasphere   *J. D. Menietti, J. L. Burch, R. L. Williams, D. L. Gallagher, and J. H. Waite, Jr.*   172

V. ACTIVE PROCESSES

The Neutral Lithium Velocity Distribution of an AMPTE Solar Wind Release as Inferred From Lithium Ion Measurements on the UKS Spacecraft   *S. C. Chapman, A. D. Johnstone, and S. W. H. Cowley*   179

Lithium Tracer Ion Energization Observed at AMPTE-UKS   *A. D. Johnstone, A. J. Coates, M. F. Smith, and D. J. Rodgers*   186

ELF Waves and Ion Resonances Produced by an Electron Beam Emitting Rocket in the Ionosphere   *J. R. Winckler, Y. Abe, and K. N. Erickson*   191

Argon Ions Injected Parallel and Perpendicular to the Magnetic Field   *R. E. Erlandson, L. J. Cahill, Jr., C. Pollock, R. L. Arnoldy, J. LaBelle, P. M. Kintner, and T. E. Moore*   201

A Comparison of Plasma Waves Produced by Ion Accelerators in the *F*-Region Ionosphere   *P. M. Kintner, J. LaBelle, W. Scales, R. Erlandson, and L. J. Cahill, Jr.*   206

VI. LABORATORY PROCESSES

Ion Acceleration in Laboratory Plasmas *Reiner L. Stenzel* 211

Ion Acceleration in Laboratory Plasmas *Noah Hershkowitz* 224

Laboratory Simulation of Magnetospheric Plasma Phenomena Using Laser Induced Fluorescence as a Diagnostic *N. Rynn* 235

Laboratory Simulation of Ion Acceleration in the Presence of Lower Hybrid Waves *R. McWilliams, R. Koslover, H. Boehmer, and N. Rynn* 245

Laboratory Experiments on Plasma Expansion *Chung Chan* 249

Ion Acceleration: A Phenomenon Characteristic of the "Expansion of Plasma Into a Vacuum" *Uri Samir, K. H. Wright, Jr., and N. H. Stone* 254

VII. MICROSCOPIC PROCESSES

Ion Acceleration by Wave-Particle Interaction *Robert L. Lysak* 261

Ion Heating in the Cusp *M. Hudson and I. Roth* 271

Plasma Simulation of Ion Acceleration by Lower Hybrid Waves in the Suprauroral Region *John M. Retterer, Tom Chang, and J. R. Jasperse* 282

Analytic Ion Conics in the Magnetosphere *G. B. Crew, Tom Chang, J. M. Retterer, and J. R. Jasperse* 286

Parametric Processes of Lower Hybrid Waves in Multicomponent Auroral Plasmas *Hannu E. J. Koskinen* 291

A New Mechanism for Excitation of Waves in a Magnetoplasma, I, Linear Theory *G. Ganguli, Y. C. Lee, and P. J. Palmadesso* 297

A New Mechanism for Excitation of Waves in a Magnetoplasma, II, Wave-Particle and Nonlinear Aspects *P. Palmadesso, G. Ganguli, and Y. C. Lee* 301

Heating of Light Ions in the Presence of a Large Amplitude Heavy Ion Cyclotron Wave *K.-I. Nishikawa and H. Okuda* 307

Linear Effects of Varying Ion Composition on Perpendicularly-Driven Flute Modes: Comparison to Rocket Observation *David N. Walker* 311

The Direct Production of Ion Conics by Plasma Double Layers *Joseph E. Borovsky and Glenn Joyce* 317

On the Spatial Scale Size of Oblique Double Layers *W. Lennartsson* 323

Double Layers in Linearly Stable Plasma *Robert H. Berman, David J. Tetreault, and Thomas H. Dupree* 328

Effects of Warm Streaming Electrons on Electrostatic Shock Solutions *E. Witt and M. Hudson* 334

Numerical Simulations of Auroral Plasma Processes: Ion Beams and Conics   *Nagendra Singh, H. Thiemann, and R. W. Schunk*   340

Numerical Simulations of Auroral Plasma Processes: Electric Fields   *Nagendra Singh, H. Thiemann, and R. W. Schunk*   343

VIII. MACROSCOPIC PROCESSES

Macroscopic Ion Acceleration Associated With the Formation of the Ring Current in the Earth's Magnetosphere   *B. H. Mauk and C.-I. Meng*   351

Ion Acceleration in Expanding Ionospheric Plasmas   *Nagendra Singh and R. W. Schunk*   362

Time-Dependent Numerical Simulation of Hot Ion Outflow From the Polar Ionosphere   *T. I. Gombosi, T. E. Cravens, A. F. Nagy, and J. II. Waite, Jr.*   366

IX. SUMMARY

Impulsive Ion Acceleration in Earth's Outer Magnetosphere   *D. N. Baker and R. D. Belian*   375

Experimental Identification of Electrostatic Plasma Waves Within Ion Conic Acceleration Regions   *P. M. Kintner*   384

A Digest and Comprehensive Bibliography on Transverse Auroral Ion Acceleration   *D. M. Klumpar*   389

Author Index   399

PREFACE

This past onescore years had seen substantial progress in understanding the nature of the plasma domains which play critical roles in the dynamics of the ionosphere and magnetosphere: the diverse auroral plasmas including the striking inverted-V events, the polar cusp and polar wind, the plasmasphere, the bow shock, the magnetotail, the plasma sheet and its boundary and earthward extension. On the other hand, with only a few exceptions, analytic descriptions of the various plasma phenomena that are relevant to the real geometry and physics of the geoplasma environment are generally inadequate.

As one example, recent satellite and rocket measurements of ion distributions in the topside ionosphere at the polar cusp and auroral altitudes exhibited numerous examples of ion conics. These are distributions of ions flowing upward into the magnetosphere with energies greater than would be expected for ions of ionospheric origin. The name "conic" refers to the fact that these distributions are strongly peaked in pitch angle, so that ions are concentrated on a cone in velocity space, indicating some form of heating transverse to the ambient magnetic field. The discovery of conics was somewhat startling, largely because no mechanism for transversely accelerating ions to what are essentially magnetospheric energies had been anticipated. The likely agent for this acceleration is some sort of wave-particle interaction. In this picture, the ions are energized (i.e., heated) perpendicular to the geomagnetic field lines by the energy-carrying plasma waves. (Popular candidates include lower hybrid waves and electrostatic ion cyclotron waves. More recent work suggests that electromagnetic ion cyclotron waves may also play a role.) One can then account for the conic form of the distribution by recalling that the magnetic field strength decreases with altitude. Thus the adiabatic motion of the ions drifting to higher altitudes transforms the heated distribution into one that is more field-aligned, i.e., a conic. Ion acceleration through oblique electrostatic double layers has also been invoked for the explanation of transverse heating of ions. For the most part, however, self-consistent, coherent theories capable of providing definitive descriptions of this observed acceleration phenomenon have not yet emerged.

There are many other examples of ion acceleration phenomena of current interest which at present defy explanation. Ion beams of varying energies have been detected throughout the magnetosphere, and yet the energization processes of these field-aligned ion distributions are still not entirely understood. Energetic ions have also been observed in several of the the magnetosphere's boundary layers, but their origin has yet to be determined. Recently, a number of active experiments have been performed with both satellites and sounding rockets. These have resulted in a number of artificial ion acceleration events, for which detailed explanations are still being sought. In some of these cases, explanations have been proposed which unfortunately falter when subjected to detailed theoretical investigation or comparison with observational data.

In recognition of this state of affairs, an AGU Chapman conference on ion acceleration in the magnetosphere and ionosphere was held at Wellesley College in the greater Boston area from June 3 to 7, 1985. The purpose of the conference was to bring together the international community of experimentalists and theoreticians engaged in the study of various aspects of ion energization processes, to promote the interchange of ideas among these active researchers, and to attempt to achieve some sort of basic understanding of these interesting and complex geoplasma phenomena. The conference attracted over 120 scientists, with contributions from seventeen countries (Brazil, Canada, China, Denmark, Finland, France, Germany, India, Israel, Italy, Japan, Norway, Sweden, Switzerland, United Kingdom, USA and USSR). The format of the conference included invited reviews and topical lectures as well as contributed oral and poster papers. Effort was made to provide ample free time for informal discussions during the evening hours and at the poster sessions.

The idea of assembling an ion acceleration conference grew from various informal discussions among the co-conveners at the Spring and Fall AGU meetings and the Gordon conference on space plasmas. Our idea was to convene a specialty conference within a secluded and serene environment conducive to a productive mutual interchange of ideas among the participants. Differing from the Gordon conference concept, however, we believed that the proceedings of the conference, including both invited and contributed presentations, should be formally published as soon as possible as a volume of the AGU monograph series. Such a volume should be useful both to the practicing scientist as well as the student or novice who wishes to step into the wonderland of this "frontier" of scientific research. We are thus doubly indebted to the co-conveners. They joined us in the effort in making this conference a reality, and they served amiably as co-editors of this timely monograph.

This monograph contains a cross-section of the material presented at the conference. To some extent, it constitutes a snapshot of the state of the field in the mid 80's. The technical content of the conference program and consequently of this monograph was the responsibility of the program committee which determined the topics to be emphasized at the conference and the speakers to be invited to present the review and topical lectures. Members of the program committee consisted of renowned scientists of international repute in ion acceleration research and include:

R. Arnoldy of the University of New Hampshire
M. Ashour-Abdalla of the University of California, Los Angeles
R. Boström of the Uppsala Ionospheric Observatory, Sweden
C.-G. Fälthammar of the Royal Institute of Technology, Sweden
L. Frank of the University of Iowa
H. R. Balsiger of Physikalisches Institute, Switzerland
D. Gorney of the Aerospace Corporation
G. Haerendel of the Max-Planck-Institute for Extraterrestrial Physics, Germany
D. Hardy of the Air Force Geophysics Laboratory
J.P. Heppner of Goddard Space Flight Center
B.K. Hultqvist of the Kiruna Geophysical Institute, Sweden
D. Klumpar of the University of Texas, Dallas and the Lockheed Palo Alto Research Laboratory
R. Lysak of the University of Minnesota
F. Mozer of the University of California, Berkeley
H. Okuda of the Princeton Plasma Physics Laboratory

R. Olsen of Marshall Space Flight Center and the University of Alabama
S. Ossakow of the Naval Research Laboratory
R. Schunk of Utah State University
R. Sharp of the Lockheed Palo Alto Research Laboratory
E. Shelley of the Lockheed Palo Alto Research Laboratory
E. Wescott of the University of Alaska
D. Williams of the Johns Hopkins University Applied Physics Laboratory
D. Winningham of the Southwest Research Institute
D. Young of the Los Alamos National Laboratory

and the co-conveners. The editors are most grateful for their contributions to this joint endeavor.

In accordance with AGU policy for its monograph series, each submitted manuscript received two "peer" reviews. We have grouped the accepted papers under several headings which follow essentially the same scheme that was adopted for the technical sessions of the conference. Thus Section I contains the two lead-off review articles provided by J.M. Cornwall and D.T. Young. These give an overview of the status of the theoretical and experimental aspects of ion acceleration processes in the magnetosphere and ionosphere at the time of the conference. Recognizing the enormity of the task of preparing such reviews, we are most appreciative of their contributions.

The current interest in high latitude ion acceleration processes is reflected by the large number of contributions in Section II. These papers discuss both perpendicular and parallel ion acceleration as may be deduced from observations of ion conics and beams made over a wide range of altitudes. In addition to the rather extensive data base of events that can be subjected to statistical scrutiny, there are also a number of new types of events which have been identified. Section III contains two excellent review articles on boundary layer ion acceleration activity by T.E. Eastman, et al., which provides a succinct study of ion distribution functions in the plasma sheet boundary layer, and B. Hultqvist, et al., which discusses ion interactions at the magnetopause. The next section contains an assortment of ion acceleration processes occurring in the equatorial regions of the magnetosphere. Recently, much progress has been made in our understanding of the ionosphere/magnetosphere system through the use of active experiments. A number of interesting presentations of results from the AMPTE experiment as well as active experiments carried out from sounding rockets are contained in Section V. Of course, ion acceleration is not unique to space physics. Section VI contains a number of reports of laboratory experiments that may have relevance to the physics of space plasmas. In particular, R.L. Stenzel discusses the acceleration of ions observed in a beautiful series of experiments designed to reveal the secrets of magnetic reconnection in a neutral sheet, N. Hershkowitz describes a number of experiments that have been performed in a variety of machines to simulate problems relevant to space applications, and N. Rynn describes results obtained with a new laser flourescence technique which is proving its worth as a diagnostic of the ion population in a number of ion acceleration experiments. Section VII contains a number of theoretical studies that have been performed with the goal of elucidating some of the microscopic details of ion acceleration. There is an excellent tutorial article by R.L. Lysak on the subject of wave particle interactions, which probably are ultimately responsible for most of the accelerated ions reported, e.g., in Section II. There is also a paper by M.K. Hudson and I. Roth summarizing their work on ion heating in the cusp by lower hybrid waves. Additional contributions are concerned with the formation of ion conics, double layers and explanations for some of the turbulent wave activity that has been observed. Penultimately, B.H. Mauk and C. Meng demonstrate in Section VIII that the formation of the ring current continues to be an active field of research. The final section consists of articles of a summary nature, and begins with a report by D.N. Baker and R.D. Belian on the impulsive acceleration of ions to high energies in the magnetosphere. P.M. Kintner describes some of the general experimental considerations and problems associated with the definitive identification of plasma waves in the ionosphere, particularly in regimes where ion conics are observed. Finally, D.M. Klumpar provides a synthesis of the current comprehension of transversely accelerated ions and an extensive bibliography on auroral ion acceleration.

In addition to the topical or summary papers mentioned above, there were similar unpublished conference presentations made by R.L. Arnoldy, M. Ashour-Abdalla, D.J. Gorney, G. Haerendel, P. Palmadesso, A. Roux, E.G. Shelley, M. Yamada, and A.Y. Wong. We found their lectures most instructive and are grateful for their efforts. We also thank R.W. Schunk and J.D. Winningham for preparing talks which at the last moment they were prevented from delivering. Thanks are due to G.H. Pettengill, Director of the Center for Space Research at MIT who, as local host, welcomed the conference participants. We also wish to thank R.G. Johnson of the White House Science Office for his introductory remarks at the opening session of the conference. We are most grateful to our skillfull moderators, J.R. Jasperse, L.A. Frank, M. Schulz, T. Moore, A.T.Y. Lui, P.M. Kintner, R. McWilliams, S. Ossakow, M.K. Hudson and B. Hultqvist, who served as session chairpersons, kept the conference on schedule and guided us through many stimulating discussions.

The local program committee was responsible for all of the details of the organization and daily operation of the conference. It also arranged the contributed and invited papers into the conference program. In addition to these responsibilities, we are grateful to J.R. Jasperse for his input in formulating the basic structure of the conference and for serving as secretary during the various organizational meetings of the program committee. J.M. Retterer ably assisted in many aspects of the daily operation of the conference and presided over one of the poster sessions. G.B. Crew assisted in every facet of the actual organization and operation of the conference and served as assistant editor of the monograph. It is impossible to detail all the help that he has rendered to the ultimate success of the conference and the monograph. Without his continued support, neither the conference nor the monograph could have been a success.

We are indebted to members of the AGU and Wellesley staff, including B. Weaver, F. Spilhaus and L. Knight, for their assistance in the organization and execution of the daily operation of the conference, and J. Holoviak for arranging the AGU endorsement of the monograph. We would especially like to thank R. Albert, who formed the indispensable link between the conference staff and AGU and put in four strenuous days at the conference site helping out with the registration and a spectrum of other difficult chores. Thanks are also due to C. Bravo who was most helpful in assisting us to get the original Chapman conference concept approved at AGU headquarters. We are most appreciative to Janet Jasperse for lending her artistic talent to the logo that appeared on the cover of the conference program and also on the list of conference participants. J. Wagner helped in many ways to get the initial organization process of the conference started. Major thanks are due to J. Buchholz, who served as conference and editorial assistant and efficiently handled all of the correspondence and files throughout the history of the conference and the editorial process of the monograph. Her cheerful disposition added a constant touch of pleasantry to the otherwise arduous atmosphere of monograph editing. We thank the authors who responded efficiently to the stringent production timetable in addition to providing the monograph with so many timely and interesting papers. The editors wish to thank the referees who took considerable efforts in providing us with the thoughtful and detailed reviews. We also wish to thank our editorial advisor, D.M. Klumpar, for his amiable assistance. Lastly, thanks are due to S. Mansberg, the monograph coordinator at AGU, who was instrumental in assuring the proper preparation of the manuscripts and was the prime moving force that kept the publication schedule of the monograph almost "on time."

This conference was sponsored by AGU, MIT, the Air Force Geophysics Laboratory, Cornell University, Dartmouth College, the White House Science Office, Lockheed Palo Alto Research Laboratory and the Aerospace Corporation. We are grateful to the Air Force Geophysics Laboratory, the Lockheed Palo Alto Research Laboratory, the National Aeronautics and Space Administration and the National Science Foundation for financial support.

Tom Chang
Massachusetts Institute of Technology
Cambridge, Massachusetts
1986

Section I. INTRODUCTION

MAGNETOSPHERIC ION ACCELERATION PROCESSES

John M. Cornwall

University of California, Los Angeles, CA 90024 and
Space Sciences Laboratory, Aerospace Corporation, El Segundo, CA 90245

Abstract. Magnetospheric ions can be heated or accelerated by shocks, reconnection and neutral-sheet electric fields, convection electric fields, cross-L diffusion driven by fluctuating electric or magnetic fields, auroral parallel electric fields, and various local wave-particle processes. Many of these mechanisms are more or less well-understood when they act alone, but the major space-physics problem remaining to be solved is their mutual interaction. The coupling of different mechanisms may be elementary in principle (e.g., using shock heating to determine appropriate initial conditions for cross-L diffusion), but it is usually complex. For example, many local plasma processes have growth rates and transport coefficients largely determined by global boundary conditions set by other processes, such as convection or diffusion. In turn, global mechanisms may be strongly affected by microscopic plasma interactions, and one must discuss mutual couplings on grossly disparate space and time scales. Examples are the sequence beginning with Speiser acceleration in the neutral sheet and ending with an aurora; the coupling of turbulence and mirror forces in the aurora; and the interplay of convection/diffusion processes and wave-particle interactions in the ring current.

1. Introduction

Since ions carry the bulk of the particle energy in the Earth's magnetosphere, the processes which energize them are quite properly a subject of primary interest for space physicists. Once a given species of ions is accelerated, it may share its energy with other ion species, with neutrals, or with electrons. These energy-loss processes are themselves of the greatest interest, and are believed by many to be involved, for example, in formation of auroral arcs and suprathermal heating of ambient electrons and heavy ions in the ring current. Some of the energy thus given up by some ions may later be recovered by other ions, for example, through the formation of ion beams and conics in the auroral zone, whereby aurorally-energized electrons add energy to cold ionospheric ions.

There are two dominant sources of magnetospheric ions: the solar wind and the ionosphere. (We will not consider cosmic rays or cosmic-ray albedo neutron decay (CRAND) sources here.) In principle, they are distinguished by their ionic composition. The solar wind is predominantly H^+ with a few percent He^{++}; the ionospheric source is mostly H^+ and O^+, with some He^+. However, for helium ions of < 1 MeV energy in the ring current, charge exchange and ionization processes blur the distinction between the two sources. Before they are accelerated by the processes considered in this paper, solar wind ions have velocities of a few hundred km/sec (thus energy $\sim$ 1 keV for H^+), while ionospheric ions have energy from 0.1 to a few eV.

The main regions and processes of ion acceleration are:

1. Shock-associated acceleration in the Earth's bow shock and nearby interplanetary shocks;
2. Electrostatic and inductive acceleration in the tail and neutral sheet ($B \simeq 0$);
3. Ring-current acceleration of ions by electric fields, which can be static or fluctuating convection fields, or induction fields associated with fluctuations in $\underline{B}$;
4. Suprathermal heating of ambient plasmaspheric heavy ions by proton-generated electromagnetic cyclotron (EMC) waves;
5. Heating in $T_\perp$ of ionospheric ions from the auroral zone and possibly from the polar cap by waves (e.g., electrostatic ion cyclotron (EIC) or lower-hybrid (LH) modes) driven unstable by electron currents.
6. Electrostatic acceleration by auroral-zone or polar-cap parallel (to $\underline{B}$) potential drops, which can be ambipolar, adiabatic, or kinetic in nature.

To this list we can also add man-made scenarios in space which test the above processes in special ways, e.g., the AMPTE artificial comet releases [Haerendel, these proceedings].

Most of these processes are interrelated in a highly complex way, and the challenge for space physics of the future is to understand their mutual interaction. Ions do not merely respond to their environment, they shape that environment as well. It is no longer enough to argue, as was common in the past, that microscopic plasma physics can be worked out or computer-simulated for an essentially homogeneous infinite medium. In fact, the arena of space physics is highly inhomogeneous, and macroscopic phenomena far away from a given point can profoundly influence local wave-particle interactions. For example, in the aurora mirror forces (due to the inhomogeneity of $\underline{B}$) are critical both in setting up a parallel potential drop and in transporting ionospheric ions which are heated by EIC (or LH waves) that owe their existence to that potential drop. The actual heating rate and other characteristics of the turbulence are determined in large part by global considerations (e.g., the ion transit time along an auroral field line or across the neutral sheet). The problem here is the enormous dynamic range: physics at the ion Larmor scale ($\sim$ 1 km in the aurora) is closely connected with physics on a global scale of $\sim 10^4$ km. Another example of the interplay of global and microscopic processes in the competition between creation of ring-current proton pitch-angle anisotropy by radial diffusion and charge exchange, and anisotropy destruction by unstable generation of ion EMC modes. The outcome of this competition helps to determine the loss rate of the protons, the wave-heating rate of ambient electrons and heavy ions, and possibly the appearance of stable auroral red arcs (SAR arcs) at mid-latitudes as a result of electron heating. A final example, related to the above: Evidently the convection electric field ($E_\perp$, for short) has much to do with inward transport and acceleration of ions which initially were on high L shells. But a sufficiently strong ring current can partially short out $E_\perp$, bringing these processes to a halt. In that case, the ring current might stop before reaching the high-density plasmasphere, where the ion-EMC resonance condition can be satisfied, and there might be no unstable wave generation. On the other hand, only quasi-DC (periods $\gtrsim$ 1 hr) convection fields can be shorted out, and more rapidly fluctuating fields are still effective. What, then, governs the temporal fluctuations of $E_\perp$, aside from fluctuations in the driving solar wind? The answer most likely is: transport and acceleration processes in the neutral sheet and plasma sheet, which are inherently unsteady because of the very presence of the ions being transported and accelerated. These examples made it clear that, as we have said, at every stage of magnetospheric physics, ions do not merely react to their environment; they are of the greatest importance in determining that environment. Moreover, the mutual couplings of ions and their environment involve careful consideration of grossly disparate space and time scales.

A note on the references: a complete list would exceed the body of the text in length, and a great many important papers will not be found in the bibliography. Papers are cited because they are up-to-date reviews, because they were pioneering efforts, or because the author happens to be particularly familiar with them.

2. Ion Acceleration in the Bow Shock and Interplanetary Shocks

Aside from these shocks, there are slow-mode shocks associated with reconnection, which we will not discuss here (see Section 3). The field of collisionless shocks encompasses such varied and complex phenomena that anything short of a full-scale review will be woefully incomplete. For such reviews, we refer to Greenstadt et al. [1984]; Kennel et al. [1984]; Russell and Hoppe [1983]; Pesses et al. [1982]; Toptyghin [1980]; as well as the special Journal of Geophysical Research issues of June 1981 and January 1985.

Speaking in the broadest possible terms, there are two major shock types: quasi-perpendicular and quasi-parallel, with $\underline{B}$ more or less perpendicular or parallel to the shock normal. One should also distinguish subcritical and supercritical quasi-perpendicular shocks, the latter having a fast-mode Mach number M larger than about 3. In this case resistivity alone--not even anomalous resistivity--can provide sufficient dissipation to satisfy the shock jump conditions. It appears that the needed extra dissipation can be furnished by processes which reflect and accelerate ions that cannot be dealt with in MHD or multi-fluid theory.

The bow shock and interplanetary shocks are capable of accelerating ions to quite substantial energies: tens of keV/Q on up to cosmic-ray energies. The energy given up to these high-energy particles is a finite fraction of the total shock energy, and we must expect that energetic-ion acceleration in shocks cannot be treated as a test-particle process. Those shock-energized ions which become trapped on closed-field lines will be further accelerated by such processes as radial diffusion; see Section 4. It is not always easy to tell whether an accelerated ion observed in the upstream region, or foreshock, is a result of a shock-associated process or is simply the result of leakage from the magnetosphere to the upstream region [Anagnostopoulos et al., 1985]. But Scholer et al. [1981] have shown that there is some local acceleration upstream of the bow shock. This upstream region is populated with a variety of waves and energetic particles, and no less a variety of theories to explain them [see, e.g., Russell and Hoppe, 1983, and articles in special <u>Journal of Geophysical Research</u> numbers of June 1981 and January 1985]. One observes spike events, in which energetic ($\gtrsim$ 100 keV) ion intensities near the shock increase suddenly, then decay after a few minutes; steplike increases of intensity associated with the shock, which may last for hours; and energetic storm particles (ESP) events, in which over a period hours before the shock energetic particle intensities increase, then decay slowly after the shock passage [see, e.g., Tsurutani and Lin, 1985]. Also seen are

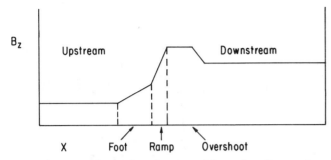

Fig. 1. A schematic of a one-dimensional quasi-perpendicular shock. The foot is formed by hot downstream ions moving to the left, whose electrostatic potential reflects upstream ions, which gain enough energy by drifting along E_y to penetrate and form the overshoot.

electron plasma waves as well as ULF and ELF waves, including Alfvén, whistler, ion acoustic, and other modes. This wave turbulence is important not only as a diagnostic of shock processes (e.g., particle beams reflected from the shock which make a given wave mode unstable), but also as a source of scattering centers for particles which can lead to multiple acceleration events for a single ion as it is scattered repeatedly in the vicinity of the shock.

The two critical issues of how a collisionless shock creates the dissipation needed for the shock to form, and how high-energy particles are accelerated in the shock, are not necessarily independent. Consider first a quasi-perpendicular supercritical shock, one in which anomalous (very possibly due to ion sound) resistivity is not adequate dissipation to sustain the shock. Leroy et al. [1982] have simulated a one-dimensional shock with a hybrid (fluid, massless electrons; particulate ions) code, adding anomalous resistivity in an ad hoc way. Results similar to their simulation are sketched in Fig. 1, with the foot being formed by left-moving downstream ions which crossed the shock (ramp). (The magnetic field prevents the electrons from crossing.) These ions produce an electrostatic potential which forms a barrier between upstream and downstream regions, and which reflects some upstream ions. These ions are accelerated by the y component of electric field associated with the flow, because they drift in the y direction in the same direction as E_y points. Eventually these ions become energetic enough to cross the potential and exit downstream (to the right on Fig. 1). They thus constitute a dissipation mechanism, removing energy from the shock flow. In one particular simulation, about 20% of the incoming ions were reflected.

A closely-related energization mechanism--called shock-drift acceleration--can multiply the energy of moderately energetic particles. Those energetic particles whose Larmor radius exceeds the shock thickness essentially conserve their magnetic moment, and thus upstream particles can be reflected from the shock if their pitch angle is large enough. It is easy to see that the effective ∇B drift of such particles always leads to acceleration in the $\underset{\sim}{V} \times \underset{\sim}{B}$ flow electric field, for fast-mode shocks (where the downstream magnetic field B_2 exceeds the upstream field B_1). For a nonrelativistic particle, the upper-limit relation between energy gain $\nabla \epsilon$ and initial energy ϵ is

$$\frac{\Delta \epsilon}{\epsilon} \leq 4 \left(\frac{B_2 - B_1}{B_1} \right) \leq 12 \qquad (2.1)$$

where the last inequality holds because $B_2 \leq 4B_1$. So particles gain energy according to what they already had, and to explain the very energetic particles seen near the Earth's bow shock with this mechanism requires it to be repeated, presumably because the newly-energized particles are reflected back to the shock by downstream turbulence one or more times. Shock-drift acceleration seems capable of explaining spike events.

Quasi-parallel shocks can support little or no shock-drift acceleration, because $\underset{\sim}{B}$ is nearly perpendicular to the shock front. The shock structure is very thick, and blends into a huge region called the foreshock, which is downstream from the line of tangency of the solar wind magnetic field on the bow shock (see Fig. 2). Near this line of tangency, field-aligned ion beams of a few keV are seen (so-called reflected ions), and further in are intermediate ion distributions with energies up to several hundred keV. Furthest in are the diffuse distributions, ring-shaped distributions of energetic ions. Energetic electrons are also seen (Fig. 2). Associated with these particles are a variety of waves, notably low-frequency ion-acoustic and whistler-mode waves as well as Alfvén waves.

The general idea behind theories of quasi-parallel shocks [e.g., Parker, 1961] is that particles can more or less freely stream across the shock front, but are subject to partial reflection by the shock. Dissipation is provided by unstable counter-streaming ion modes, e.g., firehose instability (at large downstream β) coupled with generation of Alfvén waves. Ions moving upstream and downstream can generate turbulence far from the actual shock (i.e., create the foreshock), and this turbulence may provide scattering centers for Fermi acceleration: The upstream and downstream scattering centers are moving toward each other (because the shock slows down the plasma bulk velocity as the plasma crosses the shock) and so a particle trapped between them is accelerated. Any upper limit on the acceleration comes from the global scale of the shock, which determines how long a particle can stay in the acceleration region. Such first-order Fermi acceleration may explain diffuse ion distributions (Fig. 2) and ESP events. As with shock-drift acceleration, the more energetic a particle is, the more it gains in a reflection. The es-

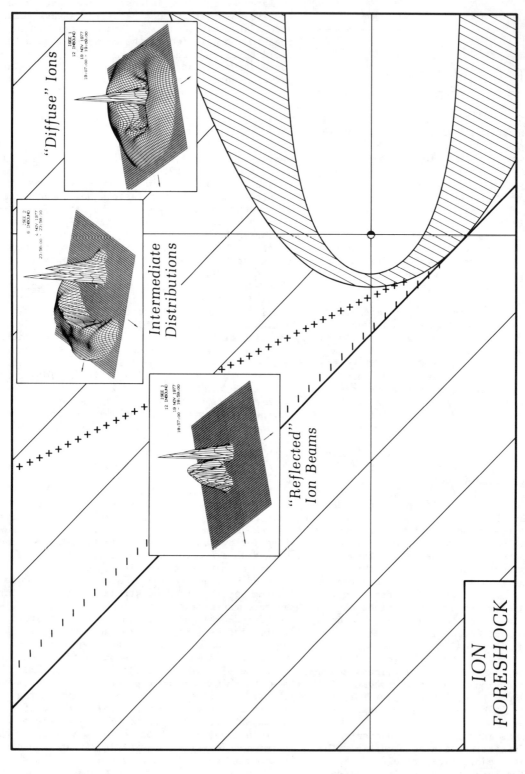

Fig. 2a. Reflected ion beams are the lower hump; the tall peak is incident solar wind. Further away from the foreshock boundary (line of tangency) the ion distribution is broader, and called intermediate; yet further away are backstreaming ring distributions, called diffuse.

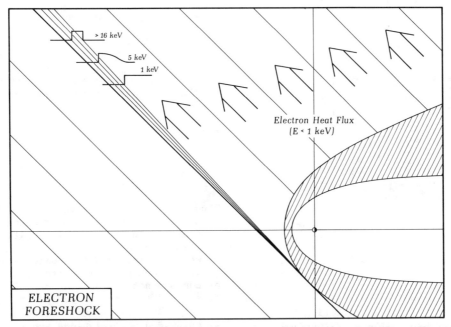

Fig. 2b. Energetic electrons seen in the foreshock turn on abruptly, and are swept away from the line of tangency by the solar wind. The foreshock also has a heat flux away from Earth, carried by ~ 1 keV electrons [from Russell and Hoppe, 1983].

Fig. 2. Schematic of ion and electron properties in the bow foreshock.

sential question that then remains is: How are particles of characteristic solar-wind energy (~ 1 keV) efficiently accelerated? Whatever the details of this process, it appears that reflected ions in the foreshock have an exponential distribution in energy per charge, with an e-folding energy of ~ 20 keV/Q. Both these energetic particles as well as the bulk solar wind can be accelerated by neutral-line reconnection processes, which we now take up.

3. Particle Acceleration and Reconnection in the Tail Neutral Sheet

Particle acceleration in a neutral sheet or near a neutral line takes place because low-energy particles are not impeded, by the small magnetic fields in the vicinity of the neutral sheet, from being accelerated by the cross-tail electric field (generally dawn-to-dusk). The process is much more effective for ions than electrons, because particles tend to gain a velocity increment independent of their mass. Our discussion will focus on tail reconnection and neutral lines; of course, reconnection can also occur at the nose and elsewhere.

There are, not unexpectedly, parallels between the neutral sheet processes and both auroral physics and collisionless-shock physics. Just as in these problems, the first attempts at understanding used MHD with resistivity added purely phenomenologically [Petschek, 1964; developments to 1975 are reviewed by Vasyliunas, 1975]. And just as in these problems, it is by no means obvious that kinetic and turbulent effects are properly modeled by setting $\underline{J} = \sigma \underline{E}$, nor is it at all clear what to choose for σ even if the structural form of this constitutive relation is correct [Lyons and Speiser, 1985]. As the MHD theories developed, people attempted to justify various candidate wave modes for anomalous resistivity, such as ion acoustic [Coroniti and Eviatar, 1977] or lower-hybrid drift (LHD) [Huba et al., 1981]. However, ion acoustic turbulence only grows in a very thin region where $B \simeq 0$, and LHD turbulence prefers $B \neq 0$ and low β: neither of these modes fills the spatial volume where dissipation is needed to drive reconvection.

Let us now turn to models invoking finite inertia, in which the acceleration of particles by the electric field in regions where $B \simeq 0$ is limited by inertial effects. Speiser [1965] first investigated single-particle acceleration in given static electric and magnetic fields, as in Fig. 3. (For a recent review, see Cowley [1980] and for examples of more elaborate numerical computations, see Lyons and Speiser, 1982; Wagner et al. 1979.) As long as $B_z > E_y$ (that is, we are not actually on the neutral line) the y component of the electric field can be transformed away in a frame moving toward the sun at velocity $U \equiv cE_yB_z^{-1}$. Ions and electrons are carried in the z-direction toward the neutral sheet by the main electric drift, and in the moving frame they are reflected

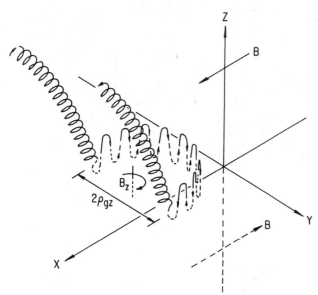

Fig. 3. Particle motion in a neutral sheet in the $E_y = 0$ frame, showing kinetic reflection of V_x by Larmor gyration through half a B_z gyroperiod [from Lyons and Speiser, 1985].

by the $\underline{B}$-field after half a gyrorotation in B_z. In the Earth's rest frame they are energized by the electric field over a distance $\sim 2R_{Lz}$, where

$$R_{Lz} = V_x \Omega_z^{-1} \simeq U \Omega_z^{-1} \quad (3.1)$$

the latter approximate equality holding for initially low-energy particles. Thus in the Earth's frame the gain in energy is $\sim 2eR_{Lz}E_y$, which is easily interpreted in the moving frame as a change in sign of V_x from $-U$ to U. In either case, the energy gain is approximately (depending on initial gyrophase) $(1/2) M(2U)^2$. Typical values of the parameters might be $E_y \sim 0.5$ mV/m, $B_z \sim 1\gamma$, so $U \sim 500$ km/sec, and a proton would gain ~ 5 keV, an electron only 3 eV. Twice the gyroradius R_{Lz} is ~ 5000 km, far less than the extent of the neutral sheet in the y direction. Ampere's law plus particle continuity specifies the electric field:

$$E_y = \frac{B_x^2}{4\pi Nde} = \frac{B_z V_a}{c} \quad (3.2)$$

where N is the particle density and [Speiser, 1970] $d \sim 2R_{Lz}$ is the spatial distance over which injected particles remain in the neutral sheet. In effect, (3.2) supplies a relation between E_y and J_y, since by Ampere's law

$$\lambda J_y = \frac{cB_x}{2\pi} \quad , \quad E_y = \frac{2\pi\lambda}{c^2} V_{Az} J_y \quad , \quad (3.3)$$

where λ is the thickness in the z-direction of the current sheet and $V_{Az}^2 = B_z^2/4$ NM. The electric field acts to accelerate both electrons and ions, thus supplying the dissipation of magnetic energy necessary to support reconnection. Note that the second equality says that the outflow velocity $2U = 2cE_y B_z^{-1}$ is twice the Alfvén velocity, as one finds in the MHD picture. In this idealized picture all the particle-energy is in parallel motion, so

$$\beta_{\parallel} = \frac{8\pi}{B_x^2} NM(V_A)^2 = 2 \quad , \quad \beta_{\perp} = 0 \quad . \quad (3.4)$$

The outgoing plasma is thus marginally firehose unstable, as also results from the MHD balance of stresses in the absence of significant gradients in x [Rich et al., 1972].

We have discussed this single-particle picture as if it took place in a steady state. This is, of course, not so, and the calculation of the influence of the accelerated particles on $\underline{E}$ and $\underline{B}$ reveals that this single-particle model essentially becomes a collisionless tearing-mode instability [Coppi et al., 1966], with particle inertia replacing or supplementing Landau damping. As particles are accelerated, the current they carry increases, and the current sheet thins; these time-dependent effects give rise to inductive electric fields ($E_y = -c^{-1} \partial A_y/\partial t$). The acceleration produced by the time-dependent fields is limited to a time $\tau \sim \Omega_z^{-1} = Mc/eB_z$ by the single-particle arguments, thus producing a current in response to $E_y(t)$ which is proportional to τ. Again, by using Ampere's law and continuity, one finds the rate γ at which the inductive fields charge to be, for small perturbations

$$\gamma \simeq \frac{\Omega_z}{\Omega_x} \frac{\overline{V}}{\lambda} \quad ; \quad \gamma^{-1} \simeq 10\text{-}100\text{'s sec} \quad (3.5)$$

where $\overline{V}$ is a characteristic particle velocity. Eventually [Coroniti, 1985] it turns out that γ is proportional to the time-dependent part of A_y, and because γ is $(\partial/\partial t)(\ell n\, A_y)$ this implies that A_y becomes singular in a finite time (explosive instability). Of course, other physical effects intervene to prevent an actual singularity, but the idea is that the inductive fields grow very rapidly to very large size. Although the dominant effect here is the current carried by accelerated particles, it may also be necessary to invoke trapped particles, carrying no current, which allow the current distribution and $\underline{B}$ to be self-consistent [Francfort and Pellat, 1976].

This inherent time dependence means that reconnection is always unsteady, even in the presence of a perfectly stationary solar-wind energy input. During the explosive phase of reconvection particles may be energized up to ~ 1 MeV or so; we have already mentioned that such energetic particles may leak out of the magnetosphere into the foreshock [Anagnostopoulos et al., 1985] and mix with shock-accelerated particles. Unsteady reconnection also means time-dependent convection, with consequences for ring-current dynamics discussed in Section 4.

Just as for collisionless shocks, we have argued that MHD supplemented with anomalous resistivity is not an adequate framework for treating neutral-line phenomena. This is not to say that turbulence is not present or is unimportant; rather, we say that any possible anomalous resistivity associated with turbulence does not adequately explain dissipation over the whole thickness of the current sheet in the vicinity of neutral-line, whereas the dissipation by drift acceleration does seem to work. MHD taken by itself is not so much incorrect as it is incomplete, and on spatial scales large compared to dissipation lengths a Petschek-like picture must emerge. In fact, the Speiser single-particle picture does fit into the general picture of MHD reconvection models, with their implicit or explicit resistive dissipation [Eastwood, 1972; Hill, 1975], but goes beyond these in providing a detailed picture of dissipation by acceleration. Coroniti [1984] has carefully discussed the failure of anomalous resistivity to explain fully either reconnection or auroras; the latter will be mentioned in Section 5.

4. Ion Acceleration and the Ring Current

Much of the basic physics of the ring current, taken as an isolated entity, was worked out long ago [for reviews, see, e.g., Schulz and Lanzerotti 1974; Cornwall and Schulz, 1979]. Important exceptions to this are new measurements of ring-current ion composition through the total range of significant energies (below hundreds of keV) [Young, these proceedings; Hamilton et al., these proceedings; see also the special issue of Geophysical Research Letters of May, 1985, discussing early AMPTE results], and observations as well as theory of heavy-ion heating by proton electromagnetic cyclotron (EMC) waves [e.g., Roux et al., 1982; Omura et al., 1985].

Of course, the ring current is not an isolated phenomenon, and we are now only beginning to solve such vital problems as the sources of the ring current, and the nature of transport processes from the sources to the current (e.g., how exactly do ions accelerated in the neutral sheet or in auroras join the bulk ring current?). Several papers in these proceedings address these questions [Eastman et al., Hultqvist; Ipavich et al.; Hones et al.].

Given that ions do reach the outer parts of the ring current, they can be accelerated by radial diffusion, or--if their energy is low enough to begin with--by quasisteady convection electric fields. These acceleration processes necessarily involve spatial transport over magnetospheric-scale distances. Inward transport, as well as charge exchange and other loss processes, tend to increase the pitch-angle anisotropy of the ions which (if the ambient plasma density is not too low) leads to EMC instability [Cornwall, 1965; Obayashi, 1965] that generates Alfvén waves and lowers the ions' average pitch angle. Energetic ions are thus scattered into the atmosphere and lost. The competition between inward transport (and anisotropy growth) and pitch-angle scattering losses determines, in principle, both the particle fluxes and wave intensities [Cornwall, 1966]. But the full problem of self-consistent particle transport, wave generation, and pitch-angle scattering has never been worked out in detail, even with simulations. Simulations tend to founder on the large spatial inhomogenities associated with the earth's magnetic field, but these are critical in every phase of the physics. It is especially important to account for the spatial variation of B along a field line when heavy-ion heating is considered [Cornwall, 1977], because equatorial proton EMC waves can cyclotron-resonate even with very low-energy heavy ions off the equator. Such resonances not only heat the heavy ions but profoundly affect the propagation of the EMC mode itself. This modification (amounting to reflection of the proton EMC waves somewhere off the equator) has been invoked by Roux et al. [1982] to explain Geos 1,2 observations [Young et al., 1981] that the right number of cold He^+ ions seems to lead to enhanced proton EMC waves and heating of the He^+ ions (Fig. 4). Additionally, magnetospheric electrons can be heated through Landau-resonant interactions with EMC waves that have (because of heavy ions or curved field lines) wave normals nearly perpendicular to $\underline{B}$ and thus a finite $E_\parallel$ [Cornwall et al., 1970; Roux et al., these proceedings]. These hot ($\geq$ few eV) electrons can be the source of stable auroral red (SAR) arcs, mid-latitude auroras that occur during storm main phase and recovery phase. A characteristic feature of all these phenomena driven by EMC waves is that they are unlikely to occur outside the plasmasphere, because the ambient plasma density there is too low to allow cyclotron resonance with the bulk of the ring current [Cornwall et al., 1970].

Let us consider radial diffusion in a bit more detail. [See Mauk and Meng, these proceedings, for a discussion of ion convection.] Radial diffusion is a process in which charged particles have a fluctuating drift velocity $\sim \underline{E} \times \underline{B}$ due to fluctuations in $\underline{E}$. These fluctuations may be an inductive response to changes in $\underline{B}$ (so-called magnetic diffusion), or direct changes in the sources of $\underline{E}$ (electrostatic diffusion). If the fluctuations occur at the drift period or multiples thereof (thus violating the third adiabatic invariant), the $\underline{E} \times \underline{B}$ velocity has a component which steadily increases or decreases L. The diffusion coefficient depends on the power spectrum of fluctuations in $\underline{E}$, and Cornwall [1972] has given the

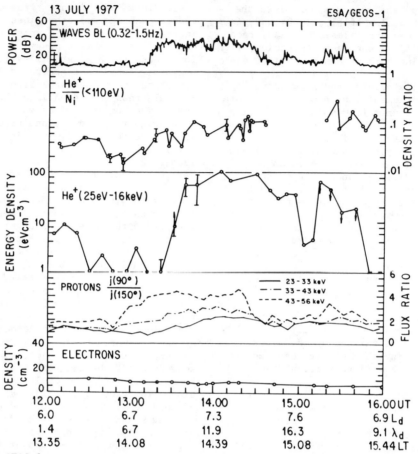

Fig. 4. GEOS-1 wave and ring-current data, showing the increase in EMC waves (top panel) and He$^+$ heating connected with an increase of energetic-proton anisotropy [from Young et al., 1981].

following expression for the electrostatic diffusion coefficient D_E based on a power spectrum coming from fluctuations with a rapid rise and exponential decay in time T, and saving only the lowest resonance:

$$D_E(\omega_D) = \frac{c^2 T L^6 <E^2>}{4 B_o^2 [1+(\omega_D T/2)^2]} \quad . \quad (4.1)$$

Here

$$\omega_D = 8 \times 10^{-4} (A/Z)(E_o/L^2) \text{ sec}^{-1} \quad (4.2)$$

is the drift frequency of a particle of charge Z, mass number A, and energy per nucleon E_o at L = 7, in keV, and $<E^2>$ is the mean square of the fluctuating electric field. For magnetic diffusion with the same sort of power spectrum of fluctuations in B (4.1) would be multiplied by ω_D^2 (and geometric factors), since $E \sim \dot{B}$. This means that for $\omega_D T \gg 1$ magnetic diffusion acts the same on all ionic species of sufficiently large energy, while electrostatic diffusion acts quite differently. Cornwall [1972] has computed radial diffusion profiles for H$^+$, He$^+$, and He^{++}, based on various boundary conditions and including charge exchange and Coulomb losses. A typical result is shown in Fig. 5, for ion energies of 500 keV/nucleon. The diffusion source is the solar wind, which has essentially no He$^+$, but note that charge exchange has produced almost as much He$^+$ as He^{++} over a wide range of L. The diffusion coefficients for these (and other) species are different according to (4.1,2), so that the study of dynamics of different ions could be, in principle, an important tool for untangling various magnetospheric processes.

Most of the ring-current protons have energy $\leq$ 100 keV. An ion of gyrofrequency Ω can resonate with a proton EMC wave at frequency ω if its energy exceeds the critical energy E_c:

$$E_c = \frac{B^2}{8\pi N} \left(\frac{1-x}{x^2}\right)\left(x - \frac{\Omega}{\Omega_p}\right)^2 \quad (4.3)$$

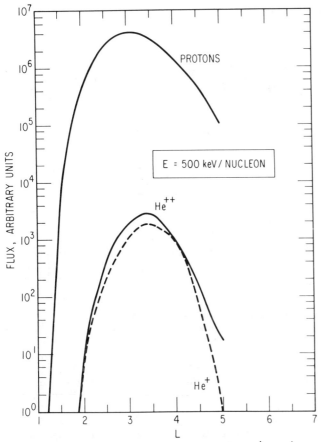

Fig. 5. Radial diffusion profiles for H^+, He^+, and He^{++}, showing the effect of charge exchange and species-dependent diffusion coefficients. The energy of all three species is 500 keV/nucleon [from Cornwall, 1971].

where $x = \omega/\Omega_p$, Ω_p is the proton gyrofrequency, and N is the cold plasma density. This equation is simply a rewriting of the cyclotron resonance condition $\omega - k_\parallel V_\parallel = \Omega$, with ω and $k_\parallel$ related by the linear dispersion relation. The normalized frequency x is determined by the energetic-proton anisotropy A ($x < A(A+1)^{-1}$), and is usually $\lesssim 1/2$. At L = 5 the magnetic energy per particle $B^2/8\pi N$ is about 120 keV/N with N in cm^{-3}, and for N < 1 (outside the plasmasphere) there are normally not enough protons above 120 keV to produce strong EMC waves. Inside the plasmasphere, however, energetic ring-current ions are in cyclotron resonance. The EMC waves thereby generated heat ambient He^+ ions, among other things, as discussed in these proceedings by Roux et al. and by Ashour-Abdalla and Coroniti. Simulations [Omura et al., 1985] and observations [Young et al., 1981; Roux et al., 1982a] show interesting nonlinear effects: He^+ gyrophase bunching and trapping in EMC wave potentials. Initially in the simulations He^+ ions are not resonant, but merely follow the oscillating

$\underset{\sim}{E}_W \times \underset{\sim}{B}$ velocity imposed by the wave electric fields $\underset{\sim}{E}_W$; at later times $V_\parallel$ grows to the point that the $\underset{\sim}{V} \times \underset{\sim}{B}_W$ force is significant, trapping takes place, and finally there is bulk heating of the low-energy ions. EMC wave saturation seems to take place by trapping, that is, the trapping frequency ω_T becomes essentially equal to the linear EMC growth rate γ:

$$\gamma \simeq \omega_T = (kV_\perp \Omega_W)^{1/2} \quad (4.4)$$

where Ω_W is the ion gyrofrequency based on the wave magnetic field B_W.

There are some important points difficult both in simulations and in analytic treatments. They have to do with inhomogeneities in B, and they affect (as we have already mentioned) the propagation of waves and the general conditions for gyroresonance, as well as the acceleration of electrons trapped in a wave propagating in an inhomogeneous medium [Roux et al., 1982b]. Another effect of inhomogeneities is the effective broadening of local cyclotron resonance: As B changes, resonance is lost, and the homogeneous EMC growth rate γ is diminished by an amount γ_{RB}, roughly given by

$$\gamma_{RB} = -\left(\frac{V_\parallel}{\pi}\frac{\partial \Omega}{\partial s}\right)^{1/2} \quad (4.5)$$

where $V_\parallel$ is the parallel velocity and Ω the gyrofrequency of the resonant particle, and s the length along the earth's field line. At the equator $\partial\Omega/\partial s$ vanishes, but off the equator—where low-energy ions can resonate with proton EMC waves generated at the equator—γ_{RB} is of order 0.1 sec^{-1}, which is not negligible compared to the linear growth rate. The challenge for future computer simulations is to take all these inhomogeneity effects into account.

5. Auroral, Polar Cap, and Cusp Region Ion Acceleration

Just as for the ring current, energetic ions furnish the basic energy source for a variety of auroral cusp phenomena, some of which involve the acceleration of cold ambient ions by waves and fields which owe their existence to the energetic ions. We restrict our explicit discussion to the auroral region. This is not the place to engage in detailed discussion of how parallel potential drops are created along auroral field lines, but it might as well be admitted that the author adheres to an adiabatic picture, in which anomalous resistivity plays a subsidiary role [Chiu and Schulz, 1978; Chiu and Cornwall, 1980; Lyons, 1980, 1981]. Anomalous resistivity is less important because it is too dissipative [Cornwall and Chiu, 1982; Coroniti, 1984]: a resistivity large enough to match observed ~ keV potential drops dissipate more power along the field line than energetic

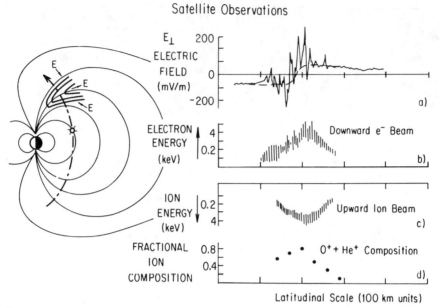

Fig. 6. Schematic view of auroral satellite observations of fields, electrons, and ions.

ions can supply. Moreover, most turbulent electrostatic modes have E_W largely perpendicular to B, so they are inefficient in impeding motion parallel to B. In the adiabatic picture, Alfvén-Fälthammar mirror forces impede electron motion along B, so they cannot quite neutralize ions which are assumed to mirror lower than the electrons would if there were no $E_{\parallel}$. In actual practice, an upward $E_{\parallel}$ can be maintained by assuming quasineutrality.

The adiabatic picture readily matches satellite observations (Fig. 6) with an appropriate choice of ionospheric and magnetospheric particle distributions, except for small-scale spatial and temporal fluctuations resulting from unstable wave modes driven by these distributions. The influence of such instabilities on ions is the subject of this section, and it is a very rich subject indeed, far too big to be reviewed in detail here. A substantial part of this conference is devoted to detailed explications of current research, so we can afford to be general here.

The electrons accelerated by auroral $E_{\parallel}$ form an upward current whose velocity is of order 3×10^9 cm/sec, and this current can drive ion acoustic waves, electrostatic ion cyclotron (EIC) waves [Kindel and Kennel, 1971], lower-hybrid (LH) waves [Chang and Coppi, 1981], and other largely electrostatic modes, as well as electromagnetic modes (whistler, AKR, etc.). As mentioned above, the electrostatic modes have $E_{\perp}/E_{\parallel} \gg 1$ (usually, $E_{\perp}/E_{\parallel} \sim (M_{ION}/M_{ELEC})^{1/2}$), and when they act directly on ions they energize their perpendicular velocity component; the result is generically termed transversely-accelerated ions (TAI). We have spoken so far of electron-driven currents in the auroral region of upward $E_{\parallel}$; there are also return currents neighboring the inverted-V auroral structure, also carried by electrons. Here there are two possibilities: the most common case is that the only available $E_{\parallel}$ is an upward ambipolar field, and the return current is carried by electrons whose drift speed is subsonic. In the second case, it appears that sometimes a good fraction of the return current is carried by electrons of energy 10's-100's of eV, presumably associated with downward $E_{\parallel}$ at altitudes above the ambipolar field ($E_{\parallel}$ can change sign when the electrons go supersonic). The source of this downward $E_{\parallel}$ is unknown. In both cases certain modes may be unstable; the common candidates are EIC and LH. Return-current EIC waves have been studied by many authors. We mention here the work of Ashour-Abdalla and Okuda [1984] which contains references to earlier work. Much work on LH heating of cold ions in auroras has been done by Chang and his collaborators [Chang and Coppi, 1981; Retterer et al., 1983, 1985]. Finally, we mention that if magnetospheric ions of several keV drive auroras as in the adiabatic model, transverse acceleration of such ions can raise their mirror points, tending to reduce the electron-ion differential pitch-angle anisotropy and thus reduce the overall parallel potential drop responsible for auroras [Cornwall and Chiu, 1982].

Evidently if ions find themselves in regions of upward E they can be transported upward, out of the ionosphere and into the ring current. Since significant $E_{\parallel}$ does not penetrate into the ionosphere (where the high density shorts it out) the question arises: how do the ions get into the region of upward $E_{\parallel}$? An attractive scenario was

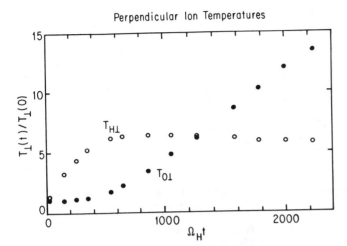

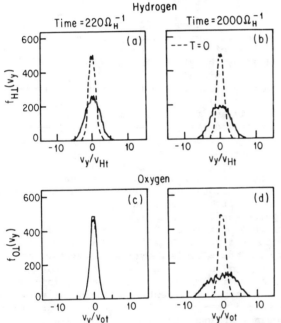

Fig. 7. Results of transverse ion heating in an ionosphere with equal amounts of H^+ and O^+. Top panel: time history of perpendicular temperature ratios. Bottom panels: ion velocity distributions at selected times [from Ashour-Abdalla and Okuda, 1984].

proposed some time ago [Ungstrup et al., 1979; Cornwall, unpublished talk at the 1978 San Francisco AGU Meeting]. First, an ionospheric ion is transversely accelerated by, e.g., EIC or LH waves, then the mirror force (∿ perpendicular energy) raises the ion to the region of large $E_\parallel$ where it is further accelerated. These energized ions can now enter the ring current and be further accelerated by convection and radial diffusion. Evidence for this two-stage process is reported by Klumpar et al. [these proceedings], but many questions remain to be answered (as raised in these proceedings) before we can say that we understand how ionospheric ions are accelerated and transported into the magnetosphere.

Can ions be transported out of the ionosphere in connection with the downward $E_\parallel$ inferred to explain return currents carried by 10-100 eV electrons? At first glance it appears that the answer is no, but in fact it is possible in conjunction with transverse ion heating [Gorney et al., 1985; Cornwall, unpublished]. With downward $E_\parallel$, ions are trapped in a potential well created by the upward mirror force and the electric field. As a result, they are continually returned to the region of transverse heating, where they eventually get enough energy to overcome the electrostatic potential barrier.

Having discussed a few of the important mechanisms for ion heating in the auroral (and cusp) regions, let us give a few examples of their implementation. For EIC return-current TAI, Fig. 7 shows some of the results of Ashour-Abdalla and Okuda [1984]. As these authors note, quasi-linear saturation of EIC growth by formation of an electron plateau is physically inappropriate, because new electrons are continually streaming into the acceleration region; this effect has been modelled in the simulation. Not modelled, however, is the inhomogeneous B-field, which strongly affects possible saturation of the EIC mode as $T_\perp/T_\parallel$ for the ions increases. The reason is that ions with large $T_\perp$ are carried upward by the mirror force and therefore cannot contribute to saturation of the waves responsible for the large $T_\perp$. As Fig. 7 shows, perpendicular temperature ratios (current temperature to initial temperature) reach ∿ 5 for H^+ and ≥ 15 for O^+, and the heating is bulk heating as opposed to a high-energy tail. It is very possible that substantially greater heating would have taken place if ions with large $T_\perp$ were allowed to be lifted out of the acceleration region.

Next we turn to LH waves. There is a very serious problem here: resonance is difficult to achieve. Accurately enough for our purposes, we write the real part of the LH dispersion relation as

$$\omega^2 \simeq \omega_{pi}^2 \left(1 + \frac{k_\parallel^2 M}{k^2 m} \right) \qquad (5.1)$$

where ω_{pi} is the ion plasma frequency, M the ion mass, and m the electron mass. We expect, after considering the imaginary part as well, that $k_\parallel/k \sim (m/M)^{1/2}$, $\omega \sim \omega_{pi}$. We also require for instability that $\omega \simeq k_\parallel u_D$ where u_D is the parallel electron drift velocity, as well as $\omega \simeq kV_{\perp i}$ ($V_{\perp i}$ is the ion perpendicular velocity). In effect, the electron drift velocity projected along $\underline{E}$ must more or less match the ion velocity, but 1 keV electrons hitting a 1 eV H^+ plasma fail by a factor of ∿ 50 to meet this condition. As a result, only high-energy ions can be accelerated unless

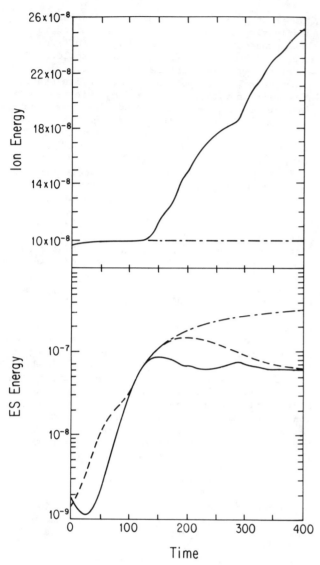

Fig. 8. Modeling results for LH ion heating; top panel is ion energy, and bottom panel is electrostatic energy. Solid lines show results with mode coupling included; other lines are results with only LH waves. The latter show essentially no heating [from Retterer et al., 1985].

there is some other process at work. Retterer et al. [1985] have proposed that the first step in the heating process involves the parametric decay of a LH mode to a lower-frequency LH mode plus a near-zero frequency oscillating two-stream (OTS) mode. This latter is caused by the ponderomotive force at the beat frequency of the parent and daughter LH waves. The growth rate of the OTS instability depends on the strength of the parent LH wave. Retterer et al. [1985] show that the phase velocity of the daughter LH mode can be small enough to resonate with the bulk of the thermal ions, thus heating them. To verify their expectations, these authors ran a one-dimensional spatially homogeneous hybrid simulation (unmagnetized particulate ions, guiding-center electron fluid) which did indeed show creation of high-energy ion tails, with velocities about three times the bulk thermal velocity. They also devised a theoretical model in which the parametric-decay mode coupling could be turned on or off. Results of their calculations with this model are shown in Fig. 8, with solid lines corresponding to mode coupling turned on, and other lines to its being turned off; in the latter case, LH turbulence forms, but there is no ion heating.

These calculations of EIC and LH ion heating are suggestive but inconclusive, for at least two reasons: they do not include the potentially all-important effects of spatial inhomogeneity of B (already mentioned), and they treat the auroral electron beam as given, without inquiring into the efforts of electrostatic turbulence on the mechanism driving auroras. Cornwall and Chiu [1982] have discussed these effects within the context of the adiabatic auroral model, and have shown that saturation of EIC turbulence is very largely determined by the global magnetospheric processes underlying auroral acceleration, and not by local plasma processes of the type usually considered (e.g., quasi-linear, ion heating, trapping, decay instabilities) for a homogeneous plasma. In the adiabatic model it is essential that ions have a more isotropic distribution in pitch angle than the electrons, or in a bi-Maxwellian, that $T_{\perp i}/T_{\parallel i} < T_{\perp e}/T_{\parallel e}$. But in the presence of EIC or LH turbulence, $T_{\perp i}$ begins to increase, which reduces the auroral potential drop and the electron beam velocity, and consequently reduces the level of turbulence. Simple arguments based on this picture show that the hot ion velocity-space diffusion coefficient D is bounded by a value $\sim V_i^3/\ell$, where V_i is a typical energetic-ion velocity and ℓ the length along the auroral field line over which there is turbulence. This sort of scaling is far different from the usual values of D inferred from quasi-linear diffusion or ion resonance broadening; these latter values are much bigger. Fig. 9 shows the result of a calculation of the parallel potential drop in the adiabatic model as modified with specified level of EIC turbulence (parametrized by D). As D is increased, the potential drop decreases, until finally no solution can be found at a value of $D \sim V_i^3/\ell$ (e.g., $V_i \sim 10^8$ cm/sec, $\ell \sim 10^9$ cm). Cornwall and Chiu's calculation relies on several approximations, and is in no sense a substitute for a full-scale simulation. But it serves to draw a moral which we have repeatedly alluded to: the microscopic plasma processes involved with ion-acceleration mechanisms are strongly coupled to global magnetospheric inhomogeneities, which in turn allow for other acceleration processes that do not rely on anomalous resistivity. It appears that this coupling of the microscopic and the macroscopic is one of the most important ingredients in

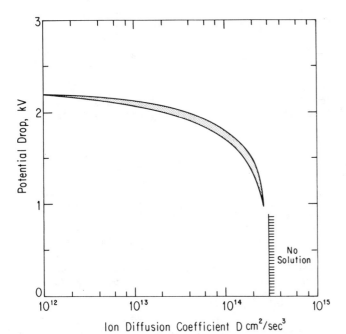

Fig. 9. Limitation of transverse ion heating by raising mirror points. As the transverse velocity diffusion coefficient D goes up, the auroral potential drop goes down, leading to an upper limit on D [from Cornwall and Chiu, 1982].

our search for a fully self-consistent magnetospheric dynamics.

Acknowledgements. This work was supported in part by the National Science Foundation.

References

Anagnostopoulos, G. C., E. T. Sarris, and S. M. Krimgis, Magnetospheric origin of energetic ($E \geq 50$ keV) ions upstream of the bow shock: The October 31, 1977, event, submitted to J. Geophys. Res., March 1985.

Ashour-Abdalla, M. and H. Okuda, Turbulent heating of heavy ions on auroral field lines, J. Geophys. Res., 89, 2235, 1984.

Chang, T. and B. Coppi, Lower hybrid acceleration and ion evolution in the suprauoral region, Geophys. Res. Lett., 8, 1253, 1981.

Chiu, Y. T. and J. M. Cornwall, Electrostatic model of a quiet auroral arc, J. Geophys. Res., 85, 543, 1980.

Chiu, Y. T. and M. Schulz, Self-consistent particle and parallel electrostatic field distributions in the magnetospheric-ionospheric auroral region, J. Geophys. Res., 83, 629, 1978.

Coppi, B., G. Laval, and R. Pellat, Dynamics of the geomagnetic tail, Phys. Rev. Lett., 16, 1207, 1966.

Cornwall, J. M., Cyclotron instabilities and electromagnetic emission in the ULF and VLF frequency range, J. Geophys. Res., 70, 61, 1965.

Cornwall, J. M., Micropulsations and the outer radiation zone, J. Geophys. Res., 71, 2185, 1966.

Cornwall, J. M., Radial diffusion of ionized helium and protons: a probe for magnetospheric dynamics, J. Geophys. Res., 77, 1756, 1972.

Cornwall, J. M., On the role of change exchange in generating unstable waves in the ring current, J. Geophys. Res., 82, 1188, 1977.

Cornwall, J. M. and Y. T. Chiu, Ion distribution effects of turbulence on a kinetic auroral arc model, J. Geophys. Res., 87, 1517, 1982.

Cornwall, J. M., F. V. Coroniti and R. M. Thorne, Turbulent loss of ring current protons, J. Geophys. Res., 75, 4699, 1970.

Cornwall, J. M. and M. Schulz, Physics of heavy ions in the magnetosphere, in solar system plasma physics, V. III, ed. L. J. Lanzerotti, C. F. Kennel, and E. N. Parker (North-Holland, Amsterdam), 1979.

Coroniti, F. V., Space plasma turbulent dissipation: reality or myth, to be published in the proceedings of the International School for Space Simulations, held at Kauai, 1984.

Coroniti, F. V., Explosive tail reconnection: the growth and expansion phases of magnetospheric substorms, J. Geophys. Res., 90, 7427, 1985.

Coroniti, F. V. and A. Eviatar, Magnetic field reconnection in a collisionless plasma, Ap. J. Ser., 33, 189, 1977.

Cowley, S. W. H., Plasma populations in a simple open model magnetosphere, Space Sci. Rev., 26, 217, 1980.

Eastwood, J. W., Consistency of fields and particle motion in the Speiser model of the current sheet, Planet. Space Sci., 20, 1555, 1972.

Francfort, P. and R. Pellat, Magnetic merging in collisionless plasmas, Geophys. Res. Lett., 3, 433, 1976.

Gorney, D., Y. T. Chiu, and D. R. Croley, Jr., Trapping of ion conics by downward parallel electric fields, J. Geophys. Res., 90, 4205, 1985.

Greenstadt, E. W., V. Formisano, C. Goodrich, J. T. Gosling, M. Lee, M. Leroy, M. Mellott, K. Quest, A. E. Robson, P. Rodriguez, J. Scudder, J. Slavin, M. Thomsen, D. Winske, and C. S. Wu, Collisionless shock waves in the solar-terrestrial environment, in Solar-terrestrial Physics Present and Future, NASA Reference Publication 1120, 1984.

Hill, T. W., Magnetic merging in a collisionless plasma, J. Geophys. Res., 80, 4689, 1975.

Huba, J. D., N. T. Gladd, and J. F. Drake, On the role of the lower hybrid drift instability in substorm dynamics, J. Geophys. Res., 86, 5881, 1981.

Kennel, C. F., J. P. Edmiston, and T. Hada, A quarter century of collisionless shock research, in Proceedings of the AGU Chapman Conference on Collisionless Shocks in the Heliosphere, ed. R. Stone and B. Tsurutani, AGU Press, Washington, 1985.

Kindel, J. M. and C. F. Kennel, Topside current instabilities, J. Geophys. Res., 76, 3055, 1971.

Leroy, M. M., D. Winske, C. C. Goodrich, C. S. Wu, and K. Papadopoulos, The structure of perpendicular bow shocks, J. Geophys. Res., 87, 5081, 1982.

Lockwood, M., J. H. Waite, Jr., T. E. Moore, J. F. E. Johnson, and C. R. Chappell, A new source of suprathermal O^+ ions near the dayside polar cap boundary, J. Geophys. Res., 90, 4099, 1985.

Lyons, L. R., Generation of large-scale regions of auroral currents, electric potentials, and precipitation by the divergence of the convection electric field, J. Geophys. Res., 85, 17, 1980.

Lyons, L. R., Discrete aurora as the direct result of an inferred high-altitude generating potential distribution, J. Geophys. Res., 86, 1, 1981.

Lyons, L. R. and T. W. Speiser, Evidence for current-sheet acceleration in the geomagnetic tail, J. Geophys. Res., 87, 2276, 1982.

Lyons, L. R. and T. W. Speiser, Ohm's law for a current sheet, J. Geophys. Res., 90, 8543, 1985.

Obayashi, T., Hydromagnetic whistlers, J. Geophys. Res., 70, 1069, 1965.

Omura, Y., M. Ashour-Abdalla, K. Quest, and R. Gendrin, Heating of thermal helium in the equatorial magnetosphere: a simulation study, J. Geophys. Res., 90, 8281, 1985.

Parker, E. N., A quasi-linear model of plasma shock structure in a longitudinal magnetic field, J. Nucl. Energy, C2, 146, 1961.

Pesses, M. E., R. B. Decker, and T. P. Armstrong, The acceleration of charged particles in interplanetary shock waves, Space Sci. Rev., 32, 185, 1982.

Petschek, H. E., Magnetic field annihilation, in Proceedings of the AAS-NASA Symposium on the Physics of Solar Flares, NASA SP-50, p. 425, 1964.

Retterer, J. M., T. Chang, and J. R. Jasperse, Ion acceleration in the supra-auroral region: a Monte Carlo model, Geophys. Res. Lett., 10, 583, 1983.

Rich, F. J., V. M. Vasyliunas, and R. M. Wolf, On the balance of stresses in the plasma sheet, J. Geophys. Res., 77, 4670, 1972.

Roth, J. and M. K. Hudson, Lower hybrid heating of ionospheric ions due to ion ring distribution in the cusp, J. Geophys. Res., 90, 4191, 1985.

Roux, A., S. Perrat, J. L. Rauch, C. de Villedary, G. Kremser, A. Korth, and D. T. Young, Wave-particle interactions near Ω_{He^+} observed on board GEOS 1 and 2. Generation of ion cyclotron waves and heating of He^+ ions, J. Geophys. Res., 87, 8174, 1982a.

Roux, A., N. Cornilleau-Wehrlin, and J. L. Rauch, Acceleration of thermal electrons by ICW's propagating in a multicomponent magnetospheric plasma, J. Geophys. Res., 89, 2267, 1982b.

Russell, C. T. and M. M. Hoppe, Upstream waves and particles, Space Sci. Rev., 34, 155, 1983.

Scholer, M., D. Hovestadt, F. M. Ipavich, and G. Gloeckler, Upstream energetic ions and electrons: bow-shock associated or magnetospheric origin?, J. Geophys. Res., 86, 9040, 1981.

Schulz, M. and L. J. Lanzerotti, Particle Diffusion in the Radiation Belts (Springer, Heidelberg), 1974.

Speiser, T. W., Particle trajectories in model current sheets, 1, analytical solutions, J. Geophys. Res., 70, 4219, 1965.

Tanaka, M., Simulations of heavy ion heating by electromagnetic ion cyclotron waves driven by proton temperature anisotropies, J. Geophys. Res., 90, 6459, 1985.

Toptyghin, J. N., Acceleration of particles by shocks in a cosmic plasma, Space Sci. Rev., 26, 157, 1980.

Tsurutani, B. T. and R. P. Lin, Acceleration of > 47 keV ions and > 2 keV electrons by interplanetary shocks at 1 AU, J. Geophys. Res., 90, 1, 1985.

Ungstrup, E., D. M. Klumpar, and W. J. Heikkila, Heating of ions to suprathermal energies in the topside ionosphere by electrostatic ion cyclotron waves, J. Geophys. Res., 84, 4289, 1979.

Vasyliunas, V. M., Theoretical models of magnetic field line merging 1, Rev. Geophys. Space Phys., 13, 303, 1975.

Wagner, J. S., J. R. Kan, and S.-I. Akasofu, Particle dynamics in the plasma sheet, J. Geophys. Res., 84, 891, 1979.

Young, D. T., S. Perraut, A. Roux, C. de Villedary, R. Gendrin, A. Korth, G. Kremser, and D. Jones, Wave-particle interactions near Ω_{He^+} observed on GEOS 1 and 2 1. Propagation of ion cyclotron waves in He^+-rich plasma, J. Geophys. Res., 86, 6755, 1981.

EXPERIMENTAL ASPECTS OF ION ACCELERATION IN THE EARTH'S MAGNETOSPHERE

David T. Young

Space Plasma Physics Group, Los Alamos National Laboratory, Los Alamos, NM 87545

Abstract. This paper discusses the experimental evidence for ion acceleration in the Earth's magnetosphere. Particular emphasis is placed on compositional aspects of experimental knowledge, particularly on results obtained by mass-discriminating instruments over the past decade. All magnetospheric ion populations are addressed, including the plasmasphere, the ionospheric source regions, the tail plasma sheet, ring current and radiation belt. Some steps are taken towards an empirical synthesis of ion acceleration processes and their dependence on the nature of the ion populations upon which they operate. In keeping with the tutorial nature of this lecture, a brief introductory chapter is included on the history and nature of experimental approaches to measurements of magnetospheric ion populations.

1. Introduction

Ion acceleration processes are central to our understanding of the origins of all major particle populations within the Earth's magnetosphere. Although this fact has long been appreciated, only recently have we begun to become aware of the true complexity of these processes. This awareness has grown chiefly from experimental advances made over the past 10-15 years. In particular, it is my intent in this review to highlight areas where ion composition experiments have opened new doors of perception on magnetospheric phenomena that come loosely under the heading of ion acceleration.

By now many authors have emphasized that our improved knowledge of magnetospheric composition has led to an overthrow of earlier concepts which centered on the solar wind as the origin of most hot plasma and trapped particles within the magnetosphere. Figure 1 shows the extent to which the solar and terrestrial sources of magnetospheric plasma are interwoven. Qualitatively at least, one plasma population can be thought of as feeding another in a cascade of particle and energy flow through the magnetosphere. The boxes in Figure 1, which represent populations, are arranged to emphasize the cascade-like relationship of populations and processes. From this point of view, instead of the old paradigm of a single ultimate source for ions (the solar wind), we now find that the magnetosphere itself presents a number of intermediate source regions. Each of these usually exhibits a characteristic composition of its own that results from a complex interplay of region accessibility; ion acceleration, transport and loss processes; and ultimately the intrinsic properties of the ions themselves, such as mass to charge ratio or collision cross section. For example, the ring current does not see the ionosphere or the solar wind as a direct source of ions. Ions from either the ionosphere or solar wind must first run a gauntlet of processes before entering the ring current proper and then, perhaps, passing on to the radiation belt (Figure 1). To understand the origin of the ring current as a particle population, it is therefore necessary to come to grips with the intervening resevoirs and processes through which ring current ions pass.

While ion acceleration is the main topic of this review, it is important to keep in sight the overall dynamics of the magnetosphere. This involves the flow, not just of mass (represented by ions), but also of momentum and energy (cf. Hill, 1979). For many years it was thought that all three quantities, mass, momentum and energy, flowed almost exclusively from the solar wind to the terrestrial ionosphere via the magnetosphere. We now know that a nonnegligible mass flow takes place from the ionosphere back into the magnetosphere. Moreover, we are beginning to find that this transport may act as a kind of feedback on the magnetosphere and its processing of solar wind energy input. (Electrodynamic feedback is of course an old story in the magnetosphere.) Specific examples of processes in which ionospheric particle feedback plays a role include injection of oxygen ions into the tail plasma sheet (Section 5) and ring current (Section 6) and the effects of cold helium ions in mediating wave-particle interactions. Others may come to light in the future, perhaps in this volume.

This review is organized by dividing magnetospheric particles into five populations:

- plasmasphere (~ 1 eV)
- exo-plasmasphere (1 eV – 1 keV)
- plasma sheet (1-10 keV)
- ring current (10-300 keV)
- trapped radiation belts (> 300 keV)

The reader of course realizes that these distinctions among populations are merely conveniences. The populations overlap both in space and energy: one man's plasma sheet is often another's ring current. Their usefulness is that "populations" convey to most of us a definite picture of a range of particle energies, densities, trajectories, etc. that are under discussion.

One caveat is central to any discussion of magnetospheric ion measurements, and certainly to this review: magnetospheric ion abundances depend on solar cycle. This is well-established at $1 \sim 20$ keV energies (Section 6) and is almost certainly the case at higher energies as well. Figure 2 therefore depicts the phasing of all recent magnetospheric missions with respect to the present solar cycle. Note that AMPTE, which is the first spacecraft to probe the bulk of the ion ring current, does so near solar minimum. Also note that during most of the current solar maximum there was no satellite operating at polar latitudes.

If there is a single theme to this review, it is that composition plays a key role in ion acceleration processes and hence in magnetospheric dynamics. It is by now generally accepted that "heavy ions" can no

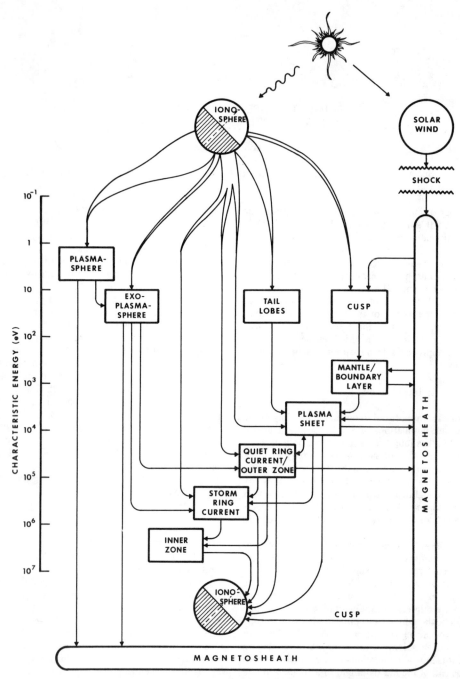

Fig. 1. Schematic relationship of magnetospheric particle populations. The ionosphere and magnetosheath are shown as both sources (top) and sinks (bottom) of particles. Approximate latitude of ionospheric sources and sinks are shown schematically (e.g. flow into the plasmasphere is from near-equatorial regions, flow into the tail lobes is polar in origin). From this diagram it can be seen that virtually all latitudes feed magnetospheric particle populations, and also act as particle sinks. The escape of energetic ions back upstream through the shock is not shown. Characteristic ion energy increases from top to bottom in the figure (scale at left) as does, roughly speaking, the lifetime of trapped populations. (Figure updated from Young, 1983b).

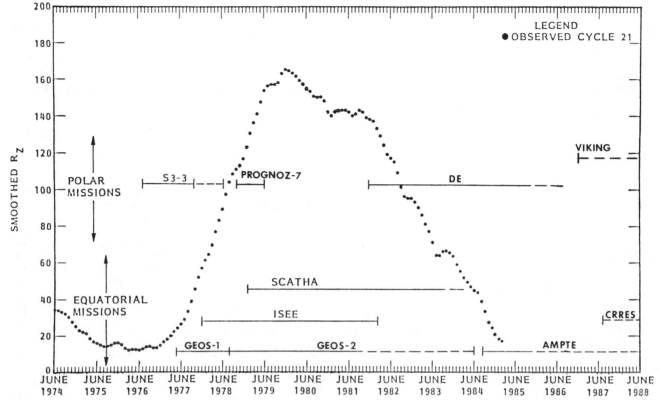

Fig. 2. Operational periods of magnetospheric spacecraft carrying ion composition experiments shown relative to the present solar cycle (using 13-month smoothed sunspot numbers, R_z). Dashed lines indicate extended (usually intermittent) or expected extension to present missions. Note that no polar spacecraft was operational for $\simeq 2.3$ years around the solar maximum period and that AMPTE/CCE ring current measurements are made near solar minimum. (Adapted from Shelley, 1985.)

longer be considered "tracer" particles. The occasion of this Chapman Conference provides an excellent opportunity to step back from our earlier pursuits and look afresh at a magnetosphere permeated by multi species plasmas—and perhaps to reflect that the magnetospheres of Jupiter and other astrophysical objects are also not likely to be made up only of "protons." Magnetospheric acceleration processes which tend to select or enrich a particular ion species can also be expected to operate in astrophysical environments with as yet unforseen consequences.

In the sections that follow I give a brief instrumentation summary followed by one section devoted to each of the five populations listed above, with one final section for the ionospheric source region. The sections are not airtight, there will be some overlap and inevitably some confusion about which phenomena belong where. But to some degree that confusion simply reflects our present, incomplete knowledge of how the magnetosphere works.

2. Instrumentation

There is a need in the growing field of ion composition studies for a thorough and critical evaluation of experimental techniques. It is beyond the scope of this review to do more than scratch the surface of this neglected but vital topic. An early review by Vasyliunas (1971) covers theoretical aspects of non-mass discriminating satellite-borne instruments. Wilken et al. (1982) and Balsiger (1982) have given useful overviews of mass analysis methods including magnetic analyzers, solid state detectors, and time-of-flight techniques. Energy-analyzing laboratory instrumentation has been discussed more rigorously by Steckelmacher (1973). What is still lacking, however, is a thorough examination of the critical tradeoffs among instrument parameters (e.g., energy range and resolution, angular range and resolution, mass range and resolution, geometric factor, signal to noise ratio, etc.) that largely determine performance. We have no good figures of merit whereby we may judge competing designs and, in comparison to photon optics, particle optics as applied to space plasma experiments exhibits a lamentable lack of rigor. A very pressing and more general issue to be considered is whether the field should invest precious resources in facility class instruments, following, for example, the lead of high energy physics and experimental astronomy. Finally, in order to sensibly discuss present day experimental results as well as future prospects, the field as a whole needs to be better informed. Such a review of instrumental techniques would at least remove some of the barriers to that discussion.

Table 1 provides a chronological sketch of instrument development in ion composition and the complementary area of ULF wave measurements. Particle instrumentation is divided into three energy regimes because different techniques tend to be applied in each. The table includes a rough classification by instrument type as well as detectable mass to ionic charge ratio (M/Q) or nuclear charge (Z).

Values of M/Q or Z included in the table are limited to those that can be resolved as reported in the literature. The appropriate comparable data for waves is, roughly speaking, the frequency threshold. References are to instrumentation papers where available, or to "first results" papers where some abbreviated description of the experimental technique can be found. Of course it would be desirable to expand this table to include ranges and resolutions for the critical tradeoff parameters mentioned above, but that far exceeds the scope of this review. Rather the reader is asked to peruse Table 1 and note the accelerating development of composition instrumentation that has taken place over the last two decades. Also note that several instrument pedigrees are evident in Table 1 (e.g., the OGO-5 magnetic spectrometer leading to RIMS on DE-1; GEOS-1 to CCE; 1969-25B to SCATHA; and IMP-7 to CCE). This shows that careful long-term instrument development by a number of specialized groups was necessary to produce the current exciting harvest of results that are the subject of this conference.

One rather subtle point that has had some impact on the course of magnetospheric physics was the early availability of composition measurements at very low ($\sim$ eV) and very high ($\sim$ MeV) energies. Both types of instrumentation were reasonably well developed in other areas of physics e.g. solid state detectors, gas proportional counters, and such devices have their origins in low- to medium-energy nuclear physics. The energy range $\sim$ 1 keV to $\sim$ 100 keV has presented the greatest technical problems for composition measurements. These were solved only with technological advances such as rare earth magnetic materials, development of space-qualified very high voltage systems ($\sim$ 30-50 kV), and very thin-walled or thin surface barrier energetic particle detectors. Still newer developments such as position-sensitive particle detection and flight-qualified 16-bit microprocessors should result in even greater advances in instrument performance. Finally, while not dealt with in this review, computer modeling of magnetospheric phenomena should be seen as another facet of experimental space physics. Its ties to the instrumentation community need to be strengthened, particularly in the area of simulations of phenomena the hardware experimentalist might expect to measure.

3. The Plasmasphere

The plasmasphere deserves attention in a review of ion acceleration processes for three reasons: first as a source of ions which may be accelerated in situ within the high altitude magnetosphere; secondly, as the background plasma that determines wave propagation and polarization characteristics critical to wave-particle interactions; and thirdly, as the medium in which more energetic particles are embedded and to which they ultimately lose energy and momentum e.g. through Coulomb collisions or pitch-angle scattering. The neutral counterpart of the plasmasphere, namely the hydrogen geocorona, is also an important loss medium (cf. Tinsley, 1976) whose characteristics, in particular long term variability, are in need of further study (Moore et al., 1985b).

During the 1960's, limited composition measurements of the near-equatorial plasmasphere by the OGO-1,-3, and -5 mass spectrometers (Table 1) all found He^+/H^+ to be < 0.01 (we adopt the convention that He^+/H^+ stands for the density ratio of He^+ to H^+) with negligible amounts of O^+ (Taylor et al., 1965; Harris et al., 1970). This assessment of the plasmasphere changed with the advent of the GEOS-1, ISEE-1 and DE-1 mass spectrometers. The latter all yielded $He^+/H^+ \sim 0.1$ as a more typical value (Young et al., 1977; Waite et al., 1984; Horwitz et al., 1984) with excursions up to 50% He^+ by number. The reason for the discrepancy between OGO and later measurements need not detain us, but it now seems well established that the newer results are indeed correct (cf. Horwitz, 1982). No sooner had the ion mass spectrometers begun to turn up large quantities of He^+, than the ULF wave experiments discovered strong emissions obviously under the control of this species (Young et al., 1981; see below). The O^+ ion is less abundant than He^+ in the

TABLE 1. Magnetospheric Spacecraft Carrying Ion Composition and ULF Wave Instrumentation

A. Cold ions (< 100 eV)

Spacecraft	Instrument Type[2]	Dates	M/Q	Reference
OGO-1	Bennett RF	1964	1,4	Taylor et al., 1965
OGO-3	Bennett RF	1966	1,4	Taylor et al., 1965
OGO-5	Magnet	1968	1,4,16	Harris & Sharp, 1969
GEOS-1	RPA+ESA+CME	1977-1979	1,2,4,8,16	Geiss et al., 1978
ISEE-1	RPA+ESA+CME	1977-	1,2,4,8,16	Baugher et al., 1980
SCATHA	RPA+magnet	1979	1,4,16	Reasoner et al., 1982
DE-1	RPA+magnet	1981-	1,2,4,8,14,16 28,30,32	Chappell et al., 1981 Craven et al., 1985

B. Hot ions (100 eV - 30 keV)

Spacecraft	Instrument Type[2]	Dates	M/Q	Reference
1969-25B	SWF+ESA	1969	1,4,16	Sharp et al., 1974
1971-92A	SWF+ESA	1971	1,2,4,16	Shelley et al., 1972
S3-3	SWF+ESA	1976-1978	1,2,4,16	Sharp et al., 1977
GEOS-1	ESA+CME	1977-1979	1,2,4,8,16	Balsiger et al., 1976
ISEE-1	ESA+CME	1977-	1,2,4,8,16	Shelley et al., 1978
GEOS-2	ESA+CME	1978-1985	1,2,4,8,16	Balsiger et al., 1976
SCATHA	SWF+ESA	1978-	1,2,4,8,16	Johnson et al., 1982
PROGNOZ 7	SWF+ESA	1978-1979	1,2,4,16	Lundin et al., 1979
DE-1	CME+ESA	1981-	1,2,4,8,16	Shelley et al., 1981
CCE	CME+ESA	1984	1,2,4,8,16	Shelley et al., 1985a

TABLE 1. Continued

C. Energetic Ions (> 30 keV)

Spacecraft	Instrument Type[2]	Dates	Z	Reference
Injun 4	SSD	1967	1,2	Krimigis & Van Allen, 1967
Injun 5	SSD	1968-1969	1,2,6-8	Van Allen et al., 1970
OV1-19	SSD	1969-1970	1,2	Blake & Paulikas, 1972
Exp. 45	SSD	1971-1974	1,2,≥4,≥9	Fritz & Williams, 1973
IMP-7	ESA+SSD	1972-1973	1,2,6-8	Fan et al., 1971
IMP-8	ESA+SSD	1973-1976	1,2,6-8	Tums et al., 1974
ATS-6	SSD	1974	1,2,6-8,>8	Fritz & Wilken, 1976
S3-2	PC+SSD	1975-1976	1,2,>4,>16	Scholer et al., 1979
ISEE-1, ISEE-3	ESA+PC+SSD	1977-	1,2,6,7,8, 10-14,≥16	Hovestadt et al., 1978
ISEE-1	TOF+SSD	1977-	1,2,6-8,>8	Williams et al., 1978
SCATHA	SSD	1978-	1,2,6-10,≥12	Blake & Fennell, 1981
CCE	ESA+TOF+SSD	1984-	1,2,6,7,8	Gloeckler et al., 1985b
CCE	TOF+SSD	1984-	1,2,6-8	McEntire et al., 1985

D. Wave analyzers (ULF range)[1]

Spacecraft	Instrument Type[2]	Frequency	Reference
Hawkeye 1	Search coil	>1 Hz	Kintner & Gurnett, 1977
Exp. 45	Search coil	>1 Hz	Taylor et al., 1975
GEOS-1,2	Search coil	>.05 Hz	Perraut et al., 1978
	Electric dipole	>.05 Hz	Perraut et al., 1978
ATS-6	Fluxgate	>.05 Hz	Mauk & McPherron, 1980
S3-3	Electric dipole	>30 Hz	Kintner et al., 1978
DE-1	Search coil	>1 Hz	Shawhan et al., 1981
	Electric dipole	>1 Hz	Shawhan et al., 1981
ISEE-1,2	Search coil	>5.6 Hz	Gurnett et al., 1978
	Electric dipole	>5.6 Hz	Gurnett et al., 1978
ISEE-1,2	Fluxgate	>16 Hz	Russell, 1978

[1] For a comparison of wave detector sensitivities see Jones (1978)
[2] RPA: Retarding Potential Analyzer, ESA: Electrostatic Analyzer, CME: Curved-plate Magnetic-Electric Filter, SSD: Solid State Detector, TOF: Time-of-flight; PC: Proportional counter.

plasmasphere. Typical values of O^+/H^+ are ~ 0.001-0.01 although at times the percentage of O^+ may be much higher in the outer plasmasphere (Horwitz et al., 1984). Rarer ion species are also present with He^{2+}, D^+, O^{2+}, N^{2+} and N^+ having been reported at abundances < 0.001. Molecular ions (N_2^+, NO^+, O_2^+) have been seen out to 3 R_E over the polar cap during an intense geomagnetic storm (Craven et al., 1985) but have not yet been found within the plasmasphere.

Ion temperatures within the inner plasmasphere are typically ~ 0.3 eV, exhibiting a slight radial gradient that leads to temperatures ~ 1 eV near the plasmapause (Comfort et al., 1985). Different ion species are in thermal equilibrium (Comfort et al., 1985), as are electrons and ions (Decreau et al., 1978).

In terms of identifying the sources of the more energetic ring current and radiation belt ions, the most important plasmaspheric feature is its compositional ordering $H^+ > He^+ > O^+$ by density. This is in distinction to the composition of auroral beams and upwelling polar cap ions (Section 8), in which $H^+ \gtrsim O^+ > He^+$ is a clearly established feature. Although such an ordering is a useful guideline for identifying the source of trapped particles at energies up to a few tens of keV, it should be kept in mind that the solar wind elemental ordering is also H > He > O and is therefore qualitatively similar to that of the plasmasphere. This is an important consideration at higher energies (> 0.1 MeV) where solid state detectors are the primary measurement tool (used alone they distinguish only nuclear, not ionic, charge). At very high energies, trapped ions gain or lose electrons through collisions and thereby work toward a charge state equilibrium (cf. Spjeldvik and Fritz, 1978). Thus initial charge states alone do not necessarily serve as viable markers of ion origins, and the plasmaspheric elemental ordering could in principle be confused with that of the solar wind. Fortunately the present sophistication of energetic ion instrumentation offers a second test of ion origin, namely the elemental C/O ratio. We return to this in Section 7.

The outer plasmasphere and the region adjacent to it (the so-called "trough") provide fertile ground for the production of plasma waves through a variety of mechanisms (cf. Shawhan, 1979). Of particular interest here are ion cyclotron waves (ICWs) just above or below the local gyrofrequencies of the two dominant cold ion species: H^+ and He^+. It has long been known that the presence of a second (or more) ion species introduces new cutoffs and resonances into the plasma wave dispersion relation (Smith and Brice, 1964).

Both propagation and polarization characteristics of ion cyclotron waves are therefore altered by the presence of He^+ (or O^+) mixed in with the cold hydrogenic background plasma found within and outside the plasmasphere. Evidence for this in the magnetosphere is found in ICWs observed in the equatorial region at $L \sim 4 - 7$ by GEOS and ATS-6. These waves display propagation and polarization characteristics clearly controlled by He^+ and O^+ (Mauk and McPherron, 1980; Young et al., 1981; Fraser and McPherron, 1982).

The ICWs observed on GEOS and ATS are thought to be generated near the equator by the pitch angle anisotropy ($T_\perp > T_\parallel$) of hot (20 $\sim$ 50 keV) protons. Ray tracing simulations by Rauch and Roux (1982) show that ICWs are reflected within $\sim 20°$ of the equatorial plane and bounce back and forth between wave "mirror points," gaining amplification near the equator on each bounce. Amplification comes about because the presence of $\sim 5\%$ He^+ enhances the ICW growth rate (Roux et al., 1982) while wave reflection occurs when the local ICW frequency matches the bi-ion hybrid frequency which is controlled, again, by the presence of cold He^+. The ICWs in turn act back on the cold He^+, and through a mechanism not yet clearly established (but possibly quasi-linear diffusion, Gendrin and Roux, 1980; or non-linear effects, Mauk et al., 1981; Roux et al., 1982), the waves strongly heat the He^+ ions transverse to the local magnetic field direction, perhaps contributing to the genesis of the exo-plasmaspheric population (Section 4).

Aside from the GEOS-1 observations directly showing He^+ heating up to ~ 100 eV, there is good circumstantial evidence for wave heating of ions based on the prevalence of so-called "pancake" or trapped pitch angle distributions (i.e. peaked at 90°) of both H^+ and He^+ (Horwitz et al., 1982; Olsen, 1981) and the resonant absorption of ICWs at both He^+ (Mauk et al., 1981) and O^+ (Fraser and McPherron, 1982) equatorial gyrofrequencies. Horwitz et al. have also reported that trapped distributions occur preferentially in the dayside magnetosphere, which is coincidentally the locus of ICW events and heated He^+ (Roux et al., 1982). These experimental results have instigated considerable theoretical work on gyroresonant wave-particle interactions and the interested reader is referred to reviews by Gendrin (1983) and papers in the present volume, as well as to theoretical work by Gomberoff and Neira (1983) and Kozyra et al. (1984). For a different view on acceleration and trapping of suprathermal ions see Curtis (1985). As Gendrin has pointed out, wave-particle interactions mediated by cold He^+ represent a kind of frictional interchange of energy between hot ions of one species (H^+) and cold ions of another (He^+) with heating effects extending as well to cold electrons (Roux et al., 1984).

In addition to wave-particle interactions, heating of the outer reaches of the plasmasphere may be a consequence of magnetic flux tube filling from the ionosphere (Banks et al., 1971; Schulz and Koons, 1972). The filling process is evident from flowing distributions of H^+ and He^+ (Sojka et al., 1983) and from the steady increase in cold isotropic plasma observed during periods of quieting magnetic activity (Horwitz et al., 1981, 1984). The outer plasmasphere is nearly always reported to be hotter (1-3 eV) than the inner core of the plasmasphere (0.5 eV) in these studies, a feature noted in earlier work (Serbu and Maier, 1970; Bezrukikh and Gringauz, 1976) and recently confirmed in detail by Comfort et al. (1985) with DE-1.

Heating of the outer plasmasphere, which may be seen as the first step in the acceleration of these ions at high altitude, has been addressed in several recent studies. Sojka et al. (1983) found that as DE-1 moved to higher L-shells, field-aligned flows of H^+ and He^+ ions changed from counterstreaming flows coming from both hemispheres, to unidirectional flows coming only from the nearest hemisphere. These observations are qualitatively in keeping with the interaction expected of counterstreaming polar wind ion beams near the equator. Such beams would be initially unstable to ion acoustic waves until sufficient density had built up at the equator (Schulz and Koons, 1972). Singh and Schunk (1983) find that suprathermal forerunner ions (~ 2.5 eV) may be initially unstable and start the process of scattering and thermalization. A colliding beam interaction has two important consequences: ion flow energy is converted to thermal energy, possibly accounting for higher temperatures in the outer plasmasphere; and the plasmasphere fills from the top (equator) down. Comfort et al. (1985) measured positive temperature gradients of ~ 0.5 K/km along flux tubes in the inner plasmasphere ($L < 4$) during magnetically quiet times. They felt this was also consistent with plasmasphere filling. Top-down filling of flux tubes with cold background plasma may also be a consideration in the wave propagation studies mentioned above since it implies that density (and composition) can vary along magnetic field lines as well as in the radial direction.

4. The Exo-Plasmasphere

As I have suggested elsewhere (Young 1983a, b) there is evidence for a suprathermal population roughly adjacent to the plasmapause, extending outward perhaps 1-2 R_E. While it could be argued that this is nothing more than the low-energy tail of the ring current or plasma sheet, the composition of this "exo-plasmasphere" is distinctly terrestrial in nature, but with a strong plasmaspheric flavor due to large He^+/H^+ and O^{++}/O^+ ratios. Typical ion energies run from ~ 10 eV (Gurnett and Frank, 1974; Horwitz et al., 1981; Sojka et al., 1983; Sojka and Wrenn, 1985) up to as high as ~ 1 keV (Balsiger et al., 1983). Pitch angle distributions range from trapped through conical (defined as having flux maxima at pitch angles $0° < \alpha < 90°$ or $90° < \alpha < 180°$) to field-aligned (Singh et al., 1982). Among the more consistent features of this plasma is that He^+, more so than H^+, tends toward trapped distributions suggestive of acceleration perpendicular to the local magnetic field (Horwitz et al., 1981; Roux et al., 1982; Sojka et al., 1983). Olsen (1981) has found a strongly trapped (within a few degrees of the equator) population that may be predominantly plasmaspheric H^+ (Quinn and Johnson, 1982). The O^+ exo-plasmaspheric component tends to be field-aligned and generally more energetic than the He^+, and is likely to be associated with direct O^+ injection from the auroral zones and subsequent trapping, rather than a purely plasmaspheric source (see Section 8, also Young, 1979; Hultqvist, 1983; Kaye et al., 1981a,b). The very interesting recent results from AMPTE/CCE are consistent with an exo-plasmaspheric source for He^+ and O^{2+} fluxes observed at the inner edge of the storm-time ring current (Shelley et al., 1985b).

A primary candidate for ion acceleration in the outer plasmasphere is wave-particle interactions (WPI) driven by the anisotropy of energetic protons (Section 3) or energetic heavy ions (Kozyra et al., 1984). In addition to heating He^+ by the former process, indirect evidence of WPI near the O^+ gyrofrequency also exists (Fraser and McPherron, 1982; Fraser, 1982) although no observations of locally heated O^+ ions have been reported. Perpendicular heating of suprathermal (not cold) H^+ by ICWs above the hydrogen gyrofrequency may also take place (Perraut et al., 1982). The ICWs in this case are generated by unstable proton ring distributions which give up energy to the less energetic (0.1-5 keV) part of the proton spectrum. In all instances of wave-particle interactions cited above, the source of proton pitch angle anisotropy is gradient and curvature drift of protons from near local midnight around to local noon. Thus the free energy for cold ion acceleration in the equatorial region is ultimately supplied by substorm particle injection or, in some cases, simply $E \times B$ convection from the tail plasma sheet.

In this section I have tried to make a case for high-altitude, near-equatorial acceleration of cold plasmaspheric ions. It seems important to make this distinction because the composition of this popula-

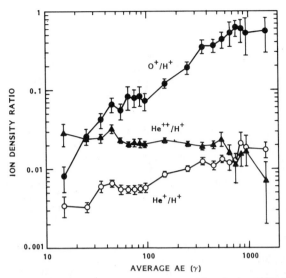

Fig. 3. Average ISEE-1 plasma sheet composition during 1978-1979 from 10-23 R_E in the energy range 0.1-16 keV/e (from Lennartsson and Shelley, 1986).

tion, which apparently contributes to formation of the innermost part of the ring current (Section 6), is different from that of directly injected auroral ions. In the exo-plasmaspheric population the presence of He$^+$ stems primarily from source composition rather than from the charge-exchange loss process which tends to increase the relative He$^+$ abundance at ring current energies (Section 6). Indeed, the two effects will augment one another, perhaps accounting for large enhancements of He$^+$ seen at the inner edge of the ring current (Balsiger et al., 1980; Lundin et al., 1980; Lennartsson et al., 1981).

5. The Plasma Sheet

No other region of the magnetosphere elicits as much controversy among experimentalists as does the magnetotail. The reasons for this seem to be its vastness, and its dynamic, time-varying nature, coupled with a lack of multiple spacecraft observations. All of these features make certain classes of measurements by a single spacecraft problematic at best. For example, controversies over spatial vs. temporal effects (e.g. plasma sheet expansion vs. flapping to name but one) are legion. And yet the magnetotail and plasma sheet are central to magnetospheric dynamics as reservoirs for the storage of nearly all energy (in the form of magnetic flux and hot plasma) and mass (ions) consumed during the substorm process.

Build-up of magnetic flux in the tail followed by its release through reconnection is the generally agreed upon paradigm for substorms (cf. Russell and McPherron, 1973; Baker et al., 1979). Through this process ion acceleration to energies as high as $\sim$ 1 MeV occurs. While it is of interest to know how these great energies come about, particularly since the typical cross-tail electric potential is only $\sim$ 50 kV, of greater importance to magnetospheric dynamics is the acceleration and earthward motion of the bulk of the plasma sheet particles which in turn feed the ring current (cf. Figure 1). Plasma sheet ion acceleration is manifested through plasma flows directed both earthward and tailward at velocities up to hundreds of km/s during substorms (Hones et al., 1974; De Coster and Frank, 1979); as plasmoids ejected from the tail, carrying with them substantial amounts of particle and field energy (cf. Hones, 1984); and as energetic ions (1 keV $\sim$ 1 MeV) accelerated along the boundary of and within the plasma sheet (cf. Scholer, 1984). All of these processes are touched on in this section, again with emphasis placed on composition aspects.

It is perhaps useful to point out that the ISEE-1 ion mass spectrometer (Table 1) is still the only mass-resolving instrument measuring bulk plasma properties to penetrate the magnetotail region beyond $\sim$ 7 R_E. ISEE-1 also carried non-mass-discriminating plasma analyzers (Eastman et al., 1984) and energetic particle instruments capable of resolving O$^+$ (Hovestadt et al., 1978). Energetic particle and plasma instruments on IMP-7 and -8 have measured characteristics of streaming magnetotail ions out to 45 R_E. Still farther down the tail (80-220 R_E), ISEE-3 has revealed the apparent solar wind character of energetic (6-150 keV/amu) ions found in plasmoids (Gloeckler et al., 1984).

In terms of particle storage times, the plasma sheet must be considered to be a temporary reservoir of magnetospheric plasma, whereas the ring current and radiation belt contain progressively longer-lived populations. Before the present era of heavy ion measurements, the plasma sheet was thought to be fed by the solar wind via the high- and low-latitude boundary layers, or by direct diffusion from the magnetosheath into the flanks of the magnetotail (cf. Cowley, 1980). Hill (1974) noted that a solar wind particle capture rate of only $\sim$ 0.1% was needed to supply the quiet plasma sheet. By the same token, the polar wind flux ($< 10^8$/cm^{2}s) could only supply $\sim$ 50% of the estimated quiet time loss rate of 5×10^{25}/s. Hill discussed other evidence which, at that time, suggested a solar wind origin for the plasma sheet. As he pointed out, it made for a tidier theory if both mass and momentum transfer requirements for the plasma sheet could be met by a single source, namely the solar wind. The most recent ISEE-1 composition measurements in the tail (reviewed below) suggest that the solar wind is indeed the primary ion source during magnetically quiet times. During disturbed times, however, the plasma sheet seems to fill with a steady stream of ionospheric material (Figure 3).

Compilations of significant amounts of ISEE-1 plasma sheet composition data have been presented by Sharp et al. (1981, 1982) and Lennartsson and Shelley (1986). With reference to Figure 3, their basic conclusions are that in magnetically quiet conditions, ions in the plasma sheet out to 23 R_E are $\sim$ 98% of solar wind origin and have been accelerated to equal energy per nucleon with H$^+$ receiving slightly more energy than He^{2+}. The average He^{2+}/H$^+$ density ratio is 0.03, somewhat less than the typical (but not simultaneously measured) solar wind value of 0.04-0.05. As magnetic activity increases (Figure 3) the O$^+$/H$^+$ density ratio increases until, as Sharp et al. infer, plasma sheet ions are $\sim$ 50% terrestrial in origin. At the same time, the amount of solar wind He^{2+} actually decreases inside 23 R_E (Lennartsson and Shelley, 1986). With increased magnetic activity, the average solar wind ion energy in the plasma sheet increases a factor of two, but in such a way that H$^+$ gets relatively more energy than He^{2+} and the two species tend toward equal energy per charge rather than equal energy per nucleon. Interestingly, Lennartsson and Shelley report that at very low levels of activity, the mean ion energy approaches solar wind values (e.g. 1 keV for H$^+$ and 4 keV for He^{2+}). Their study, covering all ISEE-1 data in 1978 and 1979, shows a definite solar cycle correlation as the average O$^+$/H$^+$ ratio increased by a factor of 3-4 with increased solar activity at all levels of magnetic activity.

More energetic ions (>20 keV) are present within the plasma sheet proper, are found in plasmoids ejected by substorm activity from the distant tail, and appear as bursts of flowing ions seen on the boundary of the plasma sheet. Within the plasma sheet, the more energetic ions extend to > 1 MeV and may appear as a distinctly non-thermal power law tail extending from the Maxwellian part of the main population distribution (Sarris et al., 1981). More complex energetic ion distributions were found in a study with ISEE-1 (Ipavich and Scholer, 1983). Ipavich et al. (1984) also report that the energetic O^+ content of the plasma sheet (as measured by O^+/H^+) increases with magnetic activity at least a factor of 10, up to O^+/H^+ ~ 0.4, qualitatively similar to the Sharp et al. (1982) results.

Magnetic activity is also associated with the occurrence of beams of energetic ions flowing earthward, tailward, or simultaneously counterstreaming along the boundary of the tail plasma sheet (Krimigis and Sarris, 1979; Mobius et al., 1980; Williams, 1981; Lui et al., 1983). Ion beams have the appearance of "bursts" in particle data and have been recorded most frequently in association with crossings of the plasma sheet boundary layer. They seem to be a consequence of substorm activity and hence reconnection (Krimigis and Sarris, 1979). (Since the plasma sheet is known to thin and then expand with substorm phases (cf. Hones, 1984), it seems more likely that the motion of the plasma sheet past a satellite is responsible for bringing the plasma sheet boundary layer into view during crossings, than is the relatively slow (~ 1 km/s) motion of the satellite in inertial space. However Eastman et al. (1984) argue that the boundary layer persists even during magnetically quiet periods ($K_p = 0$) and is a permanent feature of the tail.) The composition of these energetic particles seems to be primarily solar wind, based on He^{2+}/H^+ ratios (Mobius et al., 1980), however, Ipavich et al. (1985) report energetic O^+ streaming tailward during two substorms on March 22, 1979 (CDAW 6 analysis).

As discussed by many of the above authors and summarized recently in Baker and Fritz (1984) and Scholer (1984), energetic ion beams may be accelerated as part of the reconnection process at one or more X-type neutral lines formed during substorm activity at distances of 10 R_E to > 100 R_E tailwards from the Earth. The formation of neutral lines is not necessarily a prerequisite for acceleration, since acceleration at any region of weak normal magnetic field component also rather naturally leads to ion flows at the outer boundary of the plasma sheet (Lyons and Speiser, 1982).

Important experimental evidence for the neutral line/sheet acceleration process is discussed by Lyons and Evans (1984). Low altitude satellite data shows that precipitating electrons causing the discrete aurora are associated both temporally and spatially with energetic proton fluxes (30-800 keV) that resemble distributions seen in the high latitude boundary of the plasma sheet. The observed proton flux spectra are in agreement with characteristics computed for neutral sheet acceleration, the important implication being therefore, that discrete aurora often occur on field lines that map along the boundary layer and are connected to regions of current sheet acceleration. In addition to neutral line/sheet acceleration, rapid changes in tail magnetic fields give rise to large induction electric fields which will also accelerate ions to very high energies (Baker et al., 1979). As discussed by a number of authors (cf. Cowley, 1980; Scholer, 1984), beams of energetic ions may execute several bounces through the plasma sheet and gain energy by the Fermi mechanism. The bounce motion also leads ultimately to some ions populating the plasma sheet proper (Sarris et al., 1981).

At ion energies below 1 keV, Eastman et al. (1984) have argued that the plasma sheet boundary layer is the conduit for upflowing ion events originating in the auroral zone, and that the streaming ions eventually evolve through pitch angle scattering to populate the central plasma sheet. Evidence for scattering of upward accelerated field-aligned ions as a means of populating the plasma sheet has been advanced by Ghielmetti et al. (1979) and Sharp et al. (1981). Ghielmetti et al. noted the predominance of upward flowing over downward flowing ions at lower altitudes (< 8000 km) on magnetic field lines connecting to the low latitude boundary of the auroral zone and to the inner plasma sheet. They argued, based on spatial location and relative phase space densities, that both trapped and downflowing populations could have originated from the more commonly observed upward flowing ions. Sharp et al. (1981) used ISEE-1 observations of ionospheric ion streams in the tail to make essentially the same case. Thus the ISEE-1 and S3-3 observations show evidence for ion injection over a wide range of latitudes into the central plasma sheet, without the ions necessarily passing through the plasma sheet boundary layer which is presumably a magnetic field-aligned entity. Furthermore, upward flowing auroral ions are generally associated with inverted-V events or occasionally are found to occur over a very wide latitudinal extent (Sharp et al., 1979; Lundin et al., 1982; Quinn and Johnson, 1985; Shelley, 1985). The $E \times B$ drift and resultant dispersion of the upward flowing ions (particularly the slower O^+) would also tend to cause the beams to be injected throughout the main body of the plasma sheet, as observed by Sharp et al. (1982). Observations by the Lockheed group, therefore, seem to imply direct ionospheric injection throughout the plasma sheet and not just in the boundary layer as Eastman et al. have suggested.

In summary, the picture which emerges is that at low levels of magnetic activity solar wind populates the bulk of the plasma sheet (at least within 23 R_E of the Earth) with little additional ion acceleration. Substorm activity leads to the upward acceleration of auroral ions and scattering into the plasma sheet. The formation of neutral line(s) and enhanced reconnection and convection of mantle plasma into the plasma sheet speeds up the rate of earthward acceleration of energetic ions. Increased convection in the tail lobes together with increased reconnection should also increase the proportion of solar wind ions found in the plasma sheet, but apparently does not. (This might well be a consequence of tailward movement of the reconnecting region.) Energetic ions are observed at low altitudes in conjunction with discrete auroral forms, although not necessarily originating in the aurora, and at high altitudes as beams on the plasma sheet boundary and interior. Counterstreaming of this population is understood to come about as earthward directed ions mirror and return to the spacecraft. As substorm activity increases, the flux of accelerated particles out of the auroral zone increases still further and the plasma sheet tends to fill with ionospheric material that may ultimately alter its response to subsequent substorms (Baker et al., 1982).

Shelley (1985) estimated that the relatively small terrestrial component of the quiet time plasma sheet ($\sim 2\%$) can be easily supplied by either quiet time upward flowing auroral ions (Collin et al., 1984) or by accelerated polar cap ion outflows observed on DE-1 (Yau et al., 1984). During magnetic storms (Dst ~ -100 γ), even though the outflow of both O^+ and H^+ increases significantly, the auroral acceleration region alone is sufficient to supply the observed terrestrial content of the plasma sheet. Other transport channels (magnetotail lobe streams and magnetopause boundary layers) may, in Shelley's estimation, contribute significantly, but are not likely to be the primary source of the storm-time plasma sheet (Shelley, 1985).

One final note on significant changes in plasma sheet composition with magnetic activity is the observation that the absolute density of He^{2+} decreases with increasing activity (Lennartsson and Shelley, 1986). One might speculate that this comes about because He^{2+} is preferentially denied access to the plasma sheet inside 23 R_E (the limit of ISEE coverage) as a result of changing convection and/or acceleration patterns within the magnetotail. Increased convection might, for example, cause He^{2+} entering via the magnetopause boundary layer to be carried farther downtail before it could gain access to the plasma sheet. Tailward motion of the reconnection region in the plasma sheet could have a similar effect, as would changes in the pattern of disruption of the tail current system which feeds the ionospheric Birkeland currents (Lennartsson and Shelley, 1986; Baker et al., 1982). Baker et al. (1982) have suggested a second

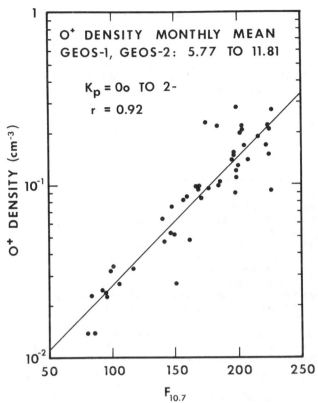

Fig. 4. Average O$^+$ density in the energy range 0.9-15.9 keV obtained near L = 6.6 by GEOS−1 and −2. Data cover the rising phase and most of the maximum of the current solar cycle (Adapted from Young et al., 1982).

species-specific effect, namely that the presence of O$^+$ in the near-Earth plasma sheet may define regions in which ion tearing mode growth rates are increased and the threshold for instability leading to substorm onset is lowered. More recent data on the distribution of O$^+$ densities in the tail are consistent with dawn-dusk asymmetries noted by Baker et al.

6. The Ring Current

Without a doubt one of the most exciting experimental results in recent annals of magnetospheric physics has been the long awaited measurements by the AMPTE/CCE experiments of ion composition in the terrestrial ring current. Ring current studies were formerly plagued by our inability to measure the composition of ions in the energy range 20-300 keV, wherein lies ~ 90% of the ring current energy density (Williams, 1980, 1983). Initial CCE results reported in the literature are limited to a single main phase magnetic storm, that of September 4-6, 1984 (Williams and Sugiura, 1985), however more extensive data are included in the present volume.

Like the plasma sheet, the Earth's ring current is a storehouse of accelerated magnetospheric particles and hence of energy. Particles in the ring current are stored by virtue of trapping in the more dipolar regions of the geomagnetic field. Because they may execute many circuits of the Earth, storage times are much longer than in the plasma sheet (days vs. hours) and this leads to their energization through radial (cross L-shell) diffusion. At the same time, particles are lost from the ring current either through charge exchange with neutral hydrogen, through pitch angle scattering into the atmosphere, or via de-trapping caused by reconfigurations of the geomagnetic field. The sources and pathways for ion entry into the ring current are better understood now as the result of CCE and earlier composition measurements, but questions remain.

The first direct measurements of ions ("protons") and electrons below 50 keV in the ring current were reported by Frank (1967, 1971) and were later extended to ~ 1 MeV by Smith and Hoffman (1973). As discussed by Lyons (1984), these and other non-composition data showed that the particle flux increases occurring during magnetic storms contributed to the main phase ring current primarily inside of L = 4. Flux increases outside this altitude were found to be no greater during main phase storms than during substorms. The inner edge of the ring current largely coincides with the plasmapause (Frank, 1971). This was predicted on theoretical grounds as the consequence of pitch angle diffusion driven by ion cyclotron wave-particle interactions (Cornwall et al., 1970). There is general agreement, however, that the primary loss mechanism for ring current ions is charge exchange with geocoronal neutral hydrogen rather than wave-induced pitch angle scattering. It should be kept in mind that both of these loss mechanisms are composition dependent, and, indeed, they may be coupled through charge exchange-induced anisotropies in the hot ion pitch angle distributions (Solomon and Picon, 1981).

The geostationary orbit (L $\simeq$ 6.6) lies near the outer portion of the quiet time ring current, in the general region where strong plasma injections are observed in the dusk to midnight local time sector. McIlwain (1972, 1974) has used ATS-5 plasma data to demonstrate that this region is literally a gateway for the passage of kilovolt plasma from the tail plasma sheet and the ionosphere (Mauk and McIlwain, 1975) into the inner magnetosphere. Injection, drift and dispersion of ionospheric material in this region has been described by, among others, Kaye et al. (1981b) and Strangeway and Johnson (1984) using S3-3 and SCATHA composition data. Particles injected near L = 6.6 are on trapped orbits under most conditions of magnetic activity, and from there begin to convect or diffuse inward into the ring current and radiation belts.

Using data from the GEOS-1 and -2 ion mass spectrometer, Young et al. (1982) showed that there was a strong solar cycle dependence in the number density of the terrestrial ions He$^+$ and O$^+$ at energies of 1-15 keV (Figure 4). Similar long term trends attributable to the solar cycle have been found in ISEE-1 composition data from the tail plasma sheet between 10 and 23 R$_E$ (Lennartsson and Shelley, 1986) and in upflowing ion events observed with DE-1 (Yau et al., 1985) and S3-3 (A. Ghielmetti, private communication, 1984).

In retrospect, the ionospheric origin of kilovolt magnetospheric ions should have lead one to expect that some form of solar cycle dependence might be observable. Responses of the upper atmosphere and F-region ionosphere to solar cycle variations in the solar EUV output are well known (cf. White, 1977). Young et al. (1982) argued that increases in solar EUV (as quantified by F$_{10.7}$, the 10.7 cm radio flux index) would increase the production rate of ionospheric species, but more importantly would raise the scale heights of both ions and neutrals. Because of its greater mass, the scale height of O$^+$ would be affected the most (note that the ordinate in Figure 4 is logarithmic) and that of H$^+$ the least, with He$^+$ falling somewhere in between. This in fact is what is observed on GEOS in terms of ion density correlations with F$_{10.7}$. Moore (1980, 1984) has examined the limitations placed on O$^+$ outflow by charge exchange of O$^+$ on neutral hydrogen. He found that acceleration of O$^+$ to supersonic speeds was sufficient to promote its escape simply by shortening its path length through the neutral exospheric hydrogen. Lockwood (1984) has discussed the question of solar cycle control of

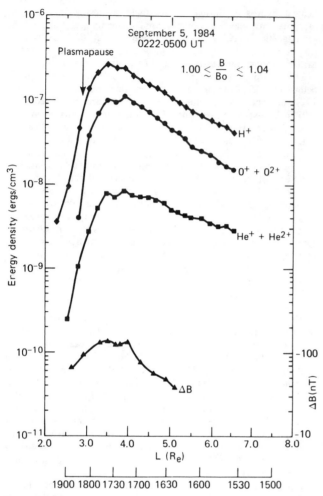

Fig. 5. AMPTE/CCE ring current composition during the magnetic storm of Sept. 4-7, 1984. Energy density is an integral over ~10 eV to ~5 MeV (from Krimigis et al., 1985).

the auroral O^+ acceleration process in some detail. This is taken up in Section 8.

Moving on to higher ion energies, in the earlier absence of direct measurements of ring current composition in the critical 20 keV-300 keV range, a number of studies were undertaken in order to infer indirectly the ring current composition during the storm recovery phase. These were based on deducing the characteristic time scales for ring current decay. Observed decay rates of 1-3 days in the storm-time Dst index (Tinsley, 1976) and particle fluxes (Lyons and Evans, 1976; Smith et al., 1981) were too slow to be accounted for by H^+ lifetimes against charge exchange, leading to the suggestion that He^+ or O^+ should be the dominant ring current ion at energies below 50 keV. This surmise was confirmed below 20 keV by GEOS-1, ISEE-1 and PROGNOZ-7 data, and seems to be consistent with available CCE results (Krimigis et al., 1985). The dominant "other" species turns out to be O^+ rather than He^+.

Published data from the AMPTE/CCE composition instruments are thusfar limited to a single storm in September, 1984. The CCE is in a 15.7 hr. orbit with apogee initially near 1300 LT and 8.8 R_E. The ring current that developed during the September storm was fairly intense (preliminary Dst = -120γ) but asymmetric (Williams and Sugiura, 1985). During the storm initial phase on Sept. 4, K_p reached 7o and 8-, on Sept. 5 it reached 7o.

Figure 5 is a summary plot of ion energy density during the storm main phase when the CCE was inbound near local dusk. Prior to the storm the quiet time ring current energy density was dominated by H^+ with a peak at 100 to 300 keV. During the storm main phase, O^+ fluxes increased the most dramatically of all ions at energies < 300 keV although there were appreciable increases in H^+ and He^+. Gloeckler et al. (1985a) note that H^+, He^+, He^{2+} and O^+ spectra are all relatively similar below 300 keV/e with peaks at ~ 15 keV/e and ~ 150 keV/e. There is no obvious difference between the energy spectra of ions of terrestrial origin (O^+, He^+) and those of solar wind origin (He^{2+}, $[CNO]^{5+}$) when plotted at equal E/Q, suggesting primarily E/Q dependent acceleration. At energies < 17 keV Shelley and co-workers find that the H^+ and O^+ pitch angle distributions, however, show significant differences, with the H^+ exhibiting progression from isotropic toward trapped characteristics with decreasing L, and O^+ evolving from field aligned toward more isotropic distributions consistent with a field-aligned source.

The relative abundances of solar wind ions below 300 keV/e observed within the ring current (He^{2+}, $[CNO]^{5+}$, $Si^{>4+}$, and $Fe^{>9+}$) are remarkably similar to their solar wind values, suggesting little mass or charge discrimination effects in solar wind ion entry into the magnetosphere or in subsequent acceleration (Gloeckler et al., 1985a). One further note is that an integrated oxygen charge state spectrum for 1-300 keV/e shows that although charge exchange processes are at work creating O^{3+}, O^{4+} and O^{5+}, the solar wind and ionospheric sources are still clearly delineated (Kremser et al., 1985). During the storm recovery phase, O^+ above ~ 20 keV decays rapidly and an important question is whether it contributes significantly to the higher oxygen charge states at energies > 100 keV as suggested by Spjeldvik and Fritz (1978). Preliminary AMPTE/CCE data (Kremser et al., 1985) show that solar wind carbon is essentially C^{4+} — C^{6+} at L > 8 but at $4 \leq L \leq 6$ only C^{2+} and possibly C^+ are present.

Acceleration and injection of ring current ions is known to occur through several pathways, although the relative contribution and time-dependence of each remains to be elucidated. Following Williams (1983), I list four "subliminally popular" generation mechanisms for the ring current and add two more that are suggested by recent observations and theoretical work. Except for estimates provided by Hultqvist (1983) and Shelley (1985), no attempt has been made to assess quantitatively the importance of these different sources. The pathways and related acceleration processes are depicted schematically and keyed to numbers given in Figure 6.

Earthward $E \times B$ drift of plasma sheet ions driven by the cross-tail convection electric field (1). Acceleration occurs through μ and J conservation, with the side effect that initially isotropic particle distributions become anisotropic and are therefore potentially unstable to electromagnetic ion cyclotron waves (Kaye et al., 1979). In some sense this rapid inward transport by unsteady convection can be thought of as a radial diffusion process. The Rice model (Wolf et al., 1982) simulates formation of the ring current magnitude quite well by $E \times B$ convection, however the ring current ends up at the inner edge of the model plasma sheet (L ~ 5) which is not in agreement with the inner edge observed at L ~ 3. Although not touched on in this review, modeling "experiments" such as that of the Rice and, more recently, Lockheed groups (Cladis and Francis, 1985) are a much needed adjunct to in situ experiments.

Upward accelerated auroral ions (2). The particle source is the topside ionosphere and hence tends to be rich in O^+. Acceleration is effectively parallel to the magnetic field, regardless of whether it is initially transverse or parallel, due to the action of the mirror force at low altitudes. Acceleration is followed by isotropization near the

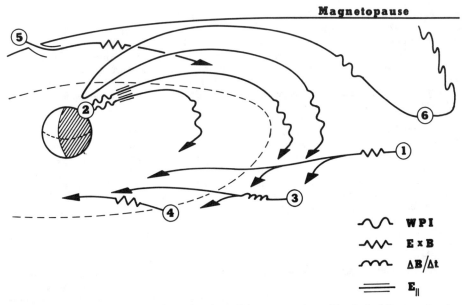

Fig. 6. Pathways for injection and acceleration of ring current ions. The dashed line represents a rough boundary between outer (quiescent) and inner (storm) ring current positions. Note that trapping of ionospheric source populations in the equatorial regions may play a large role in generating the ring current.

equatorial plane. Pitch angle diffusion towards trapped distributions is clearly an important intermediate step in populating the ring current by this mechanism.

Recently Hultqvist (1983) has examined the injection of hot O^+ fluxes into the dayside magnetosphere. He argues for acceleration over a limited altitude range in the topside ionosphere (< 1000 km) and over a broad range of local times and L-values since observed fluxes appear rapidly and are widely dispersed. Auroral ions are known to populate both the near and distant plasma sheet (Section 5) as well as the dayside ring current (Hultqvist, 1983). Newly injected field-aligned ions will tend to charge exchange with neutral hydrogen at lower altitudes, leading to alterations in composition and pitch angle distributions which may produce the O^+ conical distributions observed near the equator (Lyons and Moore, 1981). Evidence for pitch angle scattering has been obtained most recently from CCE (Shelley et al., 1985b), from SCATHA observations of near-equatorial beams (Richardson et al., 1981) as well as the work of Ghielmetti et al. (1978). Cladis and Francis (1985) have constructed a theoretical model of auroral ion transport based on Monte Carlo calculations of ion diffusion in arbitrary wave fields.

In situ acceleration at the substorm injection boundary, first proposed by McIlwain (1974) (3). Subsequent work on this model by Moore et al. (1981) has led to the suggestion that the impulsive, dispersionless nature of the substorm injection process is caused by an earthward propagating compressional wave set loose by the collapse of magnetotail field lines at the onset of rapid reconnection in the tail (cf. Baker et al., 1979). The induced electric field associated with this rapid magnetic field change is responsible both for injecting and heating plasma sheet particles. Quinn and Southwood (1982) find that impulsive increases in the convection electric field will more effectively accelerate ions with smaller rather than larger equatorial pitch angles through violation of the bounce adiabatic invariant. This process causes bouncing ion clusters (Quinn and McIlwain, 1979) observed frequently on ATS. Quinn and Johnson (1985) have identified the substorm injection boundary with intense beams from the ionosphere which are similar to the field-aligned component of "zipper" events. This process has been observed only in the outer (L > 6) ring current regions. Presumably it would move earthward with the equatorward expansion of auroral activity during main phase storms. In any case, populations injected near L = 6.6 serve as the feeder for other processes (see below) that may further accelerate ions.

Earthward adiabatic transport of a pre-existing trapped particle population (Lyons and Williams, 1980) (4). Tests of this hypothesis with Explorer 45 data showed that pre-storm distributions of ions and electrons could reproduce storm-time distributions quite well under the assumption that an azimuthal equatorial electric field of 0.3 ~ 1.0 mV/m acts over an appreciable (90° ~ 270°) range of drift longitudes. The new CCE data set offers an excellent opportunity to repeat this experiment, particularly since apogee is beyond geostationary orbit and the expected injection boundary.

Direct injection of magnetosheath plasma into the outer ring current during main phase storms (5). Injection may be into the dayside boundary layer plasma (Lundin et al., 1982) from which ions can in principle drift into the outer ring current. PROGNOZ-7 observations suggest that impulsive solar wind injections occur in a manner remarkably like the model proposed by Lemaire (1977) and Heikkila (1979). Intruded solar wind plasma seems to be necessary in order to explain the so-called "mixed regions" containing hot, non-flowing, solar wind enriched plasma observed preferentially in the dawn-side magnetosphere with GEOS (Balsiger et al., 1980) and ISEE-1 (Lennartsson et al., 1981).

Field-aligned injection of ions from the tail neutral sheet (6). As discussed in Section 5, bursts of field-aligned energetic ions in the plasma sheet boundary layer result from acceleration in the tail neutral sheet. Some of these ions (and electrons) are able to reach the auroral zone (Lyons and Evans, 1984) but in addition can also populate the main plasma sheet (Sarris et al. 1981) through the angular divergence inherent in the acceleration process, as well as through $E \times B$ convection and scattering. Injection of energetic ions directly into the outer ring current near geostationary orbit occurs during substorms (Baker et al., 1979) and there is no reason why these could not also contribute to main phase storms. Since these ions are seen at 6.6 R_E it is likely that they are also injected even deeper into the ring current by the same process, namely large scale induction electric fields.

In summary, we have identified at least six mechanisms by which the injection of ions into the inner or outer ring current has been observed to occur. As a number of authors have noted, it is unlikely that only one process is dominant; it is certainly conceivable that all processes are in part responsible for ring current generation. The task is then to study (or model) each in systematic fashion and particularly to understand how the contribution from each varies with storm phase and from storm to storm. Since some processes favor injection of a more solar wind-like component (5,6), others a pure terrestrial component (2), or still others a mixture of the two in variable proportions (1,3,4) it seems likely that the composition of the injected ring current will depend on the relative contribution of each injection mechanism. Of course changes in the source populations feeding these injection pathways are also known to occur (e.g. the solar cycle effect on O^+ and He^+) and will further complicate this picture.

7. Radiation Belt

For ions the locus of the radiation belt extends from the limit of stable trapping (L ≃ 5 on the long term) down to L ≃ 1.1 where ion losses to collisions with the atmosphere can no longer be supported. The most energetic ions (> 100 MeV) are protons originating from the cosmic ray production of neutrons in the atmosphere and their subsequent decay (CRAND). Below this energy, radiation belt ions apparently originate beyond the stable trapping region, based on evidence that they are largely solar in origin (see below). The issue is not settled, however, with recent AMPTE/CCE findings indicating strong injections of O^+ during magnetic storms at energies up to ~ 1 MeV (Krimigis et al., 1985). Acceleration to radiation belt energies is thought to occur through the process of radial diffusion driven by fluctuations in the large-scale magnetospheric electric and magnetic fields which violate the third adiabatic invariant. Since the source of these large scale fluctuations is the solar wind, either through compressions of the magnetopause boundary or through variations in the convection electric field, it is again the solar wind that is the ultimate energy source for ion acceleration. In this regard, the rapid acceleration and transport of MeV O^+ found by CCE clearly is an important issue that needs to be addressed.

Somewhat surprisingly, at these very high energies instrumentation has proved adequate to the requirements for composition measurements. For example, one key measurement is that the radiation belt C/O elemental abundance is > 0.5 (Hovestadt et al., 1978), a characteristic of only the solar wind since the ionospheric C/O ratio is < 10^{-5} (Blake, 1973). There exist sufficient heavy ion measurements such that both the origin and the relevant physical processes in the radiation belt are relatively well understood. Excellent reviews on the place of heavy ions in radiation belt physics have been given in recent years by Schulz and Lanzerotti (1974), Schulz (1975), Spjeldvik (1981) and Spjeldvik and Fritz (1983).

At first, measurements of heavy (Z ≥ 2) ions were made only at low altitudes. These showed He and CNO fluxes to be a factor ~ 10^2 less than expected, based on known solar wind abundances. The problem turned out to be very steep equatorial ion pitch angle distributions that are strongly peaked at 90°. Inner zone (L < 2) ion fluxes were found to fit a $sin^n \alpha$ function with n ≃ 8 for hydrogen and n ≃ 12 for He (Blake et al., 1973). Similarly, in the outer zone (L ≃ 3-5) n ≃ 5 for hydrogen, n ≃ 6 − 8 for He (Fritz and Williams, 1973; Blake et al., 1980; Hovestadt et al., 1981) and n ≃ 16 for C and O (Hovestadt et al., 1981). These data refer to ion energies of ≃ 0.5 MeV/nucleon. The origin of steep pitch angle distributions has been variously ascribed to the steepness of the ion energy distributions in the source region e.g. in the plasma sheet, to reduced effectiveness in pitch angle scattering of heavy ions, or to differential loss of small pitch angle particles. Ions which conserve μ and J while diffusing inward will gain most energy near 90° pitch angle and least near 0°. Thus a steep initial energy distribution, such as that found in the plasma sheet (Section 5), will contribute energetic particles primarily near 90° pitch angles. At low altitudes (L < 2) Fennell et al. (1974) stress that radial diffusion is small and pitch angle scattering favors protons over heavier ions. Finally, charge exchange and Coulomb energy loss will be greatest for small equatorial pitch angles (at least at total energies up to a few MeV) and these ions will be preferentially lost (Cowley, 1977; Lyons and Moore, 1981).

Cornwall (1972) and Schulz (1975) suggested that if He^{2+} has less than four times the temperature of H^+ in the source region (plasma sheet) this would lead to a steeper He^{2+} than H^+ energy/nucleon spectrum and thence to the observed species dependent pitch angle distributions. Recent ISEE-1 data (Lennartsson and Shelley, 1986) show, however, that $E_{He^{2+}} = 4 E_{H^+}$ to within ≃ 15% for all levels of magnetic activity except the highest, where equal ion energy per charge is more likely. The only CNO charge state measurements in the outer magnetosphere (L ≃ 6-7) which address this issue are those of AMPTE (Gloeckler et al., 1985a). These data show that during a single storm the $[CNO]^{\geq +3}$ average energy per charge was qualitatively similar for all ions, including terrestrial species. The He^{2+} and CNO group average energy per nucleon was only about 1/2 that of H^+. Thus if storm or disturbed-time injections and subsequent rapid diffusion are a major contributor to the radiation belts, as seems likely, then the steep pitch angle distributions at L < 5 may be understood in terms of the AMPTE/CCE data. The average ISEE-1 data on H^+ and He^{2+} from 10 to 23 R_E are in rough agreement with this concept for high levels of magnetic activity (storms) but not for low activity.

As mentioned earlier, composition of the radiation belts for Z ≃ 2 ions measured near the equatorial plane at L ≃ 2.5-4 strongly indicates a solar wind origin, at least at energies of 0.4-1.5 MeV/nucleon (Hovestadt et al., 1978, 1981). The ISEE-1 data also show that Ne, Si, and Mg are present at energies < 0.6 MeV/nucleon. Ions with higher rigidities apparently cannot be stably trapped due to violation of the Alfven criterion (Hovestadt et al., 1978). The latter requires that the ion gyroradius be small (< .05) relative to $\rho |\nabla B|/B$ where ρ is the particle gyroradius. Further evidence for the validity of this trapping criteria comes from an observed C/O ratio of 1 ~ 4 at energies of 0.45 to 1.3 MeV/nucleon at L ≃ 2.5-4. Since solar wind C/O values are ≃ 0.5, this enrichment of C relative to O can be understood from the fact that the O^{6+} ion has a sufficiently large gyroradius to violate the Alfven criterion outside L ≃ 3.5, whereas C^{6+} does not (Hovestadt et al., 1978).

Observations now show considerable amounts of O^+ in the outer magnetosphere and ring current at energies of 1 ~ 300 keV. One might therefore ask what becomes of so much O^+ in the inner radiation belt where values of C/O ~ 1 apparently preclude a terrestrial ion source.

One answer is clearly that rapid charge exchange helps to dispose of O^+ by creating fast neutral oxygen atoms which are immediately lost. Below ~ 500 keV, O^+ is much more likely to capture an electron than it is to lose one (Spejeldvik and Fritz, 1978). Furthermore, the injected O^+ spectrum (at equal energy/nucleon) is simply too soft and/or too field-aligned to promote very many ions into the MeV range through radial diffusion. A third important effect, noted by Cornwall (1972), is that radial diffusion driven by electric fluctuations is proportional to ion $(Q/M)^2$, hence terrestrial O^+ and O^{2+} should diffuse far slower than O^{6+} and would thereby be diminished relative to higher charge states before much acceleration occurred. Given these three mechanisms which should mitigate against O^+, it is interesting that AMPTE/CCE storm-time results show that O^+ fluxes at 300 keV/e are $> 10^3$ more intense than $[CNO]^{\geq 3+}$ fluxes (Gloeckler, et al. 1985a). This indicates that despite losses, O^+ is energized and injected rapidly enough, at least down to L = 3.7, to overcome its disadvantages relative to solar wind O^{6+}.

Explorer 45 data have yielded considerable detail on the storm-time injection and subsequent decay of radiation belt ions (see review by Spjeldvik and Fritz, 1983). By studying a series of storms in 1972 (the declining phase of Solar Cycle 20), Fritz and Spjeldvik have found, in addition to large variability from storm to storm, that rapid injection of MeV ions (1 ~ 20 MeV depending on mass) can occur down to L ≃ 2.5 during a large storm. Largest enhancements occurred at the smallest L values and were far stronger for heavier than for lighter ions. Conversely, the heavier ion fluxes decayed more quickly than did the lighter ions. Decay periods were consistent with charge exchange and Coulomb losses nearer the Earth (L < 3.5) and with radial diffusion farther out.

In summary, it is generally considered that the broad principles of radiation belt theory are reasonably well understood. Nothing in the recent experimental results would contradict such a statement, however non-steady-state conditions during rapid injection of storm-time particles are still poorly elucidated. As this brief review demonstrates, within the radiation belt a number of mass or species dependent effects are known to be important. Chief among these are:
— species and origin (terrestrial or solar wind) dependent energy spectra and pitch angle spectra occur in the radiation belt source regions,
— species dependent trapping takes place in the case of high vs. low rigidity ions,
— species dependent radial diffusion results from electric (but not magnetic) fluctuations (Cornwall, 1972), and
— loss processes are virtually entirely species dependent (charge exchange, Coulomb collisions, resonant wave-particle interactions).

Although no single theoretical or modeling effort currently addresses all of the above effects, it is possible that progress will be made, for example, under the aegis of the CRRES/SPACERAD program (Gussenhoven et al., 1985).

8. Ionospheric Acceleration Processes

Mass spectrometers on S3-3 made the first observations of upward acceleration of kilovolt H^+ and O^+ ions over the auroral zones (Shelley et al., 1976). This was quickly followed by the identification of conical, rather than exclusively field-aligned, ion distributions, which imply a strong acceleration process acting transverse to the magnetic field (Sharp et al., 1977). There is also evidence for ion (and electron) acceleration through magnetic field-aligned electric potential drops associated largely with "inverted-V" structures (Mizera and Fennell, 1977; Hultqvist and Borg, 1978; Sharp et al., 1979; Collin et al., 1981; Fennell, 1984; Klumpar et al., 1984).

Acceleration via field-aligned potential drops seems to be relatively well understood, however the jury is still out on the nature of the transverse acceleration processes (cf. Fennell, 1984). Evidence has been presented in a few studies that transverse and field-aligned acceleration processes are related in a causative sense, as intuition suggests they might be if both are manifestations of the auroral process (Klumpar et al., 1984; Heelis et al., 1984; Kintner et al., 1979). Moore et al. (1985a) report that strong transverse O^+ heating, and a net upward heat flux carried by O^+, are located at the edges of and below regions of parallel ion acceleration. Statistical studies, on the other hand, have tended to show that field-aligned beams and conics (conics are the high altitude manifestation of transversely accelerated ions due to the action of the mirror force and conservation of the first invariant) have different spatial distributions (compare Ghielmetti et al., 1978 and Gorney et al., 1981) and therefore might not necessarily be the product of the same underlying phenomena. A more recent and intensive statistical study with DE-1 (Yau et al., 1984) is largely inconclusive on this subject except to show that high altitude conics are probably derived from lower altitude transverse acceleration. Of course transverse acceleration at low altitudes produces an effective parallel acceleration through the action of the mirror force, but the real question is whether regions of transverse acceleration are intimately related to field-aligned potential drops and, ultimately, to the auroral process itself. Observations of transverse ion heating within regions of field-aligned currents, of intense ion cyclotron and lower hybrid (LH) wave activity, and of so-called electrostatic shocks (Heelis et al., 1984; Mizera et al., 1981) help reinforce the consensus that upward acceleration of ions is indeed just one more aspect of the auroral process. There are, however, several candidate mechanisms for transverse ion acceleration—the question is which one(s) does nature choose and how do they fit into a consistent scheme of auroral physics.

At low altitudes (< 1000 km) the total observed ion acceleration is rather small although this really must be judged in terms of the roughly 100-fold gain over very low initial ionospheric temperatures (Ashour-Abdalla et al., 1981). As a number of authors have noted, one problem with wave acceleration is that the heated ions are soon moved out of the acceleration region by the mirror force, thereby limiting their energy gain. Observations at middle altitudes (1000-5000 km) typically find that upward accelerated ions are seen in association with intense fluxes of precipitating electrons (not all studies have had access to magnetometer data, hence electron fluxes are often taken as an indicator of field-aligned currents) and occur in regions of low ambient plasma density (Klumpar, 1979). Theoretical studies have suggested that one agent for transverse acceleration is electrostatic ion cyclotron (EIC) waves (Ungstrup et al., 1979; Dusenbery and Lyons, 1981; Ashour-Abdalla and Okuda, 1983). On the other hand, Chang and Coppi (1981) have pointed out that intense lower hybrid (LH) waves are more typically observed in this middle altitude range and will also accelerate ions transversely. The EIC waves take their free energy from drifting ionosopheric electrons moving either down (Ungstrup et al., 1979) or up (Dusenbery and Lyons, 1981) the auroral field lines. Simulations in which cold drifting electrons provide the free energy have shown that oxygen cyclotron waves should also become unstable and preferentially heat O^+ transverse to the magnetic field (Ashour-Abdalla et al., 1981; Ashour-Abdalla and Okuda, 1983). Unfortunately, oxygen cyclotron waves have not yet been reported and are inherently more difficult to observe than hydrogen waves. Chang and Coppi (1981) and Retterer et al. (1983) have argued that LH waves are seen at lower altitudes and are more intense than EIC waves. The LH waves are driven by electrons precipitating through the cold background plasma at altitudes of 1000 ≤ 5000 km, and are sufficiently intense to accelerate ions to keV energies. Following Chang and Coppi, it seems possible that LH waves are the low-altitude feeder for ions that are further accelerated at higher altitudes by EIC waves.

At higher altitudes (> 5000 km) intense electrostatic-hydrogen cyclotron waves are found frequently in association with more energetic transverse ion heating (Kintner et al., 1979). In addition to waves, localized regions of abrupt reversals and enhancements in the transverse auroral electric field occur and have been termed electrostatic shocks (Mozer et al., 1977). These are found at altitudes of 1 R_E out to a few R_E, are generally embedded in inverted-V structures, and are associated with upward accelerated ions and electrostatic ion cyclotron waves as noted above. Both O^+ and H^+ ions which have received strong transverse heating ($T_\perp \sim 60$ eV) have been observed (Sharp et al., 1983). Borovsky (1984) has made a case for transverse ion acceleration in oblique (to the magnetic field) double layers. This can produce a number of the observed ion acceleration features and would resemble the electric field signatures of electrostatic shocks. Acceleration in oblique double layers can in principle provide a large amount of transverse energy necessary, for instance, to fulfill the observational requirements of Collin et al. (1981). The latter find that kilovolt upflowing oxygen is on the average about twice as energetic as H^+ and that about half of its total energy must have come from acceleration perpendicular to the magnetic field.

A related subject that has received much attention of late is suprathermal ion escape (in some cases this is referred to as "upwelling") primarily of O^+, at energies <100 eV and often ~ 10 eV, observed at high latitudes over the polar cap (Shelley et al., 1982; Gurgiolo and Burch, 1982; Waite et al., 1985), at the boundary between polar cusp and polar cap (Lockwood et al., 1985), over the auroral zones (Moore et al., 1985a) and, by inference, in the magnetotail (Sharp et al., 1981). (Note that the gravitational escape energy for O^+ is 10.5 eV.) Although it might be thought that these ions are an high energy tail on the classical polar wind, it now appears that they are heated by transverse acceleration processes similar to those outlined above. For example, Moore et al. (1985a) find at times that a significant portion of the thermal plasma is transversely accelerated, leading to large upward heat fluxes carried by O^+ ions. The heating of upwelling ions is generally inferred to take place at altitudes below where the charge exchange of O^+ with neutral hydrogen is important, otherwise the upflow of O^+ would not reach measured flux levels.

As discussed by Moore (1980, 1984) the neutral hydrogen exosphere represents a barrier to O^+ escape unless the O^+ is first accelerated to energies ~ 10 eV which, coincidentally, is the gravitational escape energy as well. Lockwood (1984) has examined this mechanism, including the effects of elastic scattering, and has showed that acceleration to ~ 10 eV is sufficient for both H^+ and O^+ to escape the ionosphere from an altitude of 600 km. He hypothesized that the topside ionosphere is unstable to ion cyclotron waves and that O^+ seen in upward flowing ions is heated initially by these waves. Some contradiction between Lockwood's results and the GEOS observations arises because the O^+ escape flux should be highest at low topside ion densities whereas the reverse seems to be true for the solar cycle effect observed by Young et al. (1982). Moore's results, which do not concern specific heating processes, agree with GEOS trends. In summary, oxygen ion escape is an important cross-disciplinary area in which further work is needed to sort out the competing effects of ionospheric chemistry, ion transport, and the plasma physical problem of ion acceleration. We are far from understanding the chemical control which the ionosphere and solar activity exercise over the makeup of magnetospheric ion populations.

Analysis of the stability of upwelling ions and other injected ionospheric populations is needed in order to ascertain under what circumstances streaming ions might generate waves that lead to ion trapping at high altitudes. I have argued in this review that ion upflow and subsequent trapping is a widespread process that completes the pathway of mass and energy transfer from the ionosphere to the equatorial magnetosphere (cf. Figure 6). This process converts parallel ion flow energy, for example via wave-particle interactions, into perpendicular thermal energy. The upflow and subsequent trapping of ions thus represents momentum and energy, as well as mass, transferred from the ionosphere to the equatorial magnetosphere rather than the other way around, which is the more conventional picture. Evidence that these processes are indeed occurring can be found in the injection of energetic auroral ions into the plasma sheet (Section 5) and dayside magnetosphere (Hultqvist, 1983).

To summarize this section, present observations seem to allow for the existence of several related and even complementary auroral ion acceleration mechanisms. Quite possibly more than one may be found to operate at one time in the region of a given auroral field line. While details may vary, all observations and theories have in common field-aligned currents as the causative agent. Current or drift driven instabilities generate waves which transversely accelerate the observed ions. Subsequent acceleration by parallel electric fields or further acceleration by waves may occur. None of the proposed mechanisms make outrageous demands on free energy sources and all reproduce some aspect of this very rich acceleration phenomena. This places stringent requirements on future experiments if one is to distinguish the different acceleration mechanisms that are proposed.

Acknowledgements. This paper is dedicated to Dick Sharp, with the hope that he may once again join us in a field of endeavor which sorely misses his talents. I would like to thank Dan Baker, Larry Lyons, Jim Burch, Bill Peterson, Tom Moore and two other referees, who brought the work on LH waves to my attention, for their helpful comments on the manuscript. I'm not sure how to express my thanks to Dick Johnson and Tom Chang for inviting me to give and write this review, so I'll leave it at that. I also thank Sharon Trujillo for careful typing of the manuscript. This work was supported by the U.S. Department of Energy.

References

Ashour-Abdalla, M., H. Okuda, and C. Z. Cheng, Acceleration of heavy ions on auroral field lines, *Geophys. Res. Lett.*, 8, 795, 1981.

Ashour-Abdalla, M., and H. Okuda, Transverse acceleration of ions on auroral field lines, in *Energetic Ion Composition in the Earth's Magnetosphere*, edited by R. G. Johnson, p. 431, Terra Scientific Publishing Co., Tokyo, 1983.

Baker, D. N., R. D. Belian, P. R. Higbie, and E. W. Hones, Jr., High-energy magnetospheric protons and their dependence on geomagnetic and interplanetary conditions, *J. Geophys. Res.*, 84, 7138, 1979.

Baker, D. N., E. W. Hones, Jr., D. T. Young, and J. Birn, The possible role of ionospheric oxygen in the initiation and development of plasma sheet instabilities, *Geophys. Res. Lett.*, 9, 1337, 1982.

Baker, D. N., and T. A. Fritz, Hot plasma and energetic particles in the Earth's outer magnetosphere: New understandings during the IMS, in *Proc. Conf. Achievements of the IMS*, ESA SP-217, p.85, European Space Agency, Paris, 1984.

Balsiger, H., Recent developments in ion mass spectrometers in the energy range below 100 keV, *Adv. Space Res.*, 1982.

Balsiger, H., P. Eberhardt, J. Geiss, A. Ghielmetti, H. P. Walker, D. T. Young, H. Loidl and H. Rosenbauer, A satellite-borne ion mass spectrometer for the energy range 0 to 16 keV, *Space Sci. Instrum.*, 2, 499, 1976.

Balsiger, H., P. Eberhardt, J. Geiss, and D. T. Young, Magnetic storm injection of 0.9 to 16- keV/e solar and terrestrial ions into the high-altitude magnetosphere, *J. Geophys. Res.*, 85, 1645, 1980.

Balsiger, H., J. Geiss, and D. T. Young, The composition of thermal and hot ions observed by the GEOS-1 and -2 spacecraft, in *Energetic Ion Composition in the Earth's Magnetosphere*, edited by R. G. Johnson, p. 195, Terra Scientific Publishing Co., Tokyo, 1983.

Banks, P. M., A. F. Nagy, and W. I. Axford, Dynamical behavior of thermal protons in the mid-latitude ionosphere and magnetosphere, *Planet. Space Sci.*, 19, 1053, 1971.

Baugher, C. R., Chappell, J. L. Horwitz, E. G. Shelley, and D. T. Young, Initial thermal plasma observations from ISEE-1, *Geophys. Res. Lett., 9*, 657, 1980.

Bezrukikh, V. V., and K. I. Gringauz, The hot zone in the outer plasmasphere of the Earth, *Planet. Space Sci., 38*, 1085, 1976.

Blake, J. B., Experimental test to determine the origin of geomagnetically trapped radiation, *J. Geophys. Res., 78*, 5822, 1973.

Blake, J. B., and G. A. Paulikas, Geomagnetically trapped alpha particles 1. Off-equator particles in the outer zone, *J. Geophys. Res., 77*, 3431, 1972.

Blake, J. B., J. F. Fennell, M. Schulz, and G. A. Paulikas, Geomagnetically trapped alpha particles 2. The inner zone, *J. Geophys. Res., 78*, 5498, 1973.

Blake, J. B., J. F. Fennell, and D. Hovestadt, Measurements of heavy ions in the low-altitude regions of the outer zone, *J. Geophys. Res., 85*, 5992, 1980.

Blake, J. B., and J. F. Fennell, Heavy ion measurements in the synchronous altitude region, *Planet. Space Sci., 29*, 1205, 1981.

Borovsky, J. E., The production of ion conics by oblique double layers, *J. Geophys. Res., 89*, 2251, 1984.

Chang, T., and B. Coppi, Lower hybrid acceleration and ion evolution in the subauroral region, *Geophys. Res. Lett., 8*, 1253, 1981.

Chappell, C. R., S. A. Fields, C. R. Baugher, J. H. Hoffman, W. B. Hanson, W. W. Wright, H. D. Hammack, G. R. Carignan, and A. F. Nagy, The retarding ion mass spectrometer on Dynamics Explorer-A, *Space Sci. Instrum., 5*, 477, 1981.

Cladis, J. B., and W. E. Francis, The polar ionosphere as a source of the storm time ring current, *J. Geophys. Res., 90*, 3465, 1985.

Collin, H. L., R. D. Sharp, E. G. Shelley, and R. G. Johnson, Some general characteristics of upflowing ion beams over the auroral zone and their relationship to auroral electrons, *J. Geophys. Res., 86*, 6820, 1981.

Collin, H. L., R. D. Sharp, and E. G. Shelley, The magnitude and composition of the outflow of energetic ions from the ionosphere, *J. Geophys. Res., 89*, 2185, 1984.

Comfort, R. H., J. H. Waite, Jr., and C. R. Chappell, Thermal ion temperatures from the retarding ion mass spectrometer on DE-1, *J. Geophys. Res., 90*, 3475, 1985.

Cornwall, J. M., F. V. Coroniti, and R. M. Thorne, Turbulent loss of ring current protons, *J. Geophys. Res., 75*, 4699, 1970.

Cornwall, J. M., Radial diffusion of ionized helium and protons: A probe for magnetospheric dynamics, *J. Geophys. Res., 77*, 1756, 1972.

Cowley, S. W. H., Pitch angle dependence of the charge-exchange lifetime of ring current ions, *Planet. Space Sci., 25*, 385, 1977.

Cowley, S. W. H., Plasma populations in a simple open model magnetosphere, *Space Sci. Rev., 26*, 217, 1980.

Craven, P. D., R. C. Olsen, C. R. Chappell, and L. Kakani, Obsevations of molecular ions in the Earth's magnetosphere, *J. Geophys. Res., 90*, 7599, 1985.

Curtis, S. A., Equatorial trapped plasmasphere ion distributions and transverse stochastic acceleration, *J. Geophys. Res., 90*, 1765, 1985.

DeCoster, R. J. and L. A. Frank, Observations pertaining to the dynamics of the plasma sheet, *J. Geophys. Res., 84*, 5099, 1979.

Decreau, P. M. E., J. Etcheto, K. Knott, A Pedersen, G. L. Wrenn, and D. T. Young, Multi-experiment determination of plasma density and temperature, *Space Sci. Rev., 22*, 633, 1978.

Dusenbery, P. B., and L. R. Lyons, Generation of ion-conic distribution by upgoing ionospheric electrons, *J. Geophys. Res., 86*, 7627, 1981.

Eastman, T. E., L. A. Frank, W. K. Peterson, and W. Lennartsson, The plasma sheet boundary layer, *J. Geophys. Res., 89*, 1553, 1984.

Fan, C. Y., G. Gloeckler, and E. Tums, An electrostatic deflection vs. energy instrument for measuring interplanetary particles in the range 0.1 to 3 MeV/charge, *Proc. 12th Int. Conf. Cosmic Rays, 4*, 1602, 1971.

Fennell, J.F., IMS contribution to the understanding of auroral precipitation, transport, and particle sources, in *Proc. Conf. Achievements of the IMS*, ESA SP-217, p. 731, European Space Agency, Paris, 1984.

Fennell, J. F., J. B. Blake, and G. A. Paulikas, Geomagnetically trapped alpha particles, 3, Low-altitude outer zone alpha-proton comparisons, *J. Geophys. Res., 79*, 521, 1974.

Frank, L. A., On the extraterrestrial ring current during geomagnetic storms, *J. Geophys. Res., 72*, 3753, 1967.

Frank, L. A., Relationship of the plasma sheet, ring current, trapping boundary, and plasmapause near the magnetic equator and local midnight, *J. Geophys. Res., 76*, 2265, 1971.

Fraser, B. J., Pc 1-2 observations of heavy ion effects by synchronous satellite ATS-6, *Planet. Space Sci., 30*, 1229, 1982.

Fraser, B. J. and R. L. McPherron, Pc 1-2 magnetic pulsation spectra and heavy ion effects at synchronous orbit: ATS 6 results, *J. Geophys. Res., 87*, 4560, 1982.

Fritz, T. A., and D. J. Williams, Initial observations of geomagnetically trapped alpha particles at the equator, *J. Geophys. Res., 78*, 4719, 1973.

Fritz, T. A., and B. Wilken, Substorm generated fluxes of heavy ions at the geostationary orbit, in *Magnetospheric Particles and Fields*, edited by B. M. McCormac, p. 171, D. Reidel, Dordrecht, Holland, 1976.

Geiss, J., H. Balsiger, P. Eberhardt, H. P. Walker, L. Weber, D. T. Young, and H. Rosenbauer, Dynamics of magnetospheric ion composition as observed by the GEOS mass spectrometer, *Space Sci. Rev., 22*, 537, 1978.

Gendrin, R., Wave particle interactions as an energy transfer mechanism between different particle species, *Space Sci. Rev., 34*, 271, 1983.

Gendrin, R., and A. Roux, Energization of helium ions by proton-induced hydromagnetic waves, *J. Geophys. Res., 85*, 4577, 1980.

Ghielmetti, A. G., R. G. Johnson, R. D. Sharp, and E. G. Shelley, The latitudinal, diurnal, and altitudinal distributions of upward flowing energetic ions of ionospheric origin, *Geophys. Res. Lett., 5*, 59, 1978.

Ghielmetti, A. G., R. D. Sharp, E. G. Shelley, and R. G. Johnson, Downward flowing ions and evidence for injection of ionospheric ions into the plasma sheet, *J. Geophys. Res., 84*, 5781, 1979.

Gloeckler, G., F. M. Ipavich, D. Hovestadt, M. Scholer, A. B. Galvin, and B. Klecker, Characteristics of suprathermal H^+ and He^{++} in plasmoids in the distant tail, *Geophys. Res. Lett., 11*, 1030, 1984.

Gloeckler, G., B. Wilken, W. Studemann, F. M. Ipavich, D. Hovestadt, D. C. Hamilton, and G. Kremser, First composition measurement of the bulk of the storm time ring current (1 to 300 keV/e) with AMPTE-CCE, *Geophys. Res. Lett., 12*, 325, 1985a.

Gloeckler, G., et al., The Charge-Energy-Mass (CHEM) spectrometer for 0.3 to 300 keV/e ions on AMPTE-CCE, *IEEE Trans. Geosci. Remote Sensing, GE-23*, 234, 1985b.

Gomberoff, L., and R. Neira, Convective growth rate of ion cyclotron waves in a H^+-He^+ and $H^+-He^+-O^+$ plasma, *J. Geophys. Res., 88*, 2170, 1983.

Gorney, D. J., A. Clarke, D. Rowley, J. Fennell, J. Luhmann, and P. Mizera, The distribution of ion beams and conics below 8000 km, *J. Geophys. Res., 86*, 83, 1981.

Gurgiolo, C., and J. L. Burch, DE 1 observations of the polar wind—A heated and unheated component, *Geophys. Res. Lett., 9*, 945, 1982.

Gurnett, D. A., and L. A. Frank, Thermal and suprathermal plasma densities in the outer magnetosphere, *J. Geophys. Res., 79*, 2355, 1974.

Gurnett, D. A., F. L. Scarf, R. W. Fredericks, and E. J. Smith, The ISEE-1 and ISEE-2 plasma wave investigation, *IEEE Trans. Geosci. Electron.*, GE-16, 225, 1978.

Gussenhoven, M. S., E. G. Mullen, and R. C. Sagalyn, CRRES/SPACERAD Experiment Descriptions, *Rep. AFGL-TR-85-0017*, Air Force Geophysics Lab., Bedford, MA, 1985.

Harris, K. K., and G. W. Sharp, OGO-V ion spectrometer, *IEEE Trans. Geosci. Electron.*, GE-7, 93, 1969.

Harris, K. K., G. W. Sharp, and C. R. Chappell, Observations of the plasmpause from OGO 5, *J. Geophys. Res.*, 75, 219, 1970.

Heelis, R. A., J. D. Winningham, M. Sugiura, and N. C. Maynard, Particle acceleration parallel and perpendicular to the magnetic field observed by DE-2, *J. Geophys. Res.*, 89, 3893, 1984.

Heikkila, W. J., Impulsive penetration and viscous interaction, in *Magnetospheric Boundary Layers*, ESA SP-148, p. 375, European Space Agency, Paris, 1979.

Hill, T. W., Origin of the plasma sheet, *Rev. Geophys. Space Phys.*, 12, 379, 1974.

Hill, T. W., Rates of mass, momentum, and energy transfer at the magnetopause, in *Magnetospheric Boundary Layers*, ESA SP-148, p. 325, European Space Agency, Paris, 1979.

Hones, E. W., Jr., Plasma sheet behavior during substorms, in *Magnetic Reconnection in Space and Laboratory Plasmas*, edited by E. W. Hones, Jr., Geophys. Monogr. Ser., Vol. 30, 178, 1984.

Hones, E. W., Jr., A. T. Y. Lui, S. J. Bame, and S. Singer, Prolonged tailward flow of plasma in the thinned plasma sheet observed at $r \sim 18 R_E$ during substorms, *J. Geophys. Res.*, 79, 1385, 1974.

Horwitz, J. L., The ionosphere as a source for magnetospheric ions, *Rev. Geophys. Space Phys.*, 20, 929, 1982.

Horwitz, J. L., C. R. Baugher, C. R. Chappell, E. G. Shelley, D. T. Young, and R. R. Anderson, ISEE 1 observations of thermal plasma in the vicinity of the plasmasphere during periods of quieting magnetic activity, *J. Geophys. Res.*, 86, 9989, 1981.

Horwitz, J. L., C. R. Baugher, C. R. Chappell, E. G. Shelley, and D. T. Young, Conical pitch angle distributions of very low-energy ion fluxes observed by ISEE 1, *J. Geophys. Res.*, 87, 2311, 1982.

Horwitz, J. L., C. R. Chappell, D. L. Reasoner, P. D. Craven, J. L. Green, and C. R. Baugher, Observations of low-energy plasma composition from the ISEE-1 and SCATHA satellites, in *Energetic Ion Composition in the Earth's Magnetosphere*, edited by R. G. Johnson, p. 263, Terra Sci. Publishing Co., Tokyo, 1983.

Horwitz, J. L., R. H. Comfort, and C. R. Chappell, Thermal ion composition measurements of the formation of the new outer plasmasphere and double plasmapause during storm recovery phase, *Geophys. Res. Lett.*, 11, 701, 1984.

Hovestadt, D., G. Gloeckler, C. Y. Fan, F. M. Ipavich, et al., The nuclear and ionic charge distribution particle experiments on the ISEE-1 and ISEE -C spacecraft, *IEEE Trans. Geosci. Electron.*, GE-16, 166, 1978.

Hovestadt, D., B. Klecker, E. Mitchell, J. F. Fennell, G. Gloeckler, and C. Y. Fan, Spatial distribution of $Z \geq 2$ ions in the outer radiation belt during quiet conditions, *Adv. Space Res.*, Vol. 1, 305, 1981.

Hultqvist, B., On the origin of hot ions in the disturbed dayside magnetosphere, *Planet. Space Sci.*, 31, 173, 1983.

Hultqvist, B., and H. Borg, Observations of energetic ions in inverted V events, *Planet. Space Sci.*, 26, 673, 1978.

Ipavich, F. M., and M. Scholer, Thermal and suprathermal protons and alpha particles in the Earth's plasma sheet, *J. Geophys. Res.*, 88, 150, 1983.

Ipavich, F. M., A. B. Galvin, G. Gloeckler, D. Hovestadt, B. Klecker, and M. Scholer, Energetic (>100 keV) O^+ ions in the plasma sheet, *Geophys. Res. Lett.*, 11, 504, 1984.

Ipavich, F. M., A. B. Galvin, M. Scholer, G. Gloeckler, D. Hovestadt, and B. Klecker, Suprathermal O^+ and H^+ ion behavior during the March 22, 1979 (CDAW 6) substorms, *J. Geophys. Res.*, 90, 1263, 1985.

Johnson, R. G., R. J. Strangeway, E. G. Shelley, J. M. Quinn, and S. M. Kaye, Hot plasma composition results from the SCATHA spacecraft, in *Energetic Ion Composition in the Earth's Magnetosphere*, edited by R. G. Johnson, p. 287, Terra Sci. Publishing Co., Tokyo, 1983.

Jones, D., Introduction to the S-300 wave experiment on board GEOS, *Space Sci. Rev.*, 22, 327, 1978.

Kaye, S. M., M. G. Kivelson, and D. J. Southwood, Evolution of ion cyclotron instability in the plasma convection system of the magnetosphere, *J. Geophys.Res.*, 84, 6397, 1979.

Kaye, S. M., R. G. Johnson, R. D. Sharp, and E. G. Shelley, Observations of transient H^+ and O^+ bursts in the equatorial magnetosphere, *J. Geophys. Res.*, 86, 1335, 1981a.

Kaye, S. M., E. G. Shelley, R. D. Sharp, and R. G. Johnson, Ion composition of zipper events, *J. Geophys. Res.*, 86, 3383, 1981b.

Kintner, P. M., and D. A. Gurnett, Observations of ion cyclotron waves within the plasmasphere by Hawkeye 1, *J. Geophys. Res.*, 82, 2314, 1977.

Kintner, P. M., M. C. Kelley, and F. S. Mozer, Electrostatic hydrogen cyclotron waves near one Earth radius altitude in the polar magnetosphere, *Geophys. Res. Lett.*, 5, 139, 1978.

Kintner, P. M., M. C. Kelley, R. D. Sharp, A. G. Ghielmetti, M. Temerin, C. Cattell, P. F. Mizera, and J. F. Fennell, Simultaneous observations of energetic (keV) upstreaming and electrostatic hydrogen cyclotron waves, *J. Geophys. Res.*, 84, 7201, 1979.

Klumpar, D. M., Transversely accelerated ions: An ionospheric source of hot magnetospheric ions, *J. Geophys. Res.*, 84, 4229, 1979.

Klumpar, D. M., W. K. Peterson, and E. G. Shelley, Direct evidence for two-stage (bimodal) acceleration of ionospheric ions, *J. Geophys. Res.*, 89, 10779, 1984.

Kozyra, J. U., T. E. Cravens, A. F. Nagy, E. G. Fontheim, and R. S. B. Ong, Effects of energetic heavy ions on electromagnetic ion cyclotron wave generation in the plasmapause region, *J. Geophys. Res.*, 89, 2217, 1984.

Kremser, G., W. Studemann, B. Wilken, G. Gloeckler, D. C. Hamilton, F. M. Ipavich, and D. Hovestadt, Charge state distributions of oxygen and carbon in the energy range 1 to 300 keV/e observed with AMPTE/CCE in the magnetosphere, AMPTE Scientific Preprint ASP 34, MPI fur Aeronomie, Katlenburg-Lindau, W. Germany, 1985.

Krimigis, S. M., and J. A. Van Allen, Geomagnetically trapped alpha particles, *J. Geophys. Res.*, 72, 5779, 1967.

Krimigis, S. M., and E. T. Sarris, Energetic particle bursts in the Earth's magnetotail, in *Dynamics of the Magnetosphere*, edited by S.-I. Akasofu, p. 599, D. Reidel, Holland, 1979.

Krimigis, S. M., G. Gloeckler, R. W. McEntire, T. A. Potemra, F. L. Scarf, and E. G. Shelley, Magnetic storm of September 4, 1984: A synthesis of ring current spectra and energy densities measured with AMPTE/CCE, *Geophys. Res. Lett.*, 12, 329, 1985.

Lemaire, J., Impulsive penetration of filamentary plasma elements into the magnetospheres of the Earth and Jupiter, *Planet. Space Sci.*, 25, 887, 1977.

Lennartsson, W., R. D. Sharp, E. G. Shelley, R. G. Johnson, and H. Balsiger, Ion composition and energy distribution during 10 magnetic storms, *J. Geophys. Res.*, 86, 4628, 1981.

Lennartsson, W., and E. G. Shelley, Survey of 0.1-16 keV/e plasma sheet ion composition, *J. Geophys. Res.*, in press, 1986.

Lockwood, M., Thermospheric control of the auroral source of O^+ ions for the magnetosphere, *J. Geophys. Res.*, 89, 301, 1984.

Lockwood, M., J. H. Waite, Jr., T. E. Moore, J. F. E. Johnson, and

C. R. Chappell, A new source of suprathermal O⁺ ions near the dayside polar cap boundary, *J. Geophys. Res., 90*, 1611, 1985.

Lui, A. T. Y., T. E. Eastman, D. J. Williams and L. A. Frank, Observations of ion streaming during substorms, *J. Geophys. Res., 88*, 7753, 1983.

Lundin, R., I. Sandahl, B. Hultqvist, A. Galeev, O. Likhin, A. Omelchenko, N. Pissarenko, O. Vaisberg, and A. Zacharov, First observations of the hot ion composition in the high latitude magnetospheric boundary layer by means of PROGNOZ-7, in *Magnetospheric Boundary Layers, ESA SP-148*, p. 91, European Space Agency, Paris, 1979.

Lundin, R., L. R. Lyons, and N. Pissarenko, Observations of the ring current composition at L < 4, *Geophys. Res. Lett., 7*, 425, 1980.

Lundin, R., B. Hultqvist, E. Dubinin, A. Zackarov, and N. Pissarenko, Observations of outflowing ion beams on auroral field lines at altitudes of many Earth radii, *Planet. Space Sci., 30*, 715, 1982.

Lyons, L. R., Ring current dynamics and plasma sheet sources, in *Proc. Conf. Achievements of the IMS*, ESA SP-217, p.233, European Space Agency, Paris, 1984.

Lyons, L. R., and D. S. Evans, The inconsistency between proton charge exchange and the observed ring current decay, *J. Geophys. Res., 81*, 6197, 1976.

Lyons, L. R., and D. J. Williams, A source for the geomagnetic storm main phase ring current, *J. Geophys. Res., 85*, 523, 1980.

Lyons, L. R., and T. E. Moore, Effects of charge exchange on the distribution of ionospheric ions trapped in the radiation belts near synchronous orbit, *J. Geophys. Res., 86*, 5885, 1981.

Lyons, L. R., and T. W. Speiser, Evidence for current sheet acceleration in the geomagnetic tail, *J. Geophys. Res., 87*, 2276, 1982.

Lyons, L. R., and D. S. Evans, An association between discrete aurora and energetic particle boundaries, *J. Geophys. Res., 89*, 2395, 1984.

McEntire, R. W., et al., The Medium Energy Particle Analyzer (MEPA) on the AMPTE/CCE spacecraft, *IEEE Trans. Geosci. Remote Sensing, GE-23*, 230, 1985.

McIlwain, C. E., Plasma convection in the vicinity of the geosynchronous orbit, in *Earth's Magnetospheric Processes*, edited by B. M. McCormac, p. 268, D. Reidel, Dordrecht, Holland, 1972.

McIlwain, C. E., Substorm injection boundaries, in *Magnetospheric Physics*, edited by B. M. McCormac, p. 143, D. Reidel, Dordrecht, Holland, 1974.

Mauk, B. H., and C. E. McIlwain, ATS-6 UCSD auroral particles experiment, *IEEE Trans. Aerospace Electron. Systems, AES-11*, 1125, 1975.

Mauk, B. H., and R. L. McPherron, An experimental test of the electromagnetic ion cyclotron instability within the Earth's magnetosphere, *Phys. Fluids, 23*, 2111, 1980.

Mauk, B. H., C. E. McIlwain, and R. L. McPherron, Helium cyclotron resonance within the Earth's magnetosphere, *Geophys. Res. Lett., 8*, 103, 1981.

Mizera, P. F., and J. F. Fennell, Signatures of electric fields from high and low altitude particle distributions, *Geophys. Res. Lett., 4*, 311, 1977.

Mizera, P. F., et al., The aurora inferred from S3-3 particles and fields, *J. Geophys. Res., 86*, 2329, 1981.

Mobius, E., F. M. Ipavich, M. Scholer, G. Gloeckler, D. Hovestadt, and B. Klecker, Observations of a nonthermal ion layer in the plasma sheet boundary during substorm recovery, *J. Geophys. Res., 85*, 5143, 1980.

Moore, T. E., Modulation of terrestrial ion escape flux composition (by low-altitude acceleration and charge exchange chemistry), *J. Geophys. Res., 85*, 2011, 1980.

Moore, T. E., Superthermal ionospheric outflows, *Rev. Geophys. Space Phys., 22*, 264, 1984.

Moore, T. E., R. L. Arnoldy, J. Feynman, and D. A. Hardy, Propagating substorm injection fronts, *J. Geophys. Res., 86*, 6713, 1981.

Moore, T. E., C. R. Chappell, M. Lockwood, and J. H. Waite, Jr., Superthermal ion signatures of auroral acceleration processes, *J. Geophys. Res., 90*, 1611, 1985a.

Moore, T. E., A. P. Biddle, J. H. Waite, and T. L. Killeen, Auroral zone effects on hydrogen geocorona structure and variability, *Planet. Space Sci., 33*, 499, 1985b.

Mozer, F. S., C. W. Carlson, M. K. Hudson, R. B. Torbert, B. Parady, J. Yatteau, and M. C. Kelley, Observations of paired electrostatic shocks in the polar magnetosphere, *Phys. Rev. Lett., 38*, 292, 1977.

Olsen, R. C., Equatorially trapped plasma populations, *J. Geophys. Res., 86*, 11235, 1981.

Perraut, S., R. Gendrin, P. Robert, A. Roux, C. De Villedary, and D. Jones, ULF waves observed with magnetic and electric sensors on GEOS-1, *Space Sci. Rev., 22*, 347, 1978.

Perraut, S., A. Roux, P. Robert, R. Gendrin, J. A. Savaud, J. M. Bosqued, G. Kremser, and A. Korth, A systematic study of ULF waves above F_{H+} from GEOS 1 and 2 measurements and their relationships with proton ring distributions, *J. Geophys. Res., 87*, 6219, 1982.

Quinn, J. M., and C. E. McIlwain, Bouncing ion clusters in the earth's magnetosphere, *J. Geophys. Res., 84*, 7365, 1979.

Quinn, J. M., and R. G. Johnson, Composition measurements of equatorially trapped ions near geosynchronous orbit, *Geophys. Res. Lett., 9*, 777, 1982.

Quinn, J. M., and D. J. Southwood, Observations of parallel ion energization in the equatorial region, *J. Geophys. Res., 87*, 10536, 1982.

Quinn, J. M., and R. G. Johnson, Observation of ionospheric source cone enhancements at the substorm injection boundary, *J. Geophys. Res., 90*, 4211, 1985.

Rauch, J. ., and A. Roux, Ray tracing of ULF waves in a multicomponent magnetospheric plasma: Consequences for the generation mechanism of ion cyclotron waves, *J. Geophys. Res., 87*, 8191, 1982.

Reasoner, D. L., C. R. Chappell, S. A. Fields, and W. J. Lewter, Light ion mass spectrometer for space plasma investigations, *Rev. Sci. Instrum., 53*, 441, 1982.

Retterer, J. M., T. Chang, and J. R. Jasperse, Ion acceleration in the subauroral region: A Monte Carlo model, *Geophys. Res. Lett., 10*, 583, 1983.

Richardson, J. D., J. F. Fennell, and D. R. Croley, Jr., Observations of field-aligned ion and electrons from SCATHA (P78-2), *J. Geophys. Res., 86*, 10105, 1981.

Roux, A., S. Perraut, J. L. Rauch, C. de Villedary, G. Kremser, A. Korth, and D. T. Young, Wave-particle interactions near Ω_{He+} observed on board GEOS 1 and 2, 2. Generation of ion cyclotron waves and heating of He⁺ ions, *J. Geophys. Res., 87*, 8174, 1982.

Roux, A., N. Cornilleau-Wehrlin, and J. L. Rauch, Acceleration of thermal electrons by ICWs propagating in a multicomponent magnetospheric plasma, *J. Geophys. Res., 89*, 2267, 1984.

Russell, C. T., The ISEE 1 and 2 fluxgate magnetometer, *IEEE Trans. Geosci. Electron., GE-16*, 239, 1978.

Russell, C. T., and R. L. McPherron, The magnetotail and substorms, *Space Sci. Rev., 15*, 205, 1973.

Sarris, E. T., S. M. Krimigis, A. T. Y. Lui, K. L. Ackerson, L. A. Frank, and D. J. Williams, Relationship between energetic particles and plasmas in the distant plasma sheet, *Geophys. Res. Lett., 8*, 349, 1981.

Scholer, M., Energetic ions and electrons and their acceleration processes in the magnetotail, in *Magnetic Reconnection in Space and Laboratory Plasmas*, edited by E. W. Hones, Jr., Geophys. Monogr. Ser., Vol. 30, 216, 1984.

Scholer, M., D. Hovestadt, G. Hartmann, J. B. Blake, J. F. Fennell, and G. Gloeckler, Low-altitude measurements of precipitating protons, alpha particles, and heavy ions during the geomagnetic storm on March 26-27, 1976, *J. Geophys. Res., 84*, 79, 1979.

Schulz, M. and H. C. Koons, Thermalization of colliding ion streams beyond the plasmapause, *J. Geophys. Res., 77*, 248, 1972.

Schulz, M., and L. J. Lanzerotti, *Particle Diffusion in the Radiation Belts*, Springer Verlag, New York, 1974.

Schulz, M., Geomagnetically trapped radiation, *Space Sci. Rev., 17*, 481, 1975.

Serbu, G. P., and E. J. R. Maier, Observations from OGO 5 of the thermal ion density and temperature within the magnetosphere, *J. Geophys. Res., 75*, 6102, 1970.

Sharp, R. D., R. G. Johnson, E. G. Shelley, and K. K. Harris, Energetic O^+ ions in the magnetosphere, *J. Geophys. Res., 79*, 1844, 1974.

Sharp, R. D., R. G. Johnson, and E. G. Shelley, Observation of an ionospheric acceleration mechanism producing energetic (keV) ions primarily normal to the geomagnetic field direction, *J. Geophys. Res., 82*, 3324, 1977.

Sharp, R. D., R. G. Johnson, and E. G. Shelley, Energetic particle measurements from within ionospheric structures responsible for auroral acceleration processes, *J. Geophys. Res., 84*, 480, 1979.

Sharp, R. D., D. L. Carr, W. K. Peterson, and E. G. Shelley, Ion streams in the magnetotail, *J. Geophys. Res., 86*, 4639, 1981.

Sharp, R. D., W. Lennartsson, W. K. Peterson, and E. G. Shelley, The origins of the plasma in the distant plasma sheet, *J. Geophys. Res., 87*, 10, 420, 1982.

Sharp, R. D., W. Lennartsson, W. K. Peterson, and E. Ungstrup, The mass dependence of wave particle interactions as observed with the ISEE-1 energetic ion mass spectrometer, *Geophys. Res. Lett., 10*, 651, 1983.

Shawhan, S. D., Magnetospheric plasma wave research 1975-1978, *Rev. Geophys. Space Phys., 17*, 705, 1979.

Shawhan, S. D., D. A. Gurnett, D. I. Odem, R. A. Helliwell, and C. G. Park, The plasma wave and quasistatic electric field instrument (PWI) for Dynamics Explorer-A, *Space Sci. Instrum., 5*, 535, 1981.

Shelley, E. G., Circulation of energetic ions of terrestrial origin in the magnetosphere, *Adv. Space Res., 5*, No. 4, 401, 1985.

Shelley, E. G., R. G. Johnson, and R. D. Sharp, Satellite observations of energetic heavy ions during a geomagnetic storm, *J. Geophys. Res., 77*, 6104, 1972.

Shelley, E. G., R. D. Sharp, and R. G. Johnson, Satellite observations of an ionospheric acceleration mechanism, *Geophys. Res. Lett., 3*, 654, 1976.

Shelley, E. G., R. D. Sharp, R. G. Johnson, J. Geiss, P. Eberhardt, H. Balsiger, G. Haerendel, and H. Rosenbauer, Plasma composition experiment on ISEE-A, *IEEE Trans. Geosci. Electron., GE-16*, 266, 1978.

Shelley, E. G., D. A. Simpson, T. C. Sanders, E. Hertzberg, H. Balsiger, and A. Ghielmetti, The energetic ion composition spectrometer (EICS) for the Dynamics Explorer-A, *Space Sci. Instrum., 5*, 443, 1981.

Shelley, E. G., W. K. Peterson, A. G. Ghielmetti, and J. Geiss, The polar ionosphere as a source of energetic magnetospheric plasma, *Geophys. Res. Lett., 9*, 941, 1982.

Shelley, E. G., A. Ghielmetti, E. Hertzberg, S. J. Battel, K. Altwegg-von Burg, and H. Balsiger, The AMPTE/CCE hot-plasma composition experiment (HPCE), *IEEE Trans. Geosci. Remote Sens., GE-23*, 241, 1985a.

Shelley, E. G., D. M. Klumpar, W. K. Peterson, A. Ghielmetti, H. Balsiger, J. Geiss and H. Rosenbauer, AMPTE/CCE observations of the plasma composition below 17 keV during the September 4, 1984 magnetic storm, *Geophys. Res. Lett., 12*, 321, 1985b.

Singh, N., W. J. Raitt, and F. Yasuhara, Low-energy ion distribution functions on a magnetically quiet day at geostationary altitude (L = 7), *J. Geophys. Res., 87*, 681, 1982.

Singh, N., and R. W. Schunk, Numerical simulations of counter-streaming plasmas and their relevance to interhemispheric flows, *J. Geophys. Res., 88*, 7867, 1983.

Smith, P. H., and R. A. Hoffman, Ring current distributions during the magnetic storms of December 16-18, 1971, *J. Geophys. Res., 78*, 4731, 1973.

Smith, P. H., N. K. Bewtra, and R. A. Hoffman, Inference of the ring current ion composition by means of charge exchange decay, *J. Geophys. Res., 86*, 3470, 1981.

Smith, R. L., and N. Brice, Propagation in multicomponent plasmas, *J. Geophys. Res., 69*, 5029, 1964.

Sojka, J. J., R. W. Shunk, J. F. E. Johnson, J. H. Waite, and C. R. Chappell, Characteristics of thermal and suprathermal ions associated with the dayside plasma trough as measured by the Dynamics Explorer Retarding Ion Mass Spectrometer, *J. Geophys. Res., 88*, 7895, 1983.

Sojka, J. J., and G. L. Wrenn, Refilling of geosynchronous flux tubes as observed at the equator by GEOS-2, *J. Geophys. Res., 90*, 6379, 1985.

Solomon, J., and O. Picon, Charge exchange and wave-particle interaction in the proton ring current, *J. Geophys. Res., 86*, 3335, 1981.

Speiser, T. W., Particle trajectories in model current sheets, 1, Analytical solutions, *J. Geophys. Res., 70*, 4219, 1965.

Spjeldvik, W. N., Transport, charge exchange and loss of energetic heavy ions in the Earth's radiation belts: Applicability and limitations of theory, *Planet. Space Sci., 29*, 1215, 1981.

Spjeldvik, W. N., and T. A. Fritz, Theory for charge states of energetic oxygen ions in the Earth's radiation belts, *J. Geophys. Res. 83*, 1583, 1978.

Spjeldvik, W. N., and T. A. Fritz, Experimental determination of geomagnetically trapped energetic heavy ion fluxes, in *Energetic Ion Composition in the Earth's Magnetosphere*, edited by R. G. Johnson, p. 369, Terra Scientific Publishing Co., Tokyo, 1983.

Steckelmacher, W., Energy analysers for charged particle beams, *J. Sci. Instrum., 6*, 1061, 1973.

Strangeway, R. J., and R. G. Johnson, Energetic ion mass composition as observed at near-geosynchronous and low altitudes during the storm period of February 21 and 22, 1979, *J. Geophys. Res., 89*, 8919, 1984.

Taylor, H. A., H. C. Brinton, and C. R. Smith, Positive ion composition in the magnetoionosphere obtained from the OGO-A satellite *J. Geophys. Res., 70*, 5769, 1965.

Taylor, H. A., H. C. Brinton, and M. W. Pharo, Contraction of the plasmasphere during geomagnetically disturbed periods, *J. Geophys. Res., 73*, 961, 1968.

Taylor, W. W. L., B. K. Parady, and L. J. Cahill, Jr., Explorer 45 observations of 1 to 30-Hz magnetic fields near the plasmapause during magnetic storms, *J. Geophys. Res., 80*, 1271, 1975.

Tinsley, B. A., Evidence that the recovery phase ring current consists of helium ions, *J. Geophys. Res., 81*, 6193, 1976.

Tums, E., G. Gloeckler, C. Y. Fan, J. Cain, and R. Sciambi, Instrument to measure energy and charge of low energy interplanetary particles, *IEEE Trans. Nucl. Sci., NS-21*, 210, 1974.

Ungstrup, E., D. M. Klumpar, and W. J. Heikkila, Heating of ions to superthermal energies in the topside ionosphere by electrostatic ion cyclotron waves, *J. Geophys. Res., 84*, 4289, 1979.

Van Allen, J. A., B. A. Randall, and S. M. Krimigis, Energetic carbon, nitrogen and oxygen nuclei in the Earth's outer radiation zone, *J. Geophys. Res., 75*, 6085, 1970.

Vasyliunas, V. M., Deep space plasma measurements, *Methods Exp. Phys., 9B*, 49, 1971.

Waite, J. H., J. L. Horwitz, and R. H. Comfort, Diffusive equilibrium distributions of He$^+$ in the plasmasphere, *Planet. Space Sci., 32*, 611, 1984.

Waite, J. H., Jr., T. Nagai, J. F. E. Johnson, C. R. Chappell, J. L. Burch, T. L. Killeen, P. B. Hays, G. R. Carignan, W. K. Peterson, and E. G. Shelley, Escape of suprathermal O$^+$ ions in the polar cap, *J. Geophys. Res., 90*, 1619, 1985.

White, O. R., *The Solar Output and Its Variation*, Colorado Associated University Press, Boulder, CO, 1977.

Wilken, B., T. A. Fritz, and W. Studemann, Experimental techniques for ion composition measurements in space, *Nuc. Instrum. Meth., 196*, 161, 1982.

Williams, D. J., Ring current composition and sources, in *Dynamics of the Magnetosphere*, edited by S.-I. Akasofu, p. 407, D. Reidel, Dordrecht, Holland, 1980.

Williams, D. J., Energetic ion beams at the edge of the plasma sheet: ISEE 1 observations plus a simple explanatory model, *J. Geophys. Res., 86*, 5507, 1981.

Williams, D. J., The Earth's ring current: causes, generation and decay, in *Progress in Solar-Terrestrial Physics*, edited by J. G. Roederer, p.223, D. Reidel, Dordrecht, Holland, 1983.

Willams, D. J., E. Keppler, T. A. Fritz, B. Wilken, and G. Wibberenz, The ISEE 1 and 2 medium energy particles experiment, *IEEE Trans. Geosci. Electron., GE-16*, 270, 1978.

Williams, D. J., and M. Sugiura, The AMPTE charge composition explorer and the 4-7 September 1984 geomagnetic storm, *Geophys. Res. Lett., 12*, 305, 1985.

Wolf, R. A., M. Harel, R. W. Spiro, G. H. Voigt, P. H. Reiff, and C. K. Chen, Computer simulation of inner magnetospheric dynamics for the magnetic storm of July 29, 1977, *J. Geophys. Res., 87*, 5949, 1982.

Yau, A. W., B. A. Whalen, W. K. Peterson, and E. G. Shelley, Distribution of upflowing ionospheric ions in the high-altitude polar cap and auroral ionosphere, *J. Geophys. Res., 89*, 5507, 1984.

Yau, A. W., P. H. Beckwith, W. K. Peterson and E. G. Shelley, Long-term (solar cycle) and seasonal variations of upflowing ionospoheric ion events at DE 1 altitudes, *J. Geophys. Res., 90*, 6395, 1985.

Young, D. T., Ion composition measurements in magnetospheric modeling, in *Quantitative Modeling of Magnetospheric Processes*, edited by W. P. Olson, *Geophys. Monogr. Ser., Vol. 21*, 340, 1979.

Young, D. T., Heavy ions in the outer magnetosphere, *J. Geophys., 52*, 167, 1983a.

Young, D. T., Near-equatorial magnetospheric particles from ~ 1 eV to ~ 1 MeV, *Rev. Geophys. Space Phys., 21*, 402, 1983b.

Young, D. T., S. Perraut, A. Roux, C. de Villedary, R. Gendrin, A. Korth, G. Kremser, and D. Jones, Wave-particle interactions near Ω_{He^+} observed on GEOS 1 and 2, 1. Propagation of ion cyclotron waves in He$^+$- rich plasma, *J. Geophys. Res., 86*, 6755, 1981.

Young, D. T., J. Geiss, H. Balsiger, P. Eberhardt, A. Ghielmetti, and H. Rosenbauer, Discovery of He^{2+} and O^{2+} ions of terrestrial origin in the outer magnetosphere, *Geophys. Res. Lett., 4*, 561, 1977.

Young, D. T., H. Balsiger, and J. Geiss, Correlations of magnetospheric ion composition with geomagnetic and solar activity, *J. Geophys. Res., 87*, 9077, 1982.

Section II. HIGH LATITUDE PROCESSES

LOW-ALTITUDE TRANSVERSE IONOSPHERIC ION ACCELERATION

A.W. Yau and B.A. Whalen

Herzberg Institute of Astrophysics, National Research Council of Canada, Ottawa, Canada K1A 0R6

P.M. Kintner

School of Electrical Engineering, Cornell University, Ithaca, NY 14853 U.S.A.

Abstract. Ions transversely accelerated to hundreds of eV have been observed at 400-600 km altitudes from a number of rocket payloads and near 1400 km altitude from Isis-2. We present preliminary results from a high-altitude ($\approx$1000 km apogee) sounding rocket experiment into the source region of a transverse ion acceleration event. The event was observed during a large auroral substorm ($\gtrsim$500 γ) and at 500-600 km altitude. Inside the event, ions accelerated up to a few hundred eV perpendicular to the local magnetic field were observed. At the bottom edge of the ion acceleration region, the ion pitch-angle distribution peaked at 90° (within 1°). The energy spectra of the accelerated ions were power-law like; the ion intensity was proportional to $E^{-\alpha}$ where α was near 2. The present observation, in combination with less detailed measurements from previous experiments, indicates that transverse ion acceleration at low altitudes is a common occurrence during auroral substorms, and accounts for intense fluxes of upflowing ionospheric ions in the active-time high-altitude ionosphere and magnetosphere.

Introduction

We report energetic particle observations in the source region of transversely accelerated ions (TAI). The TAI event was observed at 500-600 km altitude during a large auroral substorm ($\gtrsim$500 nT). Inside the event, ions of ionospheric origin were energized up to $\approx$300 eV and perpendicular to the local geomagnetic field. The observation was made from a high-altitude ($\approx$1000 km apogee) Black Brant X sounding rocket. The payload was codenamed MARIE and was a follow-on to two earlier payloads in which TAI events below the O-H charge-exchange altitude were first observed in situ [Whalen et al., 1978; Yau et al., 1983].

The two earlier TAI observations were made in the expansive phases of two auroral substorms. The first event was observed at 400-550 km altitude, and featured TAI up to $\approx$500 eV at intensities of $\approx 10^7$ $(cm^2\ s\ sr\ keV)^{-1}$. The second event was observed at $\approx$580 km altitude, near the rocket apogee, and had lower TAI intensity. The aim of the present payload was three-fold: to explore the possibility of common occurrence of TAI at low altitudes during auroral substorms (to the extent possible using sounding rockets); to study the behaviour of TAI immediately above the source region; and to obtain first energetic ion composition measurements and detailed wave measurements inside TAI source region.

Shelley et al. [1976] reported the first direct evidence of upflowing energetic (>keV) ions of ionospheric origin (UFI) from ion composition data on S3-3 near 1 R_e altitude. Sharp et al. [1977] found UFI events with upward O^+ fluxes of clearly ionospheric origin peaking in the 130°-140° pitch-angle range, and interpreted them as resulting from ions accelerated at 90° at lower altitudes (i.e., TAI), followed by adiabatic expansion away from the ionosphere.

Direct observations of TAI have been made on the Isis-1 and -2 satellites above 2700 km and at 1400 km, respectively [Klumpar, 1979], and on the two sounding rockets noted above at 400-600 km. However, the importance (i.e., occurrence frequency and intensity) of low-altitude TAI and its relationship with ion conics at higher altitudes remain to be established. In particular, since O^+ ions originating below the O-H charge-exchange altitude region (above 500 km near solar minimum and higher near solar maximum) are expected to be lost to charge-exchange with neutral hydrogen under charge-exchange equilibrium conditions, the question arises as to whether TAI below charge-exchange altitudes are indeed the source of O^+ ion conics observed at high altitudes.

Sounding rockets with $\lesssim$1000 km apogee are well suited for observing low-altitude TAI events as they make possible particle measurements of high spatial and angular resolutions, and wave measurements of limited Doppler shift and broadening, relative to orbiting satellites. The aim of this report is to present the preliminary results

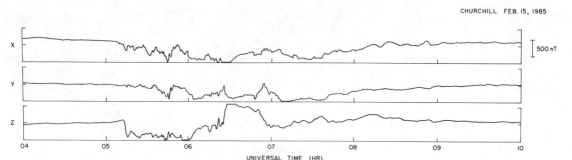

Fig. 1. Ground magnetogram at Churchill, Canada, on February 15, 1985.

of the particle measurements from the MARIE payload and to discuss their theoretical significance.

Experiment

The MARIE payload was launched on a Black Brant X rocket (BBX-01) from Churchill Research Range, Canada, on February 15, 1985, at 0525:10 UT, into the expansive phase of the second of a series of multiple substorms, and it reached an apogee of 978 km at 0534:42 UT. Figure 1 is the ground magnetogram at Churchill from 04 h to 10 h, and shows a negative bay of ≈500 nT at 0545 UT.

The payload was instrumented with a series of electrostatic-analyser electron and ion spectrometers in the 47 eV to 23.7 keV range, ion composition spectrometer in the 0.1-2 keV range, high-sensitivity fluxgate magnetometers, ion drift meters, Langmuir probes and wave receivers. Only data from the electrostatic analysers are presented in this preliminary report.

Results

Figure 2 summarises the electrostatic-analyser data in the upleg. The figure shows (a) the range of pitch-angles sampled by the ion spectrometer as a result of rocket precession and spin motion, (b) 47-366 eV ion fluxes at 0°-90°; (c) corresponding fluxes at 90°-120°; (d) 0.89-22.9 keV ion fluxes at 0°-90°; (e) corresponding fluxes at 90°-120°; and (f) 0.048-23.7 keV electron fluxes at 0°-90°. Ions between α ≈ 10° and ≈160°

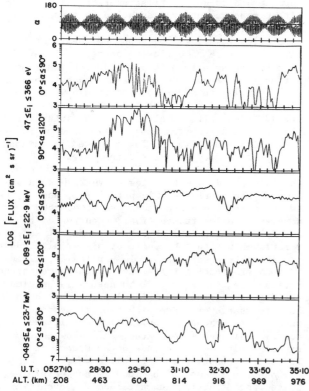

Fig. 2. BBX-01 MARIE summary data. (a) ion pitch angle; (b) 47-366 eV ion flux at 0°-90° and (c) at 90°-120°; (d) 0.89-22.9 keV ion flux at 0°-90° and (e) at 90°-120°; (f) 0.048-23.7 keV electron flux at 0°-90°.

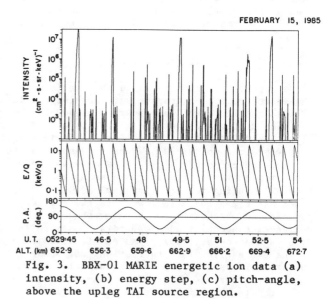

Fig. 3. BBX-01 MARIE energetic ion data (a) intensity, (b) energy step, (c) pitch-angle, above the upleg TAI source region.

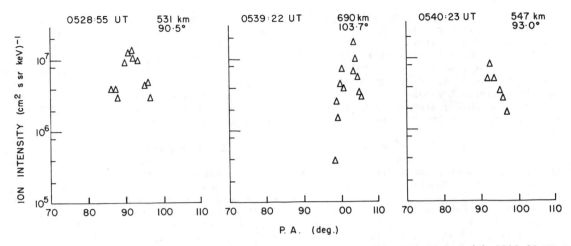

Fig. 4. Pitch-angle distribution of TAI at (a) 0528:55, (b) 0539:22 and (c) 0540:23 UT.

were sampled; α > 90° corresponds to upward moving particles. The ion spectrometer stepped through the energy ranges in panels (b-c) and (d-e) in 6 and 9 logarithmic steps, respectively. The flux values in these panels were obtained by integrating the measured intensity data and averaging the integrated data over the pitch-angle range. In general, not all of the 6 or 9 energy steps are sampled at a particular pitch-angle, nor is the full pitch-angle range sampled at a particular energy step. The aliasing in energy-angle sampling explains the apparent, irregular fluctuations in the averaged fluxes, particularly where the angular distribution is strongly peaked (see Figures 3 and 4).

Intense fluxes of transversely accelerated ions (ions peaked at ≳90° pitch angles) at 47-366 eV were present at ≈540 km altitude and above, and appeared as a >5-fold enhancement in the 90°-120° ion flux (panel c) over the downcoming flux (panel b) between 0529:00 and 0530:10. It is important to note that the ion distribution peaked at 90° (within a degree) near 0529:00 and the bottom edge of the region of TAI flux (see Figure 4a below), and that it peaked at larger angles (∼100°) at higher altitudes (above the source region; see Figure 3 below). No corresponding enhancement was present at keV energy (panels d and e). The 90°-120° flux at 0.89-22.9 keV was very comparable to, and slightly lower than, the 0°-90° flux throughout the upleg. Likewise, the 0°-90° and 90°-120° fluxes at 47-366 eV (panels b and c) were also comparable at other times in the upleg.

The intense fluxes of hundred-eV TAI were again encountered in the downleg at 694-529 km altitude, and appeared as a >5-fold enhancement in the 90°-120° ion flux at 47-366 eV over the 0°-90° flux between 0539:20 and 0540:30 UT.

Figure 3 shows the measured ion intensity, energy, and pitch-angle data in high-time resolution for the period from 0529:45 to 0529:54, when the payload was at 653-673 km altitude, about 130-150 km above the bottom edge of the TAI region. The figure shows that enhancement in ion intensity was observed only when the spectrometer was stepping through the hundred-eV steps and sampling at greater than 90° pitch-angle. Note in particular the two enhancements near 0529:52.5 (670 km altitude) that peaked near 100° and varied significantly in peak intensity. Assuming adiabatic invariance, the distribution maps down to a source altitude of ≈530 km, consistent with the observation of 90°-peaked distribution at that altitude. The variation may be temporal in origin but may also be due to azimuthal asymmetry in the intensity distribution.

Figure 4 shows the angular distribution of the TAI at (a) 0528:55 UT and 531 km upleg, when intense TAI fluxes were first encountered; (b) at 0539:22 UT and 690 km downleg; and (c) at 0540:23 UT and 547 km downleg, near the bottom edge of the downleg TAI event. In (a), the distribution

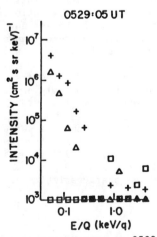

Fig. 5. Ion energy spectra at 0529:05 UT. Squares: 30°-60°; triangles: 60°-90°; pluses: 90°-120°.

peaked at 90.5°, suggesting that the payload was very near or inside the source region of transverse ion acceleration at this time. In contrast, the distribution in (b) peaked at 104°, suggesting that the payload was immediately above the TAI region at this time. Assuming adiabatic invariance, the distribution maps down to a source altitude of ≈550 km, suggesting that the TAI region is located near 550 km, consistent with the distribution in (c). Note that all three distributions were narrow. Distribution (a) had a FWHM of ≈5°, which is comparable to the instrument angular response, and suggests that the TAI region was limited to 100 km or less in altitudinal extent.

Figure 5 shows the measured ion energy spectra at 0529:05 UT, inside the upleg TAI source region at 30°-60° (squares), 60°-90° (triangles) and 90°-120° (pluses). The 90°-120° spectrum was power-law like, $j(E) \propto E^{-2.3}$, and a factor of 3-5 above the 60°-90° spectrum, at all energies up to ≈400 eV. Note that at keV energy, the 90°-120° intensity was lower than the 30°-60° and 60°-90° intensities, indicating that the keV ions are of magnetospheric origins.

Summary and Discussions

Figures 2-5 show energetic ion results from the MARIE payload, in and above the source region of a low-altitude (500-600 km) transverse ion acceleration event during a large auroral substorm. Ions accelerated up to a few hundred eV and perpendicular to the local magnetic field (within 1°) were observed inside the event. The intensity spectrum of the accelerated ions was power-law like, $j(E) \propto E^{-\alpha}$ where α varied between 2 and 2.5; α ≈ 2.3 in Figure 5.

The present observation, together with two earlier, less detailed observations of TAI, which were also made in the expansive phases of auroral substorms and at 400-600 km altitude, establishes a number of important characteristics of low-altitude transverse ion acceleration events:

1. TAI events at low altitudes (down to 400 km, and below the O-H charge-exchange altitude region) appear to be a frequent occurrence during auroral substorms.
2. Inside low-altitude TAI regions, ions are accelerated perpendicularly up to a few hundred eV and $\gtrsim 10^7$ $(cm^2\ s\ sr\ keV)^{-1}$ intensity.
3. The energization process is perpendicular to the local magnetic field; the ion pitch-angle distribution peaks at 90° to within 1°, and has angular width of 10° or less.

These observed characteristics place very serious constraints on possible theoretical energization mechanisms. A detailed examination of the various proposed mechanisms using the particle data above and other data (wave, composition, and field data) in the flight is currently in progress.

The above characteristics also underline the importance of TAI events below the O-H charge-exchange region as a source of ion conics at higher altitudes. Moore [1980] examined the fate of moderately energized (to ≈10 eV) oxygen ions travelling upward through the O^+-H charge-exchange altitude region; he found that the loss rate of these ions due to collisional charge-exchange in the O-H charge-exchange region is small compared with the equilibrium rate, and that a substantial fraction of them may reach high altitude. Hultqvist [1983] examined ion compositon data from a storm-time Prognoz-7 pass through the dayside magnetosphere from the point of view of required cold plasma source density, and inferred that the observed energetic (<16 keV) O^+ ions must originate from a narrow altitude region below 1000 km. The present observation provides direct evidence in support of Hultqvist's conclusion.

References

Hultqvist, B., On the origin of the hot ions in the disturbed dayside magnetosphere, Planet. Space Sci., 31, 173-184, 1983.

Klumpar, D.M., Transversely accelerated ions: An ionospheric source of hot magnetospheric ions, J. Geophys. Res., 84, 4229-4237, 1979.

Moore, T.E., Modulation of terrestrial ion escape flux composition (by low-altitude acceleration and charge exchange chemistry), J. Geophys. Res., 85, 2011-2016, 1980.

Sharp, R.D., R.G. Johnson, and E.G. Shelley, Observations of an ionospheric aceleration mechanism producing energetic (keV) ions primarily normal to the geomagnetic field direction, J. Geophys. Res., 82, 3324-3328, 1977.

Shelley, E.G., R.D. Sharp, and R.G. Johnson, Satellite observations of an ionospheric acceleration mechanism, Geophys. Res. Lett., 3, 654-657, 1976.

Whalen, B.A., W. Bernstein, and P.W. Daly, Low altitude acceleration of ionospheric ions, Geophys. Res. Lett., 5, 55-58, 1978.

Yau, A.W., B.A. Whalen, A.G. McNamara, P.J. Kellogg, and W. Bernstein, Particle and wave observations of low-altitude ionospheric ion acceleration events, J. Geophys. Res., 88, 341-355, 1983.

TRANSVERSE AURORAL ION ENERGIZATION OBSERVED ON DE-1 WITH SIMULTANEOUS PLASMA WAVE AND ION COMPOSITION MEASUREMENTS

W.K. Peterson and E.G. Shelley

Lockheed Palo Alto Research Laboratories
Palo Alto. California 94304

S.A. Boardsen and D.A. Gurnett

The University of Iowa. Iowa City, Iowa 52242

Abstract. The abundance of oxygen ions observed flowing into the magnetosphere cannot be explained by a single-step, parallel acceleration mechanism. Some transverse energization of ionospheric ions is required and has been observed from ionospheric altitudes to the plasma lobes in the earth's magnetotail. Progress in understanding the nature of the various transverse energization mechanisms has been slow because of the relative lack of examples with sufficiently resolved (in time, energy, and frequency) particle and wave data.

Starting in early 1984. the Dynamics Explorer-1 satellite systematically acquired a coordinated set of high time resolution plasma and plasma wave observations from the earth's auroral zone. We have selected several intervals from the 0 to 10 kHz wideband data from the Dynamics Explorer-1 satellite with intense low frequency emissions and evidence of harmonic structure and have examined in detail the high resolution ion data obtained simultaneously. In this report we present detailed data from events in the cusp and evening auroral zone and comment briefly on how these data can be used to more fully understand auroral acceleration processes.

Introduction

Energetic (> 100 eV) ionospheric ions are observed in abundance in the earth's magnetosphere. Johnson [1983] has compiled a set of papers reviewing the work of the last decade in magnetospheric ion composition, which has shown that the ionosphere and solar wind make approximately equal contributions to the ambient magnetospheric plasma. Shelley [1985]. in a survey of the available data on the circulation of energetic ionospheric plasma in the magnetosphere. concluded that during magnetically active times the primary source of ionospheric ions in the magnetosphere is the auroral zone. Shelley also noted that the relatively minor terrestrial component of the plasma sheet during magnetically quiet periods could be supplied from either the auroral zones or polar caps. Hultqvist [1983] concluded from the dynamic changes observed in the energetic ionospheric plasma component in the magnetosphere that the only possible sources were at ionospheric altitudes (i.e., less than 1000 km). Hultqvist noted that a transverse energization or acceleration mechanism is probably required to extract ions from these low altitudes. Transverse acceleration of plasma is not limited, however, to ionospheric altitudes; it has been observed in all regions of the magnetosphere (e.g., Johnson, 1983).

The systematic theoretical investigation of the coupling of the magnetospheric and ionospheric plasmas was started with the the work of Kindel and Kennel [1971], who explored a class of plasma instabilities in the ionosphere which were driven by field-aligned currents. Kindel and Kennel pointed out that these instabilities lead to transverse energization of ions. Progress in plasma wave theory in ionospheric-magnetospheric coupling has been relatively slow due to the limited capabilities of instruments in providing the detailed, high resolution, in situ diagnostics required to test the theories. Our current understanding of the many different kinds of magnetospheric acceleration processes is reviewed in this volume by Cornwall [1986]. One of the goals of the the Dynamics Explorer program is to provide detailed, high resolution diagnostics of the plasma environment in the auroral acceleration region. The purpose of this paper is to describe a data set assembled from Dynamics Explorer-1 and illustrate its relevance to studies of the auroral acceleration processes.

Data

Many of the instruments on the DE-1 satellite have the capability to cover relatively wide ranges with high spectral resolution. These instruments are programmed to optimize the combination of spectral range, spectral resolution and time resolution for specific scientific studies. A program to acquire a systematic and extensive collection of low frequency (0 to 10 kHz) wideband data from the DE-1 Plasma Wave Instrument [Shawhan et al., 1981] and high time resolution ion composition data from the Energetic Ion Composition Spectrometer [Shelley et al., 1981] in the auroral zones was started in January 1984. The analysis of this coordinated data set was be-

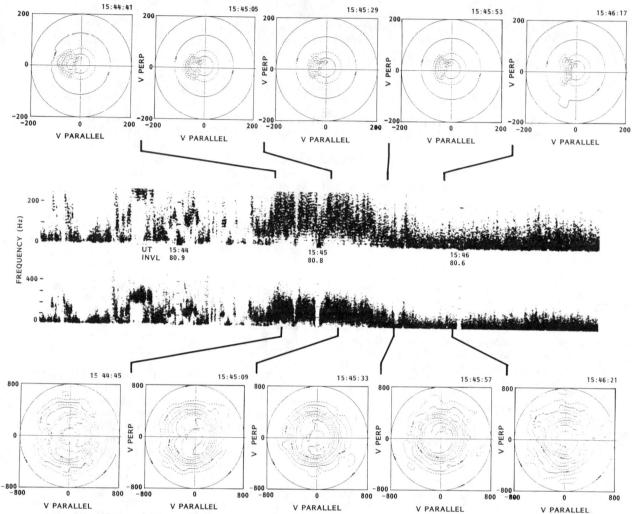

Fig. 1 High time resolution plasma wave (center panels) and ion composition data from an interval on March 15, 1984. See text for a description of the format.

gun using analog wideband data because digital ion and plasma wave data were available for only a small number of events. Kinter et al., [1979] have shown, however, that plasma wave emissions near multiples of the hydrogen cyclotron frequency are, at times, associated with transverse ion energization. We have scanned the first ~300 days of the Plasma Wave Instrument (PWI) wideband data set for intense low frequency emission events with evidence of harmonic structure lasting tens of seconds and then obtained high time resolution Energetic Ion Composition Spectrometer (EICS) data and digital swept frequency receiver (SFR) data from the PWI instrument.

Approximately 200 data acquisition intervals constituting over 300 hours of wideband data were scanned. Thirty-one events lasting on the order of tens of seconds or more were identified from the wideband data. No harmonic structures spaced at the oxygen gyrofrequency were found in this initial visual survey. More sensitive, and therefore more time consuming, methods will be used to search intervals identified in this inital survey for oxygen cyclotron waves. The 300 days scanned included complete local time coverage, but the 31 intense events were observed primarily in the evening auroral zone (17) and in the cusp region (7). The remaining events were in the midnight sector (5) or polar cap (2). One event was observed below 2000 km and two events were observed above 19,000 km.

Energetic ion composition and SFR (swept frequency receiver, Shawhan et al., 1981) data for 10 of these events were available at the time this report was prepared. The SFR data for these events shows considerable power at low frequencies (less than 1 kHz) from the electric dipole antenna and very little power from the magnetic loop antenna, indicating that the low frequency plasma waves observed are electrostatic.

We have selected two representative events, one from the cusp and one from the evening auroral zone, to present here.

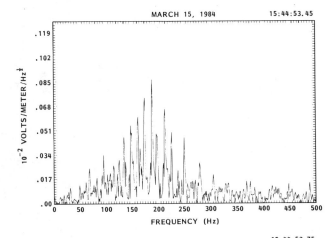

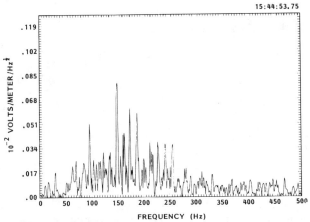

Fig. 2. Electric field strength vs. frequency from the wideband receiver for two intervals on March 15, 1984. Note that the absolute values indicated are accurate only within a factor of 4.

A Magnetospheric Cusp Crossing.

The cusp crossing selected is from March 15, 1984, when the DE-1 satellite was at an altitude of 19,500 km and 10:55 magnetic local time. An overview of the time dependence of the low frequency emissions for this interval is shown in the center panels of Figure 1 which displays the wave intensity for two frequency ranges 0 to 500 and 0 to 250 Hz encoded in gray scale as a function of frequency and time for the period 15:44 to 15:46. Black indicates the highest intensity. Universal time (UT) and invariant latitude (INVL) are indicated at one-minute intervals between the two center panels. Intense, banded, low frequency emissions are seen from ~15:44:30 to ~15:45:30 and intermittently before 15:44:30 in the 0 to 250 Hz (top) spectrogram. Figure 2 shows the electric field strength as a function of frequency (from 0 to 500 Hz) from the wideband receiver for two times during this interval. The intense, banded, low frequency emissions near multiples of the hydrogen gyrofrequency (~13 Hz) are clearly visible in this high resolution data. The lower hybrid frequency was estimated to be in the range 240 to 300 Hz between 15:44 and 15:46. This estimate was based on plasma density measurements derived from the upper cutoff of the whistler mode radiation (Ann Persoon, private communication) and relative H^- and O^+ densities determined from the energetic ion composition spectrometer. The calculation of the lower hybrid frequency was based on cold plasma theory [Stix, 1962] and assumed propagation perpendicular to the magnetic field.

High time resolution, mass analyzed ion data are presented above and below the central spectrograms in Figure 1 in the form of contours of constant phase space density. The end of the time interval over which the data for each contour plot were acquired is indicated by the heavy solid lines. Oxygen data (above) and hydrogen data (below) were obtained in alternate satellite spin periods (6 seconds). Twenty-four seconds or four complete spin periods were required to cover the full energy-angle range for both hydrogen and oxygen ions.

The contour lines are spaced at half decade intervals and are plotted in a coordinate system aligned with the local magnetic field. Velocities parallel (horizontal axis) and perpendicular (vertical axis) to the magnetic field for ± 200 km/sec for oxygen and ± 800 km/sec for hydrogen are displayed. The oxygen ions shown in the top panel

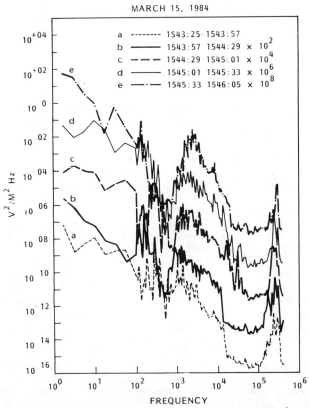

Fig. 3. Stacked plot of power spectral density as a function of frequency observed in the electric antenna (in units of v^2/m^2-Hz) from the SFR for a cusp crossing on March 15, 1984. The frequency resolution is 30% below and 1% above 100 Hz.

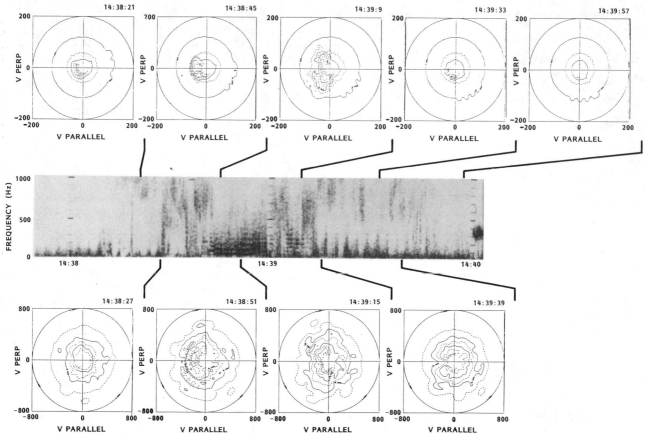

Fig. 4. High time resolution plasma wave and ion composition data from an interval on January 4, 1984. The format is similar to that of Figure 1.

are flowing from the ionosphere below. The contour plots are produced by contouring the surface defined by the larger of the one-count response of the instrument or the actual measurement at each sample point. The 'real' distribution then 'pokes through' the one-count-per-sample surface: thus the full circles and large arcs centered on the origin in the oxygen contour plots (top) are from the one-count-per-sample surface.

The sequence of hydrogen contour plots in the bottom panel of Figure 1 shows the evolution in energy and angle of injected magnetosheath H$^+$ ions which is the result of the combined effects of convection electric fields and spreading in pitch angle of the downward flowing cusp ion beam because of the magnetic mirror force. These effects have been described in detail by Peterson [1985], Roth and Hudson [1985], and Burch et al. [1982]. The hydrogen data for the period ending at 15:44:45 show the collimated, down flowing, high energy magnetosheath ions typically observed on the edge of the cusp. In the next two accumulation intervals, the distribution retains a peak in velocity which increases in angular extent. At the end of the interval shown in Figure 1, the hydrogen distributions are essentially isotropic. Since latitude is decreasing with time during this interval, the convection direction is sunward. Gorney [1983] and Roth and Hudson [1985] have pointed out that hydrogen ion distributions such as those for the intervals ending at 15:45:09 and 15:45:33 are un-

stable and could be the source of plasma waves of the type observed in the wideband data at this time.

The character of the upflowing oxygen distribution also changes over the time interval shown in Figure 1 from approximately isotropic in a frame of reference moving with the bulk flow velocity up the magnetic field line to a distribution which, for the interval ending at 15:45:29, has more thermal energy perpendicular to the magnetic field than parallel to it. Ungstrup et al. [1986] have also presented data from the ISEE-1 satellite suggesting transverse energization in the moving frame of reference of one of the plasma ion components.

Ions are expected to interact only with low (less than 1 kHz) plasma waves, but the higher frequency components can provide additional information about the local plasma environment. Figure 3 is a stacked plot of power spectral density as a function of frequency observed in the electric antenna from the swept frequency receiver. With the exception of the auroral kilometric radiation above 10^5 Hz, the cusp SFR spectra in Figure 3 are similar to those presented by Lin et al. (1984) obtained from DE-1 on magnetic field lines populated with magnetosheath ions. The power spectral density in the 100 to 200 Hz range measured in the magnetic loop antenna for the same intervals presented in Figure 3 is approximately 100 times lower than that from the electric dipole antenna, indicating that the low frequency plasma waves shown in Figure

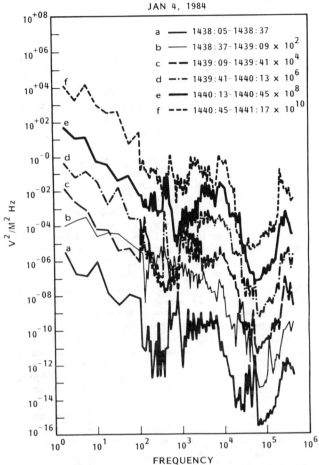

Fig. 5. Stacked plot of power spectral density as a function of frequency from the SFR for an evening auroral zone crossing on January 4, 1984. The format is similar to that of Figure 3.

2 are electrostatic. In fact below ~10^5 Hz, most of the power spectral density is seen in the electric antenna.

An Evening Auroral Zone Crossing.

Figure 4 presents high time resolution EICS and PWI data in a format similar to that of Figure 1 for an interval on January 4, 1984, when the satellite was at an altitude of 10,000 km near the poleward edge of the energetic ion precipitation zone at 17:40 magnetic local time and an invariant latitude of 78 degrees, moving equatorward. Intense bands of wave emissions spaced at the local hydrogen gyrofrequency (~60 Hz) below about 400 Hz are clearly visible from 14:38:40 to 14:39:10 in the central panel. Because of uncertainties in determining the plasma density (Ann Persoon, private communication) and the indication that most of the ion density is from ions with energies below the 10 eV lower limit selected for the EICS instrument at this time, we can only establish a broad range for the lower hybrid frequency. For equal hydrogen and oxygen densities the lower hybrid frequency would be ~725 Hz. If only 10% of the density is oxygen and the rest hydrogen, then the lower hybrid frequency would be ~1 kHz.

The contours of hydrogen phase space density in the lower panel show isotropic distributions for the the intervals ending at 14:38:27 and 14:39:39 and higher mean energies transverse to the magnetic field direction during the period of intense wave emissions near 14:39.

While there is clear evidence for upward flowing oxygen ions above the EICS 10 eV cut-off energy from 14:38:21 through 14:39:09, it is only for the interval ending at 14:39:09 that the data are adequate to provide meaningful phase space density contours. As noted above, the nearly circular contours centered on the origin are the contours of the one-count-per-sample instrument response. The series of oxygen contours in the top panel thus suggest that a low intensity upflowing oxygen ion beam exists until at least 14:39:33 which has its maximum energy and/or intensity during the measurement cycle ending at 14:39:09. Oxygen contours for the interval ending at 14:39:09 show that most, but not all, of the ion energy above the instrument threshold is perpendicular to the local magnetic field. Because of the wide energy width of the lowest energy channel, the magnitude of this parallel energy is not as well known as the magnitude of the perpendicular energy. Conservation of the first adiabatic invariant and the relative magnitudes of the parallel and perpendicular energies leads to the conclusion that this distribution is consistent with the transverse energization of the oxygen ions occurring between 1000 km below the spacecraft and locally.

Figure 5 is a stacked plot of power spectral density as a function of frequency from the electric antenna of the SFR. Data from the magnetic loop antenna do not show intense power spectral density below ~10^5 Hz. We conclude that the 0-1 kHz waves displayed in Figure 4 are electrostatic and that the plasma waves above ~1 kHz are typical of those seen in the high altitude auroral zone from DE-1 [Gurnett et al., 1983].

Discussion

Stix [1962], in the introduction to his book on plasma physics, noted that a plasma is distinguished from other states of matter primarily by its collective properties. Transverse energization of ions by plasma waves and growth of plasma waves at the expense of particle energy are two of the central topics of space plasma physics. The detailed investigation of plasma processes associated with perpendicular ion heating in existing data sets has proven to be difficult, primarily because of the small number of well documented examples (see, for example, Kintner and Gorney, 1984).

Starting in 1984, the Dynamics Explorer-1 satellite systematically acquired a coordinated set of high time resolution plasma and plasma wave observations from the auroral zone. We have begun a program to use this data set to study the plasma processes associated with perpendicular ion heating. The relatively small scale nature of perpendicular ion energization, the vast amount of data to be searched, the established association of plasma wave emissions near multiples of the hydrogen cyclotron frequency with transverse ion energization (e.g., Kinter et al., 1979), and the ready availability of wide band plasma wave data lead us to limit our initial investigation to intervals with intense low frequency (less than 1 kHz) emissions with evidence of harmonic structure lasting tens of seconds. When digital ion data are available, they will also be searched for regions of transverse ion energization and compared with simultaneous plasma wave data. The present results, while preliminary, do provide some new insights into the study of collective plasma interactions

We have presented above two representative examples obtained from the Dynamics Explorer-1 satellite which show an increase in the transverse (to the magnetic field) ion energy in association with intense low frequency plasma waves with harmonic structure. The Plasma Wave Instrument digital swept frequency receiver data and the Energetic Ion Composition Spectrometer data for both examples presented were typical of other cusp and auroral zone intervals encountered by the DE-1 satellite.

There have been previous reports of simultaneous measurements of plasma waves and ion distribution functions (e.g., Kintner, 1980; Kintner and Gorney, 1984; or Yau et al., 1986); however, the data from Dynamics Explorer extend the energy range for ion species identification to the thermal range and extend the altitude range sampled to over 22,000 km. In addition to the data presented here, data from two other Dynamics Explorer-1 instruments, a fluxgate magnetometer [Farthing et al., 1981] and a second generation cold plasma analyzer [Chappell et al. 1981], are also available for most of the intervals considered. Analysis of these data are in progress. Unfortunately, no electron plasma data are available for the period of this study.

The data presented above for the cusp crossing of March 15, 1984, are qualitatively similar to the first order analysis of the energy balance between incoming magnetosheath and accelerated ionospheric cusp ion distributions and plasma waves using published data from S3-3 and DE-1 by Roth and Hudson [1985]. A complete analysis of the full DE-1 data set for several cusp crossings, including the thermal plasma composition and field-aligned currents, will provide the in situ diagnostics for a more detailed investigation of the energy balance in this and similar events. Presentation of such an analysis is beyond the scope of this first report.

One of the more important tools in the study of the collective behavior of plasma particles and waves is the identification of the mode(s) of the plasma waves. For example, the difference in the mathematical descriptions of electrostatic ion-cyclotron and lower hybrid waves are in the approximations used to account for electron motions (e.g., Chang and Coppi, 1981). Roth and Hudson [1985] have noted that plasma wave data similar to that shown in Figures 1-3 indicate that the lower hybrid description of electron motion must be considered in models of cusp plasma.

We note also that the O^+ distribution in Figure 1 for the period ending at 15:45:29 is not what is generally referred to as a conic distribution (see, for example, Sharp et al., 1977 and Gorney et al., 1981) because the distribution shows a considerable velocity parallel to the magnetic field. In fact, the O^+ distribution in Figure 1 for the measurement interval ending at 15:45:29 can be approximately represented by a bi-maxwellian distribution with temperatures of 180 and 250 eV parallel and perpendicular to the local magnetic field and a bulk flow velocity of 27 km/sec up the magnetic field line. Klumpar et al. [1984] have presented bi-modal O^+ distributions which they describe as the result of both parallel and perpendicular energization occurring on the same magnetic field line. In the examples presented by Klumpar et al., the O^+ distributions were 'folded' toward the magnetic field direction. This folding was interpreted as the result of the gradient in magnetic field strength between the region where the ions acquired their transverse energy and the satellite observation position. The O^+ distribution for the interval ending at 15:45:29 does not show evidence of 'folding' toward the magnetic field direction, suggesting that the region of transverse O^+ energization is quite near to the satellite.

The data from the auroral zone presented in Figures 4 and 5 show the simultaneous increase in transverse energy in both H^+ and O^+ plasma components in the presence of plasma wave emissions at multiples of the hydrogen gyrofrequency. This is not the only such example found in the DE-1 data set to date. However, as noted above, the O^+ contours in Figure 4 are consistent with the oxygen having acquired most of its transverse energy below the satellite position. It is hoped that detailed examination of the field aligned currents inferred from the magnetometer data and thermal plasma data from the RIMS instrument will resolve some of the ambiguities and lead to a more complete understanding of the plasma processes associated with perpendicular ion heating.

Lennartson [1983], in a systematic listing of the known or postulated ion acceleration mechanisms operating in the auroral regions, concludes with the remark: "Even though certain mechanisms have received more attention than the others in the literature, it is probably fair to say that the 'basic' mechanisms have yet to be agreed upon." It is clear that the high resolution data and large number of events in the Dynamics Explorer data set introduced here will provide a good body of data to test the limits and establish the relative importance of these auroral acceleration mechanisms.

Acknowledgements. The authors wish to thank Ann Persoon for providing density data from the Plasma Wave Instrument. This research was supported in part by NASA under contract NAS5-28710 and grants NAG5-310 and NGL-16-001-043.

References

Burch, J.L., P.H. Reiff, R.A. Heelis, J.D. Winningham, W.B. Hanson, C. Gurgiolo, J.D. Menietti, R.A. Hoffman, and J.N. Barfield, Plasma injection and transport in the mid-altitude polar cusp, Geophys. Res. Lett., 9, 921, 1982.

Chang, T. and B. Coppi, Geophys. Res. Lett., 8, 1253, 1981.

Chappell, C.R., S.A. Fields, C.R. Baugher, J.H. Hoffman, W.B. Hanson, W.W. Right, H.D. Hammack, G.R. Carignan, and A.F. Nagy, The retarding ion mass spectrometer on Dynamics Explorer-A, Space Sci. Instru., 5, 477, 1981.

Cornwall, J.M., Magnetospheric Ion Acceleration Processes, this volume, 1986.

Farthing, W.H., M. Sugiura, B.G. Ledley, and L.J. Cahill, Jr., Magnetic field observations on DE-A and -B, Space Sci. Instrum., 5, 551, 1981.

Gorney, D.J., A. Clarke, D. Croley, J. Fennell, J. Luhmann, and P. Mizera, The distribution of ion beams and conics below 8000 km, J. Geophys Res., 86, 83, 1981.

Gorney, D.J., An alternative interpretation of ion ring distributions observed by the S3-3 satellite, Geophys. Res. Lett., 10, 417, 1983.

Gurnett, D.A., S.D. Shawhan, and R.R. Shaw, Auroral hiss, Z mode radiation, and auroral kilometric radiation in the polar magnetosphere: DE-1 observations, J. Geophys. Res., 88, 329, 1983.

Hultqvist, B., On the origin of the hot ions in the disturbed dayside magnetosphere, Planet. Space Sci., 31, 173, 1983.

Johnson, R.G., Editor, Energetic Ion Composition in the Earth's Magnetosphere, Terra Scientific Publishing Co., Tokyo, 1983.

Kindel, J.M. and C.F. Kennel, Topside current instabilities, J. Geophys. Res., 76, 3055, 1971.

Kintner, P.M., M.C. Kelley, R.D. Sharp, A.G. Ghielmetti, M. Temerin, C. Catell, P.F. Mizera, and J.F. Fennel, Simultaneous observations of energetic (keV) upstreaming ions and electrostatic hydrogen cyclotron waves, J. Geophys. Res. 84,, 7201, 1979.

Kintner, P.M., On the distinction between electrostatic ion cyclotron waves and ion cyclotron harmonic waves, Geophys. Res. Lett., 7, 585, 1980.

Kintner, P.M. and D.J. Gorney, A search for the plasma processes associated with perpendicular ion heating, J. Geophys. Res., 89, 937, 1984.

Klumpar, D.M., W.K. Peterson, and E.G. Shelley, Direct evidence for two-stage (bimodal) acceleration of ionospheric ions, J. Geophys. Res., 89, 10779, 1984.

Lennartsson, W., Ion acceleration mechanisms in the auroral regions General principles, in Energetic Ion Composition in the Earth's Magnetosphere, R. G. Johnson, Editor, Terra Scientific Publishing Company, Tokyo, 1983.

Lin, C.S., J.L. Burch, S.D. Shawhan and D.A. Gurnett, Correlation of auroral hiss and upward electron beams near the polar cusp, J. Geophys. Res., 89, 925, 1984.

Peterson, W.K., Ion injection and acceleration in the polar cusp, in The Polar Cusp, J.A. Holtet and A. Egeland eds., Reidel, p. 67, 1985.

Roth, I. and M.K. Hudson, Lower hybrid heating of ionospheric ions due to ion ring distributions in the cusp, J. Geophys. Res., 90, 4191, 1985.

Shawhan, S.D., D.A. Gurnett, D.L. Odem, R.A. Helliwell and C.G. Park, The plasma wave and quasi-static electric field instrument (PWI) for Dynamics Explorer-A, Space Sci. Instrum., 5, 535, 1981.

Sharp, R.D., R.G. Johnson, and E.G. Shelley, Observation of an ionospheric acceleration mechanism producing energetic (keV) ions primarily normal to the geomagnetic field direction, J. Geophys. Res., 82, 3324, 1977.

Shelley, E.G., D.A. Simpson, T.C. Sanders, E. Hertzberg, H. Balsiger, and A. Ghielmetti, The energetic ion composition spectrometer (EICS) for the Dynamics Explorer-A, Space Sci. Instrum., 5, 443, 1981.

Shelley, E.G., Circulation of energetic ions of terrestrial origin in the magnetosphere, Adv. Space Res., 4, 1985.

Stix, T.H., The Theory of Plasma Waves, McGraw-Hill, New York, 1962.

Ungstrup E., W. Lennartsson, C.A. Cattell, R.R. Anderson, R.J. Fitzenreiter, G. Gustafsson, and C.-G. Falthammar, Particle and wave observations in the auroral acceleration region, this volume, 1986.

Yau, A.W., B.A. Whalen, and P.M. Kintner, Low-altitude transverse ionospheric ion acceleration, this volume, 1986.

OBSERVATIONS OF COHERENT TRANSVERSE ION ACCELERATION

T. E. Moore, J. H. Waite, Jr., M. Lockwood,[1] and C. R. Chappell

Space Science Laboratory, NASA Marshall Space Flight Center, Huntsville, Alabama 35812

Transverse heating of the low-energy core ions of the high topside ionosphere to temperatures on the order of 10^5 K is frequency observed by the retarding ion mass spectrometer on the Dynamics Explorer 1 spacecraft. In a small fraction of the observations, torus or ring-shaped O^+ distribution functions are observed, indicating a coherent acceleration to energies on the order of 10 eV in the ion plasma bulk frame of reference. The tori have radii on the order of 10 km/s, are oriented with their axes parallel to the local magnetic field direction, and in general have substantial convective and field-aligned bulk flow velocity components, in addition to spacecraft ram motion. Significant anisotropies occur outside the torus shaped core, with conical tails extending well beyond the 50-eV limit of RIMS observations, carrying an upward ion heat flux. It is suggested that the mechanism which tends to produce ion torus distributions is much more widespread than the tori themselves, but that the tori are observed only when the local conditions (e.g., electron temperature) inhibit wave growth and consequent velocity space diffusion which would otherwise fill the torus center and transversely heat the distribution tail. Such coherent acceleration seems to require a mechanism different from those usually invoked in connection with transverse ion heating. Possible mechanisms include supersonic ion-neutral relative convection or perpendicular electric fields with structure on the order of the ion gyroradius.

Introduction

It has been known for some time that topside ionospheric ions are often accelerated or heated in the direction transverse to the local magnetic field [Sharp et al., 1977; Whalen et al., 1978; Klumpar, 1979; Moore, 1984] in association with auroral processes. The mechanism of this process has remained something of a mystery, however, with no clear consensus appearing to date. Two principal mechanisms have been proposed to account for the transverse heating: ion cyclotron resonant interactions with waves driven by auroral processes such as field-aligned currents, and interactions of ions with small scale auroral potential structures, e.g., oblique double layers.

It has been determined that this heating occurs frequently on the edges of and below auroral potential structures located in the vicinity of 1 R_E altitude, as reported by Moore et al. [1985]. It also occurs almost continuously in the dayside auroral zone above 2000 km altitude, as reported by Lockwood et al. [1985a]. A few observations have been obtained of transverse heating at rocket altitudes [Whalen et al., 1978; Yau et al., 1983], where it appears to produce a hot superthermal tail rather than bulk heating. Transverse heating leads to the production of significant fluxes of heavy ionospheric ions having energies greater than that required for terrestrial escape, and therefore significantly influences the composition of the ionospheric outflow. The purpose of this paper is to report observations of coherent transverse acceleration (as distinct from transverse heating) of ionospheric ions in the nightside auroral zone. The ion data are from the Dynamics Explorer (DE) 1 retarding ion mass spectrometer (RIMS) which resolves coarse features of the ion species distribution functions in the energy range from spacecraft potential to 50 eV.

Observations

With an upper energy limit of 50 eV, RIMS simultaneously observes the thermal ionosphere and more energetic or superthermal features which appear within it, especially in the auroral zone. Such features have been discussed by Moore et al. [1985], who provided representative examples of the data, and by Lockwood et al. [1985a] as transversely accelerated core ions or TACIs. The latter, statistical study of these events showed that they are relatively rare, with greatest frequency in the nightside auroral zone. Spin-time spectrograms of these events, examples of which were published in the studies referred to above, are characterized by an abrupt bifurcation of the otherwise ramped integral ion flux into two peaks which are approximately symmetric with respect to the local magnetic field direction. Bearing in mind that the peaks represent integral flux of ions from spacecraft potential to 50 eV, the two angular peaks give the appearance of two separate plasma beams in relative motion. Moore et al. [1985] argued that the integral flux observations for this type of event indicate a toroidal distribution function (a ring-shaped distribution).

These events have been further analyzed exploiting the differential response (in energy) of the high mass channel of RIMS. Using the RIMS response determined from laboratory calibration, we have interpreted the RIMS ion flux measurements as phase space densities. The high mass channel, used for O^+ measurements, is sufficiently narrow in energy response (nominally 7 to 14 eV FWHM, depending on the center energy) to differentially resolve coarse features of the ion distribution function. In practice, due to the appreciable ram energy (about 5 eV for O^+), features with dimensions larger than 1 eV are resolved. The RIMS radial head, which scans the spin plane, has an angular response perpendicular to the spin plane which is approximately 40 degrees wide FWHM, while the spin has been sectored into bins of 20 degrees width. Further details of the RIMS response characteristics may be found in Chappell et al. [1981] and Olsen et al. [1985].

The full two-dimensional distribution function obtained from this procedure is plotted at several times during the representative event in Figure 1. Here the velocity axes represent the spacecraft frame of reference, with the horizontal axis oriented parallel to the spacecraft velocity vector and the vertical axis oriented outward with respect to the Earth. The direction of the local magnetic field is indicated by a line through the origin. The phase space density is indicated as a function of the two velocity coordinates by both line and tone contouring, with the corresponding values given at the right.

---

[1] Presently at Rutherford Appleton Laboratory, Didcot, Oxfordshire, England.

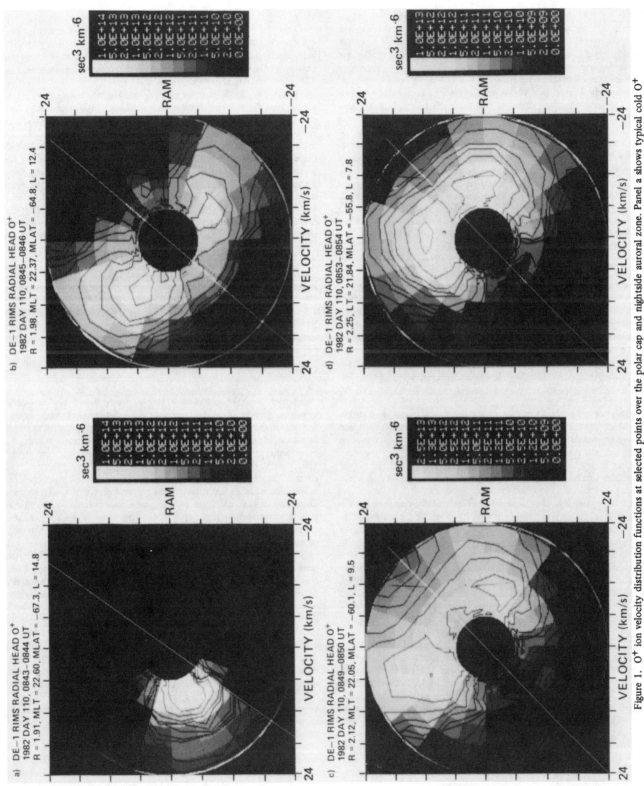

Figure 1. O$^+$ ion velocity distribution functions at selected points over the polar cap and nightside auroral zone. Panel a shows typical cold O$^+$ plasma over the polar cap region. Panels b-d show the abrupt appearance and subsequent evolution of a coherent transverse acceleration event.

The first panel, Figure 1a, is useful for orientation to this type of data display. Here DE 1 is passing antisunward across the polar cap at an altitude of 0.91 R_E, near midnight local time. The observed plasma approximates the instrument impulse response function, i.e., the response to an unresolved excitation. While the cold ionospheric O^+ observed probably has a temperature on the order of a few tenths of an electron volt, the apparent temperature is approximately 0.8 eV. This is essentially the minimum resolvable temperature for RIMS at an orbital speed of approximately 7 km/s, so that this is an upper limit on the true temperature. The bulk velocity relative to the spacecraft indicates comparable contributions of convection and downward field-aligned flow. If the plasma were stationary with respect to the Earth, the distribution centroid would be displaced by 7 km/s to the left of the origin by the spacecraft motion. The downward field-aligned flow is frequently observed in heavier ions within the polar cap, and represents the transient response of the ionosphere to heating in the dayside auroral zone. The ionosphere expands outward as it convects through the dayside auroral zone and is heated, then sinks back toward Earth as it convects antisunward across the polar cap toward the nightside auroral oval, forming what has been termed the cleft ion fountain [Lockwood et al., 1985b; Horwitz et al., 1985].

The low speed limit of the data is not an instrumental cutoff, but rather represents the lowest center energy (3.5 eV) of the lowest energy bandpass, which is 7 eV wide FWHM. The energy sweep is exponential, so that there are many steps at low energy even though fine detail cannot be resolved there. It has been assumed in plotting this data that the effect of spacecraft potential is to shift the energy associated with a particular measurement of phase space density, without change in phase space density magnitude, a procedure which can be defended on the basis of Liouville's theorem. In this particular case, it has also been assumed that a spacecraft potential is in fact zero, though a different potential would not qualitatively modify the plots, but would shift the inner plot contours radially toward (negative potential) or away from (positive potential) the velocity origin.

In the second panel, Figure 1b, we see the change in the ion distribution which occurs when DE 1 passes into the auroral oval on this pass. Subsequent panels display the space-time evolution of the distribution function as the spacecraft passes equatorward of the polar cap boundary and rises somewhat in altitude. Orbital parameters are listed above each frame. The splitting of the rammed integral flux into two distinct peaks is clear.

It is also clear that the distribution is dramatically energized, in the sense of occupying a larger region of velocity space, by the process active here. The greatest energization of the initially cold distribution is in the direction perpendicular to the local magnetic field. More importantly, it is clear that two distinct maxima exist, and the contours show a relative minimum between them. That this is indeed the case becomes unquestionable in the third panel, Figure 1c. In view of the expected symmetry about the magnetic field direction, it appears that the observed distribution function is a spin plane slice of the full distribution, which in turn is a figure of revolution about the magnetic field direction, i.e., an ion ring or torus.

The ion ring is initially observed to have little parallel bulk velocity, a radius of 12 km/s, and a "thickness" of approximately 5 km/s. It is evident that a roughly gyrotropic perpendicular hot tail exists, which evolves in time to become a "conic" distribution rooted in the ring distribution. The ring evolves in space and perhaps time, acquiring a significant upward parallel bulk velocity, on the order of 10 km/s in Figure 1c. Note as well in Figure 1c that the cone angle defined by the contours near 24 km/s (50 eV) is considerably smaller than the cone angle defined by the centroid of the ring itself, even though the ring has acquired an appreciable parallel velocity by this time. This type of behavior adds new dimensions to the concept of pitch angle "focusing" [e.g., Gorney et al., 1981] under first adiabatic invariant conservation. Of course, we are observing ions here which are significantly influenced by gravity as well as the magnetic mirror force, so that the change of pitch angle with altitude is reduced for the low-energy ions which lose parallel energy in overcoming the gravitational potential.

Figure 1d shows that the ring feature fades as the spacecraft moves equatorward and upward in altitude. The resulting distribution is virtually indistinguishable from that seen frequently in dayside auroral oval ion heating events, which have been termed upwelling ion events [Moore et al., 1985; Lockwood et al., 1985a; Moore et al., 1986].

Figures 2a-d show plots of phase space density versus energy representing cuts through the distributions of Figures 1a-d, respectively. Two cuts are shown in each frame. The parallel cut passes through the distribution in the direction parallel to the magnetic field, intersecting the maximum of phase space density. The "perpendicular" cut passes through the distribution perpendicular to the local magnetic field intersecting the maximum of phase space density, but passes outward from the ring along the conic "ridge" of the distribution, i.e., along the direction of minimum slope of the phase space density or maximum e-folding energy. This choice of direction for the "perpendicular" cut is thought to best reflect the character of the conical wings of the ring distribution. Energy is measured relative to the plasma frame of reference. These plots have the advantage of quantitatively exhibiting the slopes of the distributions, which may be characterized by e-folding energies analogous to temperatures, shown in parentheses. Having phase space densities and characteristic energies it is straightforward to estimate densities, which may also be computed by direct numerical integration of the data. Calculated O^+ densities for panels a-d are 69, 167, 61, and 35 cm^{-3}, respectively. Comparable densities of H^+ are also present during this period, leading to total electron densities of a few hundred per cubic centimeter at the peak of the event.

Discussion

We have seen from these observations that the high ionosphere ion distribution function can become ring or torus shaped due to auroral processes. These ion distributions should be distinguished from another type of ion distribution which has been referred to as ring-shaped in recent literature [Roth and Hudson, 1985]. The latter is formed as energetic magnetosheath ions penetrate deep into the magnetospheric cusp region. The resulting distribution is perhaps better described as shell-shaped, but contains regions in velocity space with positive distribution function slope, in common with the ring distributions reported here. The most important distinction is that the rings reported here have formed in the "background" ionospheric plasma itself, at densities of a few hundred per cubic centimeter, and not in the more energetic tail of an otherwise cold background plasma.

The existence of such ring shaped distributions implies an acceleration mechanism capable of imparting a preferred perpendicular energy to all ions within this region of space. Such a mechanism is fundamentally different from a heating mechanism, such as quasilinear diffusion, which increases the average energy of the ion distribution without creating a depleted region of velocity space near the bulk frame of the plasma. In addition to the obvious difference in ion distribution function shape, a coherent transverse acceleration process implies an altogether different chain of cause and effect from that usually invoked to account for transverse ion heating and conic formation. This is because ion ring distributions are unstable to the low-frequency ion waves to which conic generation is usually attributed. This raises the possibility that the waves are a by-product of the coherent acceleration process, and that transversely heated ion distributions are formed by these waves as a consequence of the diffusion of ring distributions. This diffusion would clearly scatter ions to higher as well as lower perpendicular energies in the process of smoothing the ring toward bi-Maxwellian form.

Moreover, the observation that ion conic formation is associated with the edges of auroral structures and with field-aligned electron streaming [Moore et al., 1985; Collin et al., 1982; Arnoldy et al., 1985] may merit

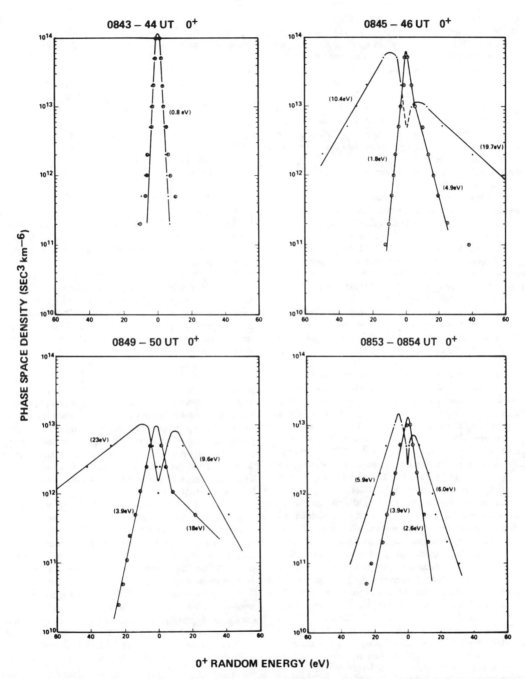

Figure 2. Plots of phase space density as a function of energy measured relative to the apparent plasma bulk frame of reference, corresponding to cuts through the two-dimensional distributions of Figure 1. Each plot shows one cut parallel to the local magnetic field direction and passing through the maximum (circled data points). A second cut (dotted data points) passes perpendicular to the magnetic field direction through the two maxima in panels b-d, then outward along the "ridge" defining the conical hot tail, i.e., along the path of minimum slope or maximum e-folding energy. Energies to right and left of center represent upward and downward parallel energy and right and left perpendicular energy, respectively, with reference to Figure 1.

reinterpretation in light of this scenario. Low-frequency waves with predominantly perpendicular wave vectors, capable of heating ions, have been attributed to these electron streams [Dusenbery and Lyons, 1981; Chang and Coppi, 1981]. The flow of energy suggested in this scenario is from the process driving field-aligned currents, and presumably the electron streaming, to the waves and thence on to ion transverse heating. The presence of ion ring distributions virtually requires that the energy flow be in the opposite direction, the waves being driven by the ions, and dissipating in randomization of the ion energy toward bi-Maxwellian distributions and ion conics, and in the parallel heating of ambient electrons.

St-Maurice [1978] has considered the instability of ion ring distributions and found from linear growth rate calculations that the threshold for instability is significantly affected by the degree of coupling of wave energy to the electron parallel motions by Landau damping. This result has two implications for the present discussion.

First, the linear instability threshold represents a measure of how extreme ion ring distributions can become before waves limit the further departure from Maxwellian. Landau damping of the waves by the electrons raises the threshold, permitting more observable ring distributions. The degree of electron Landau damping is strongly dependent upon the ratio of electron temperature to ion mean energy. The unstable waves have wave vectors so close to perpendicular that their parallel phase speeds are faster than electron thermal speeds unless the electrons are even hotter than the heated ions. Therefore, Landau damping is appreciable only when the electrons are hot relative to the ions, in which case, the instability threshold is elevated, permitting the formation of very distinct rings. Hence the ambient electron conditions have a strong effect on the observability of ion ring distributions. Although electron observations are not available for comparison here, enhanced electron temperatures may be expected in this region due to auroral electron precipitation.

Second, when electron Landau damping is strong, an appreciable fraction of energy of the ion ring formation mechanism should be dissipated in parallel heating of the electrons. Thus, observations of bi-directional parallel streaming of electrons may represent the result of coherent transverse ion acceleration, rather than the cause of transverse ion heating. It should be noted that ion conic formation is associated with the edges of auroral arcs [Moore et al., 1985], as are observations of bi-directional parallel electron streaming [Sharp et al., 1977; Collin et al., 1982] at satellite altitudes, and downward parallel electron streaming at rocket altitudes [Arnoldy et al., 1985]. This is suggestive of a coherent ion acceleration process operative above rocket altitudes.

An actual mechanism for the formation of ring distributions of the type discussed here has been addressed by the work of St-Maurice and Schunk [1979, and references therein]. They show how ion pickup by a convection electric field of sufficient magnitude, subsequent to collisions with neutral atoms or other scattering centers of appropriate density, leads to the natural formation of toroidal distributions. The basic requirements are a convection speed which exceeds the ion thermal speed, and a collision frequency ordering with the ion-ion collision frequency much smaller than the ion-neutral (or other collision center) collision frequency, which is in turn much smaller than the ion gyrofrequency. Those authors concluded that this mechanism is applicable to conditions in the topside F region ionosphere, but chiefly below 400 km. Though this appears inconsistent with our observations at about 6000 km, it remains the only analytic description of ion ring formation, and appears to require only an appropriate collisional medium to work. Since the ion gyrofrequency is much greater than the ion-ion collision frequency everywhere above the F region, the required medium must interact with the ions more strongly than they interact with themselves, and have a significantly different effective frame of reference.

Another mechanism is suggested by observations near the Earth's bow shock [Sckopke et al., 1983]. Partial reflection of solar wind ions by the electrostatic potential of the shock appears to result for certain configurations of the magnetic field in the formation of torus-shaped halos around the decelerated solar wind ions downstream of the shock. This suggests that electrostatic structures (e.g., oblique double layers) on the scale of the ionospheric ion gyroradius may be capable of producing ion ring distribution such as we observe. Borovsky [1984] and Yang and Kan [1983] have argued that ion conic distributions will be formed by oblique double layers. It does not seem unreasonable that ring distributions might be formed as well. As pointed out by Greenspan [1984], gyrophase bunching should be expected in this mechanism, and this is suggested by the asymmetry of our observed rings (see figures). However, a mathematical description which would identify appropriate conditions for ring or conic formation does not appear to exist for this mechanism.

Conclusions

The transverse heating of the low-energy ionospheric ions known to occur in the topside ionosphere at times appears in the form of coherent transverse acceleration, forming ion ring distributions having radii on the order of 10 km/s, corresponding to 10 eV O^+. The rings are oriented with their axes parallel to the local magnetic field, and in general have substantial convective and field-aligned bulk flow velocities. Significant anisotropies exist outside the ring-shaped core, with conical hot tails extending well beyond the 50-eV limit of RIMS observations, carrying an upward ion heat flux.

Based upon theoretical ideas concerning ion ring distributions, their observability should be strongly dependent upon the ratio of ambient electron temperature to the ion mean energy, due to the effect of electron Landau damping on the ion ring instability. This suggests that the mechanism which leads to ion ring distribution formation is much more widespread than the ring observations would suggest, and that typical conditions lead to an ion distribution which is only weakly ring-shaped and nearer to bi-Maxwellian, as in upwelling ion transverse heating events.

Coherent acceleration capable of producing ion ring distributions provides an alternative account of the low-frequency ion waves observed in association with the auroral zone transverse ion heating. Ion rings are themselves unstable to such waves, and contain ample free energy to drive them to thereby develop hot transverse tails or conics, and to produce parallel heating of ambient electrons by Landau damping of the nearly perpendicularly propagating waves. Such a scenario represents a significant departure from the more conventional view that ion heating is derived from the electrodynamics of field-aligned currents via waves driven unstable by electron streaming.

Encounters between ionospheric ions and small scale discrete structures such as oblique double layers might be expected to produce ion ring distributions, but only if they have scales smaller than the ring ion gyroradius, only a few hundred meters for the O^+ ions reported here. However, such structures contain very strong convective channels which may contribute to ion ring formation by the St-Maurice and Schunk mechanism, even when they have large scales in relation to the ring ion gyroradius.

A simple mechanism exists for producing ion rings based upon very rapid convective motions of the ionospheric plasma, with the convection electric field providing the energy. The convective velocity relative to scattering centers must exceed the thermal speed of the unaccelerated ionospheric ions. The scattering centers must have an appropriate density and cross section. The scattering could be provided by neutral atoms, though this seems unlikely at the altitudes of a few to several thousand kilometers where ion rings are being observed.

Acknowledgments. We are indebted to the engineering and science staff of the University of Texas at Dallas and to the RIMS team at the Marshall Space Flight Center. Assistance with data reduction software was provided by the staff of the Boeing Corporation. Spacecraft magnetic aspect was provided by M. Sugiura. Support for M. Lockwood was provided by the National Research Council under their Resident Research Associate program. The authors have benefitted from stimulating discussions with J.-P. St-Maurice and W. K. Peterson and from the use of computing and network facilities provided by the Data System Technology Program and the Space Physics Analysis Network at MSFC.

References

Arnoldy, R. L., T. E. Moore, and L. J. Cahill, Jr., Low-altitude field-aligned electrons, *J. Geophys. Res., 90,* 8445-8551, 1985.

Borovsky, J. E., The production of ion conics by oblique double layers, *J. Geophys. Res., 89,* 2251-2266, 1984.

Chappell, C. R., S. A. Fields, C. R. Baugher, J. H. Hoffman, W. B. Hanson, W. W. Wright, and H. D. Hammack, The retarding ion mass spectrometer on Dynamics Explorer-A, *Space Sci. Instrum., 5,* 477-491, 1981.

Chang, T., and B. Coppi, Lower hybrid acceleration and ion evolution in the suproauroral region, *Geophys. Res. Lett., 8,* 1253, 1981.

Collin, H. L., R. D. Sharp, and E. G. Shelley, The occurrence and characteristics of electron beams over the polar region, *J. Geophys. Res., 87,* 7504-7511, 1982.

Dusenbery, P. B., and L. R. Lyons, Generation of ion-conic distribution by upgoing ionospheric electrons, *J. Geophys. Res., 86,* 7627-7638, 1981.

Gorney, D. J., A. Clarke, D. Croley, J. Fennell, J. Luhmann, and P. Mizera, The distribution of ion beams and conics below 8000 km, *J. Geophys. Res., 86,* 83-89, 1981.

Greenspan, M. E., Effects of oblique double layers on upgoing ion pitch angle and gyrophase, *J. Geophys. Res., 89,* 2842-2848, 1984.

Horwitz, J. L., and M. Lockwood, The cleft ion fountain: A two dimensional kinetic model, *J. Geophys. Res., 90,* 9749-9762, 1985.

Klumpar, D. M., Transversely accelerated ions: An ionospheric source of hot magnetospheric ions, *J. Geophys. Res., 84,* 4229, 1979.

Lockwood, M., J. H. Waite, Jr., T. E. Moore, J. F. E. Johnson, and C. R. Chappell, A new source of suprathermal O^+ ions near the dayside polar cap boundary, *J. Geophys. Res., 90,* 4099-4116, 1985a.

Lockwood, M., M. O. Chandler, J. L. Horwitz, J. H. Waite, Jr., T. E. Moore, C. R. Chappell, The cleft ion fountain, *J. Geophys Res., 90,* 9736-9748, 1985b.

Moore, T. E., Superthermal ionospheric outflows, *Rev. Geophys. Space Phys., 22,* 264-274, 1984.

Moore, T. E., C. R. Chappell, M. Lockwood, and J. H. Waite, Jr., Superthermal ion signatures of auroral acceleration processes, *J. Geophys. Res., 90,* 1611-1618, 1985.

Moore, T. E., M. Lockwood, M. O. Chandler, J. H. Waite, Jr., C. R. Chappell, A. Persoon, and M. Sugiura, Upwelling O^+ ion source characteristics, *J. Geophys. Res.,* in press, 1986.

Olsen, R. C., R. H. Comfort, M. O. Chandler, T. E. Moore, J. H. Waite, Jr., D. L. Reasoner, and A. P. Biddle, DE 1 RIMS operational characteristics, *NASA Tech. Memo. TM-86527,* 1985.

Roth, I., and M. K. Hudson, Lower hybrid heating of ionospheric ions due to ion ring distributions in the cusp, *J. Geophys. Res., 90,* 4191-4203, 1985.

Sckopke, N., G. Paschmann, S. J. Bame, J. T. Gosling, and C. T. Russell, Evolution of ion distributions across the nearly perpendicular bow shock: Specularly and non-specularly reflected-gyrating ions, *J. Geophys. Res., 88,* 6121-6136, 1983.

Sharp, R. D., R. G. Johnson, and E. G. Shelley, Observation of an ionospheric acceleration mechanism producing energetic (keV) ions primarily normal to the magnetic field direction, *J. Geophys. Res., 82,* 3324-3328, 1977.

St-Maurice, J.-P., On a mechanism for the formation of VLF electrostatic emissions in the high latitude F region, *Planet. Space Sci., 26,* 801-816, 1978.

St-Maurice, J.-P., and R. W. Schunk, Ion velocity distributions in the high-latitude ionosphere, *Rev. Geophys. Space Phys., 17,* 99, 1979.

Whalen, B. A., W. Bernstein, and P. W. Daley, Low altitude acceleration of ionospheric ions, *Geophys. Res. Lett., 5,* 55, 1978.

Yang, W. H., and J. R. Kan, Generation of conic ions by auroral electric fields, *J. Geophys. Res., 88,* 465, 1983.

Yau, A. W., B. A. Whalen, A. G. McNamara, P. J. Kellog, and W. Bernstein, Particle and wave observations of low-altitude ionospheric ion acceleration events, *J. Geophys. Res., 88,* 341-355, 1983.

TRANSPORT OF ACCELERATED LOW-ENERGY IONS IN THE POLAR MAGNETOSPHERE

J. L. Horwitz,[1] M. Lockwood,[2] J. H. Waite, Jr.,[3] T. E. Moore,[3] C. R. Chappell,[3] and M. O. Chandler[1]

Recent satellite observations of low-energy (0-50 eV) ionospheric ions in the polar cap magnetosphere suggest that these ions are injected at the dayside cleft topside ionosphere. The ions are transported through the polar magnetosphere chiefly through the combination of their energy- and mass-dependent field-aligned motion and the energy- and mass-independent convection. Using a two-dimensional kinetic model, several interesting consequences of this ion flow from a narrow cleft source have been simulated and observed. These include: (1) the Kp/convection-dependent filling of the polar magnetosphere with ionospheric heavy ions, in which these ions are "blown" further into the polar cap magnetosphere from the cleft during high Kp/convection; (2) the mass- and energy-dependent dispersion of these ions, as in a kind of "geomagnetic spectrometer"; (3) the creation of "supersonic" ion outflows as a natural velocity-filter effect of this geomagnetic spectrometer; and (4) the "parabolic flow" of gravitationally bound heavy ions from the cleft ionosphere resulting in downward flow into the polar cap.

Introduction

With the realization that plasma of ionospheric origin is an important and frequently dominant portion of the plasma population throughout the magnetosphere [e.g., Horwitz, 1982; Young, 1986], there has been high interest in the properties and phenomena associated with upflowing ionospheric ions in the polar regions. Equally important, however, are considerations of the transport of these ions into the high-altitude magnetosphere. In particular, a major revision in our picture of the polar cap magnetosphere is under way as it has recently been revealed by Dynamics Explorer (DE) 1 observations [Lockwood et al., 1985a,b; Waite et al., 1985] that the dominant source of at least heavy ionospheric ions (e.g., O^+) appears to be not the broad polar cap, but rather a relatively confined region associated with the cleft ionosphere.

It appears, then, that many of the important features of the ions in the polar cap magnetosphere may not be revealed by steady-state one-dimensional models, at least in the case of the heavy ions. Models and observations must consider the two- and three-dimensional nature of the polar ion transport. In this report, we describe a new, two-dimensional kinetic model describing the transport of ionospheric ions into the polar magnetosphere and compare some simulation results from this model with recent observations of low-energy (0-50 eV) ions from DE 1.

Calculation Techniques and Assumptions

The calculation of ion trajectories, distribution functions, and ion bulk parameters is described in Horwitz [1984] and Horwitz and Lockwood

---

[1]Department of Physics, The University of Alabama in Huntsville, Huntsville, Alabama 35899
[2]Rutherford Appleton Laboratory, Chilton Didcot, OX11, OQX, England
[3]Space Science Laboratory, NASA Marshall Space Flight Center, Huntsville, Alabama 35812

[1985], and will be summarized here. The ion trajectories are two-dimensional trajectories appropriate to convection in the noon-midnight meridian. The magnetospheric magnetic field model is that of Luhmann and Friesen [1979], which consists of a dipole and thin current sheet to produce the tail. The transverse velocity is the convection velocity, $\underline{v}_\perp = \underline{E} \times \underline{B}/B^2$, where $\underline{E}$ is obtained by specifying $E_{\perp,i}$ at 300 km altitude at the geomagnetic pole and mapping its magnitude as $B^{1/2}$ elsewhere throughout the magnetosphere [Horwitz, 1984]. The parallel motion is obtained by integrating the parallel force equation, with magnetic gradient force, gravity, and parallel electric field terms. The parallel electric field is given by specifying $E_{\parallel,i}$ at the ionosphere and allowing it to vary with geocentric distance as $1/r^2$ [see Horwitz and Lockwood, 1985, for discussion]. Thus, the guiding center equations of motion which are integrated are:

$$m \frac{dv_\parallel}{dt} = m\underline{g} \cdot \hat{b} - \mu \frac{\partial B}{\partial s} + qE_{\parallel,i} \left(\frac{r_i}{r}\right)^2 \hat{r} \cdot \hat{b} \qquad (1)$$

$$v_\perp = \frac{\underline{E} \times \underline{B}}{B^2} \qquad (2)$$

where m,q = ion mass and charge; g = gravitational acceleration; μ = magnetic moment = $\epsilon_{\perp,gc}/B$, with $\epsilon_{\perp,gc}$ meaning the perpendicular kinetic energy in the $\underline{E} \times \underline{B}/B^2$ drift rest frame; $\partial B/\partial s$ = gradient of magnetic field magnitude along B; $\hat{b},\hat{r}$ = unit vectors along $\underline{B}$ and radially outward, respectively; and r_i,r the geocentric radial distance to 300 km altitude and to the ion's position, respectively. The trajectories discussed here would be substantially modified in the presence of larger parallel electric fields as indicated by Heppner et al. [1981] and Winningham and Gurgiolo [1982], as well as wave-particle interactions.

The evaluation of the distribution functions in the magnetosphere is based on Liouville's theorem and the ion trajectory code. An array of energies and pitch angles specifies the phase space at a given location over which we wish to define the distribution function. For each point in this phase space we integrate the ion trajectory backward in time. If the trajectory does not intersect the specified source region (see below) the distribution function value is zero at that energy and pitch angle. If such intersection does occur, the distribution value is that of the source distribution at the mapped energy and pitch angle.

The source distribution is taken to be the upgoing half of a bi-Maxwellian with specified density and parallel and perpendicular energies within a specified latitude range taken to be representative of the cleft region. The densities and thermal energies used in this report are in the range of cleft source region measurements obtained by Moore et al. [1986] and Waite et al. [1986].

Having obtained the ion distribution function, bulk parameters are obtained as various moments. Here we calculate the density, parallel average energy, and parallel flux moments at a spatial grid for the two-dimensional cross-section of the polar magnetosphere and contour these

HORWITZ ET AL. 57

parameters to indicate their spatial distribution in the polar magnetosphere.

Ion Observations

The low-energy ion measurements in this paper are from the retarding ion mass spectrometer (RIMS) experiment on DE 1, which was launched into an elliptical polar orbit, with 4.6 R_E geocentric distance apogee, in August 1981. RIMS measures the energy, angle, and pitch angle characteristics of ions in the energy 0-50 eV. A description of the instrument is given by Chappell et al. [1981].

Results

We will highlight some of the many interesting results obtained for transport of low-energy ionospheric ions through the polar magnetosphere. Figure 1 shows data and ion trajectories illustrating the concept of the "geomagnetic spectrometer." The top panel shows the peak energies of field-aligned flows for H^+, He^+, and O^+ detected as DE 1 moved poleward over the cleft and onto polar cap field lines. The first detection of each ion was sequential according to the ion mass, with O^+ observed most poleward of the cleft. Also, the upper cutoff energy in the spectrograms for each ion generally declines with the poleward movement. In the middle part of Figure 1, we have taken representative measured ion mass/energy/location combinations and integrated the trajectories backward. The ions all map back to a common narrow source that is in fact consistent with an upward flow and heating disturbance detected nearly simultaneously with the DE 2 satellite. Hence, this event illustrates the classic dispersion of ions of different masses and energies from a narrow source, controlled by the field-aligned motion and transverse convection, that we refer to as the geomagnetic spectrometer.

If the source for at least heavy ions in the polar magnetosphere is the narrow cleft topside ionosphere, then the strength of the convection electric field will determine the degree to which the polar cap is populated with these ions, in essence by "blowing" these cleft ions into the central and nightside polar cap magnetosphere. Figures 2 and 3 display simulations and observations to illustrate this concept. In Figure 2, we show contours of density, parallel average energy, and parallel flux in two columns in which source and other conditions are similar except that the convection electric field in the ionosphere is 20 mV/m in the left column case and 80 mV/m in the right column. Focusing on the densities in the two top panels, we see that in the low convection case, the O^+ is largely confined to the dayside near the source, whereas in the high convection (right column) case, large densities are seen in the central and nightside polar cap near 2-3 R_E geocentric distance. In Figure 3, we show statistical plots of DE 1 orbital segments in which the O^+ density essentially exceeded 1 O^+/cc, projected onto the GSM X-Z plane and organized into Kp range bins. It is seen that for low Kp, the segments of detectable O^+ tend to be short and confined to the dayside, whereas for the high Kp case, they extend into the central and nightside polar cap. Since the convection strength varies with Kp, these observations are consistent with the enhanced convection blowing of O^+ into these regions illustrated in Figure 2, with further effects probably also associated with increased ion source fluxes during high Kp.

The character of the ion distribution functions resulting from transport from a narrow source is also of interest. In Figure 4 are plotted distribution functions associated with the right column (high convection case) of Figure 3 for three spatial locations. The left-most distribution depicts the upgoing half of the bi-Maxwellian specified in the middle of the source location. The middle panel distribution depicts a conical or "bowl-shaped" distribution located slightly above and poleward of the source location associated chiefly with adiabatic folding of the distribution toward the magnetic field direction. The right-most panel depicts the distribution at

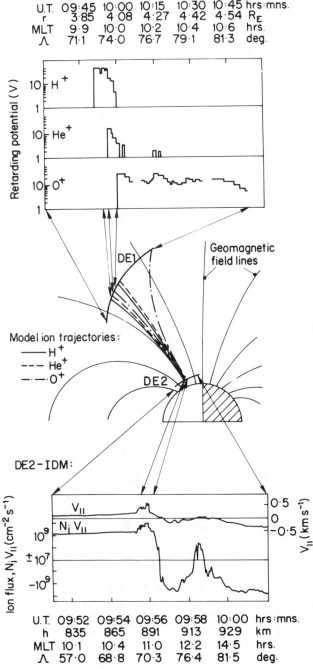

Figure 1. DE 1 low-energy ion data illustrating the geomagnetic spectrometer. The top panels show peak energies versus time for field-aligned flows of H^+, He^+, and O^+ from E = 0-50 eV ions. The variation of these fluxes is consistent with mass and energy dispersion of cleft origin ions, as indicated by the trajectory tracings of these trajectories back to a narrow source in the middle panel. The bottom panel displays near-simultaneous DE 2 data indicating ionospheric heating and flows in the locations indicated by the backwards trajectory tracings of the higher-altitude DE 1 ion flows. (From Lockwood et al., 1985c; copyright Nature, 1985)

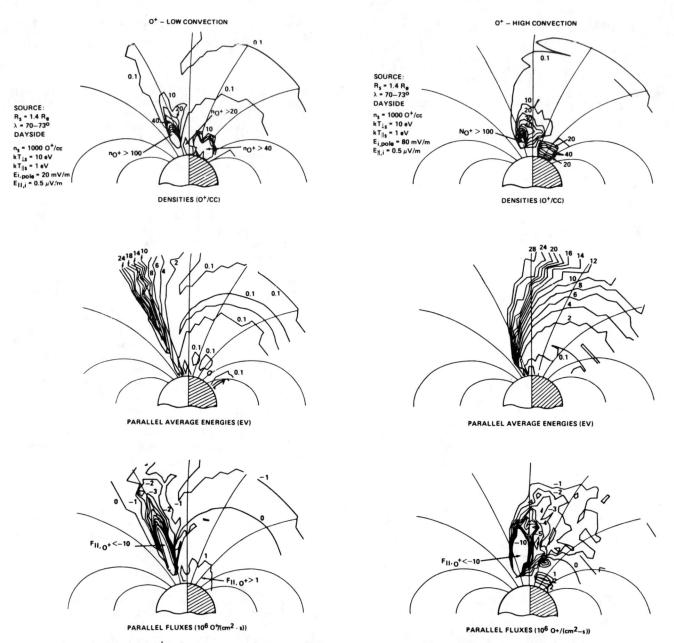

Figure 2. Calculated O^+ bulk parameters of density, parallel average energy and parallel flux (negative is upward) for conditions describing the characteristics of the cleft ion fountain which, at the ionosphere, is 20 mV/m on the left and 80 mV/m on the right. The only difference between the left and right columns is in the convection electric field. Principal features illustrated are the enhanced filling of the polar magnetosphere during increased convection (contrasting the left and right top panels of density), and the corresponding dispersion in parallel average energy seen in the contrast of the middle panels of these two columns.

a somewhat more distant location. Only a small region of velocity space is occupied around a centroid of about 10 km/sec upward flow velocity. This is a result of the velocity-filter effect in adiabatic flow from a narrow source region. Horwitz et al. [1985] have shown that the parallel Mach number for such conditions is given essentially as the ratio of the invariant latitude/horizontal distance of the location to the source over the width of the source, under certain simplifying conditions. This mechanism can explain the "supersonic" O^+ outflows originally reported by Waite et al.

[1985], provided there is an ion heating mechanism in the cleft region.

A final phenomenon of interest seen in the data and modeling is parabolic flow of heavy ionospheric ions from the cleft to the polar cap. In this case, heated but gravitationally bound ions flow initially upward out of the cleft ionosphere but fall downward after convecting into the polar cap. The small amounts of positive (downward) fluxes seen in the nightside portions of the low polar cap in Figure 2 result from this effect; these downward fluxes are more intense when source energies are low and

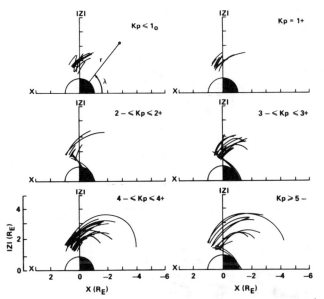

Figure 3. DE 1 orbital segments in the GSM X-Z plane for which the O^+ ions were observed (>1 O^+/cc), in Kp-ordered bins. The increased extension of these segments toward the nightside with Kp may be consistent with enhanced convection "blowing" of the O^+ from the cleft source.

hence more of the source population is gravitationally bound. In Figure 5, DE/RIMS data showing this effect are displayed. The four panels depict the time variation of upward parallel velocities for four ionospheric ions for an orbital segment somewhat like that shown in Figure 1, in which the spacecraft passed over the cleft ionosphere and into the polar cap. The analysis to obtain these parallel velocities requires an assumption about the convection velocity; in Figure 5 we show several curves for different convection values assumed in the data analysis, expressed in terms of the mapped ionospheric electric field. For assumed ionospheric electric fields less than about 60 mV/m, the heavy ions have downward parallel components beyond about 1236 UT. At this time, the DE 2 satellite measured a convection electric field of about 40 mV/m (R. A. Heelis, private communication, 1985), so it is reasonable to presume that these correspond to downward flows of heavy ionospheric ions in the polar cap, and further that this event was consistent with parabolic flow of the ions from the cleft topside ionosphere.

Conclusions

In summary, we have simulated and observed many interesting aspects of low-energy ion transport in the polar magnetosphere from the narrow cleft topside ionosphere source, arising from a heating mechanism. These aspects include: (1) the Kp/convection-dependent filling of the polar magnetosphere with ionospheric heavy ions due to "blowing" from the cleft source; (2) the mass and energy dispersion of cleft ionospheric ions due the combination of field-aligned motion and convection — the "geomagnetic spectrometer"; (3) the formation of "supersonic" ion flows as the natural result of adiabatic flow from a narrow source; and (4) the parabolic flow of gravitationally bound heavy ions from the cleft to the polar cap. There is every reason to expect that further exciting and fundamental results on the nature of this transport of low-energy ions will emerge as the simulations and observational studies proceed.

Acknowledgments. This research was supported in part at The University of Alabama in Huntsville by NASA contract NAS8-33982 and NSF grant ATM-8503102. Support for M. Lockwood came from the National Research Council under the Resident Research Associate program while he was at MSFC. We are also grateful to the RIMS team at MSFC and UAH and to the programming staff at the Boeing Corporation for assistance with the data reduction software. We would like to thank the Space Physics Analysis Network (SPAN) for use of computing and networking facilities.

References

Chappell, C. R., S. A. Fields, C. R. Baugher, J. H. Hoffman, W. B. Hanson, W. W. Wright, and H. D. Hammack, The retarding ion mass spectrometer on Dynamics Explorer-A, *Space Sci. Instrum.,* 5, 477, 1981.

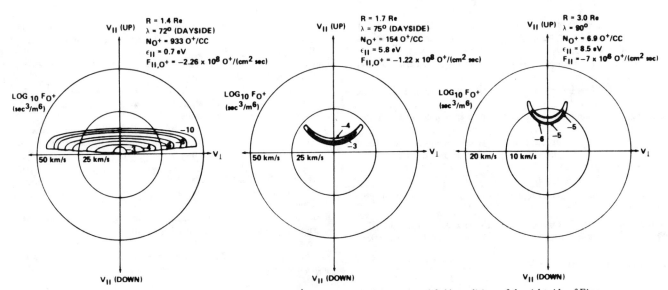

Figure 4. Example calculated velocity distributions of O^+ appropriate to the source and field conditions of the right side of Figure 2. These distributions illustrate, respectively, the distribution at the specified source location, the formation of a conic or bowl distribution above and poleward of the source, and a narrow ("supersonic") distribution more distant from the source.

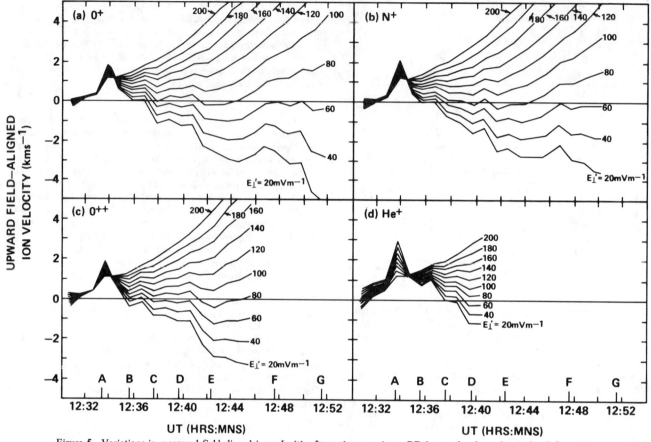

Figure 5. Variations in measured field-aligned ion velocities for various species as DE 1 moved poleward over the cleft and into the polar cap. For ionospheric electric fields of less than 60 mV/m used in the analysis to obtain the parallel velocities, the flows are consistent with parabolic flow of heated ionospheric ions from the cleft to the polar cap. Data at the times A-G were used by Lockwood et al. [1985b] to trace ions back to a cleft source [see Lockwood et al., 1985b].

Heppner, J. P., M. L. Miller, M. B. Pongratz, G. M. Smith, L. L. Smith, S. B. Mende and N. R. Nath, The Cameo Barium releases: E$_\parallel$ fields over the polar cap. *J. Geophys. Res., 86,* 3519, 1981.

Horwitz, J. L., The ionosphere as a source for magnetospheric ions, *Rev. Geophys. Space Phys., 20,* 929, 1982.

Horwitz, J. L., Features of ion trajectories in the polar magnetosphere, *Geophys. Res. Lett., 11,* 1111, 1984.

Horwitz, J. L., and M. Lockwood, The cleft ion fountain: A two-dimensional kinetic model, *J. Geophys. Res., 90,* 9749, 1985.

Horwitz, J. L., J. H. Waite, Jr., and T. E. Moore, Supersonic ion outflows in the polar magnetosphere via the geomagnetic spectrometer. *Geophys. Res. Lett., 12,* 757, 1985.

Lockwood, M., J. H. Waite, Jr., T. E. Moore, J. F. E. Johnson, and C. R. Chappell, A new source of suprathermal O$^+$ ions near the dayside polar cap boundary, *J. Geophys. Res., 90,* 4099, 1985a.

Lockwood, M., M. O. Chandler, J. L. Horwitz, J. H. Waite, Jr., T. E. Moore, and C. R. Chappell, The cleft ion fountain, *J. Geophys. Res., 90,* 9736, 1985b.

Lockwood, M., T. E. Moore, J. H. Waite, Jr., C. R. Chappell, J. L. Horwitz, and R. A. Heelis, The geomagnetic mass spectrometer — mass and energy dispersions of ionospheric ion flows into the magnetosphere, *Nature, 316,* 612, 1985c.

Luhmann, J. G., and L. M. Friesen, A simple model of the magnetosphere, *J. Geophys. Res., 84,* 4405, 1979.

Moore, T. E., M. Lockwood, M. O. Chandler, J. H. Waite Jr., C. R. Chappell, A. Persoon, and M. Sugiura, Upwelling O$^+$ ion source characteristics, *J. Geophys. Res.,* in press, 1986.

Waite, J. H., Jr., T. Nagai, J. F. E. Johnson, C. R. Chappell, J. L. Burch, T. L. Killeen, P. B. Hays, G. R. Carignan, W. K. Peterson, and E. G. Shelley, Escape of suprathermal O$^+$ ions in the polar cap, *J. Geophys. Res., 90,* 1619, 1985.

Waite, J. H., Jr., T. E. Moore, M. O. Chandler, M. Lockwood, A. Persoon, and M. Sugiura, Ion energization in upwelling ion events, this monograph, 1986.

Winningham, J. D., and C. Gurgiolo, DE-2 photoelectron measurements consistent with a large scale parallel electric field over the polar cap, *Geophys. Res. Lett., 9,* 977, 1982.

Young, D. T., Experimental evidence for ion acceleration processes in the magnetosphere, this monograph, 1986.

ION ENERGIZATION IN UPWELLING ION EVENTS

J. H. Waite, Jr.,[1] T. E. Moore,[1] M. O. Chandler,[2] M. Lockwood,[3] A Persoon,[4] and M. Suguira[5]

A source of H^+, He^+, O^+, and N^+ outflow from the ionosphere has been identified near the polar cusp/cleft using the Dynamics Explorer/ retarding ion mass spectrometer data set. This ion outflow termed "upwelling ions" is characterized by large outfluxes of H^+ and O^+ ions (10^8 to 10^9 cm^{-2} s^{-1}) and high transverse ion temperatures (~10 eV). This paper reports on the associated particle and field characteristics of one such upwelling ion event on March 12, 1982. Field-aligned currents and strong E x B convection channels are associated with the event as well as strong broadband plasma wave emission. One or all of these sources may play an important role in the ion energization in this region.

Introduction

The Dynamics Explorer/retarding ion mass spectrometer (DE/RIMS) has observed a myriad of low-energy ion outflows from the high-latitude ionosphere including the polar wind [Sojka et al., 1983; Nagai et al., 1984], polar cap O^+ outflow [Waite et al., 1985], auroral ion fountains [Moore et al., 1985], and an interesting new source of low-energy ion outflow in the dayside cusp/cleft called upwelling ion events [Lockwood et al., 1985]. Upwelling ion events are characterized by a localized transverse heating of ionospheric H^+, He^+, O^+, and N^+ ions that produces a substantial outflow of plasma that is transported into the polar magnetosphere. Such cusp ion outflow events are thought to be the low-energy extension of ion conics seen in the cusp by energetic ion mass spectrometers on earlier satellites [Shelley, 1979] and are also thought to be similar to the observed ion outflows of Gurgiolo and Burch [1982]. The roughly equal heating of the various ion species results in field-aligned flow velocities for the ions which are inversely proportional to the square root of their mass. The resulting velocity filter effect of solar-driven E x B ion convection provides a mass dispersion of ionospheric plasma in the polar cap that we call the Geomagnetic Mass Spectrometer [Waite et al., 1986; Moore et al., 1985]. The ultimate result of upwelling ion events is the creation of relatively high density O^+ and H^+ field-aligned flows over the polar cap with energies from 2 eV to >100 eV which have reported earlier by Shelley et al. [1982] and Waite et al. [1985].

The purpose of this paper is to present some of the source characteristics of the upwelling ion region using the complement of neutral and ion particle, and electromagnetic wave experiments onboard the DE 1 spacecraft. These measurements indicate a region near the dayside polar cap where H^+ and O^+ ions have been transversely heated to ~7 eV and are flowing out of the ionosphere with a flux of $>10^8$ cm^{-2} s^{-1} (at 1.8 R_E geocentric distance). The persistence of this source region from a statistical point of view equates to a source strength of $>10^{25}$ O^+ ions s^{-1} feeding the polar cap magnetosphere and eventually, through convective transport, the magnetotail and plasmasheet [Lockwood et al., 1985; Horwitz, 1984].

Observations

The data used in this study were obtained using instruments onboard the DE 1 spacecraft. Two separate data sets are reported in the paper. The first data set consists of a statistical study of O^+ ion outflow using over 2 years of data from the DE/RIMS, a medium resolution retarding potential analyzer/ion mass spectrometer covering the energy range from 0 to 50 eV and the mass range from 1 to 40 AMU. This data set is used to establish the occurrence probability, in time and space, of upwelling ion events. The second data set is comprised of measurements by all wave and particle instruments onboard DE 1 during an upwelling ion event on March 12, 1982. Further information on the instruments onboard the DE 1 spacecraft can be found in the special DE issue of *Space Science Instrumentation,* Volume 5, 1981.

The Upwelling Ion Region

The observations used in the statistical study consisted of all available DE/RIMS measurements of O^+ below 3 Earth radii geocentric distance over the 2-year period from October 19, 1981, to October 18, 1983; a total of 14,278 1-minute averaged samples. The O^+ flows were classified, according to the O^+ spin angle distribution, as having one of the following forms: (1) asymmetric spin angle distribution and upward flow (upwelling ions); (2) symmetric spin angle distribution, with the mean value shifted from the spacecraft ram direction towards the upward field-aligned direction (field-aligned upflow); (3) upgoing transversely accelerated core ions (double peaked, spin angle distributions), or (4) no detectable ion flow [Lockwood et al., 1985]. However, in this paper we are only interested in the upwelling ions.

Upwelling ions are easily characterized in the DE/RIMS data from their appearance in a spin phase angle versus time spectrogram such as that shown in Figure 1. This multi-panel spectrogram shows the radial head data for O^+ in the top two panels in a spin phase angle versus time format and the + and -Z head data (axial, transverse to line-of-flight), respectively, in the bottom two panels using a retarding potential versus time spectrogram format. The spin axis of DE 1 is normal to the spacecraft orbit and allows the detector, located in the spin plane (the "radial" head), to sample the full pitch angle range of the ions once per spin. The + and -Z heads are located antiparallel and parallel to the spin angular momentum vector, respectively. The radial head data, panels 1 and 2, are plotted as a function of the spin phase angle with respect to the satellite orbital direction on the vertical axis, time being on the horizontal axis. The upward and downward field-aligned directions are marked on the plot by the dashed and dotted white traces, respectively. Note the sudden

---

[1] Space Science Laboratory, NASA Marshall Space Flight Center, Huntsville, Alabama 35812.
[2] The University of Alabama in Huntsville, Alabama 35899.
[3] Rutherford Appleton Laboratory, Didcot, Oxfordshire, OX11, OQX, England.
[4] University of Iowa, Iowa City, Iowa 52242.
[5] NASA Goddard Space Flight Center, Greenbelt, Maryland 20771.

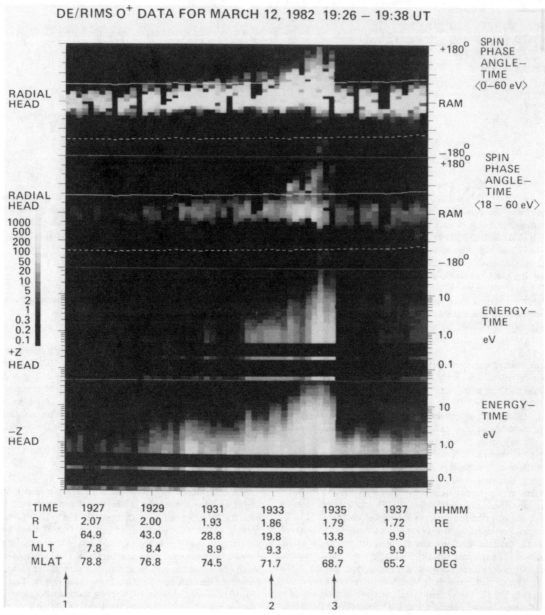

Figure 1. DE 1 RIMS spectrograms from the radial and axial heads for the time period of the upwelling ion event 1925 to 1938 UT. Integral fluxes are coded according to the grey scale at the right of the figure. The top two panels show spin-time spectrograms of O^+ from the radial head averaged over two different energy ranges, and the bottom two panels show retarding potential-time spectrograms of O^+ from the axial head.

excursion of the O^+ ion flux toward the upward field-aligned direction at 1928 UT with a dramatic equatorward cutoff at 1935 UT. These are the boundaries of the upwelling ion event denoted by points 1 and 3, respectively, in Figures 1, 3, and 5, and defined according to the criteria of Lockwood et al. [1985]. Point 2 represents the poleward boundary and point 3 the equatorward boundary of the current sheets (see also Figure 3) which are correlated almost one for one with upwelling ion events [Lockwood et al., 1985]. The top panel shows average spin curves for all O^+ ions from 0 to 60 eV, and panel 2 shows spin curves for O^+ ions between 18 and 60 eV. Note that a conical ion signature emerges at the higher energies. The bottom two panels indicate the energy distribution of the plasma, from 0 to 50 eV, transverse to the orbital direction and to the magnetic field direction. The peak in the O^+ ion flux and associated energy distribution from 1934 to 1935 UT indicates that transverse heating is taking place in this upwelling ion event.

The asymmetric spin distribution seen in the above example for O^+ is always present in all ion species measured (H^+, He^+, O^+, O^{++}, N^+). These events have been statistically categorized in the 2-year DE/RIMS data set described above. The results of this survey [Lockwood et al., 1985] are shown in Figure 2—an occurrence frequency plot for O^+ upwelling ion events with occurrence frequencies plotted as a function of invariant latitude and magnetic local time. Note the localized nature of

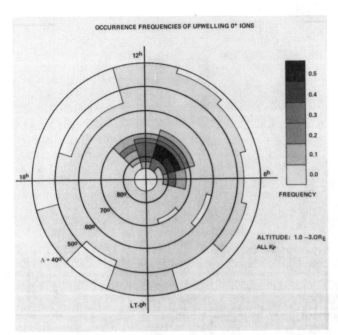

Figure 2. A polar dial plot showing the occurrence probability of upwelling ion events in the DE/RIMS data between October 19, 1981, and October 18, 1983.

the upwelling ion region, a band near the dayside polar cap boundary skewed noticeably toward the morning sector and most probably associated with the polar cusp/cleft. At the peak of the occurrence plot, near 74° invariant latitude and 10 hours magnetic local time, the probability for observing an upwelling ion event between 0.4 and 1.2 R_E altitude is essentially unity.

Source Characteristics of the Upwelling Ion Region

Outfluxes of H^+ and O^+ ions between 10^8 and 10^9 cm^{-2} s^{-1} are consistently seen in upwelling ion regions near the polar cusp/cleft. These ions show indications of significant transverse ion heating. Therefore, it is interesting to examine the characteristics of particles, electric and magnetic fields, and plasma waves in this region to get some insight into the ion energization responsible for the outflow. The data set is taken from an event over the northern polar cap on March 12, 1982, at approximately 19:35 UT.

Magnetic field signature. Lockwood et al. [1985] showed that a strong correlation existed between the appearance of upwelling ion signatures and the occurrence of a signature in DE 1 magnetometer data which is indicative of field-aligned currents. Analysis of the DE 1 magnetometer data for the March 12 event provided current densities, deduced under an assumption of infinite current sheets, that are plotted in the upper panel of Figure 3. The B component of the magnetic field, as compared to the MAGSAT model of Earth's internal field, is shown in the lower panel. The field-aligned currents are divided into two regions: a poleward region of predominantly downward current (region 1) and an equatorward region of predominantly upward current (region 2) with the boundary being near invariant latitude 75.7° or at 19:35:06 in time. The timing of the magnetic perturbation indicates that the equatorward edge of the upwelling ion event was coincident with the equatorward edge of the upgoing current sheet; however, the event extended throughout both the upward and downward directed current sheets.

Interpretation of PWI spectrogram. Figure 4 is a frequency-time spec-

trogram from the electric field amplitude measurements of the plasma wave instrument (PWI) for the period of this upwelling ion event. Several plasma wave modes commonly observed by DE 1 are illustrated in this spectrogram. The intense emissions above 130 kHz from 1835 to 1852 UT are auroral kilometric radiation. The emissions ranging from 1 to 40 kHz (1835-1901 UT) which have a well-defined upper frequency cutoff well below the electron cyclotron frequency represent the poleward expansion of the characteristic funnel-shaped nightside auroral hiss emissions. The less intense emissions above the auroral hiss which have well-defined cutoff at the electron cyclotron frequency are broadband Z-mode emissions (1840-1850 UT). Cusp-associated auroral hiss emissions of relatively low intensity are found below 100 kHz during the satellite's cusp pass in the northern hemisphere (1925-1935 UT) and again up to frequencies above 400 kHz during the cusp pass in the southern hemisphere (2014-2017 UT). Plasmaspheric hiss emissions occur below 50 kHz at 1935 to 1950 UT just equatorward of the cusp-associated auroral hiss emissions and again in the southern hemisphere below 10 kHz at 2007 to 2014 UT.

The cusp-associated auroral hiss emissions which occur during the time of this upwelling ion event (1925-1935 UT) cut off at frequencies below the electron cyclotron frequency. Cold plasma theory predicts that, in low density regions of the magnetosphere where the electron plasma frequency is less than the electron cyclotron frequency, these whistler mode emissions will be driven into resonance at the electron plasma frequency. Using a technique for deriving the electron density from the upper frequency cutoff of the auroral hiss emissions (described in detail in Persoon et al. [1983]), a density profile is obtained during the time of the upwelling ion event. Calculated densities increase from 100 ± 45 cm^{-3} at 1930 UT to 430 ± 194 cm^{-3} at 1934 UT. Limitations of the technique have been discussed fully in Persoon et al. The large uncertainty in the density determination is primarily due to the low intensity of the auroral hiss emissions with respect to the background noise level near resonance, making it impossible to locate the upper frequency cutoff to an accuracy greater than ±20%. This uncertainty in locating the upper frequency cutoff corresponds to an uncertainty in the density determination of ±45%. The resulting densities are, however, consistent with the RIMS ion density

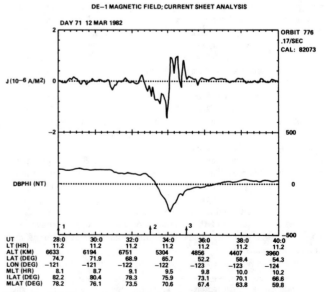

Figure 3. DE 1 magnetometer signature for the upwelling event showing interpretation in terms of two adjacent field-aligned current sheets having current densities of approximately 1 microamp/m^2 each.

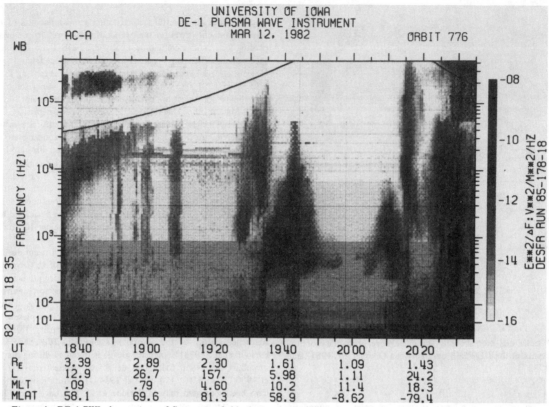

Figure 4. DE 1 PWI observations of fluctuating fields during the upwelling event. Note particularly the low-frequency emissions associated with the peak of the event, appearing from 1 Hz to well above the proton gyrofrequency. The black trace marks the electron gyrofrequency.

analysis. The density profile indicates that abrupt enhancements in the plasma density occur at 1930 UT and 1934 UT in connection with the ion heating observed by RIMS.

From 1933 to 1935 UT, intense low-frequency emissions can be seen extending from the lowest frequency of 1 Hz to well above the proton cyclotron frequency at 140 Hz. Similar broadband electric field noise has been found to occur at low altitudes over both the evening and morning auroral regions. The noise is believed to be generated in a source region above 2 R_E and has an average Poynting flux ($\sim 2 \times 10^{-2}$ erg cm^{-2} s^{-1}) directed downward toward Earth [Gurnett et al., 1984]. Gurnett et al. have estimated the wave energy to be very large for these emissions providing a possible source of energy for the O^+ ions.

Convection electric field. A polar plot of the derived horizontal drift vectors from the quasi-static electric field detector on the DE plasma wave instrument is shown in Figure 5. One of the most striking characteristics of the upwelling ion event is the particularly strong channel, or jet, of plasma flow, approaching 10 km s^{-1} (corresponding to an electric field of nearly 100 mV m^{-1}) at 1934 UT, which is coincident with the current sheet interface and the upwelling ion event. This may be an important source of energization for the event since the thermal velocity of the observed heated ions is of the same order of magnitude as the convection jet speed, and both are much larger than the thermal speed of the surrounding ionosphere.

Plasma observations. The low-energy ion observations of the DE/RIMS instrument have been described earlier and are shown in Figure 1. The O^+ data indicate a cold, rammed ionospheric flow of O^+ poleward and equatorward of the upwelling event with a marked field-aligned departure during the event (from 1932 to 1935 UT). The asymmetric O^+ spin distribution during this time indicates an outflow of O^+ exceeding 10^8 cm^{-2} s^{-1} at an altitude of 1.8 R_E ($\sim 10^9$ cm^{-2} s^{-1} when mapped down to ionospheric altitudes) and a transverse heating of the plasma to several electron volts ($T_{\parallel} \sim 3$ eV, $T_{\perp} \sim 7$ eV). H^+ ions also show signs of transverse heating and increased fluxes ($>10^8$ cm^{-2} s^{-1} at 1.8 R_E) in the upwelling ion region. However, two differences between O^+ and H^+ are evident. First, the upwelling ion region for H^+ is embedded in a region of H^+ polar wind outflow ($>10^7$ cm^{-2} s^{-1} at 1.8 R_E) both poleward and equatorward of the region. Furthermore, in the upwelling region the H^+ low-energy ion population shows a highly non-Maxwellian distribution, unlike O^+, which would seem to indicate some mixing of ionospheric and magnetosheath plasma in this region. The He^+ ion population has a much smaller flux ($\sim 10^6$ cm^{-2} s^{-1} at 1.8 R_E) and is not as hot as the O^+ ion population during the event (~ 4.5 eV).

The density of both H^+ and O^+ during the event has been determined using the DE/RIMS measurements [Moore et al., 1986] and is shown in the upper half of Figure 6. Also plotted for comparison is the total plasma density calculated from the DE/PWI using the upper frequency auroral hiss cutoff. The agreement is good except near the peak of the event where the DE/RIMS calculation tends to underestimate the density due to the strong transverse heating. This heating causes much of the distribution to be transverse to the magnetic field and spin plane and, hence, out of the view of the radial head. Densities at the peak of the event reach almost 400 cm^{-3}, quite high for an altitude of 4500 km in the polar cap. The O^+ transverse temperature during the event is shown in the bottom half of Figure 6. The temperatures vary from 0.6 eV at 1926 UT in the polar cap to 7.2 eV at the peak of the event near 1935 UT. H^+ shows a similar transverse heating, but also shows a non-Maxwellian high energy tail during the event.

Indeed, the DE/energetic ion composition spectrometer measures

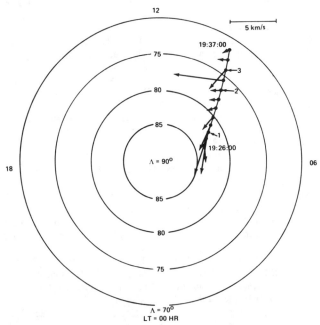

Figure 5. A polar plot of the horizontal ion drift velocities calculated from the DE 1 dc electric field data. Note the antisunward flow evolving into an intense eastward crosswind jet at the time of the upwelling ion event.

what appears to be a tail of the upwelling O^+ distribution; however, due to the low characteristic energy of the O^+ (by energetic plasma standards), no energetic O^+ temperatures can be derived. The energetic H^+, however, shows features consistent with injection of magnetosheath plasma into the magnetosphere on closed field lines [Moore et al., 1986], but a detailed study of the energetic plasma characteristics of such events has not yet been completed.

Discussion and Conclusions

In this paper, we have indicated the existence of a large upwelling of low-energy ions near the dayside polar cap boundary that serves as a persistent source of plasma for the polar cap magnetosphere ($>10^{25}$ O^+ and H^+ ions s^{-1}). We have examined the characteristics of this upwelling region using instruments onboard the DE 1 spacecraft and found the most notable features of such events to include transverse ion heating to temperatures of order 10^5 K (7.2 eV), large outward flow of the heavy ion O^+, enhanced outflow of light ions H^+ and He^+, moderately strong field-aligned current sheets of both senses, an associated intense eastward convection channel, and strong emission of waves in the frequency range near and below the proton gyrofrequency.

Although recent models have stressed the role of hot electrons in increasing ion outflow [Barakat and Schunk, 1983], the observations reported in this paper clearly indicate that ion heating plays a major role in the enhanced outflow of heavy ions in the upwelling region. Such large ion outfluxes, resulting from increased ion heating have been theoretically demonstrated using the ion transport model of Gombosi et al. [1985, 1986]. The model produced transient O^+ outfluxes of $>10^9$ cm^{-2} s^{-1} [see Gombosi et al., this monograph] sufficient to explain the observed ion outfluxes. However, more detailed modeling (using calculated rather than parameterized ion heating) is needed to fully understand the phenomenon. The source of this ion heating is in fact one of the most fascinating questions surrounding this phenomenon. Over 6×10^{-3} erg cm^{-2} s^{-1} are required to heat the ions to the temperatures and densities observed during the March 12, 1982, event.

As potential sources of energy for the ion heating, we have strong field-aligned currents (which have been correlated with upwelling ion events on a sound statistical basis [Lockwood et al., 1985]), strong perpendicular electric fields associated with a plasma convection jet observed during this event, and various wave emissions that may have propagated from a distant generation site or that may have been locally generated by plasma current or flow instabilities that are associated with the chosen event, but have also been observed in the cusp region in past studies [Gurnett et al., 1984]. We feel that the correspondence between the O^+ heating, its characteristic random energy (10 eV), and the flow energy of the convection jet for O^+ is highly suggestive of a collisional or joule heating mechanism. However, it is not clear how such a mechanism can work at an altitude where there are ordinarily few collisions on time scales of interest or how the energy could be transported up from lower altitudes where such heating is strong.

Acknowledgments. We are indebted to the engineering and science staff of the University of Texas at Dallas and to the RIMS team at Marshall

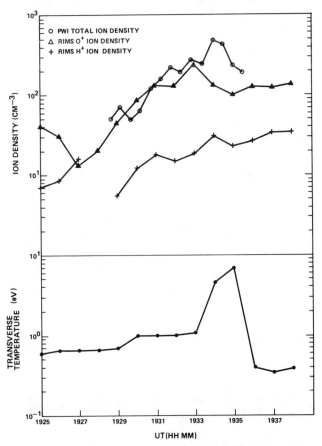

Figure 6. The top panel shows the PWI and RIMS derived total plasma density. The second panel indicates the transverse O^+ ion temperature during the event derived from the -Z head using a drifting Maxwellian fit to the data.

Space Flight Center. Assistance with data reduction software provided by the staff of the Boeing Corporation and hardware provided by the MSFC/SPAN network is gratefully acknowledged. Support for M. Lockwood and partial support for M. Chandler was provided by the National Research Council under the Resident Research Associateship program. We further thank W. K. Peterson and E. G. Shelley of the DE/EICS science team for providing us with energetic ion composition spectrometer data used in this study.

References

Barakat, A. R., and R. W. Schunk, O^+ ions in the polar wind, *J. Geophys. Res., 88,* 7887-7894, 1983.

Gombosi, T., T. E. Cravens, and A. F. Nagy, A time-dependent theoretical model of the polar wind: Preliminary results, *Geophys. Res. Lett., 12,* 167-170, 1985.

Gombosi, T. I., T. E. Cravens, A. F. Nagy, and J. H. Waite, Jr., Time-dependent ion outflow calculations, this monograph, 1986.

Gurgiolo, C., and J. L. Burch, DE-1 observations of the polar wind — A heated and an unheated component, *Geophys. Res. Lett., 9,* 945-948, 1982.

Gurnett, D. A., R. L. Huff, J. D. Menietti, J. L. Burch, J. D. Winningham, and S. D. Shawhan, Correlated low-frequency electric and magnetic noise along the auroral field lines, *J. Geophys. Res., 89,* 8971-8985, 1984.

Horwitz, J. L., Features of ion trajectories in the polar magnetosphere, *Geophys. Res. Lett., 11,* 1111, 1984.

Lockwood, M., J. H. Waite, Jr., T. E. Moore, J. F. E. Johnson, and C. R. Chappell, A new source of suprathermal O^+ ions near the dayside polar cap boundary, *J. Geophys. Res., 90,* 4099-4116, 1985.

Moore, T. E., C. R. Chappell, M. Lockwood, and J. H. Waite, Jr., Suprathermal ion signatures of auroral acceleration processes, *J. Geophys. Res., 90,* 1611-1618, 1985.

Moore, T. E., M. Lockwood, M. O. Chandler, J. H. Waite, Jr., A. Persoon, and M. Sugiura, Upwelling O^+ ion source characteristics, *J. Geophys. Res.,* in press, 1986.

Nagai, T., J. H. Waite, Jr., J. L. Green, C. R. Chappell, R. C. Olsen, and R. H. Comfort, First measurements of supersonic polar wind in the polar magnetosphere, *Geophys. Res. Lett., 11,* 669-672, 1984.

Persoon, A. M., D. A. Gurnett, and S. D. Shawhan, Polar cap electron densities from DE 1 plasma wave observations, *J. Geophys. Res., 88,* 10,123-10,136, 1983.

Shelley, E. G., Ion composition in the dayside cusp, in Proceedings of Magnetospheric Boundary Layers Conference, *Eur. Space Agency Spec. Publ. ESA SP-148,* 1979.

Shelley, E. G., W. K. Peterson, A. G. Ghielmetti, and J. Geiss, The polar ionosphere as a source of energetic magnetospheric plasma, *Geophys. Res. Lett., 9,* 941-944, 1982.

Sojka, J. J., R. W. Schunk, J. F. E. Johnson, J. H. Waite, and C. R. Chappell, Characteristics of thermal and suprathermal ions associated with the dayside plasma trough as measured by the Dynamics Explorer Retarding Ion Mass Spectrometer, *J. Geophys. Res., 88,* 7895-7911, 1983.

Waite, J. H., Jr., T. Nagai, J. F. E. Johnson, C. R. Chappell, J. L. Burch, T. L. Killeen, P. B. Hays, G. R. Carignan, W. K. Peterson, and E. G. Shelley, Escape of suprathermal O^+ ions in the polar cap, *J. Geophys. Res., 90,* 1619-1630, 1985.

Waite, J. H., Jr., M. Lockwood, T. E. Moore, M. O. Chandler, J. L. Horwitz, and C. R. Chappell, Solar wind control of the geomagnetic mass spectrometer, in *Solar-Wind Magnetosphere Coupling,* edited by Y. Kamide and J. A. Slavin, Terra-Reidel, Kyoto, in press, 1986.

OBSERVATIONS OF TRANSVERSE AND PARALLEL ACCELERATION OF TERRESTRIAL IONS AT HIGH LATITUDES

H. L. Collin, E. G. Shelley, A. G. Ghielmetti, R. D. Sharp

Lockheed Palo Alto Research Laboratory, Palo Alto, California 94304

Abstract. Previous studies of upflowing ions indicate that ion beams acquire their energy by a process which results in the heavier ions gaining additional transverse energy. Data from S3-3 are presented which provide evidence that ion beams can be the result of a two step acceleration process in which ion conics formed at lower altitudes are then accelerated through a potential drop. These ion conics, generated in regions below such potential drops and presumably within upward current regions, are produced by a mechanism which provides more energy to heavier ions.

Introduction

The Lockheed ion mass spectrometer on the S3-3 satellite has provided numerous observations of upward flowing terrestrial ions in the energy range 0.5 - 16 keV. These have been observed in the altitude range of 2000 km to 8000 km over the auroral regions. H^+ and O^+ are the most common species with He^+ occurring occasionally [Ghielmetti et al., 1978]. Ions have been observed which have undergone acceleration primarily transverse to the geomagnetic field direction (ion conics) [Sharp et al., 1977] or primarily parallel to the field direction (ion beams) [Shelley et al., 1976].

The ion conics indicate the dominance of a transverse acceleration process. They typically have falling energy spectra [Ungstrup et al., 1979] and are most commonly seen at energies below 500eV, the threshold of the ion mass spectrometer [Gorney et al., 1981]. They are frequently observed by S3-3 at altitudes between 2000 and 8000 km. In contrast the ion beams rarely occur below 5000 km and typically have distribution functions which are field aligned and are peaked at energies of a few keV [Shelley et al., 1976]. Associated with the ion beams are signatures in the electron distribution functions which indicate the existence of potential drops below, and sometimes above, the satellite.

These ion beam observations suggest that they acquire their energy by falling through the potential drop. However, such an explanation is too simple. The O^+ component of a beam generally has more energy than the H^+ component and is less closely field aligned than the H^+ [Collin et al., 1981]. Observations at higher altitudes [Lundin et al., 1982] show that the He^+ component also has more energy than the H^+. This implies that beams are not only accelerated parallel to the field direction, but are also accelerated transversely, and that the transverse acceleration process is mass dependent.

One explanation of the beams' transverse component of energy is that conics from lower altitudes were accelerated through a potential drop to form the beams [Chiu et al., 1983; Klumpar et al., 1984] and that the original conics had been produced by an acceleration process which accelerated heavy ions more strongly.

Ion Beams and Potential Drops

The data discussed here were acquired by the Lockheed ion mass spectrometer [Sharp et al., 1979] on the S3-3 satellite. The magnitudes of the potential drops below the satellite can be determined by measurements of the widening of the electrons' loss-cones. For a review of the effects of potential drops on electron distributions see Kaufmann [1984]. The method adopted here was similar to that used by Collin et al. [1984]. A composite function was used to represent the observed electron distribution. This consisted of two polynomials, one to represent the mirroring electron flux and another to represent the loss cone. The function was fitted to the observed electron distribution using a nonlinear least squares method and the 50represent its width. From this the potential drop can be found [Kaufmann, 1984] provided that the satellite altitude and the electron energy is also known. Within or below a potential drop the electron distribution comprises the accelerated primary component and lower energy electrons which are trapped below the potential [Kaufmann 1984]. Only those energies which were above the peak in the distribution and so could be relied upon to be from the primary distribution were used. Since the energy channels were very broad [Sharp et al., 1979] an effective energy was determined for each by fitting to an accelerated Maxwellian distribution and finding for each channel the energy weighted by the Maxwellian. The results obtained by this method showed good agreement with those obtained by a more detailed analysis [Sharp et al., 1979]. In order to determine the uncertainty in the evaluation of a potential drop a period of data was selected during which there was no evidence of a potential drop and electrons of all energies had loss cones of the same width. When the potential drops in this period were evaluated they were found, as expected, to be close to zero. Their scatter gave a measure of the uncertainty of the evaluation. This uncertainty was included with the

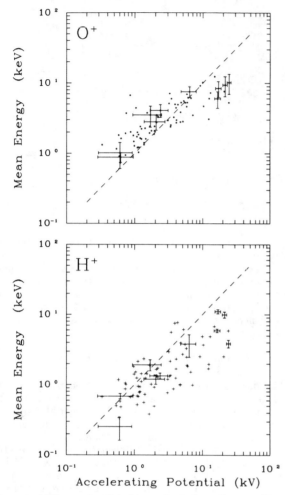

Fig. 1. The potential drop, Φ, below the satellite as determined from electron loss-cone measurements and the corresponding ion mean energies, $<E>$. The error bars show 1σ confidence levels for a representative sample of points. The dashed lines indicate $<E> = e\Phi$.

statistical errors when computing the total uncertainty. The estimated potential drop was rejected if count rates were low or if good fits could not be obtained. The ion mass spectrometer made measurements at three energies simultaneously [Sharp et al., 1979]. The mean energy, $<E>$, of the ions in the beam was characterized by the differential flux weighted mean of the three energies. See Ghielmetti et al. [1986] for details.

When the energies of the ion beams and the corresponding potential drops were compared they showed good statistical correlation (Figure 1). However, the energy of the H^+ tended to be slightly less than that corresponding to the potential drop, whereas the energy of the O^+ tended to be higher. This is consistent with the heavier O^+ having acquired additional energy.

Ion Conics and Potential Drops

The energies of most of the ion conics observed by S3-3 were too low to permit the distribution functions of their constituent ion species to be characterized by the ion mass spectrometer. However, a few multi-component conics were energetic enough to be examined in detail [Collin and Johnson, 1985]. One of these events was observed at an altitude of 3780 km over the evening sector auroral zone, Figure 2 and Figures 3a and 3d. During the four spins in which the ion conic was observed the associated electron distributions in the highest energy spectrometer, 7.3 - 24 keV, showed minima at pitch angles of 90°. These indicated that the electrons had been accelerated through a potential drop, Φ, not far above the satellite [Kaufmann, 1984]. The 90° minima could also be seen in the 1.6 - 5.0 keV spectrometer, but were much less pronounced. This implied that most of the electrons in this energy range were approximately isotropic, probably secondaries trapped below the potential drop, and that only those near the upper end of the range had been accelerated down thorough the potential drop. Since only the accelerated electrons would have exhibited the 90° minima $e\Phi$ was probably close to this spectrometer's upper energy limit of 5.0keV. The electrons did not show enhanced loss-cones at this time. This indicates that there was little or no potential drop below the satellite. The conic was composed of H^+ and He^+ with very little O^+ and showed marked mass dependent energy differences. While the H^+ was very soft and not detectable above 1.5 keV, the He^+ was much harder with substantial fluxes throughout the range of the spectrometer, 0.5 - 16 keV.

The existence of a potential drop above the conic indicates that the two step acceleration process suggested by Chiu et al. [1983] and Klumpar et al. [1984] was operating in this event. In order to determine what S3-3 would have seen above the potential drop the observed conic distributions were mapped through 5kV and through a change in B corresponding to moving from 3780 km to 8000 km in a process analogous to that used by Klumpar et al. [1984]. No attempt was made to simulate the possible effects of any non-adiabatic process. The resulting distributions, Figures 3b and 3e, were more energetic and the pitch angle widths of the conics were much reduced and somewhat energy dependent, similar to the distributions described by Klumpar et al. [1984] and Ghielmetti et al. [1986]. The ion mass spectrometer's angular acceptance range was approximately 6° full width and its pitch angle sampling rate for a given mass was about 15° and thus it could not resolve the details of such narrow distributions during a single spin. In order to simulate what S3-3 would have observed had it passed over this event at the greater altitude of 8000 km the ion mass spectrometer sampling was applied to the accelerated conic distributions, the dots in Figure 3 show the points sampled, the details of the conical distributions were not completely resolved and the 'observed' distribution appeared more beam-like, Figures 3c and 3f. On the basis of these distributions, the mean energy of He^+ component of the beam was computed to be 7.0keV and that of the H^+ was 5.8keV. $E_{He}/E_H \approx 1.2$. The pitch-angle width of the He^+ would have been 22°, somewhat greater than that of the H^+, 18°. These values are comparable to the typical values reported in statistical studies of the ion beams. Collin et al. [1981] found $E_O/E_H \approx 1.7$ and mean pitch angle widths of 15° for H^+ and 22° for O^+. This correspondence supports the hypothesis that the two stage acceleration model was operating in this event.

Five other conics permitted detailed examination by the ion mass spectrometer [Collin and Johnson 1985]. In none of these cases did the electrons show the signatures

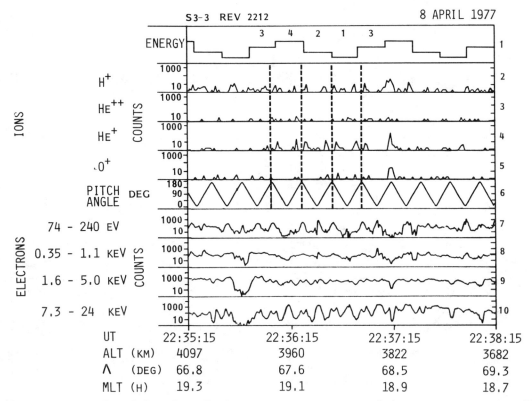

Fig. 2. Survey plot of data from the ion mass spectrometers and electron spectrometers. The abscissa indicates the universal time, altitude, invariant latitude and magnetic local time at which the data were acquired. Panel 1 shows the mass spectrometers' sequence of energy steps. Panels 2-5 show the counts of ions summed once per second from selected channels from all three ion mass spectrometers. Panel 6 indicates the pitch angle of the look direction of all the spectrometers and panels 7-10 display the counts per 0.5s for the electron spectrometers. The broken lines indicate the satellite spins during which H^+ and He^+ conics were seen.

of potential drops above the satellite at the time the conics were observed, although there was evidence of potential drops in adjacent regions in some cases. In none of these cases, when the conics were composed of only H^+ and O^+, was there evidence of significant difference between the energies of the components.

None of the other conics observed by the ion mass spectrometer were suitable for detailed study, but 8 additional cases, making a total of 14, were found where the energies and intensities were sufficient to give a rough indication of the relative hardness of the energy spectra of the ion species comprising the conics. In 3 of these cases the electrons indicated a potential drop above the satellite. In all three of these cases the higher mass ion appeared to be the more energetic. In the remaining 11 cases, in which there was no clear evidence of a potential drop, there were three cases in which the higher mass ion appeared to be more energetic and two in which the lighter ion appeared more energetic. In the remaining cases the ions' energies appeared to be independent of mass.

Discussion

The excess of energy of the O^+ observed by S3-3 is in contrast to the events studied by Reiff et al. [1986]. Their observations from DE-1 and DE-2 showed ion beams all of whose components had approximately equal average energies per ion and these energies were close to that which the beams could have acquired by falling through the simultaneously measured potential drop. Reiff et al. [1986] deduced that although they had become thermalized by a wave instability mechanism within or above the acceleration region the ions had not gained significant energy except from the potential drop.

The excess energy of the O^+ observed on S3-3 may also be explained, at least in part, by wave instability mechanisms. For example, Kaufmann and Ludlow [1986] and Bergmann and Lotko [1986] have shown that under suitable conditions a two stream instability can transfer energy from the H^+ to the O^+ component. However the operation of one mechanism does not necessarily eliminate the possibility that other mechanisms may be operative, at least on occasions. The April 8th event demonstrates that ions do sometimes become substantially energized below the main potential drop acceleration region.

The available mass resolved observations of conics with good energy and temporal resolution are too limited to allow firm conclusions to be drawn, but there are some indications that in regions below upward directed potential drops ion conics are generated by mechanisms which provide more energy to heavier ions. Outside these regions the conics are more often generated by mechanisms

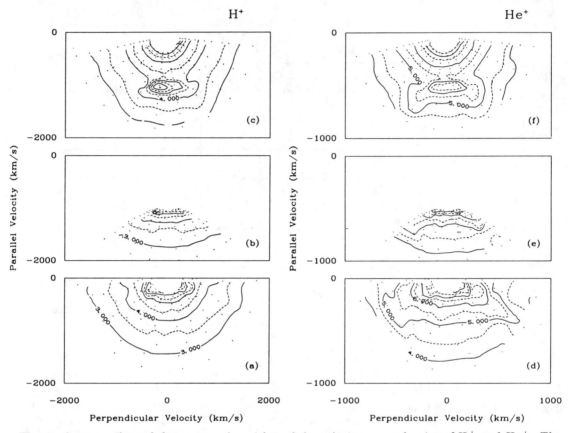

Fig. 3. Contour plots of the common logarithm of the velocity space density of H^+ and He^+. The dots indicate the locations of the data points. Panels a and d correspond to the observed conics which are displayed in Figure 2. Panels b and e show these distributions after passing through a potential drop and into a region of reduced **B**. Panels c and f show how these modified distributions would appear when viewed with the ion mass spectrometer's sampling resolution. The contours which indicate an isotropic distribution in some of the plots represent the 1/2 count surface and are included to indicate the threshold of sensitivity.

which appear to be independent of mass. The possibility that more than one mechanism may be responsible for the production of ion conics has been discussed previously by Cattell [1984] who pointed out that several of the suggested mechanisms have spatial distributions which are different from those of observed conics and so could not account for all conics.

Conclusions

Previous ion beam studies indicated that the energy of ion beams is the result of a process, or processes, which not only gives the ions parallel energy, but which results in the heavier ions gaining a larger proportion of transverse energy. In some cases a potential drop and wave instabilities mechanisms can adequately account for the distibution functions of the beams. The present study provides evidence for the two step acceleration process [Chiu et al., 1983; Klumpar et al., 1984] in which ion conics are first formed at lower altitudes and are afterwards accelerated upward through a potential drop. In the regions below such potential drops, and presumably upward currents, the original ion conics are generated by a mechanism which provides more energy to heavier ions while there are indications that outside these regions the mechanisms which generate ion conics may be independent of mass.

Acknowledgments. The Lockheed experiment and data aquisition was supported by the Office of Naval Research under contract N00014-78-C-0479. This work was supported by the National Science Foundation under grant ATM-8317710, by NASA under contract NASW-3395 and by the Lockheed Independent Research Program.

References

Bergmann, R. and W. Lotko, Transition to unstable flow in parallel electric fields, *J. Geophys. Res.*, in press, 1986.

Cattell, C. A., Associations of field-aligned currents with small-scale auroral phenomena, in *Magnetospheric Currents*, editor T. A. Potemra, A.G.U.,Washington, 304-314, 1984.

Chiu, Y. T., J. M. Cornwall, J. F. Fennell, D. J. Gorney and P. F. Mizera, Auroral plasmas in the evening sector: satellite observations and theoretical interpretations, *Space Sci. Rev.*, *35*, 211, 1983.

Collin, H. L., R. D. Sharp, E. G. Shelley and R. G. Johnson, Some general characteristics of upflowing ion beams and their relationship to auroral electrons, *J. Geophys. Res.*, *86*, 6820, 1981.

Collin, H. L., R. D. Sharp and E. G. Shelley, The magnitude and composition of the outflow of energetic ions from the ionosphere, *J. Geophys. Res.*, *89*, 2185, 1984.

Collin, H. L. and R. G. Johnson, Some mass dependent features of energetic ion conics over the auroral regions, *J. Geophys. Res.*, *90*, 9911, 1985.

Ghielmetti, A. G., R. G. Johnson, R. D. Sharp and E. G. Shelley, The latitudinal, diurnal and altitudinal distributions of upward flowing energetic ions of ionospheric origin, *Geophys. Res. Lett.*, *5*, 59, 1978.

Ghielmetti, A. G., E. G. Shelley, H. L. Collin and R. D. Sharp, Ion specific differences in energetic field aligned upflowing ions at $1R_E$, *this volume*, 1986.

Gorney, D. J., A. Clarke, D. Croley, J. F. Fennell, J. Luhmann and P. F. Mizera, The distribution of ion beams and conics below 8000 km, *J. Geophys. Res.*, *86*, 83, 1981.

Kaufmann, R. L., What auroral electron and ion beams tell us about magnetosphere-ionosphere coupling, *Space Sci. Rev.*, *37*, 313, 1984.

Kaufmann, R. L. and G. R. Ludlow, Interaction of H^+ and O^+ beams: Observations at 2 and 3 R_E, *this volume*, 1986.

Klumpar, D. M., W. K. Peterson and E. G. Shelley, Direct evidence for two-stage (bimodal) acceleration of ionospheric ions, *J. Geophys. Res.*, *89*, 10779, 1984.

Lundin, R., B. Hultqvist, E. Dubinin, A. Zackarov and N. Pissarenko, Observations of outflowing ion beams on auroral field lines at altitudes of many earth radii, *Planet. Space Sci.*, *30*, 715, 1982.

R. D. Sharp, R. G. Johnson and E. G. Shelley, Observations of an ionospheric acceleration mechanism producing energetic (keV) ions primarily normal to the geomagnetic field direction, *J. Geophys. Res.*, *82*, 3224, 1977.

R. D. Sharp, R. G. Johnson and E. G. Shelley, Energetic particle measurements from within ionospheric structures responsible for auroral acceleration processes, *J. Geophys. Res.*, *84*, 480, 1979.

E. G. Shelley, R. D. Sharp and R. G. Johnson, Satellite observations of an ionospheric acceleration mechanism, *Geophys. Res. Lett.*, *3*, 654, 1976.

Ungstrup, E., D. M. Klumpar and W. J. Heikkila, Heating of ions to suprathermal energies in the topside ionosphere by electrostatic ion cyclotron waves, *J. Geophys. Res.*, *84*, 4289, 1979.

ACCELERATED AURORAL AND POLAR-CAP IONS: OUTFLOW AT DE-1 ALTITUDES

A.W. Yau

Herzberg Institute of Astrophysics, National Research Council of Canada, Ottawa, Canada K1A 0R6

E.G. Shelley and W.K. Peterson

Lockheed Palo Alto Research Laboratory, Palo Alto, CA 94304 U.S.A.

Abstract. Data from the Dynamics Explorer-1 Energetic Ion Composition Spectrometer (DE-1 EICS) in the period from September 1981 to May 1984 were used to determine the mass composition, magnitude, magnetic activity dependence, long-term variations, and topology (MLT-invariant latitude distribution) of energetic (0.01-17 keV/e) terrestrial ion outflow. The period coincided with the declining phase of the current solar cycle (cycle 21). At both magnetically quiet and active times, the O^+ outflow rate exhibited long-term variations which correlated with the declining solar radio flux. Overall, the O^+ outflow rate in the 1981-82 period was a factor of 2 larger than the 1983-84 rate. Any corresponding variation in the H^+ outflow rate, if present, was much smaller and not statistically significant. The O^+ ion outflow rate increased exponentially with the Kp index, the rate at very disturbed times (Kp $\geq$ 6) being 3×10^{26} ions s^{-1} at solar maximum and a factor of 30 larger than the quiet-time (Kp = 0) value. The increase in the H^+ outflow rate with Kp was more modest, the disturbed time (Kp $\geq$ 6) rate being 7×10^{25} ions s^{-1} and a factor of 5 larger than the quiet-time value. The results point to the importance of perpendicular ion acceleration processes at low altitudes.

Introduction

We present results from a statistical study based on data from the Energetic Ion Composition Spectrometer (EICS) on Dynamics Explorer-1 (DE-1), in which we determined the mass composition, magnitude, magnetic activity dependence, long-term variations, and topology of energetic (0.1-17 keV) terrestrial ion outflow. These results establish the quantitative significance of upflowing auroral and polar cap ionospheric ions (UFI) as a source of magnetospheric plasma, and reveal a number of important aspects concerning ionospheric ion acceleration processes.

The morphological characteristics of upflowing ions have been the subject of a number of studies using data on S3-3 below 8000 km [Ghielmetti et al., 1978; Gorney et al., 1981; Collin et al., 1981, 1984; Ghielmetti and Johnson, 1983] and data on DE-1 up to 23000 km [Yau et al., 1984]. The present study is a follow-on to the Yau et al. study which was focussed on the occurrence frequency distribution of upflowing ions at 8000-23300 km altitude in the auroral zone and the polar cap; the data base used in the present study is an extension of that of Yau et al.

Data

Since its launch on August 3, 1981, DE-1 has provided a unique and comprehensive data set for studying ionospheric ion acceleration. The DE-1 data set is unique in three respects: the energy range of its ion composition measurements, its period of operation relative to the solar cycle, and its orbital coverage. On DE-1, the Retarding Ion Mass Spectrometer (RIMS) measured ions in the $\lesssim$50 eV/e range, while the Energetic Ion Composition Spectrometer (EICS) measured ions up to 17 keV/e. Together, the two instruments cover the full energy range of energetic UFI. DE-1 was launched near the peak of the current solar cycle (Cycle 21), and the DE data set covers the late solar-maximum and early declining phase of the solar cycle. The DE-1 orbit has an apogee of 4.6 R_e geocentric, an inclination of 90°, and an orbit-plane local-time drift period of 12 months and a line-of-apsides drift period of 18 months.

The DE-1 orbit is extremely useful for UFI studies because not only does it cover the entire auroral zone and polar cap up to 4.6 R_e at all local times, but it also enables data sampling in all seasons of the year and repeated sampling of a given season in successive years. This is illustrated in Figure 1, in which the geographic latitude and local time of the DE-1 apogee are shown as a function of time from September 1981 to August 1984. The different circle types at the top of the figure denote the different sea-

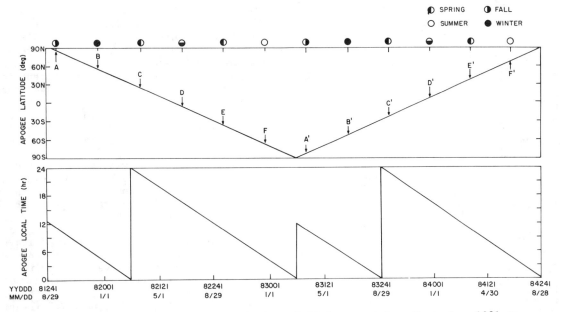

Fig. 1. Geographic latitude and local time of DE-1 apogee from September 1981 to August 1984. The identifiers indicate season of apogee data. A and A': fall equinox; B and B': winter solstice; C, C', E and E': spring equinox; F and F': summer solstice; D and D': winter in one hemisphere and summer in the other (equatorial apogee).

sons. The figure shows that the apogee latitude and local time in a given season in the first orbital cycle (the first 18 months of the mission) are revisited in the same season in the second orbital cycle. Identifiers A and A' show, for example, apogee coverage at the local-noon polar cap (83°N latitude and 1100 LT) at the northern fall equinox of 1981, and again at the southern fall equinox in 1983. Other identifier pairs show corresponding revisits of apogee latitude and local time at spring equinox (C and C', E and E'), summer solstice (F and F') and winter solstice (B and B').

The data base used in this study was acquired between September 15, 1981 and May 31, 1984. It consists of 96-s averaged integral ion fluxes of H^+ and O^+ in three energy intervals (0.01-1, 1-4, 4-17 keV) and nine pitch-angle bins. Data coverage was fairly complete at high altitude (above 16000 km, within ≈1 R_e of DE-1 apogee) for all magnetic local times and invariant latitudes above 56° (auroral and polar-cap latitudes), and for all four seasons of the year. Details of the data base were described in Yau et al. [1985].

Figure 2 shows an example of UFI event during magnetically active times. The six spectrograms show H^+ and O^+ ion fluxes at 0.01-1, 1-4 and 4-17 keV, respectively, as a function of pitch-angle and time. Very intense (>10^7 cm^{-2} s^{-1} sr^{-1}) O^+ UFI fluxes at 0.01-1 keV were observed throughout the southern dayside auroral zone and appear as enhancement in the spectrogram at pitch angles which decrease with increasing altitude, from

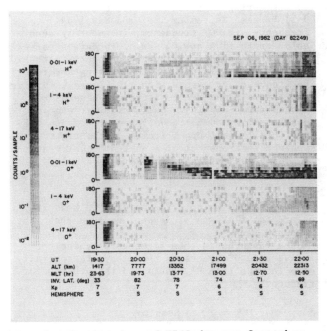

Fig. 2. Spectrograms of EICS data on September 6, 1982 between 1930 and 2210 UT, when Kp was 6-7. Data are 96-s averages and binned into nine pitch-angle bins. Ordinate of spectrograms is pitch angle; 0° is away from the ionosphere in the southern hemisphere. Note the upflowing O^+ ion beams in the 0.01-1 keV channel in the 0°-20° bin between 2100 and 2210 UT.

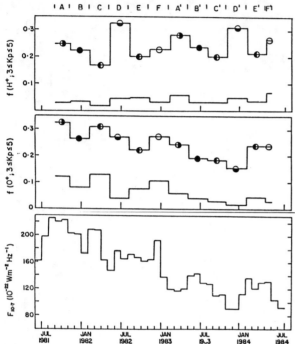

Fig. 3. Averaged occurrence frequency of UFI events during magnetically active times (Kp = 3-5) between September 1981 and May 1984. (a) H+ frequency, (b) O+ frequency, (c) monthly mean solar radio flux at 10.7 cm. Histograms marked by circles indicate frequency of events with >10^6 cm^{-2} s^{-1} sr^{-1} fluxes; unmarked histograms refer to frequency of events with >10^7 fluxes. Different circle types denote different seasons of apogee data sampling. Solid circles: winter; open circles: summer; semi-open circles on the left: fall; semi-open circles on the right: spring; semi-open circles on top: equal winter-summer (equatorial apogee) coverage.

≈90° at ≈13000 km to <20° above 17500 km. The upward ion flux was at least 2 decades above the downward flux.

The features in this figure are typical of active-time O+ events in that they are generally more intense than H+ events and that they occur frequently.

Results

Figure 3, adapted from Yau et al. [1985], shows the averaged occurrence frequencies of H+ and O+ upflowing ions below 1 keV at moderately active times (3- $\leq$ Kp $\leq$ 5+) in the twelve 91-day periods (seasons) identified in Figure 1. In panels a and b, the histograms marked by circles are occurrence frequencies of events with >10^6 (cm^2 s sr)$^{-1}$ fluxes; the unmarked histograms show the portion of events with >10^7 (cm^2 s sr)$^{-1}$ fluxes. Panel c shows the monthly mean 10.7 cm solar radio fluxes at 1 A.U. in the 91-day periods. The O+ UFI frequency displays a continual trend of overall decrease. In contrast, no long-term trend of decrease is apparent in H+ UFI. The September 1981 - August 1984 period coincided with the early declining phase of the present solar cycle, when $F_{10.7}$ decreased from a high of 222 (x 10^{-22} Wm^{-2} Hz^{-1}) in September 1981 to a low of 93 in November 1983. The long-term trend of decreasing O+ occurrence frequency is found to correlate with the solar radio flux, with the O+/H+ ratio peaking near solar maximum. Its short-term variation also appears to track the corresponding variation in the solar flux.

In both H+ and O+, the outflow was dominated by (over 90 percent) <1 keV ions. Figure 4 shows the active-time (Kp = 3-5) upward O+ ion flux <J_n> at 0.01-1 keV and 1000 km reference altitude, as a function of invariant latitude for the four MLT sectors. Triangles denote the 81-82 data, circles the 83-84 data. In both periods, the upward ion flux peaked near 78° invariant in

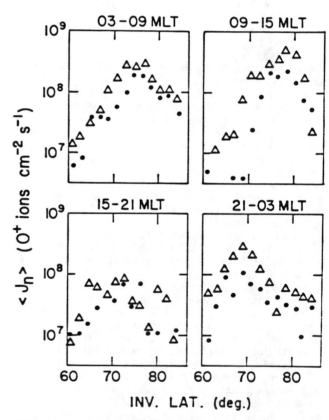

Fig. 4. Active-time (Kp = 3-5) upward O+ ion fluxes at 0.01-1 keV normalised to a reference altitude of 1000 km, as a function of invariant latitude at different MLT sectors. Triangles: September 1981 - January 1983 data; dots: February 1983 - May 1984 data.

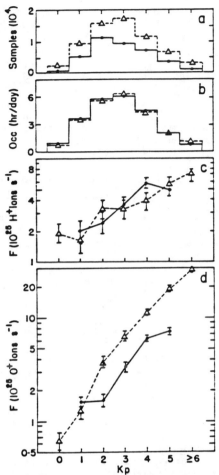

Fig. 5. H$^+$ and O$^+$ outflow rates at 0.01-1 keV as a function of Kp. Triangles: September 1981 - January 1983 data; dots: February 1983 - May 1984 data. (a) data sampling distribution, (b) Kp occurrence frequency distribution, (c) H$^+$, and (d) O$^+$ outflow rates.

the noon sector. In the midnight sector, the flux peaked at lower invariant latitude (near 70°) and was smaller by a factor of 2. The corresponding quiet-time (Kp = 0-2) distribution was qualitatively similar but smaller in magnitude. Overall, the flux was above a factor of 2 higher in 81-82 than in 83-84.

Figure 5 displays the 0.01-1 keV H$^+$ and O$^+$ outflow rates in the 81-82 (triangles) and 83-84 (circles) periods as a function of Kp. In addition it shows the distributions of data sampling and Kp occurrence. In both periods, data sampling was very adequate for the intermediate Kp conditions (Kp between 1 and 5), and was in proportion to the Kp occurrence distribution. Also, the Kp occurrence distributions in the two periods are very similar. In 81-82, Kp was ≤2, 3 to 5, and ≥6 in 41, 54 and 5 percent of the time. The data distribution in the three Kp ranges was 42, 53 and 5 percent, respectively.

In 83-84, the corresponding Kp occurrence distribution was 43, 53 and 4 percent, and the data distribution was 45, 51 and 4 percent, respectively. The apparent lack of dependence of Kp on solar cycle was puzzling in view of the fact that large magnetic storms (Dst << -100 nT) were limited to the near solar maximum period, and that the Kp index is a composite indicator of both magnetic storms and substorms.

For a given Kp condition, the O$^+$ outflow rate was higher in 81-82 than in 83-84 by a factor of 2. In contrast, the H$^+$ outflow rates in the two periods were equal within statistical error. The higher O$^+$ rate in 81-82 is attributed to the long-term variations of O$^+$ UFI occurrence which are correlated with variations in the solar flux (see Figure 3).

The O$^+$ outflow rate increased exponentially with Kp by a factor of ≈30 from ≈1 x 10^{25} ions s^{-1} at Kp = 0 to ≈3 x 10^{26} ions s^{-1} at Kp ≥ 6 in 81-82. Empirically, F(Kp) ∝ exp(0.56 Kp).

Summary and Discussions

At both magnetically quiet and active times, the outflow rate of upflowing O$^+$ ionosphere ions exhibits appreciable long-term variations which are correlated with the solar flux. In comparison, the variation in the H$^+$ outflow rate, if present, is much smaller and not statistically significant. The O$^+$ outflow rate is highly dependent on magnetic activity, as measured by the Kp index, and increases exponentially by a factor of 30 from Kp = 0 to Kp = 6. The corresponding increase of the H$^+$ rate with Kp is much more modest, being a factor of 5.

The qualitative similarity between the geomagnetic and solar activity dependences of UFI may be understood in terms of atmospheric and ionospheric scale height considerations. At times of increased geomagnetic activity, ion and neutral temperatures rise, as a result of ionospheric and atmospheric heating. The resulting increase in scale height effectively lifts O$^+$ to higher - ion acceleration - altitudes, thereby facilitating O$^+$ ion acceleration. The increased scale height also shifts the O$^+$-H charge exchange to higher altitude, and effectively reduces the overall probability of removal of low-altitude O$^+$ by charge-exchange, thereby increasing their probability of survival to (and hence occurrence at) high altitude. At the same time, the increase in scale height has little influence on the light H$^+$ ions from the topside ionosphere. The net result is an increase in intense O$^+$ UFI events and overall O$^+$ abundance. Likewise, at times of increased solar EUV flux (which is correlated with the F$_{10.7}$ radio flux), increased atmospheric heating occurs, resulting in increase in scale heights at the lower ionosphere and ultimately increase in O$^+$ UFI abundance at high altitude.

On the basis of PROGNOZ-7 data, Hultqvist [1983] provided convincing arguments that the energetic O$^+$ ions observed at PROGNOZ-7 altitude originate from a narrow altitude region in the lower ionosphere (below 1000 km). Moore [1980, 1984] examined the effect of O^+-H charge exchange on accelerated O^+ flux from low altitude, and its dependence on neutral atmospheric composition and temperature. Moore concluded that perpendicular energization of several eV below the cross-over altitude would be necessary for the upgoing O^+ ions to overcome charge-exchange and to reach high altitude, and that the O^+ outflow should increase with increased neutral atmospheric temperature. Perpendicular ion acceleration at low altitudes (below 1000 km) has been observed previously on the ISIS-1 and -2 spacecraft [Klumpar, 1979] and on sounding rockets [Whalen et al., 1978; Yau et al., 1983]. The present result is consistent with the argument of Hultqvist [1983] and points to the importance of low-altitude ion acceleration processes in understanding ionospheric ion acceleration at high altitudes.

Young et al. [1982] found marked O^+ density increase in the (outer) ring current with geomagnetic activity; the increase in the H^+ density was small. Young et al. found that $O^+/H^+ \propto \exp(0.17 \, K_p)$ for 0.9-16 keV ions in the ring current. The corresponding ratio in UFI outflow, from Figure 5 above, is proportional to $\exp(0.30 \, K_p)$. Since the K_p index is a quasi-logarithmic measure of magnetic disturbance, the exponential increase of the O^+/H^+ ion composition ratio in both UFI and ring current in this dependence reflects not only the source and sink relationship between the two on a long time-scale, i.e., that the ring current is the storage depot of the outflowing ions, but also that the ring current accommodates on a relatively short time scale the changing dynamics of the auroral ion outflow.

In conclusion, the acceleration of O^+ UFI is modulated by the atmospheric scale height, which increases with increasing solar activity (EUV flux). As a result, O^+ UFI occurrence frequency and outflow rate increase at solar maximum. In contrast, there is no significant solar-cycle variations in H^+ UFI occurrence and outflow. UFI is O^+-dominant near solar maximum and comparable in H^+ and O^+ near solar minimum.

Acknowledgement. This research was supported in part by NASA under contract NAS5-28710.

References

Collin, H.L., R.D. Sharp, E.G. Shelley, and R.G. Johnson, Some general characteristics of upflowing ion beams over the auroral zone and their relationship to auroral electrons, J. Geophys. Res., 86, 6820-6826, 1981.

Collin, H.L., R.D. Sharp, and E.G. Shelley, The magnitude and composition of the outflow of energetic ions from the ionosphere, J. Geophys. Res., 89, 2185-2194, 1984.

Ghielmetti, A.G., R.G. Johnson, R.D. Sharp, and E.G. Shelley, The latitudinal, diurnal, and altitudinal distributions of upward flowing energetic ions of ionospheric origin, Geophys. Res. Lett., 5, 59-62, 1978.

Ghielmetti, A.G., and R.G. Johnson, Variations in the occurrence frequency of UFI events from July 1976 to April 1979, Trans. Amer. Geophys. Union EOS 64, 807, 1983.

Gorney, D.J., A. Clarke, D. Croley, J.F. Fennell, J. Luhmann, and P. Mizera, The distribution of ion beams and conics below 8000 km, J. Geophys. Res., 86, 83-89, 1981.

Hultqvist, B., On the origin of the hot ions in the disturbed dayside magnetosphere, Planet. Space Sci., 31, 173-184, 1983.

Klumpar, D.M., Transversely accelerated ions: An ionospheric source of hot magnetospheric ions, J. Geophys. Res., 84, 4229-4237, 1979.

Moore, T.E., Modulation of terrestrial ion escape flux composition (by low-altitude acceleration and charge exchange chemistry), J. Geophys. Res., 85, 2011-2016, 1980.

Moore, T.E., Superthermal ionospheric outflows, Rev. Geophys. Sp. Phys., 22, 264-274, 1984.

Whalen, B.A., W. Bernstein, and P.W. Daly, Low altitude acceleration of ionospheric ions, Geophys. Res. Lett., 5, 55-58, 1978.

Yau, A.W., B.A. Whalen, A.G. McNamara, P.J. Kellogg, and W. Bernstein, Particle and wave observations of low-altitude ionospheric ion acceleration events, J. Geophys. Res., 88, 341-355, 1983.

Yau, A.W., B.A. Whalen, W.K. Peterson, and E.G. Shelley, Distribution of upflowing ionospheric ions in the high-altitude polar cap and auroral ionosphere, J. Geophys. Res., 89, 5507-5522, 1984.

Yau, A.W., P.H. Beckwith, W.K. Peterson, and E.G. Shelley, Long-term (solar-cycle) and seasonal variations of upflowing ionospheric ion events at DE-1 altitudes, J. Geophys. Res., 90, 6395-6407, 1985.

Young, D.T., H. Balsiger, and J. Geiss, Correlations of magnetospheric ion composition with geomagnetic and solar activity, J. Geophys. Res., 87, 9077-9096, 1982.

ION SPECIFIC DIFFERENCES IN ENERGETIC FIELD ALIGNED UPFLOWING IONS AT 1 R_E

A.G. Ghielmetti, E.G. Shelley, H.L. Collin, R.D. Sharp

Lockheed Palo Alto Research Laboratory, Palo Alto, California 94304

Abstract. Measurements of energetic (0.5 to 16 keV) upward flowing ion distributions above the auroral regions within broad structures associated with parallel electric fields below the satellite are presented. The pitch angle distributions of the two major ion species (H^+ and O^+) were both field aligned as expected. However, the angular widths of the distributions and their dependence on energy were inconsistent with purely parallel acceleration: The H^+ distributions became more strongly field aligned with increasing energy, while the O^+ distributions became broader. The H^+ ion distributions were characterized by a high parallel (~0.6 keV) and low perpendicular (~0.1 keV) temperature. In contrast, the O^+ transverse and parallel temperatures were comparable (~1 keV). Within these acceleration regions, the O^+ ions were on the average a factor of 2 to 3 more energetic than H^+. The mean energies of the H^+ ions were substantially less than the potential drops below the satellite, while the O^+ ion energy corresponded more closely to the potential. Candidate interpretations include preferential perpendicular heating of the O^+ component coupled with an H^+ source- and electric field region extended in altitude, and/or parallel energy loss processes acting primarily on the H^+ beam.

Introduction

Within the topside auroral ionosphere (from a few 100 km upward), ambient ionospheric ions are commonly accelerated to energies ranging from a few eV to tens of keV. The resultant upward flowing ions exhibit either field aligned (beams) [Shelley et al., 1976; Mizera and Fennell 1977]. or conical pitch angle distributions (conics) [Sharp et al., 1977]. These observations imply the existence of two principally different classes of acceleration processes: one acting primarily parallel, the other transverse to the magnetic field direction.

Processes proposed to account for ion conics include wave particle interactions and perpendicular electric field structures with scale sizes of the order of or less than the ion gyroradius. Ion beams may have been generated by parallel electric fields but may, in principle, also have evolved from conical distributions through adiabatic transport from lower altitudes. However, since the majority of the energetic field aligned upward flowing ion events seen at S3-3 originated above 5000 km altitude, this second interepretation can be excluded for the majority of events [Lysak et al., 1979; Ghielmetti et al., 1978; Gorney et al., 1981]. For recent summaries covering the topics of ion acceleration and comprehensive lists of references, see Ashour-Abdalla and Okuda [1983], Lennartsson [1983], and Sharp et al. [1983].

Mass selective effects are a feature of many of the proposed acceleration mechanisms, and hence detailed studies of ion distributions may provide valuable information on the nature of the participating processes. Morphological studies of auroral upward flowing ions [Ghielmetti et al., 1978; Collin et al., 1981; Sharp et al., 1983] have indeed provided evidence that the two primary ion species (H^+ and O^+) have on the average different energy and pitch angle distributions. It is not immediately apparent from these, however, whether the two ion species were affected differently by the same processes or whether different processes were acting in spatially separate regions.

The present work describes the mass dependent features in upward flowing ions (UFI) observed within spatially wide acceleration structures that show consistent evidence for parallel electric fields below the satellite. For this purpose, near simultaneous measurements of the energy- and pitch angle-distributions of upflowing H^+ and O^+ ions are examined and intercompared. This limited initial study focuses on three broad parallel electric field structures in which both ion species were present.

Experiment

The data used in this study were acquired by the Lockheed plasma composition experiment aboard the polar orbiting S3-3 satellite. Spin axis and experiment view directions were oriented such that complete pitch angle scans to within about 5° of the magnetic field direction were obtained every spin period. The three ion mass spectrometers were operated at exponentially spaced energy levels and, hence, simultaneously acquired a 3-point energy spectrum once per second. Approximately once every spin period, the energy levels were stepped to one of four values, thus covering the energy range from 0.5 to 16 keV in 12 steps once every 64 sec. Since a complete mass scan from 1 through 32 AMU/charge required 1 second, the two major ion species H^+ and O^+ were always sampled within ~0.5 sec (9° of rotation) of each other. In addition, four broad band electron spectrometers provided contiguous coverage of the energy range from 70 eV to 24 keV, with a temporal resolution of 0.5 seconds. The

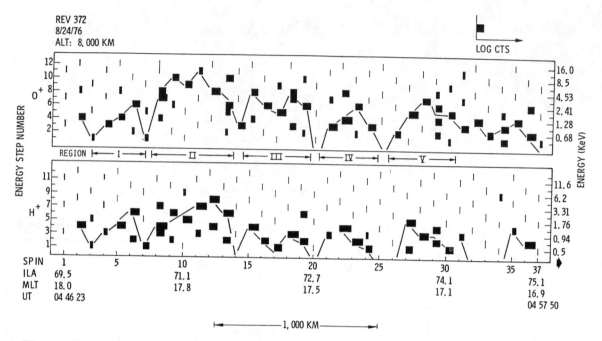

Fig. 1. Energy versus time representation of upward flowing ion flux intensities for auroral zone pass on August 24, 1976. Count rates were averaged over a 30° upward pitch angle interval and are encoded by a black bar extending to the right side.

angular acceptances of both types of instruments were ±3° FW. More detailed descriptions of the instrumentation are available from Sharp et al. [1979].

Data Selection

The spin period of the satellite (~19 sec) corresponds at apogee to a horizontal distance of ~80 km. The resulting spatial and temporal resolution of the measurements is therefore inadequate to resolve any of the small scale features associated with auroras. To be able to recognize characteristic features in the pitch angle and energy distributions and to differentiate these from spatial and temporal variations, it is necessary to combine measurements from several spins. This study will consequently be limited to spatially wide acceleration structures. In addition, we require that the differential energy fluxes of both H^+ and O^+ UFI peak well within the instrument energy range (0.5 to 16 KeV). To limit the study to parallel electric field acceleration regions, the simultaneous presence of widened upward loss cones in the electron fluxes is required.

About 50 acceleratation structures that meet the above criteria have been identified in a noncomprehensive survey. The example selected for detailed presentation is of somewhat broader extent than usual but exhibits many of the typical features.

Results

The event occurred on August 24, 1976, near the peak of a small magnetic storm (peak Dst = -45γ, and Kp = 5+). At that time the spacecraft was crossing the northern auroral zone near the dusk LT meridian at an altitude of about 8000 km.

Overview. Figure 1 gives an overview of the upward flowing ion observations. In this energy versus time representation, the intensity is encoded by means of a black bar with the magnitude plotted parallel to the time axis. The size of a black bar is proportional to the log of the count rate (~ differential energy flux), and its origin is aligned with the location in time corresponding to the upward pitch angle direction. Each data point represents an unweighted average of the countrate over a ±30° pitch angle interval centered about the upward direction. Any contributions from isotropic fluxes were subtracted. As noted above the instrument performed simultaneous measurements at three different energies. Occasionally, when the energy–stepping occured within the upward source cone, additional sampling points became available. In Figure 1, only energy steps actually measured are displayed, and steps at which there was no response are drawn as a thin vertical line. As a guide to the eye, the energy steps at which the count rates peaked were connected by straight line segments.

As seen from Figure 1 regions of UFI fluxes extended nearly contiguously from 69.7° to 75° invariant latitude (ILA). During most of this period both ion species (H^+ and O^+) were present with substantial fluxes that peaked at energies between 0.5 and 10 keV. Several structures can be discerned from Figure 1, each characterized by an increase followed by a decrease in the energy at which the peak fluxes occured (marked regions I-V in Figure 1). These modulations are consistently seen in both species and are suggestive of ion inverted V signatures.

Examination of the electron fluxes from this experi-

Table 1. Average properties of upward flowing ions and electric fields
within acceleration regions.

Region	Energy [keV] H$^+$	O$^+$	Energy Ratio [O$^+$ / H$^+$]	Potential Drop kV	Energy/(Pot drop) [H$^+$ / ϕ]	[O$^+$ / ϕ]
II	2.6 ± .2	4.8 ± .6	2.4 ± .4	5.4 ± .9	0.47 ± .05	0.98 ± .17
III	1.1 ± .1	2.5 ± .2	3.0 ± .4	2.1 ± .3	0.39 ± .04	1.24 ± .15
V	1.1 ± .1	1.9 ± .2	2.1 ± .3	1.7 ± .4	0.58 ± .07	1.05 ± .12

ment revealed the presence of enhanced upward loss cones within each of the ion structures described above. Hence, except for spins during which UFI were not observed, a parallel potential drop was consistently present beneath the satellite [Cladis and Sharp, 1979]. Within the same regions, the electron fluxes exhibited pitch angle modulation signatures [Sharp et al., 1979] that are indicative of parallel electric fields above the satellite. To further verify the above observations we examined the energy-time spectrogram derived from the Aerospace ion and electron detectors for this pass. This experiment [Mizera and Fennell, 1977], which provided more detailed energy resolution, shows several inverted V signatures in the electron fluxes that coincided with regions II, III, and V. Similar correlations between the electron and UFI fluxes are described in Sharp et al. [1979]. We conclude, therefore,

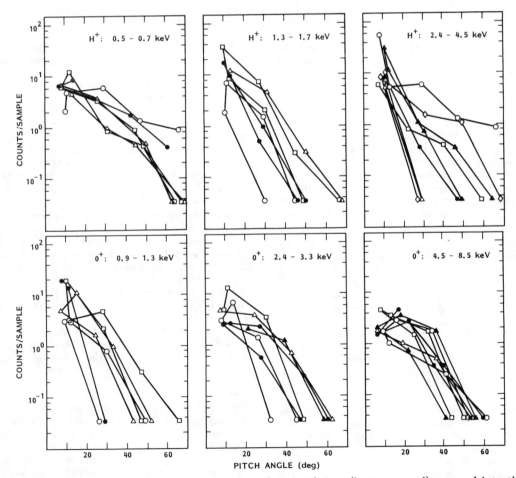

Fig. 2. Pitch angle distributions of H$^+$ (top panel) and O$^+$ ions (bottom panel) grouped into three energy ranges for acceleration regions II and III. Inbound and outbound legs are plotted along the same axis. Different symbols indicate data obtained on different spins.

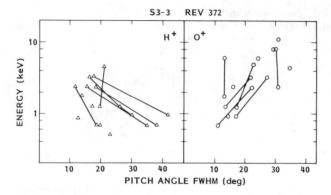

Fig. 3. Full width at half intensity of pitch angle distributions observed in regions II and III as a function of energy. Data points determined from simultaneous measurements are connected by a straight line.

that parallel electric fields were present both below and above the spacecraft throughout most of the ion acceleration structures of Figure 1.

Mean ion energy. The most striking feature in Figure 1 is the systematic difference in the energy of the upward flowing H^+ and O^+ ions at which the count rates ($\sim$ differential energy flux) peaked. In order to quantify this difference, we derive a number flux weighted average energy for each species and for each upward pitch angle interval ($\pm 30°$). From these, a ratio of the mean energies is then formed which may serve as a measure of the difference in energy. To verify the accuracy and usefullness of averages obtained from mere 3-point differential energy flux measurements, we have simulated the instrument response to a set of hypothetical ion flux distributions and applied the mean energy algorithms used above to the resultant count rates. The resultant averages were then compared to the actual distribution means. These model calculations indicate that resonably good estimates for the mean energy are obtained as long as the temperature is ≥ 0.3 keV (stationary Maxwellian) or $\geq 0.1 \times |\vec{E}|$ (drift) where $|\vec{E}|$ (drift) ≥ 0.5 keV (convecting Maxwellian). As seen from Figure 4, these conditions were generally satisfied during this event. Depending on the particular set of energy steps in use, the values obtained either under- or overestimate the true mean. However, by averaging the results from all four combinations of energy steps, these differences are reduced substantially. Hence, in order to minimize these systematic uncertainties and to smooth out temporal/spatial variations, the single spin based ratios of the mean energies were then averaged (relative error weighted) over the acceleration structures identified in Figure 1. The final uncertainties in these values are estimated to be less than 25%.

An implicit condition for obtaining meaningful values for the average ion energy is that the bulk of the ion distribution fall within the measurement range of the instrument. This was not the case in regions I and IV, where the inferred potentials below the satellite were generally near the lower instrument threshold (0.5 keV), and where UFI were primarily observed in the lowest energy channels. Consequently, further discussion will be limited to the three major acceleration structures.

Table 1 summarizes the average properties of the particles and potentials within these acceleration structures. As evident from this, the [O^+/H^+] energy ratios were significantly above unity, ranging from about 2 to 3. The most energetic upward flowing ions were seen to occur in region II. In regions III and V, the H^+ ion energies were close to the lower instrument threshold (0.5 keV) and, as a result, the calculated mean values are probably overestimates. This suggests that the actual [O^+/H^+] energy ratios in regions III and V were somewhat higher than given in Table 1.

Angular distributions. Figure 2 displays the pitch angle distributions observed in regions II and III. They have been assembled according to their energy into a low, medium, and high energy range. Although there is considerable scatter between curves from different spin periods, one can nevertheless recognize some qualitative differences in the degree of field alignment. Both the H^+ and the O^+ distributions appeared to be field aligned (beams). The H^+ distributions (Figure 2a) are flatter at low energies and more highly field aligned at higher energies. The shoulder at pitch angles $> 30°$ at the highest energies is caused by the presence of an isotropic H^+ component. In contrast, the $O+$ distributions are more field aligned at low energies and substantially wider at the higher energies. There is some evidence in the energetic O^+ distributions for a plateau or a decrease at pitch angles $< 20°$. However, sampling resolution and counting statistics were insufficient to unambiguously identify a possible conical component.

Energy-pitch angle dependence. These above systematic energy dependences of the pitch angle distributions are further illustrated in Figure 3. In this, the FWHM of the H^+ and O^+ distributions are plotted as a function of energy. Values obtained from simultaneous measurements (during the same spin period) are connected by straight lines. Figure 3 more clearly shows the decrease in the angular width of the H^+ distributions and the widening of the O^+ distributions. It is evident from this figure that these tendencies are is indeed real features of the "instantaneous" pitch angle distributions and not simply the result of sampling effects.

Magnitude of parallel potential. The magnitude of the parallel potential difference between the spacecraft and the ionosphere was determined from the electron loss cone enhancement. A detailed discussion of the methods applied is given in Sharp et al. [1979] and Cladis and Sharp [1979]. Estimates were derived for each half of the loss cone from each of the three more energetic electron detectors, from which an average was formed for that particular spin. A relative error weighted average of the ratio of the mean ion energy to the potential difference below was then obtained for each region and ion species. These average ratios together with the average magnitude of the potential below the spacecraft are included in Table 1. One sees from this that the ratios were substantially less than unity for H^+ and near or slightly above unity for O^+ in all three structures. The uncertainties in the estimates of the potential differences may be somewhat larger than indicated since unknown effects such as pitch angle scattering may in principle have introduced additional scatter and biases. However, since the values obtained from all four electron detectors were usually in good agreement, these effects are insufficient to account for the low relative

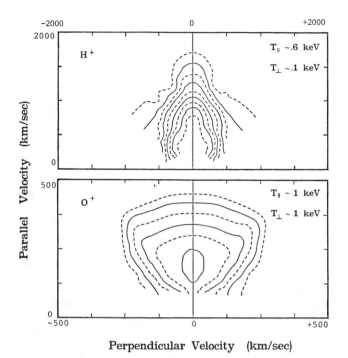

Fig. 4. Phase space density contours of H^+ and O^+ ion beams constructed from smoothed energy and angular spectras measured within acceleration region II.

H^+ energy. Note that the mean ion energies do not take into account the details of the pitch angle distributions. Proper weighting by solid angle would tend to further lower the H^+ but increase the O^+ mean energy. The O^+ relative energies are therefore consistent with being somewhat larger than unity.

Summary and Conclusions

We have examined the upward flowing ion signatures from three broad acceleration structures that showed evidence for parallel electric fields below the satellite. Within all three regions, the distributions exhibited evidence for complex mass dependent acceleration effects. The O^+ ions were consistently more energetic than H^+, and when averaged over individual structures, the O^+ mean energies were a factor of two to three higher. It is not clear at the present time how representative these examples are. Previous statistical studies have yielded typical energy ratios of about 1.7 [Collin et al.; 1981, Sharp et al., 1983]. However, these studies applied different selection criteria and did not average over wide acceleration regions.

In order to qualitatively illustrate what the different energy dependences of the H^+ and O^+ UFI distributions imply in velocity space, we provide in Figure 4 a set of phase space density contours. Note that these were not derived from individually measured data points, but were constructed from a fit to the parallel energy distributions observed in region II, and from smoothed pitch angle distributions (Figure 2) with energy dependent FWHMs derived from Figure 3. Included in Figure 4 are the estimates for the characteristic temperatures inferred from the energy and pitch angle spectra. The elongated contours representing the H^+ ion beam imply a population with a relatively high parallel (~ 0.6 keV) and low perpendicular temperature (~ 0.1 keV). In contrast, the O^+ contours became progressively wider with increasing velocity and represent a population with high parallel and perpendicular temperatures.

The high perpendicular temperature of the O^+ ions evidently requires a transverse heating process (to $T \sim 1.5$ keV). If an ion conic had resulted from this transverse heating near 5000 km altitude (the lower border of energetic ion acceleration [Ghielmetti et al., 1978; Gorney et al., 1981]) and was subsequently accelerated at a higher altitude by the parallel electric field (5 kV, Table 1) of region II (Figure 1), it would give rise to the distribution of Figure 4 at 8000 km altitude. Hence, the O^+ ion beam observations are consistent with a two-stage acceleration process similar to the one described by Klumpar et al. [1984] and Collin et al. [1986]. However, the evidence presented here is not sufficient to uniquely identify this particular sequence of processes.

In a scenario where the O^+ ions were accelerated through the full parallel potential difference after having been heated transversly, the ratio of the mean energy to the potential drop is expected to be substantially higher than unity [see Kaufmann et al., 1976]. Since this ratio was near unity in the examples presented here, one must further assume that the O^+ ion beam had lost a part of its streaming energy.

A principal difference between the H^+ and O^+ ion populations is the temperature anisotropy (Figure 4). In the case of H^+, this implies that the high parallel temperature did not originate through perpendicular heating. This suggests that the ion beam was thermalized primarily in the parallel velocity direction, perhaps via an ion beam instability [Kintner et al., 1979; Bergmann and Lotko, 1986; Kaufmann and Ludlow, 1986] that also led to a net loss of parallel energy. Evidence for similar effects have been presented by Reiff et al. [1986]. Alternate interpretations include time varying electric fields and/or an H^+ source region and electric field region that are distributed in altitude [Kaufmann et al., 1976]. These models predict a lower energy for the H^+ population relative to both the O^+ ions and the potential drop.

The observed ion energies and temperatures imply preferential acceleration of the heavier ion component. This is predicted by wave particle interaction theories [Ashour-Abdalla and Okuda, 1983] and by structured perpendicular electric fields [Lennartsson et al., 1983; Yang and Kan, 1983]. The results presented here provide new constraints on the auroral acceleration processes. Taken alone, however, they are not adequate to unambiguously confirm any one of the above suggested interpretations.

Acknowledgments. The authors wish to express their appreciation to Dr. Fennell, Aerospace Corp. for providing their unpublished data. The Lockheed experiment and data acquisition was supported by the Office of Naval Research under contract N00014-78-C-0479. This work was supported by the National Science Foundation by Grant ATM-8317710, by NASA under contract NASW-3395 and by the Lockheed Independent Research Program.

References

Ashour-Abdalla, M. and H. Okuda, Transverse acceleration of ions on auroral field lines, in *Energetic Ion Composition in the Earth's Magnetosphere*, Ed. R.G. Johnson, Terra Publishing Co., D. Reidel, Tokyo, 43-72, 1983.

Bergmann R. and Lotko W., Transition to unstable ion flow in parallel electric fields, submitted to *J. Geophys. Res.*, 1985.

Cladis, J.B. and R.D. Sharp, Scale of electric field along magnetic field in an inverted V event, *J. Geophys. Res., 84*, 6564, 1979.

Collin, H.L., R.D. Sharp, E.G. Shelley, and R.G. Johnson, Some general characteristics of upflowing ion beams over the auroral zone and their relationship to auroral electrons, *J. Geophys. Res., 86*, 6820, 1981.

Collin, H.L., E.G. Shelley, A.G. Ghielmetti, and R.D. Sharp, Observations of transverse and parallel acceleration of terrestrial ions at high latitudes, [this issue].

Ghielmetti, A.G., R.G. Johnson, R.D. Sharp, and E.G. Shelley, The latitudinal, diurnal, and altitudinal distributions of upward flowing energetic ions of ionospheric origin, *Geophys. Res. Lett., 5*, 59, 1978.

Gorney, D.J., A. Clarke, D. Croley, J. Fennell, J. Luhmann, and P. Mizera, The distributions of ion beams and conics below 8000 Km, *J. Geophys. Res., 86*, 83, 1981

Kaufmann, R.L., D.N. Walker, and R.L. Arnoldy, Acceleration of auroral electrons in parallel electric fields, *J. Geophys. Res., 81*, 1673, 1976.

Kaufmann, R.L. and G.R. Ludlow, Interaction of H^+ and O^+ beams: observations at 2 and 3 Re, [this issue].

Kintner, P.M., M.C. Kelley, R.D. Sharp, A.G. Ghielmetti, M. Temerin, C. Cattell, P.F. Mizera, and J.F. Fennell, Simultaneous observations of energetic (keV) upstreaming ions and electrostatic hydrogen cyclotron waves, *J. Geophys. Res., 84*, 7201, 1979.

Klumpar, D.M., W.K. Peterson, and E.G. Shelley, Direct evidence for two-stage (bimodal) acceleration of ionospheric ions, *J. Geophys. Res., 89*, 10779, 1984.

Lennartsson, W., Ion acceleration mechanisms in the auroral regions: general principles, in *Energetic Ion Composition in the Earth's Magnetosphere*, Ed. R.G. Johnson, Terra Publishing Co., D. Reidel, Tokyo, 23-41, 1983.

Lysak, R.L., M.K. Hudson, and M. Temerin, Ion heating by strong electrostatic ion cyclotron turbulence, *J. Geophys. Res., 85*, 678, 1980.

Mizera, P.F. and J.F. Fennell, Signatures of electric fields from high and low altitude particle distributions, *Geophys. Res. Lett., 4*, 311, 1977.

Reiff, P.H., H.L. Collin, E.G. Shelley, J.L. Burch, and J.D. Winningham, Heating of upflowing ionospheric ions on auroral field lines, [this issue].

Sharp, R.D., R.G. Johnson, and E.G. Shelley, Observations of an ionospheric acceleration mechanism producing energetic (keV) ions primarily normal to the geomagnetic field direction, *J. Geophys. Res., 82*, 3324, 1977.

Sharp, R.D., R.G. Johnson, and E.G. Shelley, Energetic particle measurements from within ionospheric structures responsible for auroral acceleration processes, *J. Geophys. Res., 84*, 480, 1979.

Sharp, R.D., A.G. Ghielmetti, R.G. Johnson, and E.G Shelley, Hot plasma composition results from the S3-3 spacecraft, in *Energetic Ion Composition in the Earth's Magnetosphere*, Ed. R.G. Johnson, Terra Publishing Co., D. Reidel, Tokyo, 167-193, 1983.

Shelley, E.G., R.D. Sharp, and R.G. Johnson, Satellite observations of an ionospheric acceleration mechanism, *Geophys. Res. Lett., 3*, 654, 1976.

Yang, W.H. and J.R. Kan, Generation of conic ions by auroral electric fields, *J. Geophys. Res., 88*, 465, 1983.

HEATING OF UPFLOWING IONOSPHERIC IONS ON AURORAL FIELD LINES

P. H. Reiff,[1] H. L. Collin,[2] E. G. Shelley,[2] J. L. Burch,[3] and J. D. Winningham,[3]

Abstract. The two coplanar Dynamics Explorer (DE) spacecraft provide a unique opportunity to test for the effects of electric fields aligned parallel to magnetic field lines by sampling, nearly simultaneously, the velocity distribution functions of ions and electrons at two points on the magnetic field lines: DE 1 at high altitudes (9000-15000 km) and DE 2 at low altitudes (400-800 km). The upflowing ion distribution typically can be characterized by a sharp peak and a falloff at high energies of the form exp-$\{(E-E_p)/E_0\}$, with E_p being the peak energy and E_0 a characteristic energy. This is the function that one expects if a Maxwellian of thermal energy E_0 is accelerated upwards by a parallel electric field with $e\Phi=E_p$. The acceleration mechanism cannot be a simple parallel electric field, however, for two reasons: first, the characteristic energy E_0 is considerably larger than the ionospheric thermal energy (and is typically hundreds of eV - 20-30% of E_p); and secondly, because the energy of the peak E_p is typically 30-50% smaller than would be expected from the potential differences $e\Phi$ inferred by two other independent techniques (precipitating electrons at DE 2 and widening of the loss cone for electrons at DE 1). The distribution does appear to be consistent with an ionospheric source, heated within (or above) the acceleration region, since the ion average energy is comparable to $e\Phi$.

Introduction

In a related paper [Reiff et al., manuscript in preparation, 1986] we have shown that Dynamics Explorer 1 and 2 particle measurements taken concurrently at ~12000 km and at ~700 km along auroral field lines are consistent with the presence of an upward-directed electric field aligned along the magnetic field at altitudes between the two spacecraft.

High-resolution measurements of the ion and electron velocity distributions obtained simultaneously at high and low altitudes on the same auroral field line allow the determination of three simultaneous and independent estimates of the field-aligned potential drop along the field line between the two spacecraft. The orbits of DE 1 and DE 2 traverse the upper and lower parts of the potential structures that are inferred from inverted-V electron signatures [Gurnett, 1972]. The peak energy of the inverted-V electron distribution as observed at low altitudes (DE 2) provides an estimate of the field-aligned potential drop above DE 2, Φ_{TOT}, [Evans, 1974] (see Figure 1 and the text below). Meanwhile DE 1 provides two measures of the potential drop occurring below that satellite, Φ_{MID}. One measure is provided by the energy of the peak of the upgoing ion beam velocity space distribution, E_{peak}, (see Figures 2 and 3 and the text below), and the other Φ_{MID} is provided by the enlargement of the atmospheric loss cones of the upcoming electrons [Knight, 1973; Cladis and

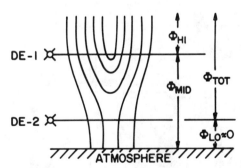

Fig. 1. Schematic of the DE 1 and DE 2 orbits and the potential drops defined from the measured particle distribution. The voltage Φ_{HI} is estimated from precipitating electrons at DE 1 and is typically small for the orbits presented here. Φ_{MID} is estimated from the enlarged loss cone of upgoing electrons at DE 1, and Φ_{TOT} is estimated from the energy of the precipitating electrons at DE 2.

---

[1]Dept. of Space Physics and Astronomy, Rice University, Houston, Texas 77251
[2]Lockheed Palo Alto Research Laboratory, Palo Alto, California 94304
[3]Southwest Research Institute, San Antonio, Texas 78284

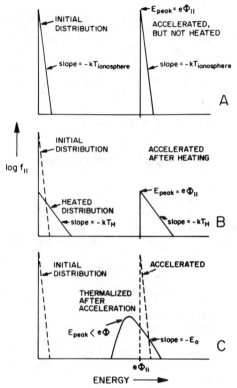

Fig. 2. (a) Schematic distribution functions $f(\alpha = 180°)$ expected if the upgoing ionospheric ions are accelerated by a parallel potential drop Φ. (b) Same, but now they are heated before being accelerated. The peak should still occur at $E = e\Phi$. (c) Distribution expected if the cold accelerated distribution thermalized after acceleration. The peak will occur at an energy $< e\Phi$.

Sharp, 1979; Mizera et al., 1981]. The absence of upgoing accelerated ions at DE 2 indicates that the field-aligned potential drop below DE 2, Φ_{LO}, is negligible, so that we may set $\Phi_{TOT} = \Phi_{MID} + \Phi_{HI}$, where Φ_{HI} is the potential difference above DE 1. The downcoming electrons observed at DE 1 provide this measure of Φ_{HI}; since this is generally small, we may usually take $\Phi_{TOT} \sim \Phi_{MID}$. We thus have three independent estimates of the field-aligned potential drop between the two satellites. The general agreement between these three estimates supports the hypothesis that a quasistatic field-aligned potential drop is a fundamental ingredient of the auroral acceleration process, while discrepancies of detail provide important clues to the nature of associated processes such as ion and electron heating.

The data from four auroral conjunction events (Figure 4a-d) show a remarkable agreement between the parallel potential drops estimated from the electron measurements Φ_{TOT} and Φ_{MID}. The potential difference inferred from the energy of the peak of the upflowing ion distributions, E_{peak}, however, is frequently 30% (and in one case 70%) less than that inferred from the electron distributions. This is a puzzle, especially in the light of earlier work using S3-3 data [Collin and Sharp, 1984] which indicated that the upflowing ion energy was consistent with the potential inferred from the concurrently measured electron loss cone measurements. However, they did not determine a potential from the peak in the velocity space distribution as we have attempted to do here. Rather they integrated the distribution over both pitch angle and energy to get the mean energy per ion which they found to be consistent with the potential inferred from electron loss-cone

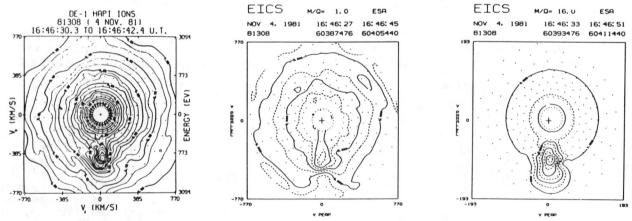

Fig. 3. (a) HAPI total-ion velocity-space contour plot. The upgoing ion beam associated with the conjunction arc is observed as the "bullet" in the lower center of the plot, offset from the origin by the velocity gained in the acceleration process - here roughly 400 km/s. EICS velocity-space contours for hydrogen (b) and oxygen (c). The velocity gained in the acceleration process is roughly four times larger for hydrogen than oxygen, consistent with the existence of an upward parallel electric field below the spacecraft.

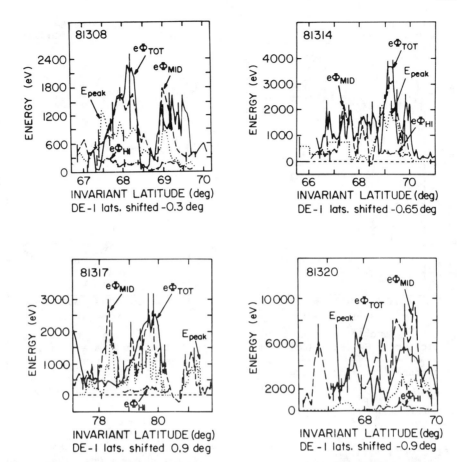

Fig. 4. (a) Summary of accelerating potentials estimated in the three independent ways: by the energy of precipitating electrons at low altitudes (eΦ_{TOT}), by the enhancement of the loss cone of electrons at high altitudes (eΦ_{MID}), and by the energy of upflowing ions at high altitudes (E_{peak}). Since, in this event, part of the acceleration region occurs above DE 1 altitudes ($\Phi_{HI} > 0$), Φ_{TOT} should be compared to the sum of Φ_{MID} and Φ_{HI}. The DE-1 data is shifted by -.3 degrees of invariant latitude. Comparison with the plasma sheet fluxes indicates that -.1 degrees of latitude shift may be attributed to magnetic field mapping errors; the remaining -.2 degrees of latitude shift may indicate a time variation or an arc motion. (b) Same as (a), but for event 2. The DE 1 data is shifted in invariant latitude by -.6 degrees, of which -.2 may be attributed to mapping errors. (c) Same as (a), but for event 3; shifted by .9 degrees. This large latitude shift is likely caused by the patchy nature of the arc and the fact that the two spacecraft, at slightly different local times, were not traveling transverse to it. (d) Same as (a), but for event 4; shifted by -.9 degrees.

measurements. As we shall show below, the mean energy is the best measure of the parallel potential drop, since it is valid whether or not the ions are thermalized after traversing the potential drop.

Figure 5 shows two examples of upgoing ion distributions measured by HAPI (i.e., not separated by mass). The distributions, taken at pitch angles ~180°, are well described by a distribution of the form

$$f(\alpha \sim 180°) = (\text{const}) \exp - \{E - E_{peak}\}/E_o\}$$

which is the form that one would expect by Liouville's theorem if a Maxwellian distribution of temperature E_o is accelerated through an electric field of potential drop E_{peak}. However, it is clear that the characteristic energies E_o are considerably larger than ionospheric temperatures (which are typically tenths of an eV). The ion heating observed in the second example (Figure 4b) is even more dramatic, with characteristic energies reaching several hundred eV.

Characteristics of the Ion Heating

If the original unaccelerated ion distribution was Maxwellian, with thermal energy kT_i, then after acceleration through a potential

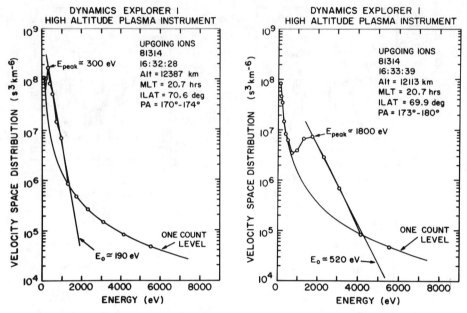

Fig. 5. Distribution functions for upflowing ($\alpha = 180°$) ions at two times during event 2. The ion characteristic energy E_0 (estimated from the slope of the line) is a substantial fraction of the accelerating potential (estimated from the peak of the distribution function); the best-fit Maxwellian temperature (not shown) is $\approx E_0/4$.

drop Φ, $\Phi \gtrsim$ 1kV, the mean energy per ion, $\langle E_i \rangle$, would be:

$$\langle E_i \rangle = e\Phi + 2kT_i \qquad (1)$$

and the energy of the peak of the velocity space density, E_{peak}, would be equal to $e\Phi$ since only the upward moving part of the original distribution would have been accelerated (Figure 2a) (see [Kaufmann et al., 1976]). Had the ions been first heated to well above ionospheric temperatures, $kT_H \gg 1$ eV, the peak would still occur at $e\Phi$ and the mean energy after acceleration would be $e\Phi + 2kT_H$, measurably higher than $e\Phi$ (Figure 2b). However, if the original ions had temperatures typical of the ionosphere, $kT_i \lesssim 1$ eV, then the mean energy after acceleration would be very close to $e\Phi$:

$$\langle E_i \rangle = e\Phi \qquad (2)$$

If such an accelerated cold ion distribution (Figure 2c, right) were thermalized within or above the electric field region, its final velocity space distribution should approximate a drifting Maxwellian of thermal energy kT_f. In that case, the mean energy per ion would be:

$$\langle E_i \rangle = E_{peak} + 5/2 \, kT_f \qquad (3)$$

where E_{peak} corresponds to $1/2 \, mV^2$, where V is the drift velocity. If the heating occurred without additional energy being acquired (or lost), we have from (2) and (3):

$$e\Phi = E_{peak} + 5/2 \, kT_f \qquad (4)$$

with E_{peak} considerably less than $e\Phi$ if the ions have been substantially heated. If the thermalization mechanism was waves generated from the unstable ion distribution, then one would expect the average ion energy would be less than $e\Phi$, since the ions would have lost energy to the waves.

The actual measured distributions (Figures 5 and 6) can be fit reasonably well with either flowing Maxwellians

$$f = (\text{const}) \exp - \{1/2 \, M(V-V_0)^2/kT_f\} \qquad (5)$$

or accelerated distributions

$$f = (\text{const}) \exp - \{(E - E_{peak})/E_0\}. \qquad (6)$$

It should be emphasized that the best-fit characteristic energy E_0 is, in general, several times larger than the best-fit thermal energy kT_f. For example, in Figure 6a, the oxygen distribution can be approximated by an accelerated distribution with $E_{peak} \sim 1000$ eV and $E_0 = 230$ eV, or as a Maxwellian (Figure 6e) with $n_0 \approx 0.8$ cm³, $V_0 \approx 100$ km/s, $kT_0 \approx 45$ eV. For Hydrogen the Maxwellian values are $n_H \approx 0.15$, $V_H \approx 330$, $kT_H \approx 40$. In Figure 6b, the Oxygen, Helium and Hydrogen characteristic energies are shown as 450, 520 and 200eV, respectively. In Figure 6f, the corresponding Maxwellian values are $n_0 \approx 0.75$, $V_0 \approx 140$, $kT_0 \approx 80$; $n_{He} \approx 0.4$, $V_{He} \approx 240$, $kT_{He} \approx 150$; and $n_H \approx 0.04$, $V_H \approx 580$, $kT_H \approx 40$.

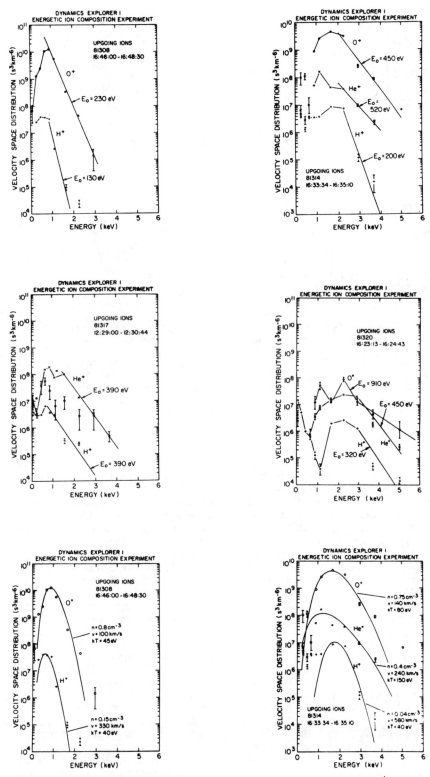

Fig. 6. (a) EICS upgoing ion distribution functions for event 1. O^+ values are shown as o's; H^+ values as +'s. (b-d) same as (a), but now also He^+ data are included (shown as x's). The characteristic energies E_O are estimated from the slopes of the line at high energies. (e-f) same data as (a) and (b), but now Maxwellian fits to the data are shown.

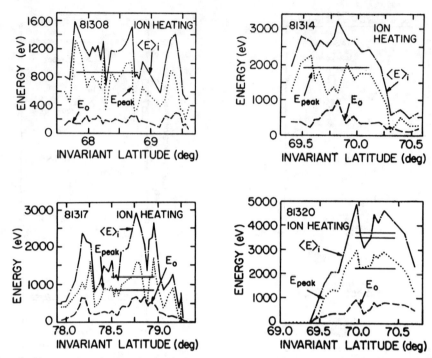

Fig. 7. Plot of the peak of the ion distribution E_{peak}, the ion characteristic energy E_O and the ion average energy $\langle E_i \rangle = E_{peak} + 2 E_O$ for the four events studied. The horizontal bars show the independently-determined average energies calculated by integrating the EICS data over energy and pitch angle. When more than one bar is shown, (c) and (d), the average energies differ slightly among the species. In (c), the Helium energy is larger than that of H^+ and O^+; in (d), He^+ is largest, then O^+, then H^+.

For ease in computation, we have used the accelerated form (Eqn. 6) to calculate the average ion energy $\langle E_i \rangle = E_p + 2E_O$ (Figure 7). This will, in general, be larger than the average energy from the Maxwellian distribution $\langle E_i \rangle = E_p + 5/2\ kT$, and reflects the substantial "high energy tail" which is typically observed, plus the fact that it does not include any fluxes below the peak. We have used the HAPI ion data to calculate $\langle E_i \rangle$ because of its finer time resolution than the EICS.

The horizontal bars in Figure 7 are the integrated average energies from the EICS data (from the spectra plotted in Figure 6). We see that the EICS and HAPI data are complementary: The HAPI yielding better time resolution and the EICS allowing a separate determination of the characteristics of each species. It is apparent that the integrated value of the average energy is closer to the Maxwellian form (Eqn. 5) than the accelerated form (Eqn. 6). For example, for the data shown in Figure 6f, the Maxwellian estimates of the average energies are 1.87, 1.58, and 1.85 keV for Hydrogen, Helium, and Oxygen, respectively. The corresponding integrated values are 1.92, 1.85, and 1.86 keV.

Figure 8 shows the comparison of the estimated $\langle E_i \rangle$ (from Eqn. 6) to the two electron measurements $e\Phi_{TOT}$ and $e\Phi_{MID}$. Now, the agreement with the other two (electron) potential measurements, although still not perfect (Figure 8a-d), is within the combined error bars of the measurements, except for event 4. Use of the more accurate form (Eqn. 5), however, would make the ion energies still somewhat less than the electron energies.

There are two obvious candidates for a wave instability mechanism to heat the ions. (Collisional heating is insignificant in the altitude range of interest.) One is a current-driven instability, drawing free energy from the flow velocity difference between the electrons and the ions. To search for evidence of this mechanism, we plot in Figure 9a the high-altitude total-ion characteristic energy E_O versus the concurrently-measured high-altitude electron current (data from all four events are combined in this plot). We see only a weak correlation ($r = 0.12$). The second mechanism is a two-stream instability, drawing free energy from the difference in flow velocity between the two major ion species. To search for evidence of that mechanism, we plot in Figure 9c the ion characteristic energy versus ΔV_i, where ΔV_i is the velocity difference that would have occurred between the hydrogen and oxygen beams had the ions received the beam energy inferred from the electron loss cone measurements. That is, $\Delta V_i =$

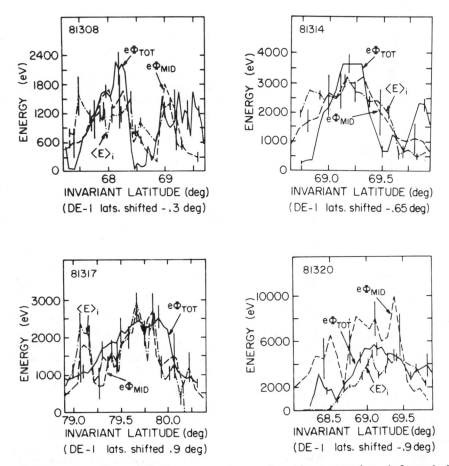

Fig. 8. Plot of the comparison of the auroral accelerating energies inferred in the three independent ways: from the energy of the precipitating electrons at low altitudes $e\Phi_{TOT}$, from the enhancement in the loss cone of electrons at high altitudes $e\Phi_{MID}$, and from the average parallel energy of upflowing ions measured at high altitudes $\langle E_i \rangle$. Now the agreement is extremely good, well within errors of the measurements except for the last case.

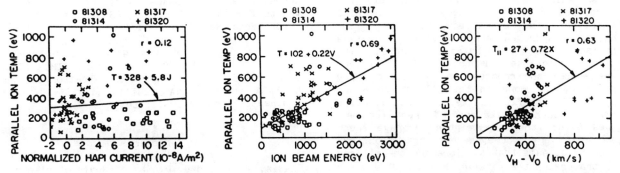

Fig. 9. (a) Scatter plot of high-altitude upflowing ion parallel characteristic energies versus concurrently-measured high-altitude upward electron currents. Typically no upflowing ions were observed whenever the electron current was zero or downward. (b) Scatter plot of ion characteristic energies versus the measured ion beam energy. A good correlation was found with the thermal energy 22% of the beam energy (r = 0.69). (c) Scatter plot of ion characteristic energies versus the difference between the inferred hydrogen and oxygen beam velocities.

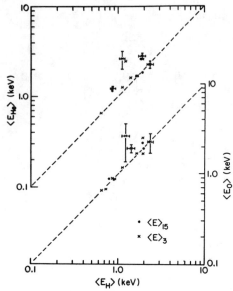

Fig. 10. Comparison of the flux-weighted average energies for He^+/H^+ (top) and O^+/H^+ (bottom). Although the characteristic energies can be very different, the flux-weighted average energies are quite comparable.

10.4 km/s $(e\Phi_{MID})^{1/2}$. This comparison shows a reasonable correlation (r = 0.63), although the best correlation is found between the ion characteristic energy and the ion beam energy (r = 0.69; Figure 9b).

If the heating mechanism is the two-stream instability, one would expect that the O^+ ions would gain energy and the H^+ ions would lose energy; i.e., the characteristic energy of O^+ should be larger than that of the H^+ ions. This appears to be the case in three of the four events studied (Figure 6 a-c). A prediction of an equilibrium distribution, however, would require a detailed kinetic simulation including the effects of all ion and electron species.

One can compare the flux-weighted average energies which are determined separately for each species by a full integration of the EICS data (Figure 10). Shown above is a comparison of helium to hydrogen energies; shown below is a comparison of oxygen to hydrogen energies. The dots are calculated using the full fifteen energy steps; the crosses are calculated using only 3, as was done in the Collin and Sharp study of S3-3 ion fluxes. We see that, although the E_o's of each species are different (compare Figures 6a-d), the flux-weighted average energies are in very good agreement, again confirming our hypothesis that the ions are thermalized after traversing the parallel electric field (or at least within it).

Conclusions

Comparison of particle distribution functions observed nearly simultaneously above and below an auroral acceleration region show encouraging agreement between the parallel electrostatic potential difference calculated from three independent observations: the peak energy of the precipitating electrons observed at low altitude, the widening of the loss cone of mirrored electrons observed at high altitude, and the average energy of upgoing ions observed at high altitude. The fact that the average energy of the ions, and not the energy of the peak of the distribution function, correlates best with the electron-determined estimates of the potential difference implies that the ions are heated as they traverse the acceleration region or above it. Characteristic energies attained by the upflowing ionospheric ions are typically 200-300 eV and have been observed as large as 1 keV, these thermal energies being typically 25% of the electron beam energy. Best-fit Maxwellian temperatures, however, are roughly one-fourth as large as the characteristic energies.

The fact that the ion characteristic energy correlates with the velocity difference between the upflowing hydrogen and ion beams, but not with the electron current, suggests that an ion two-stream instability may be responsible for the ion heating. It is now established that a solar-cycle effect is apparent in the composition of upgoing ion beams [Yau et al., 1985], with heavy ions being more prevalent during times of high solar activity. Since our data were taken close to solar maximum, the flux of upgoing heavy ions may be unusually large and the heating rates therefore considerably larger than would be observed near solar minimum.

Acknowledgments

The authors acknowledge helpful discussions with R. A. Bergmann, T. W. Hill, W. K. Peterson, and C. Goertz. Additional assistance in data analysis was provided by M. Friedman, R. A. Frahm, K. Birkelbaum, and R. E. Brazile. This research was supported in part by the National Aeronautics and Space Administration under grants NGR-44-006-137 and NAS5-302 and by the Atmospheric Sciences Division of the National Science Foundation under grant ATM 85-18710 at Rice University, by the NASA under contract NAS5-28710 and the NSF under grant ATM83-17710 at Lockheed Palo Alto Research Laboratories, and by NASA contracts NAS5-28711 and NAS5-28712 at Southwest Research Institute.

References

Cladis, J. B. and R. D. Sharp, Scale of electric field along magnetic field in an inverted V event, J. Geophys. Res., 84, 6564-6572, 1979.

Collin, H. L. and R. D. Sharp, The relationship between the energies of upflowing ions and the measured accelerating potential, EOS, Trans AGU, 65, 257, 1984.

Evans, D. S., Precipitating electron fluxes formed by a magnetic field aligned potential difference, J. Geophys. Res., 79, 2853-2858, 1974.

Gurnett, D. A., Electric fields and plasma observations in the magnetosphere, in Critical Problems of Magnetospheric Physics, ed. E. R. Dyer, p. 123, National Academy of Sciences, Washington, D.C., 1972.

Kaufmann, R. L., D. N. Walker, and R. L. Arnoldy, Acceleration of auroral electrons in parallel electric fields, J. Geophys. Res., 81, 1673-1682, 1976.

Knight, S., Parallel electric fields, Planet. Space Sci., 21, 741-750, 1973.

Mizera, P. F., J. F. Fennell, D. R. Croley, Jr., A. L. Vampola, F. S. Mozer, R. B. Torbert, M. Temerin, R. Lysak, M. Hudson, C. A. Cattell, R. G. Johnson, R. D. Sharp, A. Ghielmetti, and P. M. Kintner, The aurora inferred from S3-3 particles and fields, J. Geophys. Res., 86, 2329-2340, 1981.

Yau, A. W., P. H. Beckwith, W. K. Peterson, and E. G. Shelley, Long-term (solar cycle) and seasonal variations of upflowing ionospheric ion events at DE-1 altitudes, J. Geophys. Res., 90, 6395-6407, 1985.

INTERACTION OF H$^+$ AND O$^+$ BEAMS: OBSERVATIONS AT 2 AND 3 R$_E$

R. L. Kaufmann and G. R. Ludlow

Department of Physics, University of New Hampshire, Durham, NH 03824

Abstract. Since H$^+$ and O$^+$ upgoing auroral ion beams frequently are accelerated to comparable energies, the H$^+$ beam streaming velocity is several times higher than the O$^+$ beam velocity. Both S3-3 and DE-1 data show effects of the interaction between these two beams. Each species is heated, and there is a net transfer of energy from H$^+$ to O$^+$ so that an extended O$^+$ high energy tail is formed. Linear stability analysis, using the measured ion distribution functions, identified two plasma waves which can be driven by the beams. It is suggested that this mechanism could transfer energy from a lighter to a heavier species in a number of space plasmas.

Observations

We began this work by looking for evidence that upgoing H$^+$ and O$^+$ beams interact so that energy is transferred from H$^+$ to O$^+$ ions [Kaufmann, 1984]. Figures 1a and 1b are H$^+$ and O$^+$ distribution functions from the Lockheed ion spectrometer on the S3-3 satellite at 2R$_E$ geocentric [Shelley, et al., 1976]. This orbit was unusual because the O$^+$ beam peaked at an energy (W$_p$ = 1.9 keV) that was much higher than that of the H$^+$ beam (W$_p \le$ 840 eV). In fact, the H$^+$ peak was at such a low energy that it was not resolved clearly on this orbit. The detector's limited energy range produced difficulties when the two distribution functions were added

$$F(v) = f_{H+}(v) + (m_{H+}/m_{O+})f_{O+}(v) \quad (1)$$

as in Figure 1c. The factor m_{H+}/m_{O+} is included in (1) because we will be evaluating the linear dielectric function [e.g., Dusenbery and Kaufmann, 1980]

$$\varepsilon(\vec{k},\omega) = 1 + \sum_{p=-\infty}^{\infty} \sum_{\gamma} \frac{4\pi n_\gamma z_\gamma^2 e^2}{k^2 m_\gamma} \int d\vec{v} \left(\frac{J_p^2}{\omega - k_\| v_\| - p\omega_{c\gamma}} \right)$$

$$\cdot \left(\frac{p\omega_{c\gamma}}{v_\perp} \frac{\partial f_\gamma}{\partial v_\perp} + k_\| \frac{\partial f_\gamma}{\partial v_\|} \right) \quad (2)$$

in order to study the stability of measured distribution functions. In the above expression $\omega_{c\gamma}$ is the cyclotron frequency of the γ'th species (negative for electrons), z_γ is the charge, and J$_p$ is the Bessel function of order p with argument $k_\perp v_\perp / |\omega_{c\gamma}|$. Since the contribution of each species in (2) is inversely proportional to its mass, the combined F(v) in (1) can be treated as an equivalent H$^+$ distribution function. When the H$^+$ measurements were extrapolated down to energies near the O$^+$ peak, the H$^+$ contribution dominated so that double peaking was not evident in data from this orbit. Because of these instrumental limitations, double peaking was seen on only 2 of the 6 S3-3 orbits studied. However, all orbits showed the principal effect for which we were looking. In each case, the O$^+$ fluxes had a more extended high energy tail than the H$^+$, indicating that this portion of the O$^+$ distribution function was selectively energized.

The rather unusual circumstance that H$^+$ and O$^+$ beams peaked at different energies on this orbit suggests that a detector with good energy resolution but no mass resolution may see a double peak. Figure 1d shows what happens when the above data from the two species are combined according to energy and count rate in order to simulate a detector with no mass resolution. The resulting F(E) is plotted using the velocity of a H$^+$ ion at the measured energy

$$F(E) = f_{H+}(E) + (m_{H+}/m_{O+})^2 f_{O+}(E) \quad (3)$$

No extrapolation is required here because H$^+$ and O$^+$ fluxes are added at the same energy rather than the same velocity as in (1). The H$^+$ dominates and O$^+$ simply broadens the peak in Figure 1d when seen at the available energy resolution. The beam plasma was 80% H$^+$ on this orbit.

Figure 2 shows the 4 distribution functions made by the Aerospace detector, which had no mass resolution, during the same time interval [Mizera and Fennell, 1977]. Even though the energy resolution near the peak is better, no evidence is found of a second peak. The presence of two species simply generates what appears to be a peak that is extended along the v$_\|$-axis. Comparison of the two detectors (Figures 1d and 2)

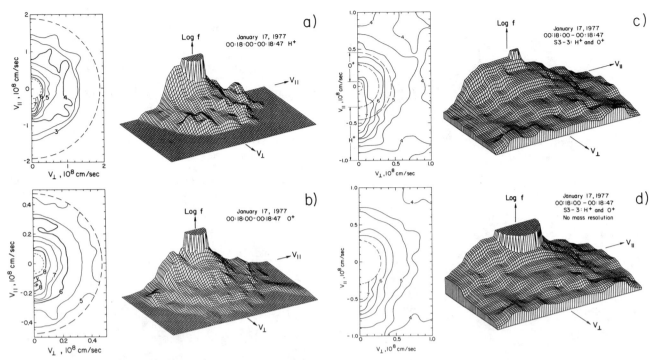

Fig. 1. a,b) H^+ and O^+ distribution functions seen by S3-3. Upgoing ion beams appear on the lower half of the contour plots and the left side of the three-dimensional plots. Dashed lines and the flat-topped pillars show limits of the detector's energy range. Most of the O^+ plot is simply a ½ count artificial base level that was added to all measurements. The O^+ count rates are significantly above this base level only near the upgoing beam velocity. c) H^+ and O^+ measurements are combined according to (1). Fluxes must be extrapolated beyond the detector's energy range, indicated for each species on the contour plot, in order to generate this panel. d) Measurements are combined according to (3) to simulate a detector with no mass resolution.

shows qualitatively similar structures, and about a factor of 3 disagreement in absolute flux levels. One conclusion from this study of the two detectors is that it is difficult to find evidence for double peaking in combined H^+ and O^+ distribution functions with detectors which have no mass resolution. The presence of two species with slightly different energy spectra will tend to produce peaks which appear to be stretched out along the $v_\parallel$ axis.

Figure 3 shows data from the Lockheed ion mass spectrometer on DE-1 [Shelley et al., 1981] at a radial distance of 3.3 R_E. The energy resolution and range are sufficient to show the two peaks clearly on the combined (eqn. 1) distribution function plot (Figure 3c). The lower velocity peak is mostly O^+ and the higher velocity peak is mostly H^+. If the valley between the peaks fills in through wave-particle interactions, the lower energy side of the H^+ peak and the higher energy side of the O^+ peak are expected to broaden. These effects can be seen in Figures 3a and 3b. The apparent double peaking of O^+ in Figure 3b is produced by time changes, and is not present in other sampling intervals on this orbit.

To summarize the observations, every orbit that was examined both from S3-3 at 2 R_E and DE-1 at higher altitude provided evidence for an interaction between H^+ and O^+ components of upgoing ion beams. The most obvious feature involves parallel acceleration of O^+ ions in this species' high energy tail. This is a natural consequence of quasilinear diffusion and other wave-particle interactions which tend to fill in the valley between the H^+ and O^+ peaks. Distribution functions from the two species usually were found to peak at approximately the same energy, but on some occasions (e.g. Figure 1) the O^+ peaks at much higher energy. Both species are much hotter than is typical of the ionosphere, which is the presumed plasma source.

Analysis

Since the DE-1 data is the most detailed available, our analysis started using Maxwellian fits to the distribution function in Figure 3. We do not yet have complete information from other DE-1 detectors, so had to guess at some parameters that were used in this preliminary

94 H$^+$ AND O$^+$ BEAM INTERACTIONS

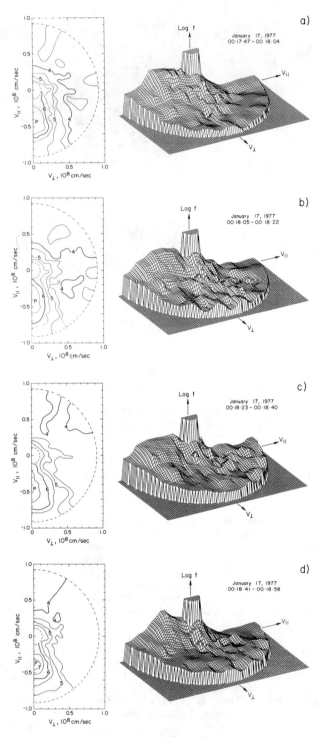

Fig. 2. Similar to Fig. 1. These measurements were made by the S3-3 ion detector which had better temporal resolution and lower energy limits, but no mass resolution.

analysis. For example, we assumed that no cold background plasma is present within the ion beams. This assumption was motivated by the suggestion that any cold ions generated well above the ionosphere in these beam regions will be accelerated up and become part of the beam [Kaufmann and Kintner, 1984].

Figure 4 shows one interaction that could produce quasilinear diffusion. The upper right, lower left, and lower right panels show H$^+$, O$^+$, and electron contributions to the real and imaginary parts of the linear dielectric function (2), while the upper left panel combines all species. Frequencies are shown in a coordinate system fixed with respect to the earth. The unstable wave, marked by the large dot where ε_R = 0, involves an O$^+$ ion acoustic wave propagating faster (relative to the earth) than the O$^+$ beam and a H$^+$ ion acoustic wave propagating slower than the H$^+$ beam (propagating down toward the

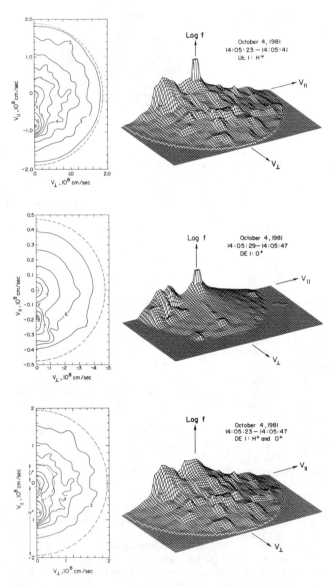

Fig. 3. Similar to Fig. 1. Measurements from DE-1 which had more extended energy coverage. The O$^+$ count rate was significant only near the upgoing beam velocity.

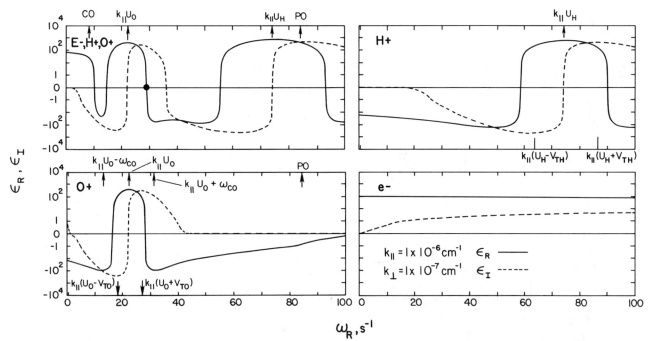

Fig. 4. Real (solid line) and imaginary (dashed line) parts of the linear dielectric function. The plasma parameters used were; e^-: $N = 0.12$ cm^{-3}, $T = 2000$ eV; H^+: $N = 0.052$ cm^{-3}, $T_\parallel = 150$ eV, $T_\perp = 105$ eV, $u_H = 7.4 \times 10^7$ cm/s; O^+: $N = 0.065$ cm^{-3}, $T_\parallel = 330$ eV, $T_\perp = 140$ eV, $u_O = 2.2 \times 10^7$ cm/s. The symbols PO, CO, and V_{TO} refer to the O^+ plasma frequency, cyclotron frequency, and parallel thermal velocity. The growing wave is indicated by the solid circle.

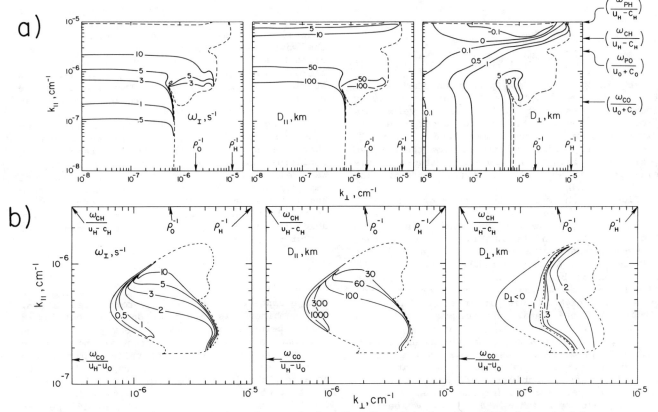

Fig. 5. Growth rates (ω_I) and growth lengths ($D_\parallel$, $D_\perp$) for the two plasma models studied. The symbols ρ_H and c_H refer to the H^+ cyclotron radius and sound speed.

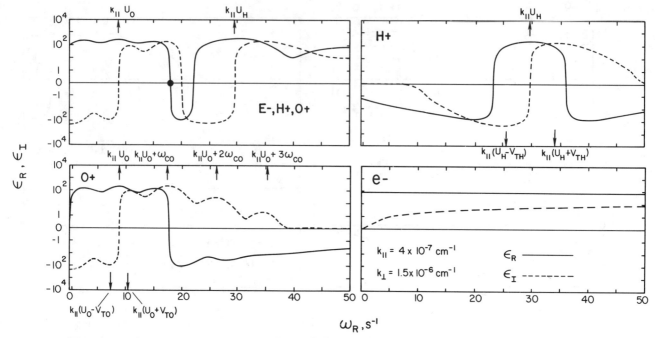

Fig. 6. Similar to Figure 4 except that T_e = 1000 eV, and the plot is for a different k.

earth as seen by an observer moving with the H^+ beam). These waves interact strongly because the growing mode has a phase velocity, $v_p \simeq u_H + c_O \simeq u_H - c_H$; the u's are beam velocities and the c's are sound speeds

$$c_i^2 = \kappa T_e n_i / m_i n_e \qquad (4)$$

A comparison of the two panels on the left of Figure 4 shows that the wave phase velocity (heavy dot) is about 1.5 O^+ thermal velocities from the O^+ beam peak. The two upper panels in Figure 4 show that this same phase velocity is about 4.5 H^+ thermal velocities from the H^+ beam peak. As a result, interactions between particles and this wave should be more effective at heating O^+ than H^+.

Figure 5a shows that these waves grow ($\omega_I > 0$) over a large range of k-space, with maximum growth rates at small propagation angles. The two right hand panels in Figure 5a show that growth lengths become short near $k_{\parallel} = 10^{-5}$ cm^{-1} (wavelength = 6 km), but the minimum growth length cannot be determined because our linear calculation become inaccurate when growth rates become so high. This interaction has been studied in more detail by Bergmann and Lotko [1986].

In order to find a region in which the Maxwellian fits produced a growing wave in this mode, we had to introduce an electron temperature of 2000 eV, which probably exceeds the actual T_e. We have not yet searched for the lowest T_e which will support such waves. However, we decided first to try looking at other wave interactions. The one in Figure 6 appeared with T_e = 1000 eV, which probably is closer to the actual electron temperature. Figure 5b shows that this mode grows only at relatively large propagation angles, and that growth lengths can be 10's of km near $k_{\parallel} = 10^{-6}$ cm^{-1}. At the $\vec{k}$ shown in Figure 6, the wave phase velocity is about 6 O^+ thermal velocities from the O^+ beam peak and about 3 H^+ thermal velocities from the H^+ peak. This wave at the wave number used for Figure 6 therefore would be less effective at heating the O^+ beam than the wave in Figure 4.

To summarize the preliminary analyses, we found that low frequency ion waves can grow with short enough growth lengths to reach large amplitudes before leaving the ion beam region. We are not yet sure which wave mode is most important or if more than one wave is unstable in the actual plasma. These calculations will be improved as soon as we have collected all available information on electron temperature and background density.

Conclusions

Ion distribution functions have been examined over a range of altitudes on auroral field lines. The combined distribution function contains a valley between the H^+ and O^+ beam peaks. These two components appear to interact strongly and to have partially filled in the valley through wave particle interactions. We suspect that the observed distribution function has evolved to the point at which it is only weakly unstable (growth lengths comparable to the thickness of the ion beam region). The net results of this process are to heat both species, to transfer parallel energy from H^+ to O^+, and to produce an extended O^+ high energy tail. This interaction therefore may provide a second mechanism, in addition to transverse heating, that could make O^+ systematically more energetic than H^+.

Finally, we would like to suggest that similar interactions could be important in the ionosphere, where very cold low energy upgoing flows of H^+ and O^+ may exist. If so, this will provide a mechanism to enhance the outflow of O^+ at the bottom of the major parallel acceleration region. For example, a quick preliminary calculation showed that the instability in Figure 4 is present if 1000 °K H^+ and O^+ beams with densities of 5×10^3 cm^{-3} are accelerated to a streaming energy of 1 eV, provided the electrons are hot enough so that ion acoustic waves exist. Low altitude (1000 to 2000 km) measurements of ionospheric O^+ and H^+ distribution functions, when combined according to (1) should show double peaking and evidence for a filling in of the valley between the peaks (similar to Figure 3) if this interaction is important. The proposed mechanism also should be considered in other space plasma regions (e.g. the plasma sheet boundary layer) where two or more species are present but are found to have different bulk streaming velocities.

Acknowledgements. The Lockheed S3-3 and DE-1 data were provided by H.L. Collin and W.K. Peterson. The Aerospace S3-3 data were supplied by D.J. Gorney. This work was supported by the National Science Foundation under grant ATM-84-00784.

References

Bergmann, R., and W. Lotko, Transition to unstable ion flow in parallel electric fields, J. Geophys. Res., in press, 1986.

Dusenbery, P.B. and R.L. Kaufmann, Properties of the longitudinal dielectric function: An application to the auroral plasma, J. Geophys. Res., 86, 5969-5976, 1980.

Kaufmann, R.L., Two-component upgoing ion beams, EOS Trans. Am. Geophys. Union, 65, 1056, 1984.

Kaufmann, R.L. and P.M. Kintner, Upgoing ion beams 2. Fluid analysis and magnetosphere-ionosphere coupling, J. Geophys. Res., 89, 2195-2210, 1984.

Mizera, P.F. and J.F. Fennell, Signatures of electric fields from high and low altitude particles distributions, Geophys. Res. Lett., 4, 311-314, 1977.

Shelley, E.G., R.D. Sharp and R.G. Johnson, Satellite observations of an ionospheric acceleration mechanism, Geophys. Res. Lett., 3, 654-656, 1976.

Shelley, E.G., D.A. Simpson, T.C. Sanders, E. Hertzberg, H. Balsiger and A. Ghielmetti, The energetic ion composition spectrometer (EICS) for the dynamics explorer-A, Space Sci. Inst., 5, 443-454, 1981.

BANDED ION MORPHOLOGY: MAIN AND RECOVERY STORM PHASES

R. A. Frahm and P. H. Reiff

Department of Space Physics and Astronomy, Rice University, Houston, TX 77251

J. D. Winningham and J. L. Burch

Department of Space Sciences, Southwest Research Institute, San Antonio, TX 78284

Abstract. Ion bands appear in a spectrogram display as a continuous line of enhanced ion energy flux, whose median energy increases as the satellite travels poleward in the low and mid-altitude magnetosphere. These ion bands occur with highest energy flux at zero degrees pitch angle. Ion bands similar to those described previously have been investigated using data from the Low Altitude Plasma Instrument (LAPI) and High Altitude Plasma Instrument (HAPI) flown on the Dynamics Explorer (DE) satellites. The purpose of this paper is to present the results of a statistical study of band occurrence and to present and describe the current models for band formation. The morphology of ion bands has been examined for main and recovery storm phases covering the period from September 1981 to December 1981. Bands are more likely to be seen during the main phase of magnetic storms than during recovery phase. Bands are more prevalent in the evening sector and occur at higher invariant latitudes ($\approx 5°$) than those in the pre-noon sector. Two current models have been proposed to describe bands or band-like signatures in ion spectrograms. The first is a time-of-flight effect as in the bouncing ion clusters (seen at geosynchronous orbit). The second is convective dispersion, where ions from the opposite hemisphere's ionosphere experience significant motion perpendicular to magnetic field lines and become dispersed in latitude as they travel parallel to a magnetic field line. The data tend to favor convective dispersion, although time-of-flight effects can also be seen.

Introduction

Winningham et al. [1984] have defined the ion band as the appearance on an energy-time spectrogram of a continuous line of enhanced energy flux which monotonically decreases in energy with decreasing latitude. Peak energy flux is observed at 0° pitch angle, and the energy flux decreases as the pitch angle increases. Bands have been reported as a night side phenomenon. An example of such a band is shown in the High Altitude Plasma Instrument (HAPI) spectrogram of Figure 1 from September 26, 1981. This plot is a collection of energy sweeps near 180° pitch angle for the ions from sensor 1 on the upper panel and sweeps near 0° pitch angle for the ions from sensor 1 on the lower panel. A distinct band can be seen in the ions of 0° pitch angle beginning at 14:29 UT (≈ 1000 eV) to 14:41 UT (≈ 8 eV). This is an example of a main storm phase, night side, northern hemisphere, HAPI band. Under the convective dispersion model of Winningham et al., the up-going ions (180° pitch angle) from 14:16 UT to 14:24 UT would be possible sources for bands that would form in the opposite (southern) hemisphere.

High and low altitude plasma data have been examined from the Dynamics Explorer (DE) satellites. The high altitude plasma data came from the HAPI, which was flown on the Dynamics Explorer-1 spacecraft. This spacecraft is a polar-orbiting satellite that is spin stabilized, with its spin axis perpendicular to the orbit plane, at 1 spin every 6 seconds. DE-1 has an apogee of $\approx 29,000$ km and perigee altitude of ≈ 650 km. HAPI look directions lie in a plane parallel to the spin axis, which is perpendicular to the orbit plane, at 0°, ± 12.5°, and ± 45° from the spacecraft equator. Placement of the HAPI on DE-1 may be found in Hoffman et al., 1981. For more information on the HAPI, the reader is referred to Burch et al., 1981 or Frahm, 1984.

The low altitude plasma data came from the Low Altitude Plasma Instrument (LAPI) which was flown on the Dynamics Explorer-2 spacecraft. This spacecraft is polar orbiting (apogee altitude, ≈ 1300 km; perigee altitude, ≈ 300 km); and normally spins once per orbital revolution. LAPI comprises 15 analyzers which look at ± 0°, 7.5°, 15°, 30°, 45°, 60°, 97.5°, 105°, 112.5°, 135°, 165°, 172.5°, and ± 180° to the fiducial line on a scan platform. The scan platform lies in a

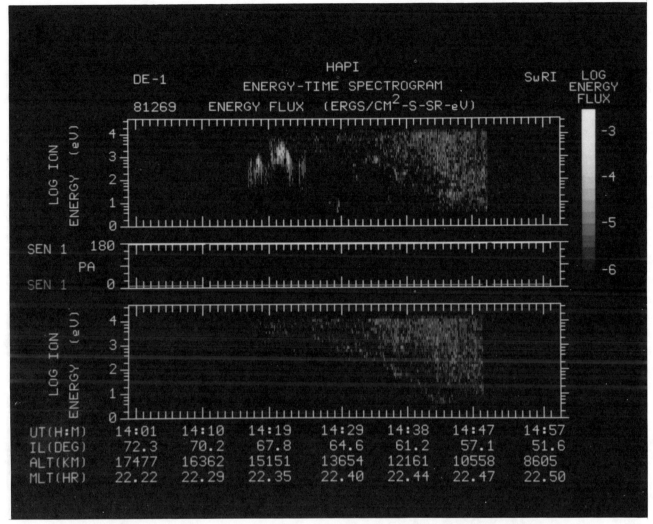

Fig. 1. HAPI banded ion spectrogram. Universal time (UT) is plotted on the horizontal axis. On the vertical axis of the upper and lower panel is plotted the log of the particle energy (in eV). Energy flux (ergs/cm²-s-sr-eV) is shaded using the scale at the upper right. Plotted on the center panel is the pitch angle (PA). This HAPI plot is a collection of energy sweeps near 180° PA for the ions from sensor 1 on the upper panel and sweeps near 0° for the ions from sensor 1 on the lower panel. After each time interval, the time (UT), invariant latitude (IL), altitude (ALT), and magnetic local time (MLT) are given. A distinct band can be seen in the ions of 0° PA beginning at 14:29 UT (≈1000 eV) to 14:41 UT (≈8 eV). This is an example of a main storm phase, night side, northern hemisphere, HAPI band. Under the convective dispersion hypothesis, the up-going ions (180° PA) from 14:16 UT to 14:24 UT would be possible sources for bands that would form in the opposite (southern) hemisphere. This pass occurred on day 81269.

plane parallel to the orbit plane and is typically slaved to the magnetic field direction. Thus, the mounting angles are approximately equal to the local pitch angle. For more information on the LAPI, refer to Winningham et al., 1981 or Frahm, 1984.

In this paper, we present the results of a storm phase morphological study of ion bands seen by the HAPI and LAPI. This study has revealed several occurrences of day side ion bands, an example of which is shown in the low altitude plasma data of Figure 2 from October 15, 1981. Plotted are the electrons from sensor 24 on the upper panel and ions from sensor 25 on the lower panel, both at about 15° pitch angle. A distinct band can be seen in the ions beginning at

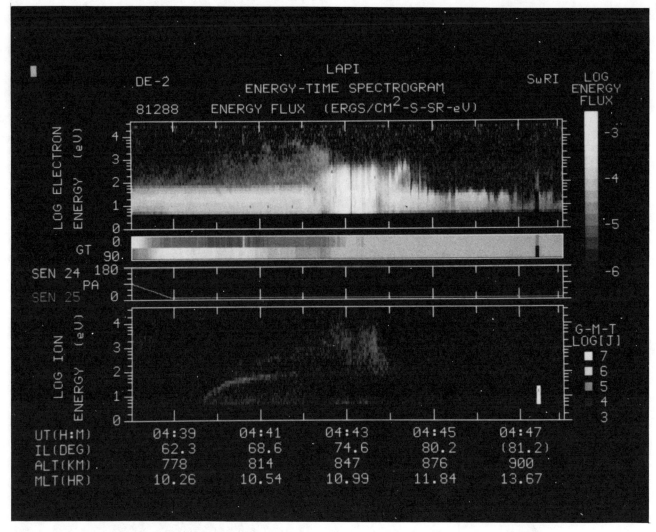

Fig. 2. LAPI banded ion spectrogram. This spectrogram is very similar in format to the HAPI spectrogram of Figure 1. A fourth panel has been added (upper center) to display LAPI's Geiger-Muller tubes (GT). Number flux (#/cm²-s-sr-eV) for the GT has been shaded using the scale at the lower right. Plotted here are the electrons from sensor 24 on the upper panel and ions from sensor 25 on the lower panel, both at about 15° PA. A distinct band can be seen in the ions beginning at 04:39:45 UT (≈8 eV) and extending to 04:41:30 UT (≈70 eV). This is an example of a recovery storm phase, day side, northern hemisphere, LAPI band. This pass occurred on day 81288.

04:39:45 UT (≈8 eV) and extending to 04:41:30 UT (≈70 eV). This is an example of a recovery storm phase, day side, northern hemisphere, LAPI band. In addition, two models are reviewed that describe how velocity distribution can produce bands or similar features on ion spectrograms: the time-of-flight model [Klumpar et al., 1983] and the convective dispersion model [Winningham et al., 1984]. The two models are not mutually exclusive, but represent two ends of a spectrum of circumstances that can form ion dispersion.

Morphological Results

Using the Dst index to identify storm phase, nine magnetic storms were selected from September, 1981 to December, 1981. Only simple, large storms were chosen--those with a Dst index minimum ≈-50 γ or less, and fluctuations within the storm of less than ≈20% of the peak storm intensity. Storm times were selected because the increase in magnetospheric activity leads to an enhancement in the occurrence of both inverted

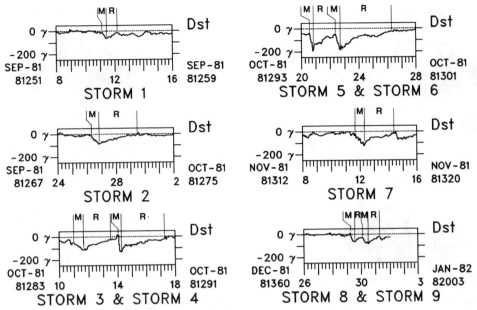

Fig. 3. Storms used in this study. Dst is plotted and identified for each storm. Storm phases are indicated with an "M" for main storm phase and "R" for recovery storm phase. Storms were identified from September 1981 to December 1981. Both HAPI and LAPI data were used for the first seven storms: storms 8 and 9 include LAPI data, only.

"V" events and magnetospheric convection (suggested to be, respectively, the ion source and mechanism of convective dispersion). It is also likely that time-of-flight dispersions will become more frequent because of the rapid time changes, particularly substorm ion injections. HAPI and LAPI data were available for the first seven storms, but only the LAPI data were available for the last two storms. The hourly Dst indices during each of the storm times identified in this study are plotted in Figure 3.

Ion band occurrence results from this study are summarized in Figure 4. Figure 4a contains our results from main storm phase data; Figure 4b contains our results from recovery storm phase data. Trajectories of both satellites, both northern and southern hemisphere, and for all storms are taken into account in these summary plots. The plots of Figure 4 do not include any latitude below 50°. All data below 50° have also been examined, but no bands were observed.

The satellites were not operating during the entire duration of the storms. Figures 4a and 4b show that there is more satellite coverage during recovery phase than main phase (Number of Passes code in Figure 4) (for a detailed table, see Frahm, 1986). Comparison of these figures also shows that the fraction which contain bands (Passes Which Show Bands to Total Passes code in Figure 4) is larger for main phase than for recovery phase by a factor of 2 or 3. It is significant that not all magnetic local times (MLT) have been covered by the satellites during the time of this study. At most, twelve of the twenty four MLT hours have been covered adequately at latitudes of band formation. Recovery phase trajectories heavily dominate the 6 H to 12 H and 18 H to 0 H sectors.

Comparison of the invariant latitude-magnetic local time distributions in Figure 4 shows that bands occur at larger invariant latitudes during recovery phase than during the main phase. Day side events tend to occur slightly poleward (≈5°) of night side events for each storm phase. The plots show that band precipitation is limited to latitudes above 50° invariant, and are not detected within the polar cap (despite extensive coverage). The majority of bands are detected at the equatorward edge of the auroral zone in the diffuse aurora (defined by precipitation). In a few cases, a band has been seen to extend across most of the auroral zone (including the discrete zone), and a few of these bands appear to contain gaps at locations of discrete electron features. Bands have been seen to have a concave downward (Figure 2), linear (Figure 1), or concave upward structure on the spectrogram. Sometimes, a band exhibits a structure which is a concave upward-linear combination or concave downward-linear combination. This structure shape must be caused by variations in the magnitude and direction of the convection electric field. The median energy of the ions within the band is always observed to decrease with decreasing latitude. This is

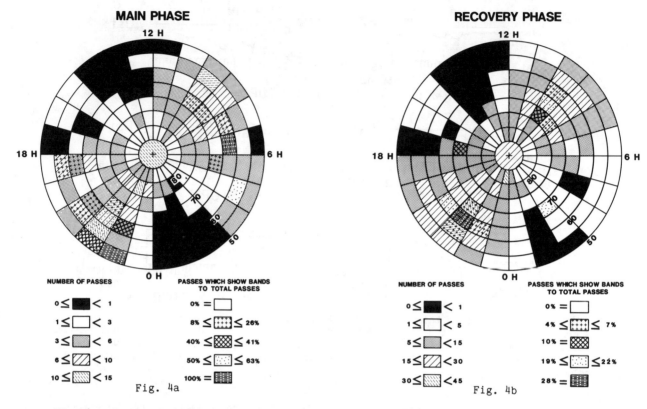

Fig. 4. Storm morphology. These polar projections are split into bins of every hour in MLT and every 5° in IL between 50° and 85°. All MLT with IL above 85° are taken as a single bin. The areas are shaded in two categories: by the total number of satellite passes (Number of Passes code) and by the fraction of those satellite passes which contain bands to the total number of satellite passes (Passes Which Show Bands to Total Passes code). Each area is coded with one shade from each category (each area has two shades concurrently). (a) These main phase passes include data from all storms for both HAPI and LAPI, and for northern and southern hemisphere. (b) These recovery phase passes include data from all storms from both HAPI and LAPI, and for northern and southern hemisphere.

independent of whether the satellite is moving equatorward or poleward. We have observed cases of bands (night sector in the southern and day sector in the northern hemisphere) where the satellites are traveling poleward which show the lowest energy ions arriving first. This type of dispersion is readily explained by the convective dispersion model and is not compatible with the time-of-flight model unless the ion source is moving in a certain way.

All bands exhibit their strongest energy flux at 0° pitch angle. The majority of LAPI bands have a discernible signal (although it is usually extremely weak and highly suspect) in the 97.5° sensor, but is nonexistent in the 105° sensor. A few LAPI bands exhibit sightly abnormal pitch angle dependences: some exhibit an extremely weak signal at 112.5°, but not in greater pitch angle sensors, while others are not detectable in sensors with pitch angles greater than 45°. Bands seen in HAPI have a higher energy flux at larger pitch angles and their signal disappears by 145° pitch angle. The maximum energy flux is usually less in those bands observed by HAPI than by LAPI, although the band energy flux is greater for HAPI at pitch angles greater than 60°.

Models

Two processes have previously been discussed to explain the occurrence of band-like ion signatures on auroral field lines. The first is parallel velocity dispersion based on the Quinn and McIlwain [1979] model used to explain certain features seen in geosynchronous orbit data; namely, ion clusters bouncing between hemispheres. This model was adapted by Klumpar et al. [1983] to explain oxygen ion dispersion signatures seen at intermediate magnetic latitudes. It will be referred to as the time-of-flight hypothesis. A model to explain ion

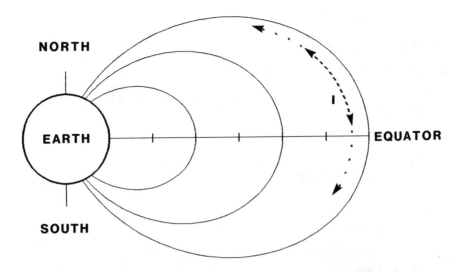

Fig. 5a. Time-of-flight hypothesis. Schematic view in a meridian plane of the Quinn and McIlwain bouncing ion cluster model. This schematic attempts to show ion cluster motion after injection. The source location is labeled by I. As the ion cluster moves along the magnetic field line, the high energy particles outrun the low energy particles.

bands was discussed by Winningham et al. [1984] and was based on low and mid-altitude data from the HAPI and LAPI. In this model, the ions are hypothesized as coming from the opposite hemisphere, accelerated upwards by electric fields in an inverted "V", and dispersing in energy perpendicular to the magnetic field. It is also possible that the ion source is in the same hemisphere as these measurements if the ions had bounced one or more times. It will be referred to as the convective dispersion hypothesis. The dispersion is similar in form (but opposite in direction) to that which occurs in cusp injection events [Burch et al., 1982, and references therein].

The model described by Quinn and McIlwain is shown schematically in Figure 5. Clusters (or groups) of ions are created on a magnetic field line during a time period that is short compared to the ion bounce time, Figure 5a [see Quinn and Southwood, 1982, for possible source mechanisms]. Once formed, they travel along the magnetic field line on which they were created, bouncing between hemispheres as they go. As the ion clusters travel along their path, the ions with higher energy (higher velocity) outrun the ions of lower energy (lower velocity) at the same pitch angle. If one observes an ion cluster at a single pitch angle and from a fixed point in space, one observes higher energy ions first. As time progresses, successively lower and lower energy ions are detected, ie. dispersion in energy occurs parallel to the magnetic field. Figure 5b shows an example of how an ion spectrogram should appear to an observer at a constant latitude (typical of geosynchronous orbit) (multiple bounces are not shown for clarity). Here it should be stressed that the observation is at a fixed pitch angle and at constant latitude where the only variable is time. This model has been very successful in explaining spectrograms observed by the auroral particles experiment on ATS-6 [the University of California at San Diego auroral particles experiment on ATS-6 is described by Mauk and McIlwain, 1975]. Ion clusters may be created anywhere along the magnetic field line, including the ionosphere and equator, and at any magnetic local time.

The observations discussed by Klumpar et al., were taken on L-shells comparable to those sampled by ATS-6, but at intermediate magnetic latitudes ($\approx 35°$); and consequently, farther from the equatorial plane. The ion species being detected were directly identified by an ion composition spectrometer [Energetic Ion Composition Spectrometer (EICS) is described by Shelley et al., 1981] which sampled ion flux versus energy and pitch angle. They observed variations in oxygen ion flux as a function of both energy and pitch angle that they interpreted to be the result of time-of-flight dispersion along the flux tube from an initial, transient

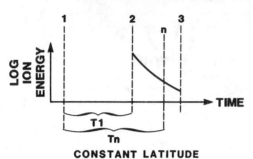

**ENERGY-TIME SPECTROGRAM
(FIXED PITCH ANGLE)**

Fig. 5b. Time-of-flight hypothesis. An idealized spectrogram for this model shows that when the source injection occurs, there is a time delay, T_1, before the spacecraft detects those ions which were injected by the source. The satellite will continue to detect ions of successively lower energy, n, until the lowest energy ions pass the satellite. Note that, for clarity, multiple bands are not shown.

event. Their interpretation employs many of the same features as that of Quinn and McIlwain and extends the treatment to a broad range of pitch angles and to observing points substantially off the equatorial plane. The ions observed at the spacecraft at any given time are those that have common travel times from the source. Since travel time is the length of the ion's spiral path around and along the magnetic field divided by its velocity, there will be a unique set of energies and arrival pitch angles that are populated at a given instant. This produces a saw-toothed pattern on an energy-time spectrogram with median energy decreasing as a function of time as shown in Figure 5c. This pattern is a temporal effect. It should be independent of the direction of motion of the satellite.

The convective dispersion model is shown schematically in Figure 6. Under this description, the source of ion bands is the up-going ions associated with inverted "V"'s that occur at low to mid altitudes (Figure 6a). Here, parallel electric fields accelerate electrons downward and ions upward [Reiff et al., this volume]. As the ions travel along their path, the convection electric field has a force component which causes an inward drift acting throughout the ion's bounce path. This force displaces different energy ions differently. High energy ions are fast and spend less time under the influence of this force. Their path is therefore slightly different from the magnetic field line that the source was located on. Because of this difference, any conjugate parallel electric field in the opposite hemisphere will not reflect the ions from the source hemisphere; and thus, the ions may precipitate at low altitudes. The convection field displaces lower energy particles even farther from the source field line, because they are traveling slower and they spend more time under the influence of this force. Thus, they are convected inward more than higher energy ions and will be observed at lower latitudes; ie., dispersion in energy occurs perpendicular to the magnetic field in addition to the time-of-flight effects along a field line.

Figure 6b shows a schematic spectrogram produced by the convective dispersion hypothesis (multiple bounces are not shown for clarity). This pattern is a spatial effect, and thus, should be dependent on satellite direction. This figure assumes that an inverted "V" is occurring in both hemispheres, simultaneously on the same magnetic field line; although the convective dispersion hypothesis does not require conjugate aurora. The time-of-flight effect would produce a series of angular "V"'s in a spin-time spectrogram. This morphology is observed by both HAPI and LAPI.

Model Differences: Time-of-Flight
with Respect to Convective Dispersion

The apparent difference between these models arises in part because they emphasize two

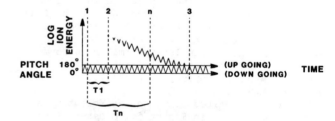

**ENERGY-TIME SPECTROGRAM
(SPINNING PITCH ANGLE)**

Fig. 5c. Pseudo-spectrogram from the time-of-flight hypothesis. Here the observation point is at intermediate magnetic latitudes and the detectors are spinning through the magnetic field. The satellite is at 1 when source injection occurs. At some time later, T_1, the ions will begin to envelope the spacecraft. The satellite will continue to detect ions, n, until the ions have passed the satellite or the satellite passes out of the populated flux tube and particles are no longer detected.

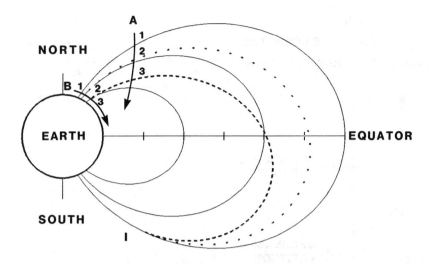

Fig. 6a. Convective dispersion hypothesis. Schematic view in a meridian plane of band formation with the convective dispersion hypothesis. The source location is labeled by I, and the approximate paths of DE-1 and DE-2 are shown.

different aspects of particle motion in the magnetosphere. The time-of-flight model, driven by observations made at nearly constant L-values, stresses the temporal evolution of the ion distribution. The convective dispersion model, driven by observations at low and mid-altitudes where the satellites are moving approximately normal to L-shells, considers the spatial dispersion produced by cross field drift. The real differences between the two models concerns the source location, and the spatial and/or temporal distributions of the ion source. In the convective dispersion model the source is taken to be a latitudinally restricted, longitudinally extended, and temporally stable feature. No a priori assumption of source location is made in the time-of-flight model; rather, the data allows the source location to be determined. In the time-of-flight model, the convection electric field is not strong enough to produce significant motion of ions across magnetic field lines, and the resulting dispersion is viewed as a result of a temporally and/or spatially localized source. The fact that a given dispersion is observed at mid-altitudes by HAPI and at low altitudes by LAPI at a slightly different time suggests that the source is steady state, and thus, convective.

Conclusion

As can be seen from Figure 4, bands are seen above 50° IL in the auroral zone, but do not occur over the pole. Bands are more frequently observed during main storm phase rather than during recovery storm phase. It is possible that this finding may be due to an increase in the ion "inverted-V" source during the main phase. In addition, during the main phase, storm compression is known to deplete shielding, which permits the convection electric field to penetrate to low latitudes [Wolf et al., 1982]. Night side bands are detected more frequently than day side bands, and are typically observed at lower invariant latitudes (on the average ≈5°).

Band-like structures in ion energy-time spectrograms can be produced as a consequence of the natural motion of heavy ions in the magnetosphere. Two models have been described as explanations of the observed band features. The

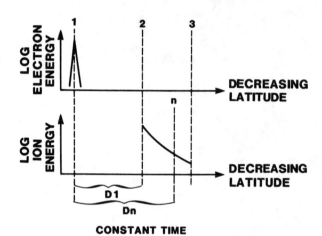

ENERGY LATITUDE SPECTROGRAM

Fig. 6b. Convective dispersion hypothesis. An idealized spectrogram of precipitating ions for the convective dispersion hypothesis. The satellite encounters 1 where source injection occurs. At some distance later, D_1, the satellite will pass into the ions from that source. The satellite will continue to detect ions, n, until it passes through that precipitation and particles are no longer detected. Note that multiple bands are not shown for clarity of the pattern that will be produced.

time-of-flight hypothesis considers primarily ion dispersion along a magnetic field line from a source located anywhere along that magnetic field line and does not address the production of latitude dispersion at low altitudes. The convective dispersion hypothesis addresses the possibility of convective ion drift perpendicular to magnetic field lines (due to the observed convection electric field) as the ions travel parallel to the magnetic field. In a time-of-flight dispersion, the energy decreases as time increases; in convective dispersion, the energy decreases as latitude decreases (if the flow is toward the equator).

The convective dispersion model best explains the band pattern observed here because the bands appear to be stable and have been observed with the same energy-latitude dependence no matter which direction the satellites are moving. Variations within this pattern can be easily accounted for by a slight change in the convection electric field which is also observed. The time-of-flight model describes a similar dispersion pattern which has been used to explain oxygen ion signatures seen by EICS. There is little or no doubt that the true motion of heavy ions in the magnetosphere involves both dispersion along the field lines and convection across the magnetic field. Both processes must be included in any attempt to describe the circulation of ions injected into the magnetosphere from any source.

Acknowledgements. We are indebted to David Klumpar for his many comments and criticisms of this paper. We would also like to acknowledge M. Sugiura for providing the Dst values used in Figure 3. This research was supported by the NASA Graduate Student Researchers Program and NASA grant NGR-44-006-137 at Rice University; and AFGL contract FY7121-84-N-0006, NASA contracts NAS5-28711 and NAS5-28712 at Southwest Research Institute.

Thanks to J. M. Quinn and J. R. Sharber for their comments on this paper.

References

Burch, J. L., J. D. Winningham, V. A. Blevins, N. Eaker, W. C. Gibson, and R. A. Hoffman, High-Altitude Plasma Instrument for Dynamics Explorer-A, Space Sci. Instrum., 5, 455-463, 1981.

Burch, J. L., P. H. Reiff, R. A. Heelis, J. D. Winningham, W. B. Hanson, C. Gurgiolo, J. D. Menietti, R. A. Hoffman, and J. N. Barfield, Plasma injection and transport in the mid-altitude polar cusp, Geophy. Res. Lett., 9, 921-924, 1982.

Frahm, R. A., Cusp particle detection and ion source oscillations, M.S. thesis, Rice Univ., Houston, TX, 1984.

Frahm, R. A., Banded Ion Morphology, Ph.D. thesis, Rice Univ., Houston, TX, 1986.

Hoffman, R. A., G. D. Hogan, and R. C. Maehl, Dynamics Explorer spacecraft and ground operations systems, Space Sci. Instrum., 5, 349-367, 1981.

Klumpar, D. M., W. K. Peterson, E. G. Shelley, and J. M. Quinn, Localized magnetospheric ion injection outside the cusp, EOS Trans. AGU, 64, 297, 1983.

Mauk, B. H., and C. E. McIlwain, ATS-6 UCSD Auroral Particles Experiment, IEEE Trans. Aerosp. Electron. Syst., 11, 1125-1130, 1975.

Quinn, J. M., and C. E. McIlwain, Bouncing ion clusters in the earth's magnetosphere, J. Geophys. Res., 84, 7365-7370, 1979.

Quinn, J. M., and D. J. Southwood, Observations of parallel ion energization in the equatorial region, J. Geophys. Res., 87, 10536-10540, 1982.

Reiff, P. H., H. L. Collin, E. G. Shelley, J. L. Burch, and J. D. Winningham, Heating of upflowing ionospheric ions on auroral field lines, this volume.

Shelley, E. G., D. A. Simpson, T. C. Sanders, E. Hertzberg, H. Balsiger, and A. Ghielmetti, The Energetic Ion Composition Spectrometer (EICS) for the Dynamics Explorer-A, Space Sci. Instrum., 5, 443-454, 1981.

Winningham, J. D., J. L. Burch, N. Eaker, V. A. Blevins, and R. A. Hoffman, The Low Altitude Plasma Instrument (LAPI), Space Sci. Instrum., 5, 465-475, 1981.

Winningham, J. D., J. L. Burch, and R. A. Frahm, Bands of ions and angular V's: a conjugate manifestation of ionospheric ion acceleration, J. Geophys. Res., 89, 1749-1754, 1984.

Wolf, R. A., M. Harel, R. W. Spiro, G.-H. Voigt, P. H. Reiff, and C.-K. Chen, Computer simulation of inner magnetospheric dynamics for the magnetic storm of July 29, 1977, J. Geophys. Res., 87, 5949-5962, 1982.

THE CONDUCTANCE OF AURORAL MAGNETIC FIELD LINES

D. R. Weimer

Regis College Research Center, 235 Wellesley St., Weston, MA 02193

D. A. Gurnett and C. K. Goertz

Department of Physics and Astronomy, University of Iowa, Iowa City, IA 52242

Abstract. Recent results from the Dynamics Explorer satellites have indicated that in the auroral zone a linear relationship exists between the field aligned current density and the potential drop parallel to the magnetic field lines. Evidence for this "Ohm's law" relationship was found in the mapping of perpendicular electric fields and field-aligned currents between high and low altitudes. The mapping depends on the perpendicular wavelength of the electric field variations. A scale length in the mapping formula is determined by the ratio of the parallel field line conductance and the ionospheric Pedersen conductance. The wavelength and the conductivity ratio also control the relationship between the perpendicular electric and magnetic fields at high altitudes.

We show here that at the short-wavelength limit the ionospheric conductivity is no longer important in the relationship between the north-south electric field and the east-west magnetic field at high altitudes (i.e., above the parallel potential drop). At the short-wavelength limit the relationship takes on a simple form: The integral of the perpendicular electric field results in a potential profile which, according to the linear theory, is proportional to the current density. Assuming that the currents are in the form of "infinite sheets" orientated east-west, the second integral of the electric field is proportional to the magnetic field.

High time-resolution data from the DE-1 satellite are shown here for two events with very large electric fields which reversed directions within a short distance. The results agree very well with the linear theory. The field line conductance is determined to be of the order of 10^{-9} mho/m^2. The same conductance appears to be valid for both upward and downward currents. Ions are accelerated from the ionosphere to magnetosphere by the potential drops in regions of upward current.

Introduction

A considerable body of scientific evidence exists which indicates that there are electric fields parallel to the magnetic field lines in the auroral zone. These parallel electric fields accelerate electrons into the ionosphere to form the aurora, and they also accelerate ions in the opposite direction. It is not so well known exactly how the parallel electric fields are formed. Several mechanisms have been suggested, such as the magnetic mirror force, double layers, anomalous resistivity, and hydromagnetic waves. These theories are discussed in the reviews by Shawhan et al. [1978] and Stern [1983].

In order to determine which mechanism is responsible for the parallel electric fields it is useful to study the current-voltage relationship along the magnetic field lines. Lyons et al. [1979] and Menietti and Burch [1981] had found evidence for a linear "Ohm's law" relationship between the field-aligned current and potential drop, based on measurements of precipitating electron energy fluxes at the ionosphere. The "Ohm's law" relationship was also verified in a recent study of the perpendicular electric fields measured nearly simultaneously by the two Dynamics Explorer spacecraft near magnetic conjunctions in the auroral zone [Weimer et al., 1985]. Parallel field line conductances of the order of 10^{-8} mho/m^2 were measured.

The study by Weimer et al. [1985] had been limited to electric field and current structures with spatial widths greater than 20 km at the base of the field lines. Naturally, the question arises about whether or not the same linear relationship still holds for very narrow current sheets. The 20 km limit was imposed by the measurement of the electric field on the high altitude DE-1 satellite. The electric field in the orbit plane of DE-1 is measured with a single, rotating double-probe. A static electric field appears in the "raw" data as a sine wave with a six second period. In the previous study the electric fields were derived from a least square fit of the raw data to a sine wave. This procedure works extremely well as long as the electric fields change on a time scale greater than the satellite spin period, but rapid, short-wavelength variations are filtered out. In order to obtain the electric field with a better spatial resolution, the sine wave modulation can be removed from the data in a more direct manner. In this paper we will show electric and magnetic field measurements with the highest possible resolution. The data shows that an "Ohm's law" is still valid for current structures just a few km across.

Theory

In a paper by Smiddy et al. [1980] it is shown that the height-integrated ionospheric Pedersen conductivity (Σ_p) is an important factor in the relationship between the field-aligned currents and the perpendicular electric fields. Just above the ionosphere the relationship is:

$$j_\parallel = -\frac{dE_x}{dx} \Sigma_p \qquad (1)$$

Here we use a coordinate system in which the Z axis is upward along the magnetic field line, X is southward, and Y is eastward. The field-aligned currents are not measured directly but are inferred from the derivative of the east-west magnetic field component measured on the satellite while moving in the north-south direction through sheets of current, which are assumed to be very long in the east-west and up-down directions. The magnetic and electric fields are related to each other according to

$$\Delta B_y = -\mu_o \Sigma_p E_x \qquad (2)$$

where ΔB_y is the difference between the measured magnetic field and the earth's dipole field. Equation (2) has been confirmed by a number of measurements of magnetic and electric fields just above the ionosphere [Smiddy et al. 1980; Sugiura, 1984]. At higher altitudes Equation (2) may not be valid due to the possible existence of an electric potential drop parallel to the magnetic field ($V_\parallel$). The "Ohm's law" we are investigating assumes that the field-aligned current density at the ionosphere and the parallel potential drop are related by

$$j_\parallel = -a V_\parallel \qquad (3)$$

where "a" is the field line conductance. The sign convention is such that the current is positive (upward) when the potential drop from low to high altitude is negative. The physical justification for Equation (3) has been discussed by Knight [1973], Lyons et al. [1979], Fridman and Lemaire [1980], and Chiu et al. [1981].

It is shown by Weimer et al. [1985] that with the definition of a constant inverse scale length or "critical wavenumber",

$$\frac{a}{\Sigma_p} = k_o^2 \qquad (4)$$

the Fourier transforms of the high-altitude quantities $j_\parallel$, B_y, and E_x have a relationship which depends on their spatial wavenumber, k:

$$\tilde{j}_\parallel = \frac{-ik}{(k/k_o)^2 + 1} \Sigma_p \tilde{E}_x \qquad (5)$$

$$\mu_o \tilde{j}_\parallel = ik \tilde{B}_y \qquad (6)$$

The tildas indicates that the quantities are transformed from a spatial domain to a wavenumber domain. The spatial variations are presumed to be in the north-south (x) direction.

With a field line conductance of 10^{-8} mho/m^2 and a Pedersen conductance of 5 to 10 mho, the inverse of the critical wavenumber, k_o^{-1}, will be in the range of 22 to 32 km. The wavelength $\lambda_o = 2\pi/k_o$ is 140 to 200 km. At the limit of a very small wavelength and large wavenumber, where $(k/k_o)^2 \gg 1$, Equation (5) can be simplified to

$$\tilde{j}_\parallel = -(1/k) a \tilde{E}_x \qquad (7)$$

From Equation (3) we see that

$$-ik\tilde{V}_\parallel = \tilde{E}_x \qquad (8a)$$

or in real space,

$$-\frac{dV_\parallel}{dx} = E_x \qquad (8b)$$

where E_x is the measured electric field at the satellite altitude. Therefore, at very small wavelengths the ionospheric conductivity does not influence the high altitude electric field, and there is a very simple relationship between the north-south electric field and the parallel potential drop. By integrating E_x one can obtain a "potential profile" ($V_\parallel$ as a function of x), which should be proportional to the current density if the "Ohm's law" of equation (3) is correct. Integration of E_x a second time (with multiplication by 'a' and μ_o) should result in ΔB_y. This prediction can be tested by comparing E_x and ΔB_y in cases where the short-wavelength variations are large in comparison to the long-wavelength variations.

Observations

An example of high-resolution electric field data from the 200 meter double probe on DE-1 [Shawhan et al., 1981] is shown in the top panel

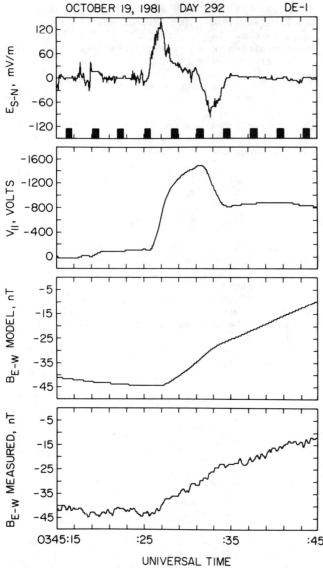

Fig. 1. Top graph is north-south electric field measured in the orbit plane of DE-1 on day 292 of 1981. The dark blocks above the time axis indicate where data gaps have been filled in. The second graph shows the potential obtained by integrating this electric field, and the third graph shows a model of the east-west magnetic field obtained with a second integration. At the bottom is shown the measured magnetic field.

of Figure 1. The "raw" electric field (which is spin modulated) is digitized at a rate of 16 samples per second. The electric field in Figure 1 has had the spin modulation removed by a technique in which each measurement is divided by the sine of the angle between the double-probe antenna and the magnetic field in the spin plane. This works well when the "spin phase angle" is large, but when the antenna is nearly parallel to the magnetic field there can be problems: A small error in the phase angle results in a distorted electric field, which goes to infinity when the angle is zero. To eliminate this effect gaps are introduced in the de-spun data whenever the magnitude of the phase angle is less than 30 degrees. The subsequent steps in the data analysis, however, require a continuous electric field, so the gaps are filled in with values which smoothly connect the data on both sides of the gaps. The resulting plot shows a large amplitude electric field which points southward for 6.0 seconds then reverses to a northward direction for 3.5 second. At the time of the measurement the velocity component of DE-1 perpendicular to the magnetic field was 4.77 km/sec (southward), thus the total width of this structure is 45 km. These fields were measured at an altitude of 10,500 km and invariant latitude of 69.4°, so the 45 km "maps" to a width of 9.5 km at the base of the magnetic field line.

Integration of the electric field plot in Figure 1 (also multiplied by the appropriate time and velocity constants) results in the electric potential function which is shown in the second plot from the top in the same figure. Multiplication of this potential by a conductance results in a value for the local current density; integration a second time results in the east-west deviation of the magnetic field shown in the third plot. The bottom plot in Figure 1 shows the actual value of the magnetic field measured by the magnetometer on DE-1 [Farthing et al., 1981].

As can be seen there is a very good agreement between the integrated and measured values. To obtain this agreement a trial-and-error adjustment of two parameters was necessary, namely a voltage offset and the field-line conductance. The potential in Figure 1 is shown to start at zero. A constant voltage may be added to compensate for the actual initial value.

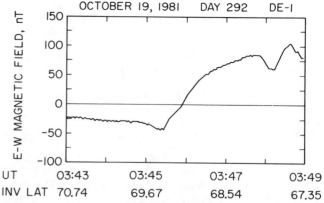

Fig. 2. East-west magnetic field component from the magnetometer on DE-1 from 3:43 to 3:49 UT on day 292, 1981. The earth's dipole field has been subtracted from the measured values.

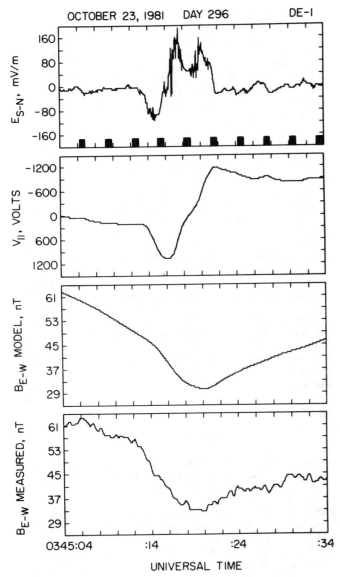

the following upward current. A narrow but large magnitude upward current sheet is located right on the boundary. This peak in the upward current occurs precisely at the point where the parallel potential is most negative.

The addition of a constant to the parallel potential corresponds to the addition of a constant slope in the model magnetic field. The magnitudes of the relative changes in the magnetic field are determined by the value chosen for the field line conductance. In this case a conductance of $4 \cdot 10^{-10}$ mho/m^2 yielded the best match between the model and measured magnetic field. At the point where the potential drop is -1300 volts the local current density is $5.2 \cdot 10^{-7}$ A/m^2. Due to convergence of the magnetic flux tubes both the current density and the east-west magnetic field increase towards lower altitudes. The "mapping" factor for both of these quantities from a high-altitude location to the base of the magnetic field line is $L^{3/2} \cos^3 \lambda_m$, where λ_m is the magnetic latitude. For the present case the mapping factor is 4.32, so the "normalized" field line conductance is $1.7 \cdot 10^{-9}$ mho/m^2. This is the ratio between the current density at the base of the field line ($2.2 \cdot 10^{-6}$ A/m^2) and the total potential drop along the field line ($V_\parallel$=1.3 kV).

Another case is shown in Figure 3. This data is from day 296 of 1981. The format is the same as in Figure 1. This event has some similarities to the previous case, yet there are important differences. Figure 4 shows that, like before, the electric field "spike" occured on the southward edge of a region of downward current. But this time the downward current is much larger in magnitude and confined to a relatively narrow region, embedded within a larger region of upward current. Whereas in the previous case most of the magnetic field variations were due to upward currents, in this second example the downward currents are equally important.

As before, different parameters were tested in order to find a model magnetic field which best

Fig. 3. Electric and magnetic fields measured with DE-1 on day 296, 1981. The format is the same as in Figure 1. The third graph from the top was obtained by twice integrating the measured electric field, while the measured magnetic field is shown at the bottom.

An offset of +200 volts yielded the best result. This positive potential at the high altitude location corresponds to a downward current. The justification for this assumption is given in Figure 2, which shows the measured east-west magnetic field for a much longer period of time. Prior to the electric field spike at 3:45:25 UT there was a gradual negative (westward) slope in the magnetic field, which corresponds to a downward current. The large electric field occurs at the boundary between the downward and

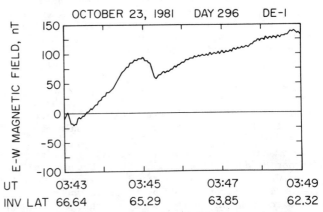

Fig. 4. East-west magnetic field measured with DE-1 from 3:43 to 3:49 UT on day 296, 1981. The geomagnetic field has been removed from the data.

matched the measured field. In this case the most reasonable results were obtained with the addition of a +3 mV/m offset to the electric field before doing the first integration. This compensated for a negative, large scale convection electric field which would be linearly proportional to ΔB according to Equation 2. A +500 volt offset was added to the potential before doing the second integration. This is justifiable on the basis that the region of downward current had been penetrated a few seconds before the starting time of Figure 3. As before, a conductance of $4 \cdot 10^{-10}$ mho/m^2 was used to generate the model magnetic field. The DE-1 satellite was at an altitude of 9000 km and an invariant latitude of 65.1°, so the current projection factor to the field line base is 3.75. This results in a normalized field line conductance of $1.5 \cdot 10^{-9}$ mho/m^2.

Discussion

The evidence which has been shown here provides a convincing case for the validity of Equation 3 for current structures with small spatial widths. The linear relationship between current density and potential is the simplest and, as far as we know, only explanation for the magnetic field signatures which are observed concurrent with large-amplitude electric fields. In general, the large amplitude electric fields occur on the boundary between upward and downward currents, where the high-altitude potential must reverse signs within a short distance.

A certain amount of adjustment of the starting potential and field-aligned conductance was required to get the good agreements shown here. But the adjustments were within reason and relatively minor. The second integral would not have given the observed agreement if the "Ohm's law" was not a good approximation. Indeed, given the gaps in the measured electric field data it is amazing that the integrated values come as close as they do to matching the measured magnetic fields, since the integrations cause errors to accumulate forward in time.

The field line conductances measured with this technique generally are close to 10^{-9} mho/m^2. The question remains: which of the proposed mechanisms for creating a parallel potential drop along magnetic field lines can produce a conductance of this value? It is important to note that the conductance is the same for both upward and downward currents. Although a better model for the data in Figure 3 maybe could be obtained by using different conductances for the upward and downward current regions, the values can not be much different. This fact tends to eliminate the magnetic mirror force as a cause of the parallel potential drop, since it works in one direction only. It also seems that the kinetic theory yields smaller values for 'a' [Fridman and Lemaire, 1980] although the difference may be within limits of the model. In both examples shown here intense, short-wavelength "turbulence" in the electric field is present where the currents are most intense, suggesting that large amplitude plasma waves may be present. Thus anomalous resistivity [Lysak and Dum, 1983] may be important. We emphasize again that the results presented here are restricted to the domain of small perpendicular scale lengths.

The data suggests that there is a current generator in the magnetosphere above the observation point--this current generator may have an associated electric field, depending on the source impedance [Lysak, 1985]. The Kelvin Helmholtz instability may be the source of these currents (P. F. Bythrow, unpublished manuscript, 1985). It appears that the currents produce the parallel potential drops, perhaps by the formation of double layers or anomalous resistivity. We note that the "Ohm's law" also suggests that $\vec{J} \cdot \vec{E} > 0$ above the ionosphere and that electromagnetic energy is dissipated through ion acceleration in the magnetosphere.

Acknowledgments. The authors thank M. Sugiura for providing data from the magnetometer instrument on DE-1. The research at the University of Iowa was supported by NASA through grant NAG5-310 from Goddard Space Flight Center and grants NGL-16-001-043 and NGL-16-001-002 from NASA Headquarters. The work at Regis College was supported by USAF contract F19628-C-84-0126.

References

Chiu, Y. T., A. L. Newman, and J. M. Cornwall, On the structure and mapping of auroral electrostatic potentials, J. Geophys. Res., 86, 10029-10037, 1981.

Farthing, W. H., M. Sugiura, B. G. Ledley, and L. J. Cahill, Magneytic field observations on DE-A and -B, Space Sci. Instrum., 5, 551-560, 1981.

Fridman, M. and J. Lemaire, Relationship between auroral electron fluxes and field aligned electric potential differences, J. Geophys. Res., 85, 664-670, 1980.

Knight, S., Parallel electric fields, Planet. Space Sci., 21, 741-750, 1973.

Lyons, L. R., D. S. Evans, and R. Lundin, An observed relation between magnetic field-aligned electric fields and downward electron energy fluxes in the vicinity of auroral forms, J. Geophys. Res., 84, 457-461, 1979.

Lysak, R. L. and C. T. Dum, Dynamics of magnetosphere-ionosphere coupling including turbulent transport, J. Geophys. Res., 88, 365-380, 1983.

Lysak, R. L., Auroral electrodynamics with current and voltage generators, J. Geophys. Res., 90, 4178-4190, 1985.

Menietti, J. D., and J. L. Burch, A satellite investigation of energy flux and inferred potential drop in auroral electron energy spectra, Geophys. Res. Lett., 8, 1095-1098, 1981.

Shawhan, S. D., C.-G. Falthammer, and L. P. Block, On the nature of large auroral zone electric fields at 1-R_E altitude, J. Geophys. Res., 83, 1049-1054, 1978.

Shawhan, S. D., D. A. Gurnett, D. A. Odem, R. A. Helliwell, and C. G. Park, The plasma wave instrument and quasi-static electric field instrument (PWI) for Dynamics Explorer-A, Space Sci. Instrum., 5, 535-550, 1981.

Smiddy, M., W. J. Burke, M. C. Kelley, N. A. Saflekos, M. S. Gussenhoven, D. A. Hardy, and F. J. Rich, Effects of high-latitude conductivity on observed convection electric fields and Birkeland currents, J. Geophys. Res., 85, 6811-6817, 1980.

Stern, D. P., Electric currents and voltage drops along auroral field lines, Space Sci. Rev., 34, 317-325, 1983.

Sugiura, M., A fundamental magnetosphere-ionosphere coupling mode involving field-aligned currents as deduced from DE-2 observations, Geophys. Res. Lett., 11, 877-880, 1984.

Weimer, D. R., C. K. Goertz, D. A. Gurnett, N. C. Maynard, and J. L. Burch, Auroral zone electric fields from DE-1 and -2 at magnetic conjunctions, J. Geophys. Res., 90, 7479-7494, 1985.

Section III. PLASMA SHEET AND BOUNDARY LAYER PROCESSES

VELOCITY DISTRIBUTIONS OF ION BEAMS IN THE PLASMA SHEET BOUNDARY LAYER

T. E. Eastman, R. J. DeCoster, and L. A. Frank

Department of Physics and Astronomy, The University of Iowa, Iowa City, Iowa 52242

Abstract. Field-aligned ion beams commonly occur in the plasma sheet boundary layer. With sufficient energy and angular resolution, plasma measurements of ion beam velocity distributions there can be used as a diagnostic tool to evaluate ion acceleration processes. Model fits to data are shown for adiabatic deformation of a flowing Maxwellian, acceleration due to field-aligned potentials, and current-sheet acceleration. Isodensity contours of the velocity distributions typically show a closer spacing within the low-energy portion of the ion beams that is not consistent with adiabatic deformation alone acting upon an initially isotropic distribution function. Detailed comparisons of acceleration models with the overall shapes of the observed beam distributions indicate that the ion beams have been subjected to some potential drop which may be field-aligned (DeCoster and Frank, 1979) or perhaps cross-tail as for the current-sheet acceleration process (Lyons and Speiser, 1982). Measurements of velocity dispersion effects have also proven to be a useful diagnostic tool although propagation effects are difficult to separate from signatures of the initial acceleration process. Ion acceleration processes in the magnetotail are primarily associated with the plasma sheet boundary layer. This observational result has important implications for magnetotail dynamics.

Introduction

Figure 1 illustrates various transport regions of the earth's magnetosphere. The magnetospheric boundary layer, which is adjacent to and earthward of the magnetopause, projects down along field lines to the particle cusp region. The high-latitude portion of the aurora at all other local times is the low-altitude projection of the plasma sheet boundary layer (Lyons and Evans, 1984). Recent studies have demonstrated the importance of these boundary layers for plasma transport and acceleration within the magnetosphere (Lundin and Dubinin, 1985; Eastman et al., 1985).

In this review, we will present some recent satellite observations obtained within the transition region of the plasma sheet from the near-isotropic distributions of the central plasma sheet to the very low density conditions typical of the lobe region which often includes low-energy ion streaming (Sharp et al., 1981). This transition region is most clearly distinguished by high-speed ion beams such as those described by Williams (1981). Figure 2 shows the contrast in ion velocity distributions that is typical for this transition region, or plasma sheet boundary layer, and for the central plasma sheet and lobe regions. The two regions of the plasma sheet are most clearly distinguished by the ratio of drift to thermal speed. Within the central plasma sheet this ratio is always small whereas it is typically large within the plasma sheet boundary layer for one or more components of the velocity distributions. Another distinctive feature of the plasma sheet boundary layer is that the several components of the ion velocity distributions shown in Figure 2 commonly occur together, frequently including counter-streaming high-speed ion beams. These multi-component distributions limit the utility of working solely with plasma moments parameters because velocity and temperature estimates thus obtained depend on what component(s) are referred to. The central plasma sheet usually has only one high-temperature ion component which can be fitted approximately by a Maxwellian with a high-energy (> 50 keV) tail (Sarris et al., 1981). Thus it is appropriate to characterize this plasma by its various moments. In contrast, when considering the multiple components of the ion velocity distributions in the plasma sheet boundary layer, misleading moments parameters will frequently be calculated, e.g., high "temperatures" will be obtained especially in the presence of counter-streaming ion beams. In terms of density, velocity and temperature, the plasma sheet boundary layer can, in this manner, closely mimic the central plasma sheet although their velocity distributions may be significantly different (Eastman et al., 1984).

Recently deployed plasma instruments are beginning to provide the energy and angular resolution needed to use the profiles of beam

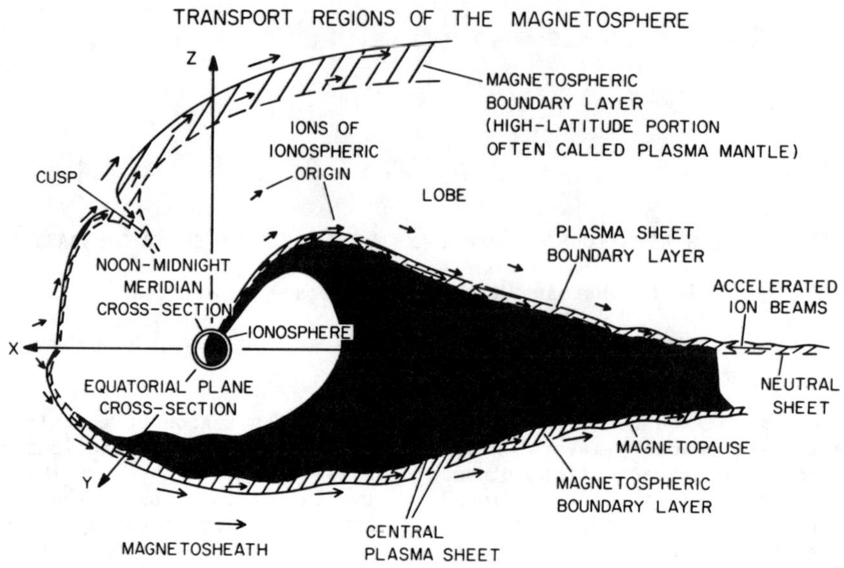

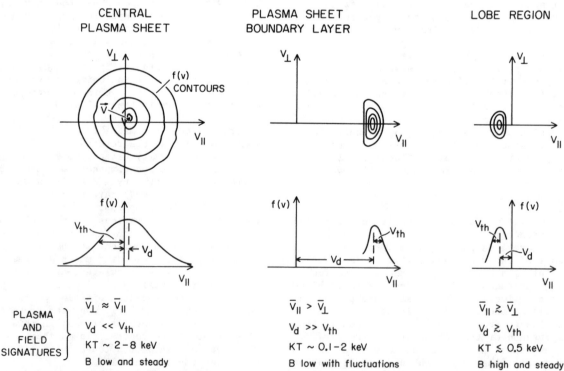

Fig. 1. Plasma regimes of the earth's magnetosphere as identified on the basis of plasma observations. The arrows denote typical bulk flow velocities which have their largest gradients near the magnetospheric boundary layers. In the equatorial plane cross-section, the tail size does not generally decrease with increasing distance as implied by this figure.

Fig. 2. Typical ion velocity distributions are shown as observed within the three primary regions of the earth's magnetotail (not including the magnetospheric boundary layer). The upper row of figures show contour plots of phase-space density in $V_\perp - V_\parallel$ space and the bottom row of figures show the corresponding cuts of these distributions taken at $V_\perp = 0$.

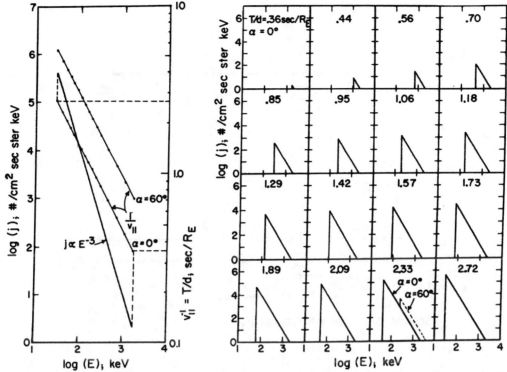

Fig. 3. Model calculations are shown here which illustrate the appearance and evolution of 'peaked' energy spectra observed by the ISEE 1 medium-energy particle instrument. A representative E^{-3} source was used and the left panel shows the source spectra with normalized propagation time curves, $v_\parallel^{-1} = T/d$. The sixteen panels on the right show the time evolution in calculated instrument response for ions at a pitch angle = 0 based on propagation from an E^{-3} source an arbitrary distance, d, away. Actual arrival times for these spectra are equal to T/d multiplied by the appropriate source distance, d. Williams (1981) used this technique, along with detector sampling characteristic and variable source parameters, to explain observed ion beam characterics observed in the plasma sheet boundary layer. (from Williams, 1981)

components within the plasma sheet boundary layer as a diagnostic for assessing source regions and acceleration processes. One early success in this endeavor was accomplished by Williams (1981) using data from the ISEE medium-energy particle experiment. Ion beams were observed to be streaming, both simultaneously and in time sequence, at various pitch angles and especially parallel and antiparallel to B. Figure 3 shows part of the modelling performed by Williams using only single-particle motion conserving the first adiabatic invariant $\mu = E_\perp/B$. The appearance and evolution of "peaked" energy spectra are shown based on using propagation effects from a representative E^{-3} source. At later times, ions of successively lower energies (or higher pitch angle) will arrive at the spacecraft. One event studied in detail clearly demonstrated that the ISEE observations could be explained by the assumption of a temporally variable source located ~ 80-100 Re from the earth in the antisolar direction. The frequent absence of earthward-directed or tailward-directed ion beam components could be explained by expected particle drift motions during their propagation to the near-earth mirror point and back to the vicinity 0f the spacecraft. This modeling shows how the evolution of $f(\vec{v})$ depends on spatial location and time variations of the source and spacecraft location with respect to the source. In this review, we will not attempt to provide a comprehensive summary of the many research efforts now in progress that focus on the plasma sheet boundary layer. The number of such efforts in progress probably outnumber the total number published to date. The basic theme of this review is that the measurement of particle velocity distributions within the plasma sheet boundary layer provides important diagnostic information on source and propagation characteristics and, most acceleration processes.

A New Diagnostic Tool

The measurement of particle velocity distributions, especially in the inner

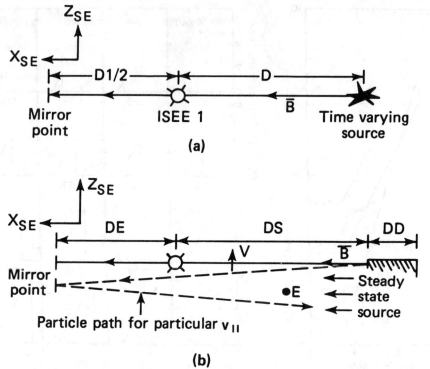

Fig. 4. Schematic diagrams of source geometry used in modeling ISEE medium energy particle instrument responses for (a) time-varying and (b) steady-states sources. (from Williams and Speiser, 1984).

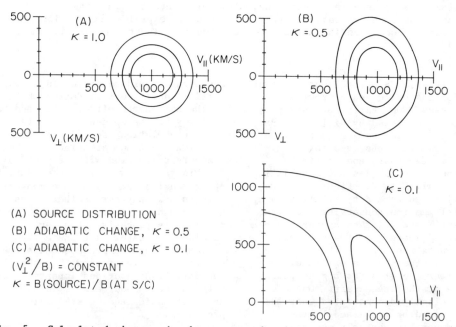

Fig. 5. Calculated changes in phase-space density contours using adiabatic deformation applied to an initial flowing Maxwellian distribution. Note how the crossings of the $V_\parallel$ axis remain constant as the distribution deforms with increasing values of the adiabatic change parameter.

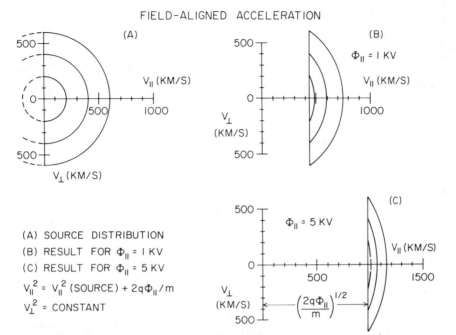

Fig. 6. Calculated changes in phase-space density contours using field-aligned acceleration applied to an initial stationary Maxwellian distribution. Note how the low-energy side of the distribution is truncated at a minimum speed associated with the applied potential.

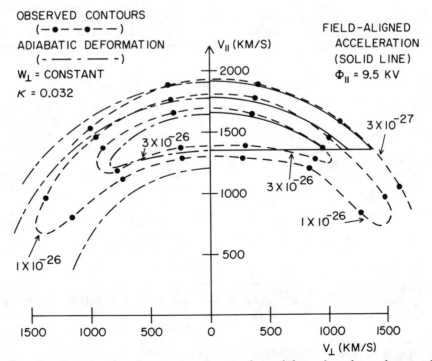

Fig. 7. Phase-space density contours are plotted here based on observations by the University of Iowa LEPEDEA plasma instrument on board ISEE 1. These contours are for an earthward-travelling high-speed ion beam observed in the plasma sheet boundary layer. Optimal model comparisons with field-aligned acceleration and adiabatic deformation are shown.

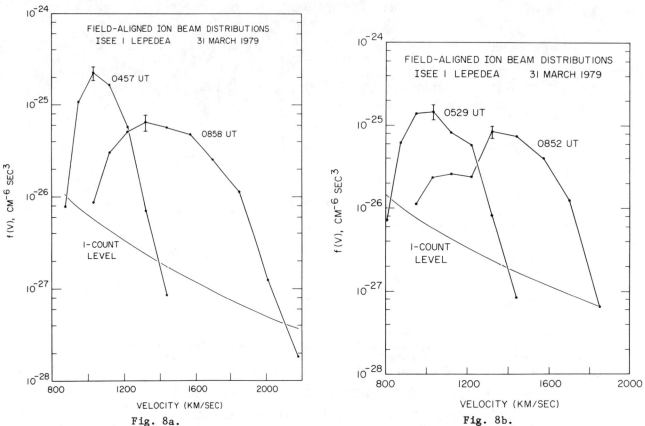

Fig. 8. a, b. Energy spectra for four field-aligned ion beams are shown which were sampled by the ISEE 1 LEPEDEA on March 31, 1979. These spectral samples resolve the low-energy portion of the ion beams and show that it typically has a steeper slope than the high-energy portion of the beams.

magnetosphere, has often been used before as a diagnostic tool for inferring source and propagation characteristics. For example, equatorial ptich-angle distributions have been used to test the role of diffusion for ring current ions (Williams and Lyons, 1974). We are now beginning to use measurements of full three-dimensional velocity distributions. Combined with the modelling of source distributions and propagation characteristics, various acceleration processes can be tested. This is an important new diagnostic tool for determining the dominant acceleration processes in the earth's magnetotail.

In addition to the propagation effects mentioned in the introduction, it is also important to distinguish time-varying sources from steady-state sources. Williams and Speiser (1984) have used their ISEE data to resolve these two distinct possibilities. In Figure 4, the geometry for these two source types is illustrated. Particle velocities, pitch angle, source distance and $\vec{E} \times \vec{B}$ drift all enter into the calculation which becomes relatively complicated even for single-particle motions.

Closely analogous to many observations of ions > 24 keV in the plasma sheet boundary layer, the models for both source types show the evolution from monoenergetic field-aligned ion beams to smoke rings in both the earthward-directed and tailward-directed beam components. Williams and Speiser find that single instruments from single spacecraft are not able to uniquely distinguish space and time variations. Further, adding complex boundary features such as parallel and transverse waves makes this distinction of space and time variations difficult even for two-point observations.

The smoke-ring effect that is so commonly observed at higher energies is usually not observed in the plasma energy range up to about 20 keV based on high-time resolution angular scans using the University of Iowa LEPEDEA instrument on ISEE 1. There are also many events during which the ion beams maintain their identity for 1-5 minutes or more. Samples of such cases that are stable for more than one instrument cycle avoid time aliasing of the data and allow for detailed measurements of the three-dimensional velocity distributions. These

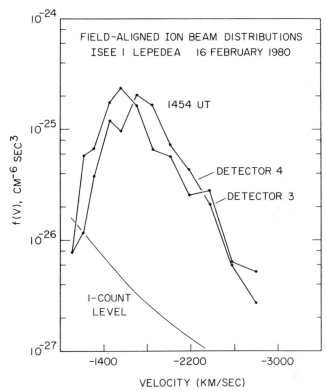

Fig. 8c. Energy spectra for a tailward-directed high-speed ion beam observed on February 16, 1980. The low-energy portion of the beam is again steeper than the high-energy portion. Samples of the beam are shown from both detector 3 (whose view angle is centered 35 degrees above the ecliptic) and the equatorial plane detector 4.

measurements provide additional diagnostic informtaion concerning possible acceleration processes.

Ion Beam Acceleration Processes

Pre-acceleration + adiabatic deformation:

For a flowing Maxwellian distribution created within some nonadiabatic acceleration region (involving reconnection ?) in the distant magnetotail, adiabatic deformation will cause the source ions to increase their pitch angle so as to preserve the first adiabatic invariant $\mu = E_\perp/B$. This deformation process is shown in Figure 5 for two values of the adaibatic change factor, $K = B$ (source)/B (at the spacecraft).

Field-aligned acceleration:

An alternative acceleration process is field-aligned acceleration where the source ions are subjected to a potential drop Φ. As shown in Figure 6, such direct acceleration results in a truncated low-energy cutoff at a speed of $(2q\Phi/m)^{1/2}$. Phase-space density contours for the remaining distribution are segments of concentric circles centered on the $V_\perp, V_\parallel = 0,0$ origin.

Model comparisons:

Model comparisons for adiabatic deformation and field-aligned acceleration have been made by DeCoster and Frank (1979) and Birn et al. (1981). An example of such a comparison is presented in Figure 7 which shows phase-space density contours for a sunward-directed ion beam sampled by the ISEE-1 LEPEDEA instrument. Model contours are also shown for both models described above. Contours for the field-aligned acceleration model are shown on the right side of Figure 7 and contours for the adiabatic deformation model are shown on the left side. The best instrument resolution is on the high-energy portion of the beam and here the contours are closely matched with the field-aligned acceleration model. In the low-energy portion of the beam at high pitch angles, both models diverge from the observations. Birn and colleagues (1981) argue in favor of the adiabatic deformation model based on model comparisons with data obtained by the LANL/MPI plasma analyzers on the ISEE spacecraft. Their contours in the high-energy portion of the ion beams also show good correlations with the field-aligned acceleration model. However, their comparison technique weighted model-observation contour differences over all energies and, based on differences at low energies, they argued in favor of the adiabatic deformation model.

The crucial difference in the conflicting results described above is that the field-aligned acceleration model predicts a steeper slope on the low-energy portion of the ion beams. In order to better resolve this difference in slope, we have selected some examples of ion beams which remain stable for two or more instrument cycles where each cycle takes 128 seconds to cover all energies. However, typical ion beam samples covering ten energy steps or less are fully sampled in velocity space within 40 seconds. Figure 8 shows some of these examples sampled at high-data rate in the high-energy mode of the plasma instrument. This provides more than twice the energy resolution available for previous published studies. The beam profiles shown in Figure 8a,b are for earthward-directed beams in the plasma sheet boundary layer. These cases all clearly show a steeper slope on the low-energy portion of the ion beams. As shown above, adiabatic deformation does not change the spacing of these contours relative to the accelerated source population. For an initial flowing Maxwellian distribution which is isotropic in the rest frame, the contour spacing should remain equal along the $V_\parallel$ axis. We conclude that our observations support an acceleration process that can lead to closer spacing of phase space density contours in the low-energy portion of the ion beams. Ion beams travelling antisunward are also observed within the plasma sheet boundary layer. One example is shown in Figure 8c. Once again, the low-energy portion of the ion beams is steeper than the high-energy portion similar to

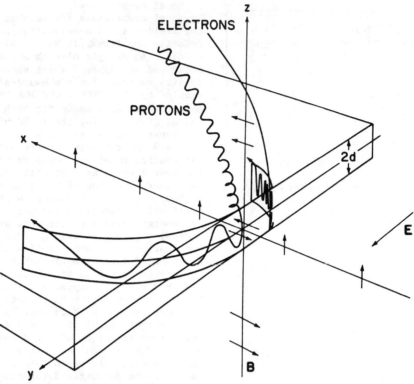

Fig. 9. Ion and electron motion in a model current sheet is illustrated here. The current sheet has a small magnetic field, B_z, normal to the sheet and a guiding center drift along the $E \times B_z$ direction adds to the particle motions which involve a meandering about the neutral sheet due the changing sign of B_x and the associated Lorentz force. During its simultaneous displacement along the direction of the applied electric field, the particles pick up energy from this cross-tail electric field. (adapted from Speiser, 1965, and Lyons and Williams, 1984)

the earthward-directed ion beams. These tailward-directed ion beams are likely the result of mirroring near earth of ion beams that were initially accelerated tailward of the spacecraft (Birn et al., 1981).

Current-sheet acceleration:

Before most of the relevant spacecraft observations were available, Speiser (1965; 1967) demonstrated the probable importance of current-sheet acceleration for space plasmas. Recent modeling and comparisons with observations by Lyons and Speiser (1982) and Speiser and Lyons (1984) strongly support the importance of the current-sheet acceleration process. Figure 8 illustrates particle motion in an idealized current sheet where the usual guiding center theory, conserving the first two adiabatic invariants, is invalid because significant magnetic field changes occur in distances less than a gyroradius. Particles entering the current sheet perform "Speiser-type" meandering trajectories about the midplane of the sheet while gaining the energy associated with the cross-tail potential. Lyons and Speiser (1982) showed how source distributions like those observed in the magnetospheric boundary layer (mantle or low-latitude boundary layer) can result in accelerated ion beams similar to those observed within the plasma sheet boundary layer. Current-sheet acceleration also leads to phase-space density contours with a steeper slope on the low-energy side of the accelerated ion beams as shown in the bottom of Figure 10. This figure also compares all of the ion beam acceleration processes discussed above. Our results indicate that adiabatic deformation of initially flowing, Maxwellian distributions is not adequate by itself to explain observed ion beam velocity distributions. However, adiabatic deformation effects must occur in the magnetized plasma and should be incorporated in any modeling effort in addition to field-aligned acceleration (third example in Figure 10) or any other potential acceleration process. Viable acceleration processes must also produce a steeper slope on the low-energy side of the ion beam contours as shown for both field-aligned acceleration and current sheet acceleration.

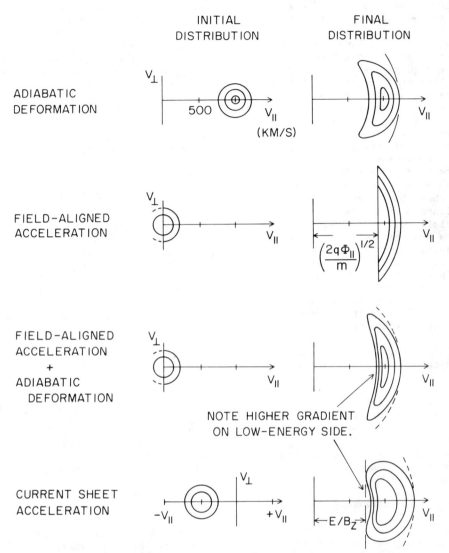

Fig. 10. A comparison of various ion beam acceleration processes showing sample initial and final velocity distributions. Field-aligned acceleration and current sheet acceleration are shown in the second and fourth rows. Both of these processes can have superimposed adiabatic deformation effects (as shown in row 3) or adiabatic deformation may be the dominant effect as illustrated in the top row.

Conclusions

As we have shown, ion beam velocity distributions provide an important diagnostic tool for inferring source, propagation and acceleration characteristics for particle beams observed within the plasma sheet boundary layer. The use of this diagnostic tool is potentially crucial, for example, for assessing the relative importance of reconnection for magnetotail processes. Current-sheet acceleration is closely related to reconnection as shown by Cowley (1980). However, one important distinction is that reconnection necessarily involves a change in field topology and current-sheet acceleration does not.

It is also important to pursue alternative acceleration processes. For example, Merlino, Cartier and D'Angelo (1985) have performed laboratory experiments showing that electrostatic ion-cyclotron (EIC) wave heating cab produce ion beams and conics in a diverging field geometry. Similar conditions are available in the earth's magnetosphere and further work is needed to evaluate the applicability of this processes. Another alternative process is that of plasma expansion into a vacuum (C. Goertz, private communication). Gurevich et al. (1966), Singh and Schunk (1982), and others have shown that when a charge neutral plasma is allowed to expand into a vacuum, the high mobility of electrons results in a parallel electric field which slows down the electrons and accelerates the ions. Recent numerical simulations show that significant beam densities exist for beams moving at about 3-7 times the sound speed or 10-50 times kT_e. An excellent review article by Samir et al. (1983) provides an up-to-date summary of developments concerning the expansion of a plasma into a vacuum. This vacuum expansion process could accelerate these ions recently injected into the low-latitude boundary layer from the magnetosheath. Field lines which map back to this source region are contained in the plasma sheet boundary layer. Earthward-directed ion beams can also occur which have mirrored near earth resulting in tailward-directed ion beams as well as counter-streaming ion velocity distributions with their superimposed drifts and all the complexities of the multiple-component ion distributions observed in the plasma sheet boundary layer.

This review of ion acceleration in the plasma sheet boundary layer has not addressed all of the many possible propagation and acceleration processes that may be applicable. However, the in-situ observation of particle velocity distributions provides a powerful diagnostic tool for inferring source, propagation and acceleration characteristics. There are many unresolved problems and many worthy tests of the best available models and observations are not yet made.

Acknowledgements. Valuable suggestions for this study were provided by Dr. L. Lyons of the Aerospace Corporation and Prof. C. Goertz of the University of Iowa. We gratefully acknowledge data reduction assistance provided by K. Duenser and M. J. Jacobs of the University of Iowa. This research was supported in part by the National Aeronautics and Space Administration under contracts NAS5-28700, grants NAG5-295 and NGL 16-001-002, and the Office of Naval Research under contract N00014-85-K-0404.

References

Birn, J., T. G. Forbes, E. W. Hones, Jr., and S. J. Bame, On the velocity distribution of ion jets during substorm recovery, J. Geophys. Res., 86, 9001, 1981.

DeCoster, R. J., and L. A. Frank, Observations pertaining to the dynamics of the plasma sheet, J. Geophys. Res., 84, 5099, 1979.

Cowley, S. W. H., Plasma populations in a simple open model magnetosphere, Space Sci. Rev., 26, 217, 1980.

Eastman, T. E., L. A. Frank, W. K. Peterson, and W. Lennartsson, The plasma sheet boundary layer, J. Geophys. Res., 89, 1553, 1984.

Eastman, T. E., L. A. Frank, and C. Y. Huang, The boundary layers as the primary transport regions of the earth's magnetotail, accepted for publication in J. Geophys. Res., 1985.

Gurevich, A. V., L. V. Paryiskaya, and L. P. Pitaevsky, Self-similar motion of rarefied plasma, Sov. Phys. JETP, Engl. Transl., 22, 449, 1966.

Lundin, R., and E. Dubinin, Solar wind energy transfer regions inside the dayside magnetopause: Accelerated heavy ions as tracers for MHD-processes in the dayside boundary layer, Plan. Sp. Sci., 33, 891, 1985.

Lyons, L. R., and T. W. Speiser, Evidence for current sheet acceleration in the geomagnetic tail, J. Geophys. Res., 87, 2276, 1982.

Lyons, L. R., and D. S. Evans, An association between discrete aurora and energetic particle boundaries, J. Geophys. Res., 89, 2395, 1984.

Lyons, L. R., and D. J. Williams, Quantitative aspects of magnetospheric phyiscs, D. Reidel Pub. Co., Dordrecht, Holland, 1984.

Merlino, R. L., S. L. Cartier, and N. D'Angelo, Laboratory experiments on electrostatic ion-cyclotron waves in a diverging magnetic field, submitted to J. Geophys. Res., 1985.

Sarris, E. T., S. M. Krimigis, A. T. Y. Lui, K. L. Ackerson, L. A. Frank, and D. J. Williams, Relationship between energetic particles and plasmas in the distant plasma sheet, Geophys. Res. Lett., 8, 349, 1981.

Sharp, R. D., D. L. Carr, W. K. Peterson, and E. G. Shelley, Ion streams in the magnetotail, J. Geophys. Res., 86, 4639, 1981.

Singh, N., and R. W. Schunk, Numerical calculations relevant to the initial expansion of the polar wind, J. Geophys. Res, 87, 9154, 1982.

Speiser, T. W., Particle trajectories in model current sheets, 1. Analytical solutions, J. Geophys. Res., 70, 4219, 1965.

Speiser, T. W., Particle trajectories in model current sheets, 2, Applications to auroras using a geomagnetic tail model, J. Geophys. Res., 72, 3919, 1967.

Speiser, T. W., and L. R. Lyons, Comparison of an analytical approximation for particle motion in a current sheet with precise numerical calculations, J. Geophys. Res., 89, 147, 1984.

Williams, D. J., Energetic ion beams at the edge of the plasma sheet: ISEE 1 observations plus a simple explanatory model, J. Geophys. Res., 86, 5507, 1981.

Williams, D. J., and L. R. Lyons, The proton ring current and its interaction with the plasmapause: Storm recovery phase, J. Geophys. Res., 79, 4195, 1974.

Williams, D. J., and T. W. Speiser, Sources for energetic ions at the plasma sheet boundary: Time varying or steady state?, J. Geophys. Res., 89, 8877, 1984.

ION INTERACTIONS IN THE MAGNETOSPHERIC BOUNDARY LAYER

B. Hultqvist, R. Lundin, and K. Stasiewicz

Kiruna Geophysical Institute, P.O. Box 704, S-981 27 Kiruna, Sweden

Abstract. New observational results obtained by means of PROGNOZ-7 in the dayside boundary layer of the Earth's magnetophere are described and their interpretations are discussed. The main results are as follows:
1) Magnetosheath plasma elements penetrate into the boundary layer where they act as generators of an electric field.
2) Different ion species may have strongly different flow vectors in the boundary layer.
3) Cold ionospheric ions are accelerated in the generator regions of the boundary layer perpendicularly to the magnetic field lines to the $\underline{E}x\underline{B}/B^2$ velocity; i.e. all originally cold ions, independent of mass, obtain the same drift velocity. They also obtain the same thermal velocity.
4) The higher the local magnetospheric plasma density is in the boundary layer the less is the perpendicular velocity difference between the local plasma and the penetrating plasma.

Introduction

In this report some new observations of ion interaction and acceleration made in the dayside magnetopause boundary layer by means of PROGNOZ-7 will be summarized and their relations to basic physical processes will be reviewed. Some of the results have been published recently [Lundin 1984, Lundin and Dubinin, 1984, 1985], others have been submitted for publication [Stasiewicz et al. 1985] and still others are under preparation. PROGNOZ-7 measured the fluxes, energy spectra and flow of the four major ion species in the boundary layer, H^+, O^+, He^+ and He^{2+}, with a higher temporal resolution than has been done before, during a small number of passes through the dayside boundary layer in which the ion-mass-separating plasma experiment was provided with a telemetry bit-rate of above 4kb/s. In those few passes it was for the first time possible to determine individual flow vectors for the different major ion species in the boundary layer with a reasonably good time resolution. All results reviewed here are based on this high bit-rate data from the dayside boundary layer equatorward of the cusp (low-latitude boundary layer and entry layer). The reader is referred to Lundin et al. [1982] for a description of the instrument and to the original papers by Lundin [1984] and Lundin and Dubinin [1984, 1985] for references and a general literature summary about boundary layer investigations.

In this review we will limit ourselves to the presentation of a few new kinds of observation in the boundary layer and their interpretations, namely
- observations of penetration of magnetosheath plasma elements into the boundary layer;
- observations of different flow vectors for different ion species;
- on perpendicular acceleration and "heating" of cold terrestrial ions in the boundary layer; and
- on the dependence of the interaction between penetrating magnetosheath plasma elements (injection regions) and "local" magnetospheric plasma on the local plasma density.

The observations reported appear generally to be in agreement with a "viscous" type of interaction between the solar wind plasma and the magnetosphere at the magnetopause, with the boundary layer being a braking region for streaming magnetosheath plasma, in which particle kinetic energy is converted into electromagnetic energy (generator, with $\underline{E}\cdot\underline{J}<0$).

Plasma structure

An example of high time resolution data taken by means of the mass-separating plasma experiment on PROGNOZ-7 is shown in Figure 1. The spacecraft crossed the magnetopause about 1 hour and 20 minutes before magnetic noon at a GSM latitude of 49.7° and passed more or less radially through the dayside magnetosphere towards the low altitude perigee. Shown in Figure 1 is only the first fifth of

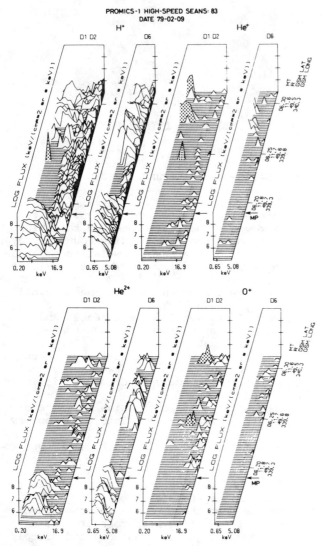

Fig. 1. Energy spectra as function of time for H^+, He^{2+}, He^+ and O^+ ions during the inbound magnetopause (marked MP) crossing by PROGNOZ-7 on 9 February 1979. The D1D2 panels show data from the perpendicular ion mass spectrometers, which scan the GSE YZ-plane, and the D6 panels show data from the sunward oriented spectrometer. Time (MT = UT + 3h) and orbital parameters are given along the inclined coordinate axis [after Lundin, 1984].

the period of passing through the boundary layer. The staggered energy spectra in Figure 1 were taken about every ten seconds for the four major ion species. There were three ion mass spectrometers onboard together with some particle spectrometers without mass separation. Two of the ion mass spectrometers measured perpendicularly to the Sun-Earth line and the third one covered a region fairly close to the sunward direction (the spin axis of PROGNOZ-7 was continously pointing towards the Sun within a few degrees). A description of the experiment has been given by Lundin et al. [1982].

The magnetopause was crossed at 0619 MT (MT=UT+3 hours). Outside the magnetopause one can see in Figure 1 the typical magnetosheath plasma with "soft" ion spectra, composed of H^+ and He^{2+} and without any measurable amounts of O^+ ions. Just inside the magnetopause the spacecraft passed for about five minutes through a typical magnetospheric plasma much more energetic than the magnetosheath one. After that a mixture of magnetosheath and magnetosphere types of plasma was observed, covering the entire energy range of the ion mass spectrometers (0.2-17 keV) and containing measurable amounts of ions of ionospheric origin. In some limited regions the He^+ and O^+ fluxes were comparable with the H^+ and He^{2+} fluxes. These larger fluxes of terrestrial ions had very narrow energy spectra and were found at low energies (see the marked peaks at ~0624, 0628 and 0631 MT). In one case also the terrestrial H^+ peak can readily be identified in Figure 1 (at 0624 MT). Besides these low energy peaks there was a low flux of O^+ ions, primarily at higher energies, everywhere inside the magnetopause.

It should be noted that the He^{2+} ions were seen only with the spectrometer measuring close to the sunward direction in the region of mixed magnetosheath and magnetosphere plasmas well inside the magnetopause, whereas in the magnetosheath they were detected also with the perpendicular detectors. At 0624 MT the situation was for a brief period similar for the H^+ ions. Contrary to the antisunward alignment of the solar wind ions in the boundary layer is the detection of narrow banded low energy terrestrial ions only with the perpendicular spectrometers.

As mentioned earlier, Figure 1 contains only about one fifth of the boundary layer data obtained in the passage. There were several more regions of the kind shown in Figure 1 with energetic magnetospheric plasma intermingled with regions of mixed magnetosheath and magnetosphere ions, including ions of ionospheric origin. The boundary layer plasma thus appears to be composed of clumps of plasmas of magnetospheric and magnetosheath origin and also mixed magnetosheath and magnetosphere plasmas. It can be concluded that the mixed regions in the boundary layer do not represent excursion of the spacecraft into the magnetosheath; the properties of the plasma and of the magnetic field (not shown in Figure 1) are too different from those outside the magnetopause. They rather look like regions of penetration of magnetosheath plasma into the boundary layer.

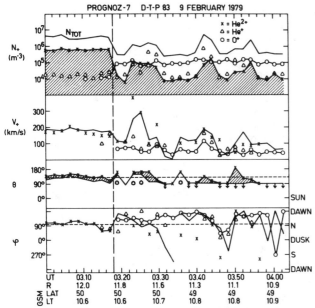

Fig. 2. Density and flow-velocities during an inbound magnetopause crossing by PROGNOZ-7. The magnetopause is marked by the vertical dashed line. The upper panel shows the ion densities deduced from the ion composition spectrometers for He^{2+}, He^+ and O^+. N_{TOT} gives the total ion density. The second, third and fourth panels show the magnitude and direction of the ion flow for H^+ (marked by solid curve), He^{2+}, He^+ and O^+ (marked by symbols). The straight dashed line in the θ-panel marks the average magnetopause tangent direction and the corresponding straight line in the ϕ panel marks the direction of the magnetopause normal [after Lundin, 1984].

Such a conclusion was drawn also by Lundin and Aparicio [1982] from observations in the plasma mantle. We will discuss these penetration regions further in later sections. A more detailed description of the observations of the penetration regions can be found in Lundin and Dubinin [1984].

Different flow velocities of different ion species

The first evaluations of individual flow vectors for the different ion species in the boundary layer from the high bit-rate data taken by means of the PROGNOZ-7 experiment showed to our surprise large differences between the flow vectors of the various ion species. These differences have been found to occur primarily in the regions of mixed magnetosheath and magnetosphere plasmas described in the previous section and may amount to 200 km/s in magnitude and 90° in direction between solar wind ions (H^+, He^{2+}) and those

of ionospheric origin (He^+, O^+). They have been seen in all the high bitrate PROGNOZ-7 passes through the dayside boundary layer, but as the number of such passes is small we cannot really present statistical results. The phenomenon is, however, certainly of common occurrence in the dayside boundary layer.

Data from the entire passage through the boundary layer, in which the data in Figure 1 were obtained, are shown in Figure 2. The uppermost frame contains the ion densities determined by means of the ion mass spectrometers (individual symbols) as well as with the total ion spectrometer without mass separation, assuming ions to be H^+ (solid curve N_{TOT}). The other three frames give the magnitude and direction of the flow vectors, $\underline{V}$. Figure 3 shows the magnitude of $-\underline{V} \times \underline{B}$, the equivalent electric field, and its direction in the three uppermost panels. The last panel gives the pressure contributions of the plasma and the magnetic field as well as their sum (P_{TOT}).

Both Figures 2 and 3 demonstrate that there are large differences between flow vectors mainly in the regions within the

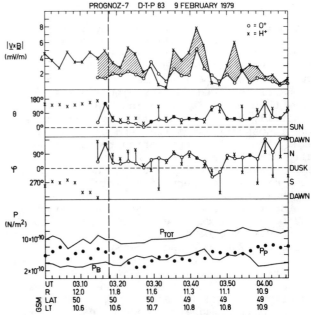

Fig. 3. Magnitude and direction of the $\underline{V} \times \underline{B}$ term for H^+ and O^+ during the PROGNOZ-7 magnetopause crossing contained in Figure 1 and Figure 2. The shaded area indicates an excess force due to magnetosheath ions. The two panels below give the direction in polar coordinates in the satellite despun frame of reference (approximately ecliptic coordinates). The bottom panel shows the plasma pressure (dots), magnetic pressure (P_B) and total pressure (P_{TOT}) [after Lundin, 1984].

TABLE 1. Some Representative Boundary Layer Parameters From Three Prognoz-7 Magnetopause Passages

PARAMETER			LLBL/EL	EXTERIOR CUSP	PLASMA MANTLE
H^+	N	(m^{-3})	$(3-10) \times 10^6$	$(3-30) \times 10^6$	$(2-10) \times 10^6$
	$T_\perp$	(eV)	500-7000	400-800	500-800
	d_p	(km)	300-3000	300-3000	300-3000
O^+	N	(m^{-3})	8×10^4	5×10^4	3×10^4
	$T_\perp$	(eV)	3500	4500	2500
	ΔT	(eV)	0-1500	0-2000	0-900
	B	(nT)	30-40	45-70	30-60
	d_B	(km)	≥ 900	≥ 1500	≥ 600
	$\Delta V_\perp$	(km/s)	0-170	0-200	0-100
Inferred Values	$(\Delta V)_{\nabla p}$	(km/s)	≤ 300	≤ 70	≤ 70
	$(\Delta V)_a$	(km/s)	≤ 50	≤ 30	≤ 50

Comments: $\Delta T = (T_\parallel - T_\perp)_{H^+} - (T_\parallel - T_\perp)_{O^+}$
LLBL/EL means low latitude boundary layer/entry layer.
The B-values given are two minute averages.

boundary layer where the density is enhanced due to large contributions of solar wind plasma, i.e. the penetration regions described in the previous section. Large differences are, however, observed also outside the regions of enhanced density. The figures also show that the solar wind ions (H^+ and He^{2+}) generally have a higher flow velocity than the ionospheric ions (He^+ and O^+). The solar wind ions in the penetration regions generally have the largest flow component in the antisunward direction whereas the ionospheric ions mainly flow perpendicularly to the Sun-Earth line.

Lundin and Dubinin [1984] have estimated the maximum velocity error that is due to count rates to be some 20% for a typical flow speed of ~ 250 km/s and a density of 0.05 cm^{-3}. Flow speeds above some 25 km/s may be determined from the PROGNOZ-7 data.

Obviously, the measurement results shown in Figures 2 and 3 are in disagreement with the so called magnetohydrodynamical approximation, $\underline{E} + \underline{V} \times \underline{B} = 0$, generally applied in magnetospheric physics. A closer investigation of the fluid description of a collisionless hot plasma, based on either Vlasov or single-particle theory, shows that the expression for the macroscopic ion flow velocity of an ion species (with mass m, number density n, and electric charge q) has the following general form:

$$\underline{V} = \underline{V}_o + \frac{B}{B^2} \times [\frac{\nabla p_\perp}{qn} + \frac{p_\parallel - p_\perp}{qn}(\hat{b} \cdot \nabla)\hat{b} + \frac{m}{q}\frac{d\underline{V}_o}{dt}] \quad (1)$$

where

$$\underline{V}_o = V_\parallel \hat{b} + \underline{E} \times \underline{B}/B^2, \quad \hat{b} = \underline{B}/B, \quad d/dt = \partial/\partial t + (\underline{V}_o \cdot \nabla);$$

$$p_\perp = nm \langle v_\perp^2 \rangle / 2 \quad \text{and} \quad p_\parallel = nm(\langle v_\parallel^2 \rangle - V_\parallel^2)$$

(Parker, 1957; Volkov, 1966).

From equation (1) it can be seen that the perpendicular flow velocity is equal to the $\underline{E} \times \underline{B}$ drift velocity only when all the other terms in (1) are much smaller than the $\underline{E} \times \underline{B}$ drift, i.e. in an isotropic and homogeneous plasma, which certainly is not characteristic for the boundary layer.

The difference between perpendicular flow velocity components of two ion species may thus be written [Stasiewicz et al., 1986]

$$(\underline{V}_1 - \underline{V}_2)_\perp = (\Delta \underline{V})_{\nabla p} + (\Delta \underline{V})_a + (\Delta \underline{V})_i \quad (2)$$

with

$$(\Delta \underline{V})_{\nabla p} = \frac{\underline{B}}{B^2} \times (\frac{\nabla p_{\perp 1}}{q_1 n_1} - \frac{\nabla p_{\perp 2}}{q_2 n_2}) \quad (3a)$$

$$(\Delta \underline{V})_a = (\frac{\underline{B} \times \nabla B}{B^3} + \frac{\mu_o \underline{J}_\perp}{B^2})[(\frac{p_\parallel - p_\perp}{qn})_1 - (\frac{p_\parallel - p_\perp}{qn})_2] \quad (3b)$$

$$(\Delta \underline{V})_i = \frac{\underline{B}}{B^2} \times (\frac{m_1}{q_1}\frac{d\underline{V}_{o1}}{dt} - \frac{m_2}{q_2}\frac{d\underline{V}_{o2}}{dt}). \quad (3c)$$

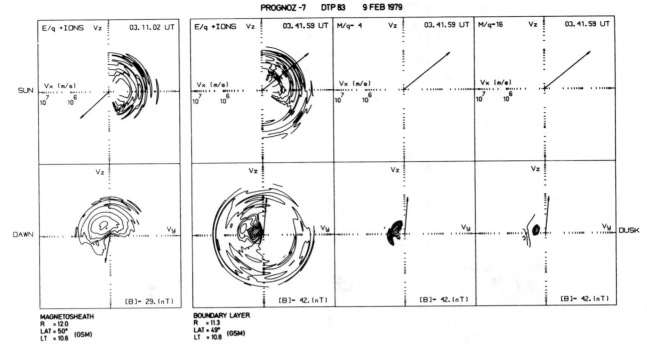

Fig. 4. Flux contour plots for ions in the XZ (upper panels) and YZ (lower panels) velocity planes. The frame of reference is the geocentric solar-ecliptic coordinate system. The magnetic field orientation and magnitude is symbolized by the arrow in each panel, and the magnetic induction is given in the bottom panels. The flux contours represent logarithmic levels of spectral energy fluxes (4.6×10^4, 1×10^5, 2.3×10^5, 4.6×10^5, 1×10^6 .... keV cm^{-2} s^{-1} sr^{-1} keV^{-1}). The leftmost panels (03.11.02 UT) represent "typical" energy per charge (E/q) ion flux contours in the magnetosheath. The other group of panels gives an example of a boundary layer magnetosheath injection structure (at 03.41.59 UT), the ones to the left as measured by the E/q-ion spectrometers and the ones to the right as measured by the ion composition spectrometers (He$^+$ and O$^+$) [after Lundin and Dubinin, 1985].

We see from (3a-c) that the difference between the flow vectors of two ion species (1 and 2) is composed of three components, one of which is due to different perpendicular pressure gradients (3a), another to different pressure anisotropies in the presence of a magnetic field gradient and/or perpendicular current (3b), and the third to different inertial effects (3c).

All these terms may play significant roles in the magnetopause boundary layer. The magnitudes of $(\Delta V)_{\nabla p}$ and $(\Delta V)_a$ have been estimated from PROGNOZ-7 data obtained in the dayside boundary layer (LLBL and near the cusp) as well as in the plasma mantle. The results are shown in Table 1. The third line from the bottom of the table contains the observed ranges of flow speed differences between H$^+$ and O$^+$ ions. The two last lines give the estimated values of $(\Delta V)_{\nabla p}$ and $(\Delta V)_a$. It can be seen that the order of magnitude of the inferred values is similar to the observed ones and the inertial term is still not taken into account. We therefore conclude that different flow vectors of different ion species should be the normal situation to observe in the boundary layer rather than the exception.

Acceleration of cold ionospheric ions in the boundary layer

The kind of narrow peaks of ionospheric ions at low energies marked in Figure 1 show up as a population near the origin (hatched) in Figure 4 in a contour diagram for the flow velocity projection onto the x-z and y-z coordinate planes of the geocentric solar-ecliptic coordinate system. The contours represent constant energy fluxes and they are logarithmically spaced ($5 \cdot 10^4$, $1 \cdot 10^5$, $2.3 \cdot 10^5$, $4.6 \cdot 10^5$, $1 \cdot 10^6$ ... keV cm^{-2} s^{-1} sr^{-1} keV^{-1}). The lefthand diagrams represent magnetosheath measurements and the righthand three frames were taken in the boundary layer passage for which data are shown in Figures 1-3. The first one of these was obtained from data from the total ion spectrometer. Another

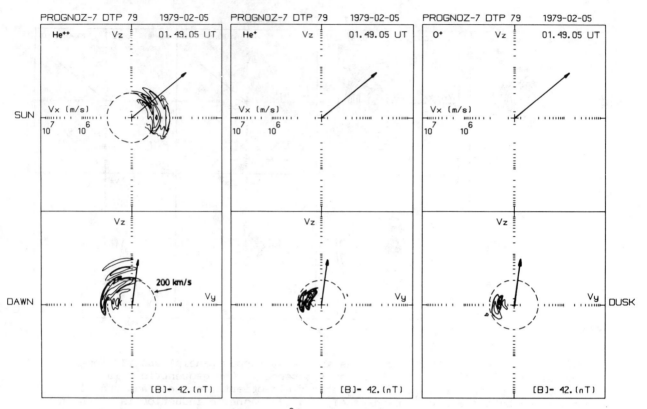

Fig. 5. Flux contour plots for He^{2+}, He^+ and O^+ ions in the dayside boundary layer ($R=11R_E$, $LT=10.8h$, GSM $LAT=49°$). Con-stant spectral energy fluxes $keV/(cm^2 \cdot s \cdot sr \cdot keV)$ are shown in logarithmic scale (4.6×10^4, 10^5, 2.3×10^5...) in the XZ and YZ GSE planes. The magnetic field magnitude and orientation is given by an arrow in each panel [after Stasiewicz et al., 1986].

example is illustrated in Figure 5. The data were obtained in the dayside boundary layer ($R=11\ R_e$, $LT=10.8$, GSM $LAT=49°$). The left-hand set of contours represent the solar wind ion flow and the other two magnetospheric ion flow.

Both Figure 4 and Figure 5 demonstrate the important result that the ionospheric ions in the acceleration process are given a flow velocity perpendicular to the magnetic field direction, whereas the solar wind ions have a substantial flow component along the magnetic field lines. It should be noted that the strict perpendicularity of the acceleration of the cold ions with respect to the magnetic field lines is not consistent with what is expected to be seen in the vicinity of a reconnection region.

We can also see in Figures 4-5 that the magnitude of the flow velocities of the terrestrial ions is about the same for O^+ and He^+. This latter result is more clearly demonstrated in Figure 6 where the perpendicular beam velocities of O^+ and He^+ are plotted and compared with the straight line on which the observational points would fall if the measured flow speeds of O^+ and He^+ were exactly equal. As can be seen the observational points all fall so close to the straight line that we may conclude, considering the limited accuracy of the measured flow vectors [see Lundin and Dubinin, 1985 for more details], that the measurements are consistent with the ionospheric ions achieving the same velocity increase in the acceleration process irrespective of ion mass.

An investigation of the ion velocity component perpendicular to the macroscopic flow vector has also demonstrated an independence of mass. This is demonstrated in Figure 7, where the peak "thermal" velocities perpendicular to the beam velocity vector are plotted.

Figure 8, finally, shows that the beam (the macroscopic flow perpendicularly to the magnetic field lines) has a factor of 2-4 higher perpendicular flow velocity than the "thermal" velocity (perpendicular to the beam direction).

A general conclusion that can be drawn on the basis of the data in Figures 6-8 is thus that the acceleration process working on cold

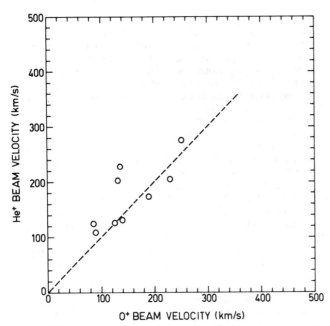

Fig. 6. O^+ beam velocity plotted versus He^+ beam velocity for 9 cases where both O^+ and He^+ ions had clearly distinguishable energy peaks. Dashed line indicates equal velocity for O^+ and He^+ [after Lundin and Dubinin, 1985].

ions in the dayside boundary layer provide all cold ions with roughly the same velocity increase irrespective of ion mass. For further details and discussion see the original paper by Lundin and Dubinin [1985].

An obvious physical mechanism that gives all plasma constituents the same velocity is the application of an electric field and the associated acceleration of the plasma to the ExB/B^2 velocity. Lundin and Dubinin [1985] have interpreted the observations described as an effect due to the electric field generated by the solar wind plasma elements penetrating into the boundary layer with an excess momentum and/or pressure gradients and pressure anisotropies.

The observed relationship between beam velocity and "thermal" velocity may then be understood as due to the penetrating solar wind plasma having a flow velocity component along the magnetopause which is 2-4 times larger than the component in the antisunward direction. To these flow velocity components correspond electric field components perpendicular to the respective velocity components. It should be noted that even if the inward flow of the magnetosheath plasma element turns completely to be along the magnetopause after some time inside the magnetopause, the inward flow energy is kept as gyration energy by ions. Depending on the gyrophase of the individual ions when the electric field component driving the ions inward disappears, the ions will have a distribution of velocities which is kept as gyromotion and which makes up the "thermal" spread observed. In the same way the main beam velocity is transformed into gyration velocity of the individual ions if it decreases or disappears. In other words: the ions keep the speed and energy they have when the driving field disappears and the motion becomes a gyro motion in the absence of an electric field. This mechanism thus has the important characteristic that even a transient electric field, generated by e.g. penetrating solar wind plasma elements, gives a "permanent" increase of the gyration energy of individual "cold" ions, and the effect of several such more or less transient events add in "pumping up" the magnetic moment of the originally cold ions. In terms of energy this process due to fluctuating electric fields, produced as suggested above or otherwise, favours the heavy ions, as the energy increase and corresponding magnetic moment increase due to equal velocity increase is proportional to the mass.

Lundin (personal communication) has pointed out that the perpendicular electric field generated in the magnetospheric boundary layer is expected to contribute to increasing the magnetic moment of the cold ions far away from the boundary layer along the

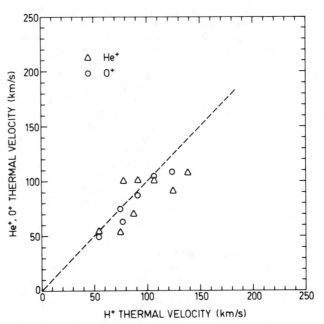

Fig. 7. H^+ thermal velocity plotted versus He^+ and O^+ thermal velocity for the 14 cases. Dashed line indicates equal thermal velocities [after Lundin and Dubinin, 1985].

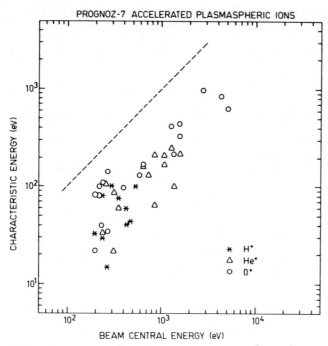

Fig. 8. A scatterplot of the beam [peak] energy versus characteristic energy of low temperature ion beams in the boundary layer, as observed from PROGNOZ-7. The dashed line indicates that the beam energy equals the characteristic energy [after Lundin and Dubinin, 1985].

magnetic field lines, perhaps even down to the uppermost ionosphere, by mapping of the electric field to low altitudes and favouring in this process the heavy O^+ ions in the way described in the previous paragraph. The mirror force acting on the increased magnetic moment brings the ions upward into the magnetosphere. This process thus may (contribute to) explain the astonishing favoring of heavy ionospheric ions that appears to characterise the ionosphere as a plasma source for the magnetosphere. A more detailed study of the mechanism is under way. Other species dependent and altitude dependent mechanisms have also been suggested to explain this phenomenon [see e.g. Hultqvist, 1982 and 1983 for discussion and references].

Dependence of the interaction in the boundary layer on local plasma density

When the density of cold local ions in the boundary layer is very low, it is natural to look upon them as test particles which are affected by the electric field generated by the penetrating magnetosheath plasma. The generator region is part of an electric circuit comprising also the field-aligned currents and the ionospheric resistor [see e.g. Lundin, 1984]. If the local plasma density is comparable with the penetrating plasma density it is expected to affect the entire process in the boundary layer. Lundin and Dubinin [1985] have investigated the dependence of the interaction in the boundary layer on the local plasma density in the same high-bitrate PROGNOZ-7 data, which has been analysed from different points of view in the previous sections. Their main result is shown in Figure 9, where along the vertical axis the difference between $\varepsilon_\perp(H^+)=|\underline{V}x\underline{B}|$ for H^+ ions, assumed to be composed primarily of solar wind protons, and $\varepsilon_\perp(O^+)=|\underline{V}x\underline{B}|$ for local ions, definitely of ionospheric origin, divided with $\varepsilon_\perp(H^+)$, is plotted. Along the horizontal axis is shown the ratio of the total ion density (penetrating + local) divided with the ion density of magnetosheath origin. The cold local plasma density has been estimated by assuming that it contains 14% He^+ ions, as found to be typical in the outer plasmasphere [e.g. Young et al., 1977] and observed also by PROGNOZ-7 in the boundary layer [Lundin and Dubinin, 1985]. The oxygen ion density, $n(O^+)$, is the magnetospheric O^+ ion contribution which is predominantly due to ions which are not cold. The magnetospheric H^+ ions are included in $n(H^+)$. Filled symbols represent data points for which the statistical error in determining the He^+ density is less than 50%. A power law fit to the 17 most significant data points (filled symbols) is represented by the dashed curve $(R_\varepsilon=1.35 \cdot R_n^{-2.58})$ for which the correlation coefficient is 0.88.

Figure 9 clearly demonstrates that the local magnetospheric plasma in the penetration regions strongly affects the energy transfer process. It shows, for instance, that when the local plasma density is comparable with or larger than the injected plasma density, the local plasma takes up most of the excess momentum and both populations have nearly the same drift velocity. Conversely, a low local plasma density is associated with a low exchange of momentum and energy. We may expect that then most of the excess momentum of the penetrating magnetosheath plasma is used to drive dynamo current along magnetic field lines through the ionosphere and that the local plasma merely convects with a velocity given by the electric field generated by the penetrating magnetosheath plasma (the test particle case).

All the observations reported here thus indicate a braking of the solar wind plasma in the boundary layer and associated perpendicular acceleration of the local magnetospheric plasma to the $\underline{E}x\underline{B}/B^2$ convection velocity in the penetration regions. As mentioned earlier this kind of situation has been observed in all high-bitrate passes of PROGNOZ-

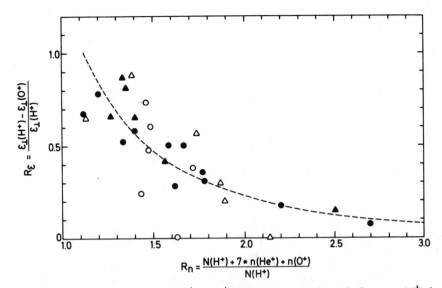

Fig. 9. Difference between $\varepsilon_\perp = |-V \times B|$ for magnetosheath plasma (H^+ ions assumed to be mostly of solar wind origin) and magnetospheric local ions (O^+) as a function of the total plasma density (composed of magnetosheath plasma and cold and hot local plasma) divided by the magnetosheath plasma density in the penetration region [after Lundin and Dubinin, 1985].

7 through the dayside boundary layer, which indicates that it is commonly occurring.

Acknowledgement. The PROMICS 1 experiment from which most of the data discussed in this review were obtained was a joint USSR-Swedish experiment flown on PROGNOZ 7. The principal experimenters on the Soviet side were N. Pissarenko and A. Zakharov, whose important contributions to the success of the experiment is gratefully acknowledged. The experiment was supported in part by the Swedish Board for Space Activities.

References

Hultqvist, B., Recent progress in the understanding of the ion composition in the magnetosphere and some major question marks, Rev. Geophys. Space Phys., 20, 589, 1982.

Hultqvist, B., On the origin of the hot ions in the disturbed dayside magnetosphere, Planet. Space Sci., 31, 173, 1983.

Lundin, R., and B. Aparicio, Observations of penetrated solar wind plasma elements in the plasma mantle, Planet. Space Sci., 30, 81, 1982.

Lundin, R., B. Hultqvist, M. Pissarenko and A. Zakharov, The plasma mantle: composition and other characteristics observed by means of the PROGNOS-7 satellite, Space Sci. Rev., 31, 247, 1982.

Lundin, R., Solar wind energy transfer regions inside the dayside magnetopause-II. Evidence for an MHD generator process, Planet. Space Sci., 32, 757, 1984.

Lundin, R., and E. Dubinin, Solar wind energy transfer regions inside the dayside magnetopause-I. Evidence for magnetosheath plasma penetration, Planet. Space Sci., 32, 745, 1984.

Lundin, R., and E.M. Dubinin, Solar wind energy transfer regions inside the dayside magnetopause-III. Accelerated heavy ions as tracers for MHD-processes in the dayside boundary layer, Planet. Space Sci., 33, 891, 1985.

Parker, E.N., Newtonian development of the dynamical properties of ionized gases of low density, Phys. Rev., 107, 924, 1957.

Stasiewicz. K., R. Lundin, and B. Hultqvist, On the interpretation of different flow vectors of different ion species in the magnetospheric boundary layer, submitted for publication to J. Geophys. Res., 1986.

Volkov. T.E., Hydrodynamic description of a collisionless plasma, in Review of Plasma Physics (Ed. M. A. Leontovich), Vol. 4, pp 1-21, Consultants Bureau, New York, 1966.

Young. D.T., J. Geiss, H. Balsinger, P. Eberhardt, and A. Ghielmetti, Discovery of He^{2+} ions of terrestrial origin in the outer magnetosphere, Geophys. Res. Lett., 4, 561, 1977.

ION ACCELERATION DURING STEADY-STATE RECONNECTION AT THE DAYSIDE MAGNETOPAUSE

A.D. Johnstone[1], D.J. Rodgers[1], A.J. Coates[1], M.F. Smith[1] and D.J. Southwood[2]

Abstract. During encounters with the dayside magnetopause, accelerated ions were observed with high time resolution by the AMPTE UK spacecraft. The flow direction of these ions was closely coupled to the magnetic field direction. The accelerated ions were observed for minutes and were accompanied by an outward flow of magnetospheric ions.

Introduction

Ever since Dungey (1961) suggested that geomagnetic and interplanetary fields could become connected at the Earth's magnetopause it has been realised that ions would be accelerated in the process (Levy et al 1964). It was not until the two ISEE spacecraft were launched together in 1978 that the accelerated particles were directly observed (Paschmann et al 1979). Although several cases of quasi-steady reconnection were found on the ISEE passes through the magnetopause (Sonnerup et al 1981), it was not detected on most of the crossings. Further evidence for reconnected field lines was provided by observations of energetic magnetospheric particles in the magnetosheath. The pitch angle distributions were anisotropic with respect to the magnetic field in a direction indicating streaming out of the magnetosphere (Scholer et al., 1981; Sonnerup et al., 1981).

When the three AMPTE spacecraft were launched together in August 1984, two of the spacecraft, UKS and IRM were placed in close proximity to each other in an orbit which took them regularly through the magnetopause and bow shock. Both carried three-dimensional plasma analysers with a time resolution of one revolution of the spin-stabilised spacecraft (5 secs for UKS) which enabled them to observe the complete distribution of positive ions while the spacecraft passed through the structure of the magnetopause. We present here results from the UKS on one crossing of the magnetopause when the effects of reconnection were observed. In the data the distribution of magnetosheath ions, which are accelerated as they cross the magnetopause, can be distinguished from the distribution of trapped ions inside the magnetosphere, which are observed to leak out of the magnetosphere on the open field lines.

Instrumentation

The AMPTE-UKS spacecraft carried a magnetometer, 3-d analysers for positive ions and electrons, a particle correlator and a plasma wave instrument with electric and magnetic sensors (Ward et al 1985). In this presentation data from the magnetometer and the positive ion

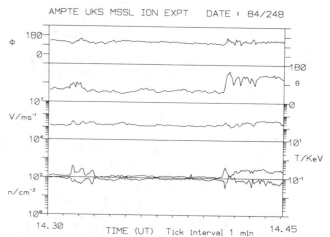

Fig. 1. The density n, temperature T and flow velocity V, θ, Φ (GSE coordinates) of the total population during the inbound magnetopause crossing of 4 September 1984. There is an encounter with a flux transfer event at 1431UT before the spacecraft crosses the magnetopause for the first time at 1442. It is seen most clearly in this plot as a change in the flow direction from northward to southward. The ion flow speed exceeded the magnetosheath speed throughout the period from 1442UT to 1446UT.

---

[1] Mullard Space Science Laboratory, University College London, Holmbury St Mary, Dorking, Surrey, U.K.

[2] Blackett Laboratory, Imperial College of Science and Technology, London SW7 2AZ, U.K.

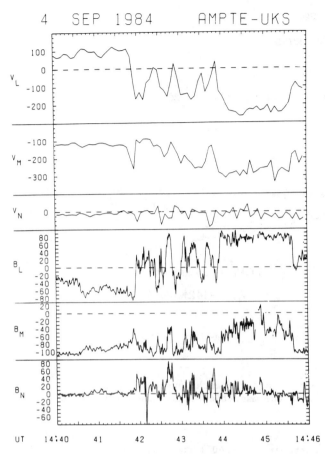

Fig. 2. A comparison between the flow velocity and magnetic field in the boundary normal system. The N direction, calculated as the direction of minimum variance, should be normal to the magnetopause. The velocity and magnetic field are correlated for both L and M components.

analyser are used. The magnetometer transmits 16 vector measurements per second with an overall accuracy of 0.1 nT (Southwood et al 1985).

The positive ion analysers consist of electrostatic energy analysers with a combined field of view of 180° with respect to the spin axis of the spacecraft. As the spacecraft rotates the analysers sweep through the energy range from 20keV/q down to 10eV/q precisely 16 times each revolution. The data are transmitted as three-dimensional matrices, in energy, polar angle, and azimuthal angle.

The intrinsic resolution of the instrument of 120 energies, 8 polar angles and 16 azimuthal angles is reduced by combining adjacent elements of the matrix to fit the telemetry rate available. During the pass described here the transmitted resolution was 60 contiguous energy bands ($\Delta E/E = 14\%$), 4 polar angle sectors (45°) and 8 azimuthal sectors (45°) (Coates et al 1985). The bulk parameters, density, flow velocity and temperature are calculated on the ground by integrals over the distribution. If there are two distinct populations as here, then bulk parameters can be obtained for both populations simultaneously by setting suitable limits for the two integrations after inspecting the distribution.

Observations

A magnetic storm began with a storm sudden commencement at 0746UT on 4 September 1984. At this time the AMPTE-UKS was in the solar wind at a geocentric distance of $12R_e$ on the inbound leg of its orbit at a local time of approximately

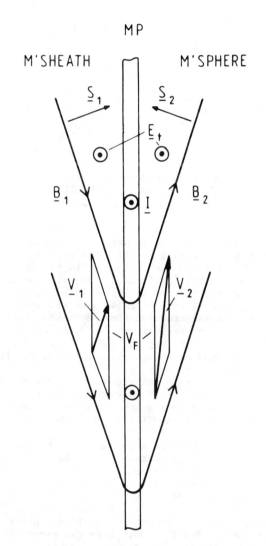

Fig. 3. Geometric construction of plasma flow acceleration in the de Hoffman-Teller frame which moves along the magnetopause with speed V_F. The flow vectors shown are those of inbound magnetosheath ions. Magnetospheric ions flow outward in a similar way.

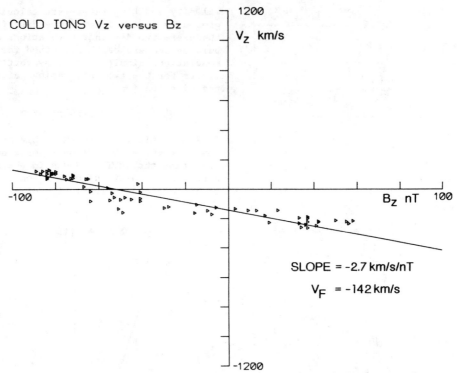

Fig. 4. A scatter plot of V_z (GSE coordinates) of magnetosheath ions calculated separately against B_z. The flow is anti-parallel to the magnetic field; southward inside the magnetopause, and northward outside.

1500 and south of the geomagnetic equator. The bowshock was crossed at 1320UT at a geocentric distance of $9.95R_e$, well inside the average position at that local time. The magnetopause was first encountered at 1442UT at a distance of $8R_e$ and a local time of 1534. The UKS spacecraft detected accelerated ion flows from then until the instruments were switched off for operational reasons at 1446UT, but the IRM (G. Paschmann private communication) continued to observe them for another 30 minutes.

The ensuing magnetic storm, which reached a maximum depression of 120nT in the equatorial field at the Earth's surface has been studied in detail by the AMPTE-CCE spacecraft (eg. Williams & Sugiura 1985).

An overview of the magnetopause crossing, as seen by the ion experiment on UKS is shown in figure 1. At the beginning, the spacecraft is in the magnetosheath where it encounters a flux transfer event between 1431UT and 1432UT. The accompanying bipolar signature in the normal component of the magnetic field is not presented here. At 1442UT UKS encounters the magnetopause for the first time and a complex series of crossings follows as the magnetopause moves backwards and forwards over the spacecraft.

During these crossings, seen most clearly in figure 1 as a change in the flow direction from northward ($\theta \sim 60°$) to southward ($\theta \sim 120°$), the flow speed increases from the 140km/s observed in the magnetosheath to as much as 400km/s. Throughout the period from 1442UT to 1446UT the ions are accelerated above magnetosheath speeds. At the same time the density decreases slightly and the temperature increases. In figure 2 the components of the flow velocity are compared with the magnetic field. The coordinate system for this plot is the boundary normal system with the N axis directed along the outward normal to the magnetopause (Russell and Elphic 1978). Here the magnetopause crossings are readily identified in the usual way by the change in sign of the B_L component. There is a clear correlation between the velocity components and the corresponding magnetic field component, which suggests that the tangential stress balance relation (Sonnerup et al 1981) is obeyed and that therefore the acceleration is due to magnetic reconnection across the magnetopause. Before applying this relationship we first note that two distinct particle populations are apparent in the data matrices from the instrument. The first, and dominant population numerically, is the magnetosheath population which is accelerated in crossing the magnetopause. The second population consists of trapped particles from the magnetosphere which should be able to leak out if

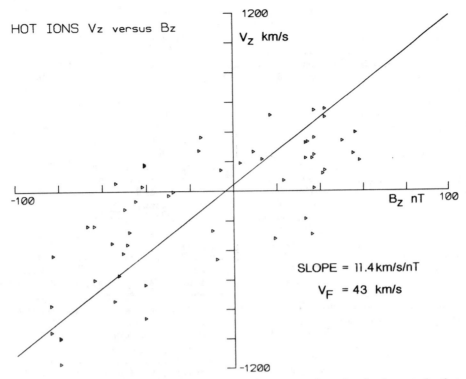

Fig. 5. A scatter plot of V_z (GSE coordinates) of magnetospheric ions calculated separately against B_z. The flow is parallel to the magnetic field; northward inside the magnetopause, and southward outside.

on open field lines. This population is more energetic than the first with few particles detected below 5 KeV. Since the magnetosheath distribution does not extend above 5 KeV in energy the two populations can be separated. Accordingly the bulk parameters have been calculated for this second population alone. The density ranges between $0.1 cm^{-3}$ and $0.3 cm^{-3}$ compared with the density of the first population in the range $60 cm^{-3}$ to $100 cm^{-3}$.

Analysis

The flow of plasma through an open magnetopause is shown in figure 3 (Paschmann 1984). In this diagram magnetosheath ions are shown crossing into the magnetosphere. An equivalent diagram can be drawn for magnetospheric ions moving in the opposite direction. The frame of reference is moving along the magnetopause at a speed V_F, which can be thought of as the field line speed (Cowley 1982). In this frame the ions are moving parallel to the magnetic field. If the magnetopause is a rotational discontinuity, as it should be near a reconnection point, then the field aligned velocity $V = B/\sqrt{\mu_o \rho}$ (Paschmann et al 1979). In the stationary reference frame, essentially the spacecraft frame, the total particle velocity is obtained by adding the field-line velocity V to the field-aligned velocity V_F

$$V_1 = V_F \pm B/\sqrt{\mu_o \rho} \tag{1}$$

The sign to be taken depends on the direction of the normal component of magnetic field B_n. This in turn depends on whether the observation is made north or south of the reconnection line. If south, B_n is outward, and the sign is negative. This relationship provides a test for the existence of a rotational discontinuity, and hence, an open magnetopause. Equation (1) expresses, tangential stress balance at the discontinuity in a simplified form since, for example, pressure anisotropy has been ignored, (Paschmann et al 1979). Simultaneous observations of B, V and ρ can be tested for this relationship in a scatter plot of $B/\sqrt{\rho}$ against V. Since it is a vector relationship it applies to each cartesian component. In figure 4 the test has been applied to the Z component in GSE coordinates, with the further simplification that variations in density have been ignored. Even so, the linear dependence between B_z and V_z is apparent.

The line, a least-squares fit has a negative slope implying that the observation is made south of the reconnection line. This is consistent with the south-eastward flow given in figure 2. The field line velocity component V_{FZ} is given by the intercept of the line on the V_z axis. It is equal to 142km/s in a southward direction. From the slope the average density can be computed. The value obtained, 65cm^{-3}, is consistent with the direct measurements shown in figure 1.

We have made a similar plot (figure 5) for the V_z component of the magnetospheric ion population. In this case it is not to be construed as a test of tangential stress balance because the population is a minor constituent. Again it shows a linear dependence between V_z and B_z but, because the density is low and the thermal velocity is high, there is a lot more statistical scatter in the points. The slope has the opposite sign showing that the magnetospheric ions are passing through the magnetopause in the opposite direction. The slope does not have the significance of being proportional to the square root of the density because stress balance is not involved. We are using the magnetic field component as an indicator of position. The best-fit line does not pass through the same value of V_{FZ} on the velocity axis. However, as the thermal velocity is much higher because of the high average energy of the ions, and hence the scatter of the points is large, the result is not inconsistent with the same value of V_{FZ}, as in figure 4. What this plot essentially shows is that the magnetospheric ions are moving northward inside the magnetopause and southward outside. This is consistent with the magnetic configuration of an open magnetopause south of the reconnection line.

Conclusions

The plasma analysers on the two high altitude AMPTE spacecraft combined, for the first time, high time resolution (5 secs) with complete 3-dimensional coverage. This enabled them to resolve 3-dimensional flow structure in the magnetopause. Data from the UKS ion experiment for a notable example of steady state reconnection have been presented here. The condition for tangential stress balance in a rotational discontinuity is obeyed, confirming that the accelerated ion flows observed are the consequence of magnetic reconnection.

The south-eastward direction of the accelerated flow, and the outward direction of the normal component of B deduced from the sign of the slope of the scatterplot of B_z against V_z, both show that the observations were made south of the reconnection line. Simultaneously, magnetospheric ions were detected moving outward through the magnetopause.

The improved coverage and resolution, provided by the instrument on the AMPTE-UKS spacecraft should eventually enable a much more detailed study of the conservation relations during the crossing and provide much stricter tests of the theory of magnetic reconnection.

References

Coates, A.J., J.A. Bowles, R.A. Gowen, B.K. Hancock, A.D. Johnstone, S.J. Kellock AMPTE-UKS 3-dimensional ion experiment, IEEE Trans. Geosci. and Rem. Sens., GE-23, 287, 1985.

Cowley, S.W.H. The causes of convection in the Earth's magnetosphere: A review of developments during the IMS, Rev. Geophys. Space Phys., 20, 531, 1982.

Dungey, J.W. Interplanetary magnetic field and the auroral zones, Phys. Rev Lett. 6, 47, 1961.

Levy, R.H., H.E. Petschek, G.L. Siscoe, Aerodynamic aspects of the magnetospheric flow, AIAA Journal, 2, 2065, 1964.

Paschmann, G. Plasma and particle observations at the magnetopause: implications for reconnection in Magnetic Reconnection in Space and Laboratory Plasmas ed Edward W Hones Jr., p114, American Geophysical Union, Washington DC, 1984.

Paschmann, G, B.U.O. Sonnerup, I. Papmastorakis, N. Sckopke, G. Haerendel, S.J. Bame, J.R. Asbridge, J.T. Gosling, C.T. Russell, R.C. Elphic, Plasma acceleration at the Earth's magnetopause: evidence for reconnection, Nature, 282, 243, 1979.

Russell, C.T. and R.C. Elphic, Initial ISEE magnetometer results: magnetopause observations, Space Sci. Rev. 22, 681, 1978.

Sonnerup, B.U.O., G. Paschmann, I. Papamastorakis, N. Sckopke, G. Haerendel, S.J.Bame, J.R. Asbridge, J.T. Gosling, C.T. Russell, Evidence for magnetic field reconnection at the Earth's magnetopause, J. Geophys. Res. 86, 1009, 1981.

Scholer, M, F.M. Ipavich, G. Gloeckler, D. Hovestadt, B. Klecker, Leakage of magnetospheric ions into the magnetosheath along reconnected field lines at the dayside magnetopause. J. Geophys. Res. 86, 1299, 1981

Southwood, D.J., C. Mier-Jedrzejowicz, C.T. Russell, The fluxgate magnetometer for the AMPTE-UKS sub-satellite, IEEE Trans. Geosci. and Rem. Sens. GE-23, 301, 1985.

Ward, A.K., D.A. Bryant, T. Edwards, D.J. Parker, A. O'Hea, T.J. Patrick, P.H. Sheather, K.P. Barnsdale, A.M. Cruise, The AMPTE-UKS spacecraft, IEEE Trans. Geosci and Rem Sens. GE-23, 202, 1985.

Williams, D.J. and M.Sugiura, The AMPTE Charge Composition Explorer and the 4-7 September 1984 Geomagnetic storm. Geophys. Rev. Lett. 12, 305, 1985.

THE EFFECT OF PLASMA SHEET THICKNESS ON ION ACCELERATION NEAR A MAGNETIC NEUTRAL LINE

R. F. Martin, Jr.[1]

NOAA Space Environment Laboratory, Boulder, Colorado 80303

Abstract. The motion of a charged particle in a magnetic neutral line field and a perpendicular electric field is calculated. Full numerical calculations are performed for parameters relevant to the magnetotail. We consider an acceleration region of length L and height T embedded in the plasma sheet. It is found that for a thin plasma sheet (T=1000 km) the motion is relatively simple and particles can be energized significantly. For a thick plasma sheet (T=20,000 km) the motion can be very complicated and particles can actually lose energy. Applications to the magnetotail are discussed.

Introduction

Acceleration of particles by magnetic reconnection at a magnetic neutral line has long been proposed as an important energization mechanism in the geomagnetic tail. Numerous authors have studied the problem by calculating orbits of test particles in prescribed fields meant to model the expected or observed field structure in the tail. This single particle approach has been remarkably successful in predicting the properties of the low density plasma in the magnetosphere. To date, most studies in the magnetotail have dealt with a magnetic field reversal geometry, often with a small uniform, normal field added, and a uniform, cross-tail electric field. [see Speiser and Lyons, 1984 and Wagner et al., 1979, and references therein]. These fields are designed to model the actual field conditions downstream from the presumed neutral line. Particle motion in the near vicinity of the neutral line itself has received comparatively little attention. Rusbridge [1971] has calculated orbits numerically for a simple magnetic x-line configuration. The perpendicular electric field was added by Sonnerup [1971], who assumed the current sheet adiabatic invariant [Speiser 1970] to hold if the x-line angle is very small. Stern [1979] derived a potential function and used it to study the qualitative behavior of particle orbits. In this paper the dynamics of charged particles in the magnetic neutral line plus perpendicular electric field geometry will be further studied. In particular, we are interested in the dependence of particle energization on the thickness of the plasma sheet in which the assumed neutral point acceleration region is embedded.

Basic Model

The following magnetic field will be assumed:

$$\underline{B} = B_o (z \hat{x} + \delta x \hat{z})/D \qquad (1)$$

where B_o and D are reference field strength and distance scales, respectively, and the carats denote unit vectors. Standard magnetospheric coordinates in which the x-axis points sunward, are used. A uniform cross-tail electric field will also be assumed in the y-direction, with magnitude $E = \varepsilon B_o$.

The magnitude of B increases with distance, so the model fields can clearly be applicable only locally in the tail. We therefore define a local "acceleration region" to be a rectangle of height T in the z-direction (essentially the current sheet thickness) and length $L = T/\delta^{1/2}$ in the x-direction. The neutral line lies at the center of this box, which is to be considered embedded in the plasma sheet. This acceleration region is then essentially the same as Coroniti's [1985] "reconnection region" when one is considering a near earth neutral line.

In order to get meaningful numbers out of the calculation requires realistic numbers to be put in, and here one is hampered by the lack of direct observations of neutral points in the magnetosphere. In particular, the parameter δ and the size of the acceleration region are difficult to pin down. Two dimensional neutral points are "observed", however, in numerical simulations of the tearing mode instability. Ugai [1984] and Hayashi and Sato [1979] present

---

[1] National Academy of Sciences, National Research Council Resident Research Associate

results of local MHD simulations. From figures in both references, a value of $\delta = 0.03$ seems to be reasonable. The thickness of the "X" region is approximately one third to half of their current sheet thickness. Global MHD simulations also result in the formation of neutral point similar in form locally to the fields used in this paper [Brecht et al., 1982; Lyon et al., 1981]. Although the resolution of these simulations is not high, the value of δ quoted above is not unreasonable.

In this paper, then, the value $\delta = 0.03$ will be taken as a rough estimate. The size of the "X" region seems to be rather large in the global simulations: on the order of several earth radii. For this study, two values of sheet thickness will be used: 1000 km for a relatively thin current sheet [Lyons and Speiser 1984, Rothwell and Yates, 1979], and 20,000 km for a thick sheet, to correspond to the global simulations mentioned above.

Values for the parameters B_o and ε are still needed. Taking $B(x = 0, z = T/2) \approx 20$ nT, as a representative value for the edge of the current sheet, defines B_o. A typical value of the cross tail electric field is $E_o \approx 3 \times 10^{-4}$ V/m, which will define ε. The value of ε depends weakly on T, but for simplicity, $\varepsilon \approx 10^{-5}$ will be used for both values of T.

Particle Dynamics

Choosing the length scale (in Gaussian units) to be $D = c/\omega_c$, where $\omega_c = qB_o/mc$, q is the charge of the particle, m is its mass, and c is the speed of light, yields the following simple form for the equations of motion:

$$\dot{v}_x = s\, \delta x v_y$$
$$\dot{v}_y = s\,[\varepsilon + (z v_z - \delta x v_x)] \quad (2)$$
$$\dot{v}_z = s\, z v_y$$

Here, s is the sign of the charge, the dot represents time differentiation, and lengths have been made dimensionless with D, times with ω_c^{-1}, and fields with B_o.

The equations of motion (2) have immediate integrals of y-momentum:

$$v_y = s[\varepsilon t + (z^2 - \delta x^2)/2] + P_y \quad (3)$$

and energy:

$$v_x^2 + v_y^2 + v_z^2 = 2\varepsilon y + E_3 \quad (4)$$

where P_y and E_3 are constants. Combining the y-momentum integral (3) with the equations of motion (2), the problem can be reduced to a set of two-dimensional, non-autonomous equations in the z and x variables.

The resulting equations are useable for analytic and numerical study of this dynamical system. Due to space limitations, detailed analytic results will be presented elsewhere, and only a rough summary is presented here. In fact, it can be shown that this dynamical system is probably nonintegrable, in the sense that no integrals of motion other than (3) and (4) exist. Despite this, the qualitative behavior can be studied using an effective potential function. The motion can be visualized as a particle moving in a two-dimensional, slowly varying effective potential [Stern 1979; Martin, in preparation]. The allowed regions of motion are "valleys" which parallel the asymptotic x-lines. The width of the valleys decreases with distance from the neutral point, so particles will eventually mirror, and cannot escape to infinity. The potential valleys "convect" at the E x B velocity. For a particle started at z = T/2 near x = 0, for example, this convection is downward toward the neutral line initially, then outward along the x-axis. Thus, all such particles have access to the neutral region (as long as $\varepsilon \neq 0$), and hence will undergo nonadiabatic motion. To the extent that particles are tied to a particular valley, this picture begins to look like the fluid picture of field lines convecting inward above and below the neutral point and outward to the left and right. However, the particles are not in general tied to the convecting valley, but their motion alternated erratically between crossing between valleys and mirroring in a single valley (and of course the field lines are fixed here). Thus, although the gross motion does follow this convection, the underlying dynamics are quite a bit more complicated (and interesting).

The trajectory is thus a complex combination of what Sonnerup [1971] called "meandering" and "noncrossing" orbits. The special case when a particle has zero initial velocity in the x-direction is equivalent to that of a magnetic field reversal layer with no normal field. This limit has been studied by Speiser [1965], and Rothwell and Yates [1979], and has a closed form solution.

Numerical Calculations Relevant to the Magnetotail

Protons were started at the edge of the acceleration region rectangle at initial height z_o = T/2 for various values of x_o, initial energy, and pitch angle. Particles were considered to have exited from the acceleration region when the absolute value of their x-coordinate is more than a gyroradius greater than L/2. For the higher energies, particles with initial pitch angles less than about 45° often satisfied this criterion before they even entered the acceleration region rectangle. This is because they follow field lines above the X-lines down a potential valley. Although they will eventually mirror and return to enter the acceleration region, this type of orbit seems inconsistent with a more realistic magnetotail model in which such particles would likely reenter the tail lobes and drift back down into the plasma sheet

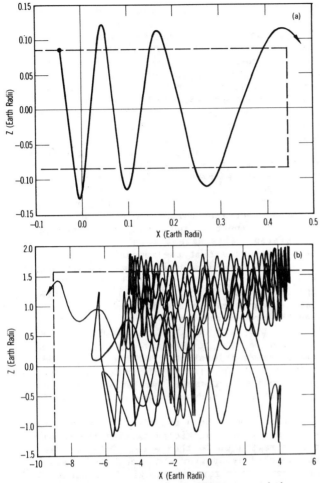

Fig. 1. Proton orbits for initial x-position $x_0 = -0.01L$; (a) thin sheet (T = 1000 km), 5 keV initial energy, 80° initial pitch angle; (b) thick sheet (T = 20,000 km), 100 keV initial energy, 70° initial pitch angle. Dashed lines show the boundary of the acceleration region.

downstream of the neutral line region. Hence, the present simplified model cannot adequately deal with these orbits and they are not considered in the following.

The basic motion is as discussed above, but the details vary considerably depending on the initial parameters and sheet thickness. Each time the particle visits the central region where E is significant relative to B, it is accelerated. The time spent there generally decreases on each visit. When the motion is adiabatic (in regions where B is dominant), the particle is decelerated, as the gradient drift is against the electric field. In the transitional region, where B is still dominant but the motion is nonadiabatic, energy can be gained or lost. Thus, the net energy gain for a given trajectory depends on the relative amount of time spent in each region, while the particle is within the L by T acceleration region.

Consider now the effect of varying the thickness T. Figure 1 shows representative orbits for the two sheet thicknesses considered. It can be seen that for a thin current sheet, the particle exits the acceleration region before any mirroring can occur. On the other hand, for the thicker sheet the particle can go down one or another valley and get mirrored several times before exiting from the acceleration region. This qualitative behavior is independent of initial conditions over a wide range (as verified by calculating orbits for several hundred initial conditions) and has an important effect on the energy gained by the particle. For the thin sheet, the particle spends a relatively large amount of time in the region where E dominates, and can therefore gain a significant amount of energy. For the thicker sheet the particle loses energy on each excursion down a valley, and combining this with a relatively long initial adiabatic drift toward the center, the net effect can actually be an energy loss.

This is shown graphically in Figure 2 for

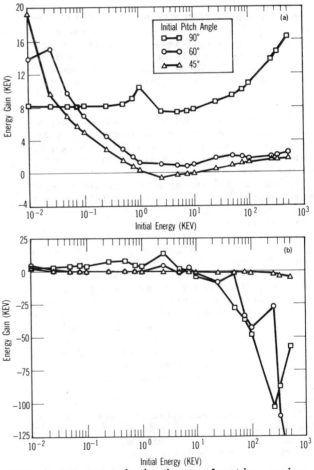

Fig. 2. Energy gain in the acceleration region vs. initial energy for the same initial x-position as in Figure 1; (a) thin sheet; (b) thick sheet.

particles with initial pitch angles of 90°, 60°, and 45°, and initial x-position near the center of the box. It can be seen from the figure that for initial energies greater than about 1 keV, the thin sheet is a more efficient accelerator of protons. For the thick sheet, all initial pitch angle particles are decelerated above 5 keV input energy.

It is interesting to compare these results with those of Speiser and Lyons [1984], who consider a field reversal geometry with a constant normal field. Using a value for this constant normal field equal to the average normal field within the acceleration region used here, one obtains energy gains of about 15 keV for the thin sheet and 1 keV for the thick sheet (for 90° initial pitch angle). Thus, the thin sheet values are not that different, while the thick sheet neutral point region is a more efficient accelerator for low energy particles, and less efficient than the constant normal field model for high energy particles.

Conclusions

We have shown that a thin plasma sheet containing an X-type neutral line acceleration region can be an efficient accelerator of protons. Contrary to expectations from fluid theory, however, thicker plasma sheets can either accelerate or decelerate protons. The results of the numerical calculations can be summarized as follows:

1) The energy gained by a singly charged ion while passing through the acceleration region depends on sheet thickness, initial energy, initial pitch angle, initial x-position, and particle mass.

2) Low initial energy particles are always energized; higher pitch angle particles tend to gain greater energy.

3) Higher initial energy particles usually gain energy in the thin sheet but lose energy in the thicker sheet. Higher pitch angles tend to lose more energy.

4) Scaling the particle mass by a factor α scales both the dimensionless sheet thickness and the ratio ϵ by a factor $\alpha^{-1/2}$. The dimensionless energy would then scale with a factor of α^{-1}. Thus, a 160 keV oxygen ion in a 10,000 km sheet would have the same orbit as a 10 keV proton in a 2500 km sheet with one quarter the electric field. Mass scaling for the field reversal geometry has been previously calculated by Rothwell and Yates [1985] and Baker, et al. [1982].

5) The largest excursion for accelerated particles in the cross-tail y-direction was about 10^5 km, or less than one half of the width of a 40 earth radius wide plasma sheet. Some decelerated particles, however, could reach the flanks of the plasma sheet.

Rothwell and Yates [1979], for a current reversal geometry with no electric field, noted that a thin sheet is required to obtain orbits which cross the z = 0 plane. The present case, although more complicated, is analogous. Within the acceleration region, the orbits in the thin sheet are mainly of a crossing variety (e.g. Figure 1). Since such an orbit samples the strong E field near z = 0 it gains significant energy. The thick sheet orbits are partly of this type, but also consist of the noncrossing excursions down the valleys and the initial noncrossing drift, both of which cause a decrease in kinetic energy. Thus, during substorm onset when the plasma sheet thins, one expects more accelerated particles, which seems consistent with observation.

Property (4) above shows that as the sheet thickness increases during substorm recovery, one should expect to see higher energy particles decelerated. This agrees with observations in the tail showing that the intensity of high energy (130 keV) ions decays more rapidly during substorm recovery than that of lower energy ions [Ipavich et al, 1985]. Their data also suggests that oxygen intensity may decay more rapidly than the proton intensity. From the scaling given in property (4) it might seem that this would not be true here, since the oxygen effectively sees a thin sheet. But the energy and electric field scaling modify this, so the net result is not obvious and further work on comparison of the theory and observation is indicated.

It should also be noted that in a more realistic tail model, particles exiting the acceleration region would then enter a region like that modeled by Speiser and Lyons [1984] and and others. Thus further energization would occur until the particle is ejected from the current sheet. Order of magnitude calculations indicate that this further energization can compensate for the energy loss to high energy particles in the thick sheet, for some parameter values. Nonetheless, the total energy gain is still greater for the thin sheet.

Finally, recall that the dynamical system represented by equations (2) is most likely nonintegrable. This leads one to ask the question as to whether the system is chaotic. Calculations of the Lyapunov exponent [for a definition see Henon, 1983] do indicate chaotic motion, which is most pronounced for the thicker current sheets considered here. The effects of this stochasticity will be discussed in a separate paper.

Acknowledgements. The author gratefully acknowledges discussions with T.W. Speiser and H.H. Sauer and thanks F. Ipavich for pointing out the tail observations.

References

Baker, D.N., E.W. Hones, Jr., D.T. Young, and J. Birn, The possible role of ionospheric oxygen in the initiation and development of plasma sheet instabilities, Geophys. Res. Lett., 9, 1337-1340, 1982.

Brecht, S.H., J.G. Lyon, J.A. Fedder, and K. Hain, A time dependent three-dimensional simulation of the earth's magnetosphere: reconnection events, J. Geophys. Res., 87, 6098-6108, 1982.

Coroniti, F.V., Explosive tail reconnection: the growth and expansion phases of magnetospheric substorms, J. Geophys. Res., 90, 7427-7447, 1985.

Hayashi, T., and T. Sato, Externally driven magnetic reconnection and a powerful magnetic energy converter, Phys. Fluids, 22, 1189-1202, 1979.

Henon, M., Numerical exploration of Hamiltonian systems, Choatic Behavior in Deterministic Systems., Les Houches session XXXVI, 1981, North-Holland Publ., 55-170, 1983.

Ipavich, F.M., A.B. Galvin, M. Scholer, G. Gloeckler, D. Hovestadt, B. Klecker, Suprathermal O^+ and H^+ ion behavior during the March 22, 1979 (CDAW 6), substorms, J. Geophys. Res., 90, 1263-1272, 1985.

Lyon, J.A., S.H. Brecht, J.D. Huba, J.A. Fedder, P.J. Palmadesso, Computer simulation of a geomagnetic substorm, Phys. Rev. Lett., 46, 1038, 1981.

Lyons, L.R., and T.W. Speiser, Evidence for Current sheet acceleration in the geomagnetic tail, J. Geophys. Res., 87, 2276-2286, 1982.

Rothwell, P.L., and G.K. Yates, Active experiments and single ion motion in the magnetotail, in Active Experiments in Space, ESA-195, pp. 341-345, 1985.

Rothwell, P.L., and G.K. Yates, A dynamical model for the onset of magnetospheric substorms, in Dynamics of the Magnetosphere, D. Reidel Publ. Co., ed. S.-I. Akasofu, 497-518, 1979.

Rusbridge, M.G., Non-adiabatic charged particle motion near a magnetic field zero line, Plasma Phys., 13, 977-987, 1971.

Sonnerup, B.U.O., Adiabatic particle orbits in a magnetic null sheet, J. Geophys. Res., 76, 8211-8221, 1971.

Speiser, T.W., Particle trajectories in model current sheets, 1, analytical solutions, J. Geophys. Res., 70, 4219-4226, 1965.

Speiser, T.W., Conductivity without collisions or noise, Planet. Space Sci., 18, 613-622, 1970.

Speiser, T.W., and Lyons, L.R., Comparison of an analytical approximation for particle motion in a current sheet with precise numerical calculations, J. Geophys. Res., 89, 147-158, 1984.

Stern, D.P., The role of O-type neutral lines in magnetic merging during substorms and solar flares, J. Geophys. Res., 84, 63-71, 1979.

Wagner, J.S., J.R. Kan, and S.-I. Akasofu, Particle dynamic in the plasma sheet, J. Geophys. Res., 84, 1979.

Ugai, M., Self-consistent development of fast magnetic reconnection with anomalous plasma resistivity, Plasma Phys. Controlled Fusion, 26, 1549-1563, 1984.

Section IV. EQUATORIAL REGION PROCESSES

ACCELERATION OF ENERGETIC OXYGEN ($E > 137$ KEV) IN THE STORM-TIME RING CURRENT

A. T. Y. Lui, R. W. McEntire, S. M. Krimigis, and E. P. Keath

The Johns Hopkins University, Applied Physics Laboratory, Laurel, Maryland 20707

Abstract

Acceleration of energetic oxygen ($E > 137$ keV) in the storm-time ring current region during the geomagnetic storm of September 4-7, 1985 is investigated with measurements from the Medium Energy Particle Analyzer (MEPA) on the Charge Composition Explorer (CCE). Large intensity increases of oxygen ions, up to a factor of 2×10^3, were seen at $L \approx 2.5$ to 7.0 during the main phase of the geomagnetic storm. Analysis shows that the intensity enhancement of the locally mirroring oxygen ions at $L \approx 3.5$ to 5.5 is more than expected from betatron acceleration of the prestorm local population. However, an inward radial displacement by approximately $1.5 R_e$ of the pre-storm energetic oxygen population accompanied by betatron acceleration can account for the intensity increase at this L range. The result also indicates a new finding that an additional oxygen ion source or acceleration mechanism exists at $L \approx 6.5$ to 8.

Introduction

Although many spacecraft have surveyed the trapped particle and ring current populations in the past and extensive research on these particles has been conducted (Davis and Williamson, 1963; Smith and Hoffman, 1973; Spjeldvik, 1983; Johnson, 1983; Lyons and Williams, 1984, and references therein), the Charge Composition Explorer (CCE) of the Active Magnetospheric Particle Tracer Explorers (AMPTE) mission carries for the first time a comprehensive set of particle detectors that allows direct measurement of the composition of the main ring current population (Krimigis et al., 1982). A concerted effort has been made by the CCE investigators and others to examine the storm-time ring current characteristics for the magnetic storm of September 4-7, 1984 (see the special issue of *Geophysical Research Letters*, Vol. 12, No. 5, 1985). From these studies, the ring current energy density for that particular storm is found to reside mainly in protons and the contribution from oxygen ions is not more than 30% (Krimigis et al., 1985; Gloeckler et al., 1985). Furthermore, the electrical current density is deduced to be largely from the pressure gradient of the ring current protons (McEntire et al., 1985a).

This paper reports further analysis of the magnetic storm period in the high energy portion of the population ($E > 56$ keV for protons, $E > 72$ keV for helium, and $E > 137$ keV for the carbon-nitrogen-oxygen (CNO) group) measured by the Medium Energy Particle Analyzer (MEPA) on CCE. From pulse height analysis (PHA) data for this period, the CNO channels are found to be dominated by oxygen ions. Therefore, the ion intensity for CNO is referred to hereafter as the oxygen ion intensity. The largest increase in particle intensity for MEPA is observed in the rate channel for oxygen ions at 137 to 365 keV during the main phase of the magnetic storm (McEntire et al., 1985a). The purpose of this paper is to investigate the evolution of the large increases of oxygen ions by examining their phase space density variations during the course of the storm.

CCE Orbit and MEPA

The CCE was placed in a near-equatorial orbit with an inclination angle of 4.8°, an apogee of $8.8 R_e$, a perigee of 1108 km, and an orbital period of 15.7 hr. This orbital configuration permits the CCE to sample the near-equatorial particle intensity of L-shells from $L = 1.1$ to 9 in less than 8 hr. The MEPA contains a thin-foil solid-state-detector telescope that determines incident ion mass by measuring the time of flight (TOF) and total energy. Each event recorded by MEPA is eligible for 8-bit PHA of the logarithms of its TOF and energy signal, yielding a partition of 256 energy channels. The angular distribution of the ions is sampled in 32 sectors perpendicular to the spacecraft spin axis. The counts in each sector are accumulated and read out individually at one to four spin intervals (about 6 to 24 s), depending on the channel. Further details on the MEPA instrument are given in McEntire et al. (1985b).

Radial Profiles of Particle Intensities

The top panel of Figure 1 shows the D_{st} values during the magnetic storm interval. In this storm, three instances of ring current enhancements are noted, the first at approximately 1600 UT on September 4, the second at approximately 0200 UT on September 5, and the third at approximately 1200 UT on September 5 (Williams and Sugiura, 1985). There are two extrema in the D_{st} values, one at the time of the first ring current enhancement and the other at 0700 to 0800 UT on September 5. Below the D_{st} panel are radial profiles of spin-averaged particle intensities obtained from 10 spacecraft passes — eight passes during the storm interval and two prestorm passes for comparison. The radial profiles obtained in the inbound (outbound) passes are given at the left (right). An order of magnitude increase in proton intensity (56 to 190 keV) is seen in the L range of 2.5 to 4 in the dusk sector. In the pre-noon sector, no substantial change is found until the early recovery phase of the storm. The time evolution of the helium intensity is very similar, but the increase is larger. For the oxygen ions, the intensity increase is seen from $L = 2.5$ to 7 with an enhancement factor of more than 10^3 at $L = 3.5$ to 4.0.

Temporal Evolution of Phase Space Density at $L = 4$

It is possible that the increase in oxygen ion intensity may be due to betatron acceleration of the local population in association with changes in the local magnetic field strength. To explore this possibility, we restrict ourselves to consider only particles at 90° pitch angle for simplicity. The evaluation of acceleration mechanisms for particles with

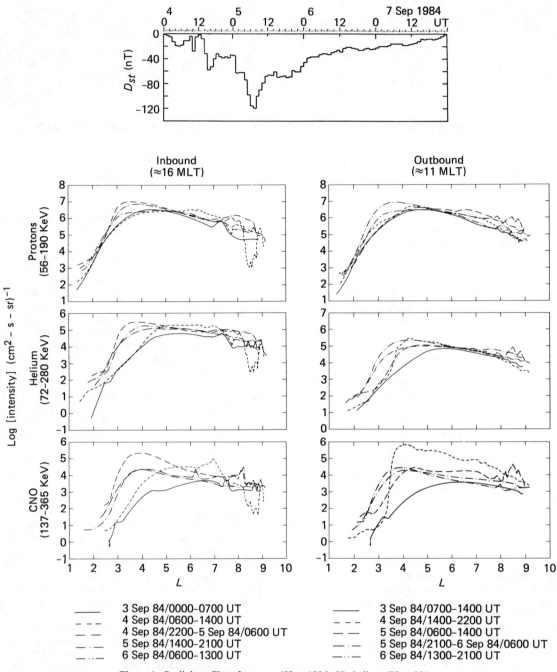

Figure 1. Radial profiles of protons (59 to 190 keV), helium (72 to 280 keV), and CNO (137 to 365 keV) during the magnetic storm of September 4-7, 1985, (McEntire et al., 1985a).

other pitch angles (e.g., Fermi acceleration) requires a realistic magnetic field model capable of reproducing the changes in the field configuration during a geomagnetic storm and will not be attempted here. Since the spacecraft trajectory is near the magnetic equator ($B/B_0 = 1.0$ to 1.04), we are essentially examining the intensity variations of the equatorially mirroring particles. For betatron acceleration, in which the magnetic moment is conserved, the equatorially mirroring particles should show no increase in their phase space density (PSD) as a func-

tion of their magnetic moment. More precisely, this is strictly true if the particle trajectories were followed and loss processes can be neglected. However, the present measurements may be regarded as "following the trajectories of equatorially mirroring particles" if the prestorm ring current population is azimuthally symmetric. Figure 2 shows the result of the analysis. The local magnetic field values have been kindly provided by Drs. T. A. Potemra and L. J. Zanetti. The oxygen ion intensity is obtained by combining the intensity measured in the TOF

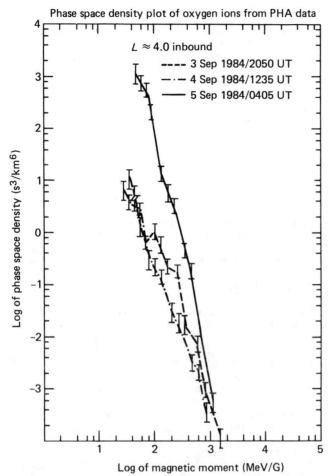

Figure 2. Phase space density of oxygen ions at $L = 4$ during three passes.

Radial Profiles of Phase Space Density

In order to study the radial transport as a mechanism for PSD increase at $L = 4$, we have extracted the PSD at 300 MeV/gauss from $L = 3$ to 9 at 0.5 L intervals. Figure 3 shows the resulting radial profile for the September 3 prestorm pass (the dashed curve), and for the September 4 pass during the main phase (the solid curve). The prestorm radial profile indicates that the PSD is relatively constant at a high value outside of $L = 6$ and decreases by about a factor of 20 from $L = 6$ to 5; the decrease is presumably associated with the plasmapause where strong pitch angle diffusion and significant loss can take place.

There is a significant increase in the storm-time PSD from $L = 3.5$ to 5.5. It is important to point out that if the prestorm radial profile is shifted inward by about 1.5 L, there is very good agreement between the two profiles at the gradient of the PSD and further out. The close agreement suggests that the PSD increase between $L = 3.5$ to 5.5 can be accounted for by an inward displacement of the pre-existing trapped particle population by approximately 1.5 R_e. The inward displacement may be achieved with a stochastic radial diffusion process or an enhanced global convection. If the latter is the case, then the 1.5 R_e displacement in an interval of 4 to 8 hr requires an average inward transport speed of 0.4 to 0.8 km/s and an average convection electric field of the order of 0.1 to 0.01 mV/m, which are reasonable values. It should be pointed out that the result of ion acceleration in the storm-time ring current region arising from inward displacement of the preexisting population has been discussed by several researchers (Söraas and Davis, un-

oxygen rate channels and the PHA data. Counts from 14 PHA energy channels are summed to provide good counting statistics for each energy interval shown. During the first storm pass on September 4, the PSD shows no significant change in comparison with the prestorm values over the range of magnetic moment covered by MEPA. However, in the subsequent inbound pass on September 5, the PSD shows a significant increase. Progressively larger enhancements are found at lower values of the magnetic moment or energy. This result indicates that the increase in the oxygen ion intensity in the locally mirroring population cannot be due to simply the betatron acceleration of the prestorm local population.

There are at least two possible ways to account for the increase in the PSD. One is by pitch angle diffusion of particles originating at smaller pitch angles. This includes the ionosphere as a possible immediate source because ionospheric ions which initially populate small pitch angles in the equatorial region could diffuse in pitch angle space to 90°. Since the pitch angle distribution at $L = 4$ is a pancake, with a maximum at 90°, the pitch angle diffusion process will tend to depopulate the 90° angle particles and thus this first possibility is unlikely. The second possibility is by particles transported radially and is examined in the next section. It should be noted that the oxygen flux enhancement for this storm period at all L values consists primarily of O^+ (Gloeckler et al., 1985; Krimigis et al., 1985) rather than higher charge states, implying an ultimate ionospheric source.

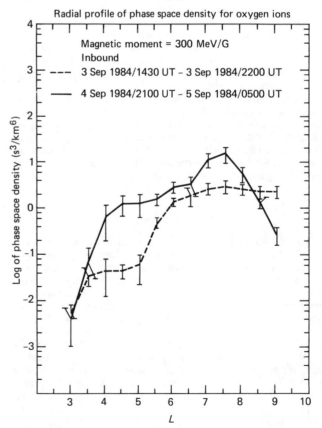

Figure 3. Radial profile of the phase space density at 300 MeV/gauss for the prestorm pass and the pass during the main phase of the magnetic storm.

published reports, 1968, 1969; Burns and Krimigis, 1972; and Lyons and Williams, 1980). The last authors reached this conclusion by comparing measured ion energy spectra with the theoretically predicted spectra assuming 1 to 3 R_e inward displacement during two geomagnetic storms. This paper differs from their earlier work in three aspects. First, oxygen ions are studied here whereas ion fluxes (nondiscriminating for different species) were used in the previous work. Second, the radial profiles of the PSD at a given magnetic moment are examined. Third, measurements are made out to $L \approx 9$, extending the earlier measurements (limited to $L \leq 4$; for Lyons and Williams, 1980). The extension to larger L values shows the departure from this simple inward displacement picture at $L = 6.5$ to 8.

Another significant increase in the PSD profile is between $L = 6.5$ to 8. There are three possible explanations for this increase. One possibility is that energetic oxygen ions are fed from the ionosphere along magnetic field lines to the equatorial region. The ionospheric oxygen ions, starting at small pitch angles in the equatorial region, could increase their pitch angles via some pitch-angle scattering process. We note that the angular distributions of these ions at this L value range (corresponding to the auroral latitudes) are cigar shaped (i.e., peaked at small pitch angles) and thus are consistent with this hypothesis. However, ion acceleration near the auroral ionosphere is usually of the order of a few keV and another process is required to bring the oxygen ions to hundreds of keV. The second possibility is due to an inward radial diffusion of PSD outside the CCE orbit, i.e., beyond $L = 9$. This appears unlikely because the relatively constant PSD at $L > 6$ for the prestorm radial profile suggests that the PSD outside of $L = 9$ is probably not higher than that at $L = 9$. The third possibility is acceleration associated with dipolarization of magnetic field configuration during the geomagnetic storm. Recently, Mauk and Meng (1985) have modeled the effect of this field dipolarization to show that this process can lead to large acceleration of particles with small pitch angles which could then be pitch-angle scattered to 90°. However, this acceleration process does not appear to be very efficient for particles with high initial energies.

Another point worth mentioning is that the storm-time radial profile starts with a lower PSD at $L = 9$. This may be attributed to particle loss due to inward displacement of the magnetopause during the early phase of the magnetic storm and particles escaping in the magnetosheath on the dayside. Consistent with this reasoning is the observation that the dayside magnetopause is found to pass inside the CCE orbit during the early phase of this geomagnetic storm (Shelley et al., 1985).

Summary and Conclusions

The intensity enhancement of oxygen ions in the storm-time ring current for the magnetic storm of September 4-7, 1985 has been investigated. It is found that the observed increase in oxygen ion intensity at $L = 3.5$ to 5.5 is not due to simple betatron acceleration of the locally mirroring oxygen population. However, an inward displacement of the pre-storm oxygen population accompanied by betatron acceleration can account for the observed oxygen ion enhancement at this L range ($L = 3.5$ to 5.5), in agreement with the result of Lyons and Williams (1980). In addition, there is an indication of a source for oxygen ions at 90° pitch angle present in the region at $L = 6.5$ to 8. Three candidates for the source are ionospheric acceleration with subsequent pitch angle diffusion, inward radial transport of an enhanced phase space density region outside the CCE orbit ($L = 9$), and acceleration associated with dipolarization of the magnetic field configuration during the geomagnetic storm accompanied by subsequent pitch angle diffusion. It appears that none of the three candidates is entirely satisfactory for the increased phase space density at $L = 6.5$ to 8. Therefore, it cannot be ruled out that an unknown mechanism other than these three candidates may be responsible.

Acknowledgment. This work has been supported by NASA under Task I of U. S. Navy Contract N00024-85-C-5301.

References

Burns, A. L., and S. M. Krimigis, Changes in the distribution of low-energy trapped protons associated with the April 17, 1965 magnetic storm, *J. Geophys. Res., 77*, 112, 1972.

Davis, L. R., and J. M. Williamson, Low-energy trapped protons, *Space Res., 3*, 365, 1963.

Gloeckler, G., B. Wilken, W. Studemann, F. M. Ipavich, D. Hovestadt, D. C. Hamilton, and G. Kremser, First composition measurement of the bulk of the storm-time ring current (1 to 300 keV/e) with AMPTE/CCE, *Geophys. Res. Lett., 12*, 325, 1985.

Johnson, R. G., *Energetic Ion Composition of the Earth's Magnetosphere,* D. Reidel Publishing Co., Dordrecht, 1983.

Krimigis, S. M., G. Haerendel, R. W. McEntire, G. Paschmann, and D. A. Bryant, The active magnetospheric particle tracer explorers program, *EOS, 63*, 843, 1982.

Krimigis, S. M., G. Gloeckler, R. W. McEntire, T. A. Potemra, F. L. Scarf, and E. G. Shelley, Magnetic storm of September 4, 1984: a synthesis of ring current spectra and energy densities measured with AMPTE/CCE, *Geophys. Res. Lett., 12*, 329, 1985.

Mauk, B. H., and C.-I. Meng, Macroscopic ion acceleration associated with the formation of the ring current in the earth's magnetosphere, submitted to *Proceedings of the the Chapman Conference on Ion Acceleration,* 1985.

McEntire, R. W., A. T. Y. Lui, S. M. Krimigis, and E. P. Keath, AMPTE/CCE energetic particle composition measurements during the September 4, 1984 magnetic storm, *Geophys. Res. Lett., 12*, 317, 1985a.

McEntire, R. W., E. P. Keath, D. E. Forth, A. T. Y. Lui, and S. M. Krimigis, The Medium Energy Particle Analyzer (MEPA) on the AMPTE/CCE spacecraft, *IEEE Trans. Geoscience and Remote Sensing, GE-23*, 230, 1985b.

Lyons, L. R., and D. J. Williams, A source for the geomagnetic storm main phase ring current, *J. Geophys. Res. 85*, 523, 1980.

Lyons, L. R., and D. J. Williams, *Quantitative Aspects of Magnetospheric Physics,* Reidel Publ. Co., 1984.

Shelley, E. G., D. M. Klumpar, W. K. Peterson, A. Ghielmetti, H. Balsiger, J. Geiss, and H. Rosenbauer, AMPTE/CCE Observations of the plasma composition below 17 keV during the September 4, 1984 magnetic storm, *Geophys. Res. Lett., 12*, 321, 1985.

Smith, P. H., and R. A. Hoffman, Ring current particle distributions during the magnetic storms of December 16-18, 1971, *J. Geophys. Res., 78*, 4731, 1973.

Spjeldvik, W. N., Ionic composition of the earth's radiation belts, *J. Geophys., 52*, 215, 1983.

Williams, D. J. and M. Sugiura, The AMPTE Charge Composition Explorer and the 4-7 September 1984 Geomagnetic storm, *Geophys. Res. Lett., 12*, 305, 1985.

ON THE LOSS OF O^+ IONS (< 17 keV/e) IN THE RING CURRENT DURING THE RECOVERY PHASE OF A STORM

J.B. Cladis and O.W. Lennartsson

Lockheed Palo Alto Research Laboratory, Palo Alto, California

Abstract. Under favorable circumstances some trapped ions that drift toward the west may be observed twice along their drift paths by the ISEE-1 satellite; once while the satellite is inbound and again while it is outbound. Since, in the absence of diffusion and loss, the phase space densities of the ions at the two observation points should be the same, the drift paths may be identified by comparing the measured phase space densities along the outbound and inbound legs. The H^+ and O^+ measurements made on the satellite during the recovery phase of the storm of December 11, 1977, were examined in this manner. Using the inbound measurements as initial conditions, ion drift paths were calculated for a dipole magnetic field containing a corotation electric field and a convection field. It was found that for a convection electric field appropriate for $Kp = 2$, the phase space densities of H^+ at the two measurement points along the drift paths were approximately the same. However, the phase space densities of O^+ ions, which should have followed the same adiabatic drift paths, were much lower along the outbound leg. The implied loss of O^+ ions along these westward drift paths is attributed to a bounce–resonance interaction of the ions with standing Alfvén waves that were simultaneously observed with five satellites. The observations also revealed that the phase space densities of the eastward drifting H^+ ions (energies at which the corotation drift exceeds the curvature-gradient drift) were about the same in the L shells at the two measurement points. Again, however, the corresponding O^+ phase space densities were lower along the outbound leg. These low-energy ion observations are attributed to a decrease with time of the phase space density of the O^+ ions (but not the H^+ ions) in the magnetotail during the convection process.

Introduction

During magnetic storms, high fluxes of O^+ and H^+ ions are injected into the ring current region [e.g., Balsiger et al., 1980; Krimigis et al., 1985]. The measurements of Balsiger et al. [1979] of ions in the energy-per-charge range 0.9–16 keV/e reveal that at such times the number density of O^+ quite often exceeds the H^+ density and the O^+ density in the dayside magnetosphere is higher than it is on the nightside. After a storm at $L > 5$, the O^+ to H^+ density ratio in this energy interval decreases rapidly, with the O^+ density decreasing faster—and the H^+ density decreasing slower—than the charge-exchange loss rates.

In the study discussed here, the behavior of these ions during the recovery phase of a magnetic storm was examined by comparing the ion distribution functions measured with the ISEE-1 satellite in the magnetosphere along its inbound and outbound legs. The results indicate that the difference in the day–night O^+ densities and the decrease of the O^+/H^+ density ratio are due principally to the convection process which continually replaces the O^+ in the outer ring current with an O^+ phase space density that is monotonically decreasing in the magnetotail during the recovery phase. The phase space density of H^+ that is convected inward is relatively constant. Moreover, a correlation of the ISEE-1 O^+ data with multisatellite measurements of the electric and magnetic field fluctuations implies that the ions which drifted toward the west in the pre-noon local time sector were lost through a bounce resonance interaction with standing Alfvén waves.

Spacecraft Measurements

During the magnetic storm of December 11, 1977, the D_{st} index decreased sharply from about 0 at 0020 UT to about -100 nT at 0800 UT and further decreased to about -125 nT at 1200 UT before entering the recovery phase. The three hourly K_p indices beginning at 0000 UT were 4, 5-, 5+, 5+, 5+, 4-, and 3. About 15 hours after the onset of the storm, the ISEE-1 satellite entered the magnetosphere and observed the injected H^+ and O^+ ions along the path (L value versus magnetic local time) shown in Figure 1. The ions were measured in the energy-per-charge range 0.01–17.4 keV/e with the onboard mass spectrometers [Shelley et al., 1978]. Some of the measurements made during this storm were discussed by Lennartsson et al. [1979]. The phase space densities of the H^+ and O^+ ions in the L intervals 8.40–8.63, 7.01–7.61, and 5.84–6.93 while the satellite was inbound and outbound are shown in Figure 2. The local pitch angles of the ions were in the range 45–135°, corresponding to equatorial pitch-angle ranges of 41.9–72.4° and 107.6–138.1° on the inbound path, and 32.3–49.1° and 130.9–147.7° on the outbound path. The counting statistics of the outbound O^+ measurements were poor in comparison to those of the other measurements because the background was high.

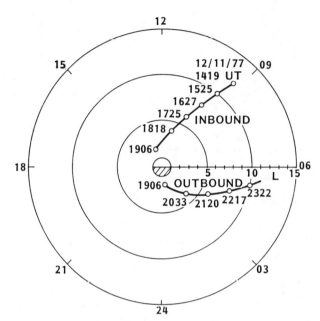

Fig. 1. ISEE-1 satellite trajectory (L versus magnetic local time) through magnetosphere during recovery phase of storm of December 11, 1977.

comparable to the O^+ flux. Accordingly, in order to improve the statistical accuracy, the counting rates were averaged over broader energy intervals. The energies over which the counting rates were averaged and the resulting standard deviation of the measurements based on counting statistics alone are indicated by the bars in parts d–f of Figure 2.

Simultaneously, magnetic pulsations were measured in the pre-noon local time sector near the synchronous orbit with five satellites: GEOS-2, SMS-2, ATS-6, ISEE-1, and ISEE-2 [Singer et al., 1979]. In addition, the corresponding electric field fluctuations were measured with spherical double probes [Mozer et al., 1978] on the ISEE-1 satellite. The electric field measurements made while the satellite was inbound were provided to us by Mozer [personal communication, 1984] and are shown in Figure 3. Singer et al. [1979] reported that the nature of these electric and magnetic field variations was consistent with the presence of standing Alfvén waves. They also reported that the period of these waves was consistent with the period inferred from the plasma composition measured with the ISEE-1 satellite. Another feature of the observations is that the period of the pulsations measured by the satellites decreased toward earlier local times.

Analysis

Loss of Westward Drifting O^+ Ions. If ion motions are adiabatic, then in the absence of loss, the phase space densities of the ions should be constant along their trajectories. Hence, if some of the westward-drifting ions observed along the inbound leg again reached the satellite on its outbound leg, the phase space densities at the two observation points should be the same. In order to test this condition, bounce-averaged drift paths of ions in a dipole magnetic field with superimposed corotation and convection electric fields were computed as described by Cladis and Francis [1985] to determine whether such drift paths could be found, and if so, at what points the drift paths intersected the outbound leg. The recent Kp-dependent convection electric field patterns constructed by Baumjohann et al. [1985] from data obtained with the GEOS-2 electron gun experiment were used for these calculations.

Connecting drift paths were found for convection electric fields ranging from zero to the values given by Baumjohann et al. for Kp = 2. During the course of the measurements, the Kp index was higher than 2; it was decreasing from 4^- to 3. Nevertheless, the drift motions computed with the convection field for Kp = 2 are regarded to be appropriate because the actual magnetic field intensity in the region under consideration was higher than that given by the dipole field.

The two arrows at the abscissæ of the graphs in Figure 2 denote energies of ions that have critical drift paths when the convection field is neglected. The equatorial pitch angle of the ions was taken to be 45.5°, which is at the center of the overlapping pitch angle intervals of the ions measured along the inbound and outbound legs. In each graph, the lower denoted energy is that of an ion that has a stationary drift path; i.e., its corotation drift is equal and opposite to its curvature-gradient drift. Higher-energy ions drift toward the west and lower-energy ions drift toward the east. An ion with the energy denoted by the second arrow has a drift path that connects the two satellite positions at the mean value of the L interval shown at the top of the graph. At these energies, the phase space densities of the ions at the two measurement points should be the same for a negligible convection field.

As shown in Figure 2, the H^+ densities at these energies are higher on the outbound leg (broken-line curves in panels a – c) than on the inbound leg (solid-line curves) by about a factor of 2, while the O^+ densities at these energies are lower on the outbound leg (data points in panels d – f) than on the inbound leg (solid-line curves) by nearly an order of magnitude. Both these H^+ and O^+ density differences probably would have been somewhat larger if the spectrometers viewed the same pitch-angle intervals along the outbound and inbound legs because the distribution functions of H^+ generally decrease—and those of O^+ increase—toward lower pitch angles. This effect of the pitch angle differences, however, tends to be reduced by the effect of L-shell splitting. For H^+ better agreement (to within a factor of 1.5) was reached when the convection field for Kp = 2 was used. In this case, for example, an ion with an energy of 18.4 keV at the point L = 6.3, LT = 152° of the inbound leg drifted in 4.86 hours to the point L = 8.11, LT = 73° of the outbound leg where its energy was 11.3 keV. Further agreement probably cannot be expected because of the limitations of the magnetic and convection electric field models. However, the discrepancy in the O^+ densities remains large, implying a rapid loss mechanism that acts selectively on the O^+ ions along their drift paths.

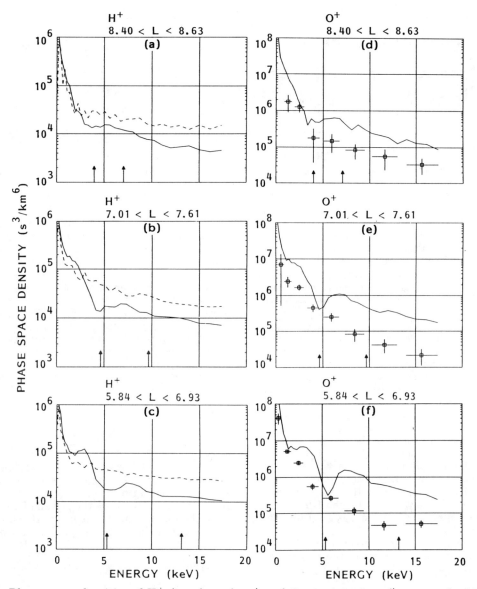

Fig. 2. Phase space densities of H^+ (graphs a, b, c,) and O^+ (graphs d, e, f) measured while the satellite was inbound (solid lines) and outbound (broken lines in a, b, c; data points in d, e, f). The L value intervals are given at the tops of the graphs.

The electric field fluctuations measured on the ISEE-1 satellite (Figure 3) have dominant frequency components near the bounce frequencies of the westward drifting O^+ ions. Figure 4 shows the power spectrum of the E_y component from 1610 UT (L = 8.5) to 1712 UT (L = 6.3), and Figure 5 shows the bounce times versus energy of H^+ and O^+ ions in a dipole magnetic field for the extreme parameters: L = 8.5, $\alpha_0 = 42°$ and L = 6.3, $\alpha_0 = 90°$. The horizontal broken lines in Figure 5 are the wave periods at one-fourth the height of the principal power peak, near 2.8 mHz. Note that the dipole-field bounce times of O^+ ions from about 4 to 50 keV are within the periods of the principal power peak. The bounce periods of the westward drifting H^+ ions are far removed from the wave periods. It appears, therefore, that the loss of westward drifting O^+ ions is through a bounce-resonance interaction with the standing Alfvén waves.

Loss of Eastward Drifting Ions. Note in Figure 2 that the O^+ densities along the outbound leg are lower than those along the inbound leg at all energies, not only at the high energies discussed above, whereas the corresponding densities of H^+ are approximately the same at low energies. As discussed by Lennartsson and Shelley [1985], the average value of the O^+ number-density

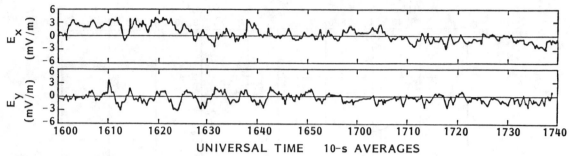

Fig. 3. E_x and E_y components (GSE coordinates) of electric field fluctuation measured on ISEE-1 satellite [Mozer et al., 1978].

in the magnetotail (10 to 23 R_E) decreases appreciably as the Kp and AE indices decrease (about an order of magnitude over the full ranges of these indices) while the average H^+ number density increases somewhat as these indices decrease. We wondered, therefore, whether the O^+ densities were lower along the outbound leg because they were lower in the magnetotail before they were convected inward to the outer ring current. To investigate this possibility, the drift paths of the ions measured along the inbound and outbound legs were computed backward in time to determine the times (hence the Kp values) at which the ions were in the magnetotail. Since the convection field patterns of Baumjohn et al. [1985] are strictly valid only near the synchronous orbit, the Volland-Stern convection potential with the shielding factor, $\gamma = 2$, was used for these calculations (see Cladis and Francis, 1985). However, at $L \leq 7$ the pattern was rotated toward earlier local times by 1.5 hours to simulate the orientation of the patterns found by Baumjohann et al. [1985]. The rotation angle was reduced smoothly to zero from $L = 7$ to $L = 10$. The coefficient of the potential was related to the Kp index, as it varied with time during the storm, through the expression given by Maynard and Chen [1975]. The trajectories were terminated when the ions reached 12 R_E in the magnetotail.

The results of these calculations revealed that the ions detected along the inbound leg were in the magnetotail (at 12 R_E) at the time when the Kp index was about 5^+, and that the ions detected along the outbound leg with initial energies less than about 12 keV were in the magnetotail at least six hours later in time when the Kp index was about 3. Accordingly, these results verify that the loss of O^+ implied by the data, at least at energies less than about 12 keV, may be attributed to the convection process that continually replaces O^+ in the outer ring current during the recovery phase of a storm with a lower concentration of O^+ from the magnetotail. Ions initially on the outbound leg with energies higher than about 12 keV drifted backward in time toward the east, crossing the inbound leg and going around the earth, before reaching the tail. These ions were, therefore, in the magnetotail earlier than ions with similar energies observed along the inbound leg.

Discussion

Both loss mechanisms discussed here indicate that the number density of O^+ should decrease progressively toward earlier local times in the sector between the inbound and outbound legs. Since the period of a standing wave is inversely proportional to the mean Alfvén velocity along a magnetic field line, hence approximately proportional to

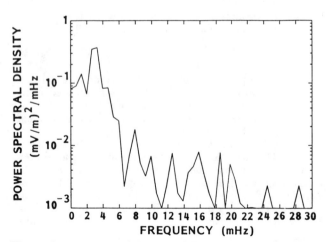

Fig. 4. Power Spectral density of E_y component of electric field fluctuations from 1610 to 1712 UT.

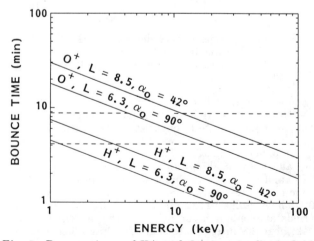

Fig. 5. Bounce times of H^+ and O^+ ions in dipole field as function of energy for L values and pitch angles given in figure. The dominant periods of the electric field fluctuations are within the horizontal broken lines.

the ion mass density at the equator, the progressive loss of O^+ toward earlier local times appears to be responsible for the multisatellite observations of the decrease of the standing-wave period in that direction.

The role of convection discussed here on the variation of the ion densities beyond the plasmapause offers an explanation of the spatial and temporal variations of O^+ and H^+ reported by Balsiger et al. [1980]. The day-night asymmetry of the O^+ number density—as well as the loss of O^+ —during the recovery phase of a storm could be due to the inward convection of an O^+ phase space density that decreases monotonically in the magnetotail. The H^+ number density remains relatively constant because the phase space density of H^+ convected inward from the tail is more nearly the same.

Conclusions

This study indicates that the phase space density of O^+ ions in the magnetotail steadily decreased with time during the recovery phase of the storm of December 11, 1977, while the density of the H^+ remained approximately the same. The observed loss of low energy O^+ ions in the outer ring current is attributed to the inward convection of the decreasing O^+ density in the magnetotail. A loss of low energy H^+ ions in the outer ring current was not observed because it was continually replenished by the convection process. It is suggested that such differences in the inward convection of O^+ and H^+ could account for several of the observations reported by Balsiger et al. [1980], viz. the higher O^+ density in the dayside magnetosphere than on the nightside, the enhanced loss rate of O^+ (faster than the charge exchange rate) at $L > 5$, and the low loss-rate of H^+ (lower than the charge-exchange rate).

The decrease in the period of the standing Alfvén waves toward earlier local times observed by five satellites (including the ISEE-1 satellite) is consistent with the progressive decrease in the O^+ density from the dayside to the nightside magnetosphere.

Evidence is also presented that the rapid loss of the westward drifting ions was through a bounce-resonance interaction of the ions with the standing Alfvén waves.

Acknowledgments. We are grateful to Y.T. Chiu, E.G. Shelley, and J.M. Quinn for helpful discussions in the course of this work. This effort was supported by the Lockheed Independent Research program and by NASA under contracts NAS5-28413 and NAS5-28702.

References

Balsiger, H., P. Eberhardt, J. Geiss, and D.T. Young, Magnetic storm injection of 0.9- to 16-keV/e solar and terrestrial ions into the high-altitude magnetosphere, *J. Geophys. Res.*, 85, pp. 1645–1662, 1980.

Baumjohann, W., G. Haerendel, and F. Melzner, Magnetospheric convection observed between 0600 and 2100 LT: Variations with Kp, *J. Geophys. Res.*, 90, pp. 393–398, 1985.

Cladis, J.B. and W.E. Francis, The polar ionosphere as a source of the storm time ring current, *J. Geophys. Res.*, 90, pp. 3465–3473, 1985.

Krimigis, S.M., G. Gloeckler, R.W. McEntire, T.A. Potemra, F.L. Scarf, and E.G. Shelley, Magnetic storm of September 4, 1984: A synthesis of ring current spectra and energy densities measured with AMPTE CCE, *Geophys. Res. Lett.*, 12, pp. 329–332, 1985.

Lennartsson, W. and E.G. Shelley, Survey of 0.1 to 16 keV/e plasma sheet ion composition, *J. Geophys. Res.*, in press, 1986.

Lennartsson, W., E.G. Shelley, R.D. Sharp, R.G. Johnson, and H. Balsiger, Some initial ISEE-1 results on the ring current composition and dynamics during the magnetic storm of December 11, 1977, *Geophys. Res. Lett.*, 6, pp. 483–486, 1979.

Maynard, N.C. and J.J. Chen, Isolated cold plasma regions: observations and their relation to possible production mechanisms, *J. Geophys. Res*, 80, pp. 1009–1011, 1975.

Mozer, F.S., R.B. Torbert, U.V. Fahleson, C.G. Falthammar, A. Gonfalone, and A. Pedersen, Measurements of quasi-static and low-frequency electric fields with spherical double probes on the ISEE-1 spacecraft, *IEEE Trans on Geoscience Electronics*, GE-16 (3), pp. 258–261, 1978.

Shelley, E.G., R.D. Sharp, R.G. Johnson, J. Geiss, P. Eberhardt, H. Balsiger, G. Haerendel, and H. Rosenbauer, Plasma composition experiment on ISEE-A, *IEEE Trans. on Geoscience Electronics*, GE-16 (3), pp. 266–270, 1978.

Singer, H.J., C.T. Russell, M.G. Kivelson, T.A. Fritz, and O.W. Lennartsson, Satellite observations of the spatial extent and structure of Pc 3, 4, 5 pulsations near the magnetospheric equator, *Geophys. Res. Lett.*, 6, pp. 889–892, 1979.

EIGENFUNCTION METHODS IN MAGNETOSPHERIC RADIAL-DIFFUSION THEORY

Michael Schulz

Space Sciences Laboratory, The Aerospace Corporation, El Segundo, California 90245

Abstract. Complete sets of orthonormal basis functions constructed according to a generalization of the quantum-mechanical WKB approximation can be used to generate a nearly-diagonal matrix representation of the radial-transport operator for ring-current ions in the presence of radial diffusion and charge exchange. The resulting eigenfunctions (constructed by weighting the basis functions in proportion to the respective components of the eigenvectors of the matrix representation) and eigenvalues provide a spatial and temporal description of the evolving phase-space density during and following a magnetospheric disturbance (e.g., a magnetic storm). A linear superposition of the basis functions can also be used to eliminate any discrepancy between the steady-state solution of the transport equation and the appropriate WKB approximation of this steady-state solution.

Introduction

If energy degradation (Coulomb loss) and pitch angle diffusion can be neglected, then the radial transport of magnetospheric ring-current ions is governed by an equation of the form

$$\partial \bar{f}/\partial t = L^2 (\partial/\partial L)[(D_{LL}/L^2)(\partial \bar{f}/\partial L)] - (\bar{f}/\tau_q)$$
$$= (\partial/\partial \Phi)[D_{\Phi\Phi}(\partial \bar{f}/\partial \Phi)] - (\bar{f}/\tau_q), \quad (1)$$

where $\bar{f}$ is the phase-space density at fixed M and J (first two adiabatic invariants), D_{LL} is the diffusion coefficient for transport in L (dimensionless shell parameter), Φ is the third adiabatic invariant (inversely proportional to L), and τ_q is the ionic lifetime against charge exchange. Jentsch [1984] has described a method for obtaining approximate steady-state solutions of this equation when D_{LL} is exactly proportional to a fixed power (β) of L, i.e., when $D_{\Phi\Phi}$ is exactly proportional to a fixed power ($4-\beta$) of Φ. The purpose of the present work is to describe an alternative method for obtaining time-dependent as well as steady-state solutions of (1) while permitting the dependence of D_{LL} upon L (or equivalently, of $D_{\Phi\Phi}$ upon Φ) to deviate somewhat from a strict power law.

The alternative method is highly advantageous because even the simplest dynamical models for magnetospheric radial diffusion lead to diffusion coefficients D_{LL} that deviate in fact from strict power laws except in certain limits. For example, the standard model [e.g., Cornwall, 1972; Schulz, 1983] for charged-particle diffusion in a dipolar magnetic field leads to a diffusion coefficient of the form

$$D_{LL} \approx \frac{1 \times 10^{-10} L^{10} \text{ day}^{-1}}{(M/y^2 \gamma Z M_0)^2 [2D(y)/T(y)]^2 + 10^{-6} L^4}$$
$$+ 7 \times 10^{-9} [Q(y)/180 D(y)]^2 L^{10} \text{ day}^{-1}, \quad (2)$$

where y is the sine of the equatorial pitch angle α_0, γ is the ratio of relativistic mass m to rest mass m_0, Z is the integer that specifies charge state, and $M_0 \equiv 1$ GeV/gauss. The auxiliary functions $Q(y)$, $D(y)$, and $T(y)$ in (2) are well approximated [Schulz and Lanzerotti, 1974, pp. 20, 21, 44; Davidson, 1976] by the algebraic expressions

$$Q(y) \approx -27.12667 - 45.39913 y^4 + 5.88256 y^8, \quad (3a)$$

$$D(y) \approx 0.4600577 + 0.1066154 y^{3/4} - 0.1997662 y, \quad (3b)$$

and

$$T(y) \approx 1.3801730 - 0.6396925 y^{3/4}. \quad (3c)$$

A further complication is that y and γ in (2) typically vary with L at fixed M and J. The variation of y is given approximately [Chen and Stern, 1975] by

$$y^{-2} \approx 1 + 1.38048 X - 0.030425 X^{4/3}$$
$$+ 0.10066 X^{5/3} + [X/2T(0)]^2, \quad (4)$$

where $X \equiv (La/8m_0 \mu M)^{1/2} J$, a is the radius of the earth, and μ is the earth's magnetic moment. The variation of γ with L is given by

$$\gamma^2 = 1 + (2\mu M/L^3 a^3 y^2 m_0 c^2), \quad (5)$$

where c is the speed of light. The limiting cases $X = 0$ ($J = 0$) and $X = \infty$ ($M = 0$) correspond to $y = 1$ and $y = 0$, respectively, but $X = \infty$ implies $M/y^2 = J^2La/32\mu m_0[T(0)]^2$ upon evaluation of the indeterminate form.

Transformations

The factors $[2D(y)/y^2T(y)]^2$ and $[Q(y)/D(y)]^2$ in (2) vary approximately as powers of L [Schulz and Lanzerotti, 1974, pp. 91, 93]. The exponents of L are given by

$$2L(\partial/\partial L)\{\ln[D(y)/y^2T(y)]\}_{M,J}$$
$$= \frac{Y(y)}{24D(y)}\left[10 - \left[1 + \frac{yT'(y)}{T(y)}\right]\frac{Y(y)}{T(y)}\right] = \begin{cases} 0, & y=1 \\ 2, & y=0 \end{cases} \quad (6a)$$

and

$$2L(\partial/\partial L)\{\ln[Q(y)/D(y)]\}_{M,J}$$
$$= \left[\frac{yD'(y)}{2D(y)} - \frac{yQ'(y)}{2Q(y)}\right]\frac{Y(y)}{T(y)} = \begin{cases} 0, & y=1 \\ 0, & y=0 \end{cases} \quad (6b)$$

respectively, where [Schulz and Lanzerotti, 1974, pp. 20, 21]

$$\left(\frac{\partial \ln y}{\partial \ln L}\right)_{M,J} = -\frac{Y(y)}{4T(y)} = \begin{cases} 0, & y=1 \\ -1/2, & y=0 \end{cases} \quad (7a)$$

and

$$Y(y) = 2y\int_y^1 (y')^{-2}T(y')\, dy'$$
$$= 6[T(y) - 2D(y)] = \begin{cases} 0, & y=1 \\ 2T(0), & y=0 \end{cases} \quad (7b)$$

The exponents of L implied by (6) remain approximately (but only approximately) independent of L for fixed (but nonvanishing) M and J. They remain strictly independent of L only for $y = 1$ and for $y = 0$, and even in these cases the form of D_{LL} specified by (2) is not strictly proportional to a fixed power of L, e.g., proportional to L^β with β a function of M and J only.

The transformation $z = (\beta - 3)\ln L$ proposed by Jentsch [1984] brings (1) into an equation of the form

$$\partial w/\partial t = (\beta-3)^2 L^{-2} D_{LL}[(\partial^2/\partial z^2) - (1/4)]w - (w/\tau_q), \quad (8)$$

where $w = L^{(\beta-3)/2}\bar{f}$, if $D_{LL} \propto L^\beta$ for fixed β. The transformation

$$\zeta = \int_L^\infty (L')^2 D_{L'L'}^{-1}\, dL' = L^2(d\Phi/dL)\int_0^\Phi D_{\Phi'\Phi'}^{-1}\, d\Phi' \quad (9)$$

introduced in the present work brings (1) into an equation of the form

$$\partial \bar{f}/\partial t = (L^4/D_{LL})(\partial^2 \bar{f}/\partial \zeta^2) - (\bar{f}/\tau_q) \quad (10)$$

without recourse to the assumption that $D_{LL} \propto L^\beta$ for fixed β. The factor $L^2(d\Phi/dL)$ in (9) is a constant, since $\Phi \propto 1/L$, and the factor L^4/D_{LL} in (10) is proportional to $D_{\Phi\Phi}^{-1}$ for the same reason. The fixed limit of integration $L' = \infty$ ($\Phi' = 0$) in (9) lies outside the domain of validity of (1) and thus requires the integrand of (9) to be evaluated by analytical extrapolation for $L' > L_1$ (see below). However, it follows from (9) that $\zeta = (\beta-3)^{-1}L^3 D_{LL}^{-1} \propto L^{3-\beta}$ if $D_{LL} \propto L^\beta$ for some fixed $\beta > 3$, and no other fixed limit of integration in (9) would lead to such a simple form for ζ. The form of (10) suggests a time-dependent solution

$$\bar{f}(L,t) = \bar{f}_\infty(L) + \sum_{n=0}^\infty a_n(t)g_n(L) \quad (11)$$

in which $\bar{f}_\infty(L)$ is the steady-state solution of (10) and the $g_n(L)$ are the eigenfunctions of the operator $\Lambda \equiv -(L^4/D_{LL})(\partial^2/\partial\zeta^2) + (1/\tau_q)$, corresponding (respectively) to the eigenvalues λ_n. The expansion coefficients $a_n(t)$ are thus given by $a_n(t) = a_n(0)\exp(-\lambda_n t)$ if the transport coefficients and boundary conditions are time independent for $t > 0$.

Steady State

The steady-state solution $\bar{f}_\infty(L)$ thus satisfies, according to (10), the equation

$$(d^2\bar{f}_\infty/d\zeta^2) - (D_{LL}/L^4\tau_q)\bar{f}_\infty = 0, \quad (12)$$

subject to the inner boundary condition that $\bar{f}_\infty(L_0) = 0$ at the top of the atmosphere and the outer boundary condition that $\bar{f}_\infty(L_1)$ correspond to the phase-space density at the inner edge of the plasma sheet. The exact solution of (12) is expressible [cf. Schiff, 1955, p. 187; Walt, 1970, p. 414] in terms of modified Bessel functions of fractional order if $D_{LL}/L^4\tau_q$ is exactly proportional to a fixed power (p) of ζ. This fact suggests a modified WKB approximation of the form $\bar{f}_\infty(L) \approx \hat{f}_\infty(L)$ for the steady-state solution of (1) and (10), where

$$\hat{f}_\infty(L) \equiv (\hat{\theta}/\hat{\theta}_1)^{1/2}(L/L_1)(\tau_q/D_{LL})_1^{-1/4}(\tau_q/D_{LL})^{1/4}$$
$$\times \frac{I_\nu(\hat{\theta})K_\nu(\hat{\theta}_0) - K_\nu(\hat{\theta})I_\nu(\hat{\theta}_0)}{I_\nu(\hat{\theta}_1)K_\nu(\hat{\theta}_0) - K_\nu(\hat{\theta}_1)I_\nu(\hat{\theta}_0)}\bar{f}_\infty(L_1) \quad (13)$$

and

$$\hat{\theta} \equiv \int_0^\zeta (D_{L'L'})^{1/2}[(L')^4\tau_q(L')]^{-1/2}d\zeta'$$
$$= \int_L^\infty [\tau_q(L')D_{L'L'}]^{-1/2}dL'$$
$$= \int_0^\Phi [\tau_q(\Phi')D_{\Phi'\Phi'}]^{-1/2}d\Phi'. \quad (14)$$

The name "modified WKB approximation" is suggested by the appearance of modified (rather than ordinary) Bessel functions in (13) as a consequence of the negative (minus) sign in (12). The optimal order ν of the modified Bessel functions $I_\nu(\hat{\theta})$ and $K_\nu(\hat{\theta})$ in (13) is given by $\nu = 1/(\bar{p}+2)$, where $\bar{p}$ is a representative value of

$$\hat{p} \equiv (d \ln D_{\Phi\Phi}/d \ln \zeta) - (d \ln \tau_q/d \ln \zeta)$$

$$= (d \ln D_{LL}/d \ln \zeta) - 4(d \ln L/d \ln \zeta)$$

$$- (d \ln \tau_q/d \ln \zeta) \quad (15)$$

within the interval $L_0 \leq L \leq L_1$. For $D_{LL} \propto L^\beta$ and $\tau_q \propto L^\gamma$ exactly (i.e., with fixed β and γ) one obtains $\hat{\theta} = 2(\beta+\gamma-2)^{-1} L \tau_q^{-1/2} D_{LL}^{-1/2}$, which is proportional to the power $(2-\beta-\gamma)/2$ of L. One further obtains $\hat{p} = (\beta-\gamma-4)/(3-\beta)$ and $\nu = (\beta-3)/(\beta+\gamma-2)$, and in this case $\bar{f}_\infty(L)$ is given exactly by (13) if $\beta > 3$ and $\beta + \gamma > 2$. The subscripts 0 and 1 in (13) denote evaluation at $L = L_0$ and $L = L_1$, respectively. The limit $\tau_q(L) \to \infty$ in (12) yields $\bar{f}_\infty(L) = [(\zeta_0 - \zeta)/(\zeta_0 - \zeta_1)]\bar{f}_\infty(L_1)$ exactly. Expansion of the modified Bessel functions in (13) for small argument yields this same result for $\bar{f}_\infty(L)$ if β and γ are fixed (i.e., independent of L).

The above development of a modified WKB approximation for $\bar{f}_\infty(L)$ is somewhat reminiscent of the Green-Liouville solutions described by Jentsch [1984]. His Green-Liouville solutions involved hyperbolic-sine (sinh) functions, which are in fact proportional to modified Bessel functions of order $\nu = 1/2$. It seems that (13) is the appropriate generalization of the procedure described by Jentsch [1984] to situations for which $\bar{p} \neq 0$.

Eigenfunctions and Eigenvalues

The development of time-dependent solutions of (1) is facilitated by the eigenfunction expansion shown in (11), where

$$L^2(d/dL)[(D_{LL}/L^2)(dg_n/dL)] - \tau_q^{-1} g_n(L) + \lambda_n g_n(L)$$

$$= (d/d\Phi)[D_{\Phi\Phi}(dg_n/d\Phi)] - \tau_q^{-1} g_n(L) + \lambda_n g_n(L) = 0 \quad (16)$$

The eigenfunctions $g_n(L)$ are required to vanish both at $L = L_0$ and at $L = L_1$. Eigenfunctions corresponding to distinct eigenvalues λ_n and λ_m are necessarily orthogonal in the sense that

$$\int_{L_0}^{L_1} L^{-2} g_n(L) g_m(L) \, dL = \delta_{nm} \equiv \begin{cases} 0, & n \neq m \\ 1, & n = m \end{cases} \quad (17)$$

Given a complete set $\{\bar{g}_n(L)\}$ of orthonormal basis functions satisfying (17) and the boundary conditions $\bar{g}_n(L_0) = \bar{g}_n(L_1) = 0$, the required eigenfunctions $g_n(L)$ and eigenvalues λ_n can be obtained by diagonalizing the matrix representation

$$\Lambda_{nm} = \int_{L_0}^{L_1} L^{-2} \bar{g}_n'(L) D_{LL} \bar{g}_m'(L) \, dL$$

$$+ \int_{L_0}^{L_1} L^{-2} \bar{g}_n(L) \tau_q^{-1}(L) \bar{g}_m(L) \, dL \quad (18)$$

of the transport operator $\Lambda \equiv -(\partial/\partial\Phi)[D_{\Phi\Phi}(\partial/\partial\Phi)] + (1/\tau_q) = -L^2(\partial/\partial L)[(D_{LL}/L^2)(\partial/\partial L)] + (1/\tau_q)$. An optimal set $\{\bar{g}_n(L)\}$ of basis functions would be one that can be constructed by means of a fairly simple prescription, but one that makes the off-diagonal elements of Λ_{nm} especially small in absolute value.

It is evident from (18) that all the eigenvalues of Λ_{nm} are positive. This situation corresponds, of course, to temporal decay of the expansion coefficients $a_n(t)$ in (11). Moreover, if the eigenvalues are ordered (as usual) so that $0 < \lambda_0 < \lambda_1 < \lambda_2 < \ldots$, then it follows from (18) that (with increasing n) radial diffusion becomes increasingly important (compared to charge exchange) for the determination of λ_n. The presence of derivatives of the $\bar{g}_n(L)$ in the first term (but not in the second term) on the right-hand side of (18) assures this. Except for the term $\tau_q^{-1} \bar{g}_n(L)$, which has no counterpart in their paper, the eigenvalue equation specified by (16) is identical in form to the one for which Schulz and Boucher [1984] successfully constructed an optimal set of orthonormal basis functions by means of a variant of the WKB approximation. A further variant of that construction is required here, since the boundary conditions of the present radial-diffusion problem differ from the boundary conditions appropriate to the pitch-angle diffusion problem treated by Schulz and Boucher [1984].

The analogous construction appropriate to the radial-diffusion problem yields orthonormal basis functions of the form

$$\bar{g}_n(L) = (2\theta_n/\theta_n^0)^{1/2} D_{LL}^{-1/4} L [\int_{L_0}^{\infty} D_{L'L'}^{-1/2} \, dL']^{-1/2} C_\nu^*(\theta_n)$$

$$\times \{[C_{\nu\pm1}(\theta_n^0)]^2 - [\alpha C_{\nu\pm1}(\alpha\theta_n^0)]^2\}^{-1/2}, \quad (19)$$

where

$$C_\nu^*(\theta_n) \equiv J_\nu(\theta_n) Y_\nu(\alpha\theta_n^0) - J_\nu(\alpha\theta_n^0) Y_\nu(\theta_n) \quad (20a)$$

and

$$C_{\nu\pm1}(\theta_n) \equiv J_{\nu\pm1}(\theta_n) Y_\nu(\alpha\theta_n^0) - J_\nu(\alpha\theta_n^0) Y_{\nu\pm1}(\theta_n). \quad (20b)$$

The argument θ_n of the ordinary Bessel functions $J_\nu(\theta_n)$ and $Y_\nu(\theta_n)$ in (19) and (20) is given by

$$\theta_n \equiv \theta_n^0 \int_L^{\infty} D_{L'L'}^{-1/2} \, dL' \div \int_{L_0}^{\infty} D_{L'L'}^{-1/2} \, dL', \quad (21)$$

where θ_n^0 is the nth positive root ($n = 0,1,2,...$) of the equation

$$G_\nu^*(\theta) \equiv J_\nu(\theta)Y_\nu(\alpha\theta) - J_\nu(\alpha\theta)Y_\nu(\theta) = 0, \quad (22)$$

and where

$$\alpha \equiv \int_{L_1}^\infty D_{LL}^{-1/2} dL \div \int_{L_0}^\infty D_{LL}^{-1/2} dL < 1. \quad (23)$$

Since $C_\nu^*(\theta_n^0) = C_\nu^*(\alpha\theta_n^0) = 0$ for each value of n, it thus follows from (19)-(23) that $\bar{g}_n(L_0) = \bar{g}_n(L_1) = 0$, as is required. The normalization and mutual orthogonality [in the sense of (17)] of the basis functions $\bar{g}_n(L)$ specified by (19) can be verified by using θ_n as the variable of integration and invoking certain indefinite integrals evaluated by Watson [1944, pp. 148-149]. Since $C_{\nu+1}(\theta_n^0) = -C_{\nu-1}(\theta_n^0)$ and $C_{\nu+1}(\alpha\theta_n^0) = -C_{\nu-1}(\alpha\theta_n^0)$, the evaluation of (19) is unambiguous despite the available choice of sign in (20b). The optimal order ν of the ordinary Bessel functions in (20a) is given by $\nu = 1/(\bar{p}+2)$, where $\bar{p}$ is a representative value of

$$p \equiv (d \ln D_{LL}/d \ln \zeta) - 4(d \ln L/d \ln \zeta)$$
$$= (d \ln D_{\phi\phi}/d \ln \zeta) \quad (24)$$

within the interval $L_0 \leq L \leq L_1$ ($\theta_n^0 \geq \theta_n \geq \alpha\theta_n^0$). For $D_{LL} \propto L^\beta$ exactly (i.e., with fixed $\beta > 3$) one obtains $\theta_n = (L_0/L)^{(\beta-2)/2}\theta_n^0$, $\alpha = (L_0/L_1)^{(\beta-2)/2}$, $p = (4-\beta)/(\beta-3)$, and $\nu = (\beta-3)/(\beta-2)$. The basis functions specified by (19) should lead to a nearly diagonal matrix representation Λ_{nm}, as defined by (18), of the transport operator Λ. In other words, the diagonal element Λ_{nn} (at least for $n \geq 4$) should greatly exceed the absolute value of each off-diagonal element Λ_{nm} ($=\Lambda_{mn}$) in the same row or column of the matrix. This major benefit of the WKB construction of basis functions enables the eigenvalues and eigenvectors of Λ_{nm} (and therefore the eigenvalues and eigenfunctions of the transport operator Λ) to be evaluated by means of a rapidly convergent perturbation theory. The formal results [Schulz and Boucher, 1984] are

$$\lambda_n \approx \Lambda_{nn} - \sum_{k \neq n} \frac{\Lambda_{nk}\Lambda_{kn}}{\Lambda_{kk} - \Lambda_{nn}} \quad (25a)$$

and

$$g_n(L) = U_{nn}\bar{g}_n(L) + \sum_{k \neq n} U_{kn}\bar{g}_k(L), \quad (25b)$$

where

$$\frac{U_{kn}}{U_{nn}} \approx \frac{1}{\Lambda_{nn} - \Lambda_{kk}} \left[\Lambda_{kn} + \sum_{j \neq k,n} \frac{\Lambda_{kj}\Lambda_{jn}}{\Lambda_{nn} - \Lambda_{jj}} \right] \quad (25c)$$

for $k \neq n$ and

$$U_{nn} = \left[1 + \sum_{k \neq n} (U_{kn}/U_{nn})^2 \right]^{-1/2} \quad (25d)$$

to assure the unitarity of the transformation from the $\{\bar{g}_n(L)\}$ to the $\{g_n(L)\}$, i.e., to assure that the $\{g_n(L)\}$ are likewise normalized in accordance with (17).

A further use of the orthonormal basis functions $\bar{g}_n(L)$ specified by (19) is to eliminate altogether the presumably small discrepancy between the exact steady-state solution $\bar{f}_\infty(L)$ of (1) and the approximate steady-state solution $\hat{f}_\infty(L)$ given by (13) and based on the modified WKB approximation. This can be done by formally expanding the discrepancy as a general linear superposition of the $\bar{g}_m(L)$ and inserting the formal expansion, viz.,

$$\bar{f}_\infty(L) = \hat{f}_\infty(L) - \sum_{m=0}^\infty A_m \bar{g}_m(L), \quad (26)$$

into (1) for $\partial\bar{f}/\partial t = 0$. The result (after the usual straightforward steps) is a set of coupled linear equations given by

$$\sum_{m=0}^\infty \Lambda_{nm} A_m = \int_{L_0}^{L_1} L^{-2}\bar{g}_n'(L) D_{LL}(d\hat{f}_\infty/dL) \, dL + \int_{L_0}^{L_1} L^{-2}\bar{g}_n(L)\tau_q^{-1}(L)\hat{f}_\infty(L) \, dL \quad (27)$$

for the expansion coefficients A_m that should be inserted in (26). The solution of (27) is numerically well-determined, since the matrix Λ_{nm} given by (18) is supposed to be nearly diagonal when the basis functions $\bar{g}_n(L)$ are constructed in accordance with (19)-(24).

Applications

The decomposition of $\bar{f}(L,t) - \bar{f}_\infty(L)$ into eigenfunctions of the transport operator $\Lambda \equiv \tau_q^{-1}(L) - L^2(\partial/\partial L)[(D_{LL}/L^2)(\partial/\partial L)]$ provides an efficient means of describing the evolution of $\bar{f}(L,t)$ following a sudden change $\Delta\bar{f}(L_1)$ in the value of $\bar{f}$ at the outer boundary ($L = L_1$) of the ring current, i.e., at the inner edge of the plasma sheet. The initial expansion coefficients $a_n(0)$ in (11) for this scenario should be roughly proportional to $(n+1)^{-1}\Delta\bar{f}(L_1)$ if experience with analogous problems in mathematical physics is a valid guide. The infinite series in (11) should thus converge for $L_0 \leq L < L_1$, but numerical evaluation of $\bar{f}(L,t)$ for early times ($0 < \lambda_0 t \ll 1$) after the sudden change in $\bar{f}(L_1,t)$ will presumably require the retention of very many terms in the series. However, the spectrum of Λ is discrete, and the eigenvalues λ_n for sufficiently large n can be estimated as

$$\lambda_n \sim (n+1)^2 \pi^2 [\int_{L_0}^{L_1} D_{LL}^{-1/2} dL]^{-2}$$
$$+ [\int_{L_0}^{L_1} D_{LL}^{-1/2} dL]^{-1} \int_{L_0}^{L_1} D_{LL}^{-1/2} \tau_q^{-1}(L) \, dL \quad (28)$$

by asymptotically expanding the Bessel functions that appear in (19) and (20). Thus, the expansion coefficients $a_n(t) = a_n(0)\exp(-\lambda_n t)$ in (11) tend to decay more and more rapidly with increasing n, and so the numerical evaluation of $\bar{f}(L,t)$ will presumably require only a few terms of the infinite series for $\lambda_0 t \gtrsim 1$. The orthonormal basis functions $\bar{g}_n(L)$ specified by (19)-(23) certainly form a complete set, since their construction is analogous to the construction of basis functions for an infinitely deep potential well in quantum mechanics [e.g., Schiff, 1955, pp. 34-36].

A very minor change in terminology allows the present method to accommodate also the case in which $\bar{f}(L_1,t)$ varies smoothly rather than discontinuously in time. In this case the solution of (12) specified by (26) is characterized as the quasi-static solution $\bar{f}_\infty(L,t)$, and this is rather well approximated by the function $\hat{f}_\infty(L,t)$ specified by (13). Since $\hat{f}_\infty(L,t)$ is constructed so as to satisfy the fixed boundary condition $\bar{f}(L_0,t) = 0$ and the time-dependent boundary condition given as $\bar{f}(L_1,t)$, the difference between $\bar{f}(L,t)$ and $\bar{f}_\infty(L,t)$ can still be expanded in eigenfunctions $g_n(L)$ that vanish at $L = L_0$ and at $L = L_1$. Numerical evaluation of $\bar{f}(L,t)$ from (11) should require fewer terms from the infinite series if $\bar{f}(L_1,t)$ varies smoothly (rather than abruptly or discontinuously) with time so as to achieve the same $\Delta \bar{f}(L_1)$. If D_{LL} were multiplied by some time-dependent (but L-independent) factor, then the basis functions specified by (19)-(24) would remain unchanged. However, the eigenvalues and eigenvectors of (18) would then vary with time, as would the quasi-static solutions specified by (13) and (26). Coupled differential equations would then relate the expansion coefficients $a_n(t)$ in (11) to each other, but the eigenfunction expansion would remain valid in concept. The neglect of charge exchange, achieved by taking the limit $\tau_q(L) \to \infty$ in (18) and (27), would mitigate these complications. In this limit the eigenfunctions of Λ and the quasi-static solution $\bar{f}_\infty(L,t)$ would remain invariant to the multiplication of D_{LL} by a time-dependent (but L independent) factor. The eigenvalues λ_n would then be multiplied by this same time-dependent factor, but the differential equations for the expansion coefficients $a_n(t)$ in (11) would no longer be coupled to each other.

The deadline and page limit for papers to appear in this volume prevent the inclusion of numerical results illustrating the usefulness of analytical methods described above for solving problems in radial-diffusion theory. However, the numerical results of Schulz and Boucher [1984], showing (for example) that the off diagonal elements of Λ_{nm} are consistently smaller (by one to several orders of magnitude) in absolute value than the corresponding diagonal elements when analogous methods are applied to a prototypical pitch-angle diffusion problem, suggest that the present approach will be found highly advantageous when applied numerically to radial diffusion problems as well. Moreover, the ease with which charge exchange can be incorporated, as in (18), into the radial-diffusion problem suggests that charge exchange could similarly be handled together with pitch-angle diffusion in problems that require this, e.g., in studies of the evolving pitch-angle distribution of ring-current ions after charge exchange has made the equatorial pitch-angle distribution anisotropic enough to generate electromagnetic ion-cyclotron waves [cf. Cornwall, 1977]. For this latter application the basis functions for pitch-angle diffusion could be constructed according to the prescription of Schulz and Boucher [1984]. A long-range goal is to treat the simultaneous occurrence of radial diffusion and pitch-angle diffusion, in which case the L-dependent eigenvalues of the pitch-angle diffusion operator will presumably enter the mathematical description of radial transport [cf. Walt, 1970] in somewhat the same way that the charge-exchange rate $\tau_q^{-1}(L)$ enters (1). However, the bimodal (radial/pitch-angle) diffusion problem is complicated by the absence of a kinematical quantity that both modes of diffusion simultaneously conserve, and so the solution is presumably much more elusive for the bimodal diffusion problem than for the isolated radial diffusion problem treated here.

Acknowledgments. Portions of this work were supported by the NASA Solar-Terrestrial Theory Program (STTP) under contract NASW-3839. Other portions of this work were supported by the Air Force Weapons Laboratory (AFWL) and by the U. S. Air Force Systems Command's Space Division (SD) under contract F04701-84-C-0085. The author thanks Dr. Alice L. Newman for correcting a numerical error that had appeared in equation (19) in the original manuscript and for suggesting a clarifying remark regarding this equation.

References

Chen, A. J., and D. P. Stern, Adiabatic Hamiltonian of charged-particle motion in a dipole field, J. Geophys. Res., 80, 690-693, 1975.

Cornwall, J. M., Radial diffusion of ionized helium and protons: A probe for magnetospheric dynamics, J. Geophys. Res., 77, 1756-1770, 1972.

Cornwall, J. M., On the role of charge exchange in generating unstable waves in the ring current, J. Geophys. Res., 82, 1188-1196, 1977.

Davidson, G. T., An improved empirical description of the bounce motion of trapped particles, J. Geophys. Res., 81, 4029-4030, 1976.

Jentsch, V., The radial distribution of radiation belt protons: Approximate solution of the steady-state transport equation at arbitrary pitch angle, J. Geophys. Res., 89, 1527-1539, 1984.

Schiff, L. I., Quantum Mechanics, McGraw-Hill, New York, 1955.

Schulz, M., Principles of magnetospheric ion composition, in Energetic Ion Composition in the Earth's Magnetosphere, edited by R. G. Johnson, pp. 1-21, Terra Sci. Publ. Co., Tokyo, 1983.

Schulz, M., and D. J. Boucher, Jr., Orthogonal basis functions for pitch-angle diffusion theory, in Physics of Space Plasmas (1982-4), edited by J. Belcher, H. Bridge, T. Chang, B. Coppi, and J. R. Jasperse, pp. 159-168, Scientific Publishers, Inc., Cambridge, Mass., 1984.

Schulz, M., and L. J. Lanzerotti, Particle Diffusion in the Radiation Belts, Springer, Heidelberg, 1974.

Walt, M., Radial diffusion of trapped particles, in Particles and Fields in the Magnetosphere, edited by B. M. McCormac, pp. 410-415, Reidel, Dordrecht, 1970.

Watson, G. N., A Treatise on the Theory of Bessel Functions, Cambridge Univ. Press, Cambridge, England, 1944.

IONOSPHERIC ION STREAMS AT ALTITUDES BELOW 14 RE

Stefano Orsini and Maurizio Candidi

Istituto di Fisica dello Spazio Interplanetario, CNR, C. P. 27, 00044 Frascati (Roma), Italy

Abstract. The properties of ionospheric ion streams flowing away from Earth into the magnetotail are studied between 2 Re and 14 Re. General agreement is found between these observations and those made already at lower and larger distances. It is shown that at altitudes lower than 10 Re the flow of these ions is dominated by the dynamics of the source region; sometimes the satellite location is well suited to determine the flow pattern of the ions and the source altitude. The data presented here indicate that at higher distances from the earth, the ExB/B^2 filter organizes the streams according to their parallel velocity, so that no source properties can be derived by looking at single cases.

Introduction

The outflux of ions from the high latitude ionosphere and their streaming down the tail magnetic field lines has been studied in detail (Sharp et al., 1981; Candidi et al., 1982; Orsini et al., 1982). The effect of the magnetospheric ExB/B^2 drift, proposed by Rosenbauer et al. (1975) as the cause of the plasma mantle structure, has been shown to cause these ions to enter the plasma sheet boundary layer and be there mixed with the plasma sheet population (Orsini et al., 1984). Theoretical models have been proposed to demonstrate that the ionospheric plasma can be responsible for refilling the plasma sheet of the plasma it looses by other processes (Pilipp and Morfill, 1978) and studies of the relative importance of the ionospheric contribution with respect to the solar wind source have been published (Sharp et al., 1982).

These studies have generally been concentrated to high altitudes, above 10 Re, or to distances very close to the source, below 2 Re by Gorney et al. (1981) and Lockwood et al. (1985) and between 2 and 4 Re by Yau et al. (1984). Here we use data from the ISEE-2 EGD positive ion experiment to extend the observations to the intermediate range.

Experiments

The data used in this paper were obtained by several instruments. The technical details for each of them can be found in the literature, as referred in the following. The main body of data has been taken from the ISEE-2 positive ion experiment EGD (Bonifazi et al., 1978); magnetic field data from the ISEE satellites have been provided by the on board magnetometers (Russell, 1978).

Data analysis and discussion

Ion interaction with the plasma sheet

During the time interval from 0730 UT to 1530 UT of April 25, 1979, day 115, the ISEE-1 and -2 satellites were outbound between 2.7 and 14 Re; the satellites explored the northern near-Earth magnetotail on the dusk side, moving from 1600 to 2200 GSM local time at a GSM latitude of 35 to 40 degrees.

Figure 1 shows the parameters derived from the low energy ion distribution, as observed by the ISEE-2 EGD experiment in this event. For a discussion of the algorithm used in computing the plasma parameters see Orsini et al. (1985). These parameters are computed under the assumption that these ions be ionospheric oxygen; none of the inferences will be affected by the real identification of the ions, since we will look at relative variations only and not at the absolute value of the numbers. The three upper panels show the magnetic field at ISEE-2, elevation, azimuth and total intensity (from the top down); these show decreasing total field as we move from 2.7 Re to 14 Re, and zero elevation as is appropriate for the high latitude nightside field lines, while azimuth changes from -90 to zero degrees as the satellite moves from the Northern hemisphere dusk side to the midnight region. At the bottom of the figure the UT time and position of the satellite in GSM coordinates are given.

While the satellite moves from low to high altitude the average density decreases from values in the range .1 to 1 per cubic centimeter to values in the range .01 to .1 per cubic centimeter; at the same time there is a trend in the thermal velocity component parallel to the magnetic field that is higher before 0946 UT and decreases after that time. This corresponds to a

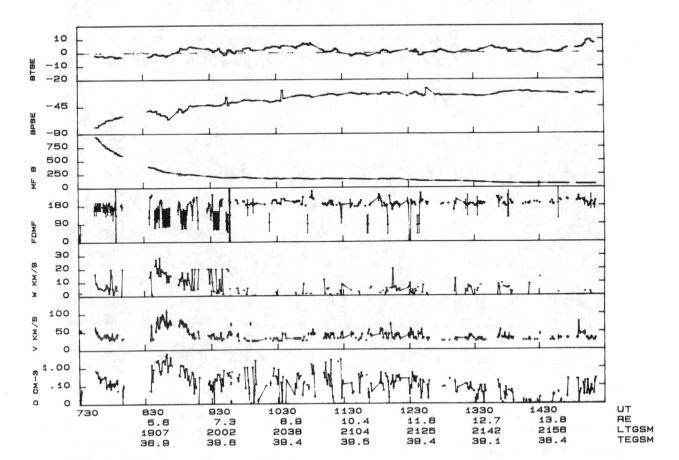

Fig. 1. The parameters of the streaming ions as computed from the EGD data (assuming O+) for day 115, 1979, 0730-1530 UT. From the bottom: density (cm-3), flow velocity (Km/s), thermal velocity component parallel to the ambient magnetic field (Km/s), flow direction with respect to the magnetic field GSE XY projection. The top three panels show the ISEE-2 GSE polar coordinates of the magnetic field (absolute value, longitude and latitude). The ISEE-2 satellite GSM position (local time, distance and latitude) is indicated at the bottom. The large error bars in panel 4 are due to the ion stream being detected in the wide angular sector.

transition from the plasma sheet into the Northern lobe, that is apparent in the ISEE-2 EGD experiment data (not shown here), in which the isotropic population above 1 keV/q disappears. The temperature of the streaming ions is higher in the (near Earth) plasma sheet than in the neighboring lobe as was already noticed at higher altitudes (R larger than 10 Re) in a statistical study by Sharp et al. (1981).

This transition may occur abruptly, like for instance on day 74, 1978 (March 15), between 0141 and 0154 UT. In such cases it is more apparent on a different display, as in figure 2, than in plots like figure 1. The data shown in figure 2 represent 9 time-contiguous full scans, each taken over 96 seconds, of the number fluxes as a function of energy and flow direction. When the isotropic signal above 2 keV/q is present, indicating the plasma sheet plasma, the energy spread of the anisotropic ions is much larger than at times when they are observed in the lobes.

This thermalization could be attributed to the interaction of these anisotropic stream with the plasma sheet energetic population. The fact that this occurs at distances very close to the Earth, together with the observation of the same effect at higher distances (Orsini et al., 1984; Orsini et al., 1985), demonstrates that the plasma sheet is supplied with ionospheric ions at all distances from the Earth, both in the near Earth high latitude extensions of the plasma sheet and beyond 10 Re in the plasma sheet proper. The ionospheric ions are thermalized at their input into the plasma sheet at any position along the tail, and

166 ION STREAMS AT LOW ALTITUDES

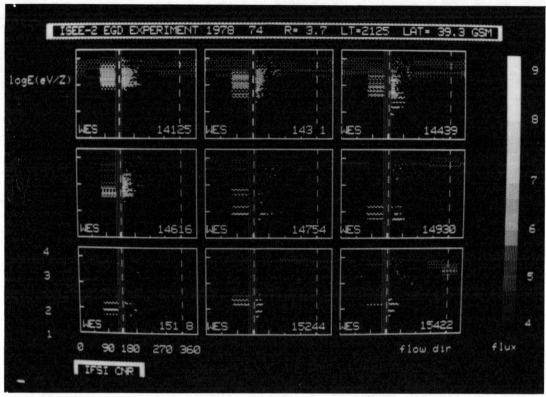

Fig. 2. Sequence of nine time-contiguous scans over 96 seconds of the number fluxes (cm-2 s-1 sr-1 (keV/q)-1) as a function of energy per charge and flow direction. The ISEE-2 EGD experiment data are shown on day 74, 1978, 0141-0154 UT. The gray tones are proportional to the log of the number fluxes according to the scale on the right side. The broken lines are the GSE XY directions of the ISEE-2 magnetic field. On the top side the ISEE-2 distance from the Earth, the GSM local time and elevation are listed (R, LT, LAT), these correspond to the position at the measurement cycle shown in the top left panel. At the bottom right of each panel the starting UT time is shown. The code WES at the bottom left indicates that the instrument is measuring in the whole energy range (from 55 eV/q to 11 keV/q).

become a part of the plasma sheet itself, following its general convection pattern. As a consequence of that, these streaming particles would be able to change "en route" the plasma sheet density, temperature and composition, also at higher latitudes, in the near Earth plasma sheet. An alternative scenario of these observations could be that these ions flow inside the plasma sheet boundary layer without penetrating into the plasma sheet; in this case a complicate mechanism able to trap these ions in the plasma sheet boundary layer should be invoked, once they drift there from the lobe region.

The ExB/B² filter effect

On day 158, 1979, between 0400 and 0630 UT, in an inbound orbit, the ISEE 2 satellite descends from 8.1 to 4.1 Re, in the 2200 to 2400 GSM local time sector, at a GSM latitude of -8 to -21 degrees; it is at all times in the Southern near Earth plasma sheet, which is detected in the energy channels above 1 keV/q by the EGD experiment.

Figure 3 shows the same parameters as in figure 1 for this period. Again we see relatively high temperatures parallel to the magnetic field in the streaming ions; moreover a slight trend shows increasing parallel temperatures with decreasing altitude. This can be interpreted as an effect of the magnetospheric ExB/ filter; as we approach the Earth, the source cannot any more be considered localized, and a wider range of ion velocities gain access to the same position in the near Earth magnetotail from different parts of the source region through different paths in the magnetospheric filter; this results in a higher computed temperature parallel to the magnetic field.

The same is seen on an individual case on day 86, 1979 (March 27) by comparing two single 96 seconds full energy scans, at 1552 UT, at 5.3 Re,

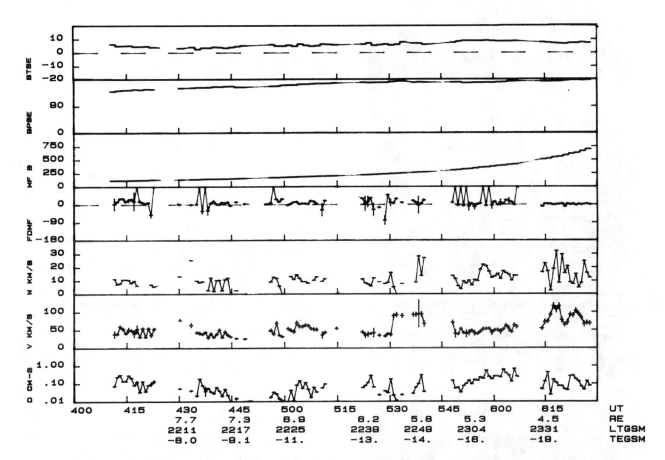

Fig. 3. Plot of the stream parameters (assuming O+) as computed from the ISEE-2 EGD experiment data for day 158, 1979, 0400-0630 UT. See figure 1 for description.

and later on at 1718 UT, at 7.6 Re, as shown in figure 4. In the upper panel, at 5.3 Re, the low energy stream is spread out between 1.0 keV/q and 8.2 keV/q (i.e. between 440 and 1250 km/s, for a proton equivalent velocity), while in the lower panel, at 7.6 Re, the spread is lower, between 400 eV/q and 1.5 keV/q (270-540 km/s). The peak energy is at 610 eV/q (340 km/s) in this last case. Other cases can be found in which the peak energy for the low energy ions is higher.

Streams close to the source region

Figure 5 shows data acquired on day 69, 1979, (March 10), between 2110 and 2123 UT, when ISEE 2 was around 2.9 Re, at 2033 GSM local time, 51 degrees above the GSM equatorial plane. The plasma sheet ions are seen at all times and result in the isotropic signal at or above 1 keV/q. At lower energy throughout this period ionospheric ions are flowing roughly along the magnetic field and away from the Earth. In this event the stream number flux, energy per charge and thermal spread strongly fluctuate during the whole period. These fluctuations can be attributed to variability of the source. The magnetospheric electric field scales with distance as $1/r$, while the magnetic field scales as $1/r^3$, such that the drift velocity, whose absolute value is E/B, scales as r^2, i.e. at lower altitudes the magnetospheric velocity filter is less and less effective. At 2.9 Re, that is relatively close to Earth on the scale of the distances covered by our previous studies, the combined effect of source extension and low drift velocity results in the observed spread in energy. It reflects in this case the energy distribution at the source unaffected by the filter that is weak at this distance and has not yet had the time to disperse the particles.

Another peculiar feature of these low altitude observations is shown in figure 6, day 129, 1978 (May 9), 0135 to 0150 UT, at 4.4 Re, 1710 GSM local time, and 44.5 degrees above the GSM equatorial plane. Streaming ions are observed along the magnetic field lines and towards the Earth, appearing first at relatively high energy,

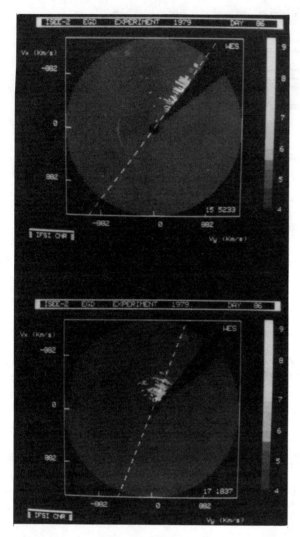

Fig. 4. Angular distributions of two full 96s energy scans as measured by the ISEE-2 EGD experiment on day 86, 1979, at 1552 UT (upper panel) and at 1718 UT (lower panel). The data are presented in a coordinate system GSE Vx versus GSE Vy. The gray tones are proportional to the log of the number fluxes (cm-2 s-1 sr-1 (keV/q)-1) according to the scales on the right of the single panels. The fan-shaped area on the upper right is the blind sector. The broken lines are the ISEE-2 magnetic field GSE X Y projections.

1.1 keV/q, and with decreasing energy as time progresses, down to the lowest energy channels of the experiment, 55 eV/q, and disappearing (panels 1 to 7; signal in the right side). After a certain delay, ions appear flowing away from Earth with energies that again decrease from 1.3 keV/q to 85 eV/q (panels 3 to 8; signal in the left side). This is the typical signature of energy-time dispersion of an ion stream that is injected at a certain distance from the observation point; the faster ions will arrive to the satellite in a shorter time than the slower ones; the ions will then flow down the magnetic field to their mirror point and will be reflected to be observed once more at the same location again with the higher energy ions appearing before the lower energy ions. The time delay between the appearance of the ions in one direction and their return in the opposite direction implies a total travel path of 2 to 3 Re; this seems rather short for ions observed at 4.4 Re altitude; still it is not inconsistent with the distances we are dealing with (especially taking into account the 96 seconds of uncertainty involved by the energy cycling of the instrument). There seems to be an increase in the energy of the ions when they appear after mirroring, which is evident in the maximum energy of the stream that changes from 1.1 keV/q to 1.3 keV/q; this feature should be studied on a larger number of cases to be able to understand whether this energy increase is consistent or it is just a statistical fluctuation observed in this case.

Figure 7 shows evidence for still another feature that is displayed by the ionospheric ions as they proceed from their source region to the high altitude magnetotail. The upper panel shows data on day 115, 1979 (April 25), 0514 UT, at 4.2 Re, 0153 GSM local time, when the satellite is at -24 degrees GSM latitude. The low energy ions are conically distributed around the magnetic field with a pitch angle of 12+2 degrees. Later on, at 0550 UT (middle panel), when the satellite has moved closer to the Earth, at 2.75 Re, 0400 GSM local time and -40 degrees GSM latitude, the ions are observed, in the same energy range with a different pitch angle of 47+12 degrees. Still later at 0602 UT (bottom panel), at 2.25 Re, 0422 GSM local time and -42 degrees GSM latitude, the pitch angle has increased up to 73+13 degrees. At lower altitude the pitch angle distribution of the ions shows up as a larger spread around the direction of the magnetic field than at higher distance; from these pitch angle values, the altitude of the source can be computed; the data at 0514, 0550 and 0602 UT, imply respectively an altitude of 1.2 to 1.8 Re, 1.8 to 2.4 Re, and 2.0 to 2.2 Re; in this last case the satellite is very close to the source and the pitch angle is very close to 90 degrees. During these 52 minutes ISEE-2 is moving in local time between 0153 and 0422, i.e. more than 35 degrees in longitude, and still the source seems to be uniformly emitting the ions that are observed in their adiabatic motion away from the Earth.

Conclusions

The ISEE-2 EGD experiment data presented in this paper show that, when moving away from the source region in the ionosphere into the tail, the velocity distribution of the ion streams is determined by the effect of the ExB/B convection,

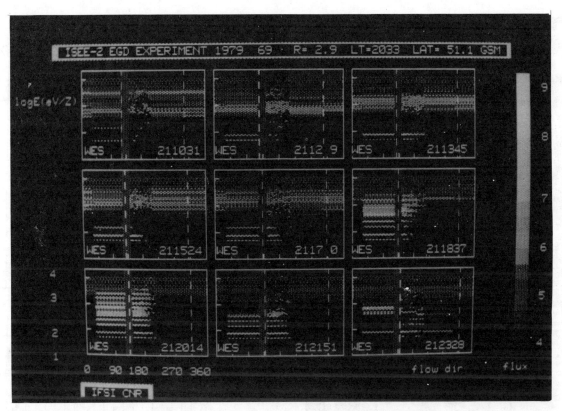

Fig. 5. Same as figure 2, for day 69, 1979, 0141-0154 UT.

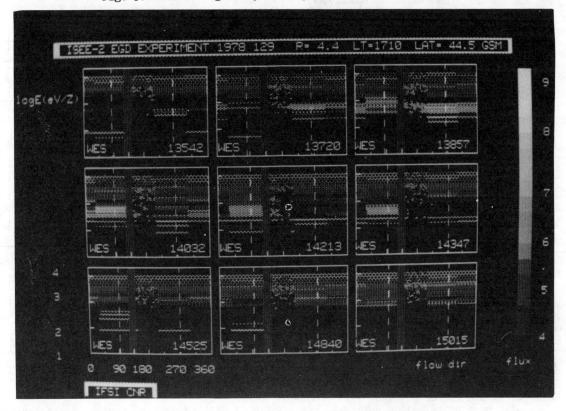

Fig. 6. Same as figure 2, for day 129, 1978, 0135-0150 UT.

170 ION STREAMS AT LOW ALTITUDES

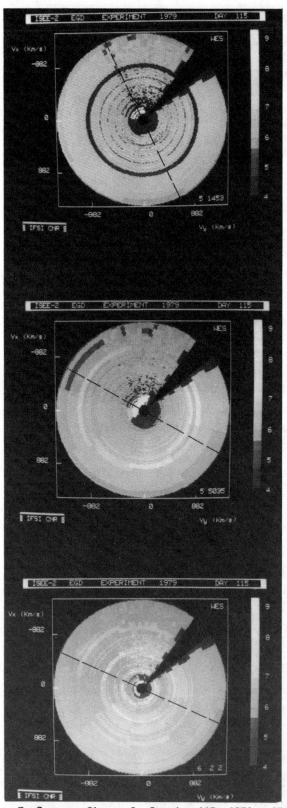

Fig. 7, Same as figure 2, for day 115, 1979, 0514 UT (upper panel), 0550 UT (middle panel) and 0602 UT (lower panel).

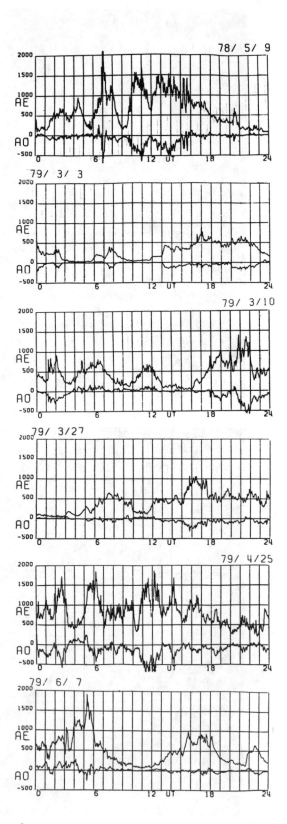

Fig. 8. AE index plots for the events presented in the text (from Kamei and Maeda).

combined with the adiabatic flow along the magnetic field. The properties of the source can only be inferred if this effect is carefully taken into account, and not just by looking at the ion parameters. This effect is reduced at closer distances and there the streams properties have stronger resemblance with the ion properties at the source. This shows up the observation of a larger energy spread, i.e. a larger stream temperature parallel to the magnetic field, when the satellite is closer to the source, and at times in the mixture of species that emerge together with wide energy distributions. Far away the velocity filter separates in energy the various populations and the local temperature of each of them decreases, so that the mixture of species is resolved also without a mass discriminator. Close to the source we notice correlation of the stream properties to AE, as on day 69, 1979 (Figure 5), when the variability of AE between 2100 and 2120 UT (Figure 8, increase from 700 nT to 1300 nT) corresponds to a sharp increase of the stream density, energy and temperature parallel to the magnetic field. Away from the Earth instead large fluctuations of AE do not show up in modulations of the stream properties; the parameters observed there are mainly dictated by the filter geometry (see for example, in Figure 1, day 115, 1979 when, in the lobe the stream parameters do not reflect the fluctuations of AE, Figure 8).

The interaction of the ionospheric ion streams with the plasma sheet plasma is shown to take place near the Earth as well as it has been previously shown at higher altitudes, so that we can conclude that the enrichment of the plasma sheet with ionospheric ions takes place at all distances. In the case shown in figure 7 it seems that this interaction starts occurring at altitudes which are very close to the source regions of the streams. In the same event the evidence of the angular spread of the ions which is increasing with decreasing altitude directly confirms that a simple adiabatic expansion, in spite of other effects, governs the flow of the streams at altitudes greater than 2 Re.

Acknowledgements. We thank C. T. Russell for providing the ISEE magnetic field data.

References

Bonifazi C., P. Cerulli-Irelli, A. Egidi, V. Formisano and G. Moreno, The EGD experiment positive ion experiment on the ISEE-B satellite, IEEE Trans. Geosci. Electron., GE-16, 243, 1978

Candidi M., S. Orsini and V. Formisano, The properties of ionospheric O+ ions as observed in the magnetotail boundary layer and northern plasma lobe, J. Geophys. Res., 87, 9097, 1982.

Candidi M., S. Orsini and A. Ghielmetti, Observation of multiple ion streams in the magnetotail low latitude boundary layer. Evidence for a double H+ population, J. Geophys. Res., 89, 2180, 1984.

Gorney, D. J., A. Clarke, D. Croley, J. F. Fennell, J. Luhmann, and P. F. Mizera, The distribution of ion beams and conics below 8000 km, J. Geophys. Res., 86, 83, 1981.

Kamei, T. and H. Maeda, World Data Center C2 for Geomagnetism, Data Books (AE), Faculty of Science, Tokyo University.

Orsini S., M. Candidi, H. Balsiger and A. Ghielmetti, Ionospheric ions in the near earth geomagnetic tail plasma lobes, Geophys. Res. Lett., 9, 163, 1982.

Orsini S., M. Candidi, V. Formisano, H. Balsiger, A. Ghielmetti and K. W. Ogilvie, The structure of the plasma sheet - lobe boundary in the earth's magnetotail, J. Geophys. Res., 89, 1573, 1984.

Orsini S., E. Amata, M. Candidi, H. Balsiger, M. Stokholm, C. Huang, W. Lennartsson, and P.-A. Lindqvist, Cold streams of ionospheric oxygen in the plasma sheet during the CDAW-6 event of March 22, 1979, J. Geophys. Res., 90, 4091, 1985.

Pilipp W. G. and G. Morfill, The formation of the plasma sheet resulting from plasma mantle dynamics, J. Geophys. Res., 83, 5670, 1978.

Rosenbauer H., H. Grunewaldt, M. D. Montgomery, G. Paschmann and N. Sckopke, HEOS 2 plasma observations in the distant polar magnetosphere, J. Geophys. Res., 80, 2723, 1975.

Russell C. T., The ISEE-1 and -2 fluxgate magnetometers, IEEE Tran Geosci. Elect., GE-16, 239, 1978.

Sharp R. D., D. L. Carr, W. K. Peterson and E. G. Shelley, Ion streams in the magnetotail, J. Geophys. Res., 86, 4639, 1981.

Sharp R. D., W. Lennartsson, W. K. Peterson and E. G. Shelley, The origins of the plasma in the distant plasma sheet, J. Geophys. Res., 87, 10420, 1982.

Yau, A. W., B. A. Whalen, W. K. Peterson and E. G. Shelley, Distribution of Upflowing Ionospheric Ions in the High-Altitude Polar Cap and Auroral Ionosphere, J. Geophys. Res., 89, 5507, 1984.

STATISTICAL STUDY OF ENHANCED ION FLUXES IN THE OUTER PLASMASPHERE

J. D. Menietti, J. L. Burch, R. L. Williams

Department of Space Sciences, Southwest Research Institute, San Antonio, TX 78284

D. L. Gallagher, J. H. Waite, Jr.

Space Science Laboratory, NASA Marshall Space Flight Center, Huntsville, AL 35812

Abstract. Statistical studies of outer plasmaspheric ions in the northern hemisphere have been made utilizing the High Altitude Plasma Instrument (HAPI) on board the Dynamics Explorer-1 satellite. The data were collected during equinox and winter seasons and during a period of solar maximum activity conditions. The data include approximately 40 dayside and over 50 nightside plasmaspheric passes covering a range of magnetic activities ($0 < Kp < 7$). A total of six magnetic storms and recovery periods and a number of quiet times are included in the sampling. The range of magnetic local times on the dayside is from about 6 hours to 12 hours, while the nightside range is from about 18 hours to 23 hours. Our results indicate a clear enhancement in the low energy ($5\ eV < E < 30\ eV$) number flux during periods of large magnetic activity in both the dayside and nightside outer plasmasphere (the inner plasmasphere was not observed). The dayside plasmaspheric fluxes were predominately upward (anti-parallel to $\bar{B}$) while the nightside plasmaspheric fluxes were predominately downward (parallel to $\bar{B}$). The net number fluxes sometimes reached a value of over $10^8\ cm^{-2}\ sec^{-1}$ (assuming H^+ as the predominate species). The largest flows up the field line occur in the outer plasmasphere and decrease in the plasma trough. The ion temperatures in the outer plasmasphere were typically lower than those in the plasma trough and auroral regions. Since the largest flows both parallel and anti-parallel to $\bar{B}$ are observed at periods of high magnetic activity, enhanced outer plasmaspheric fluxes may be due to ionospheric ions expanding into depleted plasmaspheric flux tubes. The nightside fluxes may be due to expansion of the ionosphere in the magnetic conjugate hemisphere.

Introduction

Ion outflow from the F-region into the dayside plasmasphere is a maximum at sunspot minimum when the fluxes can reach 10^8 ions $cm^{-2}\ sec^{-1}$ (Evans, 1975). Incoherent radar scatter measurements made at Millstone Hill (L=3.2) from altitudes less than 1000 km have indicated that dayside H^+ outflow is continuous, but nightside H^+ outflows diminish in the evening sector and become downward fluxes in the midnight to dawn sector (Evans and Holt, 1978). As shown by Arecibo incoherent radar scattering measurements of Vickrey et al. [1979] (L=1.4), large dayside H^+ and O^+ outflows and nightside H^+ and O^+ downward flows exist. Since not all of the ions flowing out of the dayside ionosphere are replenished by downward flows after sunset, much of the flow reaches the magnetosphere and takes part in plasmaspheric refilling and polar cap outflows. During the recovery phase plasmaspheric flux tubes are observed to refill and the plasmasphere to expand (e.g. reviews by Horwitz, 1982 and Moore, 1984). Foster et al. [1978] have shown that the H^+ outflow in the plasmapause decreases with latitude. Recently, however, Menietti et al. [1985] and Jorjio et al. [1985] have reported low energy upward flowing O^+ ions in the dayside plasmasphere. The latter authors have proposed that azimuthal electric fields, which penetrate the nightside plasmasphere at magnetically active times, inject suprathermal ions. Jorjio et al. postulate these ions then corotate to be observed on the dayside associated with ELF emissions, the source of possible heating.

In this paper we examine ion flows in the outer plasmasphere at high altitude as measured by the High Altitude Plasma Instrument (HAPI) on board the Dynamics Explorer-1 (DE-1) satellite. The data is confined to the dayside morning sector and the nightside evening sector. This is a statistical study of ion flows earlier reported on two passes by Menietti et al. [1985]. The present observations indicate that enhanced ion flows into the dayside outer plasmasphere and out of the nightside outer plasmasphere are observed associated with magnetically active periods.

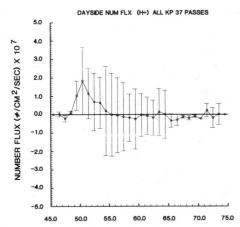

Fig. 1a. Net dayside number flux versus Λ for 37 different passes of DE-1. The data has been averaged in latitude bins of width 1°.

Orbit and Instrumentation

The DE-1 satellite is in a polar orbit (inclination near 90°) with a north polar apogee of about 25,000 km and a south polar perigee of about 675 km. For the period covered by the present analysis, DE-1 was increasing in altitude over the dayside plasmasphere and decreasing in altitude over the nightside plasmasphere. In order to prevent instrument damage, the High Altitude Plasma Instrument (HAPI) was not operated in the inner plasmasphere. Typically the lowest northern hemisphere invariant latitude observed by HAPI was about 50°.

The High Altitude Plasma Instrument consists of five electrostatic analyzers mounted in a fan-shaped angular array lying in a plane containing the spacecraft spin axis. The 0° detector is oriented at a right angle with respect to the spacecraft spin axis, while the other four detectors are oriented at ±12° and ±45° with respect to the 0° detector. The total acceptance angle of each detector is 5° by 20°. Electrons and ions are measured simultaneously at a sampling rate 64 sec^{-1} and in the energy range of 5 eV to 32 keV with an energy resolution ($\Delta E/E$) of 32%. The HAPI typically scans one complete energy sweep each one-half second and sweeps all pitch angles approximately once each six-second spin period. The HAPI instrument has been described in detail in previous publications (Burch et al., 1981).

Results and Calculations

Ion flow velocities were calculated from the HAPI particle data by evaluating moments of the velocity-space distribution function. The spacecraft velocity, as determined from the orbit/attitude data has been subtracted from the results. The number flux was evaluated by taking the product of the number density and the calculated flow velocity. This technique was chosen rather than direct integration of the calculated distribution function in order to eliminate the effects of high density ram ions at low latitudes within the plasmasphere. The ion species assumed for the calculations was H^+. It is possible that some of the flux was due to heavier species, including He^+ and O^+ (as will be discussed later). Heavier species would have the primary effect of decreasing the calculated flow velocity and net number flux by a factor of the square root of the atomic mass.

The moment calculations were performed as in Menietti et al. [1985], by first sorting the particle data by energy step and pitch angle and then numerically evaluating the zeroth and first moment integrals. The energy range of the calculations was 5 eV < E < 32 keV. The number fluxes for approximately 40 dayside and over 50 nightside outer plasmaspheric passes have been

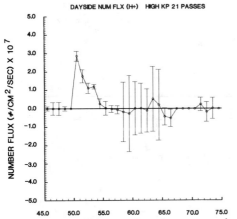

Fig. 1b. Same as 1a except only passes in the magnetic activity range $4 \leq Kp \leq 7$ have been included.

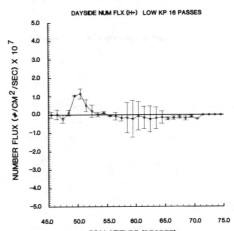

Fig. 1c. Only passes for $0 \leq Kp \leq 3$ are included.

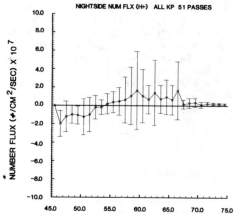

Fig. 2a. Net nightside number flux versus Λ for 51 passes of DE-1. Format is the same as in Figure 1.

evaluated by integrating the HAPI ion data for each two-spin period. The data were collected during equinox and winter seasons and during a period of solar maximum activity conditions. Each satellite pass was different, but the data included the outer plasmasphere and continued at least into the plasma trough region. Many passes extend to the auroral zone, but the dayside cusp region was not included in the sampling, so the "cleft ion fountain" as discussed by Horwitz and Lockwood [1985] was not observed.

Figure 1 displays a summary plot of the results showing averaged, net number fluxes versus invariant latitude. The top panel (Figure 1a) contains data for all levels of magnetic activity; the middle panel (Figure 1b) contains passes for magnetic index in the range $0 \leq Kp \leq 3$; and the bottom panel (Figure 1c) includes passes for the range $4 \leq Kp \leq 7$. The data has been sorted in invariant latitude bins of width 1° and then averaged within each bin. The error bars have a length of twice the probable error,

P.E., where P.E. = 0.6745σ, and σ = standard deviation. The latitude bin size, 1°, was chosen to provide good resolution and good statistics, with the number of points in each bin varying from typically 30 to over 400. In these plots and all others, the term "upward" refers to anti-field-aligned flux, and "downward" indicates field-aligned flux (indicated as negative values).

It is clear from Figure 1a that the averaged, net fluxes are distinctly upward at the lowest latitudes, with a peak of almost 2×10^7 $(cm^2 sec)^{-1}$ at about Λ = 50.5°. Detailed spectrograms from HAPI and, in many cases, the Retarding Ion Mass Spectrometer (RIMS), as well, indicated the region of net upward fluxes was the dayside northern hemisphere plasmasphere. At higher latitudes the averaged, net fluxes appeared to be primarily isotropic or slightly downward in the dayside auroral zone (equatorward of the cusp). The magnitude of the non-averaged net upward number fluxes ranged to over $10^8 cm^{-2} sec^{-1}$. We believe the near-zero net flux values at the lowest invariant latitudes (for all the panels of Figures 1 and 2) are indicative of the energy of the plasmaspheric ions falling below the lowest energy cutoff of the HAPI detector, and may not represent an actual flux decrease. In comparing the last two panels of Figure 1 it is clear that while the general characteristics of the fluxes for Λ > 55° were similar, the peak averaged, net upward plasmaspheric fluxes (near Λ = 50.5° for both panels) was almost three times larger for those passes observed on magnetically active days.

The plot format for nightside passes (Figure 2) is the same as the dayside, except the scale is twice that of Figure 1. We have omitted passes which included the polar cap regions. Most nightside passes included a portion of the nightside auroral zone, the plasma trough, and the nightside plasmasphere. In the top panel of Figure 2, for all levels of magnetic activity, distinctive net downward averaged fluxes were

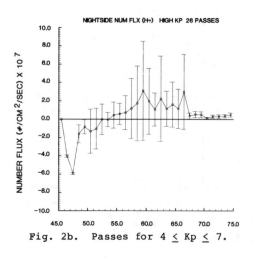

Fig. 2b. Passes for $4 \leq Kp \leq 7$.

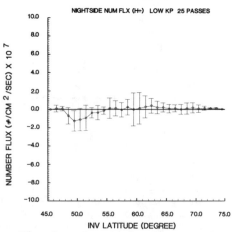

Fig. 2c. Passes for $0 \leq Kp \leq 3$.

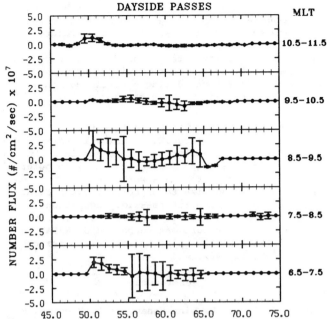

Fig. 3. A stacked-panel plot displaying averaged dayside number fluxes in invariant latitude bins (1° wide) for five different ranges of magnetic local time.

seen at the lowest latitudes with no obvious peak, but approaching -2×10^7 $(cm^2 sec)^{-1}$ at $\Lambda = 46.5°$. Examination of the detailed spectrograms revealed the downward net fluxes were seen in the nightside outer plasmasphere. At higher latitudes, poleward of $\Lambda \sim 54°$, in the plasma trough and auroral zone region, pronounced net, averaged upward fluxes, approaching the same magnitude, were detected. The net, non-averaged plasmaspheric fluxes ranged to over 5×10^7 $(cm^2 sec)^{-1}$, while those in the auroral zone region ranged to greater than $10^8 cm^{-2} sec^{-1}$.

Figures 2b and 2c are analogous to Figures 1b and 1c and show net, averaged nightside fluxes for passes with high and low magnetic activity, respectively. In Figure 2b the net, averaged downward plasmaspheric fluxes for $\Lambda < \sim 54°$ reached a peak value of almost $-6 \times 10^7 (cm^2 sec)^{-1}$ at about $\Lambda \sim 48.5°$. The net, averaged upward fluxes for $\Lambda > 54°$, in the plasma trough and auroral zone, approached 3×10^7 $(cm^2 sec)^{-1}$. In contrast, Figure 2c for the low Kp passes indicates net, averaged plasmaspheric fluxes which peaked at about ¼ the value of Figure 2b and the peak was shifted to higher invariant latitudes ($\Lambda \sim 49.5°$). The net, averaged fluxes for $\Lambda > 54°$ were much lower for the passes of lower magnetic activity.

The net, averaged number fluxes for all periods of magnetic activity for the dayside (Figure 3) and the nightside (Figure 4) are displayed in stacked-panel plots sorted as a function of magnetic local time (MLT). The data for each panel of Figures 3 and 4 has again been averaged in bins 1° wide as in Figures 1 and 2. The conclusions drawn from Figures 1 and 2 are reinforced by Figures 3 and 4, i.e., there was enhanced, net upward dayside outer plasmaspheric flux and enhanced, net downward nightside outer plasmaspheric flux. There was no clear dependence on MLT for the dayside passes, but the nightside passes seem to indicate more low-latitude net, averaged downward fluxes for the early-evening sectors (18-21 hrs. MLT).

We believe that the large fluxes observed by HAPI in the outer dayside and nightside plasmasphere are partially due to O^+ or another heavy ion species. We have examined the data from the Retarding Ion Mass Spectrometer (RIMS), also on board DE-1, for 16 of the passes of this study. O^+ was being monitored by RIMS in 8 of the 16 cases, and in 3 of these 8 cases the O^+ flux is believed to have been the dominant flux for energies, E, greater than 5 eV, the low energy threshold of HAPI. If H^+, He^+ and O^+ were observed by RIMS for E > 5 eV, H^+ was usually the dominant species followed by He^+ and then O^+. Detailed analysis of the RIMS data remains to be performed.

The ion temperatures were also determined for each of the passes. These calculations were performed crudely by determining the slope of the velocity space distribution function versus energy assuming a single-species (H^+) Maxwellian distribution. We found the ion temperatures were

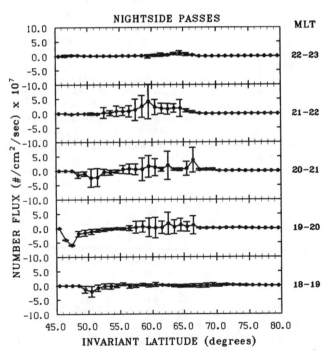

Fig. 4. A stacked-panel plot displaying averaged nightside number fluxes in invariant latitude bins (1° wide) for five different ranges of magnetic local time.

typically about 100 eV in the plasma trough region and tended to decrease dramatically in the outer plasmasphere where the net upward fluxes were observed for the dayside passes. On the nightside the data were similar, but the nightside auroral zone ions were hotter with temperatures of often several hundred eV. The rapid decrease in ion temperature was much more distinct for the dayside outer plasmasphere on the magnetically active passes.

Summary and Discussion

The statistical study presented indicates that the observation of enhanced net upward-flowing dayside and enhanced net downward-flowing nightside ions in the outer plasmasphere is a persistent feature during periods of intensified magnetic activity. Vickrey et al. [1979] have reported radar scattering measurements at Arecibo for one winter and one summer day which agree qualitatively with our results. Those authors have explained the dayside observation of outflowing H^+ and O^+ as due to the thermal expansion of the ionosphere. It is possible that this flux survives the $O^+ - H^+$ charge exchange region to be observed at DE-1 altitudes, but the region of any possible ion acceleration is not known. In the nightside plasmasphere these authors observed downward flowing O^+ and H^+ and argued this flow was due to the drop of the F-layer produced by thermal cooling and the depletion of O^+ at lower altitudes. While the flow velocities measured by Vickrey et al. agree with our measurements qualitatively in sign, the energies of the particles observed at low altitude (<1000 km) were less than the lower energy threshold of the ions observed at high altitude by HAPI.

In contrast to the proposal of Jorjio et al. [1985] mentioned earlier, we suggest that because of the association of the fluxes with magnetically active periods, it is possible that energization of lower-altitude, diffusing ions is taking place on the dayside along depleted plasmaspheric flux tubes. The nightside observations may be the result of these processes that have taken place a few minutes earlier in the conjugate hemisphere. For a large sample of the nightside passes, we have traced the magnetic field lines (Olsen-Pfitzer model) which intersect the nightside outer plasmasphere. For all of the nightside passes the field lines mapped to or near dayside positions. Hydrogen ions with an energy of 10 eV would require about 15 minutes (~60 minutes for O^+) to travel the length of the field line (L=3). Anti-solar convection of ~1 km/sec and corotation at the foot of the conjugate field line could explain why some nightside field lines map to "near" dayside positions.

The ion temperatures were lower in the dayside plasmasphere during active periods. This could indicate enhanced outflows of low-temperature dayside ionospheric ions. The nightside temperature results were not as definitive as the dayside, possibly due to collisional or wave-particle interactions during particle motion along the field line toward the nightside.

Acknowledgments. We would like to particularly acknowledge K. Birkelbach and S. Junge for their outstanding assistance in accessing the data and running the many computer programs necessary for this work. One of us (JDM), would also like to especially thank T. E. Moore and one of the referees for suggestions which led to improvement in the calculations of this study. This work was supported by NASA contract NAS5-28711 and by an internal research grant, SwRI 15-9408.

References

Burch, J. L., J. D. Winningham, V. A. Blevins, N. Eaker, W. C. Gibson, and R. A. Hoffman, High altitude plasma instrument for Dynamics Explorer-A, Space Sci. Instrum., 5, 455, 1981.

Evans, J. V., A study of F2 region daytime vertical ionization fluxes at Millstone Hill during 1969, Plan. Space Sci., 23, 1461, 1975.

Evans, J. V. and J. M. Holt, Nighttime proton fluxes at Millstone Hill, Plan. Space Sci., 26, 727, 1978.

Foster, J. C., C. G. Park, L. H. Brace, J. R. Burrows, J. H. Hoffman, E. Maier, J. H. Whitteker, Plasmapause signatures in the ionosphere and the magnetosphere, J. Geophys. Res., 83, 1175, 1978.

Horwitz, J. L., The ionosphere as a source for magnetospheric ions, Rev. Geophys. Space Phys., 20, 929, 1982.

Horwitz, J. L. and M. Lockwood, The cleft ion fountain: A two-dimensional kinetic model, J. Geophys. Res., 90, 9749, 1985.

Jorjio, N. V., R. A. Kovrazhkin, M. M. Mogilevsky, J. M. Bosqued, H. Reme, J. A. Sauvaud, C. Beghin, and J. L. Rauch, Detection of suprathermal ionospheric O^+ ions inside the plasmasphere, Adv. Space Res., 5, 141, 1985.

Menietti, J. D., J. D. Winningham, J. L. Burch, W. K. Peterson, J. H. Waite, Jr., and D. R. Weimer, Enhanced ion outflows measured by the DE-1 high altitude plasma instrument in the dayside plasmasphere during the recovery phase, J. Geophys. Res., 90, 1653, 1985.

Moore, T. E., Superthermal ionospheric outflows, Rev. Geophys. Space Phys., 22, 264, 1984.

Vickrey, J. F., W. E. Swartz, and D. T. Farley, Ion transport in the topside ionosphere at Arecibo, J. Geophys. Res., 84, 7307, 1979.

Section V.   ACTIVE PROCESSES

THE NEUTRAL LITHIUM VELOCITY DISTRIBUTION OF AN AMPTE SOLAR WIND RELEASE AS INFERRED FROM LITHIUM ION MEASUREMENTS ON THE UKS SPACECRAFT

S.C. Chapman*[1], A.D. Johnstone[2], S.W.H. Cowley[1]

(1) Blackett Laboratory, Imperial College of Science and Technology, LONDON, SW7 2BZ, United Kingdom.

(2) Mullard Space Science Laboratory, University College London, Holmbury St Mary, DORKING, Surrey, RH5 6NT, United Kingdom.

Abstract. As part of the AMPTE mission on 20 September 1984 a neutral lithium release was made in the quiet solar wind. The MSSL ion instrument on board the AMPTE-UKS spacecraft that was positioned ~30 km from the release centre enabled measurements of significant fluxes of lithium ions to be made for ~3 min. after the release, that is, long after the effects of the initial (~25s) local perturbation to the field and flow had died away. These lithium test ions move in cycloidal orbits in the steady ambient fields, so that measurements of their fluxes at the UKS can be used to infer the velocity distribution function of the collisionless neutral cloud over a restricted region in velocity space. These restrictions are such that the ion data cannot easily be used to infer whether or not the neutral $f(\underline{v})$ is shell-like, but do allow us to show that significant anisotropy is present.

1. Introduction

As part of the AMPTE (Active Magnetospheric Particle Tracer Explorers) mission (eg Bryant et al, 1985), two releases of neutral lithium vapour were made in the solar wind, from the German IRM (Ion Release Module). The UKS (United Kingdom Satellite) was located at a small (~30 km) separation from the IRM at the time of the releases, and has provided a set of in situ measurements complimentary to that obtained by the IRM (eg Lühr et al, 1986, Coates et al, 1986). The expansion speed of the released neutrals is small, ~3 km s⁻¹, so that they are all initially within a localised region of space, photoionising in sufficient concentrations to produce a diamagnetic cavity and a large surrounding perturbation to the ambient field and flow, extending to the UKS position. Shortly (~25s) after the release, however, the field and flow at the spacecraft have almost returned to their ambient values, as the effect of the perturbation in this region is reduced. Since the photoionisation time of lithium, ~1 hr, is much longer than the above time scale, most of the released neutrals are not associated with this initial perturbation, instead being ionised over a much larger volume to produce lithium mass densities which are small compared with the ambient mass densities. For the case of the second lithium release (on 20 September 1984) the ambient fields and flow were sufficiently steady in magnitude and direction that the test ion motion of the photoionised lithium is simple to determine. (The first release on 11 September 1984 took place in the presence of large amplitude upstream waves so that an analysis similar to that given here cannot be performed). Measurements of the differential energy flux of the lithium ions obtained by the MSSL ion instrument on board the UKS (Coates et al, 1985) can then be used to infer the velocity distribution function of the collisionless neutral cloud acting as a source for these ions. In this paper we begin in Section 2 by deriving a relationship between the lithium ion fluxes measured at the UKS and the velocity distribution function of the expanding neutral cloud, based upon certain simplifying assumptions. In Section 3 we discuss the significance of these assumptions and how they affect the results. Finally, in Section 4, the features of the neutral cloud obtained from this analysis will be presented.

2. Ideal Derivation of the Neutral Velocity Distribution

The geometry upon which this derivation is based is shown in Fig. 1. Here at the time of the release (t=0) we can assume that a collisionless neutral cloud is produced instantaneously at a single point in the solar wind, that is, at the IRM position, shown at the origin of the XYZ (parallel to GSE) axes. The number of neutrals produced in the velocity range $\underline{v}$ to $\underline{v} + d\underline{v}$ is then just $N_0 f(\underline{v}) d^3v$ where $f(\underline{v})$ is the velocity distribution to be found here and is normalised to the total number of neutrals released N_0. The neutrals expand radially along paths such as $\underline{r}$, producing a local number density at point C a time t later

$$n(\underline{r},t) = N_0 \frac{f(\underline{r}/t)}{t^3} \quad (1)$$

---

* Now at School of Mathematical Sciences, Queen Mary College, LONDON, E1 4NS, United Kingdom.

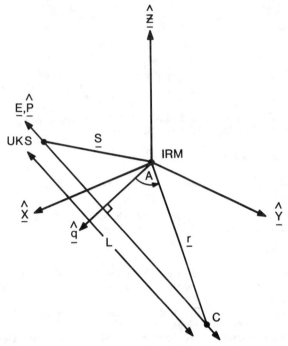

Fig. 1. The geometry of both the lithium neutral and ion paths is shown w.r.t. the spacecraft positions. The neutrals are released approximately at the IRM position, shown here at the origin of the axes drawn parallel to the XYZ GSE axes, and travel along radial paths such as $\underline{r}$. Ions seen at the UKS at a given energy were then produced from neutrals which were photoionised as they crossed the electric field vector $\underline{E}$ passing through UKS (ie at C). Since all paths $\underline{r}$ which cross $\underline{E}$ lie in the plane defined by $\underline{E}$ and the spacecraft separation vector $\underline{s}$, we define orthogonal axes $\hat{p}$ and $\hat{q}$ in this plane, with $\hat{p}$ along the $\underline{E}$ direction and $\hat{q}$ passing through the IRM position. The ion creation point C relative to the UKS and IRM positions is then defined by the angle A between the neutral path $\underline{r}$ and the $\hat{q}$ axis, and the distance L travelled by the ions along $\underline{E}$.

where we have assumed that N_0 remains approximately constant over the first ~3 min. following the release since this interval is short compared with the lithium photoionisation time $\tau_p \simeq 1$ hr. The expansion speeds of the neutrals are small (~ few km s⁻¹) compared with the typical speeds of the lithium ions seen at the UKS (~ 50-200 km s⁻¹), which in turn are small compared with the ambient solar wind flow speed of ~450 km s⁻¹. The photoionised lithium therefore is detected during the early part of its almost exact cycloidal motion, which approximates simply to acceleration along the steady ambient electric field direction ($\underline{E} = -\underline{V}_{SW} \wedge \underline{B}$). We can assume that the direction of $\underline{E}$ is constant so that the paths of the lithium ions detected at the UKS all lie along the $\underline{E}$ (or $\hat{p}$ vector) shown in Fig. 1, which passes through the UKS position (specified by separation vector $\underline{s}$ relative to the IRM). For a given ambient electric field magnitude (which is constant over the ion flight times), the distance L of the creation point C of an ion from the UKS (along $-\hat{p}$), is just given by the energy with which it arrives at the UKS $\epsilon = L E$, (ϵ will be in eV throughout this paper). The flux of ions arriving from C is proportional to the time varying local neutral number density, which from (1) is given by the velocity distribution of their source neutrals that have travelled along radial path $\underline{r}$. A measurement of the flux of ions from a given point C at a particular time t after the release is hence just equivalent to sampling $f(\underline{v})$ at neutral velocity $\underline{v} = \underline{r}/t$. The fluxes of lithium ions arriving over an interval of time from C then give a measurement of $f(\underline{v})$ over a range of speeds of neutrals that have all travelled along a particular $\underline{r}$. Finally, the range of discrete energies spanned by the ion instrument corresponds to measurements of the $f(\underline{v})$ of neutrals travelling along a range of different directions of motion $\underline{r}$ intersecting the $\hat{p}$ axis. In this way the neutral velocity distribution in a restricted slice of velocity space can be obtained from the ion data. Formally then the number of ions arriving at the detector $dN_i(\epsilon, t)$ in a given energy range ϵ to $\epsilon + d\epsilon$, in a given short sample time $d\tau$(~ 5 ms for each of the 60 energy levels of the detector) during which $n(\underline{r}, t)$ is constant, is just the number of ions created in volume $dV = dA\, d\epsilon/|\underline{E}(t)|$ (dA is the detector area projection perpendicular to $\hat{p}$) a flight time $t_A(\epsilon)$ earlier, so that

$$dN_i(\epsilon, t) = n(\underline{r},\, t - t_A)\, dV\, \frac{d\tau}{\tau_p} \qquad (2)$$

The number of arriving ions $dN_i(\epsilon, t)$ is just $dN_c(\epsilon, t)/G(\epsilon)$ where dN_c is the number of counts and $G(\epsilon)$ the energy dependent detector efficiency. Our sample of $f(\underline{v})$ is then

$$f(\underline{v}) = \frac{(t-t_A)^3}{N_0} \frac{\tau_p}{d\tau} \frac{|\underline{E}(t)|}{dA\, d\epsilon\, G(\epsilon)}\, dN_c(\epsilon, t) \qquad (3)$$

We now briefly consider the importance of departures from this ideal description.

3. Departures from Ideality

We will now briefly discuss the ways in which the actual AMPTE lithium release departs from the idealisation discussed in Section 2. We begin by listing those items which require steps to be taken to accommodate them (as has been done), and then go on to itemize those effects which may be neglected.

<u>Effects which are taken into account</u>

1) $|\underline{E}|$ varies significantly over the period during which li ions are observed, so that samples of the $f(\underline{v})$ of neutrals travelling along a fixed direction $\hat{\underline{v}} = \hat{\underline{r}}$, as required, correspond to measurements of the fluxes of ions in energy bins of the instrument which vary with time. The mean electric field $\langle E \rangle \simeq 3.92$ mV m⁻¹ is used together with the set of discrete energy bins ϵ_i of the instrument to define the fixed source locations of distance $L_i = \epsilon_i/\langle E \rangle$ from the UKS which in turn define the set of fixed directions $\hat{\underline{v}}$ along which $f(\underline{v})$ is determined.

2) Statistical fluctuation in dN_c (and hence $f(\underline{v})$) is improved by averaging ion measurements corresponding to

given source distances along $\hat{p}$ over several successive spacecraft spins.

3) Deviation of ion cycloidal trajectories from straight line orbits (along $\underline{E}$) leads to a non negligible error in the angle A (the direction of paths $\underline{r}$, see Fig. 1) of neutrals which act as a source of larger energy ions, and which will be discussed in the next section.

4) The orientation of $\underline{s}$, the UKS - IRM separation vector, is not well known at this time although probable (preliminary) GSE components of $\underline{s}$ are (16, -16, 23) km as found from ground based orbit determination (G. Spalding, private communication, 1985) with $|\underline{s}| \simeq 32$ km known by radar ranging to ± 5 km. The effect of the uncertainty in $\underline{s}$ is assessed by recalculating $f(\underline{v})$ for $\underline{s}$ orientated along the +ve and -ve directions of each of the xyz axes.

5) When operating in the mode used for the solar wind releases, the UKS ion detector does not have complete angular coverage for a given ion energy (Coates et al, 1985). Since the ion detector is spin synchronised however, the range of angles over which ions of a given energy may be detected remains constant with time. We therefore calculate the incoming angles of ions on cycloidal orbits as a function of their energy, in the time dependent $\underline{E}$ and $\underline{B}$ fields seen at this time, in order to deduce the ranges of energies with which ions arrive at angles where the detector efficiency is low (to be reported in detail elsewhere). These time dependent energy ranges $\epsilon(t)$ for the ions correspond to ranges of neutral velocities $\underline{v} = \underline{r}(t)/t$ (where $\underline{r}(t)$ is just given by $\epsilon(t)$) where measurements of $f(\underline{v})$ will be adversely affected. Appropriate estimates of these ranges of $\underline{v}$ will be given in the next section.

Effects which can be neglected

1) The releases are made from 2 canisters separated by ~ 2 km (c.f. UKS, IRM separation of ~ 30 km), such that the atoms start from a small region to a reasonable approximation.

2) The neutrals leave the canister within ~ 1/5 second, by which time the cloud is collisionless, and the canisters ignite almost simultaneously (G. Haerendel, private communication, 1985), such that all the neutrals start at nearly the same time.

3) The release time is known from magnetometer data to within 1 second, giving only a ~ 2% error in the neutral $\underline{v}$.

4) The time variation of the $\underline{E}$ direction over the time period when li ions are observed is only ~ 6° on average.

5) The time variation of $|\underline{E}|$ over the ~ few second flight times of ions is in all cases small, ~ 2%.

6) Changes in $\underline{E}$ over time scales shorter than the 10s resolution of the solar wind velocity measurement are assumed to be of order the < 10% changes in $\underline{B}$ (sampled 16 times/second).

7) Spatial variations of $\underline{E}$ due to localised flow perturbations are always on a larger scale than the ion flight paths.

8) Uncertainty in the measured $\underline{E}$ is smaller in direction than (4) and in magnitude produces a smaller error in path lengths L_i than the length widths $dL_i = d\epsilon_i/E$ themselves.

4. Results

We now present the preliminary neutral $f(\underline{v})$ that has been found for the 20 September release in the solar wind. Using the idealized geometry of Fig. 1, $f(\underline{v})$ has been derived for neutrals travelling along a set of fixed radial directions $\underline{r}$ which are constrained to lie in the $\hat{p}$, $\hat{q}$ (or $\underline{E},\underline{s}$) plane, and which are defined by the choice of $L = \epsilon_i/\langle E \rangle$. This set of radial paths $\underline{r}$ is just equivalent to a set of radial cuts in a plane in velocity space, and we begin by examining the features of $f(\underline{v})$ along an example of one of these cuts before combining the entire set. The example shown in Fig. 2 corresponds to number 23 of the 60 energy levels of the instrument, with an energy of 1160 eV (number 1 being 20 keV, and number 60 being 10 eV, the range being divided logarithmically). The position of the cut in $\underline{v}$ space depends on the orientation of $\underline{s}$ (see Fig. 1), for which we have chosen GSE components of (16, -16, 23) km. On the right hand side of the plot the energy of the centre of the 23rd bin is given, along with the angle A between the $\hat{q}$ axis and the velocity vector $\underline{v}$ (or path $\underline{r}$) of the neutrals. The radial distance of the 'source region', $|\underline{r}|$ (where ions arriving at the UKS within the appropriate energy bins were created) from the release centre is also given. During each 5s spacecraft spin a sample of the neutral f(v) along the cut is made at a particular $|\underline{v}|$ so that as well as marking the neutral speeds (in km s^{-1}) along the bottom of the plot, we have hence also at the top of the plot marked some spacecraft spin numbers, where spin number 1 would just refer to the earliest time at which the ambient field and flow are sufficiently steady that ion measurements can be used to derive values for $f(\underline{v})$. (Note that lithium ions are not seen beyond spin 33). The upper limit to the interval of velocities over which $f(\underline{v})$ is determined is given by the upper limit to the velocity range dv corresponding to this first sample and for this example is ~11.7 km s^{-1}. The normalised neutral velocity distribution funtion $f^*(\underline{v}) = N_0 G_0 f(\underline{v})/\tau_p$ has been plotted versus v in the figure and includes the normalised detector efficiency $G^*(\epsilon) = G(\epsilon)/G_0$, where at this time preliminary calibration gives $G_0 = 3.012 \times 10^{-8} m^3 s^{-2}$ ($f(\underline{v})$ has units of $2 \times 10^4 m^{-3} s^3$). A lower limit to the region of $\underline{v}$ space over which $f(\underline{v})$ may be found is given by the one count level, which is the value of $f^*(\underline{v})$ appropriate to one count being detected during each spacecraft spin ($dN_c = 1$ in equation (3)), and is marked by the dashed line on the figure. In effect then $f^*(\underline{v})$ cannot be determined in the region below this line. The statistical uncertainty in $f(\underline{v})$ is reduced by averaging over adjacent measurements made during successive spacecraft spins. The resulting loss of resolution in $\underline{v}$ is minimised by averaging over just a sufficient number of spins that each averaged measurement of f(v) represents at least 6 counts, giving a maximum uncertainty of ~40%. The actual number of counts corresponding to each of the averaged measurements has been marked on the curve of $f^*(\underline{v})$ (solid line) on Fig. 2. The new one count level that is defined by the averaging process can from simple considerations be seen to be approximately the same as that obtained for the un-averaged case. Finally we consider the effect of the incomplete angular coverage of the ion detector. For each spacecraft spin the energy

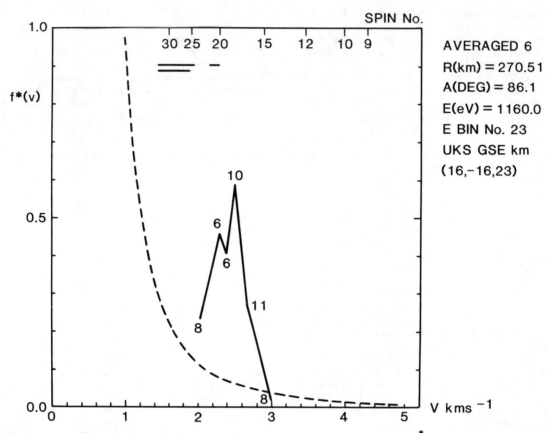

Fig. 2. A radial cut through the normalised neutral velocity distribution $f^*(\underline{v})$, in the $\underline{E}$, $\underline{s}$ plane. The location of creation point C of the ions seen at the UKS (for which neutrals moving in this particular direction act as a source) is given on the right of the plot. The normalisation of $f^*(\underline{v})$ is given in the text, and velocities are in km s^{-1}, with their equivalent spacecraft spin numbers (see text) given across the top of the plot. The solid line denotes the neutral $f^*(\underline{v})$ itself where the number of counts from which each averaged sample of $f^*(\underline{v})$ is comprised is given beside each point. The upper limit to the region within which $f(\underline{v})$ may be determined which is specified by the first measurement made in time (Spin 1) is ~11.7 km s^{-1}; the lower limit which is approximately given by the one count level (see text) is marked by the dashed line. Finally, those spacecraft spins during which ions of the particular energies used to create this plot are not expected to fall within the detector field of view are marked with the double horizontal bars across the top of the plot, and those spins during which ions are expected to arrive outside the detector FWHM are marked with single horizontal bars.

ranges over which the arrival angles of ions on cycloidal orbits lie outside the detector 'field of view' (efficiency < 10% of peak) and outside the detector FWHM (efficiency < 50% of peak) were first determined. The horizontal bars at the top of Fig. 2 indicate when this is true of the ion fluxes used to construct this plot (single bar for < 50% efficiency and double bar for < 10% efficiency). We then see that the fall in $f^*(\underline{v})$ at lower velocities corresponds to ions on cycloidal orbits arriving at angles where the detector efficiency is low. From our present analysis we can therefore only deduce that in this particular cut in $\underline{v}$ space, the neutral $f^*(\underline{v})$ is negligibly small beyond ~3 km s^{-1}, rising to ~0.5 at ~2.5 km s^{-1}, which must be regarded as a typical, rather than most probable speed.

Fig. 2 shows a cut in $f^*(\underline{v})$ obtained for particular probable orientation of the spacecraft separation vector $\underline{s}$. We now combine all 60 cuts in the $\underline{E}$, $\underline{s}$ plane in $\underline{v}$ space, calculated assuming the same $\underline{s}$ orientation, to produce a contour plot of $f^*(\underline{v})$, and this is shown in the left hand panel of Fig. 3, drawn in steps of 0.1. Some of the positions of the cuts along which $f(\underline{v})$ is determined are also shown, labelled with the number of their defining energy level $\epsilon_i = L_i E$. The neutral $f(\underline{v})$ is then determined over a region extending in angle from cut 1 to cut 60 as shown. The outer boundary to the radial extent of this region, given by the measurement made at spin 1, is marked by the solid line on the left of the diagram. At lower velocities, the contours end as $f(\underline{v})$ falls below the one count level. The centre panel of Fig. 3 shows the regions in $\underline{v}$ space where measurements of $f(\underline{v})$ are adversely affected by the

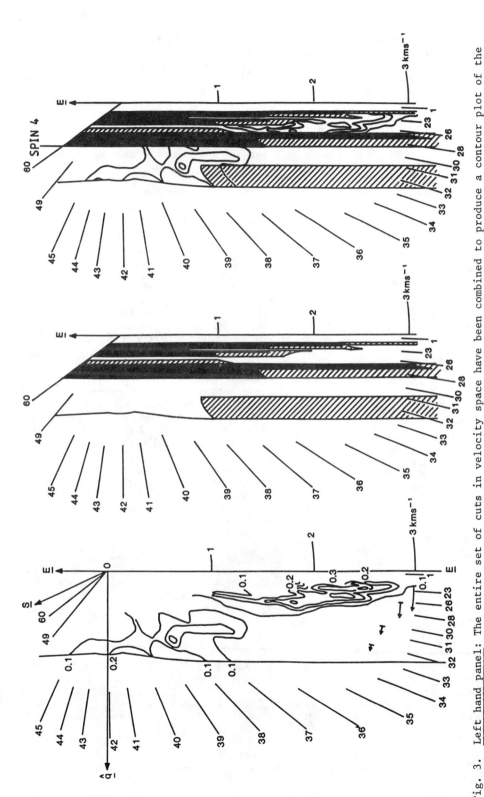

Fig. 3. Left hand panel: The entire set of cuts in velocity space have been combined to produce a contour plot of the neutral $f^*(\underline{v})$ over a limited region in the $\underline{E}$, $\underline{s}$ plane. The $\underline{E}$ and $\underline{s}$ directions are shown on the plot, along with the angular directions of some of the cuts, each of which is marked with the number of its defining energy level (see text). The contours of $f^*(\underline{v})$ are in steps of 0.1 and have been drawn up to the one count level. The neutral $f^*(\underline{v})$ can be sampled over a region in this plane extending radially up to the speeds defined by the first measurement in time used in this study (solid line) and in angle from the cut defined by the highest energy level of the instrument (No 1 equivalent to 20 keV) to the lowest level (No 60 equivalent to 10 eV). The typical systematic error introduced in the angle A of the cuts in $\underline{v}$ space by the approximation of straight line orbits for the lithium ions is given by the arrows.
Centre panel: The shaded areas on this plot (which is in the same format as the left hand panel) denote the areas over which samples of $f^*(\underline{v})$ are adversely affected by the limitations of the detector response. Measurements of the neutral $f^*(\underline{v})$ within the solid and hatched regions require ions moving on cycloidal orbits to be detected at energies which imply arrival angles where the detector efficiency is <10%, and <50% respectively. The shaded areas have been mapped to within regions in $\underline{v}$ space where the one count level becomes large.
Right hand panel: In this plot the shaded areas of the centre panel have been super-imposed on the contours of $f^*(\underline{v})$ shown in the left hand panel. The line in velocity space corresponding to neutrals which acted as a source for ions detected at the UKS during Spin 4 has also been labelled in order to allow a comparison with the results of Johnstone et al (this issue).

incomplete angular coverage of the ion detector. The regions corresponding to arrival angles of ions lying outside the detector 'field of view' and FWHM are shown by the solid and hatched regions respectively. These limitations imposed by the detector response have only been investigated for the time period up to spin 33, that is, for the region in neutral $\underline{v}$ space that extends up to the edge of the shaded area nearest the $\underline{E}$ axis on the plot, since this also extends to within the region where the one count level is large. These shaded regions are then shown superimposed on the contour plot of $f(\underline{v})$ on the right hand panel of Fig. 3. It is then clear that throughout the plot, regions corresponding to ions arriving at angles where the detector efficiency is low correspond closely to regions where $f(\underline{v})$ is small, implying that the detector response is modifying the observed neutral $f(\underline{v})$. We hence cannot, without more detailed analysis, discern whether the distribution function is shell-like, or whether it continues to have large values closer to the origin. Two features of $f(\underline{v})$ are however apparent from the figure. First, there is a clear anisotropy, with typical speeds varying with direction from ~2.5 km s^{-1} to ~1 km s^{-1} over the region where $f(\underline{v})$ may be determined. Second, where peaks in $f(\underline{v})$ can be sampled at higher speeds they are found to be a factor of ~2 larger than those measured at lower speeds. The average magnitude of $f^*(\underline{v})$ can also be used to estimate the total number of neutrals released N_O. Assuming that $f^*(\underline{v}) = 0.25$ within a sphere in velocity space of radius 2.5 km s^{-1}, and is zero elsewhere gives $N_O = 3.92 \times 10^{25}$, which is in good agreement with expectations (eg Krimigis et al, 1982).

We will now consider the effect of the uncertainty in the spacecraft orientation $\underline{s}$, and of the errors introduced by the straight line orbit approximation, on the above statements. The error introduced by the straight line orbit approximation results in an error in the angles A (from $\underline{\hat{q}}$) of the cuts in velocity space, which varies with the ion energy appropriate to each and also with the orientation of $\underline{s}$. Typical estimates of this angular error are denoted by arrows on some of the cuts where it is largest, on the left hand panel of the figure, the error always of course being in the same sense for all the cuts. It is clear that this error will not affect the overall properties of $f(\underline{v})$ that have been deduced from the contour plot. Re-calculating the contour plot of $f(\underline{v})$ for extreme orientations of $\underline{s}$ gave a variation in the range of typical speeds found for the various cuts in velocity space, from the largest range of ~2.7 km s^{-1} - 0.8 km s^{-1} ($\underline{s}$ along + Z GSE) to the smallest range of ~3.0 km s^{-1} - 2.4 km s^{-1} ($\underline{s}$ along - Z GSE).

For completeness, we will briefly compare these results with the more preliminary analysis given by Johnstone et al. (this issue) and also in more detail by Coates et al. (1986). In that analysis (e.g. their Fig. 2) the detected lithium ion energy is shown to increase linearly with time, requiring a shell-like neutral cloud with a constant expansion velocity v_{ex} ~3 km s^{-1}. For this linear fit to be appropriate, however, data is only used after 9.56.40 UT (eg Coates et al. (1986), equivalent to spin 4 here, so that the neutrals acting as a source for the detected ions have presumably reached a distance $v_{ex} t \gg |\underline{s}|$. The line in velocity space corresponding to measurements made during spin 4 has been marked on the right hand plot of Fig. 3, the data used in the preliminary analysis hence corresponds to measurements in velocity space to the right of this line, which if taken alone do not suggest a significant anisotropy. Now since measurements of the number of counts of lithium ions seen at higher energies will be strongly weighted w.r.t. those at lower energies, (from equation (3)), an analysis based simply on the number of counts, as in Johnstone et al. leads to a larger value for the typical speed of the neutral population lying to the right of the spin 4 line. Finally, from our discussion of the effect of the detector response on the neutral $f(\underline{v})$ we can see how the preliminary analysis would indicate that the neutral $f(\underline{v})$ was shell-like.

5. Conclusions

The UKS ion data obtained during the steady conditions following the second lithium release have been used to deduce the neutral velocity distribution function over a restricted region in $\underline{v}$ space. These restrictions are such that, with the present analysis it is not possible to determine whether or not the neutral $f(\underline{v})$ is shell like, but it is possible to reveal a significant anisotropy. Typical neutral speeds have been found to vary with direction over a range which may be at most ~2.7 km s^{-1} - 0.8 km s^{-1}, to at least ~3.0 km s - 2.4 km s^{-1} depending on the relative orientation of the UKS and IRM which is at present uncertain. Ion measurements from the IRM for this release have also been used to estimate expansion speeds for the neutral cloud (Paschmann et al, 1986). This measurement made from the release centre corresponds to a peak in $f(\underline{v})$ lying on the -$\underline{E}$ axis of Fig. 3, so that the maximum of ~3.8 km s^{-1} expansion speed which they derive is therefore not inconsistent with the anisotropy found here. Furthermore, for the first release, which took place during more disturbed ambient conditions, analysis of the low flux of high energy (5 - 20 keV) lithium test ions (Möbius et al, 1985) has shown that the neutrals acting as a source of these ions moved with a range of expansion speeds of 2.25 - 0.65 km s^{-1}, consistent with a neutral cloud that is anisotropic, or that is not shell-like.

<u>Acknowledgements</u>. The authors are most grateful to M.W. Dunlop, M. Six, W.A.C. Mier-Jedrzejowicz and R.P. Rijnbeek for their assistance in analysing the magnetometer data. This work was performed whilst one of the authors (S.C.C.) was supported by a UK SERC studentship.

References

Bryant, D.A., S.M. Krimigis, and G. Haerendel, Outline of the Active Magnetospheric Particle Tracer Explorers (AMPTE) Mission, <u>IEEE Trans. Geosci. Remote sensing</u>, <u>GE-23</u>, 177 1985.

Coates, A.J., J.A. Bowles, R.A. Gowen, B.K. Hancock, A.D. Johnstone, and S.J. Kellock, AMPTE-UKS three dimensional ion experiment, <u>IEEE Trans.Geosci. Remote sensing</u>, <u>GE-23</u>, 287, 1985.

Coates, A.J., A.D. Johnstone, M.F. Smith, and D.J. Rogers, AMPTE-UKS ion experiment observations of lithium releases in the solar wind, <u>J. Geophys. Res.</u>, <u>91</u>, 1331, 1986.

Johnstone A.D., A.J. Coates, M.F. Smith, D.J. Rodgers, Lithium tracer ion energisation observed at AMPTE-UKS, AGU Chapman Conference on Ion Acceleration, these prodeedings, 1986.

Krimigis, S.M., G. Haerendel, R.W. McEntire, G. Paschmann, and D.A. Bryant, The Active Magnetospheric Particle Tracer Explorers (AMPTE) program, EOS. Trans. AGU, 63, 843, 1982.

Lühr, H., D.J. Southwood, N. Klöcker, M. Acúna, B. Häusler, M.W. Dunlop, W.A.C. Mier-Jedrzejowicz, R.P. Rijnbeek, M. Six, In-situ magnetic field measurements during AMPTE solar wind Li+ releases, J. Geophys. Res., 91, 1261, 1986.

Möbius, E., D. Hovestadt, B. Klecker, M. Scholer, G. Gloeckler, F.M. Ipavich, and H. Lühr, Observation of lithium pick-up ions in the 5 to 29 keV energy range following the AMPTE solar wind releases, J. Geophys. Res., 1985 (in press).

Paschmann, G., C.W. Carlson, W. Baumjohann, D.W. Curtis, N. Sckopke, and G. Haerendel, Plasma observations on AMPTE/IRM during the lithium releases in the solar wind, J. Geophys. Res., 91, 1271, 1986.

LITHIUM TRACER ION ENERGISATION OBSERVED AT AMPTE-UKS

A.D. Johnstone, A.J. Coates, M.F. Smith & D.J. Rodgers

Mullard Space Science Laboratory,
University College London, Holmbury St Mary,
Dorking, Surrey, UK

Abstract. When AMPTE-IRM released lithium in the solar wind on 20 September 1984 the UKS spacecraft was situated approximately 33km from the release point. As the cloud of neutral lithium expanded past the UKS the first lithium ions were detected at low (~10eV/q) energies and the solar wind was seen to be decelerated. Further lithium ions were seen for approximately 4 minutes as the neutral cloud expanded further. The acceleration of the newly-born ions in the convection electric field is consistent with a shell-like expansion of the neutral cloud and can be followed until the ions reach just over twice the solar wind energy/charge.

A comparison between the momentum change in the solar wind and the momentum acquired by the freshly accelerated Lithium ions demonstrates that momentum is conserved locally and the proportion of momentum transmitted along field lines is small. Only a small faction of the energy lost by the solar wind is given to the Lithium ions. The major proportion goes into compression of the magnetic field and heating of electrons.

Introduction

When a positive ion is created in the solar wind it is accelerated by the electric field into a cycloidal orbit (fig.1). This process, sometimes called "ion pickup" is important in several space plasma regimes, for example, the pickup of ionised interstellar gas in the solar wind (Egidi et al 1984), and the mass-loading of the solar wind by cometary plasma (Johnstone 1985). The measurements of plasma, by the UKS and IRM spacecraft of the AMPTE mission, created by the release of a cloud of neutral Lithium by the IRM, enabled detailed observations of this process under controlled conditions.

Experimental Conditions

Two Lithium releases were carried out in the solar wind by the AMPTE mission on 11 September 1984 and 20 September 1984. The solar wind conditions were more favourable for the observations during the second release so only that one will be discussed here. Details of both are given by Coates et al (1986). The measurements reported here were made by the 3d-ion analyser on the UKS spacecraft (Coates et al 1985). Briefly, the instrument consists of electrostatic energy analysers covering the energy per charge range from 10 ev/q to 20 keV/q and the complete angular range from $0°$ to $180°$ with respect to the spin axis. Within one rotation of the spacecraft (5 secs) the instrument obtains a 3-dimensional energy distribution.

Expansion of the neutral lithium cloud

Neutral lithium was released from canisters deployed by the IRM spacecraft at 09:56:02 UT on 20 September 1984. The lithium cloud expanded outwards from the release site engulfing the UKS spacecraft 33kms away at approximately 09:56:15 UT and continuing, according to the observations discussed below, until it was at least 1500km in diameter, four minutes after the release.

Since the photoionisation time constant of Lithium, in the solar flux, is of the order of one hour, the production rate of ionisation in the complete cloud is essentially constant for the period of the observations. It has been measured to be 6×10^{21} per sec (G. Haerendel unpublished manuscript 1986). Once an ion has been created it is accelerated by the electric field. The trajectory of the ions, in the coordinate frame in which the electric field E_x is parallel to the x-axis and the magnetic field B_y is parallel to the y-axis, is a cycloid

$$V_x = \frac{E_x}{B_y} \sin \Omega(t-t_1)$$

(1)

$$V_z = \frac{E_x}{B_y} (1 - \cos \Omega(t-t_1))$$

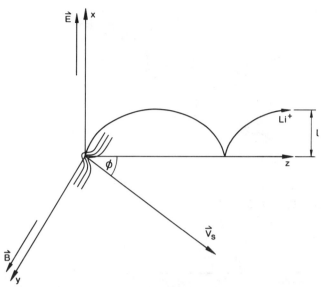

Fig. 1. Once ionised the Lithium ions follows a cycloidal trajectory in the plane perpendicular to the magnetic field as shown. The dimension across the loop, l, was approximately 6800kms in the release of 20 September 1984. The value of the angle ϕ was 17°. The distortion of the magnetic field, keeps the field vector in the yz plane, the plane containing both $\underline{B}$ and $\underline{V}$. This coordinate system is chosen for the purpose of trajectory analysis and is not aligned with the GSE system also used in this paper.

where $\Omega = (eB_y/M)$ and the atom is ionised at the origin at t_i.

The motion can be described as a guiding-centre drift at a velocity $V=E_x/B_y$ perpendicular to the magnetic field combined with a gyration around the magnetic field direction at the same velocity.

In the early part of the trajectory the motion is nearly parallel to the electric field and given approximately by

$$V_x = \frac{E_x}{B_y} \Omega(t-t_i) \qquad (2)$$

or $\frac{1}{2}MV_x^2 = e\,E_x x$

i.e. the distance travelled, x, is proportional to the energy. As the energy of the detected ions increases, the ion comes from the further away and the curvature of the trajectory becomes important.

The first Lithium ions are detected by the UKS shortly after the neutral cloud reaches the electric field line passing through the UKS at a point quite close to the UKS. As a result the ions gain only a small amount of energy before they are detected. As time progresses and the neutral cloud expands ions are created further and further away and hence gain more and more energy.

Once the cloud has expanded well past the position of UKS, the distance to the front of the cloud increases linearly with time. The variation of detected energy with time is then also approximately linear (Coates et al 1985).

$$(\tfrac{1}{2} M V^2/eE_x) = V_e(t-t_r) + r\cos\phi \qquad (3)$$

where the UKS is $\underline{r}$ from the centre of the release, $\underline{r}.\underline{E} = -rE_x\cos\phi$, and t_r is the time of the release.

Figure 2 shows the variation of the lithium ion distribution with time against a linear energy scale. The area shaded shows where the lithium count rate exceeded a threshold of 3 counts per bin. Lithium ions are eventually detected with more than twice the solar wind energy although by that time the number of counts is very small. The more darkly-shaded area is the energy bin with the highest number of counts in each spin. Two lines have been fitted to the data collected after 09:56:40 UT when the cloud has expanded well past the position of the UKS. Then the linear approximation of equation (3) is valid, the magnetic perturbation of the cloud has disappeared and the electric and magnetic fields are once more nearly constant at the pre-event levels. The upper one is fitted to the upper energy of the distribution; the lower line to the energy of the peak in the distribution. Comparing the lines with equation (3), taking $E_x = 3.8$ mv/m, (fig.3) and r = 33km gives the expansion velocity $V_e = 3.05 \pm 0.2$ km/s and $\phi = 106^\circ$.

A more detailed analysis of the expansion of the neutral lithium cloud, in which the initial velocity distribution of the Lithium atoms is derived, has been made by Chapman et al. (1986).

Apart from providing the expansion velocity of the cloud these data, on the variation of the lithium distribution with time, confirm that the Lithium ions are accelerated in the predicted manner. The direction of arrival of the ions is parallel, within the instrumental resolution, to the electric field given by

$$\underline{E} = -\underline{V} \times \underline{B} \qquad (4)$$

where the velocity and magnetic field are the values measured simultaneously in the solar wind (figure 3). Over the period that the Lithium ions are intense enough to be detected the cloud does not go far enough away for the trajectory of the ions returning to the spacecraft to be significantly deflected from the straight line parallel to the electric field. In other words, the cycloidal nature of the trajectory is not directly observed.

Effect of the Lithium cloud on the solar wind

On alternate rotations of the spacecraft the ion instrument switches to a high resolution (in energy and angle) mode centred on the expected position of the solar wind. In this mode the

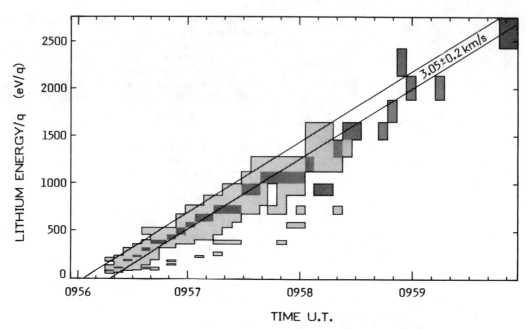

Fig. 2. Measured energy of the lithium distribution with time. The shaded areas show where lithium counts >3 per bin are detected. The energy bin corresponding to the peak of the distribution is shown by darker shading. The fitted speeds of the outer edge and the peak of the distribution are shown.

instrument can measure the density, velocity vector and temperature of the protons and alpha particles in the solar wind separately.

The speed and flow direction of the protons in the solar wind is shown in the top three panels of figure 3. The first effect of the release is detected in the measurement made at 9:56:06 UT, four seconds after the release. There is a small, but significant, deflection of the solar wind although no detectable change in the speed. In the next measurement, 10 seconds later, the speed has dropped by more than 40km/s and direction changed by approximately 10° in both θ and ϕ. Over the next thirty seconds the solar wind velocity, the electric and magnetic fields all return to the undisturbed level and the lithium ion flux drops sharply.

The behaviour of the solar wind alpha particles is similar to the protons, but they are deflected only about two-thirds as much.

The effect on the solar wind has been summarised in fig. 4 by projecting the principal vectors, $\underline{V}$ (protons and alphas), $\underline{B}$, and $\underline{E}$ into the yz plane, perpendicular to the solar direction in GSE coordinates, and giving the view from the Sun.

The diagram shows first that the variation of the magnetic field remains in the plane containing the solar wind direction and the magnetic field before the event (Luhr et al 1986). This is consistent with the notion of the field lines being draped around the cavity produced by the lithium release. The track along which the end point of the magnetic field vector moves is indicated by the arrows. The proton and alpha velocities before the event are slightly different but the vector difference between them is nearly parallel to the magnetic field (as shown by the dashed line) as it should be (Marsch et al. 1982). This provides a check on the accuracy of the velocity vector measurements. The error in the velocity components thus appears to be of the order of ± 2km/s. After the release both protons and alphas are deflected perpendicular to the magnetic field, and antiparallel to the electric field.

There are also effects on the velocity and field components parallel to the x-axis. The velocity is reduced: the electric field component, which is directed away from the Sun, in increased.

Momentum and energy conservation

In the yz plane, the momentum changes take place perpendicular to the magnetic field. The change in the solar wind momentum must balance the momentum of the Lithium ions being accelerated parallel to the electric field. We have no idea how the solar wind deflection varies across the extent of the neutral cloud but if we made the simplest assumption, that the variation is uniform across a spherical cloud whose radius is given by

$$r_c = V_e(t-t_r) \qquad (5)$$

then momentum balance is expressed by

$$(N_p V_p + 4n_a V_a) Mp V_s \pi V_e t = \alpha N_o M_L V_L \qquad (6)$$

If we calculate the velocity V_L from equation (6) then

$$V_L = 50 \text{ km/s}$$

In words, if the momentum change per second in the solar wind is shared amongst all the lithium ions created in that second then each lithium ion would acquire a velocity of 50km/s. If starting with zero velocity the energy change is 90eV. Knowing the electric field, the distance travelled by the ion in gaining this energy can be calculated. At 13 secs after release, when the magnetic field is at its maximum value (Luhr et al 1986) it is difficult to calculate. On the one hand, equation (4) is not valid, or the solar wind would not be deflected, but it can be used

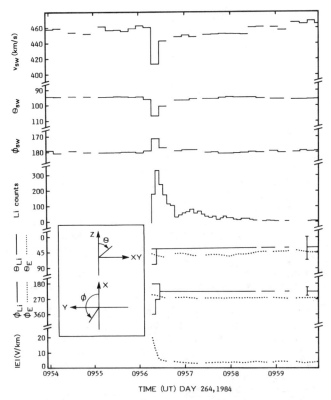

Fig. 3. Variation of the solar wind velocity and the lithium counts and directions with time for the data in Figure 2. The solar wind decelerates and deflects simultaneously with the detection of the first lithium ions at 09:56:15 UT. The electric field directions and magnitude are shown as dashed lines.

Here αN_o is the production rate of lithium ions, equal to 6×10^{21} ions/sec. In equation (6) the only unknown quantity is V_L which can therefore be obtained.

Using the values appropriate for the maximum deflection at 13 secs after the release, the total momentum change in the solar wind is 3.5 kgm/s^2. It is also possible to calculate the total momentum change per second of the complete lithium ion cloud at the same time, assuming that all the ions are following cycloidal trajectories. The component parallel to the electric field, the x direction in equation (1), is

$$\Delta M = \alpha N_o M_L (E_x/B_y) \sin \Omega(t-t_r) \quad (7)$$

$$= 30 \text{ kg m/s}^2 \quad (t-t_r = 13s)$$

The momentum change in the solar wind passing through the neutral cloud does not nearly account for the change in momentum of all the lithium ions.

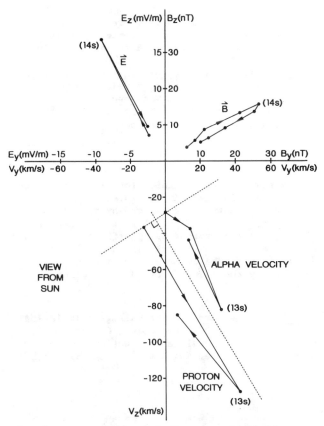

Fig. 4. A vector plot of the magnetic field, electric field and solar wind velocity during the release. Each plotted point is the end point of a vector from the origin. The succession of points connected by lines shows the time progression of the vectors during the event. The time in brackets is the time after the release. The coordinate system is GSE, not the system illustrated in figure 1. The yz plane is the plane perpendicular to the earth-sun direction, with Z perpendicular to the ecliptic in a northward direction and the view is as seen from the Sun.

to obtain an upper limit on the electric field, as in figure 4, of 19 mv/m. The lower limit, provided by the electric field in the undisturbed solar wind, is 3.8 mv/m. Thus the lithium ion should have travelled between 4.8 kms and 24 kms in becoming energised. This is comparable with, or smaller than, the dimensions of the neutral lithium cloud at that time. Momentum is apparently conserved essentially at a local level.

The energy lost by the solar wind over the cross-section of the neutral cloud is

$$\Delta E = (N_p + 4N_a) M_p V_s (V_s^2 - v^2) \pi V_e^2 t^2$$

$$= 1.35 \times 10^6 \text{ watts}$$

$$= 1400 \text{ eV per lithium ion created}$$

Only a small fraction is therefore actually going to the lithium ions. The rest goes into increasing the magnetic field and accelerating electrons (Hall et al 1986, Paschmann et al 1986). In fact the electrons may be regarded in this interaction as playing a role equivalent to neutrinos in beta-decay. They become the vehicles to remove excess energy without significantly affecting the momentum balance.

Conclusions

The newly-created lithium ions are accelerated by the electric field in the solar wind as expected. Momentum is conserved by a deflection of the solar wind antiparallel to the electric field, with the protons being deflected 1.6 times as much as the alpha particles. Most of the energy lost by the solar wind goes into compressing the magnetic field and heating electrons.

Acknowledgements. We acknowledge stimulating discussions with Drs. E. Mobius, G. Paschmann, D.A. Bryant, D.S. Hall and C. Chaloner. The Science and Engineering Research Council, UK, supported the UK participation in the AMPTE programme.

References

Chapman, S.C., Johnstone, A.D. and Cowley, S.W.H., The velocity distribution function of the neutral cloud produced by an Ampte lithium release in the solar wind, AGU Chapman Conference on Ion Acceleration, these proceedings, 1985.

Coates, A.J., Bowles, J.A., Gowen, R.A., Hancock, B.K., Johnstone, A.D. and Kellock, S.J., AMPTE-UKS Three dimensional ion experiment, IEEE Trans. Geoscience and Remote Sensing, GE-23, 287, 1985.

Coates, A.J., Johnstone, A.D., Smith, M.F. and Rodgers, D.J., AMPTE-UKS Ion Experiment observations of lithium releases in the solar wind, J. Geophys. Res. 91, 1131, 1986.

Egidi, A., Francia, P. and Villante, U. Interaction between neutral hydrogen and solar wind; spacecraft measurements of H^+ at the Earth's orbit, Geophys. Res. Lett. 11, 709, 1984.

Hall, D.S., Bryant, D.A., Chaloner, C.P., Lepine, D.R. and Bingham, R., AMPTE-UKS electron measurements during the Lithium releases of 11th and 20th September 1980, J. Geophys. Res, 91, 1321, 1986.

Johnstone, A.D., "Comets" in Solar System Magnetic Fields ed. E.R. Priest, published by D. Reidel, Dordrecht, 1985.

Luhr, H., Southwood, D.J., Klocker, N., Acuna, M., Hausler, B., Mier-Jedrzejowicz, W.A.C., Rijnbeek, R.P. and Six, M., In-situ magnetic field measurements during the AMPTE solar wind Li^+ releases, J. Geophys. Res., 91, 1261, 1986.

Marsch, E., Muhlhauser, K.H., Rosenbauer, H., Schwenn, R. and Neubauer, F., Solar wind helium ions: observations of the Helios solar probes between 0.3 and 1 AU, J. Geophys. Res., 87, 35, 1982.

Paschmann, G., Carlson, C.W., Baumjohann, W., Curtis, D.W., Sckopke, N. and Haerendel, G., Plasma observations on AMPTE/IRM for the lithium releases in the solar wind, J. Geophys. Res., 91, 1271, 1986.

ELF WAVES AND ION RESONANCES PRODUCED BY AN ELECTRON
BEAM EMITTING ROCKET IN THE IONOSPHERE

J. R. Winckler, Y. Abe and K. N. Erickson

School of Physics and Astronomy, University of Minnesota
Minneapolis, Minnesota 55455

Abstract. The injection of a powerful electron beam into the auroral ionosphere produces a myriad of large electric signals which have been studied on a free flying plasma diagnostics payload during the ECHO 6 experiment. The primary interaction of the electron beam was to produce a hot plasma region with characteristic electron energies of 100 ev. When the beam was injected in a nearly transverse spiral the most intense region extended to 60 m perpendicular to the beam but was still detectable at 110 m. Strong electric fields were observed in the hot plasma region consisting of a quasi-dc component directed towards the beam sector with large superposed low-frequency wave variations. Harmonics were seen spaced about 40 Hz near the 0+ gyro frequency with the fundamental at 75 Hz or 3/2 the 50 Hz fundamental. When the beam was injected upward parallel to B the hot plasma region and quasi-dc fields were not observed and were probably within the 40 m radius where measurements began. In this case the proton gyro wave consistently appeared with the electric vector linearly polarized in a direction transverse to the radial direction from the beam. In addition a continuous wave spectrum extended from about 200 Hz to above the system Nyquist frequency at 1250 Hz. These natural ion resonances appear to be in the electrostatic mode and may be produced by a current-driven instability as described by Kindel and Kennel. The 1 kHz accelerator pulsing frequency was also consistently observed to the largest distance surveyed of 110 m. The wave activity across the entire ELF spectrum was strongly suppressed when the system passed through field lines containing auroral activity, which together with other evidence indicates that the basic wave production mechanism was the negative potentials of the beam combined with currents flowing in the hot plasma region as well as return currents necessary for neutralizing the positive potential of the accelerator payload. Vigorous wave production in the lower hybrid and whistler mode frequency regions was observed but will not be discussed in detail.

Introduction

We have made observations which show that ionospheric ions can be excited into resonance modes and accelerated in the vicinity of a powerful electron beam injected into the ionosphere from a sounding rocket. These results present certain problems both theoretical and experimental since the basic interaction of an electron beam with a plasma has been traditionally treated on the basis of the two-stream instability and the resulting disturbances occur near the effective plasma frequency of the beam-background system (Lavergnat and Pellat, 1979). Thus plasma and electromagnetic waves have been observed and predicted in the MHz range and above for electron beams injected into the ionosphere. However, laboratory experiments have shown that lower frequency electromagnetic waves may be produced through conversion of the plasma waves to whistler mode waves within the electron beam (Stenzel and Whelan, 1982). Because our observations clearly show that the ion resonances for hydrogen, oxygen and lower frequencies are produced in the region well outside the injected beam, in a region of hot plasma also produced by the primary beam-plasma interaction, we must search for secondary processes for ELF range wave production. Since the observed waves are probably electrostatic in nature and do not propagate effectively away from the source region, one must search for a current system or other excitation source secondary to the primary injected beam from the accelerator but permeating the region of space where the observations were made. The production of electrostatic ion cyclotron waves and ion acoustic waves by broad field-aligned currents in the high ionosphere has been studied theoretically by Kindel and Kennel (1971) and their paper may provide explanations also for similar waves produced by a beam-injecting rocket system in the lower ionosphere.

The research effort consists of a number of launchings of Black Brant V and other vehicles from an auroral zone site, the latest in the series being the ECHO 6 mission sent from Poker

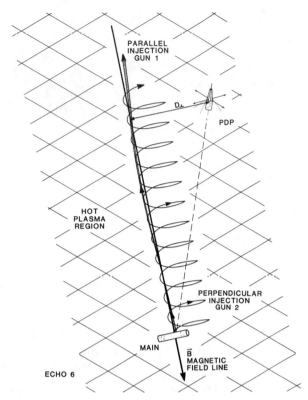

Figure 1. Geometry of the ECHO 6 Plasma Diagnostic Package with respect to the beam-inject field line. The separation speed from the accelerator was 1.5 m/s and the perpendicular distance varied from 40 to 110 m during the experiment.

Flat range near Fairbanks, Alaska. The primary mission has been to study conjugate return "echoes" from the electron beam pulses for studies of the near and distant magnetosphere. However, the injection of the beams has always yielded a variety of plasma phenomena in the near ionosphere. The use of electron beams as probes and the problems of beam stability and vehicle potentials are addressed in a recent summary which includes the ECHO 5 mission (Winckler, 1982). Although ion resonance effects and the production of a region of supra-thermal electrons by the injected beam were noticed in many of the earlier ECHO flights, our best data to date has come from the recent ECHO 6 flight launched from the Poker Flat range on March 30, 1983. This experiment included a free-flying plasma diagnostics package (PDP) equipped with two double electric probes oriented orthogonally to each other and nearly orthogonal to the magnetic field and also equipped with a Langmuir probe, ion and electron detectors and scanning photometers. The probes could study the electric field variations from DC to the 1250 Hz Nyquist frequency by digital sampling, and in addition had narrow band wave channels at 8.5 and 125 kHz.

The ELF sensitivity threshhold was about 1 mV/m which was more than adequate for beam-induced effects, but not intended for low-level natural ionospheric waves. The electric probes also yielded the ionospheric convection field and showed its variations for example over an auroral arc region, during accelerator off times.

As shown in Figure 1 the PDP moved away from the accelerator payload at a 20 degree angle to the magnetic field. When the accelerator was started at 163 s flight time the PDP was about 40 meters perpendicular distance from the magnetic field line passing through the accelerator and about 117 m above the main payload. This was completely outside the range of beam particles and none were shown by electron spectrometers on the PDP. The PDP was about 110 m perpendicular distance at accelerator termination. Electron gun 1 injected an essentially DC beam at a series of voltages up to 36 kV and 230 ma, and the beam pitch angle could be set at, for example, 0 degrees ("down" injection), 100 degrees ("out") and 180 degrees ("up", parallel to B). Gun 2 injected always at 100 degrees with an exponentially dropping energy from 40 to 10 kV each ms, giving a very large 1 kHz component in all the associated effects (luminosity, electric fields, plasma heating etc.). Electron gun waveforms and a typical power spectra from each gun in the "out" mode of the electric variations derived from the digital data are shown in Figure 2. Note the 1 kHz line in the gun 2 spectrum, not present for gun 1. Both spectra show a major decrease by about 250 s flight time, but for gun 2 the intensity recovers after 300 s near 1 kHz frequency and below 250 Hz frequency. This dropout is associated with the passage of the system across an auroral arc, and the recovery is north of the arc where auroral effects were very low. The auroral-associated wave dropout is thought to be the result of quenching of the potentials of the beam and the accelerator payload on auroral field lines, accompanied by a reduction in the extent of the hot plasma region and return currents. The PDP position, at one side of the beam field line and well above the main payload resulted in its responses being mainly due to the beam itself in the "out" and "up" modes, and very little to the return current discharge around the accelerator (main) payload. Many details of the ECHO 6 experiment have been given in several recent publications (Winckler et.al., 1984, Winckler et.al., 1985a, 1985b). The hot plasma environment and the quasi-dc potentials in the space around an electron-beam emitting system in the ionosphere have been studied in several recent experiments (Arnoldy and Winckler, 1981, Arnoldy et.al., 1985). This paper will concentrate on the production of ion resonances and ion waves by an electron beam. The rapid stochastic acceleration of ions observed near electron beam-emitting vehicles will be discussed elsewhere in this symposium (Arnoldy, 1985).

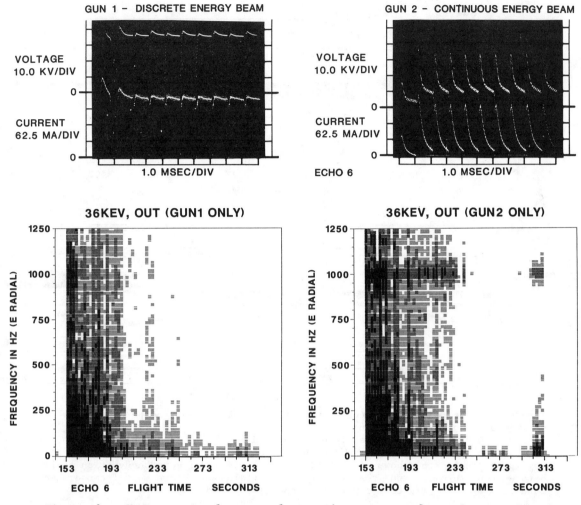

Figure 2. Upper - Accelerator voltage and current waveforms. Lower - Power spectra during the complete flight for the radial component of the disturbance fields for, left, gun 1 at 100 degree pitch angle and right, gun 2 at 100 degree pitch angle. This is an example among many spectral possibilities. Note the 1 kHz component in gun 2, which is essentially absent in gun 1 spectra.

Electric Field Observations

The electron gun injections were programmed so that when one gun was on in a steady mode for 0.5 s the other was coded in 50 ms on and off periods. This code was actually a digital word in bi-phase format to identify the exact injected pulse if conjugate echoes were detected. This pulsing format was also useful for wave studies since within the format time intervals could be found when each gun was injecting alone in 50 or 100 ms intervals and, in the case of gun 1, at a controllable pitch angle. A typical example of the electric field variations observed over a 0.6 s interval with gun coding as just described is shown in Figure 3. During this interval the PDP was located about 50 m perpendicular distance from the beam injection field line and 137 m above the accelerator. The flight system was at 200 km altitude. The vector electric fields transverse to B are plotted each 0.4 ms along the time axis. Both guns were off during the first 50 ms interval (182.320 - 182.720 s) and the background ionospheric convection field shows as a continuum of vectors directed slightly east of north (see calibration diagram to left of panels). In this example both guns were directed "out". When gun 1 was on alone it produced a similar but smaller effect, and the wave vectors partially reverted to the ionospheric field but were directed magnetically SW. A very large field was produced when gun 2 switched on again so that both guns were in the "out" format. One sees major wave fluctuations, but with the average field directed inwards toward the beam azimuthal quadrant. After 182.895 s with both

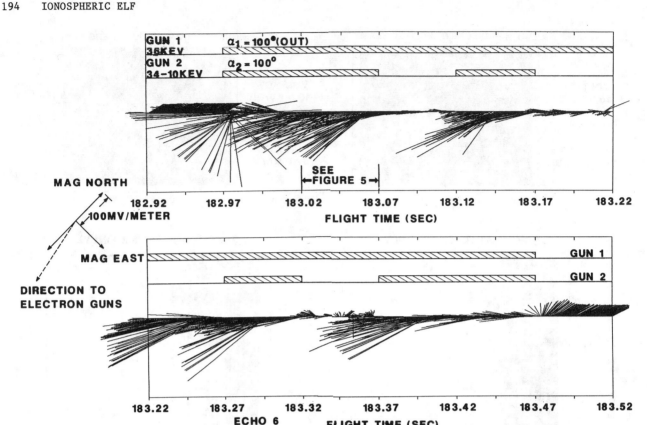

Figure 3. The electric disturbance vector in the magnetic perpendicular plane evaluated each 0.4 ms during a 0.6 s interval in which gun 1 was on steady at maximum power and 100 degree pitch angle, and gun 2 was coded on and off in 50 or 100 ms intervals (see shaded bars above each panel). A direction and magnitude scale is shown to the left. Note the NE directed convection field before gun turn-on and the large electric fields which are most intense when both guns are on. See text for further details.

guns off the field reverted to the ionospheric direction after a 10 ms decay period.

It was always observed that the greatest extent of the hot plasma region was produced when either gun 1 or gun 2 or both were directed "out", i.e. with a 100 degree pitch angle, sending the beam upwards 10 degrees from perpendicular in a spiral of about 10 m Larmour radius for the 36 kV energy. The very intense electric variations occurred when the PDP was immersed in this hot plasma region, and when the PDP passed out of this region at about 60 m distance the electric fields rapidly returned to background. We show in Figure 4 (left panels) a power spectrum of intervals selected each 10 s throughout the ECHO 6 flight when gun 1 was directed "out" and gun 2 was off, and for comparison (lower panel) a measure of the hot plasma in terms of the negative floating potential of the PDP body as described previously (Winckler et.al., 1984). It would have been preferable to have utilized the electron spectrum directly, but unfortunately the spectrometer failed early in the flight. The "out" spectrum

shows that the wave activity was concentrated below 500 Hz, and was most intense when the PDP was in the region of negative floating potential excursions as shown in the lower panel. The right panels of Figure 4 show the power spectrum selected each ten seconds for the condition that gun 1 was "up" and gun 2 off. This spectrum is almost devoid of frequencies below 250 Hz but shows a band extending to the system Nyquist frequency at 1250 Hz and also wave activity persisting to the largest distances measured at 110 m, and also clearly shows the wave dropout in the auroral arc region at 83.4 m PDP separation. The large inward-directed quasi-dc electric fields were completely absent for this "up" beam mode, and it is assumed that such fields probably existed closer to the beam (the "up" Larmour radius was only about 1 m) but lay inside the 40 m radius where measurements began.

The Excitation of Ion Resonances

To simplify the analysis we have constructed vector plots in the format of Figure 3 for 50 ms

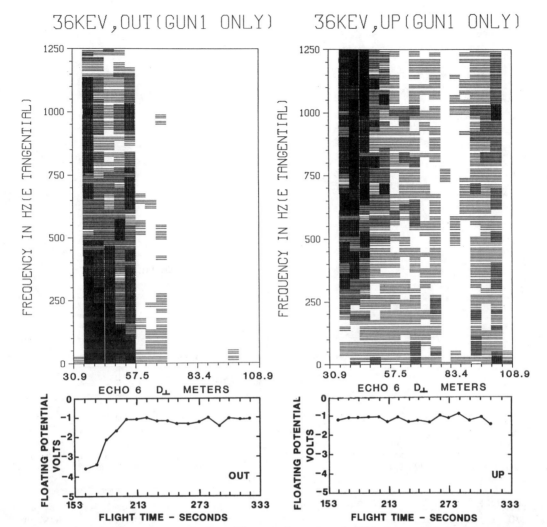

Figure 4. A comparison of the effects of an "out" (100 degree) injection, left panels, and an "up" (180 degree) injection, right panels. Note that the "out" injection produces a wave spectra of more limited radial extent and concentrated below 500 Hz. It occurs in a hot plasma region as shown by the floating potential curve (bottom). The "up" injection produces a higher frequency spectrum which extends to greater distances with only normal ionospheric floating potentials near 1 V negative, and therefore lower plasma heating over the range of perpendicular distances surveyed. The auroral absorption near 83 m is obvious.

intervals when the guns were on in a certain mode, and plotted the r sidual vector after vectorially subtracting the average vector evaluated over the 150 data samples in the interval. This procedure suppresses the VXB field, the ionospheric convection field and the quasi-dc field due to the beam and reveals the wave field. A 50 ms interval is shown in Figure 5 in which both guns were on at full power and injecting at a 100 degree pitch angle. In this case the plasma was violently heated as shown by the major rise in floating potential in the bottom traces with a maximum near 183.040 s. The wave vector in this case was rotating counter-clockwise, in the ion gyro sense, and appears to be elliptically polarized but somewhat irregular. The wave period is about 20 ms, so this is probably an O+ ion wave. We note that the subtracted average field vector (at center) is nearly 100 mV/m, and is directed radially inward towards a point in the beam quadrant as discussed earlier. Such an average electric field may produce a clockwise EXB drift of the plasma region around the beam with a resulting Doppler shift of the observed waves.

A Fourier spectrum of the O+ range waves

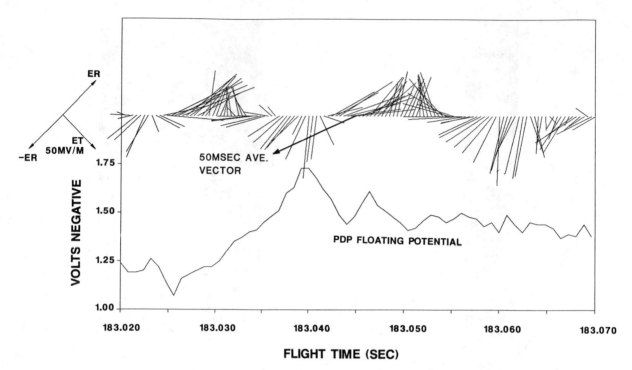

Figure 5. An example of an O+ resonance excited during maximum plasma heating with both guns injecting out (100 degree pitch angle) at maximum power. The polarization is left-handed and circular or elliptical. Note that the 50 ms average vector is nearly 100 mV/m magnitude and directed westward of the direction to the beam gyro- centers. The floating potentials show wave-associated variations.

constructed from a 0.100 s sample between 173.170 s and 173.270 s flight time when gun 1 was on alone and injecting "out" is shown in the left panel of Figure 6. At this time the PDP was about 46 m perpendicular distance from the beam and was in the inner hot plasma region as shown in Figure 4, left. This spectrum shows a series of harmonics spaced mostly at 40 Hz intervals with the "fundamental" at 75 Hz and one lower peak at 37 Hz. The harmonic spacing differs from the 50 Hz O+ gyro frequency. Nevertheless, it seems useful to associate the spectrum with the results of Kindel and Kennel (1971) for the case where the ratio Te/Ti of the electron and ion temperatures is very high (near 100) and the O+ fundamental is shifted by 3/2. It points towards the origin of the resonance in a local current source, possibly a field-aligned return current to the vehicle neutralizing the beam current. Besides the resonance peaks at the indicated frequencies in Figure 6 there appear to be two "absorption" minima at 148 Hz and 262 Hz. These distinct features seem to be superposed on a falling white noise background spectrum. It should be stressed that this spectrum is an example of many obtained over a similar frequency range for the "out" gun injections, but that the same harmonic structure is not always observed, but rather a varied and only partially identified series of resonant peaks. Prior to the ECHO 6 mission resonant oscillations in the N2+ range have been observed. Finally, we must not expect an exact comparison with the calculations of Kindell and Kennel as the ECHO series involves non-uniform plasmas with large radial gradients of density and temperature which may shift resonant frequencies and harmonic structures.

We have found that the pitch angle of the injected beam has an overwhelming effect on the ELF spectrum and the type of ion resonances observed. Figure 7, in the format of Figures 3 and 5, shows the wave vectors in a 50 ms interval during which gun 1 was in the "up" injection mode. At the start of the interval, 183.620 s, the injection was changed from "out" (100 degrees) to up (180 degrees). In the first 10 ms the inward-directed fields die away and are replaced by a growing periodic wave oscillating transverse to the radial beam direction and with a frequency near the sampling rate so that successive points are directed opposite. The hot plasma rapidly disappears also as shown by the

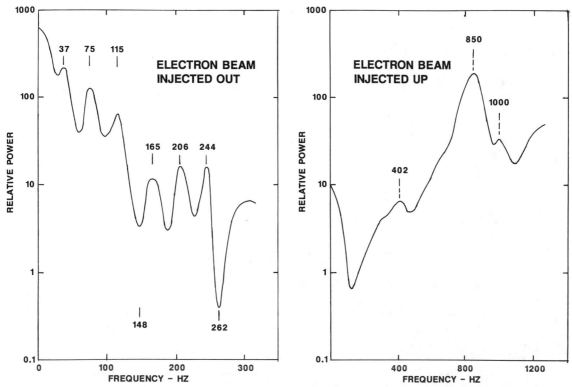

Figure 6. Power spectra for the "out" configuration derived from a 128 ms data sample (left) and for the "up" configuration from a 50 ms sample (right). Oxygen-range resonances show in the left panel while a proton resonance dominates the "up" configuration. Both spectra show some absorption features of unknown origin.

floating potential reverting to its normal 0.8 volts negative after 10 ms. The increasing wave activity is reflected in a small floating potential response towards the end of the sample. This disturbance is identifiable as a proton ion cyclotron wave, and we have observed its excitation in many of the "up" type injections. It was discussed in a previous publication (Winckler et.al., 1984). The delay in the start of this resonance must reflect the time required to set up a suitable new state of the system after terminating the "out" injection and the large hot plasma region it produces. This must be both a spatial and temporal decay, and the time constant of 30-40 ms is similar to that previously observed (Arnoldy and Winckler, 1985). The power spectrum of this resonance derived from a case where the resonance was fully developed for an entire 50 ms sample is shown in the right panel of Figure 6. The proton peak at 850 Hz is the principal feature, and is shifted 50 Hz from the 800 Hz gyro resonance. A small peak at the gun drive frequency (1000 Hz) and another at the helium gyro harmonic (400 Hz), and two absorption minima at 120 Hz and 1080 Hz complete the detectable spectral features. The proton resonance shows field strengths of 50-75 mV/m and dominates all other spectral features. This makes it possible to construct a polarization hodogram of the wave vector, reproduced here as Figure 8 (see Winckler et. al., 1984). The wave is polarized in a plane whose normal points substantially SW of the beam direction, and the wave electric field is consistent with a drift wave in the plasma flow pattern due to the observed ionospheric flow combined with the local EXB flow around the beam, as observed in the PDP reference frame. The frequency of the wave may be Doppler shifted due to the motion of the medium over the observing spacecraft.

The proton resonance was embedded in a band of excitation, and is indicated by the white dot in Figure 9. This figure summarizes all of the ECHO 6 wave data for "up" injections. The lower panel is a spectrogram in maximum detail for the entire accelerator "on" period, evaluated for the polarization transverse to the direction inward towards the beam. The top panels summarize our lower hybrid range (8.5 kHz) and whistler range (125 kHz) narrow band channels also for the "up" mode of gun 1 operating at full power, and evaluated every 10 s. The lower hybrid energy maximizes around 300 s when the system was completely on the north side of the auroral arc.

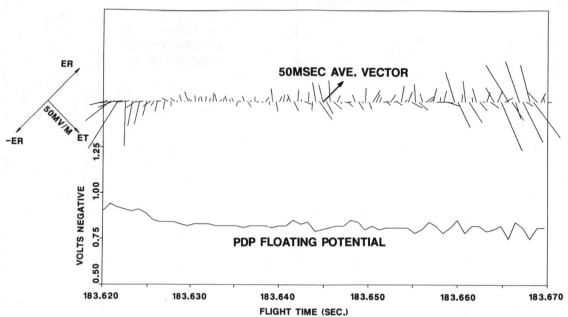

Figure 7. A 50 ms sample in the format of Figure 3 with gun 1 injecting up (180 degree pitch angle) at full power. Gun 2 had been turned off at the start of the interval but a finite time was required for the plasma to cool before the Hz proton resonance began its transverse oscillatory mode towards the right end of the sample. The lower curves are the PDP-probe single potentials, and are a measure of the PDP floating potentials and the hot plasma intensity. The latter decreases in the first 10 ms to an equilibrium value much lower than during gun 2 firing (see also Figure 5).

This was a region of enhanced effects both in the beam-produced disturbances and in the natural DC ionospheric convection electric fields as discussed by Winckler et. al., (1985b). The whistler mode channel at top also has a small maximum near 313 s, but a much larger peak at the edge of the region of intense wave production encountered at 193 s, or about 50 m perpendicular distance from the beam. No MHz range waves were measured in ECHO 6, but previous studies have shown relatively intense emissions near the local plasma frequencies due to electron beam injection (Cartwright and Kellogg, 1974, DeChambre et. al., 1980).

Summary-Discussion

We have shown that ion resonances of known ionospheric species can be excited by the injection of artificial electron beams from a sounding rocket in the auroral ionosphere.

These observations cannot be explained by the basic electron beam-plasma interaction, which occurs in the high-frequency range near the electron plasma frequency. Mode coupling, parametric decay and other similar secondary processes do not have promise to explain the large ELF effects observed.

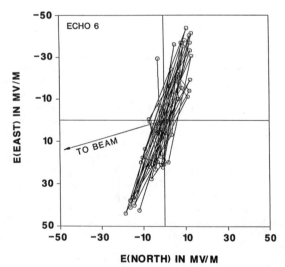

Figure 8. Polarization hodogram of the 850 Hz proton resonance. The wave field is almost linearly polarized perpendicular to the average quasi-dc electric field. The latter was generally directed inward towards the beam quadrant and presumably showed the plasma flow characteristics. Thus this resonance may be a type of drift wave.

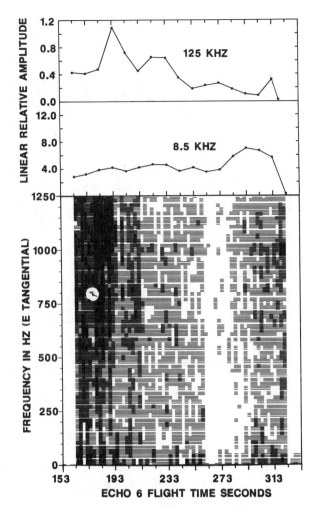

Figure 9. Summary of wave data for "up" mode injections. The spectrogram at bottom refers to the polarization component transverse to the radial direction to the beam. The white dot is the time and frequency of the proton resonance evaluated in Figure 6, right panel. The upper panels are the lower hybrid (8.5 kHz) and whistler mode (125 kHz) narrow-band channels. The lower hybrid signal maximizes on the poleward side of the auroral dropout shown in the lower spectrogram. These channels do not show the dramatic auroral attenuation of the ELF and DC ranges, indicating probably a closer association with the primary beam-plasma interaction.

The resonances are observed well outside the 10 m radius of the injected beam but in a region pervaded by hot plasma due to the beam-plasma interaction. This plasma appears with great rapidity (<1 ms) after beam turn-on out to more than 100 m. The mechanism for producing this hot plasm is not well understood.

The lower frequency range below 200 Hz shows resonances near the O+ gyro frequency which occur in an inner region of very hot plasma and strong quasi-dc electric fields which extend out to a radius of 60 m. These fields seem to be a part of the plasma flow pattern around the beam. This inner region is observed when the beam is injected near 90 degrees pitch angle and spirals past the observing payload.

The higher frequency range between 200 Hz and 1.25 kHz is preferentially excited when the beam is injected parallel to the magnetic field. In this configuration the hot plasma effects were much smaller and the quasi-dc fields were not observed. A large-amplitude proton gyro wave was repeatedly observed, as well as the 1 kHz power drive modulation and possibly some other resonances (He).

All the ELF range electric effects including the beam-generated waves as well as the quasi-dc fields, and the ionospheric convection fields, were suppressed when the system passed through field lines containing auroral precipitation above an auroral arc. In effect all transverse electric effects were suppressed because of the enhanced particle fluxes on these field lines. Conversely, in the aurora-free region on the poleward side of the arc all the transverse electric fields, both natural and artificial, became appreciably enhanced, which must be associated with a decreased natural ionospheric conductivity.

The ion resonances may have their origin in the currents flowing parallel to the magnetic field required for vehicle neutralization over a considerable cross section outside the beam, corresponding to the hot plasma region. Ion excitation by current-driven instabilities as discussed by Kindel and Kennel (1971) may be applicable to our case. In addition, the plasma flow around the beam shown by the quasi-dc fields and possible diamagnetic drift currents may generate drift instabilities which resemble the observations. All these effects are rather far removed from the primary beam-plasma interaction, but rather depend on the entire beam-emitting system and its interaction with the ionospheric flow past the system.

Some major questions for further study are: a) How is the hot plasma produced in the region exterior to the beam? b) What are the flow patterns and currents which excite the ELF range ion resonances? c) How do they depend on beam injection geometry, etc.? d) What is the exact role of the ambient plasma, its conductivity, etc.? e) Why does atomic hydrogen, despite its low abundance, produce such dominate resonance effects? and f) What are the details of the mechanism by which the aurora suppresses beam effects and what does this tell us about the physics of auroral arcs?

Acknowledgements. This work was supported by the Plasma Physics Division, NASA Headquarters, under Grant NSG 5088. The authors benefited by discussions with Profs. Robert L. Lysak and Paul J. Kellogg.

References

Arnoldy, R. L. and J. R. Winckler, The Hot Plasma Environment and Floating Potentials of a Beam-Emitting Rocket in the Ionosphere, Jour. Geophys. Res. 86, 575, 1981.

Arnoldy, R. L., Craig Pollock and J. R. Winckler, The Energization of Electrons and Ions by Electron Beams Injected in the Ionosphere, Jour. Geophys. Res. 90, 5197, 1985.

Arnoldy, R. L., Ion Effects Measured During the Injection of Electron Beams and Plasma Into the Ionosphere, Invited Paper, these proceedings.

Cartwright, D. C., and P. J. Kellogg, Observations of Radiation from an Electron Beam Artificially Injected into the Ionosphere, Jour. Geophys. Res. 79, 1439, 1974.

DeChambre, M., C. A. Gusev, Yu. V. Kushnerevsky, J. Lavergnat, R. Pellat, S. A. Pulinets, V. V. Selegel, I. A. Zhulin, High Frequency Waves During the ARAKS Experiments, Annales de Geophys. 36, 333, 1980.

Kindel, J. M., and C. F. Kennel, Topside Current Instabilities, Jour. Geophys. Res., 76, 3055, 1971.

Lavergnat, J. and R. Pellat, High-Frequency Spontaneous Emission of an Electron Beam Injected into the Ionospheric Plasma, Jour. Geophys. Res. 84, 7223-7238, 1979.

Stenzel, R. L., and D. A. Whelan, Electromagnetic Radiation from Beam-Plasma Instabilities, in Artificial Particle Beams in Space Plasma Studies, edited by B. Grandal, pp 471-480, Plenum, New York, 1982.

Winckler, J. R., The Use of Artificial Electron Beams as Probes of the Distant Magnetosphere, in Artificial Particle Beams in Space Plasma Studies, edited by B. Grandal, pp. 3-33, Plenum, New York, 1982.

Winckler, J. R., J. E. Steffen, P. R. Malcolm, K. N. Erickson, Y. Abe and R. L. Swanson, Ion Resonances and Elf Wave Production by an Electron Beam Injected into the Ionosphere: ECHO 6, Jour. Geophys. Res. 89, 7565, 1984.

Winckler, J. R., J. E. Steffen, P. R. Malcolm, K. N. Erickson, Y. Abe, and R. L. Swanson, Objectives and Design of the ECHO 6 Electron Beam Experiment; Large Ionospheric Perturbations and Energetic Particle Patterns, ARCAD Symposium, Centre d'Etudes Spatiale des Rayonnements, Toulouse, France, 22-25 May, 1984. Conference Proceedings, Jan. 1985a.

Winckler, J. R., K. N. Erickson, Y. Abe, J. E. Steffen and P. R. Malcolm ELF Wave Production by an Electron Beam emitting Rocket System and its Supression on Auroral Field Lines: Evidence for Alfven and Drift Waves, Geophys. Res. Lett. 12, 457, 1985b.

ARGON IONS INJECTED PARALLEL AND PERPENDICULAR TO THE MAGNETIC FIELD

R. E. Erlandson and L. J. Cahill, Jr.

School of Physics and Astronomy, University of Minnesota, Minneapolis, MN 55455

C. Pollock and R. L. Arnoldy

Space Science Laboratory, University of New Hampshire, Durham, NH 03824

J. LaBelle and P. M. Kintner

School of Electrical Engineering, Cornell University, Ithaca, NY 14853

T. E. Moore

Marshall Space Flight Center, Huntsville, AL 35812

Abstract. The ARCS 3 payload equipped to measure electric and magnetic field fluctuations, and charged particles was launched from Sondre Stromfjord at 0211 UT on 10 February 1985. A subpayload, carrying two argon ion accelerators, was ejected downward and parallel to the magnetic field. The spin axis of the subpayload was aligned along the field so that one accelerator ejected ions, approximately anti-parallel to the field toward the main payload and the other accelerator ejected ions approximately perpendicular to the field. In this paper we present results from our preliminary investigation of the second perpendicular and parallel ion beam injections. The investigation centers on observations from single particle ion detectors and 0 to 10 KHz electric field plasma wave receivers. An ion flux consistent with the expected dectector response due to the argon ion beam was recorded during both the perpendicular and parallel ion beam events as well as an unexpected ion flux at 90° pitch angle during the second parallel ion beam event. The waves observed during the perpendicular events were much stronger than those observed during the parallel events.

Experiment

The payload traveled southward over the auroral oval along an azimuth of 127° and reached a peak altitude of 406 km. At a time 134 s into the 600 s flight, at an altitude of 250 km, the subpayload was ejected downward at an angle of 7.5° from the magnetic field. After the first ion beam injection, as the payloads separated with a speed of 2.3 m/s, the ion accelerators operated alternately for 17 s periods with a 10 s rest period between events. During this portion of the flight the spin axis of the main payload was aligned at a 45° angle with respect to the magnetic field.

The observations described are from measurements obtained by electric field receivers and ion detectors (Figure 1). The electric field plasma wave receivers used spherical probes, separated by 3 m, on the LF E-field booms. The boom to boom DC-E electric field meter recorded electric field fluctuations in the range from 0 Hz to 625 Hz. Boom to payload electric field fluctuations in the range from 30 Hz to 10 KHz were recorded by the HF1S experiment. Two ion electrostatic analyzers sweep in acceptance energy from 25 KeV to 13 eV in a time of 5 seconds. The detectors cover all pitch angles during one spin period and do not distinguish between different ion species.

On the subpayload, two ion accelerators injected argon ions and low energy electrons (~ 1 eV) perpendicular and parallel to the magnetic field (Figure 1). In-flight monitors indicated that the argon ion accelerators operated properly and allowed estimates to be made of the total argon ion beam current and accelerating voltage. It is estimated that the total argon ion beam current is 200 $\pm$ 100 mA and the accelerating voltage of the perpendicular and parallel accelerators are 220 $\pm$ 20 V and 190 $\pm$ 20 V respectively. The maximum argon ion beam flux, estimated from the accelerating voltage and laboratory measurements of the beam, is at an

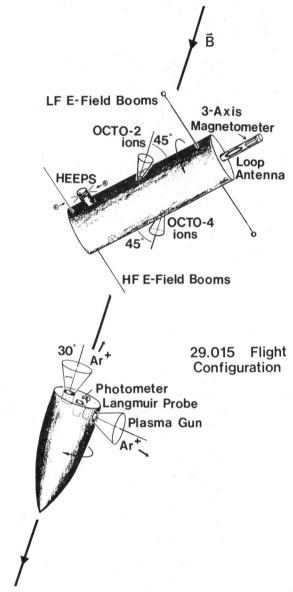

Fig. 1. ARCS 3 experimental configuration of argon plasma generators, plasma wave receivers, and particle detectors.

energy of 190 ± 20 eV and 160 ± 20 eV for the perpendicular and parallel plasma guns respectively.

Ion Observations

The ion flux at different pitch angles and energies as measured by two ion electrostatic analyzers during the second perpendicular and parallel events are presented in Figures 2 and 3. The middle channel of the figure is the energy sweep of both detectors with toggles between two different energies. The top two channels are the pitch angle and ion flux of detector OCTO-4 which looks once each spin in the direction of the subpayload. The bottom two channels are of the pitch angle and ion flux of the detector OCTO-2 which looks in the opposite direction of OCTO-4.

Several distinct peaks in the ion flux appear during the second perpendicular ion beam event (Figure 2). First, 156 eV ions are recorded at pitch angles between 90° and 100° by the OCTO-4 detector and not by the OCTO-2 detector. This is close to the expected response of the ion detectors to the perpendicular ion beam. Secondly, upflowing ions were detected at energies of 32 and 58 eV with pitch angles from 155° to 180° by the OCTO-4 detector. It is presently uncertain whether these ions are gun related or not although they appear to originate from near the subpayload. The arrows in figure 3 point to gun related features in the ion flux during the second parallel event. The peak at 156 and 202 eV with pitch angles from 155° to 180° by the OCTO-4 detectors is the expected response due to the parallel ion beam. The flux is nearly equal for both energy channels while it is expected that the ion flux at 156 eV should be greater than the flux at 202 eV since the maximum argon ion flux of the parallel gun is at an energy of 160 ± 20 eV. The other arrow points to gun related 274 eV ions at 90° to 95° pitch angles. The ion flux is similar to the flux measured during the second perpendicular event in that only one detector observes ions although at this

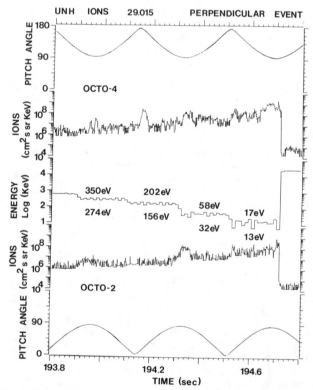

Fig. 2. Ion flux in units of $\#/cm^2$ ster s KeV during the second perpendicular ion beam event.

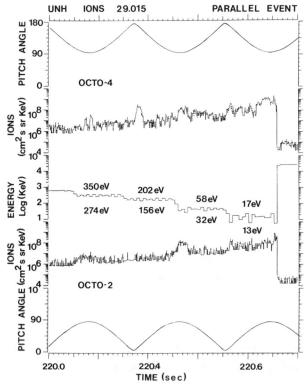

Fig. 3. Ion flux in units of #/cm² ster s KeV during the second parallel ion beam event.

time the perpendicular ion gun is off. Also, these ions are at an energy greater than the ion gun's accelerating voltage. The origin of the 274 eV ion flux at 90° pitch angle is puzzling and we speculate that they are either argon ions or ambient ions that have been transversely accelerated during the parallel ion beam injection.

Wave Observations

A spectrograph showing the time evolution of electric field fluctuations recorded by the HF1S wave receiver illustrates the different spectral features which occur during the second perpendicular and parallel ion beam events (Figure 4). The horizontal lines which continue throughout the flight are due to payload interference signals. During the perpendicular event spin modulated broadband noise and emissions near harmonics of the hydrogen gyrofrequency dominate the spectrograph whereas during the second parallel ion beam injection there was an increase in wave power between 4 to 6 KHz with a narrow emission at 5.9 KHz. The line at 5.9 KHz is at the eighth harmonic of the hydrogen gyrofrequency and is near the lower hybrid frequency. It is unclear whether the electric field fluctuations observed during the second parallel ion beam injection are generated by the parallel ion beam, the observed 274 eV ions at 90° pitch angle, or by another source of free energy.

Low frequency waves below 625 Hz were recorded

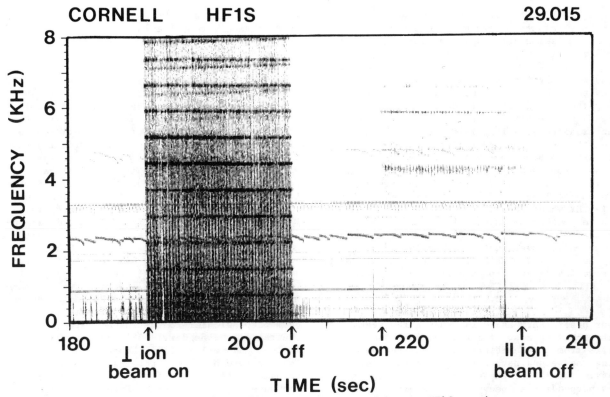

Fig. 4. Electric field fluctuations recorded by the HF1S receiver.

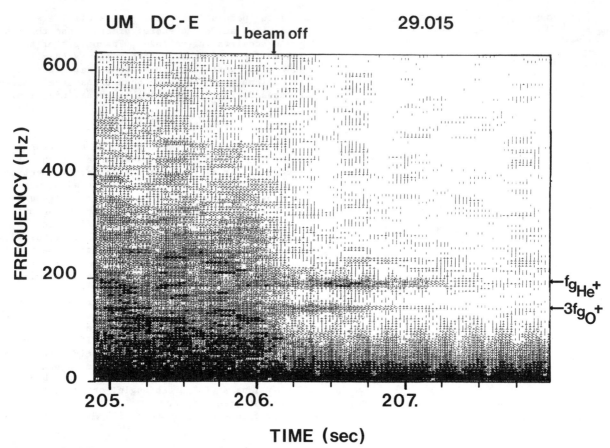

Fig. 5. Low frequency waves at the helium gyrofrequency and the third harmonic of the oxygen gyrofrequency recorded by the DC-E receiver at the end of the second perpendicular ion beam event.

by the DC-E electric field meter during the second perpendicular ion beam event whereas during the second parallel ion beam event no changes in wave activity were observed. Figure 5 is an expanded view in time of an interval at the end of the second perpendicular ion beam event. After the ion beam was turned off at 206.1 s the level of broadband noise decreased and two emission lines appeared, lasting for over 1 second. The lines are at frequencies of 140 Hz and 190 Hz which coincide with the third harmonic of the oxygen gyrofrequency (146 Hz) and the helium gyrofrequency (194 Hz) respectively. We are currently investigating emissions at the heavy ion gyrofrequencies which are found to also occur during later perpendicular ion beam events.

Summary

Our preliminary analysis, only a few weeks after receiving most of our data, included results from ion detectors and wave receivers during the second perpendicular and parallel gun events. The ARCS 3 experiment expanded on previous perpendicular heavy ion beam experiments [Haerendel and Sagdeev, 1981; Moore et al., 1982; Moore et al., 1983; Pottelette et al., 1984] by investigating both a perpendicular and parallel heavy ion beam. First, ions with energies near 274 eV were detected at 90° pitch angle during the second parallel ion beam injection may have been transversely accelerated by a mechanism which operates during the injection of parallel heavy ion beams. These ions were detected when the field line separation between payloads was approximately 50 m. Secondly, electric field fluctuations observed during the second perpendicular and parallel ion beam events differed greatly in their spectral features. The emissions at harmonics of the hydrogen gyrofrequency observed during the second perpendicular ion beam event are nearly identical to the harmonic features observed in the Porcupine experiment during the injection of a 200 eV xenon ion beam perpendicular to the magnetic field [Haerendel and Sagdeev, 1981; Jones, 1981; Kintner and Kelley, 1981; Lebreton et al., 1983]. This feature has been explained in terms of a resonant interaction of beam ions which move near the wave phase velocity by Roth et al. [1983].

Finally, waves were observed at heavy ion gyrofrequencies immediately after the second perpendicular ion beam injection. At this time the plasma wave and ion data recorded during all the perpendicular and parallel events are being investigated in more detail.

Acknowledgements. This research was supported at the University of Minnesota by NASA Grant NAG 6-11, at the University of New Hampshire by NASA Grant NAG 6-12, and at Cornell University by NASA Grant NAG 5-601.

References

Haerendel, G., and R. Z. Sagdeev, Artificial plasma jet in the ionosphere, Adv. Space Res., 1, 29, 1981.

Jones, D., Xe^+ -induced ion-cyclotron waves, Space Res., 20, 379, 1981.

Kintner, P. M. and M. C. Kelley, Ion beam produced plasma waves observed by the $\delta n/n$ plasma wave receiver during the Porcupine experiment, Adv. Space Res., 1, 107, 1981.

Lebreton, J. P., R. Pottelette, O. H. Bauer, J. M. Illiano, R. Truemann, and D. Jones, Observation of waves induced by an artificial ion beam in the ionosphere, Active Experiments in Space, ESA SP-195, 35, 1983.

Moore, T. E., R. L. Arnoldy, R. L. Kaufmann, L. J. Cahill, Jr., P. M. Kintner, and D. N. Walker, Anomalous auroral electron distributions due to an artificial ion beam in the ionospheric plasma, J. Geophys. Res., 87, 7569, 1982.

Moore, T. E., R. L. Arnoldy, L. J. Cahill, Jr., and P. M. Kintner, Plasma jet effects on the ionospheric plasma, Active Experiments in Space, ESA SP-195, 197, 1983.

Pottelette, R., J. M. Illiano, O. H. Bauer, and R. Treumann, Observation of high-frequency turbulence induced by an artificial ion beam in the ionosphere, J. Geophys. Res., 89, 2324, 1984.

Roth, I., C. W. Carlson, M. K. Hudson, and R. L. Lysak, Simulations of beam excited minor species gyroharmonics in the Porcupine experiment, J. Geophys. Res., 88, 8115, 1983.

A COMPARISON OF PLASMA WAVES PRODUCED BY ION ACCELERATORS IN THE F-REGION IONOSPHERE

P. M. Kintner, J. LaBelle, and W. Scales
School of Electrical Engineering, Cornell University, Ithaca, NY 14853

R. Erlandson and L. J. Cahill, Jr.
School of Physics and Astronomy, University of Minnesota, Minneapolis, MN 55455

Abstract. Ion beams injected into the ionosphere are known to produce waves related to the normal modes of the plasma. In this paper we examine the spectra of plasma waves produced during four sounding rocket experiments. The experimental conditions were somewhat different during each experiment. The accelerated ion was either Xe^+ or Ar^+ and the experimental geometry, described by the separation vector between the plasma wave receiver and the ion accelerator, was either parallel or perpendicular to the geomagnetic field.

Introduction

During periods of magnetic activity the earth's magnetosphere is known to fill with energetic ions in the range of 10 keV to 100 keV. In the past decade ion mass spectrometer measurements have shown that oxygen forms a substantial fraction, sometimes a majority, of the energetic ion population [Johnson, 1983]. Furthermore low altitude ion mass spectrometer measurements have shown that oxygen below 1 R_E altitude can be accelerated to several keV by two different processes - either parallel acceleration or perpendicular acceleration [Shelley et al., 1976; Sharp et al., 1977]. Hence, it appears that the earth's ionosphere substantially contributes to the earth's energetic ion population. Given this hypothesis, the questions of how ions are accelerated and how they propagate into the magnetosphere naturally arise.

These questions have been partially answered by passive observations on satellites and sounding rockets. Another approach to answering these questions is with active ion beam experiments. That is, by injecting an ion beam, the microscopic processes and ion propagation can be examined in a straightforward manner since the source of free energy is known.

Ion beams can be injected in two ways: either by an explosive shaped charge release of barium which later photoionizes or by ionizing a gas then accelerating it through an electrostatic potential. In the latter case the ion accelerator is mounted on a subpayload which is typically ejected from the instrumented payload, and subsequently the ions are accelerated toward the instruments. Only examples of plasma waves generated by ion accelerators are presented here.

Description of the Experiments

Artificial ion beams have been generated in space from several different platforms. We will discuss here 4 experiments (Porcupine and ARCS 1,2 and 3) whose operating parameters are listed in Table 1. With the exception of ARCS 1, all of the payloads consisted of a mother section with instruments and an ejected daughter section with an ion accelerator.

In all cases plasma wave activity was measured. However, in the case of ARCS 1, the waves generated could not be identified with any normal modes of a warm plasma. Consequently we will not discuss it here but a detailed examination of that flight can be found in Moore et al. [1982]. In the remaining experiments wave modes were generated which can be identified with normal modes of the plasma. We will examine these cases next with particular emphasis on

Table 1

Experiment Name	Ion Type	Ion Energy	Ion Pitch Angle	Ion Accelerator-Receiver Geometry
Porcupine	Xe^+	200 eV	$\cong 90°$	Perpendicular to $\underline{B}$
ARCS 1	Ar^+	25 eV	0-90°	Attached
ARCS 2	Ar^+	33 eV	$\cong 90°$	Parallel to $\underline{B}$
ARCS 3	Ar^+	160 eV/ 190 eV	$\cong 0°/\cong 90°$	Parallel to $\underline{B}$

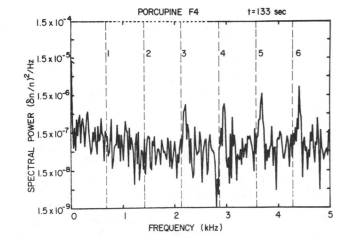

Fig. 1. Plasma wave emissions at multiples of the hydrogen cyclotron frequency produced by a Xe^+ ion beam during the Porcupine experiment.

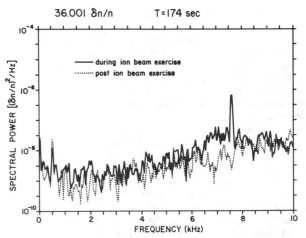

Fig. 3. Plasma wave emissions near the lower hybrid frequency produced by an Ar^+ ion beam during the ARCS 2 experiment.

repeatability of the measurements in similar physical situations.

Plasma Wave Measurements

The Porcupine sounding rocket contained four ejected subpayloads of which one was a Xe^+ ion beam. The subpayload was ejected at an angle of 52° to the magnetic field. It produced a variety of effects some of which are discussed in Haerendel and Sagdeev [1981] and Jones [1981]. During the first exercise of the Xe^+ gun emissions just above multiples of the hydrogen gyrofrequency were observed (Fig. 1). These were remarkable on several counts. First, the emissions occurred at an altitude of 200 km where H^+ should be a minor constituent of the ionosphere and second the lower order modes (n=1,2,3) were less intense than the higher order modes. Kintner and Kelley [1983] were able to explain these results by assuming that the Xe^+ beam was unmagnetized and that H^+ was 1% of the ionospheric density. In their calculation the Xe^+ beam coupled to flute mode Bernstein emissions but oxygen cyclotron damping reduced the growth rate of the lowest order modes. Malingre and Pottelette [1985] extended the calculation to show that the growth rates are ordered by the Xe^+ LHR frequency.

Since the wave generation mechanism only weakly depends on the ion beam velocity and therefore the ion mass and energy, we would expect to find similar wave activity from a perpendicular Ar^+ beam. The Bernstein modes were again observed during ARCS 3 and they were similar to those

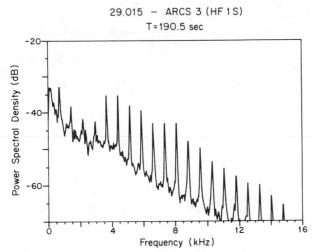

Fig. 2. Plasma wave emissions at multiples of the hydrogen cyclotron frequency produced by an Ar^+ ion beam during the ARCS 3 experiment.

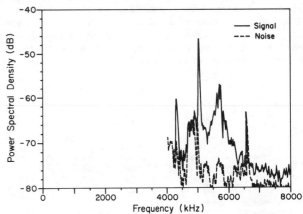

Fig. 4. Plasma wave emissions near the lower hybrid frequency produced by an Ar^+ ion beam during the ARCS 3 experiment.

observed during the Porcupine experiment. An example is shown in Figure 2. In this case the emissions at multiples of the hydrogen cyclotron frequency were produced and the lower order modes were attenuated. Inspection of the time domain data reveals that the emissions show extreme nonlinear steepening. The nonlinear steepening is responsible for the large number of harmonics generated.

At later times and larger separation distances, emissions near the O^+ lower hybrid frequency were produced by ion beams on all three flights. In the case of Porcupine a broad peak was observed at the lower hybrid frequency. However, on ARCS 2 a narrow peak was observed at the lower hybrid frequency (Fig. 3). Below the lower hybrid frequency was a broader emission spectrum with less amplitude. An explanation for waves generated near the lower hybrid frequency nonresonantly has been offered by Hudson and Roth [1984]. It was possible to establish the emission wavelength as 6 meters compared to an O^+ gyroradius of 4 meters [Kintner et al., 1984].

An example of emissions near the LHR frequency produced by the perpendicular ions on ARCS 3 is shown in Fig. 4. This wave data is extremely noisy because the experiment has been optimized to measure wavelength instead of frequency. As a result it is very sensitive to onboard interference which is labeled noise. The noise spectrum was produced from data taken just after the end of an ion beam exercise. The spectrum labeled signal was produced from data during an ion beam exercise. In this case several emissions can be identified near the lower hybrid frequency which was roughly 7 kHz. The emission contained both sharp and broad spectral peaks but at somewhat different frequencies than those present during ARCS 2.

Summary

Propagating ion beams in the lower ionosphere excited plasma waves at the normal modes of a warm plasma. They can be identified as H^+ Bernstein modes and waves near the lower hybrid resonance. Both types of plasma waves were observed during at least two experiments even though the experimental geometry was different in each case.

Acknowledgements. We would like to thank G. Haerendel for access to the Porcupine data and we would like to thank our collaborators on the ARCS 1, 2, and 3 experiments - R. Arnoldy, T. Moore, R. Kaufmann, and D. Walker. This research was supported at Cornell by NASA Grant NAG5-601 and at the University of Minnesota by NASA Grant NAG6-11.

References

Haerendel, G., and R. Z. Sagsdeev, Artificial plasma jet in the ionosphere, Adv. Space Res., 1, 29, 1981.

Hudson, M. K. and I. Roth, Thermal fluctuations from an artificial ion beam injection into the ionosphere, J. Geophys. Res., 89, 9812, 1984.

Johnson, R. G., The hot ion composition, energy, and pitch angle characteristics above the auroral zone ionosphere, High Latitude Space Plasma Physics, B. Hultqvist and T. Hagfors, eds., Plenum Publ. Co., New York, 1983b.

Jones, D., Xe^+-induced ion-cyclotron waves, Space Res., 20, 379, 1981.

Kintner, P. M. and M. C. Kelley, A perpendicular ion beam instability: Solutions to the linear dispersion relation, J. Geophys. Res., 88, 357, 1983.

Kintner, P. M., J. LaBelle, M. C. Kelley, L. J. Cahill, Jr., T. Moore, and R. Arnoldy, Interferometric phase velocity measurements, Geophys. Res. Lett., 11, 19, 1984.

Malingre, M., and R. Pottelette, Excitation of broadband electrostatic noise and of hydrogen cyclotron waves by a perpendicular ion beam in a multi-ion plasma, Geophys. Res. Lett., 12, 275, 1985.

Moore, T. E., R. L. Arnoldy, R. L. Kaufmann, L. J. Cahill, Jr., P. M. Kintner, and D. N. Walker, Anomalous auroral electron distributions due to an artificial ion beam in the ionosphere, J. Geophys. Res., 87, 7569, 1982.

Sharp, R. D., R. G. Johnson, and E. G. Shelley, Observation of an ionospheric acceleration mechanism producing energetic (keV) ions primarily normal to the geomagnetic field direction, J. Geophys. Res., 82, 3324, 1977.

Shelley, E. G., R. D. Sharp, and R. G. Johnson, Satellite observations of an ionospheric acceleration mechanism, Geophys. Res. Lett., 3, 654, 1976.

Section VI. LABORATORY PROCESSES

ION ACCELERATION IN LABORATORY PLASMAS

Reiner L. Stenzel

Department of Physics
University of California, Los Angeles, California 90024

Abstract. Acceleration and heating of ions observed in various laboratory experiments are reviewed and related to the interest in space plasma physics. Various cases of increasing complexities are studied. First, in dc electric fields such as occur in double layers and sheaths the ion acceleration is understood in terms of the electrostatic force on single particles, $\vec{f} = q\vec{E}$. Second, in time-varying electric fields the acceleration mechanism can be ion cyclotron resonance ($\omega = \omega_{ci}$), Landau resonance ($\omega = \vec{k} \cdot \vec{v}$), or non-resonant processes such as impulsive (transit time) and stochastic (strong turbulence) heating. Observations for ac ion acceleration will include minority species ion cyclotron heating, scattering of test ions by ion acoustic turbulence, and ion acceleration by the ponderomotive force of intense localized high frequency fields. In time-varying magnetic fields such as occur during magnetic reconnection both inductive and electrostatic electric fields are present in plasmas and accelerate ions. Ion jetting from a neutral sheet is observed and interpreted in fluid terms to arise from the magnetic $\vec{J} \times \vec{B}$ force or, in particle terms, to result from an electric field due to space charge separation between electrons and ions. During highly dynamic reconnection events (impulsive current disruptions) large inductive voltages (LdI/dt) are generated which are observed to form nonstationary double layers and to generate bursts of fast ions. The ejection of plasma from a disrupted current sheet steepens into a shock wave which produces anisotropic particle distributions. Space and time resolved three-dimensional velocity distribution functions are measured and analyzed with the help of computers.

Introduction

The acceleration of charged particles arises from electric and magnetic forces, $\vec{f} = q(\vec{E}+\vec{v}\times\vec{B})$. While the acceleration by electric fields leads to an energy increase magnetic fields do not energize particles ($\vec{f} \cdot \vec{v} = 0$). Time varying magnetic fields cause energization via inductive or motional electric fields. In plasmas a variety of conditions can lead to an acceleration of ions. Since the electromagnetic fields are frequently not well known the acceleration is difficult to understand. In comparison with space plasmas the situation in laboratory plasmas can be simpler since one can often isolate one problem at a time and study it under controlled conditions. Several such examples will be presented in this paper. The research on double layers is one case where laboratory plasmas have provided much insight due to detailed diagnostics, controlled parameter variations and known boundary conditions (Sato, 1982). The presently discussed experiments emphasize the physical concepts rather than the scaling aspect of the more global modeling experiments (Podgorny et al., 1980; Baum and Bratenahl, 1982).

Sheaths

A collisionless plasma usually supports strong electric fields only over distances on the order of the Debye length such as in sheaths, double layers, electrostatic waves. Sheaths play a major role in the interaction of bodies with plasmas. For example, the filling of a wake behind a spacecraft is both a problem of plasma expansion (Wright et al., 1985) and of the deflection of particles at the boundaries of the object. The latter problem has been studied in conjunction with the development of a novel ion velocity analyzer (Stenzel et al., 1982).

Fig. 1a shows the experimental set-up consisting of a double plasma device. From the source plasma an ion beam is injected into the target plasma where it may impinge on a flat disc and create a wake. With a dual energy analyzer the ion distribution is measured. One side of the analyzer contains, for purpose of comparison, a conventional retarding potential analyzer, while the other side has a novel directional analyzer. As shown in Fig. 1b the new analyzer contains a passive microchannel

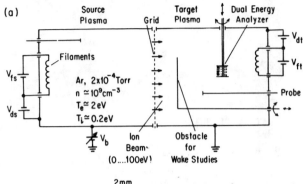

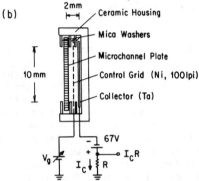

Fig. 1. a) Experimental set up for generating an ion beam incident upon an obstacle which creates a wake. b) Schematic of the directional ion velocity analyzer (from Stenzel et al., 1982).

plate instead of the first floating grid. Since the ions must pass through the long thin channels (600μ x 14μ) the analyzer has excellent angular resolution ($\Delta\theta \simeq 0.6°$). From differential flux measurements the distribution function $f(\vec{v})$ is obtained.

The properties of the unperturbed ion flow are summarized in Fig. 2 which also compares the performance of the two types of analyzers. The polar plot of the ion flux $I_c(\theta)$ shown in Fig. 2a indicates that the angular beam spread of $\Delta\theta = 3°$ is properly resolved with the directional analyzer where as the conventional analyzer trace is purely instrument broadened. Due to the collimated particle flow the directional analyzer also has a far superior velocity resolution (Fig. 2b). For a 60 eV ion beam a beam temperature $kT_b \simeq 0.0027$ eV (~31°K) is resolved. Thus, the beam is essentially monoenergetic and nondiverging. When it impinges on an object one might expect to find a long wake downstream. This is not the case.

Fig. 3 shows the ion flux measurements in the wake. The analyzer resolves two ion beams propagating at angles $\pm \theta_m$ with respect to the normal z axis. As the observation point approaches the plate the angle θ_m increases in such a fashion that all rays project to the edge of the dis . The deflected ions have essentially the same energy as the incident ones. The ions focus with a density maximum exceeding the incident beam density (Fig. 3b).

The observations lead to the conclusion that the ions are deflected by the transverse electric field concentrated in the sheath of the floating plate. The spread in angles is generated by the drop in field strength with distance from the plate. The impulsive ion deflection in the sheath is analogous to a Coulomb scattering process involving a momentum transfer but no energy exchange.

These observations can be scaled to the parameters appropriate to a low-altitude satellite. The deflection angle for the ion focus is approximately $\theta_m \simeq \tan^{-1}(5kT_e/eV_b)$. For typical values ($kT_e \simeq 0.25$ eV, ram ion energy $eV_b \simeq 0.3$ eV for H^+, 4 eV for O^+) the ions are deflected at angles $\theta_m \simeq 6°$ for H^+, 60° for O^+. Thus, the wake behind a disc-like obstacle on a spacecraft fills within tens of

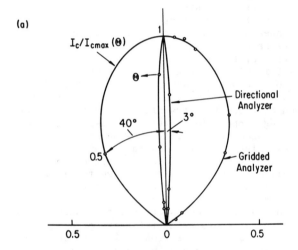

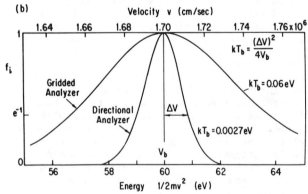

Fig. 2. Comparison between the new directional analyzer and a conventional gridded analyzer in resolving the properties of the injected ion beam. a) Normalized ion current vs. angle indicating a beam spread of $\Delta\theta \simeq 3°$. b) Normalized ion distribution functions vs. energy/velocity indicates a beam temperature $kT_b \simeq 0.0027$ eV for a 60 eV ion beam (from Stenzel et al., 1982).

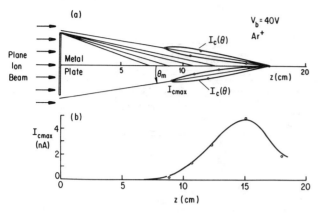

Fig. 3. Ion dynamics in the wake. a) Polar plot of ion current $I_c(\theta)$ indicates two converging ion beams at angles $\theta_m(z)$. The beam deflection occurs in the sheath surrounding the floating plate. b) Ion flux exhibits a focus in the wake (from Stenzel et al., 1982).

centimeters ($\lambda_D \simeq 1$ cm) with ions deflected by sheath electric fields at sharp boundaries. Mass separation occurs since light ions are deflected stronger than heavy ions.

Double Layers

Ions are accelerated by the electric field of double layers. Among the many different double layer experiments (Sato, 1982) we mention the V-shaped double layer in a dipole magnetic field (Stenzel et al., 1981) which resemble auroral potential structures (Akasofu, 1981). By drawing electrons to and reflecting fast ions from an electrode in a converging magnetic field (Fig. 4b) a steady-state two-dimensional potential structure $\phi(r,z)$ shown in Fig. 4a is generated. Along the symmetry axis ($r = 0$, $z \geq 0$) a parallel electric field exists as shown in Fig. 4c. This field decelerates the beam ions injected in $-z$ direction and accelerates background ions in $+z$ direction. Electrons are energized toward the positive potential side. The energy may be provided by external supplies or, in the fully magnetized case (Nakamura and Stenzel, 1982) by the injected ion beam itself.

Double layers may also be produced by inductive voltages, $V = -d\phi/dt$, in which case the potential structures become time-dependent and nonstationary (Stenzel et al., 1983). Ion bursts are generated, an example of which is shown in Fig. 4d. Two traces of ion flux vs.

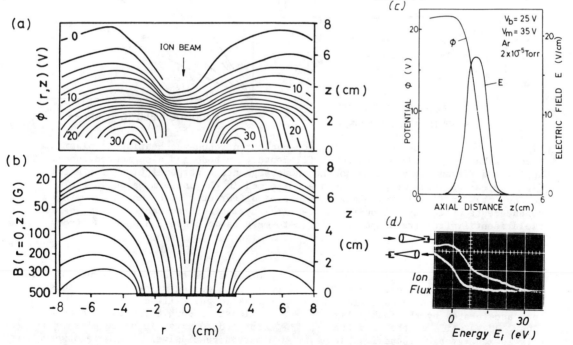

Fig. 4. a) V-shaped potential structures generated by injecting fast ions into a converging magnetic field and reflecting them at a positively biased boundary. b) Magnetic field lines. c) Axial potential and electric field of the V-shaped double layer (from Stenzel et al., 1981). d) Ion flux to a directional velocity analyzer facing toward/away from a double layer. Ions are accelerated from a nonstationary double layer produced by a disruption of a current sheet (from Stenzel et al., 1983).

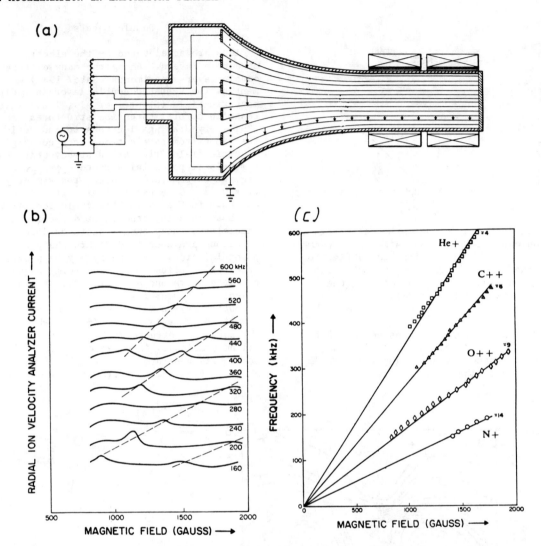

Fig. 5. a) Device for demonstrating ion cyclotron resonance heating in plasmas. A dc discharge is produced between a filamentary cathode and a segmented anode. By applying different ac voltages to the anode segments a perpendicular rf electric field is established. Experiments are performed in the uniform magnetic field region. b) Ion flux $j_\perp$ vs. magnetic field at different rf frequencies showing enhancements at ion cyclotron resonances. c) Cyclotron frequencies vs. magnetic field identify from the slope the charge-to-mass ratio hence ion species (from Stenzel, 1978).

energy are shown. When the directional analyzer points from the low potential side to the high potential side one observes a flux of energetic ions accelerated by the double layer potential ($\Delta\phi_p \simeq 30$ V) while in the opposite direction only low energy background ions are collected. It is interesting to note that the ion energy is provided by the released magnetic field energy of a plasma current system which exhibits spontaneous disruptions (see last section). The peak inductive voltages are much larger than the dc voltages which drive the steady-state currents. The same basic physical processes are thought to produce high energy particles during magnetospheric substorms (Heikkila and Pellinen, 1977).

Classical double layers have been often suspected to exist in space plasmas but not observed conclusively. Remote observations on solar flares lack the required resolution ($100\ \lambda_D \simeq 3$ cm) by 7 orders of magnitude. In-situ observations in magnetospheric plasmas could, in principle, resolve field-aligned potential structures of thickness 1 m to 1 km but conditions of large current densities ($v_d \simeq v_{th}$) are likely to appear sporadically and

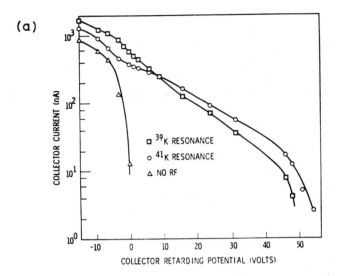

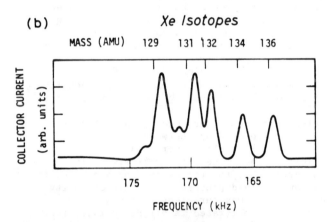

Fig. 6. a) Ion energy analyzer traces $j_\perp(V)$ showing ion cyclotron resonance heating from an initial temperature $kT_i \simeq 0.2$ eV (no rf) to $kT_i \simeq 15$ eV for selected ion species. b) Energization of ions with small mass differences are possible in uniform magnetic fields and monochromatic rf fields (from Dawson et al., 1976).

localized. Possibly, simultaneous probing with multiple detectors enhances the chance of direct observations of double layers in space.

Ion Cyclotron Resonance

The cyclotron motion of ions in a magnetic field can be resonantly excited by an ac electric field oscillating at the frequency $\omega_{ci} = qB/m_i$. An experiment which demonstrates ion energization by cyclotron heating is shown in Fig. 5 (Stenzel, 1978). A discharge plasma is generated by a filamentary cathode and a segmented anode sketched in Fig. 5a. Since the anode voltage determines the plasma potential along $\vec{B}$ a different voltage on each anode segment can be used to generate a perpendicular electric field ($\vec{E} = -\nabla_\perp \phi_p$). When the electric field oscillates at a frequency ω and the magnetic field is varied the perpendicular ion flux is resonantly enhanced as shown in Fig. 5b. A plot of the resonant frequency vs. magnetic field presented in Fig. 5c identifies from the charge-to-mass ratio ($q/m = Z e/m$) the different minor ion species added to an argon plasma.

The enhanced perpendicular ion flux is plotted in Fig. 6a vs. energy. From the slope $d(\ell n I)/dV$ one can infer perpendicular ion heating up to $kT_i \simeq 15$ eV compared to $kT_i \simeq 0.2$ eV without rf. In a highly uniform magnetic field the ion cyclotron resonances are sharply defined ($\Delta f/f < 1\%$) so that selective excitation of various isotopes is possible (Fig. 6b) and may lead to practical applications (Dawson et al., 1976).

Wave-Particle Resonance

A particle moving with velocity $\vec{v}$ in an electrostatic wave of phase velocity $\omega/\vec{k}$ experiences the wave electric field at a Doppler shifted frequency $\omega' = \omega - \vec{k} \cdot \vec{v}$. At the resonance ($\omega' = 0$) the particle interacts with the wave field like with a dc electric field which may lead to a strong acceleration. The resonant scattering and heating of a test ion beam in turbulent ion acoustic waves has been demonstrated in a laboratory experiment (Stenzel and Gekelman, 1978).

In a helium magnetoplasma column ion sound turbulence is generated by drawing field-aligned electron currents with drifts $v_d \gg c_s$ at $T_e \gg T_i$. From cross-spectral measurements the waves are found to propagate dominantly across $\vec{B}$. Typical phase fronts are shown in Fig. 7a. A beam of test argon ions is injected in the direction of wave propagation and detected downstream with a velocity analyzer. Fig. 7b shows test ion distribution functions versus energy at different levels of acoustic turbulence controlled by the anode voltage V_a. A comparison between a resonant test ion beam ($V_b = 15$ V) and a nonresonant beam ($V_b = 20$ V) is given in order to show the stronger beam broadening at wave-particle resonance. The relative beam heating $\Delta T_i/T_i$ is evaluated in Fig. 7c and plotted vs. beam energy which clearly shows the resonant enhancement when $v_b^{Ar} = c_s^{He}$. Heating rather than trapping arises due to the ion interaction with a broad spectrum of modes rather than a single wave.

Resonant interactions between waves and particles occur for both longitudinal and transverse modes ($\omega - \vec{k} \cdot \vec{v} = n\omega_c$, $n = 0, \pm 1,...$). Ions can be energized by cyclotron and Landau damping of various waves and only the simplest examples have been given here. Energization of ions by wave-particle

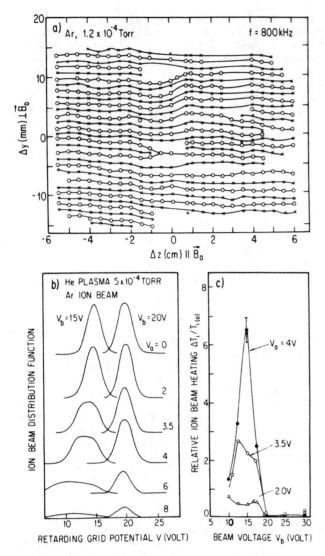

Fig. 7. a) Phase fronts of current-driven ion acoustic turbulence obtained from two-probe cross-spectral measurements. Modes propagate dominantly at $\theta = \cos^{-1}(c_s/v_d) \lesssim 90°$ with respect to the field aligned electron drift $v_d \gg c_s$. b) Test ion beam distribution $f_i(v)$ at different turbulence levels which increase with anode potential V_a. The resonant beam ($V_b = 15$ V, $v_b \simeq \omega/k$) is strongly scattered while the nonresonant ions ($V_b = 20$ V) is not broadened. c) Test ion beam temperature vs. beam energy showing a strong enhancement at wave-particle resonance, $v_b = \omega/k$ (from Stenzel and Gekelman, 1978).

interactions plays an important role in auroral physics (ion conics and beams).

Ponderomotive Force

Ion acceleration arises in many situations due to space charge electric fields which couple the electron and ion fluids. Thus, when a force is exerted on the electrons ions may also be accelerated. An example is the observation of ion energization by intense non-uniform lower hybrid waves (Gekelman and Stenzel, 1977).

In a uniform quiescent magnetoplasma ($\omega_p \simeq \omega_c$) lower hybrid waves ($\omega \simeq 50\omega_{LH}$) are excited by a circular ring antenna whose radiation pattern is that of a converging resonance cone (Fig. 8a). While the wave dispersion causes the field enhancement at the surface of the cone the geometric convergence of the energy flow causes a further intensity increase so that at the focus the wave energy density can well exceed the particle energy density ($\varepsilon_o E^2 \gg nkT$). Thus, the nonlinear effects of the ponderomotive force ($f \propto -\nabla \varepsilon_o E^2$) on the plasma can be investigated at the focus free of boundary effects which usually complicate such studies near antennas.

When an rf pulse is applied one observes within microseconds the evolution of a deep density depression in the region of strong rf fields (Fig. 8b). This rapid expulsion of plasma implies a strong ion acceleration. With an ion energy analyzer bursts of ions are detected with energies of tens of eV corresponding to the rf field energy density (Fig. 8c). Thus, during the transient stage wave energy is effectively coupled into ion kinetic energy. Subsequently a nonlinear interaction develops between the density-dependent rf field pattern and the rf field-dependent density profile. It is observed that the resonance cone breaks up into a random field pattern and that the density profile becomes turbulent. Ions are now randomly accelerated in all directions, i.e. heated, and the observed temperatures (Fig. 8d) exceed the initial values by two orders of magnitude. The relative electron heating is almost negligible ($\Delta T_e/T_e \simeq 1$). Although only the electrons respond to the high frequency electric field the ions gain most of the momentum and energy change imparted by the ponderomotive force.

Active experiments of high power rf wave excitation in space plasmas should produce similar effects as observed in the laboratory. Since typical ionospheric frequencies are lower by two orders of magnitude, a properly scaled ring exciter should be 10 m in diameter, producing a focus at a distance of 40 m, be driven at $f \simeq 100$ kHz ($f_{ce} \simeq 1.4$ MHz, $f_{pe} \simeq 3$ MHz) at power levels $P \simeq 1$ kW (scaled as $nkT \times \lambda^3$). Ion acceleration up to 50 eV would be visible at pulse lengths $t_{rf} \gtrsim 5\,f_{pi}^{-1} \simeq 75\,\mu$sec short enough to neglect the satellite motion (~ 0.5 m) compared to the focal region (~ 5m $\times$ 2m). Similar predictions can also be made for the density ducts created by high power whistler waves which lead to their self-focusing (Stenzel, 1976).

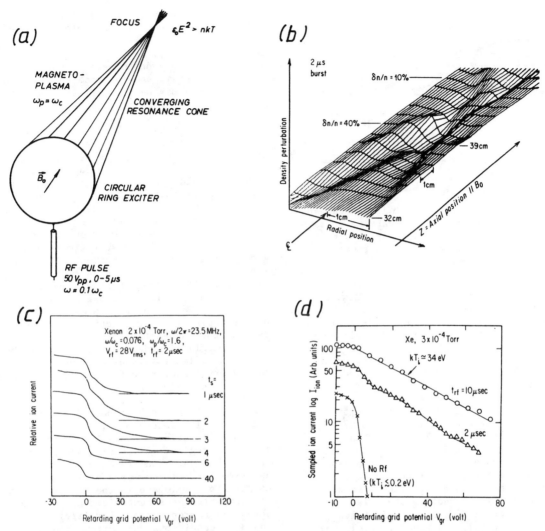

Fig. 8. a) Schematic arrangement for producing an intense localized rf electric field away from an antenna in a magnetoplasma. A circular electrostatic exciter is used to produce a converging resonance cone with a maximum field intensity at the focus. b) Density perturbation produced by the ponderomotive force of a short rf burst. c) Sampled ion energy analyzer traces showing the generation of energetic ion tails ($\lesssim$ 80 eV) due to the rf pressure exerted on the electrons and transferred via space charge fields on the ions. d) During long rf pulses (t = 10 μsec) stochastic ion acceleration results in isotropic heating to $kT_i \simeq 34$ eV (from Gekelman and Stenzel, 1977).

Magnetic Reconnection

Magnetic reconnection is of fundamental importance in space plasmas (Sonnerup, 1984). It deals with changes in magnetic field topologies and conversion of magnetic energy to particle kinetic energy. It couples nonlinear fluid and kinetic processes near magnetic null points to the global system and its boundaries. It is thought to occur spontaneously or driven, in steady-state or explosively, in many situations in space (solar flares, magnetotail, magnetopause) and in the laboratory (tokamaks, pinches). Fundamental to the reconnection problem is magnetic flux transfer by diffusion near a magnetic null region (X point or neutral sheet with at least two components of $\vec{B}$ vanishing). The flux transfer implies an electric field along the neutral line (separator) which can be taken as a measure for the reconnection rate (Vasyliunas, 1975). The reconnection electric field $\vec{E}$, the plasma current density $\vec{j}$, resistivity ρ, fluid velocity $\vec{v}$ and magnetic field $\vec{B}$ are thought to

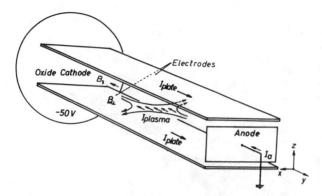

Fig. 9. Schematic picture of the UCLA reconnection experiment. A current sheet is induced in a 2m long discharge plasma by time-varying primary currents in two plate electrodes. All currents close on a coaxial cylindrical chamber wall which is omitted for purpose of clarity.

be coupled self-consistently via Ohm's law, $\vec{E} + \vec{v} \times \vec{B} = \eta \vec{j}$. Fluid models also predict dissipation of magnetic energy by Joule heating and the acceleration of ions up to the Alfvén speed (jetting).

Due to observational difficulties few of the theoretical predictions have been verified in space. Usually, magnetic fields and particle flows are detected on a spacecraft but the data obtained at one point are difficult to extrapolate to two or three dimensions. Remote observations employed for solar flares and fusion plasmas provide still less information. Thus, in order to study reconnection processes comprehensively a carefully diagnosed laboratory experiment has been constructed (Stenzel and Gekelman, 1984). A neutral sheet is established and fields ($\vec{E}$, $\vec{B}$, $\vec{j}$), particles [$f(\vec{v},\vec{r},t)$, n, T, $\vec{v}$, ρ], and waves (δn, δT, $\delta \vec{B}$) are measured in situ. In the present context, however, the processes leading to ion energization will be emphasized.

Fig. 9 is a schematic view of the experimental set up. A 2 m long discharge plasma is generated between a 1 m diam. cathode and grounded anode. The plasma of density $n \simeq 10^{12}$ cm^{-3}, temperature $kT_e \simeq 10$ $kT_i \simeq 10$ eV, and argon gas pressure 10^{-4} Torr is pulsed (5 msec on, 2 sec off). It is uniform, quiescent, nearly collisionless, highly reproducible and stable. A uniform axial dc magnetic field ($B_\parallel = 0...100$ G) may be applied. For the reconnection experiments a time varying magnetic field ($t_{rise} \simeq 100$ μsec) is applied transverse to the plasma column ($B_\perp \simeq 0...20$G). Since this field is produced by two parallel plate electrodes carrying identical axial currents its topology in vacuum contains an X-type neutral point on the axis of the device. When the space between the insulated plates is filled with plasma a secondary plasma current is induced anti-parallel to the plate currents.

It is sufficiently large (~1000 A) that it modifies the vacuum field topology strongly and a neutral sheet or magnetic island may appear self-consistently.

A few comments on the scaling of the laboratory parameters to those encountered in space plasmas may be appropriate. Density, temperature and magnetic fields are close to those encountered in solar plasmas (photosphere). While the scale length of the laboratory plasma is very small compared to that of solar plasmas it is nevertheless large compared to the smallest characteristic size of magnetic field structures, $c/\omega_{pe} \simeq 0.5$ cm. The fact that we observe current sheets of thickness $\Delta z \simeq 10$ c/ω_{pe} would suggest that those structures may also exist in solar plasmas. However, the spatial resolution of order 1 arc sec $\simeq$ 700 km in solar observations permits only global features to be observed. Thus, the laboratory plasma offers high resolution data of a magnetic neutral sheet while solar observations provide a global picture on the MHD scale. Both aspects are essential for understanding the reconnection processes. As regards the magnetotail the density is approximately 11 orders of magnitude smaller than in the laboratory case. Hence, the laboratory plasma size ($\ell_x \simeq 1$ m, $\ell_y \simeq 2.5$ m) scales to a region of $L_x \simeq 300$ km, $L_y \simeq 750$ km in the magnetotail. Current sheets of thickness $\Delta z \simeq 15$ km can be expected to exist. Again, the laboratory provides excellent spatial resolution (~1 km) but lacks the large scale MHD regions surrounding the neutral sheet. The duration of the pulsed laboratory experiment ($\tau \simeq 100$ μsec) scales ($\omega_c \tau$ = const.) to $\tau \gtrsim 10$ sec in the magnetotail which is longer than a spacecraft transit time through a current sheet ($\tau = \Delta z/v_{sc} \simeq 2$ sec). Boundary conditions and current closures are different and easier to understand in the laboratory than in the magnetosphere.

An example of the observed neutral sheet

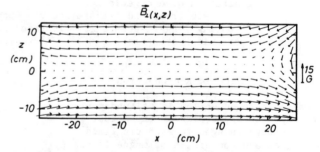

Fig. 10. Vector field of the measured magnetic field $\vec{B}_\perp = (B_x, B_z)$ in the transverse plane (x,z). A classical neutral sheet is formed in response to increasing flux generated at the plates. The observation of an electric field E_y confirms that magnetic reconnection occurs.

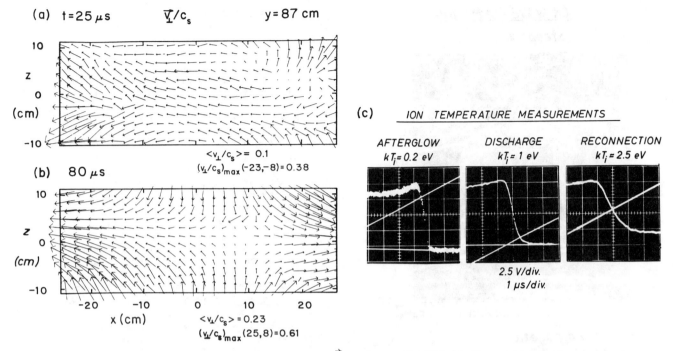

Fig. 11. a) Vector field of the ion flow $\vec{v}_\perp = (v_x, v_z)$ normalized to the sound speed $c_s = (kT_e/m_i)^{1/2}$. At early times during reconnection the unmagnetized ions respond to space charge fields associated with temperature gradients. b) Ion flow field after several Alfvén transit times ($t_A \simeq L_\perp/v_A \simeq 20$ μsec) is governed by $\vec{J} \times \vec{B}$ force exerted on electrons but transferred to ions via space charge forces (from Gekelman et al., 1982). c) Ion temperature measurements for different plasma conditions. During reconnection ions are accelerated by electric fields ($-\partial A/\partial t$, $-\nabla \phi_p$), and randomized by fluctuations in the neutral sheet.

topology is given in Fig. 10. The null region ($B_\perp = 0$, $B_\parallel = 10$ G) extends along a line so that the separatrix exhibits two contact points rather than a single one as for an X-point. Such neutral sheets arise when the induced plasma current inhibits the flux transfer across the separatrix. The flux pile-up implies that excess magnetic field energy associated with the current sheet is stored in the plasma. A disruption of the current sheet may lead to a release of this energy.

The plasma current is mainly carried by electrons. Electron currents also flow in regions of finite normal fields B_z due to drifts $\vec{E}_\perp \times \vec{B}_\perp$ associated with space charge electric fields $\vec{E}_\perp = (E_x, E_z)$. A magnetic field component $B_\parallel = B_y$ favors electron transport along the separator. The ions are essentially unmagnetized near the neutral sheet. They are directly accelerated by the inductive reconnection electric field E_y and electrostatic fields arising from charge separations. The important role of space charge fields is rarely mentioned in reconnection theories (Terasawa, 1984) or discussed in space observations where electric field measurements are very uncertain. Space charge fields arise near magnetic null regions due to the different magnetization of electrons and ions. In the x-z plane the magnetized electrons are subject to an $\vec{E}_y \times \vec{B}_x$ drift into the neutral sheet whereas the unmagnetized ions would simply follow E_y. However, space charge electric fields couple the ions to the electrons so that the ions are also accelerated in the x-z plane.

Measurements of the ion flow during reconnection have been performed (Gekelman et al., 1982). Using three orthogonal differential ion flux probes the ion drift velocity vector $\vec{v} = (v_x, v_y, v_z)$ is recorded vs. x, z and t. At two times the transverse flow $\vec{v}_\perp = (v_x, v_z)$ normalized to the sound velocity $c_s = (kT_e/m_i)^{1/2} \simeq C_A = B/(\mu_0 n m_i)^{1/2}$ is shown in Fig. 11a,b. One finds at early times a flow field governed by space charge fields associated with nonuniformities in temperature and density ($\vec{E}' = -\nabla p/n_e$). After a few Alfvén times across the device a quasi-stationary state is approached where the flow field is governed by the $\vec{J}_y \times \vec{B}_\perp$ force. This force acts on the current-carrying electrons yet is transmitted via space charge forces to the ions. The fluid is vertically compressed into

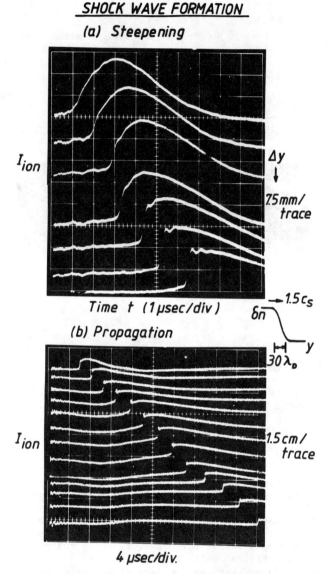

Fig. 12. The energization of plasma by the current disruptions leads to a mass ejection which steepens into an electrostatic shock wave and propagates through the ambient plasma. Probe traces show ion flux vs. time at different positions along the direction of shock propagation.

$\oint \vec{E}_i \cdot \vec{d\ell} \simeq 100$ V cannot be reduced everywhere and it is found that approximately 80 V drop off at the cathode sheath and only 20 V along the plasma column. Thus, the space charge fields have modified the net electric field distribution in an unforeseen way. The major particle acceleration takes place in the cathode sheath. The accelerated ions leave the plasma while energetic electrons are injected into the neutral sheet creating anisotropic distribution and microinstabilities (Gekelman and Stenzel, 1984). Although the actual electric field distributions may depend on the global boundaries of each problem the possibility of localized electric fields and acceleration regions exists in the nonuniform magnetosphere as well (Frank et al., 1976).

The neutral sheet exhibits a broad spectrum of fluctuations such as ion acoustic waves, whistler waves and Langmuir waves. The interaction of accelerated ions with turbulent fields leads to scattering in velocity space, i.e. heating. A comparison between the force density $\vec{f} = \vec{j} \times \vec{B} - \nabla p$ and the fluid acceleration $\rho_m d\vec{v}/dt$ has indicated the necessity to introduce an anomalous scattering process in order to account for the reduced ion acceleration (Gekelman et al., 1982). Measurements of the ion distributions in the neutral sheet indicate reasonable isotropy with ion temperatures well above those without reconnection. Fig. 11c shows the ion energy analyzer traces for three different plasma conditions. In the afterglow the ions cool due to ion-neutral collisions. During the discharge they are heated by electron-ion collisions. During reconnection they are further heated due to energy gained in electric fields and isotropized by fluctuations. The short time scale ($\Delta t \simeq 30$ μsec) for the temperature increase during reconnection excludes collisional heating ($\nu_{ei}^{-1} m_i/m_e \simeq 50$ msec) or cooling ($\nu_{in}^{-1} \simeq 0.5$ msec).

the neutral sheet and horizontally ejected at an increased speed ($v_\perp/c_s \simeq 0.6$). The ions are also accelerated along the separator but by a smaller amount ($v_\parallel/c_s \lesssim 0.2$). This points to a smaller electric field along the separator than perpendicular to it which is surprising since a large inductive field $E_y \lesssim 1$ V/cm is applied along the separator. Plasma potential measurements show that a space charge electric field $E_s = -\partial \phi_p/\partial y$ builds up opposite to the inductive electric field. However, since $\oint \vec{E}_s \cdot \vec{d\ell} = 0$ the inductive voltage $V_i = -\partial \phi_m/\partial t$

Current Disruptions and Shocks

Reconnection processes may not only occur in a driven steady way as described in the previous section but frequently appear to occur spontaneously and impulsively. Examples are flux transfer events at the magnetopause (Russel and Elphic, 1978; Cowley, 1982), explosive tearing mode instabilities (Galeev et al., 1978), and possibly the substorm event itself (Hones, 1979). Impulsive reconnection events are associated with rapid changes of currents in plasmas and may therefore also be described as current disruption events. Critical questions are the causes for current disruptions, the dynamics of the system during disruption, and the particle energization processes. The analysis of current disruptions clearly requires the knowledge of the global circuit as well as the local plasma properties

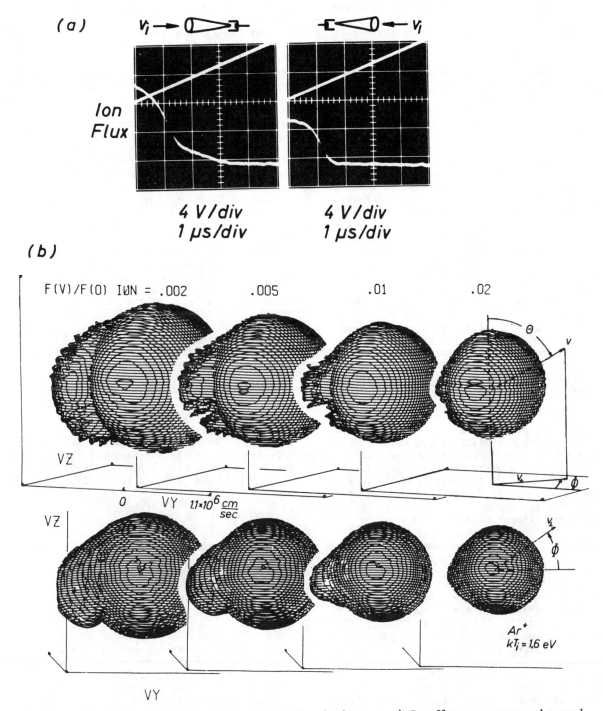

Fig. 13. Ion energization by an electrostatic shock wave. a) Ion flux vs. energy observed with a directional ion energy analyzer. After the passage of the shock a tail of energetic (4...8eV) ions are observed to follow the shock. b) Normalized ion distribution function $f_i(\vec{v})/f_i(o)$ in three-dimensional velocity space (v_x, v_y, v_z) displayed as surfaces of constant values decreasing from right to left. The distortions from spherical shape represent the tails of ions energized by the shock electric field. Top and bottom rows display the same surfaces rotated by 90° around the v_y axis.

involved in the disruption. Such comprehensive data are not available for flares or substorms. They can be obtained in laboratory plasmas but due to different geometries and boundaries the results are not necessarily applicable to other current systems. Nevertheless the understanding of one well documented case may be helpful in the study of others.

Spontaneous disruptions of the current sheet are observed when the current density is raised beyond $v_{de}/v_{the} \gtrsim 0.3$ (Stenzel et al., 1983). As described in the section on double layers the current interruption produces a large inductive voltage which drops off inside the plasma in the form of a double layer at the location of lowest conductivity. The current disrupted in the center of the original sheet is redirected to the sides. The current loss from the center of the sheet is due to plasma thinning, i.e. ion expulsion from the central region of high plasma potential. The ions are accelerated to energies corresponding to the inductive loop voltage $V = -d\phi/dt$. The magnetic energy is largely stored outside the plasma hence can be transported to the dissipation region (double layer) at the speed of light.

A fundamental question is at which speed magnetic energy can be released in a collisionless plasma (Spicer, 1982). In order to investigate this problem the external circuit inductance has been minimized and the entire current sheet is rapidly switched off with a fast magnetic switch (Stenzel and Gekelman, 1984). One might expect the excess magnetic energy associated with the current sheet to convect away within an Alfvén transit time ($L/v_A \simeq 15$ cm/5×10^5 cm/sec $\simeq 30$ μsec). Instead, a rapid ($\Delta t \lesssim 5$ μsec) dissipation of the field energy is observed. The initially laminar current sheet cascades into many filaments and loops of small scale length and random spatial shape. These appear to facilitate the dissipation process by lengthening the current paths, providing anomalous resistivity due to cross-field currents, and effectively to reduce the magnetic Reynolds number due to the smaller scale length. Although the dissipation processes are not yet known in detail, some dramatic consequences of the disruption have been observed.

The current disruption results in a potential gradient ($\Delta V \simeq 10$ V) across the magnetic switch which gives rise to an ejection of ions from the disrupted sheet into the surrounding plasma. This ion mass ejection steepens into a shock wave. Fig. 12 displays the ion flux vs. time observed at increasing distances (Δy) in the direction of propagation. The density front is found to steepen to a sharp transition ($\delta y \simeq 1$ mm $\simeq 30 \lambda_D$) and to propagate at speed $v \simeq 1.5 c_s$ as an electrostatic shock over large distances through the plasma.

Just behind this large amplitude shock the ion distribution is found to be modified. With a directional ion velocity analyzer tails in the ion distribution in the direction of shock propagation are observed. Raw data traces of ion flux vs. energy are shown in Fig. 13a while the complete distribution function is displayed in Fig. 13b. Since the function $f(v_x, v_y, v_z)$ depends on three velocity variables it is displayed as surfaces of constant value $f(\vec{v})$ = const. in 3 dimensional velocity space. Anisotropies in the distribution function become apparent at small values of the normalized function, $f(\vec{v})/f(o) \ll 1$, hence large velocities, $v > v_{th}$. Since the tail ions travel in the direction of wave propagation but are found behind the shock front they have been accelerated by the shock electric field but not reflected. The localized shock electric field ($E = -\nabla \phi$) moves rapidly compared to the thermal ions ($v_s \simeq 5 v_{th}$) hence exerts an impulse $e\int E dt \simeq \phi/v_s \lesssim mv_{thi}$ on most ions while accelerating more efficiently comoving tail ions ($\vec{v} \lesssim \vec{v}_s$) to energies $e\phi_s \lesssim kT_e$. Reflected ions have not been observed. Small amplitude shocks produce little ion acceleration and experience little damping.

Acknowledgments. The author acknowledges the fruitful collaboration with Dr. W. Gekelman in many of the reviewed experiments. Mr. J. M. Urrutia's assistance in the reconnection experiments is also greatly appreciated. This work is supported by the National Science Foundation under grants NSF ATM 84-01322 and PHY 84-10495, and AFOSR contract F19628-85-K-003.

References

Akasofu, S.-I., Auroral arcs and auroral potential structure, in Physics of Auroral Arc Formation, S.-I. Akasofu and J. R. Kan, editors, Geophys. Monograph 25, p. 1, AGU, Washington D.C., 1981.
Baum, P. J. and A. Bratenahl, The laboratory magnetosphere, Geophys. Res. Lett., 9, 435, 1982.
Cowley, S. W. H., The causes of convection in the earth's magnetosphere: A review of developments during the IMS, Rev. Geophys. Space Phys., 20, 531, 1982.
Dawson, J. M., H. C. Kim, D. Arnush, B. D. Fried, R. W. Gould, L. O. Heflinger, C. F. Kennel, T. E. Romessor, R. L. Stenzel, A. Y. Wong, and R. F. Wuerker, Isotope separation in plasmas by use of ion cyclotron resonance, Phys. Rev. Lett., 37, 1547, 1976.
Frank, L. A., K. L. Ackerson, and R. P. Lepping, On hot tenuous plasmas, fireballs, and boundary layers in the earth's magnetotail, J. Geophys. Res., 81, 5859, 1976.

Galeev, A. A., F. V. Coroniti, and M. Ashour Abdalla, Explosive tearing mode reconnection in the magnetospheric tail, Geophys. Res. Lett., 5, 707, 1978.

Gekelman, W., R. L. Stenzel and N. Wild, Magnetic field line reconnection experiments 3. Ion accelerations, flows and anomalous scattering, J. Geophys. Res., 87, 101, 1982.

Gekelman, W., and R. L. Stenzel, Magnetic field line reconnection experiments 6. Magnetic turbulence, J. Geophys. Res., 89, 2715, 1984.

Heikkila, W. J., and R. J. Pellinen, Localized induced electric fields within the magnetotail, J. Geophys. Res., 82, 1610, 1977.

Hones, E. W., Jr., Transient phenomena in the magnetotail and their relation to substorms, Space Sci. Rev., 23, 393, 1979.

Nakamura, Y., and R. L. Stenzel, in Symposium on Plasma Double Layers, Risø National Laboratory, June 16-18, 1982, Report R-472, 153-158, 1982.

Podgorny, I. M., E. M. Dubinin, and Yu. N. Potanin, On magnetic curl in front of the magnetosphere boundary, Geophys. Res. Lett., 7, 247, 1980.

Russell, C. T., and R. C. Elphic, Initial ISEE magnetometer results: Magnetopause observations, Space Sci. Rev., 22, 681, 1978.

Sato, N., Double layers in laboratory plasmas, in Symposium on Plasma Double Layers, Risø National Laboratory, June 16-18, 1982, Report R-472, 116-140, 1982.

Spicer, D. S., Magnetic energy storage and conversion in the solar atmosphere, Space Sci. Rev., 31, 351, 1982.

Stenzel, R. L., Filamentation of large amplitude whistler waves, Geophys. Res. Lett. 3, 61, 1976.

Stenzel, R. L., Method and apparatus for the electrostatic excitation of ions, United States Patent, 4,093,856, 1978.

Stenzel, R. L., and W. Gekelman, Dynamics of test particles in a turbulent magnetoplasma, Phys. Fluids, 21, 2024, 1978.

Stenzel, R. L., M. Ooyama, and Y. Nakamura, Potential double layers formed by ion beam reflection in magnetized plasmas, Phys. Fluids, 24, 708, 1981.

Stenzel, R. L., R. Williams, R. Aguero, K. Kitazaki, A. Ling, T. McDonald, and J. Spitzer, A novel directional ion energy analyzer, Rev. Sci. Instrum., 53, 1027, 1982.

Stenzel, R. L., W. Gekelman, and N. Wild, Magnetic field line reconnection experiments 5. Current disruptions and double layers, J. Geophys. Res., 88, 4793, 1983.

Stenzel, R. L., and W. Gekelman, Particle acceleration during reconnection in laboratory plasmas, Adv. Space Res., 4, 459, 1984.

Terasawa, T., On the microphysics of the magnetotail reconnection region, Adv. Space Res., 4, No. 2-3, 429, 1984.

Vasyliunas, V. M., Theoretical models of magnetic field line merging, 1. Rev. Geophys. Space Phys. 13, 303 (1975).

Wright, K. H., Jr., N. H. Stone, and U. Samir, A study of plasma expansion phenomena in laboratory generated plasma wakes: Preliminary results, J. Plasma Phys., 33, 71, 1985.

ION ACCELERATION IN LABORATORY PLASMAS

Noah Hershkowitz

Nuclear Engineering Department University of Wisconsin-Madison, Madison, WI 53706

Abstract. Recent experiments which investigate ion acceleration in "collisionless" laboratory plasmas are described. The experiments fall into two classes—those involving stationary and those involving time dependent processes. Among the stationary structures which accelerate ions are sheaths, presheaths, double layers, multiple double layers and ambipolar potentials. RF is an example of time dependent processes. Both classes of experiments are considered. Experiments with "collisionless" plasmas in multi-dipole triple plasma type devices and novel phenomena associated with the use of Ion Cyclotron Resonance Heating in the inhomogeneous magnetic field of a tandem mirror "fusion" plasma are described. New RF phenomenon include RF pitch angle scattering, RF electron pumping, and ICRF ponderomotive force.

Introduction

This paper describes two types of ion acceleration; acceleration by stationary potential structures such as sheaths, double layers and ambipolar potentials, and acceleration by time dependent phenomena in magnetic mirror B fields such as fluctuating RF fields. The latter category includes ion heating, pitch angle scattering, electron pumping, and ponderomotive force associated with waves with frequencies near the ion cyclotron frequency.

Two types of laboratory plasmas are considered. One is a low density, low temperature triple plasma device plasma [Hershkowitz et al., 1981] of the sort that is traditionally associated with basic plasma experiments in unmagnetized plasmas that simulate space plasmas. The other is a high density, high temperature, magnetized tandem mirror plasma [Hershkowitz et al., 1985a] which is usually associated with "fusion" experiments. It is worth noting however that such plasmas just barely have the minimum requirements for addressing collisionless, magnetized, space plasmas. The problem is that the Alfvén velocity scales as $B/\sqrt{nm}$, where B is the magnetic field, n the plasma density and m the ion mass. Taking a frequency as the ion cyclotron frequency gives a wavelength $\lambda = (c/e)\sqrt{\pi m/n}$. It is clear that the shortest wavelength is obtained when hydrogen is used. Even using hydrogen it is still necessary to have densities greater than 10^{12} cm^{-3} to achieve wavelengths as small as device dimensions. Thus both types of plasmas described here can be employed when space plasma properties are to be studied in the laboratory. Q-machines [Rynn, 1964] by contrast allow the study of either high density collisional magnetized plasmas or of low density magnetized plasmas for which Alfvén waves do not fit the devices.

Properties of the two types of plasmas, triple plasma (TP) device plasmas and tandem mirror plasmas are summarized in Table 1. Note that while the densities and temperatures are quite different, the relative potential steps are quite comparable. TP devices [Hershkowitz et al., 1981] consist of plasmas in three chambers that are separated by grids. Plasmas are produced by filament discharge and normally have low fractional ionization, cold ions and electron temperatures in the neighborhood of 1-3 eV. Multi-dipole surface magnetic fields are often employed in such devices to enhance the confinement of the ionizing primary electrons [Leung et al., 1975]. Tandem mirrors [Dimov et al., 1976; Fowler and Logan, 1977; Kesner et al., 1983] consist of three or more magnetic mirror cells fueled by ionization of gas puffed at their outer edges. In general, plasmas are heated by a combination of neutral beams and low and high frequency RF. The Phaedrus tandem mirror plasma makes use of electromagnetic waves near the ion cyclotron fre-

TABLE 1. Two Kinds Of Experimental Devices

Type	multi-dipole	tandem mirror
Gas	argon	hydrogen
Plasma Density (cm^{-3})	10^7-10^9	$10^{12}-10^{13}$
T_e (eV)	1-2	20
T_i (eV)	0.2	$30-10^3$
Neutral Density (cm^{-3})	10^{12}	10^{10}
B (gauss)	0-10	10^3
$e\Delta\phi/T_e$	0-18	0-2

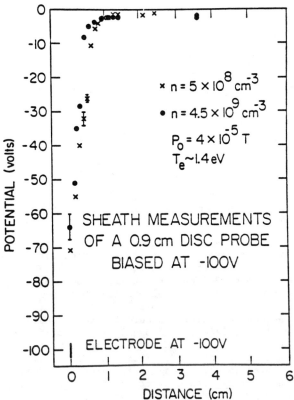

Fig. 1. Plasma potential measurement near a 0.9 cm. diameter disc probe biased at -100 V. The plasma potential was measured with an emissive probe.

quency to heat ions and electrons [Yujiri, 1982; Golovato et al., 1985]. Fractional ionization is close to 100% on axis, electron temperatures are typically 20 eV and ion temperatures range from 20 eV to 1 keV.

Stationary Plasma Potential Structures

The simplest structures found in laboratory plasmas are sheaths, the non-neutral regions associated with plasma boundaries. They are among the oldest structures identified in plasmas and have been known since Langmuir's original investigations [Tonks and Langmuir, 1929; Langmuir, 1961]. Although they are usually thought of as monotonic structures, double sheaths associated with cathodes were also investigated by Langmuir [1929]. Recent investigations have discovered double layers [Schrittweiser, 1984], which are similar to sheaths which are disconnected from plasma boundaries and a variety of multiple double layer-like structures. Although sheaths are localized, presheaths [Bohm, 1949] associated with sheaths can be as large as device dimensions. Consider Poisson's equation in one dimension

$$d^2\phi/dx^2 = -4\pi e(n_i - n_e). \quad (1)$$

Substituting $\tilde{\phi} = e\phi/T_e$, $\tilde{x} = x/\lambda_D$ where

$$\lambda_D \equiv \sqrt{T_e/4\pi n_o e^2} \text{ and } \tilde{n} = n/n_o \quad (2)$$

yields,

$$d^2\tilde{\phi}/d\tilde{x}^2 = -(\tilde{n}_i - \tilde{n}_e). \quad (3)$$

The Debye length clearly represents a characteristic length which gives a dimensionless Poisson's equation. Although the solution for sheath thickness will be given in terms of Debye lengths, it is not at all obvious that the solution will be anything like one Debye length.

It is sometimes argued that sheaths are only several Debye lengths thick but it has been known since Langmuir's original work that sheaths can be much thicker. Consider the sheath shown in Fig. 1 [Coakley, 1980] near a circular plate biased at $V_B \approx -100V$ with respect to the background plasma potential. The region in which the background plasma potential is significantly disturbed extends out to approximately 0.5 cm which is approximately 18 λ_D. Note that positive ions which fall into the sheath are accelerated toward the negatively biased plate and some eventually strike the plate with an energy equal to $-eV_B + e\phi_p$, where ϕ_p is the plasma potential far from the probe. Ions which are born within the sheath by ionization or ion-neutral charge exchange are also accelerated to the plate but with a lower energy.

Double sheaths such as those associated with space charge limited emission are well known [Langmuir, 1913, 1961]. Less well known are potential dips near positively biased plates in

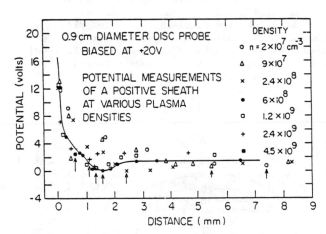

Fig. 2. Sheath measurements near a positively biased probe. Note the potential well that forms in front of the probe. The arrows indicate where the potential minimum occurred at each density.

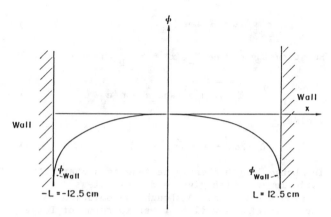

Fig. 3. Schematic of the plasma potential throughout a laboratory device. Much of the variation in potential is associated with presheaths.

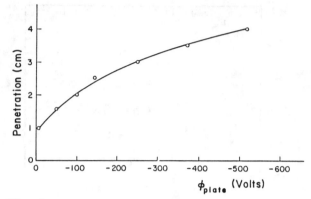

Fig. 5. Penetration of the -2 V contour into the plasma versus the plate bias voltage.

plasmas [Armstrong and Fujita, 1984]. Data are given for such a sheath in Fig. 2 [Coakley, 1980]. We have recently found that the dip is only present when there is an insulating coating on the back of the probe [Forest and Hershkowitz, 1985]. When the coating is removed, and only conductors are present, no dip is found and the plasma potential is monotonically positive in the sheath from the probe to the plasma. The insulating coating is found to float at a potential that is negative with respect to the dip potential, allowing ions electrostatically trapped in the dip to be drained off. Note that the one-dimensional ion phase space near the probe has an ion hole which accelerates and then decelerates ions before reflecting them near the probe. The position of the ion hole is determined by the Child-Langmuir condition [Langmuir, 1913] which limits the electron current to the probe to agree with the current determined by the potential dip, plasma density and electron temperature.

Most of the ion acceleration near a plasma boundary takes place in the sheath but there are also presheaths [Bohm, 1949; Emmert et al. 1980] which accelerate ions up to Bohm velocity (equal to the ion acoustic velocity) at the sheath edge. In general presheaths can extend over a significant fraction of a laboratory device as shown in Fig. 3 [Meassick et al., 1985] which was produced in a multi-dipole device which resembles one chamber of a triple plasma device.

The equipotential contours in the neighborhood of a plate (diameter = 3.5 cm) biased at -500 V are given in Fig. 4. Here the contour at -10 V extends as far as 2.5 cm from the plate. Also note, even in this unmagnetized situation, that there are very strong electric fields, and hence large ion acceleration near the edges of the biased plate. The location of the -2 V contour (really a measure of the presheath) is graphed versus plate bias potential in Fig. 5 and seen to be adequately described by the solid line given by $(L/\lambda_D) = 1.8 \sqrt{e\phi/T_e}$, the smooth curve through the data.

Laboratory double layers represent another sheath-like type of structure (see Fig. 6) [Coakley and Hershkowitz, 1979]. They consist of stationary potential steps that are disconnected

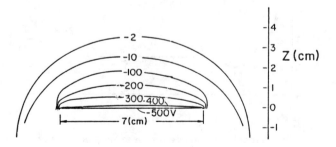

Fig. 4. Equipotential contours near a plate biased at -500 V.

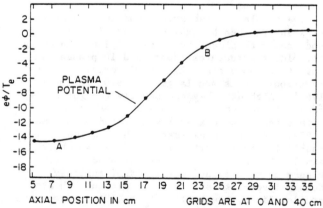

Fig. 6. A representative potential plot for a double layer as measured by an emissive probe.

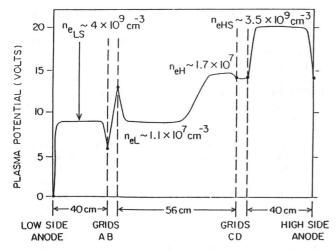

Fig. 7. Representative data for a weak double layer with a potential step equal to $e\phi/T_e \sim 5$. Note that plasma potentials and density are given throughout the TP device.

from the plasma boundary. In such structures the thickness of the transition region can often be represented as $L = a\sqrt{e\phi/T_e}$ [Torvén, 1979; Hershkowitz, N., 1985b] where a is a constant the order of 1 to 10. Similar double layers can correspond to a variety of quite different boundary conditions which are not at all apparent from data which give the potential versus spatial position. The boundary conditions are apparent in Fig. 7 which gives both the plasma potential and plasma density along the axis of the TP device for a representative double layer.

It should be noted that the high potential source is at a much higher potential than the high potential side of the double layer so that the ions from the high potential source are injected with a velocity that is far in excess of the Bohm velocity. Electrons from the low potential side are also injected with a net drift velocity. All potentials throughout the plasma are determined self-consistently and differ from those imposed on the system. So, while ions and electrons are accelerated and/or reflected at the double layer, they also have initial velocities that must be taken into account. The acceleration by the laboratory double layer is clearly not the whole story.

It is natural to assume that the potential profile between the plasma and a biased plate is monotonic because this is the simplest behavior of the plasma potential. However, there is now a considerable body of data which suggest that nonmonotonic structures can exist and are often present. For example, the sheath structure shown in Fig. 2 is nonmonotonic. Laboratory double layers also often have potential dips on the low potential side, similar to the structure shown in Fig. 8. Some of these structures depend entirely on the boundary conditions, while others may depend on turbulence. Both stationary and moving structures have been achieved in the laboratory and it should be remembered that even for stationary structures, ion and electron drifts are present at the grid boundaries and throughout the laboratory devices. The structures given in Fig. 8 are

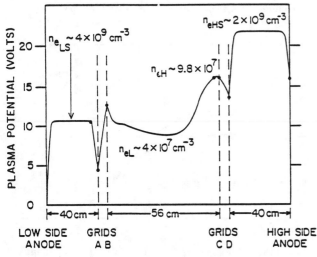

Fig. 8. Axial multiple double layer potential profile.

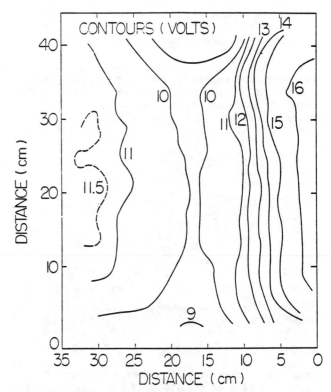

Fig. 9. Equipotential contours corresponding to Fig. 8.

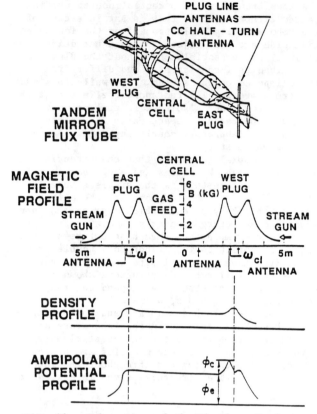

Fig. 10. Schematics of the Phaedrus Tandem Mirror device and the corresponding magnetic field, density and axial plasma potential.

ions which enter from the left. These ions are then reflected from the main double layer potential step. The ion holes reflect electron flux and for moving structures can result in a buildup of net positive charge behind the moving potential dip. This can eventually evolve into a double layer [Chan, C. et al., 1984].

The multiple double layer shown in Fig. 8 resembles in some respects very weak double layers reported in space by Temerin et al. [1982] but there are several significant differences. The potential step shown in Fig. 8 is several T_e/e in size while those seen in space were somewhat smaller. In addition, the space structures are believed to be moving. Structures showing similarities to the space double layer structures have also been seen in computer simulations.

The existence of multiple double layers with potential dips, that are stationary for many hours in laboratory plasmas, leads to an apparent problem. Why doesn't the potential dip fill up with ions and disappear? Ionization, charge exchange, and Coulomb scattering all represent possible mechanisms for filling in the dip. The answer is apparent in Fig. 9 [Hershkowitz, N. et al., 1981] which gives the the equipotential contours in the neighborhood of the multiple double layer. It is apparent that ions are not electrostatically confined in the potential dip. They can leak out at the top and the bottom. In every case where measurements have been made in the perpendicular direction, it has been found that a path was present for ions to leak out of the potential minimum present in the axial profile [Hershkowitz et al., 1985b]

RF And Magnetic Mirror Fields

Tandem mirrors make use of variety of processes to accelerate and heat ions. The simplest process is the ambipolar potential which self-consistently forms along the axis of the device to reduce electron losses in order to balance

stationary and remain so for many hours. The presence of the potential dips indicates that the plasma is self-consistently limiting the axial electron current. The dips are ion holes in phase space which accelerate and the decelerate

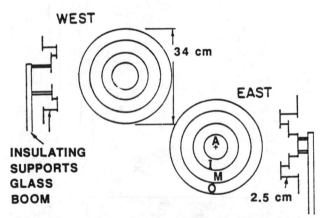

Fig. 11a. Schematic of copper end rings which were placed near the Phaedrus end walls. A, I, M, and O represent on axis, inner, middle and outer rings respectively.

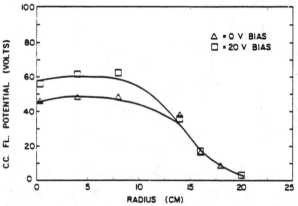

Fig. 11b. Central cell radial floating potential profiles corresponding to 0 and 20 V bias applied to the middle ring.

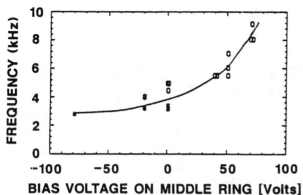

Fig. 12. Fluctuation frequency of density measured by a central cell Langmuir probe versus middle ring bias voltage.

axial ion losses. Typically the plasma potential decreases by 3-4 T_e/e between the mirror midplane and outside the mirror throat. Ions are accelerated by this potential difference in leaving the mirror cell. When collisionless plasmas are employed, off midplane perpendicular (to the magnetic field) ion velocity is converted into motion at smaller angles at the mirror midplane.

Because of the presence of a non-uniform magnetic field, magnetic mirrors provide a wide variety of possibilities for using RF to heat ions (and electrons). The Phaedrus Tandem Mirror, at the University of Wisconsin, is studying the various roles that RF can play in tandem mirrors. Phaedrus consists of a simple mirror bounded by two minimum-$|B|$ mirror cells as shown in Fig. 10. Schematics of the magnetic field, density and plasma potential profiles are also given.

Heating in each cell is provided by RF near the local ion cyclotron frequency ω_{ci} (~ 700 1Hz and 4 MHz) by line antennas in the end cells and half turn antennas in the central cell.

A variety of propagating fast magnetosonic and slow shear Alfvén waves are launched in each cell. The axial plasma potential profile consists of an ambipolar potential in the central cell mirror cell which confines central cell electrons and an ion confining potential in the west end cell. The ion confining potential has been found to be an "RF enhanced potential" that is the result of electron pumping by the RF [Hershkowitz et al., 1985c] and is not an ambipolar potential. Phaedrus plasma parameters have been summarized in Table 1.

One advantage of open magnetic field configurations in laboratory plasmas is that the radial potential profile can be modified by biasing concentric rings placed at the ends of the device (see Fig. 11a). As seen in Fig. 11b, application of a 20 V bias to the "middle" ring resulted in an increase in the potential at all flux tubes inside the biased ring. Although complete radial plasma potential control was not achieved, it was possible to alter the average radial electric field and hence the $\underline{E} \times \underline{B}$ rotation velocity. Results of such an experiment given in Fig. 12 [Severn, et al., 1984; Hershkowitz et al., 1985a] show that the rotation velocity of the lowest

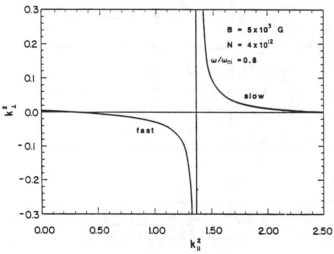

Fig. 13. Graph of $k_\perp^2$ versus $k_\parallel^2$ for $\omega/\omega_{ci} = 0.8$. This graph is based on the dispersion relation for an ∞ plasma with massless electrons.

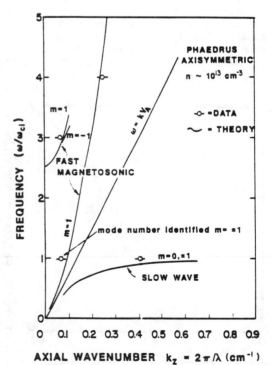

Fig. 14. Experimentally determined ω versus k in Phaedrus. Both fast and slow branches are seen. The data point at 4 ω_{ci} was identified as m = +1.

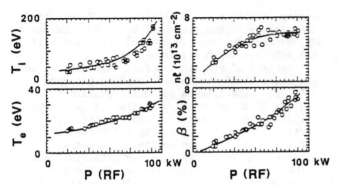

Fig. 15. Electron temperature T_e, ion temperature T_i, line density $n\ell$ and β versus coupled RF power for ponderomotive force stabilized plasmas.

frequency fluctuation could be varied from 3 to 9 kHz by changing the middle ring bias from -80 V to +70 V. The ion diamagnetic drift velocity for this configuration is in the same direction as the $\underline{E} \times \underline{B}$ drift and the frequency associated with that drift is comparable.

A graph of $k_\perp^2$ versus $k_\parallel^2$ which are the solutions to the dispersion relation for wave propagation in an infinite plasma with one ion species (hydrogen) and massless electrons is given in Fig. 13 [Fortgang, 1983]. In general there are two branches, fast and slow for frequencies less than the ion cyclotron frequency (for which the graph is drawn) and only one branch for frequencies greater than ω_{ci}. For $\omega = 0.8\,\omega_{ci}$, corresponding to the graph, it is apparent that the fast wave has $k_\perp/k_\parallel \ll 1$ and that $k_\perp$ is quite small. In the laboratory, experiments are restricted to finite radius systems (i.e., to $k_\perp > \pi/2R$) and in many cases (except at densities greater than $10^{13}\ cm^{-3}$) the $k_\perp$ determined from the infinite dispersion relation is too small to propagate. However, it has been shown [Paoloni, 1975] that if the plasma column is separated from a conducting boundary, the fast wave always propagates for azimuthal mode number with m = +1. Although this mode is right circularly polarized (in the direction of the electron rotation) it does have a significant left circularly component near the plasma surface. The experimentally measured dispersion relation measured in the Phaedrus plasma is given in Fig. 14 [Intrator, T. et al., 1984]. Both fast and slow waves were present.

Of course in space the antenna launching electromagnetic waves are not present. However, mirror loss cone distribution functions with $\partial f_\perp/\partial v_\perp > 0$ are present and these distribution functions serve as a source of free energy which can drive waves near the ion cyclotron frequency. Microinstabilities of this type were responsible for poor plasma confinement in many

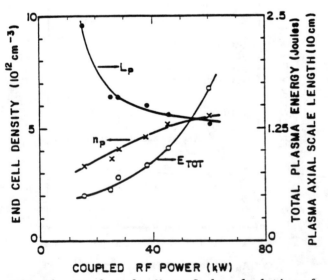

Fig. 16. Results of a Monte-Carlo calculation of the total confined energy E, plasma density n_p and plasma length L_p of the plasma trapped in the Phaedrus end cell.

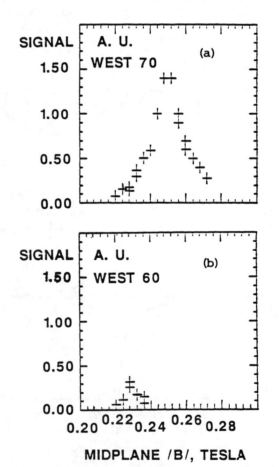

Fig. 17. SED signals at: (a) 70°, (b) 60° versus magnetic field strength at midplane.

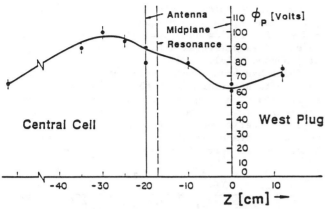

Fig. 18. Axial potential profile in the west end cell of Phaedrus, obtained with self emissive probes. Note that the RF enhanced potential peaks on the central cell side of the ion cyclotron resonance.

mirror experiments before the 2XIIB experiment [Coensgen et al., 1975] which was stabilized by filling the loss cone with a plasma stream.

Another characteristic specifically associated with magnetic mirror fields is that perpendicular ion heating at an off-midplane location is converted to parallel heating at the mirror midplane. In the Phaedrus tandem mirror RF near the ion cyclotron frequency is used to heat ions and electrons. Results are given in Fig. 15 [Golovato et al., 1985] As the RF power is increased, all parameters increase although the ion density tends to saturate. Electron heating is by Landau damping because the phase velocity of the fast magnetoacoustic wave is close to the electron thermal velocity.

It is believed that ion heating can be modelled by perpendicular ion heating as the ions pass through the ion cyclotron resonance. Fluctuating E_+ fields rotating at the ion cyclotron frequency appear as stationary DC fields to the rotating ions. Ions with appropriate phase are given increases in their perpendicular energy which can be described by [Golovato et al., 1985].

$$\Delta W_\perp = ev_\perp E_+ \tau_{eff} \cos\theta + \frac{e^2}{2m_i} \tau_{eff}^2 E_+^2,$$

$$\tau_{eff} = (2\pi/v_\parallel \frac{\partial \omega_{ci}}{\partial z})^{1/2} \quad (4)$$

It is important to note that the heating depends on τ_{eff}, the effective time the ion spends near the resonance. Small B field gradients ($\partial \omega_{ci}/\partial z$) favor ion heating because ions spend a long time near the resonance. The first term which depends on $\cos\theta$ is diffusive, both increasing and decreasing the perpendicular energy while the second term is a heating term which only results in increases in ion energy.

Golovato et al. [1985] have used Eq. (4) to model the heating and trapping in the Phaedrus mirror end cell using a Monte-Carlo code. Results are in qualitative agreement with data shown in Fig. 16. As the RF left circularly polarized E_+ field is increased, the density increases, the energy density increases and the plasma tends to shorten. The plasma shortens because ions are given perpendicular kicks near the ion cyclotron resonances and then have turning points which are increasingly closer to the resonance. This amounts to the production of ion conics at the end cell midplane by the ICRF at the ion cyclotron resonance. This phenomenon has been directly observed by Ross [1985] who used secondary emission detector (SED) arrays to examine the pitch angle distribution function of end cell ions in Phaedrus. The SED arrays respond to charge exchange neutrals which carry off the energy and direction of ions which charge exchange in the end cells. They give the signals detected by SED detectors which are pointed toward the end cell midplane at angles of 60° and 70°. Data given in Fig. 17 show that, as the strength of the axial B field is swept, the ions turning near 60° and 70° are swept past the corresponding detectors.

The end cell plasma potential in Phaedrus (shown in Fig. 18) [Hershkowitz, N. et al., 1985c] is an example of a nonambipolar potential along the B-field which can reflect central cell ions. This potential is the result of the combination of ICRF and the magnetic mirror B-field and does not appear to be associated with an increase in ion density. The potential has a relative minimum at the end cell midplane and peaks on the high B-field side of the ion cyclotron resonance (near where $\omega = \omega_{ci}$). The location of the potential peak with respect to the nonuniform B field is summarized in Fig. 19. One basic problem in understanding the interaction of ICRF with the plasma is that the wave excitation and

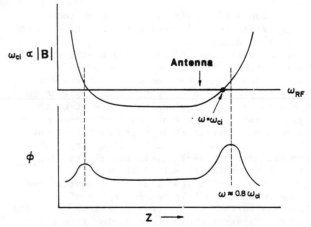

Fig. 19. The location of the ion cyclotron resonance, antenna and plasma potential peak and the magnetic field Z profile $|B| \propto \omega_{ci}$.

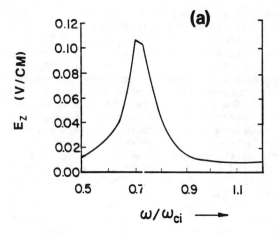

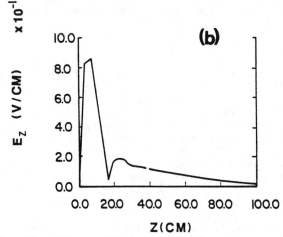

Fig. 20. Code predictions of E_z for wave propagating on the Phaedrus central cell. The code assumes a uniform B field but includes finite temperatures and radial profiles. (a) E_z versus ω/ω_{ci} at a fixed separation from the antenna. (b) E_z versus Z at $\omega/\omega_{ci} = 0.8$.

propagation varies axially and that no code currently exists for calculating the strength of the fluctuating E fields.

We can gain some understanding of the general characteristics from a code written by McVey [1984] which calculates the wave profiles for realistic plasma radial profiles and antenna configurations. Results from this code given in Fig. 20a and 20b show that the z component of the fluctuating electric field peaks near (but not at) the antenna and that for a fixed separation from the antenna the E_z field peaks when $\omega/\omega_{ci} = 0.7$ (for a particular set of plasma parameters). We believe that the potential peak is the result of a balance of pumping of electrons out of the electron potential "well" by the RF E_z against Coulomb scattering into the well [Hershkowitz et al., 1985c]. Since the Coulomb scattering rate is proportional to n^2 and the electron pumping is proportional to $n E_z^2$ where n is the plasma density, the balance requires that E_z is proportional to density. This process favors low density. Predictions of this model and experimental data are compared in Fig. 21 [Johnson et al., 1985].

There has been considerable interest in space in effects associated with the ponderomomtive force from high frequency fluctuations. The ponderomotive force associated with low frequency electromagnetic ion cyclotron waves has recently been demonstrated in laboratory plasmas. The ponderomotive force on ions can be written:

$$F_p = \frac{-e^2}{4m_i} \left(\frac{\nabla E_+^2}{\omega(\omega-\omega_{ci})} + \frac{\nabla E_-^2}{\omega(\omega+\omega_{ci})} + \frac{\nabla E_z^2}{\omega^2} \right) \quad (5)$$

This can also be written as the gradient of a ponderomotive pseudopotential. Recent experiments have demonstrated the force [Dimonte et al., 1982] and have shown that interchange stability can be achieved by taking advantage of the radial ponderomotive force [Ferron et al., 1983; Yasaka and Itatani, 1984]. In these experiments the radial ponderomotive force induced ion and electron drifts opposite to those induced by bad magnetic curvature in the transition regions at the ends of simple magnetic mirror cells. Stability was achieved when the RF frequency was greater than the ion cyclotron frequency (near the mirror midplane) and was not achieved when the RF frequency was less than the ion cyclotron frequency. Experimental data given in Fig. 22 [Ferron et al., 1983] for line density versus time exhibit this effect. The fluctuations seen when $\omega < \omega_{ci}$ were determined to be m = 1 flute modes.

The difference in the the two frequencies for the data shown in Fig. 22 was only 0.5 percent. The sensitivity to frequency depends on several effects. The most obvious is the sign change at the ion cyclotron frequency in the first term in

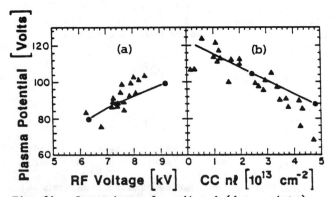

Fig. 21. Comparison of predicted (data points) and measured solid curves values of plasma potential in the Phaedrus end cell. (a) RF power is varied and density is held fixed. (b) Antenna voltage is held fixed and density is varied.

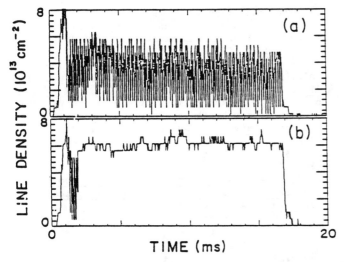

Fig. 22. Phaedrus central cell line density versus time. (a) The RF frequency ω/ω_{ci}. (b) $\omega > \omega_{ci}$.

Eq. 2. Another effect, which may be more important, is that very different radial electric field profiles are present above and below the ion cyclotron frequency because slow waves can only propagate below ω_{ci}. Yet another important effect is the ponderomotive force on electrons. In other experiments, Yasaka and Itatani [1980] have demonstrated that the azimuthal ponderomotive force associated with gradients in the RF electric fields in the in azimuthal direction can result in radial plasma convection. Taken all together, these experiments demonstrate that gradients in RF fields associated with non-uniform magnetic fields can result in significant ion and electron drifts.

Summary

This paper has presented a variety of techniques and phenomena which give ions both parallel and perpendicular energy in laboratory plasmas. In addition to heating associated with stationary potential structures, many effects associated with the combination of electromagnetic waves near the ion cyclotron frequency and magnetic mirror magnetic fields have also been presented. Several of these including drifts associated with ICRF ponderomotive forces and electron pumping by RF E_z fields have only recently been discovered. It is already known that space plasmas exhibit some of these effects. It is likely that nature has found a way to take advantage of all of them.

Acknowledgements. I would like to acknowledge the contributions of the Phaedrus Tandem Mirror Group, Dr's Tom Intrator and Dr. Chung Chan and Cary Forest to the work reported here. This work was supported by NASA Grant NAGW-275.

References

Armstrong, R.J. and Fujita, H., A negative potential dip close to an electron collecting electrode in a plasma, Proceedings, Second Symposium on Plasma Double Layers and Related Topics, Innsbruck, Austria, 194, 1984.

Bohm, D. (1949) in The Characteristics of Electrical Discharges in Magnetic Fields, ed. by Guthrie, A. and Wakerling, R.K., N.Y., McGraw Hill, p. 77.

Chan, C., Cho, M.H., Hershkowitz, N., and Intrator, T., Laboratory evidence for ion acoustic type double layers, Phys. Rev. Letts. 52, 1782, 1984.

Coakley, P. and Hershkowitz, N., Laboratory double layers, Phys. Fluids 22, 1171., 1979.

Coakley, P., Large electrostatic potential variations in laboratory plasmas, University of Iowa Ph. D. Thesis, 1980.

Coensgen, F.H., Cummins, W.F., Logan, B.G., Molvik, A.W., Maxзen, W.E., Simonen, T.C., Stallard, B.W. and Turner, M.C., Stabilization of a neutral-beam-sustained, mirror-confined plasma, Phys. Rev. Lett. 35, 1501, 1975.

Dimonte, G., Lamb, B.M. and Morales, G.J., Ponderomotive pseudopotential near gyroresonance, Phys. Rev. Lett. 48, 1352, 1982.

Dimov, G.I., Zakaidakov, V.V. and Kishinevskii, M.E., Thermonuclear confinement system with twin mirror systems, Soviet Journal of Plasma Physics 2, 326, 1976.

Emmert, G., Wieland, R., Mense, A., and Davidson, J., Electric sheath and presheath in a collisionless, finite ion temperature plasma, J. Phys. Fluids 23, 803, 1980.

Ferron, J., Hershkowitz, N., Breun, R.A., Golovato, S.N., and Goulding, R.H., RF stabilization of an axisymmetric tandem mirror, Phys. Rev. Lett. 51, 1985, 1983.

Forest, C. and Hershkowitz, N., Steady state ion pumping of a potential dip near an electron collecting anode, University of Wisconsin Phaedrus Tandem Mirror Report, PTMR85-12, 20 pp., 1985.

Fortgang, C., High power ion cyclotron resonance heating in the Wisconsin levitated octupole, Univ. of Wisconsin, Ph.D. Thesis, 1983.

Fowler, T.K. and Logan, B.G., The tandem mirror reactor, Comments on Plasma Physics and Controlled Fusion 2, 167, 1977.

Golovato, S.N., Breun, R.A., Ferron, J.R., Goulding, R.H, Hershkowitz, N., Horne, S.F., and Yujiri L., Fueling and heating of tandem Mirror end cells using rf at the ion cyclotron frequency, Phys. Fluids 28, 734, 1985.

Hershkowitz, N., Payne, G.L., Chan, C., and DeKock, J.P., Weak double layers, Plasma Phys. 23, 903, 1981.

Hershkowitz, N., Breun, R.A., Brouchous, D.A., Callen, J.D., Chan, C., Conrad, J.R., Ferron, J.R., Golovato, S.N., Goulding, R, Horne, S., Kidwell, S. Nelson, B., Persing, H., Pew, J., Ross, S., Severn, G., Sing, D., Plasma potential control and MHD stability experiments in the phaedrus tandem mirror, Proceedings Tenth International Conference on Plasma Physics and Controlled Nuclear Fusion Research, Vol. 2, IAEA, Vienna, 265, 1985a.

Hershkowitz, N., Review of recent laboratory double layer experiments, Space Sci. Rev. 41, 351, 1985b.

Hershkowitz, N., Nelson, B.A., Johnson, J., Ferron, J.R., Persing, H., Chan, C., Golovato, S.N., Callen, J.D., and Woo, J., Enhancement of plasma potential by fluctuating electric fields near the ion cyclotron frequency, Phys. Rev. Letts. 55, 947, 1985c.

Intrator, T., B. Nelson, S.N. Golovato, R. Goulding, R.A. Breun, and N. Hershkowitz, ICRF wavelength studies in phaedrus and comparison with the XANTENA code, Bull. Am. Phys. Soc. 29, 1422, 1984.

Johnson, J.W., Callen, J.D., Hershkowitz, N., Modeling ambipolar potential formation due to ICRF heating effects on electrons, Univ. of Wisc.-Madison Report 85-4, 1985.

Kesner, J., Gerver, M.J., Lane, B.G., McVey, B.D., Aamodt, R.E., Catto, P.J., D'Ipollito, D.A. and Myra, J.R., Introduction to tandem mirror physics, MIT report PFC/RR-83-35, 1983.

Langmuir, I., The effect of space charge and residual gases on thermionic currents in high vacuum, Phys. Rev. 2, 450, 1913.

Langmuir, I., The interaction of electron and positive ion space charges in cathode sheaths, Phys. Rev. 33, 954, 1929.

Langmuir, I., Collected Works of Irving Langmuir (G. Suits ed., MacMillan, NY), 1961.

Leung, K.N., Samec, T.K. and Lamm, A. (1975), Optimization of permanent magnet plasma confinement, Phys. Lett. A51, 490.

McVey, B., ICRF antenna coupling theory for a cylindrically stratified plasma, MIT Plasma Fusion Center Report PFC/RR-84-12, 1984.

Meassick, S., Cho, M.H., and Hershkowitz, N., Measurement of plasma presheath, IEEE Trans. Plasma Sci. PS-13, 115, 1985.

Paoloni, F.J. (1975), Boundary effects on m = ± 1 Alfvén in a cyclindrical, collisionless plasma, Phys. Fluids 18, 640.

Ross, S.W., The generations of sloshing ions in a tandem mirror by ion cyclotron resonance heating, Univ. of Wisconsin-Madison, Ph.D. Thesis, 1985.

Rynn, N., Improved quiescent plasma source, Rev. Sci. Instr., 35, 40, 1964.

Schrittweiser, R., Proceedings Second Symposium on Plasma Double Layers and Related Topics, Innsbruck, Austria, 411 pp., 1984.

Severn, G., Hershkowitz, N., Nelson, B.A., and Pew, J., Radial potential control in Phaedrus using end ring bias, Bull. Am Phys. Soc. 29, 1423, 1984.

Temerin M., Arny, K., Lotko, W. and Mozer, F.S., Observations of double layers and solitary waves in the auroral plasma, Phys. Rev. Letts. 48, 1175.

Tonks, L. and Langmuir, I., A general theory of the plasma of an arc, Phys. Rev. 34, 876, 1929.

Torvén, S., in Waves and Instabilities in Space Plasmas, Palmodesso, P.J. and Papadoponlos, K. (eds.), D. Reidel, Dordrecht, p. 109, 1979.

Yasaka, Y. and Itatani, R., Convective plasma loss caused by an ion-cyclotron rf field and its elimination by mode control, Phys. Rev. Letts 44, 1763, 1980.

Yasaka, Y. and Itatani, R., RF stabilization of high-density plasma in an axisymmetric mirror I: identification and stabilization of flute mode, Nuclear Fusion 24, 445, 1984.

Yujiri, L., The electron power balance in the tandem mirror phaedrus, Univ. of Wisconsin-Madison Ph.D. Thesis, 1982.

LABORATORY SIMULATION OF MAGNETOSPHERIC PLASMA PHENOMENA
USING LASER INDUCED FLUORESCENCE AS A DIAGNOSTIC

Nathan Rynn

Department of Physics, University of California, Irvine, California 92717

Abstract. There are many magnetospheric plasma effects that can be simulated in the laboratory. Among them are electrostatic ion cyclotron (EIC) waves, multiple ion effects, and lower-hybrid waves. At U.C. Irvine a powerful diagnostic tool has been developed that permits the direct observation of the ion disribution function and density as a function of time and position. The technique uses laser induced fluorescence (LIF) in a barium plasma in a Q-machine. We have been able to induce the EIC instability and to investigate a possible mechanism for auroral electron acceleration or for producing a potential drop. We have also been able to explore the effects of multiple species on ion heating. Recently, we have begun an investigation into the effects of excitation by broad-band lower-hybrid waves. We have been able to show bulk and tail heating of ions.

The Q-Machine and the Laser-Induced Fluorescence Method

At the University of California, Irvine we are engaged in a program, supported by the National Science Foundation, in which we are trying to model specific phenomena in the magnetosphere. We do not attempt to scale the magnetosphere, but seek to find a combination of parameters such that we can make experiments in the laboratory that are of interest and significance to work in the magnetosphere. Figure 1 shows a sketch of the magnetosphere areas in which we work. We are looking at the mid-region where we see electrostatic ion cylcotron waves of hydrogen ions. The rest of the area in the figure is well known to most workers. Most of our work is done in a Q-machine [Rynn, 1964] which is illustrated in Figure 2. A Q-machine is a laboratory device with the following features: (1) it uses surface, or contact ionization of a neutral vapor on a hot surface combined with thermionic emission of electrons to form the plasma; (2) the neutral vapor is delivered to the hot plate by an effuser; (3) the plasma is confined radially by a strong axial magnetic field; and (4) the vacuum chamber walls are cooled so as to condense out the background neutrals.

The ionization requirement is that the ionizer plate work function be greater than the ionization potential of the neutral vapor. The ionizer must also be hot enough to emit enough electrons to achieve the desired density. When these conditions are satisfied the net result is a quiescent plasma column, in our case 1.4 meters long and 5 centimeters in diameter, with equal electron and ion temperatures of the order of 0.25eV. Table 1 shows some typical parameters of the Q-machine. Table 2 shows a comparison of

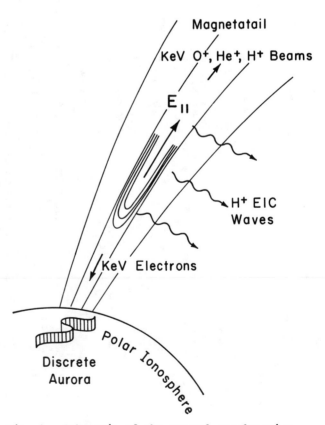

Fig. 1. Schematic of the auroral acceleration zone.

236 MAGNETOSPHERIC PLASMA SIMULATION

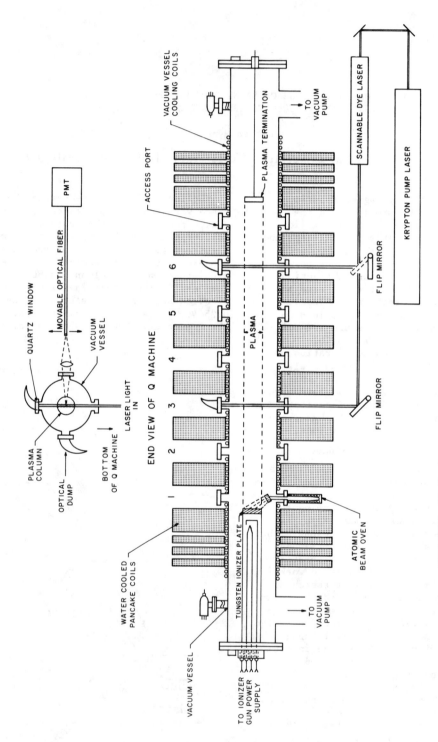

Fig. 2. Schematic of U.C. Irvine Q-machine.

TABLE 1. Typical Q-Machine Parameters*

Parameters	Description
N_{ion}	$5 \times 10^8 - 1 \times 10^{10} cm^{-3}$
B	1 – 7kG
$T_i \simeq T_e$	0.2eV
$L_{coll.}$	$\gtrsim$ 30cm
Debye length	$\lesssim$ .01cm
Gyro radii: ion	.42cm
electron	.0005cm

*Plasma: Length = 120cm, Diameter = 5cm, Current channel diameter = 1.0cm

magnetospheric parameters and Q-machine parameters. Notice that in both cases the electron gyroradius is less than the Debye length, that the ion gyroradius is greater than the electron gyroradius, and that the ratio of the current channel width to ion gyroradius in the Q-machine varies from about 2 to 7, while the arc-width to gyroradius ratio in a magnetosphere varies from 1 to 70. With these wave parameters, observed wave phenomenon observed in the magnetosphere can be modeled.

A major feature of our research involves the use of laser induced fluorescence [Hill, 1983]. We use a barium plasma with a energy level arrangement for a particular triplet of states, as illustrated in Figure 3, such that when the metastable state is excited the ground state emits photons as shown in the diagram. The laser is a dye laser that is pumped by a krypton laser and gives a very narrow line as illustrated in Figure 4. The fact that it is a narrow line allows us to scan the doppler broadened line of the barium ion, which is moving subject to the

TABLE 2. Magnetospheric and Laboratory Parameters

	Auroral Acceleration Zone	Q-Machine
Particle Parameters		
mass, μ	= 1, H^+	137, Ba^+
r_{ge}/λ_D	= 0.5	0.03
r_{gi}/r_{ge}	= 42	570
$\frac{\text{Arc width}}{r_{gi}}$	= 1 – 70	$\frac{\text{Current channel width}}{r_{gi}}$ = 2 – 7
Wave Parameters		
$e\phi/T_e \simeq 1$		1
$k_\perp/k_\parallel$	= 10	100
$\Delta\omega/\omega$	= 0.1	0.01
γ/ω	= 0.1	0.01

Optical Diagnostics

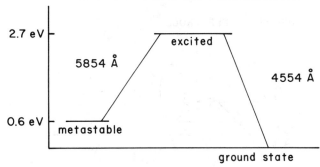

Fig. 3. Optical diagnostics.
• A cw dye laser (λ = 5854Å) excites Ba^+ ions.
• Spontaneous emission (λ = 4554Å) by excited ions is detected by a PMT.
• Signal is proportional to the density of metastable ions.

various fields that are created or exist in the plasma. The laser is tunable and scannable. Since the doppler shift is proportional to the dot product, $\vec{k}\cdot\vec{v}$, where $\vec{v}$ is the velocity of the ion, the optical signal that reaches the photomultiplier is, essentially, the ion velocity distribution along the line of sight of the laser beam. In Figure 5 we show a schematic of the diagnostic technique. The laser is beamed through the plasma and the right angle scattering is observed via a lens system that is focused on an optical fiber which sends it to the photomultiplier. The lens system itself is scannable along the laser line and the laser itself can be moved sideways so that if we want to we can scan the whole cross section of the plasma.

Large Amplitude EIC Wave Studies

A large fraction of our research in recent years has been with the use of the electrostatic

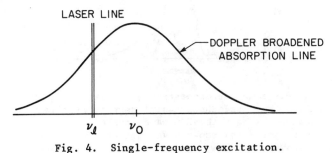

Fig. 4. Single-frequency excitation.

BASIC SCATTERING GEOMETRY

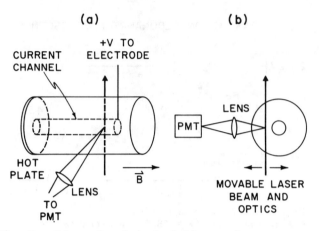

Fig. 5. Schematic of optical diagnostic system.

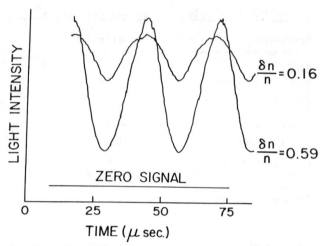

Fig. 7. Light intensity for moderately strong and very strong wave potentials.

ion cyclotron (EIC) wave as a diagnostic. It was because of the similarity between results obtained in our laboratory and some observations of EIC waves in the magnetosphere that we became involved in magnetospheric research. In this section we give some background and then describe some experiments pertinent to magnetospheric physics. The electrostatic ion cyclotron wave propagates mostly across the magnetic field with

a very small k vector along the magnetic field [Drummond and Rosenbluth, 1962]. The wave frequency is slightly higher than the ion cyclotron frequency. The EIC instability was

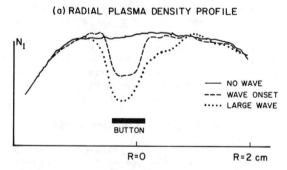

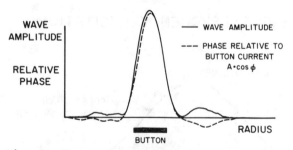

Fig. 6. Some representative data taken with the LIF system: (a) Shows radial density profiles with and without the EIC waves present. The lower figure (b) shows wave amplitude versus radius.

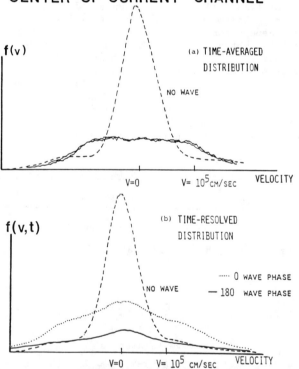

Fig. 8. (a) Time averaged ion velocity distributions with and without the wave present. (b) Time resolved ion velocity distributions showing the variation with wave phase.

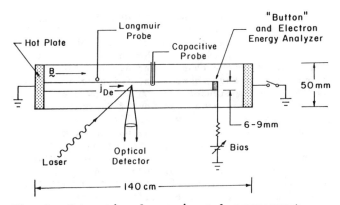

Fig. 9. Schematic of experimental arrangement for axial potential measurements.

first detected by D'Angelo and Motley (1962). A good summary is given by J. P. M. Schmitt (1972). It was first reported in the magnetosphere by Kinter (1979).

The wave is destabilized, in our device, by passing a current down a channel to a collector as illustrated in Figure 5 and in Figure 6 we

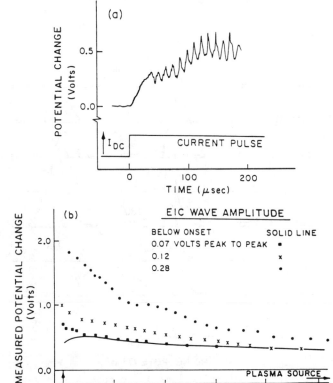

Fig. 10. (a) Current pulse applied to collector button shown EIC wave growth. (b) Measured axial potential variation.

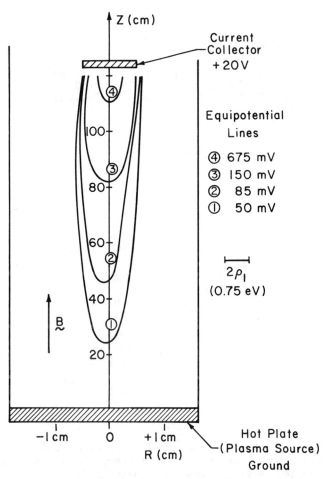

Fig. 11. Equipotential contour plot of current channel.

show a few of the results that we have been able to document using this technique. One of the most important is that just before the instability onset there is a channel of lower density in the plasma. This is due to a change in the sheath conditions at the endplates. As shown by S. von Goeler [1964] and N. Rynn [1966], these have a substantial effect on recombination at the endplates and can influence the density markedly. When the wave onset occurs the density in this channel decreases even further, due to particles being ejected by the EIC wave, so that the density in the channel in which the instability exists is lower than the surrounding plasma. The net result is, we believe, that the ion cyclotron wave which exists in this filament cannot propagate outside of it because the parameters are not right. Figure 6b shows a plot of wave amplitude versus radius along with relative phase; note that the wave amplitude is confined to the channel and that it's distribution resembles a bessel function, which one would expect with this geometry. Also shown is a standing wave pattern outside the channel

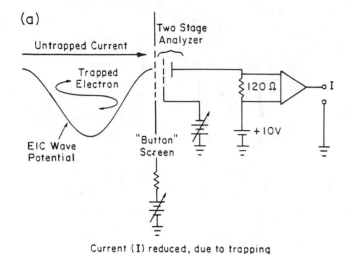

Fig. 12. Schematic of the electrostatic analyzer and a conceptualization of electron trapping at two 180° separated phases. (a) Only the untrapped distribution is sampled. (b) The entire distribution is sampled.

which indicates that there is a radial standing wave in the channel. Figure 7 shows optically received signals for just after onset of the instability and well into the instability. The main thing to notice here is the very large amplitude, on the order of 60% for $\delta n/n$, $\delta\phi/T$, indicating a large amplitude wave. In Figure 8 we show how the diagnostic allows us to see time variations. In Figure 8a, we see a time averaged distribution with and without waves, in Figure 8b we see a time resolved distribution which fluctuates with the phase of the wave.

Figure 9 shows a more detailed schematic of how we took some further measurements. In this situation we were interested in determining if we could identify a mechanism, via the variation of potential along the plasma column, that might explain the acceleration of electrons along it. The arrangement is much the same as has been described before except that we now have added a capacitive probe along with a langmuir probe. The capacitive probe is, essentially, a langmuir probe coated with glass so that it does not pick up particles directly. This means that it picks up only fluctuations and it is used to measure the plasma potential of the wave. Figure 10 shows how the plasma potential along the column is measured. In 10a we see how a current pulse was applied to the channel and how the wave grows; notice that there is a rising dc component to the wave amplitude. In 10b we show plots of the dc potential as a function of distance from the plasma source with wave amplitude as a parameter. We notice that the potential grows in amplitude as it goes toward the button. In addition, independent measurements show that the wave amplitude follows the amplitude of the potential. The important thing here is that the wave does seem to create a difference of potential as it travels along the column.

In Figure 11, we show some equi-potential plots for our plasma configuration. The potential contours are what one would expect from this particular geometry of a small button in a larger conducting cylinder. The important point is that the equi-potential plots are as expected and that we can measure them. They are reminiscent of the v-shaped curves that have been observed by satellites. This would seem to indicate that there is a very definite structure to the auroral arc. Figure 12 shows how we measured electron distributions in an attempt to

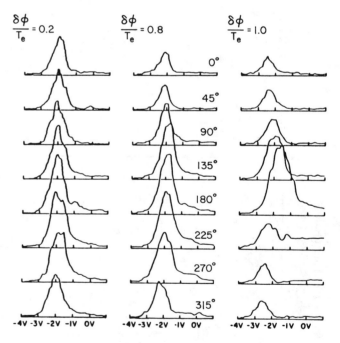

Fig. 13. The distributions are recorded at eight different phases within one barium EIC wave period and for three EIC wave saturation levels. The real electron temperature is 0.2 eV, smaller than the approximately 1.0 eV temperature indicated by the space-charge limited distributions in the figure.

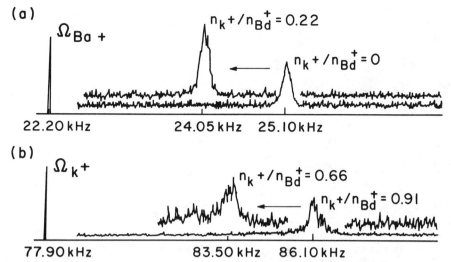

Fig. 14. EIC wave-frequency shifts at onset: (a) Shift in the barium EIC-wave frequency towards its cyclotron frequency Ω_{Ba}. (b) Shift in the potassium EIC-wave frequency towards its cyclotron frequency Ω_{K}. The bandwidth of the spectrum analyzer was 100 Hz.

show trapping. The end collector, the "button", was replaced by an electron energy analyzer, the essential features of which are shown in 12a and 12b. In 12a the particles are trapped in the wave potential and only untrapped particles go into the collector. In 12b the wave has moved such that all the trapped particles can now be deposited in the analyzer, thereby greatly modifying the velocity distribution.

In Figure 13 we show a series of velocity distributions for different amplitude fluctuations and for successive phases [Lang, 1983]. On the right hand side where we show $\delta\phi/T_e = .2$; there is not much of a modification in velocity distribution, indicating very little trapping. As we progress to $\delta\phi/T_e = .8$ to $\delta\phi/T_e = 1$, we see that as the wave fluctuates there is a gross modification in the velocity distribution. We claim, at a minimum, that this is an indication of extensive electron trapping.

This has been used as an argument to explain possible enhanced resistivity by Lang [1983], who has tried to show that if electrons are trapped and untrapped by highly non-linear waves, this would amount to a modification of the resistivity and, therefore, of electric field in the plasma column. This is a good example of what laboratory measurements can do and what ideas can be brought out that are much more difficult to do in the less accessible environment of the magnetosphere.

In Figure 14 we show the results of an

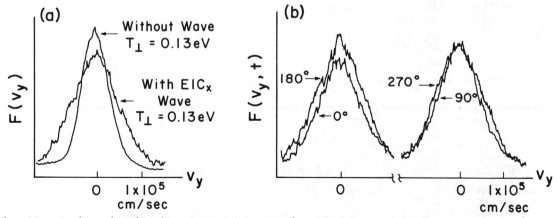

Fig. 15. Barium distributions heated by potassium EIC waves: (a) Time-averaged barium distribution function $F(V_y)$, with and without potassium-EIC waves present. (b) Time resolved $F(V_y, t)$ at four phases within a potassium EIC- wave period. $n_K/n_{Ba} = 1.2$, and no barium EIC waves were destabilized.

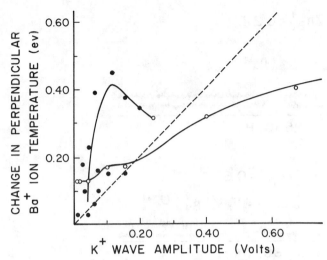

Fig. 16. Barium ion heating versus potassium EIC-wave amplitude. Optically measured barium temperatures are shown as a function of the potassium EIC wave amplitude measured with a capacitive probe. The closed circles describe heating by current-driven K^+ EICW, summarized from several different experiments. The open circles show heating by EIC waves driven externally using a variable frequency current source. The driving frequency was $f_d = 91.4$ kHz with $\Omega_K = 90.0$ kHz.

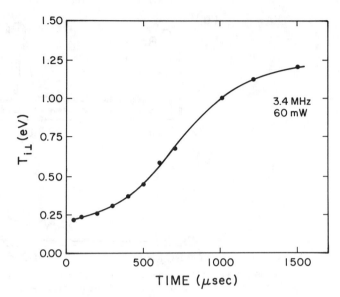

Fig. 17. Perpendicular ion temperature versus RF pulse duration. $F_{center} = 3.4$ MHz, wave power (during pulse on) = 60 mW.

experiment that attempted to show the effect of mixing two heavy ion species, in this particular case potassium and barium. In 14a we see that the shift of the frequency is toward the cyclotron frequency of barium as the ratio of potassium to barium is increased, and in 14b we see a similar effect of the shift toward the potassium cylcotron frequency as the ratio of barium to potassium is increased. This is consistent with theory [Ashour-Abdalla, 1982] and observations [Kintner, 1979], and is again indicative of the sort of control that we can get in the laboratory. We have made attempts to compare this with some simulation and this particular aspect is being continued by us at the present time.

Figure 15 shows the heating of barium ions by potassium electrostatic ion cyclotron waves. Figure 15a shows the time averaged barium distribution function with and without potassium EIC waves present. Figure 15b shows the time resolved distribution within the potassium EIC wave. Since the experimental conditions were such that no barium EIC waves were destabilized, and since both the barium and the potassium waves were in phase, it is clear that the barium was heated by the EIC waves. Effects like this are seen in the magnetosphere. In Figure 16 we show the extent of barium heating as a function of potassium EIC wave amplitude. This was done in

TABLE 3. Experimental Parameters

Characteristic	Description
Density	$9 \times 10^9 < N_i < 2 \times 10^{10}$ cm^{-3} Ba^+ Plasma
B field	3.0kG
Initial Temperature	$T_i \simeq T_e \simeq 0.2$eV
Frequency	Broadband Noise $1.5 < f_{center} < 3.9$MHz $f_{L.H.} \simeq f_{Pi} \simeq 1.8$MHz
Bandwidth	$\Delta f \simeq 1.0$MHz

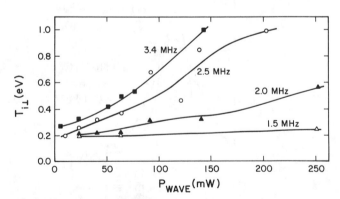

Fig. 18. Perpendicular ion temperature versus input RF power. Squares: $F_{center} = 3.4$ MHz circles, 2.5 MHz. Solid triangles, 2.0 MHz. Open triangles, 1.5 MHz.

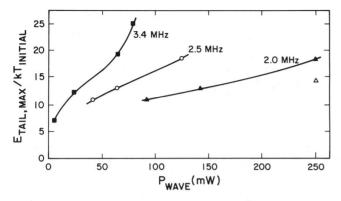

Fig. 19. Maximum observed ion energy/$KT_{initial}$ versus input RF power. Squares: F_{center} = 3.4 MHz circles, 2.5 MHz. Solid triangles, 2.0 MHz. Open triangles, 1.5 MHz.

two ways, first by applying an external signal to see what the heating would be like, and then by destabilizing the wave. The barium ion temperature was measured optically, and for the externally driven wave, one sees a definite increase in the ion temperature as the wave amplitude is increased. For the current destabilized wave, however, the heating effect goes through a maximum. This illustrates the difference between the two mechanisms and also illustrates the fact that there is a range for the destabilized case over which efficient heating occurs. This has been known to happen in other wave phenomenon in laboratory and in fusion plasmas. It is not clear from these measurements, whether we have bulk heating or tail heating orbits.

Broadband Lower Hybrid Wave Studies

The balance of this paper will be concerned with our attempts to explore the effect of lower hybrid heating on the magnetosphere. Satellite and rocket data from the supraurotal region show ion acceleration perpendicular to a geomagnetic field [Sharp, 1977]. Electric field measurements show the presence of long waves in this region [Mozer, 1980]. Chang and Coppi [1981] have shown that broadband lower hybrid waves near the lower hybrid frequency contribute to the production of the ion conics. The UCI Q-machine has been used to simulate lower hybrid perpendicular heating of ions in the supraurotal region. Laser induced fluorescence has been used for high resolution, non-perturbing measurements of the ion distribution function. The plasma consisted of a 1m long, 5cm diameter barium plasma of densities of the order of or less than 10^{10}cm contained by a 3kG main magnetic field. Waves were launched with broadband frequency spectra near the lower hybrid frequency. The waves were launched from a 12cm long cylindrical antenna which was coaxial at the plasma column and were found throughout the plasma. Substantial changes in the perpendicular ion distribution were found. Main body ion heating occurred along with tail production. Over a 10db change in input wave power we observed up to a factor of 3 enhancement in main body ion temperature.

Again the laser diagnostics were similar to what was used before, the new addition being a coaxial antenna, and an rf probe to measure the radio frequency. In Table 3 we show the experimental parameters for this particular set of investigations. Note that the lower hybrid signal was always in the form of broadband noise with the center frequency varied as shown. Figure 17 shows the effect on perpendicular ion temperature with a constant input wave power as the pulse length is increased. Note that it increases as the pulse length increases with some indications of a leveling off. This and other

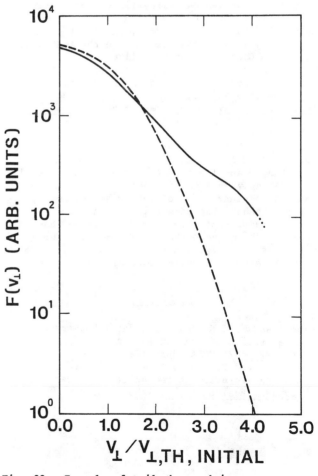

Fig. 20. Example of tail observed in perpendicular velocity distribution. Solid line: determined by laser induced fluorescence, T_I = .39 eV, F_{center} = 2.0 MHz, P_{wave} = 250 mW. Dashed line: Maxwell-Boltzmann distribution with same T_I.

phenomena are discussed more fully by McWilliams [1985] in his paper given in another part of this collection. Figure 18 shows the effect on a perpendicular ion temperature as the power is increased. Note that below the ion plasma frequency of 1.8MHz the heating is fairly flat and in fact there is not much heating at all above the original temperature of 0.2eV. Above the ion plasma frequency the heating starts to increase as the wave powers increase. Figure 19 shows the effect of the power input on tail heating. What is shown here is that the energy in the non-maxwellian tail, compared to the initial temperature, increases as the wave powers increases. Notice also that there is a marked tail-enhancement as the frequency is increased, corresponding to that which was shown in Figure 18. This means that much of the energy goes into the tail of the distribution. Figure 20 is an interesting one in that it shows a direct comparison between the tail that is excited by lower hybrid heating and the maxwellian. Note again that it is clearly demonstrated that the non-maxwellian tail is enhanced by the lower hybrid. This is in agreement with the theory of Retterer, Chang and Jasperse [1986].

Our latest results have indicated that we observed a conic in a mirror machine where we have launched this type of lower hybrid heating. We are developing tomography techniques to study these and we hope to be able to produce the instability in a mirror geometry.

This completes the survey of experiments that we have done in a Q-machine to try to model magnetospheric effects. Note that we are capable of modeling very specific phenomena so that we can obtain clear tests of theory and/or conjectures. We have the capability of varying parameters so that they bear directly on observations in the magnetosphere. We are continuing with this investigation and hope to continue to contribute to the general lore of magnetospheric physics.

References

Ashour-Abdalla, M. and H. Okuda, Transverse acceleration of ions in auroral field lines, Rep. PPG-642, UCLA, July 1982.

Chang, T. and B. Coppi, Lower hybrid acceleration and ion evolution in the supraauroral region, Geophys. Res. Lett., 8, 1253, 1981.

D'Angelo, N. and R. W. Motley, Phys. Fluids 5, 633 (1962). Also see Motley, R. W. and N. D'Angelo, Phys. Fluids, 6, 296 1963.

Drummond, W.E. and M.N. Rosenbluth, Anomalous diffusion arising from micro-instabilities in a plasma, Phys. Fluids, 5, 1507, 1962.

Hill, D.N., S. Fornaca, and M.G. Wickham, Single frequency scanning laser as a plasma diagnostic, Rev. Sci. Instrum., 54, 309, 1983.

Kinter, P.M., M.C. Kelley, R.D. Sharp, A.U. Ghielmetti, M. Temerin, C. Catell, P.M. Mizera, and J.F. Fenelli, Simultaneous observations of energetic (keV) upstreaming ions and electrostatic cyclotron waves, J. Geophys. Res., 84, 720, 1979.

Lang, A. and H. Boehmer, Electron current disruption and parallel electric fields associated with electrostatic ion cyclotron waves, J. Geophys. Res., 88, No. A7, 5564, 1983.

McWilliams, R., R. Koslover, H. Boehmer, and N. Rynn, Laboratory simulation of ion acceleration in the presence of lower hybrid waves - Proceeding of AGU Chapman Conference on Acceleration Mechanisms in the Magnetosphere, Wellesley, MA, June 3-7 1985.

Mozer, F.S., C.A. Cattell, M.K. Hudson, R.L. Lysak, M. Temerin, and R.B. Torbert, Satellite measurements and theories of low altitude auroral particle acceleration, Space Sci. Rev., 27, 155, 1980.

Retterer, J.M., T.S. Chang, and J.R. Jasperse, Ion acceleration by lower hybrid waves in the supraauroral region, J. Geophys. Res., 91, 1609, 1986.

Rynn, N., Improved quiescent plasma source, Rev. Sci. Instrum., 35, 40, 1964.

Rynn, N., Plasma column end effects, Phys. Fluids 9, 165, 1966.

Schmitt, J.P.M., Resonances of an antenna associated with the excitation of ion Bernstein modes, Phys. Fluids, 15, 2057, 1972.

Sharp, R.D., R.G. Johnson and E.G. Shelley, Observation of an ionospheric acceleration mechanism producing energetic (keV) ions primarily normal to the geomagnetic field direction, J. Geophys. Res., 82, 3324, 1977.

von Goeler, S., Influence of ion loss on diffusion of cessium plasma in the Q-machine, Phys. Fluids, 7, 463, 1964.

LABORATORY SIMULATION OF ION ACCELERATION IN THE PRESENCE OF LOWER HYBRID WAVES

R. McWilliams, R. Koslover, H. Boehmer, N. Rynn

Department of Physics, University of California, Irvine,
Irvine, California 92717

Abstract. Ion acceleration perpendicular to the geomagnetic field has been observed by satellites and rockets in the supra uroral region. Also found are broadband lower-hybrid waves, and, at higher altitudes, conical upward-flowing ion distributions. The UCI Q-machine has been used to simulate the effect of lower hybrid waves on ion acceleration. Laser induced fluorescence was used for high resolution, non-perturbing measurements of the ion velocity distribution function. The plasma consisted of a 1 m long, 5 cm diameter barium plasma of densities on the order of 10^{10} per cm^3 contained by a 3 kG magnetic field. Substantial changes in the perpendicular ion distribution were found. Main-body ion heating occurred along with non-maxwellian tail production. Over a 10 dB change in input wave power we observed up to a factor of 3 enhancement in main-body ion temperature.

Satellite and rocket data from the supra uroral region show ion acceleration perpendicular to the geomagnetic field [Sharp, 1977]. Electric field measurements show the presence of lower hybrid waves in this region [Mozer, et al, 1980 among others]. Chang and Coppi [1981] have suggested that broadband lower hybrid waves near the lower hybrid frequency contribute to the production of the ion conic distributions. Experiments at UC Irvine have been aimed at laboratory simulation of the lower magnetosphere. The purpose of these experiments is to see whether broadband lower hybrid waves can produce ion conic distributions similar to those observed in the magnetosphere.

The experiment is divided into several distinct components of which we report the initial results in this paper. Energetic, downward-flowing electrons in the suprauroral region are thought to create lower hybrid waves through instabilities. These instabilities may occasionally create ion heating and in the mirror magnetic field of the magnetosphere would help generate the observed ion conic distributions. The laboratory experiments reported here show the effect of externally launched broadband lower hybrid waves near the lower hybrid frequency on the perpendicular ion distribution function. The UCI Q-machine [Rynn, 1964] was used for these laboratory experiments. The plasma consisted of a 1 m long, 5 cm diameter barium plasma of densities of the order of 10^{10} per cm^3. Approximately uniform plasma density occurred from r = 0 to r = 1.25 cm with an approximately linear drop from r = 1.25 to r = 2.5 cm. The ion heating measurements were made at r = 0. The plasma was contained in a cylindrical geometry by a 3 kG magnetic field. The lower hybrid waves were launched with either broadband or narrowband frequency spectra near the lower hybrid frequency with control over the bandwidth, the center frequency and the amplitude of the lower hybrid waves. These waves were detected by means of a small, radially moveable coaxial rf probe with a 3 mm tip oriented along the magnetic field lines. The waves were launched with a 12 cm cylindrical antenna [Bellan, 1976] which was coaxial with the plasma column.

The waves were found throughout the plasma. Plasma density and electron temperature were inferred from Langmuir probes. Plasma density was also inferred from the propagation of higher frequency lower hybrid waves through the plasma. The ion distribution function is determined through the use of laser induced fluorescence [Hill, 1983]. This non-perturbing diagnostic consists of a laser beam passing through the plasma. The beam is from a tunable dye laser and is of sufficiently narrow line width as to excite only the ions coupling with it through the correct doppler shift, given by $\underset{\sim}{k}_L \cdot \underset{\sim}{v}_{ion} = \omega_o - \omega_L$ where ω_o is a natural transition frequency of singly ionized barium and ω_L is the laser frequency one has tuned. The distribution function of ions is then obtained by scanning the laser frequency over all of the ion population velocities. Spatial resolution of approximately 1 mm^3 is obtained by focusing detected light through a lens prior to entering a photomultiplier tube. Time resolution of this diagnostic is good down to about 10μsec. We can obtain either the perpendicular or parallel ion

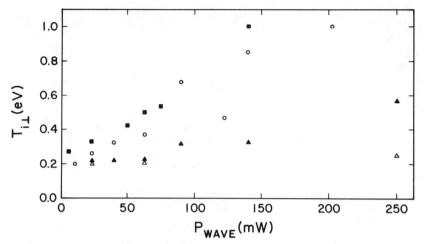

Figure 1. Perpendicular ion temperature versus input rf power. Squares: f_{center} = 3.4 MHz. Circles, 2.5 MHz. Solid triangles, 2.0 MHz. Open triangles, 1.5 MHz.

distribution functions by orienting the laser beam either perpendicular or parallel to the magnetic field.

The experiments here report measurements mainly of the perpendicular ion distribution function. Parallel distribution functions exhibited a change of 25% or less in the parallel ion temperature over the range of parameters studied here. Remarks about temperatures further in the paper then are the appropriate moment of the distribution function as if it were bi-maxwellian, and refer to the perpendicular ion distribution function. The stated temperatures are then a temperature for a maxwellian distribution fit to the data, conserving number density of the distribution. Temperature is thus an indication of main body heating which was generally close to maxwellian while the tail of the distribution could be highly non-maxwellian, as discussed below. The diagnostic, of course, gives more information than just the temperature as will be emphasized later. For this plasma the ion plasma frequency is approximately 1.8 MHz. The data reported here are using, as shown in Figure 1, center frequencies of 1.5 MHz, 2 MHz, 2.5 MHz, and 3.4 MHz with a frequency half-width of the broadband noise of .5 MHz. We have therefore launched broadband lower hybrid waves from approximately ω_{pi} to a little over $2\omega_{pi}$.

As one expects, below ω_{pi} (f_{pi} = 1.8MHz, f = 1.5MHz) there is little perpendicular ion heating observed. As we raise the frequency, approaching of the order of 2 times the plasma frequency, and still in the vicinity of the lower hybrid frequency, we see more and more ion heating as shown in Figure 1. In fact, the maximum heating observed was up to a factor of six times the initial temperature. In Figure 2 the ion temperature as a function of the pulse duration of the lower hybrid noise is reported. Initially, for very short pulses there is no ion heating. The temperature rises with time until it approaches a saturation of about five times the initial temperature for this figure. We attribute this to the plasma source being approximately 1 meter from the laser detection point in this experiment. The plasma drifting means that ions will sample the rf waves for as long as the rf is on or until the ion comes to the laser detection point, whichever comes first. The saturation which one observes here is consistent with the drift time of the ions from the plasma source to the laser detection point.

In Figure 3 we discuss some of the non-maxwellian nature of the ion distribution func-

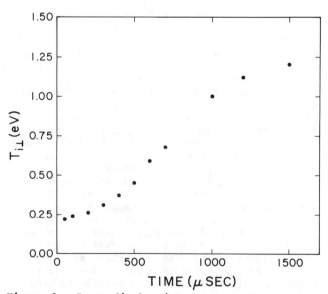

Figure 2. Perpendicular ion temperature versus rf pulse duration. f_{center} = 3.4 MHz, Wave power (during pulse on) = 60 mW.

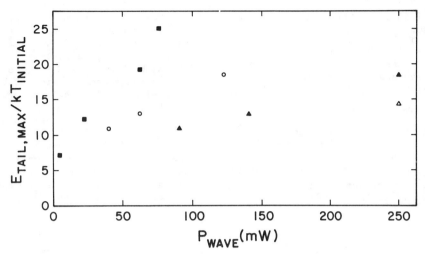

Figure 3. Maximum observed ion energy/$kT_{initial}$ versus input rf power. Squares: f_{center} = 3.4 MHz. Circles, 2.5 MHz. Solid triangles, 2.0 MHz. Open triangles, 1.5 MHz.

tions in simplified form. For different frequencies as a function of input power we plot the maximum observed tail energy of the particles which are accelerated in the perpendicular direction as compared to the initial kinetic energy of the plasma. What we speak of here is a fraction of the ions which are found to be non-maxwellian, that is, there is an enhanced tail population well above the thermal speed of the particles. We are able to detect signals as low as 2% of the peak detected laser induced fluorescence. Thus, when there are particles forming a non-maxwellian tail, they will be observed if their population is above this sensitivity level. For a non-maxwellian distribution the maximum observed tail energy corresponds to the particle energy where the laser induced signal falls to the sensitivity limit of the diagnostic. As shown, particles are observed in the tail with up to 25 times the initial thermal energy of the plasma. That is approximately the limit of resolution of our diagnostic. Figure 4 shows a complete distribution function and a comparison with the maxwellian. The solid line in the figure is the measurement of the distribution function and is not a fit through data points but is a semi-log plot of the actual output from the diagnostic. The dashed line is a maxwellian distribution fitted to the data, conserving the number of particles, so it lies slightly above the solid line at lower velocities. As one can see there is a fraction of the order of 5% to 10% of the ion population which is distinctly non-maxwellian in a tail of greater energy than a maxwellian distribution would imply.

In conclusion, broadband lower hybrid noise near the lower hybrid frequency can cause substantial perpendicular ion heating of the distribution function. The main body distributions can be found to be maxwellian while there can also be substantial non-maxwellian tails out on the distribution functions. Tail particles could be created, at least to the instrumental

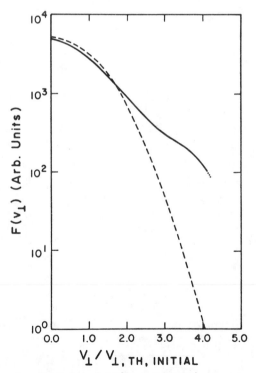

Figure 4. Example of tail observed in perpendicular velocity distribution. Solid line: determined by laser induced fluorescence, f_{center} = 2.0 MHz, P_{wave} = 250 mW. Dashed line: Maxwell-Boltzmann distribution with T_i = 0.39 eV.

detection limit, of 25 times the initial thermal energy of the plasma.

Acknowledgements. The authors would like to thank M. Okubo for assistance in operating the laser. This work was supported by NSF Grant Number ATM-8411189.

References

Bellan, P.M. and M. Porkolab, Experimental studies of lower hybrid wave propagation, Phys. Fluids, 19, 995, 1976.

Chang, T. and B. Coppi, Lower hybrid acceleration and ion evolution in the suprauroral region, Geophys. Res. Lett., 8, 1253, 1981.

Hill, D.N., S. Fornaca and M.G. Wickham, Single frequency scanning laser as a plasma diagnostic, Rev. Sci. Instrum., 54, 309, 1983.

Mozer, F.S., C.A. Cattell, M.K. Hudson, R.L. Lysak, M. Temerin and R.B. Torbert, Satellite measurements and theories of low altitude auroral particle acceleration, Space Sci. Rev., 27, 155, 1980.

Rynn, N., Improved quiescent plasma source, Rev. Sci. Instrum., 35, 40, 1964.

Sharp, R.D., R.G. Johnson and E.G. Shelley, Observation of an ionospheric acceleration mechanism producing energetic (keV) ions primarily normal to the geomagnetic field direction, J. Geophys. Res., 82, 3324, 1977.

LABORATORY EXPERIMENTS ON PLASMA EXPANSION

Chung Chan

The Center for Electromagnetics Research and Department of Electrical and
Computer Engineering, Northeastern University, Boston, MA 02115

Abstract. Laboratory experiments have been designed to study the expansion of a plasma into a vacuum, along and across magnetic field lines. It was shown that the ambipolar electric field and the existence of a moving double layer front provided very effective ion acceleration. The presence of a longitudinal magnetic field was observed to induce azimuthal E x B motion on the expanding plasma. For the expansion across magnetic field lines, a polarization electric field was developed to retard the plasma flow. The implication of these results to plasma expansion processes in space was also discussed.

Introduction

The expansion of a plasma into a vacuum or a much lower density plasma, represents an effective mechanism for ion acceleration. In collisionless plasmas with no external magnetic fields, the ambipolar electric field has been observed to accelerate ions up to MeV energies. For applications in space plasmas, the effect of the geomagnetic field becomes important and the expansion processes can be quite different under different magnetic field strengths and configurations. The examples include the releases of artificial plasma clouds in the ionosphere and magnetosphere (Wescott et al., 1980 and Wescott et al, 1985), the energization of suprathermal ions in the solar wind (Singh and Schunk, 1982), the flow of solar wind plasmas around planetary wakes (Tariq et al., 1985) and the filling-in of the wakes (Samir et al, 1983), as well as the injection and acceleration processes in the magnetotail (Eastman et al., 1985). All of the aforementioned problems required detailed understanding of the relevant plasma expansion processes along and across magnetic field lines.

In this paper, we present a brief description of our recent laboratory results on the expansion of a rarefied, collisionless plasma along and across magnetic field lines. We will mainly discuss the basic plasma processes involved and omit experimental details. The relevance of our laboratory results to the space plasma problems will also be addressed.

Self-Similar Expansion

The expansion of a collisionless plasma in one-dimension had been treated theoretically by Gurevich et al., (1966) who found the motion to be scale-invariant or self-similar. Using the ion fluid equations together with the assumption of quasi-neutrality and Boltzman electrons, the following solutions describe the evolution of the ion density n_i, ion fluid velocity v_i and the plasma potential Φ;

$$n_i = n_0 \exp[-(\frac{x}{tc_s} + 1)] \qquad (1)$$

$$v_i = c_s (\frac{x}{tc_s} + 1) \qquad (2)$$

$$\Phi = -\frac{T_e}{e}(\frac{x}{tc_s} + 1) \qquad (3)$$

where n_0, T_e and c_s are the unperturbed ion density, electron temperature and ion acoustic velocity, respectively. The self-similar variable is x/tc_s where only the ratio of the space and time variables determines the evolution processes. The self-similar solution, however, corresponds to a situation where an infinite electric field exerts infinite acceleration on the boundary ions at early times. This situation can be corrected by solving the Poisson's equation for the space charge effects at the expansion front where the electrons move ahead of the ions. As seen in the computer simulation by Crow et al,. (1975), the expansion front consisted of a region of positive space charge preceded by a region of a pure electron cloud.

Our experiments were performed in a filament discharge plasma column with the details given elsewhere (Chan et al., 1984). A plasma shutter system consisting of four electrostatic grids was used to initiate the expansion of a steady-state source plasma (with electron density $n_e \geq 10^8$ cm^{-3}, $T_e = 2$ eV and argon ion temperature $T_i \leq 0.3$ eV) into the target chamber (a vacuum with neutral pressure $P_0 \geq 10^{-5}$ Torr). Figure 1a gives the time evolution of the plasma potential profile, as

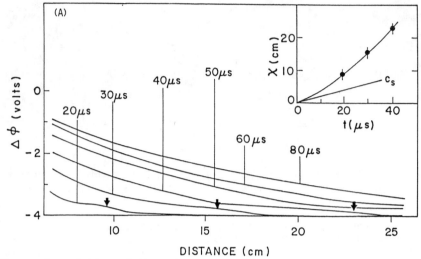

Fig. 1a. Temporal evolution of the potential profiles for the expanding plasma. The insert is the trajectory of the ion fronts whose positions are indicated by the arrows.

measured by an emissive probe, along the axis of the target chamber. The expansion was observed to take place on the ion time scale and an expanding front appeared at t = 20 μs. The ion acoustic velocity in the experiments was approximately 2 x 10^5 cm/s so that the observed ion front was supersonic. The change of curvature at the front indicated a region of ion space charge preceded by a region of excess electron space charge which corresponded to a double layer-type potential structure. At t ≥ 20 μs, expansion of the main plasma was observed. The trajectory of the ion front, as indicated by the arrows, is shown in the insert of Fig. 1a. The front is seen to accelerate to approximately 4 c_s or roughly 20 times the ion thermal velocity at t = 40 μs. The expanding plasma reached the end of the device at t = 50 μs thus limiting the maximum ion front velocity that can be obtained in our experiment. In Fig. 1b, we regraphed the plasma potential profile of Fig. 1a versus the self-similar variable x/tc_s. The potential profile possessed the self-similar form of Equation 3 (i.e. a straight line) except at early times when the expanding front was observed. In a

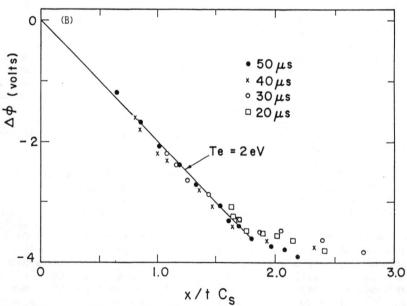

Fig. 1b. Potential profiles of Fig. 1a versus the self-similiar variable showing the dependence of Equation (3).

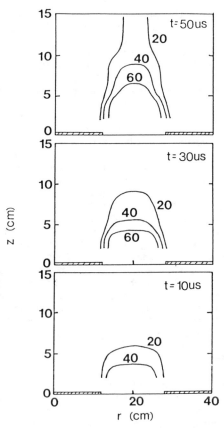

Fig. 2. Time evolving density contours for a plasma expanding in the z-direction through an aperture of 7.5 cm radius. The contours are labeled in the percentage of n_i/n_o.

recent experiment by Wright et al. (1985), the self-similar plasma expansion process was found to be the dominant wake filling mechanism in the near wake of a body which immersed in a streaming plasma. The creation and motion of an expanding ion front were also observed in that experiment.

The Effect of a Weak Longitudinal Magnetic Field

For artificial plasma cloud injection type experiments (Wescott et al., 1980), axial as well as radial expansion of a finite plasma column are of particular interest. In order to investigate the radial or the transverse motion of the expanding plasma, an aperture with a 7.5 cm radius was placed in front of the source plasma and the source plasma then expanded into the 20 cm radius target chamber. The two-dimensional equidensity contours (normalized to the unperturbed density) at various times during which the plasma was expanding in the z-direction, were shown in Fig. 2. The expansion in the axial direction was similar to the one-dimensional case, i.e. the motion was self-similar up to the expanding front. Eq. (1) and (2) can be combined to give

$$v_i = c_s \ln \frac{n_o}{n_i} \quad (4)$$

which indicates that the lower density contour expands at a higher velocity in agreement with the results shown in Fig. 2. We also noted that very little radial expansion had occurred because the main plasma flow was in the z-direction and collisional diffusion effects were negligible. However, when even a weak longitudinal magnetic field was present, the expansion process changed drastically. In Fig. 3, we showed the temporal evolution of the radial electron density profiles at a fixed z-distance with and without the presence of a uniform axial magnetic field of B = 50 G. For B = 50 G, the ratio of the electron plasma frequency to the electron cyclotron frequency was slightly less than unity near the source region with the thermal to magnetic energy ratio $\beta < 0.01$. The aperture radius was only slightly larger than the ion gyroradius so that our experimental results could be correlated with plasma expansion along a magnetic flux tube. The longitudinal magnetic field was obseved to slow down the expansion velocity axially, induce large amplitude, low frequency electrostatic oscillations in the plasma column and enhance radial expansion as indicated by the broadening of the density profiles. For a magnetic field strength of 50 G, the ion and the electron gyroradii were 7 cm and 0.06 cm, respectively. We believe that the large differences in ion-electron gyro-motion may result in a polarization electric field pointing radially inward. This electric field in turn caused an E x B motion of the plasma in the azimuthal direction. The

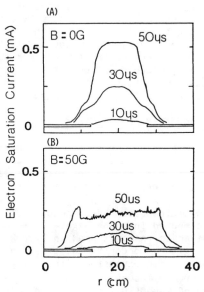

Fig. 3. Temporal evolution of the radial density profiles for the cases of (a) B = 0 and (b) a longitudinal magnetic field of B = 50 G.

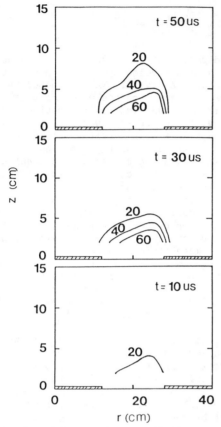

Fig. 4. Time evolving density contours for a plasma expanding across a uniform magnetic field of B = 25 G through an aperature of 7.5 cm radius.

inward-directed polarization electric field also caused the electrons to drift radially outward thus impeding the electron motion in the axial direction. As a result, the ambipolar electric field and the corresponding ion acceleration will be signicantly reduced. Thus, the axial expansion of the plasma was observed to retarded in the presence of a weak longitudinal magmetic field.

Expansion Across Magnetic Field Lines

The plasma expansion across a uniform magnetic field has been studied in the same experimental configuration as in Fig. 2 but with the magnetic field oriented perpendicular to the direction of the expansion. As shown in Fig. 4, the expansion was in the z-direction and a magnetic field of B = 25 G was directed into the page. Equidensity contours similar to the ones used in Fig. 2, were shown in Fig. 4 in order to facilitate a direct comparison between expansion without magnetic field and across magnetic field lines. With the perpendicular magnetic field, the expansion was observed to be significantly retarded (i.e. compared the n_i/n_o = 20 % contours in Fig. 2 and Fig. 4) and the plasma was essentially confined to within 10 cm from the source. There was also an apparent plasma drift to the right. If we reversed the magnetic field direction (i.e. out of the page), the plasma was observed to drift to the left instead. Since this drift appeared in both the electron and ion density profiles, it was consistent with an E x B drift with the electric field pointing towards the source (-z direction) and opposite to the direction of the ambipolar electric field. Again, this result was consistent with the observation of the polarization electric field in section III. With a gyro-radius of 14 cm, the ions can readily expand across the magnetic field lines at early times while the electrons are essentialy confined by the magnetic field to regions near the source. As such, the polarization electric field will be directed towards the source and cause an E x B drift to the right as shown in Fig. 4. In Fig. 5, we graphed the displacement of the n_i/n_o = 20 % contours in times for the two cases. Even for a weak magnetic field of 25 G, the expansion velocity was almost five times slower than the B = 0 case. Since the electrons can no longer move ahead of the ions when crossing magnetic field lines, the ambipolar acceleration of ions was not observed.

Discussion

Plasma expansion processes are expected to occur when transient density gradients are created. In this case, the inclusion of the finite density gradient scale length modifies the self-similar solution slightly but significant ambipolar ion acceleration is still expected (Feiber and Decoste, 1978). One application in space plasma problems is the refilling of the depleted plasmaspheric flux tubes after magnetic storms. Singh and Shunk (1983) have shown via numerical simulations that collisionless expansion of the conjugate ionspheric plasmas may be the dominate mechanism at the initial stage of refilling. Suprathermal forerunner ions and counter-streaming

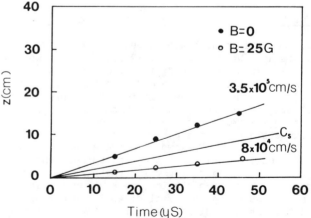

Fig. 5. The trajectories for the n_i/n_o = 20% contours of Fig. 2 and Fig. 4.

density shock fronts were observed to propagate into the "equatorial" region in that simulation. This result is similar to our result in the one-dimensional expansion experiment with B = 0. However, the expansion experiment with a longitudinal magnetic field also suggests that the effects of polarization electric field and the azithumal E x B motion may also be important along a magnetic flux tube. Another application would be in the high altitude shaped-charge plasma injection experiments (Wescott et al., 1980) where high density plasma clouds were suddenly created and expanded along and across magnetic field lines. Similar to the results present in Sections III and IV, the polarization electric field can expect to play an important role in the expansion of the barium cloud. Recent numerical simulation by Sydora et al., (1983) has, in fact, shown that the polarization electric field caused a large electron velocity shear in the azimuthal direction which, in turn, excited the Kelvin-Helmholtz instability. For some other space physics applications such as polar wind, interhemi-spheric flow and solar wind in coronal holes, the expansion will be along a strong magnetic field. For these applications, the experiments on one-dimensional expansion with B = 0 may be more relevant. However, only the expansion of a noniso-thermal plasma with $T_e/T_i \approx 10$ has, so far, been considered in our experiments. It is likely that the details of the expansion process may depend on the T_e/T_i ratio. In the experiment by Korn et al., (1970), the expansion of an isothermal plasma along a strong magnetic field was also found to be self-similar but no distinct ion front regions were observed.

Acknowledgments. This work was supported in part by the National Science Foundation.

References

Chan C., N. Hershkowitz, A. Ferreira, T. Intrator, B. Nelson and K. Lonngren., Experimental Observations of Self-similar Plasma Expansion, Phys. Fluids, 27, 266-268, 1984.

Crow J.E., P.L. Auer and J.E. Allen, The Expansion of a Plasma into a Vacuum, J. Plasma Phys., 14, 65-76, 1975.

Eastman T.E., L.A. Frank, K.I. Nishikawa, Velocity Distribution of Ion Beams in the Plasma Sheet Boundary Layer, Chapman Conference on Ion Acceleration in the Magnetosphere and Ionosphere, Wellesley, MA, 1985.

Feblber F.S. and R. Decoste, Fast Expansions of Laser-heated Plasmas, Phys. Fluids, 21, 520-522, 1978.

Gurevich A.V., L.V. Pariiskaya and L.P. Pitaerskii, Self Similar Motion of Rarefield Plasma, Sov. Phys. JETP, 22, 44-454, 1966.

Korn P., T.C. Marshall and S.P. Schlesinger, Effects of Plasma Flow on Electrostatic Disturbances in a Q Machine, Phys. Fluids, 13, 517-526, 1970.

Samir U., K.H. Wright and N.H. Stone, The Expansion of a Plasma Into a Vacuum: Basic Phenomena and Processes and Applications to Space Plasma Physics, Rev. Geophys. Space Phys., 21, 1631-1646, 1983.

Singh N. and R.W. Schunk, Numerical Calculations to the Initial Expansion of the Polar Wind, J. Geophys. Res., 87, 9154-9160, 1982.

Singh N. and R.W. Schunk, Numerical Simulations of Counterstreaming Plasma and Their Relevance to Interhemispheric Flow, Geophys. Res., 88, 7867-7877, 1983.

Sydora R.D., J.S. Wagner, L.C. Lee, E.M. Wescott and T. Tajima, Electrostatic Kelvin-Helmholz Instability in a Radially Injected Plasma Cloud, Phys. Fluids, 26, 2986-2991, 1983.

Tariq G.F., P. Armstrong and J.W. Lowry, Electrodynamic Interaction of Ganymede with the Jovian Magnetosphere and the Radial Spread of Wake Associated Disturbances, J. Geophys. Res., 90, 3995-4009, 1985.

Wescott E.M., H.C. Stenback-Nielson, T.J. Hallinan, C.S. Deehr, G.J. Romick, J.V. Olson, J.G. Roederon and R. Sydora, A High-altitude Barium Radial Injection Experiment, Geophys. Res Lett, 7, 1037-1040, 1980.

Wescott E.M., H.C. Stenback-Nielson, T. Hallinan, C. Deehr, J. Romick, J. Olson, M.C. Kelley, R. Pfaff, R.B. Torbet, P. Newell, H. Foppl, J. Fedder and H. Mitchell, Plasma-Depleted Holes, Waves and Energized Particles from High Altitude Explosive Plasma Perturbation Experiments, J. Geophys. Res., 90, 4281-4298, 1985.

Wright K.H. Jr., N.H. Stone and U. Samir, A Study of Plasma Phenomena in Laboratory Wakes: Preliminary Results, J. Plasma Phys., 33, 71-82, 1985.

ION ACCELERATION: A PHENOMENON CHARACTERISTIC OF THE "EXPANSION OF PLASMA INTO A VACUUM"

Uri Samir

Space Physics Research Laboratory, University of Michigan, Ann Arbor, Michigan 48109
and Department of Geophysics and Planetary Science, Tel-Aviv University, Tel-Aviv, Israel

K. H. Wright, Jr.

Department of Physics, University of Alabama in Huntsville, Huntsville, Alabama 35899

N. H. Stone

Space Science Laboratory, NASA Marshall Space Flight Center, Alabama 35812

The physics of the complex process by which plasma expands across a strong density gradient is briefly discussed with particular emphasis on the resulting ion acceleration. The results from a laboratory experiment that addresses the steady-state expansion of a collisionless, streaming plasma into the wake of a body are presented. The experimental results are in good qualitative agreement with theoretical predictions, thereby demonstrating the equivalence of expansion with distance in the flow direction to time in the one-dimensional theory. On this basis, the potential application of the plasma expansion process in space plasma physics is mentioned.

Introduction

It is known from theoretical studies [e.g., Gurevich et al., 1973; Crow et al., 1975; Lonngren and Hershkowitz, 1979; Singh and Schunk, 1982, 1983] that the basic phenomena involved in the "expansion of a plasma into a vacuum" are: (1) the acceleration of ions to velocities which are far above their ambient thermal velocity; (2) the creation of a rarefaction wave which propagates into the ambient plasma at the ion acoustic speed; (3) the formation of an ion front which expands into the vacuum region; and (4) the creation of strong discontinuities in the plasma parameters as well as the creation of plasma oscillations and instabilities over certain spatial regions.

Most of the experimental work to date has focused mainly on laser plasma applications. Recently, however, plasma chamber experiments were performed by Eselevich and Fainshtein [1980], Wright et al. [1985], and Raychaudhuri et al. [1986] having different applications in mind; namely, the expansion of space plasma into the wakes of bodies in the solar system.

In this paper we briefly discuss the "expansion of a plasma into a vacuum," the different mathematical approaches used to solve the problem, present some results from a recent laboratory experiment, and mention the potential for application of the plasma expansion process in space plasma physics.

Plasma Expansion: Basic Phenomena and Mathematical Approaches

The theoretical studies regarding the expansion of a plasma into a vacuum were performed either using a self-similar approach [e.g., Gurevich et al., 1973] or a self-consistent approach [e.g., Singh and Schunk, 1982].

A detailed review regarding "plasma expansion" was given by Samir et al. [1983].

The expansion is usually treated by assuming a semi-infinite plasma bounded by a diaphragm at x = 0 as depicted in Figure 1a. At a time t = 0 the diaphragm is removed and the plasma expands into the vacuum region. Figures 1b and 1c show the evolution of density and velocity, respectively, according to the self-similar solution of the hydrodynamic equations under the assumptions of quasi-neutrality and a Boltzmann relation for electrons. As the expansion begins, the electrons move ahead of the ions, due to their larger thermal velocity, developing a space charge electric field. Some of the ions are subsequently accelerated, forming a front of plasma, called the "expansion front," that moves into the vacuum. The density of ions near this front decreases with time. In addition, a region of decreased plasma density, i.e., a "rarefaction wave," moves into the ambient (undisturbed) plasma at the ion acoustic speed. Those concepts of the expansion phenomena are discussed in Gurevich et al. [1973] and Lonngren and Hershkowitz [1979] and references therein.

Figure 2 [from a kinetic solution by Gurevich et al., 1973] shows the variation of the normalized distribution functions g_1 and g_2 with the normalized velocity u for different values of the self-similar variable τ. The plasma consists of two ionic species (O^+ and H^+) and one electron distribution (one T_e) for the case $N_{10} \gg N_{20}$ where N_{10} and N_{20} are the ambient densities of the two ionic species. The functions g and u are defined:

$$g = \left(\frac{2\pi T_i}{M_i}\right)^{1/2} \cdot N_o^{-1} \cdot f \quad ; \quad u = V\left(\frac{M_i}{2kT_e}\right)^{1/2} = \frac{V}{S_o} \cdot 2^{-1/2}$$

where f is the ion distribution function; N_o is the ambient ion density; $V = V_x$ is the velocity in the direction of the X axis (see Figure 1); S_o is the ion acoustic velocity; M_i is the ionic mass; and $\tau = (x/t) \cdot (M_i/2kT_e)^{1/2}$, where t is the time.

For $\tau < 0$, the distribution function g = F(u) is the unperturbed distribution, whereas for $\tau > 1$ the distribution narrows to a delta function-like shape, which physically implies that the ions are accelerated. For a plasma with two ionic species, such as the O^+-H^+ case shown in Figure 2 where the heavy ion is dominant, the light ions are seen to be accelerated to much higher velocities than the heavy ions. Generally, the maximum acceleration of the minor ion depends on the values of M_i, Z, T_e, the ratio

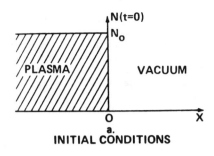

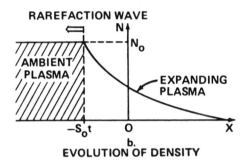

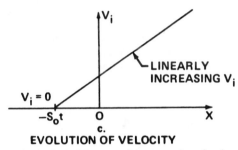

Figure 1. The basic physical features of the expansion of a plasma into a vacuum.

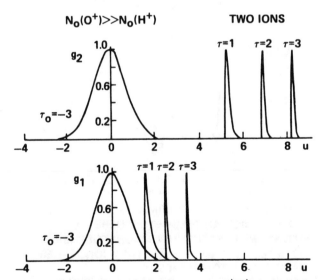

Figure 2. Examples showing the variation, in an O^+-H^+ plasma, of the normalized distribution functions (g_1, g_2) with the normalized velocity U for different values of the self-similar parameter τ. g_1 = major constituent (O^+); g_2 = minor constituent (H^+). After Gurevich et al. [1973].

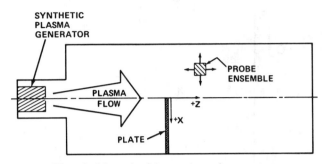

Figure 3. Schematic of the experimental arrangement.

$[R_0/\lambda_D]$ (where R_0 is a characteristic length and λ_D is the ambient Debye length), and the ratio $[N_0(O^+)/N_0(H^+)]$.

Most theoretical treatments of the plasma expansion process use the quasi-neutrality approximation with a particular variable transformation to obtain a self-similar solution; i.e., the expansion depends on the variable (x/t). Because quasi-neutrality is assumed, this solution (whether derived from a kinetic or fluid treatment) is valid only over limited ranges of x and t—the ranges in which the ions respond to the fast electrons and produce a quasi-neutral flow (see, for example, Singh and Schunk [1982]). However, Lonngren and Hershkowitz [1979] have found a variable transformation which allows a self-similar description without assuming quasi-neutrality.

Upon comparing the self-similar and self-consistent solutions one finds that some gross characteristics predicted by the self-similar approach are confirmed by the predictions of the self-consistent approach [Singh and Schunk, 1982, 1983]. In Singh and Schunk [1982, 1983], the Vlasov-Poisson equations were numerically solved. An important question which remains unsolved is the prediction of the self-similar theory regarding the existence of plasma oscillations generated by the plasma expansion process. Such oscillations were not observed in the self-consistent calculations of Singh and Schunk [1982, 1983].

Recent Results from a Laboratory Experiment

In this section we discuss some recent results from a laboratory experiment performed in a plasma chamber facility at the NASA Marshall Space

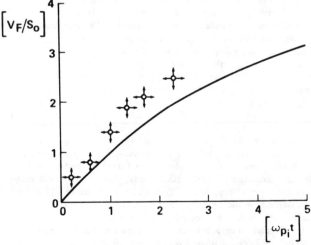

Figure 4. Variation of the normalized X-component of the ion velocity (V_F/S_0) versus normalized time ($\omega_{p_i} \cdot t$), where V_F is the velocity of the ion expansion front and $t = Z/V_{flow}$.

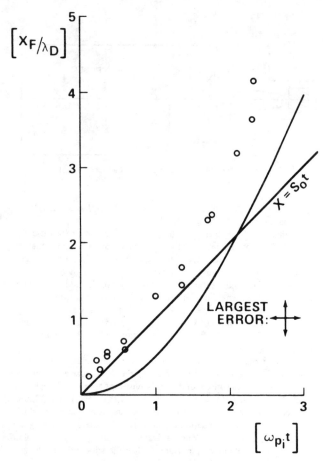

Figure 5. Variation of the normalized ion expansion front position (X_F/λ_D) versus normalized time ($\omega_{p_i} \cdot t$).

results will be relevant to the early time evolution of the expansion of plasma into the region behind the plate.

The main results from this experiment [Wright et al., 1985] show: (1) the existence of a rarefaction wave; (2) the existence of an "expansion front"; and (3) the acceleration of ions above the ambient ion acoustic speed, thus confirming some theoretical predictions [e.g., Gurevich et al., 1973; Crow et al., 1975; Singh and Schunk, 1982].

Crow et al. [1975] numerically calculated the collisionless expansion of a semi-infinite plasma under a self-consistent hydrodynamic description. They calculated the position $X_F(t)$ and the velocity $V_F(t)$ of the ion expansion front. The velocity of ions in the front was found to acheive values many times the ion acoustic speed. Approximate analytical expressions (1) and (2), given below, fit the results of Crow et al. [1975] to an accuracy of about 10% and are given in Katz et al. [1985].

$$[V_F(t)/S_o] = 2\left\{\ln(1 + \alpha\omega t) - \left(1 - \frac{0.43}{\alpha}\right)\frac{\alpha\omega t}{1 + \alpha\omega t}\right\} \quad (1)$$

$$[X_F(t)/\lambda_D] = 2\left\{(\omega t + 1/\alpha)\ln(1 + \alpha\omega t) - \omega t - \left(1 - \frac{0.43}{\alpha}\right)[\omega t - 1/\alpha \ln(1 + \alpha\omega t)]\right\} \quad (2)$$

where $\alpha = 1.6$; $X_F(t) \equiv \int V_F(t)dt$ and ω, λ_D and S_o are the ion plasma frequency, the Debye length, and the ion acoustic speed, respectively.

Figure 4 shows the normalized X-component of the ion velocity at the expansion front $[V_F/S_o]$ versus normalized time $[\omega_{p_i} \cdot t]$. The circles show the measurements and the solid line depicts the curve obtained using the approximate expression (1). As seen, the agreement between experiment and theory is qualitatively quite good.

Figure 5 shows measurements of the normalized expansion front posi-

Flight Center [Wright et al., 1985] and compare them with theoretical predictions.

The experimental arrangement is shown schematically in Figure 3. In essence, a synthetic plasma flows past a large conducting plate located $\cong 96$ cm from the source. The plate was mounted with an edge on the chamber axis and was maintained at ground potential throughout the experiment. The nominal values for the plasma properties and parameters in the laboratory experiment were: (1) drift energy, $E_i = 18$ eV; (2) ion species, N_2^+; (3) ambient density, $N_o = 1.2 \times 10^5$ cm^{-3}; (4) ambient temperature, $T_e = 0.27$ eV; (5) temperature ratio, $T_e/T_i \cong 10$; (6) ambient Debye length, $\lambda_D = 1.1$ cm; (7) ion plasma frequency, $\omega_{Pi} = 8.6 \times 10^4$ rad/sec; (8) ion acoustic speed, $S_o = 9.7 \times 10^4$ cm/sec; (9) Mach number, $M = 11.5$; (10) Φ(body) = 0; and (11) space potential, $\phi_{sp} \cong -0.5$ volts. The coordinate origin of the measurements is located at the plate edge (see Figure 3) with the positive Z axis directed downstream and the positive X axis towards the expansion region behind the plate. The probe ensemble, consisting of three instruments (a cylindrical Langmuir probe, a gridded Faraday cup, and a differential ion flux probe (DIFP)) was maneuverable in the (X,Z) plane as depicted in Figure 3. The DIFP measures the ion flow velocity in this (X,Z) plane.

The distance on the Z axis can be correlated with time by the following relation: $t = Z/V_{flow}$, where V_{flow} is the ambient plasma drift velocity. This allows comparison of the experiment with the one-dimensional, time dependent theory. For details on this variable change, see Wright et al. [1985]. A combination of plasma density and chamber length limits the experiment to an equivalent time of a few ion plasma periods. Hence, the

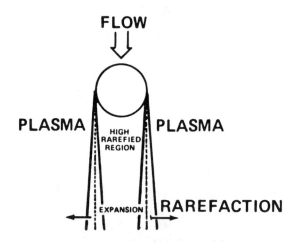

- VENUS AND MARS WITH THE SOLAR WIND.
- MOON WITH SOLAR WIND.
- COMET WITH SOLAR WIND.
- IO WITH JOVIAN MAGNETOSPHERE.
- TITAN WITH SATURNIAN MAGNETOSPHERE.
- SPACECRAFT WITH PLANETARY IONOSPHERES AND MAGNETOSPHERES.
- SPACE SHUTTLE/SPACE STATION WITH TERRESTRIAL IONOSPHERE.
- POLAR WIND EXPANSION.

Figure 6. Possible applications of the plasma expansion process in space plasma physics.

tion $[X_F/\lambda_D]$ versus normalized time $[\omega_{Pi} \cdot t]$. The straight line indicates expansion at the ion acoustic speed ($X = S_0 t$). A comparison of the experimental results with the solid curve, which represents expression (2), yields a good qualitative agreement. Hence, for both the ion expansion front velocity (X-component) and position a good qualitative agreement between the laboratory measurements and the computation of Crow et al. [1975] approximated by the analytical expressions (1) and (2) is obtained.

In summary, we have discussed the basic physical processes and phenomena involved in the expansion of a collisionless plasma into a vacuum and have shown that ion acceleration is one of the main processes involved in the expansion. We have mentioned the self-similar and the self-consistent approaches to the mathematical solution of the plasma expansion phenomena and specified the conditions for validity of the self-similar solution. We have shown via a laboratory experiment that an ion front moves into the expansion region at speeds greater than the ambient ion acoustic speed, a result which is in agreement with theoretical predictions. We submit that the process of ion acceleration via plasma expansion should not be excluded when examining ion acceleration mechanisms which operate in ionosphere-magnetosphere plasmas and particularly in high-latitude regions where a dense plasma (ionosphere) is known to expand into a less dense plasma (magnetosphere). Furthermore, as seen in Figure 6, the expansion process could also be applicable to other areas of space plasma physics.

Acknowledgments. Uri Samir acknowledges support from NASA under grant NGR-23-005-320. K. H. Wright, Jr., acknowledges support from NASA under contract NAS8-33982.

References

Crow, J. E., P. L. Auer, and J. E. Allen, The expansion of a plasma into a vacuum, *J. Plasma Phys., 14,* 65, 1975.

Eselevich, V. G., and V. G. Fainshtein, Expansion of collisionless plasma in a vacuum, *Sov. Phys. JETP Engl. Transl., 52,* 441, 1980.

Gurevich, A. V., L. V. Pariskaya, and L. P. Pitaevsky, Ion acceleration upon expansion of a rarefied plasma, *Sov. Phys. JETP Engl. Transl., 36,* 274, 1973.

Katz, I., D. E. Parks, and K. H. Wright, Jr., A model of the plasma wake generated by a large object, *IEEE Trans. Nuclear Sci., NS-32,* 4092, 1985.

Lonngren, K. E., and N. Hershkowitz, A note on plasma expansion into a vacuum, *IEEE Trans. Plasma Sci., PS-7,* 107, 1979.

Raychaudhuri, S., J. Hill, H. Y. Chang, E. K. Tsikis, and K. E. Lonngren, An experiment on the plasma expansion into a wake, *Phys. Fluids, 29,* 289, 1986.

Samir, U., K. H. Wright, Jr., and N. H. Stone, The expansion of a plasma into a vacuum: Basic phenomena and processes and applications to space plasma physics, *Rev. Geophys. Space Phys., 21,* 1631, 1983.

Singh, N., and R. W. Schunk, Numerical calculations relevant to the initial expansion of the polar wind, *J. Geophys. Res., 87,* 9154, 1982.

Singh, N., and R. W. Schunk, Expansion of a multi-ion plasma into a vacuum, *Phys. Fluids, 26,* 1123, 1983.

Wright, K. H., Jr., N. H. Stone, and U. Samir, A study of plasma expansion phenomena in laboratory wakes: Preliminary results, *J. Plasma Phys., 33,* 71, 1985.

Section VII. MICROSCOPIC PROCESSES

ION ACCELERATION BY WAVE-PARTICLE INTERACTION

Robert L. Lysak

School of Physics and Astronomy, University of Minnesota, Minneapolis, MN 55455

Abstract. Ions in the auroral zone are observed to have pitch angle distributions peaked at large angles to the magnetic field. The most likely energization mechanism for such ions is an interaction between the ions and waves at the ion gyroresonance. This paper will review the theory of perpendicular ion heating by waves, considering electrostatic ion cyclotron waves, as well as waves on the whistler resonance cone. When the wave amplitudes are weak and the spectrum broad-banded, the quasi-linear theory can predict the magnitude of heating, if the wave spectrum and distribution in space are known. For large amplitude, narrow-banded waves, stochastic heating mechanisms can rapidly accelerate ions transverse to the magnetic field. These theoretical models can be compared with results found in plasma simulations as well as with observations in space. In particular, the relative heating of different ion species will be considered and compared with observations in the auroral zone.

Introduction

In the past ten years or so it has become realized that ions of ionospheric origin are a major and sometimes the dominant constituent of the terrestrial magnetosphere. Under quiet conditions, ionospheric ions generally have energies much below escape energy and so some acceleration of these ions is necessary to account for their presence in the magnetosphere. Observations from the S3-3 satellite [Shelley et al., 1976; Sharp et al., 1977; Ghielmetti et al., 1978; Gorney et al., 1981] as well as satellites such as ISIS [Klumpar, 1979] and rocket flights [e.g., Whalen et al., 1978; Yau et al., 1983] have confirmed the existence of upflowing ions in the auroral zone and have shown that typically the ion pitch angle distribution has a peak at a pitch angle between 90° and 180°, indicating that tranverse energization of the ions has taken place. Ions that are accelerated transverse to the magnetic field at low altitudes are then accelerated upwards due to the magnetic field gradient, leading to these so-called conic distributions.

Statistical studies from satellite data show that these conic distributions occur at all local times in the auroral zone and in both upward and downward current regions. In addition the rocket data show that transverse acceleration of ions must occur at altitudes as low as 500 km [Yau et al., 1983]. Thus a mechanism which attempts to explain this acceleration must be present all around the auroral zone and must occur, at least occasionally, at ionospheric altitudes. While a number of candidate mechanisms have been proposed, it is difficult to find a single mechanism which can account for all the observations. In addition, mass spectrometer data show that oxygen ions are energized one to two times more than protons [Collin et al., 1981], so a candidate mechanism must be somewhat more efficient on heavy ions. The existence of oxygen conics at high altitudes also requires energization at low altitudes, since low energy upflowing oxygen is susceptible to charge exchange with neutral hydrogen unless it is sufficiently energized in the region in which neutral oxygen dominates neutral hydrogen [Moore, 1980].

This paper will review the theory of perpendicular ion acceleration, using both quasi-linear theory and stochastic acceleration theories. These mechanisms will then be applied to auroral zone conditions, with particular emphasis on acceleration of heavy ions and acceleration at low altitudes. Heating by waves in three frequency ranges will be considered: heating by Alfven waves at frequencies below all ion gyrofrequencies, heating by ion cyclotron waves, and heating by lower hybrid waves on the whistler resonance cone.

Theory of Ion Heating by Electrostatic Waves

A. Quasi-Linear Theory

For broad-banded, weak turbulence, wave-particle interactions may be described by a diffusion approach, in which the particles randomly walk in phase space, slowly diffusing from their unperturbed orbits. This approach may be characterized as quasi-linear since the linear particle orbits are used to calculate the non-linear effects [see, e.g., Davidson, 1972]. Such an approach is valid if the wave turbulence is weak, with wave energy density $|E|^2/8\pi nT \ll 1$, and in most cases $e\Phi/T < 1$. In addition, for a random walk approach to be valid, a particle should see a random wave field, i.e., the particle's autocorrelation time with the wave field should be shorter than the time scale of the diffusion or of the wave growth. The autocorrelation time depends on the bandwidth of the spectrum in wavenumber space and on the difference between the resonant velocity and the group speed of the wave,

$$\tau_c = 1/\Delta k_{||}(v_{res} - V_g). \quad (1)$$

Under these assumptions we can separate the distribution function into a slowly varying portion and an oscillating part, $f = \langle f \rangle + f_1$, and write the time-averaged Vlasov equation as a diffusion equation:

$$\frac{d\langle f \rangle}{dt} = \frac{\partial}{\partial \underline{v}} \cdot \underline{D} \cdot \frac{\partial \langle f \rangle}{\partial \underline{v}} \quad (2)$$

where d/dt represents the total time derivative along the unperturbed orbit and the diffusion tensor $\underline{D} \propto |E_1|^2$.

In the consideration of ion heating, the most important part of the diffusion equation is the part associated with the perpendicular velocity. Neglecting the parallel terms and using cylindrical coordinates in velocity space yields the perpendicular diffusion equation:

$$\frac{d\langle f \rangle}{dt} = \frac{1}{v_\perp} \frac{\partial}{\partial v_\perp} D_\perp(v_\perp,t) v_\perp \frac{\partial \langle f \rangle}{\partial v_\perp} \quad (3)$$

When $D_\perp$ can be expressed as a simple power law in velocity, a self-similar solution can be found to the diffusion equation [Dum, 1978]. For $D_\perp = D_0 v_\perp^{-p}$ the distribution function evolves to the "flat top" form:

$$\langle f \rangle \propto e^{-(v/v_0)^{p+2}} \quad (4)$$

where the effective thermal velocity v_0 evolves in time so that the effective perpendicular temperature satisfies:

$$\frac{d}{dt}\left[\frac{T_\perp}{T_{\perp_0}}\right]^{(p+2)/2} = (p+2)^2 D_0(t) \quad (5)$$

For perpendicular diffusion in quasi-linear theory the diffusion coefficient takes the form:

$$D_\perp = \sum_\ell \int \frac{d^3k}{(2\pi)^3} \frac{\ell^2 e^2 |E_1|^2}{m^2 \lambda^2} R(\omega - k_{||} v_{||} - \ell\Omega) J_\ell^2(\lambda) \quad (6)$$

where $\lambda = k_\perp v_\perp/\Omega$, Ω is the ion gyrofrequency, and R is a sharply peaked resonance function. In the limit where λ becomes large, the Bessel function can be expanded in an asymptotic form:

$$J_\ell(\lambda) \simeq \sqrt{2/\pi\lambda} \cos(\lambda - \frac{\pi}{2}(\ell - \frac{1}{2})) \quad (7)$$

Thus when λ is large, the cosine term averages to one half and the resulting diffusion coefficient exhibits a $v_\perp^{-3}$ dependence. This leads to a distribution function of the form $f \propto \exp(-(v_\perp/v_0)^5)$ and a heating rate which, for constant wave amplitude, goes like $T \propto t^{0.4}$. Thus slower than exponential growth is predicted for the wave amplitude. Such algebraic growth in the ion temperature was observed in simulations by Okuda and Ashour-Abdalla [1981] who also found the flattening of the ion distribution function to the self-similar form. Dusenbery and Lyons [1981] also found such flattening to occur by direct integration of the quasi-linear diffusion equation.

B. Stochastic Heating Models

When one wave mode dominates the wave field, ions may interact strongly with this mode and will therefore deviate significantly from their unperturbed trajectories. In this case ion trapping in the wave may occur. In general, the ion trapping orbits will limit the energy achievable by the ion; however, the interactions of the ions with other wave modes may allow ions to jump from one trapping orbit to another, leading to an irreversible increase in their energy. Even in the case of a single wave propagating primarily perpendicular to the magnetic field, stochastic heating can occur due to the ion interacting with the wave at more than one gyroharmonic. Stochastic heating models using lower hybrid waves were considered in a fusion context by Aamodt [1970] and Karney and Bers [1977]. The observations of narrow-banded electrostatic ion cyclotron waves by Mozer et al. [1980] motivated the application of this theory to the auroral zone by Lysak et al. [1980] and Papadopoulos et al. [1980].

The starting point for such a model is the Hamiltonian of an ion moving in a background magnetic field in the presence of an electrostatic wave field. Upon performing a canonical transformation to the action-angle variables in the background magnetic field, this Hamiltonian becomes:

$$H = \frac{p_z^2}{2m} + p_\phi \Omega + \sum_{\underline{k}} e\Phi_{\underline{k}} \cos[k_\perp(Y + r\sin\phi) + k_{||}z - \omega_{\underline{k}}t]$$

$$= \frac{p_z^2}{2m} + p_\phi \Omega + \sum_{\underline{k}} e\Phi_{\underline{k}} \sum_\ell J_\ell(k_\perp r) \cos[k_\perp Y + k_{||}z + \ell\phi - \omega_{\underline{k}}t] \quad (8)$$

where the canonical coordinates are the position z, the guiding center coordinate Y and the gyrophase ϕ and the corresponding momenta are the parallel momentum p_z, and guiding center position $m\Omega X$ and the gyromomentum p_ϕ, which is simply proportional to the magnetic moment μ of the ion. The gyroradius r is simply related to the gyromomentum with $p_\phi = m\Omega r^2/2$. The second form above comes from using a Bessel function identity to decompose the Hamiltonian in the gyroharmonics.

When one wave, say $\underline{k}_o$ is dominant, and the wave frequency is near a gyroharmonic $\omega_{\underline{k}_o} \simeq n\Omega$, the term with $\underline{k} = \underline{k}_o$ and $\ell = n$ above is most important and the Hamiltonian can be written:

$$H = \frac{p_z^2}{2m} + p_\phi \Omega + \quad (9)$$

$$e\Phi_{\underline{k}_o} J_n(k_{\perp o} r) \cos(k_{\perp o} Y + k_{\parallel o} z + n\phi - \omega_{\underline{k}_o} t)$$

$$+ e\Phi_{\underline{k}_o} \sum_{\ell \neq n} J_\ell(k_{\perp o} r) \cos(k_{\perp o} Y + k_{\parallel o} z + \ell\phi - \omega_{\underline{k}_o} t)$$

$$+ \sum_{\underline{k} \neq \underline{k}_o} e\Phi_{\underline{k}} \sum_\ell J_\ell(k_\perp r) \cos(k_\perp y + k_\parallel z + \ell\phi - \omega_{\underline{k}} t)$$

where the first two lines represent the trapping of the particle by the dominant harmonic of the dominant wave, the third line gives the interaction with other harmonics of the main wave and the final line gives the interaction with other waves in the system.

The trapping interaction is particularly simply to analyze in the case where $k_\parallel = 0$ and $\omega = n\Omega$. Then the parallel motion of the ion is decoupled and the time dependence can be removed by means of a transformation to the slowly varying gyrophase $\psi = \phi - \Omega t$, which eliminates the $p_\phi \Omega$ term from the Hamiltonian. The change in perpendicular energy is then given by:

$$\dot{p}_\phi = \frac{\partial H}{\partial \psi} = -n e\Phi_{\underline{k}_o} J_n(k_{\perp o} r) \sin(k_{\perp o} Y + n\psi) \quad (10)$$

Squaring this result and comparing with the Hamiltonian gives a harmonic oscillator equation:

$$\dot{p}_\phi^2 - n^2 e^2 \Phi_{\underline{k}_o}^2 J_n^2(k_{\perp o} r) = -H^2 \quad (11)$$

Thus the particle moves in perpendicular energy in a potential which goes like minus the square of the Bessel function with an energy which is non-positive. In this case, then, ions are trapped in potential wells which are limited by the zeroes of the Bessel function and cannot increase their energy indefinitely.

Deeply trapped ions in this potential well will undergo a trapping oscillation at a frequency which is given by $\omega_{T\perp} = A_n k_{\perp o} cE_{ko}/B_o$ where the proportionality constant A_n is given by:

$$A_n = \frac{n J_n(x_n)}{x_n} \left[1 - \frac{n^2}{x_n^2} \right]^{1/2} \quad (12)$$

where x_n is the first maxima of the nth order Bessel function. In the case of oblique propagation, this trapping frequency is modified by the parallel motion of the ions and becomes:

$$\omega_T^2 = \omega_{T\perp}^2 \mp \frac{e\Phi k_\perp^2}{m} J_n(\sqrt{2p_\phi}) \quad (13)$$

where $\omega_{T\perp}$ is the perpendicular trapping frequency given above.

When the contributions from other harmonics are included (the second line in equation (9)), the harmonics of the trapping frequency may resonate with the frequencies of these other terms, leading to the overlapping of the islands and the onset of stochastic behavior. Thus the ion may then move onto islands of increasing energy. The conditions for the onset of this stochasticity have been examined numerically by Karney [1978] and Papadopoulos et al. [1980]. It was found that the stochastic region has both a lower and an upper bound. The threshold level for stochastic onset is given by:

$$\frac{e\Phi k_\perp^2}{m\Omega^2} = \frac{\delta}{n} \frac{k_\perp r}{|J'(k_\perp r)|} \quad (14)$$

where the right-hand side is evaluated at the zero of J_n and where δ is a factor determined numerically by Karney to be about .25. For the first three harmonics the right-hand side of this equation has the values 2.38, 1.89 and 1.78, implying rather high wave amplitudes for auroral parameters. For n large, the first zero of J_n occurs for $k_\perp r \simeq n$ and $J'_n \simeq -n^{-2/3}$. This leads to an onset condition $\alpha = 0.25 n^{2/3}$ or in dimensional units, setting $n = \omega/\Omega$.

$$\frac{cE_\perp}{B} = \delta \frac{\omega}{k_\perp} \left[\frac{\Omega}{\omega} \right]^{1/3} \quad (15)$$

Note that there is a $m^{-1/3}$ dependence which ostensibly favors higher mass ions; however, it should be noted that only ions with perpendicular velocities the order of the wave phase velocity enter the stochastic region. Thus, e.g., in a hydrogen cyclotron wave, which typically has $\omega/k_\perp \simeq v_H$, the thermal speed of hydrogen, an oxygen ion would have to have an energy 16 times the hydrogen temperature to be heated stochastically.

The stochastic region extends up to a point where the condition (eq. 14) given above no longer holds. In this regime, where we can assume $k_\perp r \gg n$, the Bessel function can be expanded for a large argument and we find:

$$\alpha = \frac{\delta}{n} \left[\frac{\pi k_\perp r^3}{2} \right]^{1/2} \quad (16)$$

which allows the ion to be accelerated up to a perpendicular velocity

$$v_\perp = \left[\delta \frac{cE_\perp}{B} \right]^{2/3} \left[\frac{\omega^2}{\Omega k_\perp} \right]^{1/3} \quad (17)$$

Thus the maximum velocity goes like $m^{1/3}$ and the energy achieved by this mechanism is proportional to $m^{5/3}$.

The maximum heating rate by this mechanism may be estimated by considering the asymptotic form of the potential given by equation (12) for $k_\perp r \gg 1$. The heating rate for these ions in the tail will then be given by:

$$\frac{\partial T_\perp}{\partial t} \lesssim \left[\frac{\Omega}{k_\perp}\right]^{3/2} \left[\frac{2m}{\pi^2 T_\perp}\right]^{1/4} \quad (18)$$

so that the heating rate goes like $m^{-5/4}$. Thus the stochastic heating mechanism has mixed consequences for heavy ions. While stochastic heating of heavier ions has a lower amplitude threshold and can achieve higher energies, a pre-heating mechanism is required for these ions to enter the stochastic region and the time scale for the acceleration is slower than for the lighter ions.

Application to Auroral Zone Heating

In this section, we shall apply the theory of wave particle interactions presented above to auroral zone phenomena. We will first examine the propagation of heated ions through the auroral zone mirror field in order to clarify some of the macroscopic aspects of the auroral zone heating problem which are independent of the specific heating mechanism. Next, the relevant wave modes which can be observed in the auroral zone will be examined. Beginning at low frequencies, we will consider heating by quasi-static electric fields such as oblique electrostatic structures. Then the ion cyclotron modes which can be driven by electron drifts or by ion beams will be considered. Finally, we will discuss waves driven by the auroral electron beam, which are primarily lower hybrid waves on the whistler resonance cone but which also can include electromagnetic ion cyclotron waves propagating between the ion cyclotron frequencies of various ion species. We shall then summarize the current status of our theoretical understanding of auroral zone ion heating, with particular emphasis being placed on ion heating at low altitudes in the topside atmosphere.

A. Ion Propagation in a Mirror Field

If an ion which is in equilibrium with the gravitational and electric fields at a certain location along the field line experiences an acceleration perpendicular to the magnetic field, it will move up the field line due to the enhanced magnetic mirror force associated with its newly increased magnetic moment. Lysak et al. [1980], Chang and Coppi [1981], Gorney et al. [1985] and others have analyzed the motion of such ions in the mirror geometry. The ion equations of motion can be written in the form:

$$\dot{W}_\perp = \dot{W}_{\perp turb} - \frac{3W_\perp v_\parallel}{R} + \frac{e^2 E_\perp}{m\Omega^2} v_\parallel \frac{\partial E_\perp}{\partial z} \quad (19)$$

$$\dot{v}_\parallel = \frac{3W_\perp}{mR} + \frac{eE_\parallel}{m}$$

where the three terms in the first equation correspond to the turbulent heating, the loss of perpendicular energy due to the mirror force, and the perpendicular acceleration due to convection into a quasi-static electric field (see below), and the terms in the parallel equation describe acceleration due to the mirror force and the parallel electric field, respectively. A magnetic field with a R^{-3} radial dependence has been assumed.

In the absence of electric fields and turbulent heating, these equations describe the adiabatic mapping of ions along field lines which is used to determine the source altitude of ion conic distributions. Inclusion of the turbulent heating term was discussed by Lysak et al. [1980] and Chang and Coppi [1981] to describe heating in quasi-linear theory due to electrostatic ion cyclotron waves and lower hybrid waves, respectively. In these works, since heating occurs in an extended region along the background field, a source altitude inferred from the adiabatic assumption would only be an upper limit. A region of extended heating could moreover be distinguished from more localized heating by the sharpness of the pitch angle distribution. Retterer et al. [1983] have generalized the above works by performing test particle calculations where the assumptions made in previous works, e.g., quasi-linear heating rates assuming Maxwellian distributions, could be relaxed to obtain simulated particle distributions which can then be compared with space observations.

To compare heating rates such as those calculated in the previous section of this review with typical energies of observed ions, it is instructive to consider only the first term on the right hand side of equations (19) and (20). This approach, neglecting the effect of electric fields and also neglecting the loss of perpendicular energy due to the mirror force, allows a simple form for the energy achieved to be calculated, which is valid as long as the region in which the heating takes place is localized along the field line. If we assume the ion energy increases as $W_\perp = W_{\perp_o}(1 + \alpha t^\gamma)$, the energy achieved in a heating region of length L will be given by:

$$\frac{W_\perp}{W_{\perp_o}} = \left[\frac{mR}{3} \alpha^{2/\gamma} (\gamma+1)(\gamma+2)L\right]^{\gamma/(\gamma+2)} \quad (20)$$

Special cases include quasi-linear heating with an assumed Maxwellian distribution, which gives γ

= 0.5, self-similar quasi-linear heating, which modifies this to $\gamma = 0.4$, and the maximum stochastic heating rate giving $\gamma = 0.8$. Note that the propagation effect itself favors the heavy ions, since, for the same heating rate, a heavier ion will spend a longer time in the heating region. The mass dependence goes as $m^{\gamma/(\gamma + 2)}$, where the exponent ranges from 0.11 to 0.29 for $\gamma = 0.4$ to 0.8, leading to an enhancement in the ion energy of a factor of 1.36 to 2.21 for oxygen compared to hydrogen. It is interesting to note that ion observations in the auroral zone indicate an enhancement of about a factor of 2 in the ion energy [Collin et al., 1981; Sharp et al., 1983] so that these observations would be consistent with comparable intrinsic heating rates for oxygen and hydrogen combined with the propagation effect.

In the presence of upward parallel electric fields, the conic distributions are modified into more beam-like distributions, assuming the parallel energy gained is large compared to the perpendicular energy [Lysak et al., 1980]. Perpendicular heating of these beam distributions is evidenced by their observed width in velocity space perpendicular to the magnetic field, which typically implies a perpendicular temperature of about 50-100 eV [Kaufmann and Kintner, 1982]. This would indicate that the perpendicular heating took place first and then the distribution was accelerated through the parallel electric field. Evidence for this scenario was presented by Klumpar et al. [1984] in a rare case where the perpendicular and parallel accelerations were comparable, and a distribution intermediate between a conic and a beam was observed. Acceleration of ions in electrostatic fields will tend to mask differences in energy between ions of different mass since all ions will gain the same parallel energy.

The role of downward electric fields in trapping ions in the heating region was investigated by Gorney et al. [1985], who found that higher ion energies could be achieved if downward electric fields were present to offset the mirror force. Such a "pressure cooker" effect might be expected in downward current regions where such electric fields are present and where ion conics are often observed in the absence of wave turbulence intense enough to account for the ion distributions in a single pass of ion heating. Upgoing electron beams of up to 100 eV have been observed at 1400 km altitude by Klumpar and Heikkila [1982], indicating the presence of such downward electric fields. This effect would be independent of mass since the ratio of the mirror to electric forces depends only on the ion perpendicular energy and not on mass.

B. Quasi-Static Electric Fields

Large amplitude localized electric fields mainly perpendicular to the background magnetic field were observed on the S3-3 satellite [Mozer et al., 1977] and have been since seen on satellites such as S3-2, ISEE-1 and Dynamics Explorer. These fields can have magnitudes over 500 mV/m and correspond to potentials of a few kilovolts and so are attractive candidates for ion acceleration. On the other hand, as is well known, a uniform static perpendicular electric field does not lead to any ion energization, but only to the E x B drift. Time-varying or localized electric fields, however, can accelerate ions by means of the polarization drift, which occurs when ions see a changing electric field. In such a field the ion is accelerated in the direction of the changing electric field with a velocity:

$$v_\perp = \frac{mc^2}{qB^2} \frac{dE_\perp}{dt} \quad (21)$$

$$= \frac{mc^2}{qB^2} \left[\frac{\partial}{\partial t} + v_\parallel \frac{\partial}{\partial z} + v_\perp \frac{\partial}{\partial x} \right] E_\perp$$

where in the second line the convective derivative has been written explicitly. Taking these three terms in reverse order, the perpendicular part of the convective derivative will average out to zero provided the scale length of the electric field variation is large compared to the ion gyroradius. A short scale electric field may be able to accelerate particles with large gyroradii by this mechanism. However, if the scale length is short enough, even cold ions may be accelerated by this mechanism. In an electric field gradient, a particle's gyrofrequency will be changed [Cole, 1976] to:

$$\omega_c^2 = \Omega^2 - \frac{q}{m} \frac{\partial E_\perp}{\partial x} \quad (22)$$

When the gradient is steep enough, the frequency of gyration can decrease to zero and becomes imaginary, indicating that the ion gyromotion has been disrupted by the electric field gradient and ions are then free to be accelerated across the magnetic field. Evidently this can occur if

$$\frac{\partial E_\perp}{\partial x} > \frac{m\Omega^2}{q} = \frac{qB^2}{mc^2} \quad (23)$$

In a magnetic field of .1 G, an electric field gradient of 10 mV/m^2 is required to accelerate protons by this mechanism. Note here that the condition is more easily met for heavier ions than for protons. The maximum ion energy achievable by this process is the energy in the E x B drift of the ions $W = mc^2E^2/B^2$, again favoring the heavy ions.

In an oblique electrostatic field, the parallel motion of the ion can transport it into a region of increasing electric field. In crossing the structure, the ion gains an energy equal to the potential drop, but the important consideration as far as ion conic generation is concerned is the final pitch angle of the particle. Borovsky [1984] has shown that the pitch angle depends on the crossing time τ of the ion through the electrostatic field and the degree of magnetization of the ion species. For heavier ions, the magnetic field becomes less important and the velocity becomes aligned with the electric field while lighter ions which can gyrate many times in a crossing of the potential structure tend to stay aligned with the magnetic field.

Greenspan [1984] has shown by test particle models that a distribution of particles incident on an oblique structure can exhibit phase bunching upon exitting the structure. Such a model could account for observations of phase bunched ions seen at high altitudes near large electrostatic fields observed on ISEE by Cattell et al. [this meeting].

Acceleration by wave fields with finite frequency but lower than the ion gyrofrequency has been considered by Roth and Temerin [this meeting]. In their model a wave packet with low frequency and short perpendicular wavelength is able to accelerate ions efficiently. In particular, a subharmonic resonance in the ion motion occurs when the wave frequency $\omega = \Omega/2$ and strong energization of ions can take place. Thus if low frequency Alfvenic turbulence at frequencies below, say, the oxygen gyrofrequency is found at low altitudes, ions may be weakly accelerated and move up the field line until they reach the location where this resonance condition is met. At this location the ions would become more strongly heated and lead to a conic distribution in which the ion distribution would appear to have been accelerated at about the same altitude. This might explain low altitude acceleration of heavy ions since this resonance condition would be met at lower altitudes the heavier the ion.

In general electrostatic acceleration of ions can be particularly effective for heating cold, heavy ions. The difficulty comes in the relative rarity of large amplitude, short wavelength quasi-static perpendicular electric fields. Such fields have sometimes been observed at ionospheric altitudes where energization of oxygen is desired [Maynard et al., 1982], but it has been argued by Temerin et al. [1981a] that such low altitude fields are generally not electrostatic. Even at higher altitudes, such fields are in general restricted to the region 1 current system on the poleward side of the auroral zone, whereas conic ion distributions are seen throughout the auroral zone.

C. Electrostatic Ion Cyclotron Turbulence

The electrostatic ion cyclotron (EIC) instability has been perhaps the most popular candidate for auroral zone wave-particle interactions since it was identified by Kindel and Kennel [1971] as the mode most easily destabilized by field-aligned currents. Palmadesso et al. [1974] investigated the ion heating in the quasi-linear theory, but S3-3 observations showing EIC waves to be very narrow banded [Kintner et al., 1978] led to a number of investigations of EIC heating using stochastic heating theories [Lysak et al., 1980; Papadopoulos et al., 1980; Singh et al., 1981, 1983] and particle simulations [Ashour-Abdalla et al., 1981; Okuda and Ashour-Abdalla, 1981, 1983]. The major difficulty in invoking current-driven EIC waves for ion heating lies in the rarity of observations of EIC waves in all regions of the auroral zone where conic ions are seen. Kintner et al. [1979] showed that EIC observations were in fact correlated with ion beams rather than conics, occurring mainly in regions of upward field-aligned current above 5000 km. Thus low altitude energization of ions in both up and down current regions cannot be explained easily by this mode.

Theoretically, the difficulty in exciting EIC waves at low altitudes arises from the fact that the instability requires a relative drift velocity between electrons and ions which is roughly one third of the electron thermal speed. At low altitudes where the plasma density is high, this corresponds to a rather large critical current. In an attempt to determine where the excitation of EIC waves was most likely, Lysak and Hudson [1979] modeled the density profile based on S3-3 measurements of the lower hybrid frequency and found that for a given level of field-aligned current the EIC instability was most likely at altitudes over 5000 km. To excite these waves at lower altitudes would require either a density cavity extending to very low altitudes, such as has been suggested by Calvert [1981], or else very intense currents. Observations from DE have suggested that very narrow current structures with current magnitudes the order of 100 $\mu A/m^2$ may be present [Heelis et al., 1984]. Such current structures may be capable of exciting EIC waves but to date no study has been done to confirm the excitation of such turbulence at low altitudes.

The EIC mode is well suited for ion heating since the wave has a frequency and wavelength comparable to the ion gyrofrequency and gyroradius in the fundamental mode. Thus thermal ions can be easily energized by EIC waves of the same type (i.e., oxygen in oxygen waves, hydrogen in hydrogen waves), making this mechanism ideal for the first stage heating of cold ionospheric ions. Under special circumstances, heating of unlike ions can also occur. Heavy ions can be heated in a light ion cyclotron wave, but only if the heavy and light ion thermal speeds are comparable, i.e., the heavy ion must have a higher initial temperature than the light ion [Singh et al., 1983]. Nishikawa et al. [1983, 1985] have shown that indirect heating of oxygen ions by hydrogen cyclotron waves can be caused by a parametric decay to an oxygen cyclotron wave. The converse case, heating of light ions in a heavy ion cyclotron wave, might appear not to work well, but Nishikawa and Okuda [1985] have shown that under some circumstances, the oscillating drifts of the plasma in response to a heavy ion cyclotron wave can in turn excite the EIC instability of the light ion. While these mechanisms may be plausible, it appears that the EIC heating mechanism will primarily heat the ion species which is responsible for the type of EIC wave excited.

D. Lower Hybrid Resonance Cone Heating

The auroral electron beam at altitudes below the primary acceleration region can give rise to waves on the resonance cone both for lower hybrid waves [e.g., Maggs, 1976] and Alfven-ion cyclotron waves [Temerin and Lysak, 1984]. These waves can interact with ions at higher harmonics of the ion cyclotron frequency, which in many cases may be described using an unmagnetized model for the ions. This heating mechanism favors lighter and hotter ion species, since these waves have phase velocities which are generally larger than the thermal speed of the lightest ion in the system. Thus the heavier ions would require some amount of preheating before they could become resonant with these waves. In addition, the heating rate for ions in lower hybrid waves is inversely proportional to the mass, again favoring lighter ions. On the other hand the lower hybrid wave turbulence is seen throughout the auroral zone, and is often associated with ion conics [Temerin et al., 1981b]. Bite-outs in the wave spectrum at harmonics of the hydrogen cyclotron frequency have also been observed [Gorney et al., 1982], confirming the existence of the ion-wave interaction due to these waves.

Chang and Coppi [1981] and Retterer et al. [1983] have investigated lower hybrid wave heating using a test particle model to treat the motion of ions up the field line. Their results show that an extended region of lower hybrid turbulence can heat ions to kilovolt energies, starting with an initially warm distribution. Thus lower hybrid wave heating may be in large part responsible for the energies of ions observed in the auroral zone. This heating, however, must be preceded by some other energization mechanism to create a population of particles energetic enough to interact with the lower hybrid wave.

The difficulty of heating thermal ions with lower hybrid waves has led Retterer and Chang and Koskinen [this meeting] to investigate nonlinear wave-wave interactions driven by a lower hybrid wave pump. It was found by these authors that the excitation of sidebands with lower phase velocity could create a superthermal tail out of the main distribution which could then be heated directly by the lower hybrid pump.

Another source of lower hybrid heating was suggested by Roth and Hudson [1985] who invoked ion ring distributions as are seen in the cusp instead of an electron beam as the source of free energy for lower hybrid waves. A two stage heating process was observed in simulations, with a trapping regime occurring first when the most linearly unstable mode dominates followed by a diffusive regime in which quasi-linear processes occur. It was found that hydrogen and oxygen heating could occur, with oxygen heating enhanced in the case of a ring of solar wind α particles.

Summary and Conclusions

A large number of candidate mechanisms have been proposed in the last few years to account for the large perpendicular acceleration of ions in the auroral zone. While observational evidence has motivated and justified some of the models, it is fair to say that no one mechanism appears capable of producing all the desired features of ion conic distributions. It appears likely that at least a two stage process of ion heating would be required to account for the observations. The major problem appears to be accounting for the initial acceleration of cold ionospheric plasma at low altitudes to energies of tens of eV. Once such a warm ion population is available, it can move up the field line under the influence of the mirror force into the primary acceleration region, where acceleration to energies in the kilovolt range becomes possible.

Ion acceleration must take place at altitudes as low as 400 km [Yau et al., 1983]. At these altitudes one would require large current densities to explain excitation of EIC instabilities which could preheat the ambient (mostly oxygen) ions. Observations of oxygen EIC waves are difficult due to Doppler shifts, and only a few possible observations have been claimed [Bering et al., 1975; Bering, 1982; Yau et al., 1983; Kintner and Gorney, 1984]. Large amplitude quasi-static fields are sometimes observed [Maynard et al., 1982] and a scenario in which electrostatic fields extend to low altitudes in the ionosphere was outlined by Borovsky [1984]. Evidence from S3-3 indicates, however, that low altitude electric fields are rather turbulent [Temerin et al., 1981a] and so an electrostatic description appears inappropriate for these low altitude fields. Such turbulent low frequency fields may accelerate ions at half (or perhaps other fractions) of the ion gyrofrequency as proposed by Roth and Temerin [this meeting]. In this scenario ion heating may proceed at a low rate until the subharmonic resonance is reached, at which point a large increase in the perpendicular ion energy could be achieved. In a reasonably monochromatic wave field, all ions of a certain species would appear to have been accelerated at roughly the same altitude. In regions where electron beams are incident on the low altitude ionosphere, the lower hybrid parametric decay process might be able to energize cold ions. Observations of lower hybrid wave turbulence intense enough to exceed the threshold for the decay process at low altitudes have not so far existed. If any of these mechanisms are operative, they may be aided by the pressure cooker effect of Gorney et al. [1985], which may in particular aid in the production of conics in downward current regions where downward electric fields and current-driven EIC turbulence are likely to occur.

The fate of ions which escape the low altitude ionosphere will depend in large measure on the direction of the field aligned current. In regions of downward field aligned current, ion energies may remain lower since it is likely only weak EIC turbulence may be present. In such current regions, it may be the case that the strength of the downward electric field may determine the ultimate ion energy by extending the ion's residence time in the heating region. In upward current regions, lower hybrid turbulence driven by the electron beam and large quasi-static electric fields ("electrostatic shocks") will prove effective at further energization of ions which have been preheated at low altitudes. The intense EIC turbulence which appears to be driven by ion beams may also play a role in the acceleration of the ambient ions. In this regard, it is interesting to note that recent observations [Shelley, this meeting] have indicated that on the average hydrogen ions have slightly less energy than the upward potential drop while oxygen ions have slightly more, indicating that the faster hydrogen beam may excite an instability which could transfer energy to the oxygen ions. In any case, in the primary auroral acceleration region above 5000 km in upward current regions, it is difficult to distinguish observationally between various mechanisms, since EIC waves and large electrostatic fields are generally seen together with each other.

A study of ions returning to the ionosphere has indicated that downward flowing ions are generally only found at the equatorward edge of the auroral zone [Ghielmetti et al., 1979]. This would indicate that an ion's motion in the outer magnetosphere is not adiabatic and is further modified by wave-particle interactions as it propagates. It is interesting to note that both continued perpendicular heating as well as the excitation of waves at the Landau resonance by ion beams would tend to keep ions trapped near the equator, in the first case by increasing the first adiabatic invariant and in the second by decreasing the second invariant. In any case, it appears that ion energization by the wave-particle interaction is an essential feature in the population of the magnetosphere by ionospheric ions.

Acknowledgments. Much of the work presented in this review was done in collaboration with M. K. Hudson. The author would also like to acknowledge discussions with many participants at this meeting which clarified the issues presented here. The work was supported in part by NSF grants ATM-8218128 and ATM-8508949.

References

Aamodt, R. E., Nonlinear evolution of multiple-finite amplitude resonant loss cone modes, Phys. Fluids 13, 2147, 1970.

Ashour-Abdalla, M., H. Okuda, C. Z. Cheng, Acceleration of heavy ions on auroral field lines, Geophys. Res. Lett., 8, 795, 1981.

Bering, E. A., M. C. Kelley and F. S. Mozer, Observations of an intense field-aligned thermal ion flow and associated intense narrow band electric field observations, J. Geophys. Res., 80, 4612, 1975.

Bering, E. A., Apparent electrostatic ion cyclotron waves in the diffuse aurora, Geophys. Res. Lett., 10, 647, 1983.

Borovsky, J. E., The production of ion conics by oblique double layers, J. Geophys. Res., 89, 2251, 1984.

Calvert, W., The auroral plasma cavity, Geophys. Res. Lett., 8, 919, 1981.

Chang, T. and B. Coppi, Lower hybrid acceleration and ion evolution in the supraauroral region, Geophys. Res. Lett., 8, 1253, 1981.

Cole, K, Effects of crossed magnetic and (spatially dependent) electric fields on particle motion, Planet Sp. Sci., 24, 515, 1976.

Collin, H. L., R. D. Sharp, E. G. Shelley and R. G. Johnson, Some general characteristics of upward flowing ion beams over the auroral zone and their relationship to auroral electrons, J. Geophys. Res., 86, 6820, 1981.

Davidson, R. C., Methods in Nonlinear Plasma Theory, Academic, New York, 1972.

Dum, C. T., Anomalous heating by ion sound turbulence, Phys. Fluids, 21, 945, 1978.

Dusenbery, P. B. and L. R. Lyons, Generation of ion-conic distributions by downward auroral currents, J. Geophys. Res., 86, 7627, 1981.

Ghielmetti, A. G., R. G. Johnson, R. D. Sharp and E. G. Shelley, The latitudinal, diurnal and altitudinal distributions of upward flowing energetic ions of ionospheric origin, Geophys. Res. Lett., 5, 59, 1978.

Ghielmetti, A. G., R. D. Sharp, E. G. Shelley, and R. G. Johnson, Downward flowing ions and evidence for injection of ionospheric ions into the plasma sheet, J. Geophys. Res., 84, 5781, 1979.

Gorney, D. J., A. Clarke, D. R. Croley, J. F. Fennell, J. M. Luhmann, and P. F. Mizera, The distribution of ions beams and conics below 8000 km, J. Geophys. Res., 86, 83, 1981.

Gorney, D. J., S. R. Church and P. F. Mizera, On ion harmonic structure in auroral zone waves: the effect of ion conic damping of auroral hiss, J. Geophys. Res., 87, 10479, 1982.

Gorney, D. J., Y. T. Chiu and D. R. Croley, Trapping of ion conics by downward parallel electric fields, J. Geophys. Res., 90, 4205, 1985.

Greenspan, M. E., Effects of oblique double layers on upgoing ion pitch angle and gyrophase, J. Geophys. Res., 89, 2842, 1984.

Heelis, R. A., J. D. Winningham, M. Sugiura and N. C. Maynard, Particle acceleration parallel and perpendicular to the magnetic field observed by DE-2, J. Geophys. Res., 89, 2842, 1984.

Karney, C. F. F. and A. Bers, Stochastic ion heating by a perpendicularly propagating electrostatic wave, Phys. Rev. Lett., 39, 550, 1977.

Karney, C. F. F., Stochastic ion heating by a lower hybrid wave, Phys. Fluids, 21, 1584, 1978.

Kaufmann, R. L. and P. M. Kintner, Upgoing ion beams 1, microscopic analysis, J. Geophys. Res., 87, 10487, 1982.

Kindel, J. M. and C. F. Kennel, Topside current instabilities, J. Geophys. Res., 76, 3055, 1971.

Kintner, P. M., M. C. Kelley and F. S. Mozer, Electrostatic hydrogen cyclotron waves near one earth radius altitude in the polar magnetosphere, Geophys. Res. Lett., 5, 139, 1978.

Kintner, P. M., M. C. Kelley, R. D. Sharp, A. G. Ghielmetti, M. Temerin, C. A. Cattell, P. F. Mizera, Simultaneous observations of energetic (keV) upstreaming ions and EHC waves, J. Geophys. Res., 84, 7201, 1979.

Kintner, P. M. and D. J. Gorney, A search for the plasma processes associated with perpendicular ion heating, J. Geophys. Res., 89, 937, 1984.

Klumpar, D. M., Transversely accelerated ions: an ionospheric source of hot magnetospheric ions, J. Geophys. Res., 84, 4229, 1979.

Klumpar, D. M. and W. J. Heikkila, Electrons in the ionospheric source cone: evidence for runaway electrons as carriers of downward Birkeland currents, Geophys. Res. Lett., 9, 873, 1982.

Klumpar, D. M., W. K. Peterson and E. G. Shelley, Direct evidence for two stage (bimodal) acceleration of ionospheric ions, J. Geophys. Res., 89, 10779, 1984.

Lysak, R. L. and M. K. Hudson, Coherent anomalous resistivity in the region of electrostatic shocks, Geophys. Res. Lett., 6, 661, 1979.

Lysak, R. L., M. K. Hudson, and M. Temerin, Ion heating by strong electrostatic ion cyclotron wave turbulence, J. Geophys. Res., 85, 678, 1980.

Maggs, J. E., Coherent generation of VLF hiss, J. Geophys. Res., 81, 1707, 1976.

Maynard, N. C., J. P. Heppner and A. Egeland, Intense, variable electric fields at ionospheric altitudes in the high latitude regions as observed by DE-2, Geophys. Res. Lett., 9, 981, 1982.

Moore, T. E., Modulation of terrestrial ion escape flux composition (by low-altitude acceleration and charge exchange chemistry), J. Geophys. Res., 85, 2011, 1980.

Mozer, F. S., C. W. Carlson, M. K. Hudson, R. B. Torbert, B. Parady, J. Yatteau and M. C. Kelley, Observations of paired electrostatic shocks in the polar magnetosphere, Phys. Rev. Lett., 38, 292, 1977.

Mozer, F. S., C. A. Cattell, M. K. Hudson, R. L. Lysak, M. Temerin and R. B. Torbert, Satellite measurements and theories of auroral acceleration mechanisms, Space Sci. Rev., 27, 155, 1980.

Nishikawa, K-I., H. Okuda and A. Hasegawa, Heating of heavy ions on auroral field lines, Geophys. Res. Lett., 10, 553, 1983.

Nishikawa, K-I., H. Okuda and A. Hasegawa, Heating of heavy ions on auroral field lines in the presence of a large-amplitude hydrogen cyclotron wave, J. Geophys. Res., 90, 419, 1985.

Nishikawa, K-I. and H. Okuda, Heating of light ions in the presence of a large amplitude heavy ion cyclotron wave, J. Geophys. Res., 90, 2921, 1985.

Okuda, H. and M. Ashour-Abdalla, Formation of a conical distribution and intense ion heating in the presence of hydrogen cyclotron waves, Geophys. Res. Lett., 8, 811, 1981.

Okuda, H. and M. Ashour-Abdalla, Acceleration of hydrogen ions and conic formation along auroral field lines, J. Geophys. Res., 88, 889, 1983.

Palmadesso, P. J., T. P. Coffey, S. L. Ossakow and K. Papadopoulos, Topside ionospheric ion heating due to electrostatic ion cyclotron turbulence, Geophys. Res. Lett., 3, 105, 1974.

Papadopoulos, K., J. D. Gaffey and P. J. Palmadesso, Stochastic acceleration of large m/q ions by hydrogen cyclotron waves in the magnetosphere, Geophys. Res. Lett., 7, 1014, 1980.

Retterer, J. M., T. Chang and J. R. Jasperse, Ion acceleration in the supraauroral region: a Monte Carlo model, Geophys. Res. Lett., 10, 583, 1983.

Roth, I., and M. K. Hudson, Lower hybrid heating of ionospheric ions due to ion ring distributions in the cusp, J. Geophys. Res., 90, 4191, 1985.

Sharp, R. D., R. G. Johnson and E. G. Shelley, Observation of an ionospheric acceleration mechanism producing energetic (keV) ions primarily normal to the geomagnetic field direction, J. Geophys. Res., 82, 3324, 1977.

Sharp, R. D., W. Lennartsson, W. K. Peterson and E. Ungstsup, The mass dependence of wave-particle interactions observed with the ISEE-1 mass spectrometer, Geophys. Res. Lett., 10, 651, 1983.

Shelley, E. G., R. D. Sharp, and R. G. Johnson, Satellite observations of an ionospheric acceleration mechanism, Geophys. Res. Lett, 3, 654, 1976.

Singh, N., R. W. Schunk and J. J. Sojka, Energization of ionospheric ions by electrostatic hydrogen cyclotron waves, Geophys. Res. Lett., 8, 1249, 1981.

Singh, N., R. W. Schunk and J. J. Sojka, Preferential perpendicular acceleration of heavy ionospheric ions by interactions with electrostatic hydrogen cyclotron waves, J. Geophys. Res., 88, 4055, 1983.

Temerin, M., M. H. Boehm and F. S. Mozer, Paired electrostatic shocks, Geophys. Res. Lett., 8, 799, 1981a.

Temerin, M., C. A. Cattell, R. Lysak, M. Hudson, R. B. Torbert, F. S. Mozer, R. D. Sharp and P. M. Kintner, The small scale structure of electrostatic shocks, J. Geophys. Res., 86, 11278, 1981b.

Temerin, M. and R. L. Lysak, Electromagnetic ion cyclotron mode (ELF) waves generated by auroral electron precipitation, J. Geophys. Res., 89, 2849, 1984.

Whalen, B. A., W. Bernstein and D. W. Daly, Low altitude acceleration of ionospheric ions, Geophys. Res. Lett., 5, 55, 1978.

Yau, A. W., B. A. Whalen, A. G. McNamara, P. J. Kellogg and W. Bernstein, Particle and wave observations of low altitude ionospheric ion acceleration events, J. Geophys. Res., 88, 341, 1983.

Ion Heating in the Cusp

M. K. HUDSON

Department of Physics and Astronomy, Dartmouth College, Hanover, New Hampshire 03755

I. ROTH

Space Sciences Laboratory, University of California, Berkeley 94720

Abstract. Ion heating in the cusp is examined in the context of the intermixing of magnetosheath and ionospheric plasmas. Downflowing H^+ and He^{++} ring distributions observed on the DE and S3-3 satellites have been shown to be unstable to lower hybrid waves. Additional lower frequency modes are generated in a multispecies plasma, and become unstable for finite $k_\parallel$. The latter can lead to additional heating of oxygen as well as hydrogen, but hydrogen heating still predominates. This heating mechanism is more efficient than the electron beam generated case in terms of free energy conversion to background ion heating. They are similar in the nonlinear evolution of a broad spectrum of k modes extending to linearly stable phase velocities which are resonant with the background ions. Inclusion of superthermal (and background) electron dynamics determines a range of unstable propagation angles about perpendicular to the magnetic field of order ±20° for $M_H/M_e = 50$, and narrower for real mass ratios. It is concluded that measured ion distributions have sufficient energy flux to account for inferred conic heating via the mechanism studied.

Introduction

The cusp has long been recognized as a topologically unique region because it is open to plasma of both ionospheric and magnetosheath origin. Recent data from the DE1 spacecraft at high altitudes (>4 R_E geocentric) has explicitly shown this mixing in the observation of downflowing ion ring distributions of magnetosheath origin (H^+ and He^{++}) and upflowing ion conics of ionospheric origin (H^+ and O^+) [Peterson, 1984].

Surveying Aerospace ion data with no mass discrimination from the S3-3 polar orbiting satellite, Gorney [1983] noted that ion conics in the cusp are sometimes accompanied by downflowing ion ring distributions with $\partial f/\partial v_\perp > 0$. Figure 1 shows one such example. What we refer to as an ion ring in a contour plot of $f(v_\parallel, v_\perp)$ is really a half shell about the $v_\parallel$ axis which results when an ion beam at higher altitudes conserves the first adiabatic invariant flowing into a converging mirror field. The initial beam may be due to the injection of magnetosheath plasma associated with a reconnection event at the magnetopause, while some plasma of solar wind origin may come from continuous particle diffusion across closed field lines. Reiff et al. [1977] concluded that cusp particle data were more often consistent with merging, and that the $\mathbf{E} \times \mathbf{B}$ convection electric field was responsible for the observed latitudinal dispersion in particle energy. More energetic particles are observed at lower altitudes in constant altitude cusp crossings, because the convection electric field carries lower energy particles to higher latitudes during their time of flight to the ionosphere. This dispersion produces a local distribution which has decreasing energy at increasing pitch angle, as one would also expect for a beam-like distribution moving into a converging mirror field. Gorney [1983] pointed out that the ring distribution in Figure 1 had a positive derivative in $v_\perp$, while Cattell and Hudson [1982] found that this same event was accompanied by waves near the lower hybrid frequency, as shown in Figure 2.

More recent data from the Lockheed EICS instrument on DE1 is shown in Figure 3. With improved mass, energy and pitch angle resolution over that available on S3-3, it was possible to obtain phase space density contours for the different ion species (H^+, O^+ and He^{++}) separately. Figures 3a and 3b show phase space density contours for H^+ and He^{++}, with downflowing ion ring distributions in both, along with an upflowing H^+ conic. Figures 3c and 3d show H^+ and O^+ phase space density contours for a similar cusp crossing with an H^+ ring and O^+ conic present. All three species are not available simultaneously at high enough time resolution to produce contour plots, but what can be noted is that rings occur in plasma of solar wind/magnetosheath origin (He^{++} and energetic H^+), while conics appear in plasma of ionospheric origin (O^+ and less energetic H^+).

Roth and Hudson [1985] suggested that lower hybrid waves generated by the observed downflowing ion ring distributions may heat the background plasma and contribute to conic formation on cusp field lines. They carried out a series of one dimensional electrostatic particle simulations for H^+ and He^{++} rings in a background $H^+ - O^+$ plasma, with electrons forming a uniform, stationary background and $k_\parallel = 0$ assumed. We will show later that in a pure hydrogen plasma the $k_\parallel = 0$ mode is most unstable when a perpendicular ion ring is the free energy source, in contrast to electron beam driven modes which require finite $k_\parallel$. Waves excited are lower hybrid and Bernstein modes. The linear instability is due to the proximity of the lower hybrid branch to one of the Bernstein modes. As indicated in Figure 4 [Roth and Hudson, 1983], increasing the ring density distorts the Bernstein mode nearest the lower hybrid, producing a complex conjugate pair of roots, one of which is unstable.

The mode structure is more complicated for a He^{++} rather than H^+ ring because of the additional species, and further complicated still by the addition of O^+ to the background. Additional modes arise which produce additional heating over what is seen in a pure H^+ plasma. Figure 5 shows the eigenmodes seen in the simulations at a fixed k. Panel (a) shows the mode structure for a H^+ ring in a H^+ background. The lower hybrid occurs at a frequency

$$\omega_+^2 = \omega_p^2 + \Omega_H^2 \tag{1a}$$

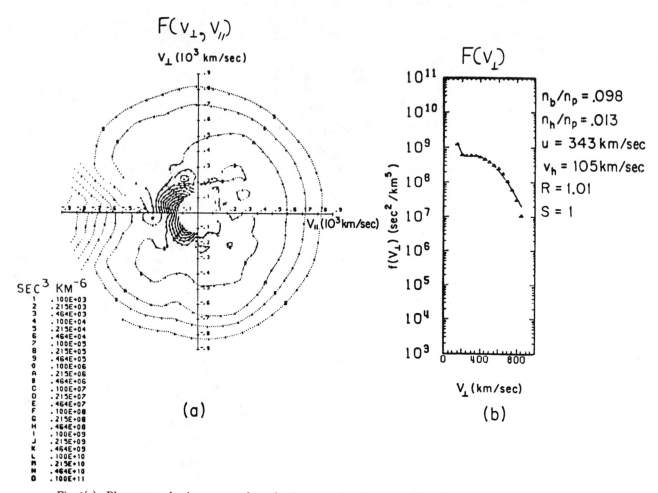

Fig. 1(a). Phase space density contours from the Aerospace ion detector on S3-3 showing downflowing ion ring and upflowing conic features. (b). Distribution function integrated over $v_\parallel$ (from Cattell and Hudson, 1982).

where ω_p is the rms plasma frequency due to all ion species present, and Ω_s is the gyrofrequency of species s. The lowest Bernstein harmonics are also evident. The addition of a minor amount of oxygen to the background in panel (b) introduces a new mode

$$\omega_-^2 \simeq \omega_O^2/(1 + \omega_H^2/\Omega_H^2 + \omega_r^2/\Omega_r^2) + \Omega_O^2 \quad (1b)$$

which is the $H^+ - O^+$ or Buschbaum hybrid resonance [Stix, 1962].

The ring term (ω_r^2/Ω_r^2) appears in the denominator when the ring is a species other than hydrogen. A He^{++} ring in a H^+ background introduces an analogous $H^+ - He^{++}$ hybrid resonance, indicated by ω_* in panel (d), where

$$\omega_*^2 = \Omega_{He}^2 + \omega_{He}^2/(1 + 4\omega_H^2/3\Omega_H^2) \quad (1c)$$

in the absence of oxygen. Temerin and Lysak [1984] have plotted

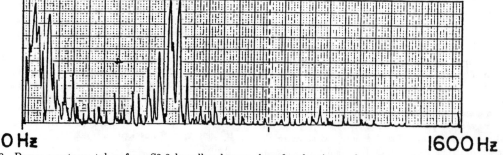

Fig. 2. Power spectrum taken from S3-3 broadband wave data for the ring and conic event analyzed by Gorney [1983] showing peaks at ω_+, $\omega_* \simeq \Omega_{He}$ and other $n\Omega_{He}$ (from Cattell and Hudson, 1982).

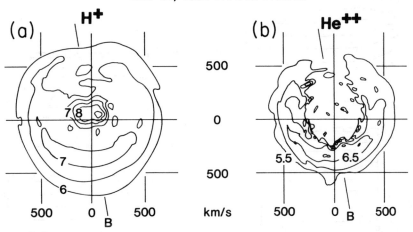
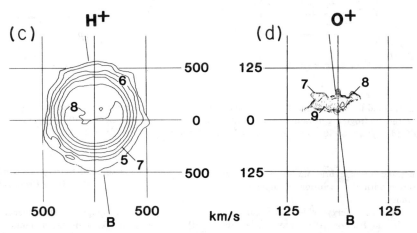

Fig. 3. Phase space density contour plots from the EICS instrument on DE1 for (a) H$^+$ and (b) He^{++} simultaneously and (c) H$^+$ and (d) O$^+$ simultaneously. The contours are log (base 10) of the phase space density in units of s^3/km^6. The data are presented in spacecraft coordinates with the direction of the magnetic field indicated. Upflowing O$^+$ (September 21) and downflowing He^{++} (October 15) are shown. Note that only H$^+$ and O$^+$ ions were sampled in September 21 case, and only H$^+$ and He^{++} ions in October 15 case (from Roth and Hudson, 1985).

the ion-ion hybrid resonance frequencies for the general case when three ion species are present. All of the above modes appear in panel (e), where a He^{++} ring is introduced into a hydrogen-oxygen background plasma. Panels (c) and (f) show that the mode structure persists for equal hydrogen and oxygen background densities. We will show later that ω_* is kinetically driven unstable by the ion ring for finite $k_\parallel$.

Figure 6 shows modes excited by a He^{++} ring in a background hydrogen-oxygen plasma summed over all k at a representative point in the system. Since this is what a probe measures, there is a good correspondence between this type of diagnostic and what is seen in the data. In particular, note the presence of peaks at the He^{++} as well as H$^+$ gyrofrequencies. Similar structure is evident in the wave data of Figure 2 [Cattell and Hudson, 1982], where Lockheed reported the simultaneous observation of He^{++} at the time of the cusp event shown in Figures 1 and 2.

Figure 7 shows the temporal evolution of the electrostatic energy summed over all k modes for the case of a He^{++} ring in a hydrogen background. Also shown is the temporal evolution of the most unstable mode with $k\rho_H = 0.5$ and a linearly stable mode with $k\rho_H = 0.25$. Most of the electrostatic energy initially resides in the most unstable mode, but the amplitude of the linearly stable mode becomes comparable at later times. Retterer et al. [1985] have suggested that nonlinear mode coupling with a low frequency quasi-mode produces a cascade of the lower hybrid daughter to higher wave numbers and smaller phase velocities for the case of electron beam generated lower hybrid waves. Such a transfer of electrostatic energy to higher k modes also occurs in

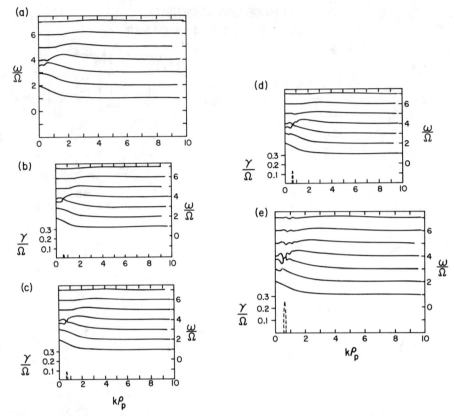

Fig. 4. Real ω/Ω (solid lines) and imaginary γ/Ω (dashed lines) parts of the complex frequency vs. $k\rho_H$, wavenumber normalized to the hydrogen gyroradius. $(\omega_H/\Omega_H)^2 = 12$; $V_r/u_H = 10$; (a) $n_r/n_H = 10^{-2}$, (b) $n_r/n_H = 1.5 \times 10^{-2}$, (c) $n_r/n_H = 2 \times 10^{-2}$, (d) $n_r/n_H = 3 \times 10^{-2}$ and (e) $n_r/n_H = 7 \times 10^{-2}$ (from Roth and Hudson, 1983).

the case of ion ring generated lower hybrid waves, and will be examined in the next section with the inclusion of superthermal background electron dynamics.

The initial burst of electrostatic energy evident in Figure 7 is characteristic of what Roth and Hudson [1985] have called the trapping phase, where energy oscillates between the particles and a single coherent wave. This behavior is particularly striking in the case of a hydrogen ring and plasma, where many trapping oscillations are seen in both electrostatic and particle kinetic energy [Roth and Hudson, 1983]. Introducing a second species, either via the ring or background, disrupts the trapping oscillations because of the appearance of new eigenmodes.

Figure 8 shows the relative heating of a hydrogen and oxygen background plasma for the case of a He^{++} ring extending to late times, $\tau \sim 6000 \, \omega_H^{-1}$. There is a broad spectrum of k modes after $100 \, \omega_H^{-1}$, so most of the heating occurs in the quasilinear regime. We see that hydrogen is heated more than oxygen, gaining most of the initial ring kinetic energy. Oxygen is, however, heated more effectively by a helium than a hydrogen ring, [cf. Roth and Hudson, 1985, Figure 12] due to the presence of lower frequency modes, particularly $\omega_* \simeq 8 \, \Omega_O$. Quasilinear heating favors the lighter mass ions, in contrast to stochastic heating processes which have been investigated for a fixed amplitude, coherent pump [Karney, 1978; Lysak et al., 1980; Papadopoulos et al., 1980; Singh et al., 1981].

Superthermal Electron Dynamics and Effects of Oblique Propagation

For oblique propagation there occur several modifications to the waves. Finite $k_\parallel$ adds the dynamics of electrons to the wave-particle interaction, modifying the eigenfrequencies and contributing to damping of waves; it also allows linear excitation at lower frequency (ω_*) due to the positive parallel slope of the ring distribution.

The He^{++} ring in our model for oblique propagation is described by a distribution function with the bulk of particles having velocities $v_x^2 + v_y^2 + v_z^2 = v^2$, and thermal spread u_r. Here $mv^2/2$ denotes the original energy of He^{++} particles as they enter onto cusp field lines. To make the analysis tractable, we shall choose the ring as

$$F_r(v) = n_r \frac{(1 - \alpha^{-1/2}\beta^{-1})^{-1}}{(\pi^{1/2} u_r)^3} \left(e^{-v^2/u_r^2} - e^{-\alpha v_\parallel^2/u_r^2 - \beta v_\perp^2/u_r^2} \right) \quad (2)$$

with $\alpha > \beta > 1$. This distribution is an anisotropic subtracted Maxwellian, cf. Cattell and Hudson [1982] for a discussion of subtracted Maxwellians.

The main effects of the superthermal electrons are in modifying the eigenmodes for oblique propagation, particularly ω_+, ω_-, and damping waves with a finite parallel component of

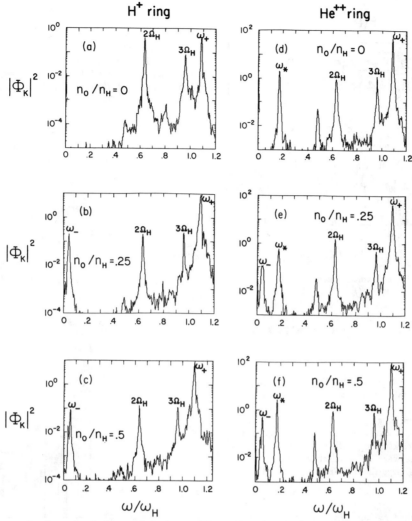

Fig. 5. Power spectra from simulations with hydrogen (Figures 5a-5c) and helium (Figures 5d-5f) ring averaged over times $0 < t\,\omega_H < 2000$. $(\omega_H/\Omega_H)^2 = 9.765$, $n_c/n_H = 0.0625$, $k\rho_H = 0.06$. Indicated are the main modes $\omega_\pm$, $n\,\Omega_H$, $m\,\Omega_O$, ω_*; (a) $n_O/n_H = 0$, (b) $n_O/n_H = 0.25$, (c) $n_O/n_H = 0.5$, (d) $n_O/n_H = 0$, (e) $n_O/n_H = 0.25$, (f) $n_O/n_H = 0.5$, in arbitrary units for comparison of relative power in different spectral peaks (from Roth and Hudson, 1985).

wavenumber. The most important eigenmodes are given for oblique propagation by

$$\omega_+^2 \simeq (\omega_H^2 + \omega_e^2 \cos^2\theta + \omega_O^2)/(1 + \omega_e^2 \sin^2\theta/\Omega_e^2) + \Omega_H^2 \quad (3a)$$

$$\omega_-^2 \simeq \frac{\omega_O^2 + \omega_e^2 \theta}{1 + (\omega_H^2/\Omega_H^2)\sin^2\theta} + \Omega_O^2 \quad (3b)$$

$$\omega_* \simeq \Omega_r \quad (3c)$$

The imaginary part of the ring susceptibility may contribute to the growth rate due to the positive slope of the ring distribution function; the electron and hydrogen/oxygen ions contribute to Landau damping. The main components of the total imaginary susceptibility around the ω_* mode become

$$-\frac{\mathrm{Im}\,\chi}{2\,\pi^{1/2}} = \frac{\omega_r^2}{k^2 u_r^2}\,\frac{\omega}{k_\| u_r} \quad (4)$$

$$\cdot \left[\alpha\,\Gamma_1\left(\frac{\Lambda_r}{\beta}\right) e^{-\Omega^2\beta/k_\|^2 u_r^2} - \Gamma_1(\Lambda_r)\,e^{-\Omega^2/k_\|^2 u_r^2}\right] \frac{\alpha}{\alpha^{1/2} - 1}$$

$$- \frac{\omega_e^2}{k^2 u_e^2}\,\frac{\Omega_r}{k_\| u_e}\,\Gamma_0(\Lambda_e)\,e^{-\frac{\Omega_r^2}{k_\|^2 u_e^2}}$$

$$- \frac{\omega_H^2}{k^2 u_H^2}\,\frac{\Omega_r}{k_\| u_H}\left[\Gamma_0(\Lambda_H) + \Gamma_1(\Lambda_H)\right] e^{-\Omega_r^2/k_\|^2 u_H^2}$$

$$- \frac{\omega_O^2}{k^2 u_O^2}\,\frac{\Omega_r}{k_\| u_O}\left[\Gamma_0(\Lambda_O)\,e^{-\omega^2/k_\|^2 u_H^2} + \Gamma_8(\Lambda_O)\,e^{-\Omega^2/k_\|^2 u_r^2}\right]$$

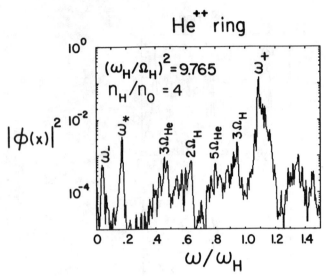

Fig. 6. Power spectrum of electrostatic potential $|\Phi|^2(x,t)$ recorded at a fixed point $x = 100\ \lambda_D$ ($L = 640\ \lambda_D$), summed over k, averaged over times $0 < t < 2000\ \omega_H^{-1}$, for a helium ring with $n_H/n_O = 4$, $n_r/n_H = 0.0625$, $(\omega_H/\Omega_H)^2 = 9.765$, in arbitrary units for comparison of relative power in different spectral peaks (from Roth and Hudson, 1985).

where $\Lambda_s = 1/2\ k_\perp^2\ u_s^2/\Omega_s^2$ and $\varsigma = \omega_* - \Omega_r$. Here we include the Cerenkov damping of all ambient populations along with the first hydrogen and eighth oxygen cyclotron damping. Cold electrons will contribute a similar damping term.

The excitation of these waves is limited to very small wavenumbers since at higher k modes, the electrons (and ambient ions) will heavily damp the waves. The electron term gives the largest damping due to its large thermal spread, hence (4) gives the marginally stable angle

$$\sin^2\theta \approx \frac{2\ \omega_e^2}{\omega_r^2}\ \frac{\Omega_r^2}{k^2\ u_e^2}\ \frac{u_r}{u_e}\ e^{-\Omega_r^2/k^2\ u_e^2\cos^2\theta} \qquad (5)$$

Setting $u_r/u_e = 1/50$, $M_H/M_e = 50$ and $k\rho_H \sim 0.06$, as obtained in the simulations for the main excited mode, the angle which solves (5) is $\theta \sim 70°$. Therefore, for the given mass ratio the linear instability at ω_* will cover the angles $\pm 20°$ off perpendicular propagation.

It was previously found that ω_+ is linearly excited by a fluid coupling of the lower hybrid to adjacent Bernstein modes for purely perpendicular propagation, neglecting electron dynamics [Roth and Hudson, 1983; 1985]. The ω_* and ω_- eigenmodes appeared for the case of a helium ring in a hydrogen/oxygen plasma because of enhancement in the thermal fluctuation level due to the ring. In the present case, the ω_* mode is linearly excited for finite but not too large $k_\parallel$ by Landau resonance with the positive slope of the helium distribution, shown in Figure 9. When $k_\parallel$ becomes too large, electron Landau damping suppresses the instability. The mode ω_- was seen but not investigated in the present study because of limitations on the time of runs including electron dynamics, and the fact that ω_* was observed to be more important than ω_- for oxygen heating in our previous study [Roth and Hudson, 1985].

Simulations

We now include a population of superthermal electrons (50 eV) in the simulations along with background ions of 1 eV. The superthermal electron density is taken to be equal to that of the background plus ring ion densities. A mass ratio of $M_H/M_e = 50$ is assumed, while $M_O/M_H = 16$ as before. A cold electron background is neglected in the runs shown, since the superthermals can

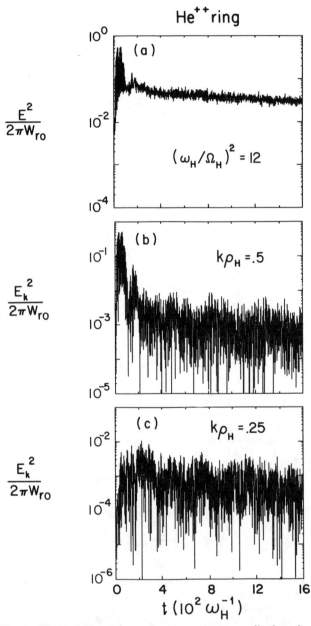

Fig. 7. Electrostatic mode energy versus time normalized to the initial ring kinetic energy, (a) total, (b) mode n = 16, and (c) n = 8. Note that most power in Figure 7a comes from Figure 7b at early times. Normalized units are the same for all three figures. $(\omega_H/\Omega_H)^2 = 12$, $n_r/n_H = 0.0625$, $n_O = 0$ (from Roth and Hudson, 1985).

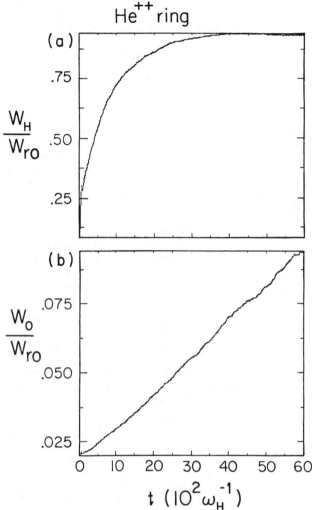

Fig. 8. Kinetic energies of (a) hydrogen and (b) oxygen for a helium ring excited case, normalized to the initial ring energy with $n_H/n_O = 4$, $n_r/n_H = 0.0625$, $(\omega_H/\Omega_H)^2 = 9.765$, $L = 320\,\lambda_D$ (from Roth and Hudson, 1985).

the total electrostatic energy at $\theta = 3°$. Figure 11 shows power spectra for two linearly unstable wavenumbers for angle $\theta = 3°$ off perpendicular propagation. The lowest mode kept in the system, $k\rho_H = 0.06$, is linearly unstable at ω_*, and ω_+ appears because of thermal fluctuation enhancement. The most unstable mode at ω_+ is $k\rho_H = 0.37$. The power spectrum for this mode shows no peak at ω_*, while ω_+ covers a broad spectrum in k.

The ω_* mode becomes kinetically unstable for finite $k_\parallel$. Its appearance in $k_\parallel = 0$ simulations is due to enhancement in the thermal fluctuation level [Roth and Hudson, 1985]. This mode was previously shown to enhance oxygen heating, since $\omega_* \simeq 8\,\Omega_O$ is gyroresonant with oxygen. Linearly, it is narrow in k for finite $k_\parallel$, appearing primarily in the lowest ($k\rho_H = 0.06$) mode.

Figures 10b and 10c show the temporal evolution of the total electrostatic energy for $\theta = 8°$ and $30°$ for comparison with Figure 10a. Linear instability and trapping oscillations are apparent in (b) and absent for (c), but the final fluctuation levels are comparable. The latter are due mainly to enhancement in the thermal fluctuation level of the plasma associated with the smaller number of particles used in the simulation than in a real plasma [Roth and Hudson, 1984, Appendix]. Figure 12 shows the kinetic energies vs. time of background hydrogen and oxygen for a long run at $\theta = 4°$. Trapping oscillations are evident in oxygen at early times. Hydrogen gains a significant fraction of the initial ring kinetic energy ($W_H/W_{ro} = 0.688$) by $\tau = 800\,\omega_H^{-1}$, comparable to that for perpendicular propagation with no electron dynamics included, cf. Figure 8 at the same time.

Figure 13 is a plot of the k spectrum for $\theta = 2°$ at early and late times in the simulation. There is a peak in the lowest k mode due to ω_* and around modes 6-8 due to ω_+. The fall off in power with k is evident at all times. The amplitude of the linearly unstable peaks is reduced and the spectrum is broader at later times.

Figure 14 plots the initial and final hydrogen and oxygen distribution functions at $\theta = 2°$. Both appear narrower at later times

have a greater effect on the modes discussed. This exaggerates the effect of the superthermals if $n_h/n_e \ll 1$, where n_h and n_e are, respectively, the superthermal and cold electron densities. We have therefore done a run with cold electrons dominant ($n_h/n_e \simeq 0.05$) as a check on our results. Full electron dynamics are included in all runs. Figure 9 shows the initial ring distribution in the simulations. The parameters of the simulations run with hot electrons only are $T_e/T_H = 50$, $T_H/T_O = 1$, $n_H/n_O = 4$, $n_H/n_{He} = 16$; $V_{He}/u_H = 10$ is the ring drift speed relative to the hydrogen thermal speed and $\Omega_H/\omega_H = 0.32$ is the cyclotron to plasma frequency ratio for hydrogen. The system length is $L = 320\,\Lambda_D$ in Debye lengths, with a grid spacing $\Delta x = 1.25\,\Lambda_D$. The time step was $\omega_e\,\Delta t = 0.14$, and the simulations were run for at least $200\,\omega_H^{-1}$, at propagation angles off perpendicular $\theta = 2°$, $3°$, $4°$, $6°$, $8°$, and $30°$.

Figure 10 shows the temporal evolution of the total electrostatic energy with superthermal electrons included for runs at $\theta = 3°$, $8°$ and $30°$. The linear growth rate is readily apparent in

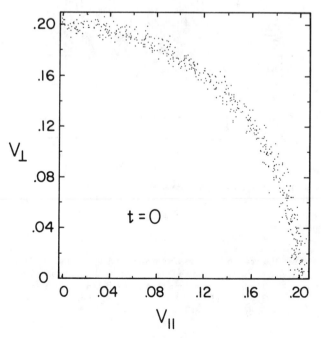

Fig. 9. Ion distribution in $v_\parallel - v_\perp$ space at the beginning of the simulations.

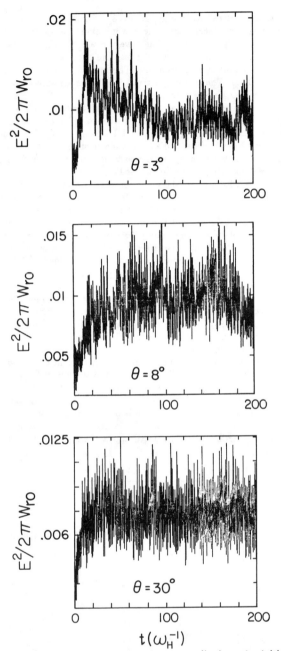

Fig. 10. Electrostatic energy vs. time normalized to the initial helium ring kinetic energy, (a) $\theta = 3°$, (b) $8°$, and (c) $30°$, including superthermal electrons. Linear instability and trapping are evident in (a).

because the scale adjusts to accomodate tail formation. In fact, hydrogen has undergone some bulk heating, with an increase from $u_H/u_{eo} = 0.02$ initially to around 0.04 and tail heating by a factor of three. Oxygen undergoes less bulk heating, but tail particles are energized by a factor of two.

Discussion

The inclusion of finite $k_\parallel$ introduces linear instability of the ω_* mode which is an eigenmode of the helium ring. This mode was previously found to enhance oxygen heating since $\omega_* \simeq 8\,\Omega_O$, and quasilinear heating rates contain $\omega - m\,\Omega_O$ denominators [Roth and Hudson, 1985]. Oxygen heating is greater in the present case because ω_* is now linearly unstable. A comparison of Figures 8 and 12 at $\tau = 800\,\omega_H^{-1}$ shows that $W_o/W_{ro} \simeq 0.028$ and 0.043, respectively, while hydrogen heating is essentially the same in the two cases. Some additional contribution to the oxygen heating comes from the trapping phase evident in Figure 12, and absent with respect to ω_* in Figure 8. Hydrogen is nonetheless heated significantly more than oxygen by the broad k spectrum of modes near the lower hybrid frequency. The latter mass dependence is characteristic of quasilinear heating processes.

The ω_* mode excited for finite $k_\parallel$ is damped by electrons, so only the lowest k modes are linearly excited since electron Landau damping is proportional to $\exp(-(\omega/k_\parallel u_e)^2)$. This factor determines an angular spread in unstable modes which cuts off around $\theta \simeq 20°$ in our simulations. The lower hybrid mode (ω_+) which contributes most to the heating of hydrogen is most unstable for $k_\parallel = 0$, where there is no electron Landau damping, since it is a nonresonant fluid instability. The effect of background and superthermal electrons on hydrogen heating is therefore minimal for this system, in contrast to the electron beam driven case which requires finite $k_\parallel$, where electron Landau and cyclotron damping could be important [Lotko and Maggs, 1979].

The evolution of the k spectrum shown in Figure 13 indicates some shift to higher k, as suggested by Retterer et al. [1985]. This has implications for ion heating since linearly unstable modes are far out in the tail of the background ion distribution. Most heating occurs after coherent trapping oscillations have ceased, and there is a broad spectrum of k modes present in the system.

Ion ring generated lower hybrid heating appears to be more efficient than the electron beam generated case. Figures 8 and 12 show that most of the ion ring energy goes into heating background hydrogen; only 1-2% goes into electrostatic wave energy, cf. Figure 10. Efficiency will be reduced if a power law vs. Maxwellian superthermal electron distribution is used, as is often measured. In the electron beam generated case Retterer et al. [1985] found a 6% conversion efficiency from electron beam to ion kinetic energy, with a comparable 1-2% in electrostatic wave energy. Given the efficiency of the ion ring generated case, the following comparison of ring and conic energetics is relevant. Figure 3a shows a hydrogen ring phase space density of 10^7 s^3/km^6 at $v = 300$ km/s. This yields an energy flux of 3.6×10^9 eV/cm^2 -s in the downflowing hydrogen ring. The upflowing hydrogen conic phase space density of 10^8 s^3/km^6 at $v = 100$ km/s yields an energy flux of 5.2×10^7 eV/cm^2 -s. There is plenty of free energy available in the hydrogen and even the helium ring, which has a lower phase space density by one and one half orders of magnitude, but a larger energy at the same velocity by a factor of 4, yielding an energy flux of 4.6×10^9 eV/cm^2 -s in Figure 3b. This exceeds typical oxygen conic energy fluxes, e.g., 3.3×10^7 eV/cm^2 -s in Figure 3d, and is greater than the hydrogen ring energy flux of 6.1×10^8 eV/cm^2 -s in Figure 3c. Since oxygen can acquire 10% of the initial helium ring kinetic energy of Figure 8, this is adequate to account for observed oxygen conic energy fluxes. Table 1 summarizes these comparisons.

We have focused on an ion heating mechanism specific to the cusp. A number of other heating mechanisms have been proposed, some applicable to the cusp as well as nightside auroral field lines. These include perpendicular acceleration of ions by electrostatic shocks [Borovsky, 1984; Greenspan, 1984], which are well correlated with conics at all local times [Redsun et al., 1984]. Ion cyclotron wave heating has been examined in the coherent [Lysak et al., 1980; Papadopoulos et al., 1980; Singh et al., 1981] and quasilinear [Lysak et al., 1980; Okuda and Ashour-Abdalla, 1981; Dusenbury and Lyons, 1981] limits. Ion cyclotron waves

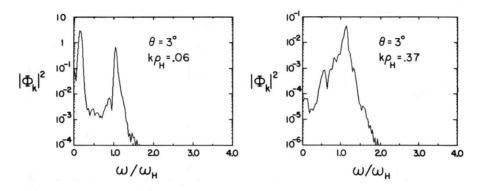

Fig. 11. Power spectra in arbitrary units for $\theta = 3°$. (a) n = 1 and (b) n = 6, with $k = 2\pi n/L$.

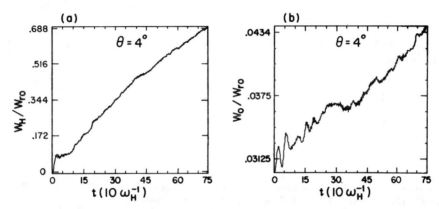

Fig. 12. Kinetic energies vs. time of (a) hydrogen and (b) oxygen for $\theta = 4°$. Note trapping oscillations in O^+ due to ω_*. Hydrogen gains a significant fraction of the initial ring kinetic energy.

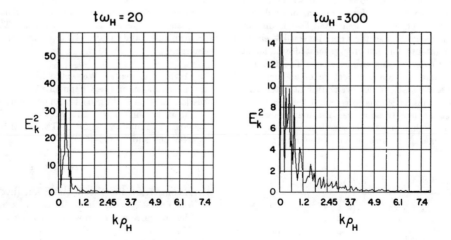

Fig. 13. Spectrum of k modes at early and late times in the same relative units for $\theta = 2°$.

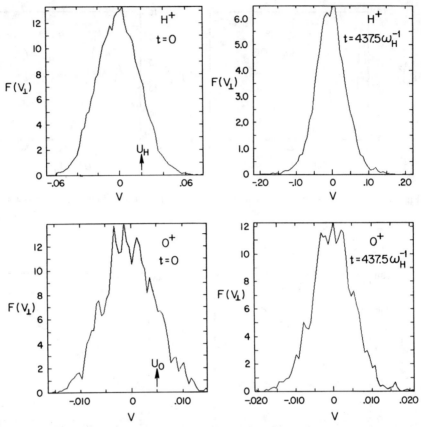

Fig. 14. Initial and final hydrogen and oxygen distribution functions $F(v_\perp)$ for $\theta = 2°$. Arrows indicate initial hydrogen and oxygen thermal speeds, to which all other speeds can be compared.

are, however, generally observed in the ion beam region above the auroral potential drop on the nightside. The mechanism closest to that studied here is heating by electron beam generated lower hybrid waves. It seems likely that lower hybrid waves are generated by different mechanisms in different regions or at different local times. While the electron beam generated case is undoubtedly most efficient in the evening sector where inverted V electron precipitation is most intense, the ion ring generated case appears likely to contribute to ion heating in the cusp. The two lower hybrid mechanisms share the common feature of an increase in power at lower, linearly stable phase velocities at later times. This increases the accessibility of ions to resonant heating.

We have not yet adequately addressed the mass dependence question. It is the case that any quasilinear heating mechanism, be it ion cyclotron waves, electron beam or ion ring generated lower hybrid waves, will preferentially accelerate hydrogen, although we have seen that oxygen is heated more by a helium than a hydrogen ring, particularly with inclusion of finite $k_\parallel$ for the ω_* mode. To the extent that oxygen conics are sometimes more energetic, particularly during disturbed times on the nightside, another mechanism(s) is required. This is not, however, a general rule, and a statistical study for the cusp has yet to emerge.

Acknowledgments. We would like to thank Tom Chang and the participants of the Chapman Conference for stimulating discussions and providing us the opportunity to review and extend our work on this topic. This work was performed under National Science Foundation grant ATM-8445010 and NASA grants NAGW-564 and NAG5-451.

References

Bergmann, Rachelle, Electrostatic ion (hydrogen) cyclotron and ion acoustic wave instabilities in regions of upward field-aligned current and upward ion beams, *J. Geophys. Res., 89,* 953, 1984.

Borovsky, J. E., The production of ion conics by oblique double layers, *J. Geophys. Res., 89,* 251, 1984.

Cattell, C. A. and M. K. Hudson, Flute mode waves near ω_{LH} excited by ion rings in velocity space, *Geophys. Res. Lett., 9,* 1167, 1982.

Table 1. Comparison of Ring and Conic Energetics

Figure	Species	$f(s^3/km^6)$	$v(km/s)$	E (eV/cm²-s)
3a	H⁺ ring	10^7	300	3.6×10^9
3a	H⁺ conic	10^8	100	5.2×10^7
3b	He⁺⁺ ring	$10^{6.5}$	300	4.6×10^9
3c	H⁺ ring	10^8	150	6.1×10^8
3d	O⁺ conic	10^9	40	3.3×10^7

Dusenbury, P. B., and L. R. Lyons, Generation of ion conic distributions by upgoing ionospheric electrons, *J. Geophys. Res., 86,* 7627, 1981.

Gorney, D. J., An alternative interpretation of ion ring distributions observed by the S3-3 satellite, *Geophys. Res. Lett, 10,* 417, 1983.

Greenspan, M., Effects of oblique double layers on upgoing pitch angle and gyrophase, *J. Geophys. Res., 89,* 2842, 1984.

Hudson, M. K. and I. Roth, Thermal fluctuations from an artificial ion beam injection into the ionosphere, *J. Geophys. Res., 89,* 9812, 1984.

Karney, C. F. F., Stochastic ion heating by a lower hybrid wave, *Phys. Fluids, 21,* 1584, 1978.

Lotko, W. and J. Maggs, Damping of electrostatic noise by warm auroral electrons, *Planet. Space Sci., 27,* 1491, 1979.

Lysak, R. L., M. K. Hudson, and M. Temerin, Ion heating by strong electrostatic ion cyclotron turbulence, *J. Geophys. Res., 85,* 678, 1980.

Okuda, H., and M. Ashour-Abdalla, Formation of a conical distribution and intense ion heating in the presence of hydrogen cyclotron waves, *Geophys. Res. Lett., 8,* 811, 1981.

Papadopoulos, K. J., D. Gaffey, Jr., and P. J. Palmadesso, Stochastic acceleration of large M/Q ions by hydrogen cyclotron waves in the magnetosphere, *Geophys. Res. Lett., 7,* 1014, 1980.

Peterson, W. K., E. G. Shelley, S. A. Boardsen, and D. A. Gurnett, Transverse auroral ion energization observed on DE1 with simultaneous plasma wave and ion composition measurements, AGU Chapman Conference on Ion Acceleration in the Ionosphere and Magnetosphere, Wellesley, MA., 1985.

Redsun, M. S., M. Temerin, and F. S. Mozer, Classification of auroral electrostatic shocks by their ion and electron associations, *J. Geophys. Res.,* in press, 1985.

Reiff, P. H., T. W. Hill, and J. L. Burch, Solar wind injection at the dayside magnetospheric cusp, *J. Geophys. Res., 82,* 479, 1977.

Retterer, J. M., T. Chang, J. Jasperse, Plasma simulation of ion acceleration by lower hybrid waves in the suprauroral region, this volume. AGU Chapman Conference on Ion Acceleration in the Ionosphere and Magnetosphere, Wellesley, MA., 1985.

Roth, I. and M. K. Hudson, Particle simulations of electrostatic emissions near the lower hybrid frequency, *J. Geophys. Res., 88,* 483, 1983.

Roth, I. and M. K. Hudson, Lower hybrid heating of ionospheric ions due to ion ring distributions in the cusp, *J. Geophys. Res., 90,* 4191, 1985.

Singh, N., R. W. Schunk, and J. J. Sojka, Energization of ions in the auroral plasma by broadband waves: Generation of ion conics, *Geophys. Res. Lett., 8,* 1249, 1981.

Stix, T. M., Theory of Plasma Waves, p. 44, McGraw-Hill, New York, 1962.

Temerin M., and R. L. Lysak, Electromagnetic ion cyclotron mode (ELF) waves generated by auroral electron precipitation, *J. Geophys. Res., 89,* 2849, 1984.

PLASMA SIMULATION OF ION ACCELERATION BY LOWER HYBRID WAVES IN THE SUPRAURORAL REGION

John M. Retterer
Boston College, Chestnut Hill, MA 02167

Tom Chang
MIT, Cambridge, MA 02139

J. R. Jasperse
AFGL, Bedford MA 01731

Abstract. The generation of lower hybrid waves below field-aligned potential drops and the effect of the resulting turbulence on the ion population in the suprauroral region are studied using particle plasma simulations. To describe the ion acceleration observed in the simulation, a theoretical model is developed using mode-mode coupling processes to generate the low-phase-velocity VLF waves with which the ions first interact. By scaling the simulation results, we show that interaction between the ions and the lower hybrid waves can account for the acceleration necessary to produce suprauroral energetic ion conic events.

Introduction

It is becoming more widely accepted that the energetic ion conics [Mizera et al., 1981] observed below field-aligned potential drops in the suprauroral region are produced as the result of ion acceleration by the VLF turbulence observed there [Mozer et al. 1980]. The turbulence is generated through the instability of the auroral electron distribution accelerated parallel to the geomagnetic field by the parallel electric field [Maggs and Lotko, 1981]. A model for the formation of ion conics in this way was proposed by Chang and Coppi [1981]. Acceleration nearly perpendicular to the field line by VLF turbulence near the lower hybrid frequency is followed by the adiabatic folding of velocities as the ions mirror and travel up the geomagnetic field line, creating the conic velocity distribution. Detailed calculations of conics using a Monte Carlo technique to model the wave-particle interaction were carried out by Retterer et al. [1983], while Crew and Chang [1985, see also this conference] have pursued analytical calculations.

Considerable uncertainty remains in the model, however, because of the difficulty in estimating the rate of the wave-particle interaction process. Empirical estimates of the velocity diffusion tensor based on observed wave amplitudes suffer because the lack of wavenumber measurements prevents us from determining the phase velocities of the waves. We cannot rely on linear calculations to give us the wave spectrum either, because there appears to be a real difficulty in linearly exciting waves of small enough phase velocity so that the resonant portion of the ambient ion distribution can account for the observed number of particles in the conics. In addition, the self-consistent evolution of the wave spectrum has been ignored in previous work.

Plasma Simulation

To address these problems, a plasma simulation was performed [Retterer et al., 1986] to provide an independent, self-consistent means of studying the generation of the turbulence and the resulting ion acceleration. The suprauroral situation was modeled by allowing a weak ($n_b/n_o \lesssim 10^{-2}$), energetic ($E_b = 1$ keV), warm ($T_b = 125$ eV) electron beam traveling along the magnetic field to destabilize a cool electron-ion plasma ($T_e = T_i = 2$ eV). We set the direction of propagation of the waves to be nearly perpendicular to the magnetic field, with $\cos^2\theta_B = m_e/m_i$ to reflect the commonly observed wave spectral peak near 1.5 times the lower hybrid resonance frequency. The velocity of the 1 keV electron beam projected onto the propagation direction is then about 32 times the initial ion thermal velocity. The phase velocities of waves excited by this beam will be far out on the tail of the ion velocity distribution, where few ions can resonantly interact with them.

Nevertheless, a finite fraction of the ions are significantly accelerated in the course of the simulation. We found that tails of energetic ions formed, emerging from both sides of the initial velocity distribution at about three times the ion thermal velocity; some ions are accelerated to velocities comparable to those of

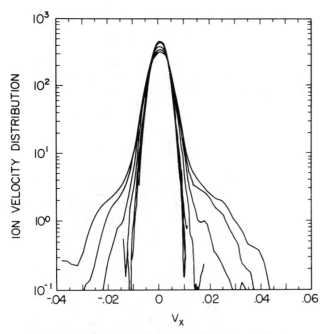

Fig. 1. The ion velocity distribution. Each curve is a plot of the ion velocity distribution at a fixed time. The times run from zero to $400/\omega_{LH}$ in increments of $40/\omega_{LH}$; the velocities are in arbitrary units in which the beam velocity is 0.08333.

the electron beam. This is illustrated in Figure 1, which presents a series of overlaid snapshots of the ion velocity distribution at different times. In addition to the tails, the core of the velocity distribution showed evidence of nonresonant heating. It can be fitted by a Maxwellian velocity distribution, in which changes in the fitted thermal velocity reflect the changes in the total wave energy. But the tails account for most of the energy transferred to the ions in the course of the instability; following wave saturation, they already contain 4% of the ions and account for half of the ion energy. In addition, electron acceleration parallel to the magnetic field (in both directions) is also observed, with the electrons reaching energies comparable to those of the ions. These accelerated electrons correspond to the Type 1 counterstreaming electrons, observed by DE-1 [Lin et al., 1982] in conjunction with ion conics.

The evolution of the fluctuation spectrum is illustrated in Figure 2. In this Figure we have plotted $|E(k,t)|^2$ as a function of k at four times, t, where $E(k,t)$ is the spatial transform of the electric field. At early times ($t \lesssim 240$), we see the linearly unstable modes (centered at k=16) emerging from the background noise. From this concentration in a relatively small region of k space, we subsequently see the wave energy redistribute itself throughout k space, through the action of mode coupling processes. The result is an approximately steady-state spectrum, achieved at $t \gtrsim 400$, which at smaller k takes the Rayleigh-Jeans form, $|E(k,t)|^2 =$ constant.

Interpretation

The interpretation of these results is clear, following from the extensive work devoted to the high-frequency analogue of the problem: electron tail formation in strong Langmuir turbulence [Kruer, 1976; Goldman, 1984; and references therein]. The intense VLF waves linearly excited by the beam parametrically decay into lower phase velocity VLF waves by coupling through nonresonant quasi-modes which are driven to finite amplitude in the turbulent state. These lower phase velocity VLF waves are then Landau damped by the plasma, accelerating the ions perpendicular to the magnetic field and the electrons (because of their restricted perpendicular mobility) parallel to the field.

Several calculations support this interpretation of our simulation. First, an analysis of the nonlinear dispersion relation [Porkolab, 1977] for the coupling of two lower hybrid waves through nonresonant quasimodes was performed. Using the amplitude and other parameters of one of the linearly excited waves in our simulation

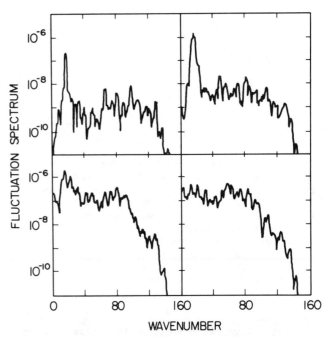

Fig. 2. The wave spectrum. In each panel is plotted $|E(k,t)|^2$ vs wavenumber k at a fixed time. In the upper row, the times are 160 and 240; in the lower row, 320 and 400. The wavenumbers are expressed in multiples of the fundamental wavenumber of the length of plasma included in the simulation.

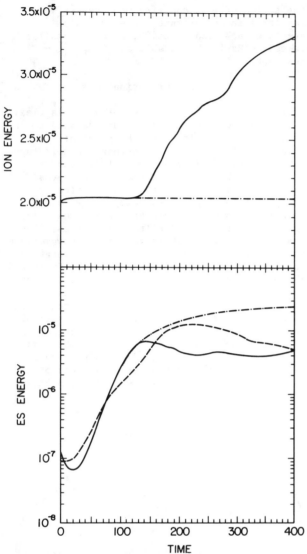

Fig. 3. The energetics of the theoretical model. The top panel gives the ion kinetic energy (in units in which the beam energy is 2.9×10^{-4}) as a function of time; the solid line gives the ion energy when mode coupling is included in the model, the dot-dash line the result without mode coupling. The bottom panel gives the electrostatic energy of the waves with the same coding of the lines as the top panel, with the addition of the dashed line showing the result from the particle simulation.

of kinetic equations, beginning with the quasi-linear equations and adding mode coupling terms to describe the nonlinear processes [Retterer et al., 1986]. Numerical solution of these equations produces an ion velocity distribution with high energy tails and a wave spectrum similar to the ones observed in the simulation. Figure 3 illustrates the energy budget of the resulting calculation, with the top panel giving the ion energy as a function of time and the lower panel the electrostatic energy of the waves. The units of energy are such that the initial beam energy density is 2.9×10^{-4}. We experimented with the model by removing the mode-coupling terms from one run to judge its effect on the evolution. The dot-dash lines are the results without the mode-coupling term; all the energy released by the beam instability remains concentrated at long wavelengths and little is shared with the ions. In contrast, with mode-coupling active most of the energy liberated is transferred to the ions, leaving much less in the waves, in agreement with the results of the particle simulation, presented here as the dashed line. In form, we see at initial times the exponential growth associated with the linear phase of the instability. Once the threshold for mode-mode coupling has been exceeded, a turbulent steady state in the waves is soon reached, while energy continues to be

as a pump wave, we calculated the phase velocities of the sideband waves excited by the three-wave coupling process. We found that these velocities agreed well with the velocities at the points where the tails emerge from the background distribution, supporting the argument that Landau damping of these sidebands accelerates the ions. We then formulated a simple set

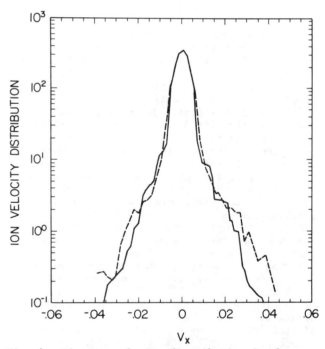

Fig. 4. The ion velocity distribution in the theoretical model. The solid line is a snapshot of the ion velocity distribution at $\omega_{LH}t = 400$. The dashed line gives the ion velocity distribution from the comparable time in the particle simulation.

transferred through the waves from the beam to the ions. In the solution without mode coupling, on the other hand, the wave energy continues to grow, although the growth rate slows as the beam velocity distribution forms a plateau at the phase velocities of the fastest-growing modes. Without mode coupling, the energy transferred to the ions is negligible.

The detailed ion velocity distribution in the mode coupling calculation shows the high-velocity tails characteristic of the particle simulation. Figure 4 shows a snapshot of the ion velocity distribution at $\omega_{LH}t=400$, illustrating the acceleration of the tails. The dashed curve here gives the comparable ion velocity distribution from the particle simulation.

We conclude from these comparisons that our theoretical model with mode-coupling does satisfactorily describe the phenomena observed in the particle simulations. While the simple resonant-quasilinear-diffusion model is adequate to describe the effect of the turbulence on the particles, a full nonlinear treatment is necessary to describe the evolution of the turbulence.

Conclusion

Scaling of our simulation results to suprauroral conditions gives results which agree well with data from observed ion conics; fraction of accelerated ions $\sim 10^{-3}$ to 10^{-2}; average energy ~ 50 eV; maximum energy ~ 1 keV. The one-dimensional model presented here does overestimate the wave electric field, predicting values of the order of 100 mV/m for the total spectrum. It does agree with the observations in demonstrating that an overwhelming portion of the energy removed from the beam goes into particle acceleration rather than waves, the proportion here being roughly ten to one. The simulation and analysis of the phenomena of lower hybrid acceleration in two or three dimensions must be postponed to a later report. As a preliminary, we note that an analysis of lower-hybrid parametric decay under suprauroral conditions has been performed by Koskinen [1985, see also this conference]. He reports that decay through nonresonant quasimodes is the dominant process and that the threshold of wave amplitude may be one mV/m or smaller, insuring that the mode-coupling processes which we have discussed can operate in the suprauroral region.

References

Chang, T., and B. Coppi, Lower hybrid acceleration and ion evolution in the suprauroral region, Geophys. Res. Lett., 8, 1253-1256, 1981.

Crew, G., and T. Chang, Asymptotic theory of ion conic distributions, Phys. Fluids, 28, 2382-2394, 1985.

Goldman, M.V., Strong turbulence of plasma waves, Rev. Mod. Phys., 56, 709-735, 1984.

Koskinen, H.E.J., Lower hybrid parametric processes on auroral field lines in the topside ionosphere, J. Geophys. Res., 90, 8361-8369, 1985.

Kruer, W.L., Saturation and nonlinear effects of parametric instability, Adv. Plasma Phys., 6, 237-269, 1976.

Lin, C.S., J.L. Burch, J.D. Winningham, and R.A. Hoffman, DE-1 observations of counterstreaming electrons at high altitudes, Geophys. Res. Lett., 9, 925-928, 1982.

Maggs, J.E. and W. Lotko, Altitude dependent model of the auroral beam and beam-generated electrostatic noise, J. Geophys. Res., 86, 3439-3447, 1981.

Mizera, P.F., J.F. Fennell, D.R. Croley, Jr., A.L. Vampola, F.S. Mozer, R.B. Torbert, M. Temerin, R.L. Lysak, M.K. Hudson, C.A. Cattell, R.G. Johnson, R.D. Sharp, P.M. Kintner, and M.C. Kelley, The aurora inferred from S3-3 particles and fields, J. Geophys. Res., 86, 2329-2339, 1981.

Mozer, F.S., C.A. Cattell, R.L. Lysak, M.K. Hudson, M. Temerin, and R.B. Torbert, Satellite measurements and theories of low altitude auroral particle acceleration mechanisms, Space Sci. Rev., 27, 155-213, 1980.

Porkolab, M., Parametric instabilities due to lower-hybrid radio frequency heating of tokamak plasmas, Phys. Fluids, 20, 2058-2075, 1977.

Retterer, J.M., T. Chang, and J.R. Jasperse, Ion acceleration in the suprauroral region: a Monte Carlo model, Geophys. Res. Lett., 10, 583-586, 1983.

Retterer, J.M., T. Chang, and J.R. Jasperse, Ion acceleration by lower hybrid waves in the suprauroral region, J. Geophys. Res., 91, 1609-1618, 1986.

ANALYTIC ION CONICS IN THE MAGNETOSPHERE

G. B. Crew, Tom Chang

M.I.T. Center for Space Research, Cambridge, MA 02139

J.M. Retterer

Space Data Analysis Laboratory, Boston College, Chestnut Hill, MA 02167

J.R. Jasperse

Air Force Geophysics Laboratory, Bedford, MA 01731

Abstract. We consider the formation of ion conics through the agency of lower hybrid wave turbulence in the Earth's magnetosphere. This process begins within a layer where significant wave-particle interactions result in a velocity space diffusion of the initially cold ion population. The conic results from subsequent adiabatic motion along geomagnetic field lines. We discuss a model for this process which permits an asymptotic determination of the ion distribution function. In particular, we assume that a suitable model for the energy spectral density of the turbulence is available. Identification of the ratio of ion thermal speed to mean wave speed as a small parameter leads to a uniformly valid asymptotic solution of the resulting quasilinear diffusion equation. These results are compared to a different approach in which a Monte Carlo procedure is used to obtain the ion distribution function. In this method, while approximations are not required in order to solve the equation, the solution is restricted by the need to include a sufficient number of particles in order to have reliable statistics. The two approaches are thus complementary, and used together can provide a wealth of information about the conic formation process. In particular we have found that conic formation is significantly affected by the presence of field-aligned potentials, as well as the spatial extent and spectrum of the turbulence.

I. Introduction

Ion conics are distributions of ions flowing upward into the magnetosphere which are strongly peaked in pitch angle and have energies significantly greater than one would expect for particles of ionospheric origin. The early observations of Sharp et. al. [1977], Ghielmetti et. al. 1978] and Klumpar [1979] were somewhat startling, largely because no mechanism for accelerating ions up to what are essentially magnetospheric energies had been anticipated. The likely agency for this acceleration is now considered to be some form of wave-particle interaction. One can then account for these distributions by coupling transverse heating at an altitude below the spacecraft with subsequent adiabatic motion in the decreasing geomagnetic field geometry. The pitch angle of the observed conic is thus determined by the ratio of the magnetic field strength at the satellite relative to that of the heating region.

Chang and Coppi [1981] have suggested that electron-beam-generated lower hybrid wave heating is responsible for the ion conics observed in the suprauroral region. Here the electron beam acts as a source of free energy which is transferred to the ions through the mediation of lower hybrid waves. In subsequent numerical work, Retterer et. al. [1983] were able to generate ion conics of the appropriate energies using a simple Monte Carlo simulation procedure. In this case, a numerical approach was essential owing to the fact that the heating region was extended over a distance on the order of an Earth radius, guaranteeing significant variations in many equilibrium quantities. On the other hand, for those events where the lower hybrid turbulence region is rather localized in altitude Crew and Chang [1985] have developed an analytic model for the conic formation process. For these events the magnetic field geometry plays no role in the heating process, thus the problem can be readily divided into two stages of heating and adiabatic motion.

The Monte Carlo procedure is quite powerful since it is sufficiently general to handle rather arbitrary wave and field geometries. However, it does take time to process each case, and the results are always limited by the statistics of the number of particles employed in the simulation. As an alternative we may use the analytic description which has the advantage of speed, but suffers from restrictions that must be imposed to retain validity.

In this paper we shall make use of the analytic approach to examine the variety of conics produced for differing spectra of lower hybrid wave turbulence. This allows us to readily examine some of the effects which, in addition to the amplitude of the lower hybrid waves, control the amount of ion heating. For example, it also depends on the ion residence time within the turbulent layer, and that in turn depends on other factors such as the field aligned potential structure. The Monte Carlo model can be used to verify these results and to determine the extent to which some of the restrictions on the analytic model can be relaxed.

In the next section we discuss briefly the assumptions required by the analytic model of Crew and Chang [1985], and summarize the principal results. The final section examines a variety of cases and draws implications for the interpretation of satellite data.

II. Analytic Model for Ion Conics

The conic formation process is essentially an evolution of a gyrotropic ion distribution that proceeds along the geomagnetic field line.

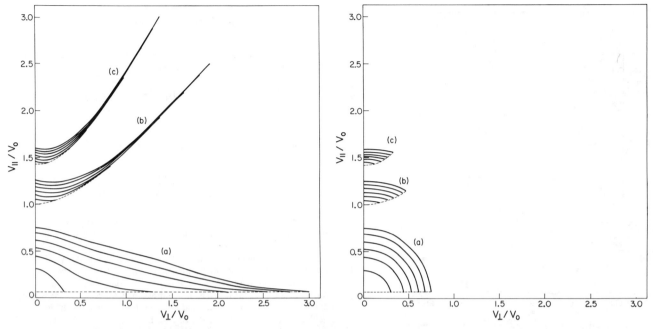

Fig. 1. This figure compares the generation of a conic which forms in the presence of turbulence (left panel) with the formation of a beam by electrostatic acceleration (right panel). Contours of constant f are separated by a factor of 10. Cases (a), (b) and (c) correspond to altitudes of 1010, 3500 and 6000 km respectively.

Thus it is natural to seek a solution of the form $f(l,v_\parallel,v_\perp)$ where the variable l is an arc length along the magnetic field line, and the velocity variable is resolved into components parallel and perpendicular to the field. One could also include dependencies on the neglected time and spatial variables, however, these are not essential to the formation of conics and are neglected here in the interest of simplicity.

The presence of lower hybrid turbulence over a range of altitudes ($-l_D \leq l \leq 0$) induces a velocity space diffusion of ions. This may be described by a diffusion coefficient

$$D = (q/m)^2 \int d\mathbf{k}\, 8\pi^2\, \delta(\omega - \mathbf{k}\cdot\mathbf{v})\, E(\mathbf{k})\mathbf{k}\,\mathbf{k}\, k^{-2} \quad (1)$$

which includes the unmagnetized response of ions resonant with the waves. Here $E(\mathbf{k})$ is the electrostatic spectral energy density of the lower hybrid waves; once this is known, the velocity dependence of the diffusion coefficient (1) can be determined. While observational data places some constraints on this quantity, our need to know the k-dependence of the spectrum forces us to look elsewhere. Specifically, we turn to local plasma simulations of the electron-beam-driven lower hybrid instability, described by Retterer et. al. [1986a, 1986b]. In these simulations both the electron beam and ion distributions can be treated kinetically, and complete spectral information on the lower hybrid waves is available. One of the important results of this work is that the wave spectrum which ultimately results departs significantly from that expected from linear analyses due to the effects of parametric and cascading processes. The parametric processes have also been discussed by Koskinen [1986].

Noting that the **k** vector of these waves is essentially orthogonal to the magnetic field, we see that the diffusion is predominantly in the perpendicular velocity variable. The appropriate equation for the evolution of f is then

$$v_\parallel \frac{\partial f}{\partial l} + \frac{qE_\parallel}{m}\frac{\partial f}{\partial v_\parallel} + \frac{v_\perp^2}{2B}\frac{dB}{dl}\left[\frac{v_\parallel}{v_\perp}\frac{\partial f}{\partial v_\perp} - \frac{\partial f}{\partial v_\parallel}\right] \quad (2)$$

$$= \frac{D_o}{v_\perp}\frac{\partial}{\partial v_\perp} v_\perp \Delta\left[\frac{v_\perp}{v_o}\right]\frac{\partial f}{\partial v_\perp},$$

where $B(l)$ is the magnetic field strength and $E_\parallel(l)$ is the parallel electric field. The diffusion coefficient has been separated into the product of a dimensional coefficient

$$D_o \equiv (2\omega_{pi}^2/\omega_o)\,(W_E\Delta_o/n_o\, m) \quad (3)$$

and a function $\Delta(R)$ (where $R \equiv v_\perp/v_o$) which describes the form of the velocity dependence of the diffusion. (Here q, m, n_o and ω_{pi}^2 are the ion charge, mass, density and plasma frequency, and ω_o and v_o are typical values of frequency and phase velocity for the waves.) In this paper we approximate this function by

$$\Delta(R) \approx \frac{R^\alpha}{1 + \eta\, R^{\alpha+3}} \quad (4)$$

where the parameters Δ, η, α, and the electrostatic energy density W_E are determined by the function $E(\mathbf{k})$. For details we refer the reader to Crew and Chang [1985]. We note, however, that $\Delta_o \sim 1$ and $\eta \sim 1/2$. The exponent α (>0) is proportional to the spectral index of the k spectrum: large values of α correspond to spectra that decay quickly at large **k**.

From Eq. (2), we note that the effects of diffusion enter the problem on the velocity scale of v_o and this quantity is typically much larger

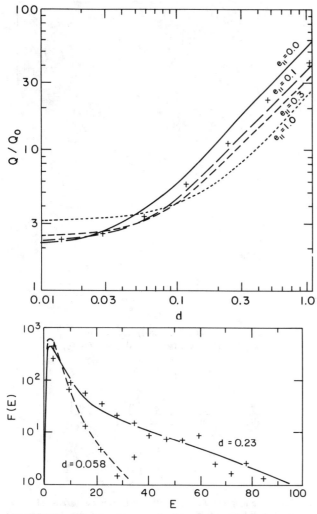

Fig. 2. This figure demonstrates the dependence of heating on the parameters d and $e_\parallel$, the former being the more pronounced. The energy flux Q carried upward by the ions is plotted in the top panel, while the energy distributions $F(E)$ (with arbitrary normalization) for two example cases are compared in the lower panel.

than the thermal velocity of the ambient ions, v_t. This fact, together with the neglect of the mirror force terms in Eq. (2) makes possible an asymptotic solution for f. The result is

$$f = f_D(l,v_\parallel,v_\perp) = \frac{n_o}{\pi^{3/2} v_t^3} \exp\left[-\frac{v_\parallel^2}{v_t^2} - \frac{e_\parallel(l+l_D)}{l_D}\right] f(t,R) \quad (5)$$

valid within the diffusion layer, i.e., for $-l_D \leq l \leq 0$. In Eq. (5) we have taken the electric field to be constant, and the function

$$f(t,R) \approx \left[\frac{P}{R}\frac{\partial P}{\partial R}\right]^{1/2} \exp\left\{-\varepsilon^{-2} P^2 [1 + 4t\Delta(P)]\right\} \quad (6)$$

describes the perpendicular velocity diffusion in terms of an implicitly defined velocity variable $P(t,R)$,

$$4t\, P\, \Delta(P)^{1/2} = \int_P^R dx\, \Delta(x)^{-1/2} \quad , \quad (7)$$

and a variable t proportional to the time it takes a particle to drift from l_D to 0. For this case of a constant electric field,

$$t(l,v_\parallel) = \frac{d}{\varepsilon^2 e_\parallel} \left\{ \frac{v_\parallel}{v_t} - \left[\left(\frac{v_\parallel}{v_t}\right)^2 - \frac{e_\parallel(l+l_D)}{l_D}\right]^{1/2} \right\} \quad (8)$$

where for convenience we have defined parameters

$$d \equiv \frac{D_o\, l_D}{v_o^2\, v_t}, \quad e_\parallel \equiv \frac{2qE_\parallel\, l_D}{mv_t^2}, \quad \varepsilon \equiv \frac{v_t}{v_o} \quad . \quad (9)$$

To understand this solution we note that the function $f(t,R)$ is the perpendicular velocity distribution of ions at a given location l having the same $v_\parallel$. This factor, which contains all of the details of the heating, is then weighted by the relative numbers of such ions. While the detailed dependence of $f(t,R)$ on its arguments is rather involved, its behavior is straightforward since it describes a diffusion in R that proceeds for a "time" t. Thus at $t=0$ it assumes the form of its "initial" condition, i.e., a Maxwellian, and for $t > 0$ this thermal "bulk" is depleted to allow the formation of a high energy "tail," the shape of which is largely determined by the function $\Delta(R)$. Clearly, larger values of t correspond to greater tail populations–more heating. From Eq. (8) we are not surprised to note that the largest values of t result from the smallest values of $v_\parallel$. Moreover, the heating monotonically increases with altitude. Additionally, we note from the definition of the parameter d in Eqs. (3) and (9) that the wave power enters in combination with the thickness of the turbulent layer, so that we would obtain the same amount of heating with half the wave power provided the layer were twice as thick.

As noted above, this solution is only asymptotically valid. In this case, it turns out that the approximation breaks down for $\varepsilon^2 t \lesssim 1$, which certainly occurs for $d \lesssim 1$. This does not have implications for the efficiency of this heating mechanism in such regimes; it merely means that expression (5) no longer provides a good description of the distribution function. We shall return to this point in the next section. There is a further limitation in that the expression (6) is a rather crude approximation when $\alpha < 2$ and $R \ll 1$. This limitation, however, is not of any great importance for the work we present here.

As noted, the distribution with the maximum heating is found at the top of the layer. Above this altitude, the particle motion is entirely adiabatic. The distribution function is then entirely expressed by

$$f(l,v_\parallel,v_\perp) = f_D(0,v_{\parallel 0}(l), v_{\perp 0}(l)) \quad , \quad (10)$$

where the trajectories $v_{\parallel 0}(l), v_{\perp 0}(l)$ follow from energy and magnetic moment conservation, and are given by

$$v_{\parallel 0}(l) = \left\{ v_\parallel^2 + v_\perp^2 [1-B(0)/B(l)] - e_\parallel v_t^2\, l/l_D \right\}^{1/2} \quad (11)$$

$$v_{\perp 0}(l) = v_\perp [B(0)/B(l)]^{1/2} \quad . \quad (12)$$

An illustrative example from Crew and Chang [1985] is shown in Figure 1 where we contrast the results obtained with and without heating. The contour plot at the bottom of each panel represents the ion distribution function which would be found at the top of a heating layer

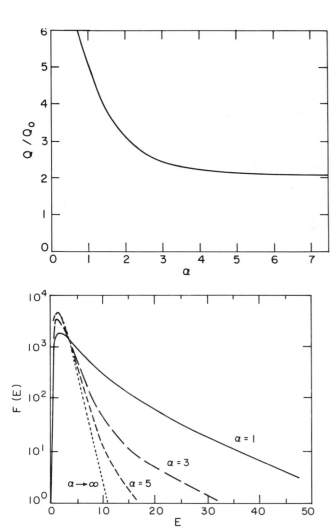

Fig. 3. Here we consider the qualitative importance of the parameter α. Again, we show the energy flux and several corresponding distributions.

10km thick in which $W_E = 10^{-13}$ erg/cm^3 ($E_{rms} = 33$ mV/m) and $\omega_{lh}/2\pi = 1.5$ kHz, which corresponds to a typical VLF hiss event reported by James [1976]. The diffusion coefficient had $\alpha = 2$, $\eta = 9/16$ and $\varepsilon = 0.2$. The upper plots in each panel represent f at higher altitudes as determined by Eqs. (10-12). While clearly the distribution in the left panel evolves into a conic, all that we obtain in the right panel (no heating) is a beam accelerated by the field-aligned electrostatic potential. We note also that since $\Delta(0) = 0$, the distributions are in fact equal at $v_\perp = 0$.

III. Results and Discussion

In this section we give some consideration to the range of possible distribution functions that may result from different values of the parameters d, $e_\parallel$ and α. The top of the heating layer is the natural place to examine these distributions, since as noted above, by this altitude the ions have gained all the energy from the waves that they can. In order to facilitate comparisons with the Monte Carlo method of generating the distribution function $f(0, v_\parallel, v_\perp)$ it is useful to introduce a reduced distribution

$$F(E) = (1/n_o) \int dv\, f(0, v_\parallel, v_\perp)\, \delta(E - v^2/v_t^2) \tag{13}$$

and the total energy flux carried by the ion distribution,

$$Q = \int dv (1/2) m(v_\parallel^2 + v_\perp^2) v_\parallel f(0, v_\parallel, v_\perp). \tag{14}$$

These quantities are readily calculated for an unheated distribution:

$$F(E) = \max\left\{\pi^{-1/2}(E^{1/2} - e_\parallel^{1/2}) \exp(-E + e_\parallel), 0\right\}, \tag{15}$$

and $Q_o \equiv n_o\, m v_t^3 / 4\pi^{1/2}$.

In Figure 2 we examine the effects on Q of varying d, for several values of $e_\parallel$. As noted above, this can be interpreted as variations in W_E, l_D, or both. Here the curves correspond to the analytic model (5), while the "+" points are calculated with the Monte Carlo code for $e_\parallel = 0.1$. As expected, the agreement is excellent for $d \to 0$, but even as $d \to 1$, the analytic values are off by at most 15% (where we recall the approximation has broken down). We also note that Q varies essentially linearly with d of order unity. The effects of changing d are also seen by comparing $F(E)$ for two cases. Here a change in d by only a factor of four (e.g. doubling E_{rms}) results in a dramatic increase in the tail ion population. The corresponding decrease in the number of bulk ions is more subtle since there are more of them.

The effects of parallel electric fields on heating is most pronounced for Q as seen in Figure 2. With little heating, the ions can gain energy only through electrostatic acceleration, thus Q increases with $e_\parallel$. On the other hand, when there is significant energy to be gained through wave-particle interactions, an increase in the parallel electric field decreases the residence time of ions in the heating layer, thus Q decreases with $e_\parallel$.

To this point, we have discussed only a single form for the diffusion coefficient, i.e., $\Delta(R)$ given by expression (4) with $\alpha = 2$. If we vary α, this corresponds to changes in the distribution of energy among the different wave modes, but with W_E held fixed. (The typical phase velocity v_o is also held fixed.) Thus in Figure 3, we see that by increasing α, the energy gained by the ions is effectively turned off, and the tail population decreases. In this limit, the amplitude of those waves which are resonant with the thermal ions is much less, hence fewer ions diffuse to higher energies. On the other hand, as α decreases, this heating process becomes much more efficient, which results in greater energy fluxes and larger tail populations.

From these results it is clear that in order to estimate the energization of conics by lower hybrid waves it is essential to know more than the total energy density of the waves. As a minimum, the thickness of the heating layer (and ideally the spatial distribution of wave energy along the field line) must be known. Furthermore, any factors which affect the ion residence time in the heating layer enter into consideration with equal importance. The average $E_\parallel$, which we have included here, is merely the most obvious factor; complicated field structures, e.g., double layers, should also be incorporated in the analysis of real events if they are likely to be present. Ideally, we should also like to know $E(\mathbf{k})$ all along the field line. In practice, this is simply not possible. However, as Figure 3 underscores, we at least need some general idea of the distribution of wave energy in $\mathbf{k}$ space. At present such information comes to us only through numerical simulations, and an understanding of the physical mechanisms which generate the waves in the first place. We have not discussed this here owing to limitations of space and the fact that more complete discussions are available elsewhere (see Retterer et. al. [1986a, 1986b] and references therein).

Finally, we note that several of the assumptions made in this analysis (e.g., the monochromaticity of the waves and the spatial homogeneity within the heating layer) can easily be relaxed to accomodate situations where more specific information is available. We do not pursue this issue here owing to the limitations of space and to the additional ambiguities introduced. However, these considerations are of importance when making a close contact with the observations, and we expect to address this in more detail elsewhere.

Acknowledgements. This work was supported by the Air Force Office of Scientific Research, and AFGL Contract Nos. F19628-83-C-0060 and F19628-86-K-0005.

References

Chang, T., and B. Coppi, Lower hybrid acceleration and ion evolution in the suprauroral region, *Geophys. Res. Lett.*, 8, 1253, 1981.

Crew, G.B., and T.S. Chang, Asymptotic theory of ion conic distributions, *Phys. Fluids.*, 28, 2382, 1985.

Ghielmetti, A.G., R.G. Johnson, R.C. Sharp, and E.G. Shelley, The latitudinal, diurnal, and altitudinal distributions of upward flowing energetic ions of ionospheric origin, *Geophys. Res. Lett.*, 5, 59, 1978.

James, H.G., VLF saucers, *J. Geophys. Res.* 81, 501, 1976.

Klumpar, D.M., Transversely accelerated ions: an ionospheric source of hot magnetospheric ions, *J. Geophys. Res.*, 84, 4429, 1979.

Koskinen, H.E.J., Parametric processes of lower hybrid waves in multicomponent auroral plasmas, this volume, 1986.

Retterer, J.M., T. Chang, and J.R. Jasperse, Ion acceleration in the suprauroral region: a Monte Carlo model, *Geophys. Res. Lett.*, 10, 583, 1983.

Retterer, J.M., T. Chang, and J.R. Jasperse, Ion acceleration by lower hybrid waves in the suprauroral region, this volume, 1986a.

Retterer, J.M., T. Chang, and J.R. Jasperse, Ion acceleration by lower hybrid waves in the suprauroral region, *J. Geophys. Res.*, 91, 1609, 1986b.

Sharp, R.D., R.G. Johnson, and E.G. Shelley, Observations of an ionospheric acceleration mechanism producing energetic (keV) ions primarily normal to the geomagnetic field direction, *J. Geophys. Res.* 82, 3324, 1977.

PARAMETRIC PROCESSES OF LOWER HYBRID WAVES IN MULTICOMPONENT AURORAL PLASMAS

Hannu E.J. Koskinen

Uppsala Ionospheric Observatory, S-755 90 Uppsala, Sweden

Abstract. Parametric instabilities of lower hybrid waves are considered in multicomponent plasmas. The plasma parameters are chosen to correspond to plasmas on auroral field lines at altitudes where acceleration of ionospheric ions perpendicular to the geomagnetic field is expected to take place. Since the wave vector matching is difficult to satisfy under such conditions, parametric coupling through nonresonant quasi-modes becomes important. At altitudes of 2000-3000 km, the threshold wave electric fields for such processes are found to be on the order of 1 mV/m. These values are within the measured amplitudes of lower hybrid waves and it is suggested that parametric processes convert the keV auroral electron beam driven lower hybrid waves into waves of lower phase velocity perpendicular to the geomagnetic field, thus producing waves that are more effective in the transverse acceleration of ions.

Introduction

The lower hybrid waves (LHW) are considered as a promising mechanism to accelerate ionospheric ions perpendicular to the geomagnetic field leading to production of ion conics. An important contribution to the development of this idea was the suggestion by Chang and Coppi [1981] that the LHW's might transfer energy from a keV auroral electron beam to ionospheric ions. An appealing feature of this mechanism is that all of the main building blocks of the model, keV auroral electrons, LHW's, and ion conics, are commonly observed on the geomagnetic field lines above the auroral region. A cartoon of this energy transfer sequence is shown in Figure 1.

The excitation and amplification of LHW's by an auroral electron beam of energies 1-10 keV was studied in a series of publications by Maggs and Lotko [Maggs, 1978; Lotko and Maggs, 1979; Maggs and Lotko, 1981]. They pointed out that convective amplification of the waves propagating inside the beam region must be considered including the damping effects of suprathermal electron populations as well as nonlocal effects of the changing magnetic field and plasma density gradients along the magnetic field. Their theoretical results are quite consistent with satellite observations [see, e.g., Gurnett and Frank, 1977; Temerin et al., 1981]. The other end of this energy transfer chain, i.e., the acceleration of ions by LWH's, was studied by Retterer et al. [1983] by simulating the quasi-linear diffusion by LHW's applying Monte Carlo technique. Recently they have conducted a more thorough simulation study of the entire energy transfer chain from an electron beam to ions through LHW's [Retterer et al., this conference].

The question, whether these two processes can be coupled together, is not as simple as it may appear. At altitudes where ion conics are produced, i.e., from 500 km to a few thousand km, the background electron and ion temperatures can be assumed to be roughly equal. By solving the linear dispersion equation for such a plasma including the keV electron beam, one finds that the perpendicular phase velocities are rather large compared with the thermal velocity of the ions, and thus only a few of the ions in the tail of the distribution function could be energized. We look for a solution to this problem by asking whether the electron beam driven LHW could be converted to a wave of a larger perpendicular wave number. The answer is "yes" if the amplitude of the LHW exceeds the threshold for a lower hybrid parametric process (LHPP). We discuss the theory of such processes and derive estimates of the threshold wave electric fields.

Wave modes

The normal wave modes of the plasma, that are important in the parametric processes of the lower hybrid waves, are lower hybrid waves (LHW), ion Bernstein waves (IBW) and electrostatic ion cyclotron waves (EIC). We also consider quasimodes which are either heavily damped normal modes, like ion cyclotron quasi-modes (ICQM) or ion acoustic quasi-modes (IAQM), or nonresonant quasi-modes (NRQM) which are not solutions to the linear dispersion equation in the absence of the pump wave.

In Figure 2 we plot some of the solutions to the full electromagnetic dispersion equation in a homogeneous anisotropic Maxwellian plasma. The

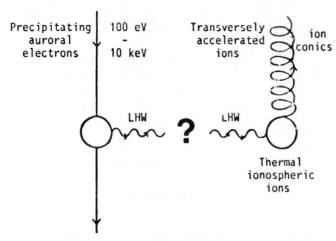

Fig 1. Energy transfer from precipitating electrons to ionospheric ions through lower hybrid waves.

solutions have been found by using the computer code WHAMP [Rönnmark, 1982; 1983]. The plasma model consists of a background magnetic field of 0.21 Gauss, and of electrons and hydrogen ions (density 3×10^9 m^{-3}, temperature 1 eV). We call this model Model A. The real parts of the solutions form "dispersion surfaces" in the $(\omega,\underline{k})$-space. In the figure, we have pointed out the LHW's, whistlers, and the lower hybrid "resonance" which all are located on the same dispersion surface. The frequency of the lower hybrid resonance is called the lower hybrid frequency and is given by

$$\omega_{LH}^2 = \frac{\sum_i \omega_{pi}^2}{1 + \frac{\omega_{pe}^2}{\Omega_e^2}} \quad (1)$$

where ω_{pi}'s are the plasma frequencies of different particle species and Ω_e the electron gyrofrequency. The LHW part of the dispersion surface is almost electrostatic and it is sufficient to study the LHW's in the electrostatic approximation. Ion Bernstein modes are found between all ion cyclotron harmonics. They were originally derived by Bernstein [1958] in the limiting case of perpendicular propagation. However, these modes are weakly damped as long as $\omega \gg k_\parallel v_{te}$ where v_{te} is the thermal velocity of electrons given by $v_{te} = \sqrt{2T_e/m_e}$, and T_e is the electron temperature. In Figure 3 we plot some of the solutions to the linear dispersion equation from a different angle so that magnitudes of the relevant frequencies and wave numbers become clearer.

The parametric dispersion equation

By a parametric process in a plasma we mean a coupling of three (or more) wave modes in an approximation where the coupling is assumed to be so weak that each of the modes may be treated in a linear approximation. We usually assume that there exists an externally supplied mode, the "pump", that is not particularly affected by the paramet-

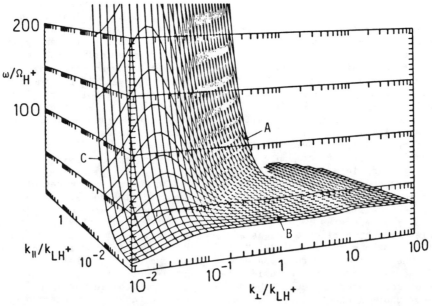

Fig 2. The dispersion surface where LHW's are located in Model A. The region of LHW's on the surface is denoted by A, the lower hybrid "resonance" by B, and the whistler mode by C. The wave number is normalized to the Larmor wave number $k_{LH^+} = 1/\rho_{H^+}$.

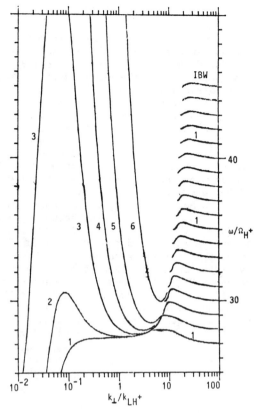

Fig 3. A part of the lower hybrid dispersion surface and some of the IBW's in Model A. The curves correspond to normalized parallel wave numbers ($k_\parallel \rho_H+$) as follows: (1) 10^{-3}, (2) 2×10^{-3}, (3) 5×10^{-3}, (4) 10^{-2}, (5) 2×10^{-2}, (6) 5×10^{-2}. All IBW's are plotted for the same parallel wave number (1).

ric process. The parametric process occurs when the pump wave beats with some (resonant or nonresonant) response of the plasma producing a sideband of the pump. There is a threshold for the process; i.e., the amplitude of the pump must be above a certain value that is defined by the damping of the sideband and of the low-frequency response, convective losses and other stabilizing mechanisms. The wave coupling must conserve energy and linear momentum. Conservation of energy is expressed in the frequency matching condition $\omega_0 = \omega_1 + \omega_2$ where ω_0 is the frequency of the pump wave and ω_1, ω_2 are the frequencies of the low-frequency response and the sideband, respectively. Similarly, conservation of momentum requires wave vector matching $\underset{\sim}{k}_0 = \underset{\sim}{k}_1 + \underset{\sim}{k}_2$. Strictly speaking, this is not quite true since, if we are considering kinetic modes, plasma particles may carry a part of the energy and momentum. In fluid-like calculations, a small mismatch can be included phenomenologically.

By lower hybrid parametric processes (LHPP) we denote parametric processes where the pump wave frequency is close to and above the lower hybrid frequency of the plasma. For LHPP's, the wave vector matching is often very difficult to satisfy if only resonant decay processes, i.e., processes where all participating modes are normal modes of the plasma, are allowed. However, the presence of the pump may create a nonresonant quasi-mode response of the plasma. These quasi-modes do not need to be real oscillations but their role is to transfer the excess of energy and momentum to the particles of the plasma.

We apply a formalism introduced by Porkolab [1977; Ono et al., 1980]. The theory is based on the dipole approximation, i.e., the wave length of the pump is assumed to be much larger than the wave lengths of the daughter waves. In processes we are studying, this often is a good approximation but its validity must be considered from case to case. We consider a plasma consisting of electrons and of a finite number of positive ion species in a homogeneous magnetic field. The particle distributions are assumed to be Maxwellian. An external uniform oscillating electric field (the pump wave) of the form $\underset{\sim}{E}_0(t,\underset{\sim}{r}) = (E_{0\perp}\hat{x} + E_{0\parallel}\hat{z}) \cos(\omega_0 t - \underset{\sim}{k}_0 \cdot \underset{\sim}{r})$ is assumed to be applied to the plasma. The coordinate system is such that the magnetic field is in the z-direction. By considering electrostatic perturbations about the (oscillating) equilibrium of this plasma, a lengthy and tedious algebraic derivation leads to the dispersion equation [Ono et. al, 1980]

$$1 + \sum_j \sum_k \sum_\sigma \frac{|\mu_j - \mu_k|^2}{8\epsilon\epsilon^\sigma}$$

$$\left\{ (\chi_j^\sigma - \chi_j)(\chi_k^\sigma - \chi_k) - \epsilon \chi_j^\sigma \chi_k^\sigma \right\} = 0 \quad (2)$$

where j, k denote the particle species, σ denotes the sidebands: + (the upper), - (the lower), the linear susceptibilities are denoted as $\chi_j = \chi_j(\omega,\underset{\sim}{k})$, $\chi_j^\pm = \chi_j(\omega \pm \omega_0, \underset{\sim}{k})$, and ϵ, ϵ^σ are the dielectric functions mode without superscripts, + or -, is the low-frequency response of the plasma. The coefficients μ_j are given by

$$\mu_j = \frac{q_j}{m_j} \left\{ \left| \frac{E_{0\parallel} k_\parallel}{\omega_0^2} + \frac{\underset{\sim}{E}_{0\perp} \cdot \underset{\sim}{k}_\perp}{\omega_0^2 - \Omega_j^2} \right|^2 + \left| \frac{(\underset{\sim}{E}_{0\perp} \cdot \underset{\sim}{k}_\perp) \cdot \Omega_j \hat{z}}{(\omega_0^2 - \Omega_j^2)\omega_0} \right|^2 \right\}^{1/2} \quad (3)$$

where Ω_j is the gyrofrequency of species j. The dispersion equation (2) has been derived in the dipole approximation ($k_0 = 0$) and the coupling is assumed to be weak ($\mu_j \ll 1$ for all species).

In many of the interesting cases of the LHPP's, $|\chi_j^\pm| \ll |\chi_j|$. Furthermore, we are interested in processes under conditions where $\mu_e \ll \mu_j$ holds (i.e., electron coupling dominates over ion coupling). These approximations simplify the parametric dispersion equation into the form

$$1 + \frac{\mu_e^2}{4\varepsilon} \chi_e \sum_i \chi_i \left(\frac{1}{\varepsilon^+} + \frac{1}{\varepsilon^-} \right) = 0 \quad (4)$$

We assume that the frequency of the low-frequency response is higher than the damping rate of the sidebands, thus the upper sideband can be neglected as off-resonant.

The (lower) sideband is a resonant mode of the plasma. Thus we may expand ε^- about $\varepsilon_R^- = 0$ (the subscripts R and I refer to real and imaginary parts, respectively, and we denote $\omega_2 = \omega_0 - \omega$):

$$\varepsilon(\omega_2) = -i(\gamma + \Gamma_2) \left| \frac{\partial \varepsilon_R}{\partial \omega} \right|_{\omega = \omega_2} \quad (5)$$

Here Re γ is the growth rate of the sideband and all linear damping mechanisms of the sideband are included in Γ_2. Im γ gives the nonlinear frequency shift which, however, is neglected in the following analysis. Thus equation (4) reduces to the form

$$\gamma + \Gamma_2 = \frac{-\mu^2}{4 \left| \frac{\partial \varepsilon_R}{\partial \omega_2} \right|} \text{Re} \left(\frac{i \chi_e(\omega_1) \sum_i \chi_i(\omega_1)}{\varepsilon(\omega_1)} \right) \quad (6)$$

Here and hereafter we denote the frequency of the low-frequency response by ω_1 in order to avoid confusion and drop the superscript "-" of the sideband. The threshold for the process is found by setting $\gamma = 0$ and solving for μ (or for the electric field). If the low-frequency response is also a resonant mode (EIC or IBW), $\varepsilon(\omega_1)$ can be expanded in the same way.

We discuss LHW's and IBW's as possible sidebands. If the sideband is a LHW, the derivative $|\partial \varepsilon_R / \partial \omega_2|$ can be written in a compact form [Koskinen, 1985]

$$\frac{\partial \varepsilon_R}{\partial \omega_2} = \frac{2}{\omega_2} \left(1 + \frac{\omega_{pe}^2}{\Omega_e^2} + \sum_i \frac{3\omega_{pi}^2}{2\omega_2^4} k_\perp^2 v_{ti}^2 \right) \quad (7)$$

where v_{ti} is the thermal velocity of ions of species i. For an IBW we find

$$\frac{\partial \varepsilon_R}{\partial \omega_2} = \sum_i \sum_{n=1}^{\infty} \frac{4n^2 \omega_{pi}^2 \omega_2}{(\omega_2^2 - n^2 \Omega_i^2)^2} \frac{\Gamma_n(b_i)}{b_i} \quad (8)$$

$\Gamma_n(b_i) = I_n(b_i) \exp(-b_i)$, I_n is the modified Bessel function of the first kind of order n, $b_i = k_\perp \rho_i / 2$, and ρ_i the ion gyroradius $\rho_i = v_{ti} / \Omega_i$.

One ion species

We compute thresholds of LHPP's in two models of plasma at an altitude about 2500 km. Model A was introduced in Figures 2 and 3. Model B is otherwise the same but the plasma density is a factor of 15 smaller.

Based on the results of the previously referred works by Maggs and Lotko and on typical satellite spectra, we assume that the peak amplitude of the pump LHW lies on the lower hybrid surface at a frequency of about $\sqrt{2} \, \omega_{LH}$. By using the program WHAMP we have examined numerically where in the wave number space an electron beam of energy of 4 keV and temperature of 1 keV excites LHW's of that frequency. In Model A, $k_\parallel \rho_i$ varies approximately between 10^{-2} and 4×10^{-2} and $k_\perp \rho_i$ between 0.4 and 2. In Model B, the corresponding intervals are from 5×10^{-3} to 2×10^{-2} and from 0.2 to 0.8. The density of the beam was assumed to be 1% of the density of the ambient plasma. From the discussion by Koskinen [1985] we conclude that wave vector matching for a resonant decay channel like LHW -> LHW + EIC can hardly be satisfied. Thus we expect that the low-frequency response is a NRQM.

In models A and B the electron coupling dominates over the ion coupling. Because we are mainly interested in pump frequencies not far from ω_{LH}, we can use the inequality $\omega_0 \ll \Omega_e$. Thus μ_e reduces to

$$\mu_e \cong \frac{e}{m_e} \left(\frac{E_{0\parallel}^2 k_\parallel^2}{\omega_0^4} + \frac{E_{0x}^2 k_y^2}{\omega_0^2 \Omega_e^2} \right)^{1/2} \quad (9)$$

where we have chosen $\underset{\sim}{E}_{0\perp} \parallel$ x-axis. The first term comes from the parallel drift and the second from the $\underset{\sim}{E} \times \underset{\sim}{B}$-drift of the background electrons due to the electric field of the pump. By some simple algebra, it can be shown that the $\underset{\sim}{E} \times \underset{\sim}{B}$-term dominates over the parallel term [Koskinen, 1985].

In many applications, our plasma models would be considered as collisionless. However, the collisionless damping rate of the LHW's and IBW's (except very close to ion cyclotron harmonics) is so small that the collisional damping determines the thresholds of the parametric processes. We estimate the collisional damping rates by using the Krook model. The application of this model for collisions of charged particles may be criticized since it does not conserve energy and momentum. However, we want to find estimates for the thresholds in space plasmas and thus the inaccuracy introduced by a non-self-consistent collision model is a minor problem. Estimates of the damping rates with sufficient accuracy are given by $\Gamma \cong \nu_e/2$ for LHW's and $\Gamma \cong \nu_i$ for IBW's [see, e.g., Kaw, 1976]. We compute the electron collision frequency ν_e according to Banks [1966] as in Koskinen [1985]. In Model A this gives $\nu_e = 0.13 \, s^{-1}$. The ion collision frequency is more difficult to determine. We estimate it as $\nu_i \cong (m/m_e)^{1/2} \nu_e$. We use the formulae given by Porkolab [1977] for the susceptibilities of the low-frequency response.

We have surveyed a wide range of "reasonable" values of variables in the (ω, k)-space by assuming that the pump frequency is $\sqrt{2}\,\omega_{LH}$. For the channel LHW -> LHW + NRQM, typical threshold electric fields are between 1 and 2 mV/m in Model A. In Model B the corresponding values are smaller by a factor of about 3. For the channel LHW -> IBW + NRQM, the threshold values vary more than in the case where the sideband is a LHW. The smallest ones are on the same order of magnitude as above but slightly higher (about by a factor of 2) in both models. Notice that, at sideband frequencies above ω_{LH}, there are frequency gaps just below the ion cyclotron harmonics.

It is interesting to compare our results with simulations by Retterer et al. [this conference]. In spite of small wave numbers of the LHW's excited by electrons, they "observed" substantial ion heating which they addressed as a consequence of a parametric process. In order to get considerable heating they needed, however, wave amplitudes of tens of mV/m. The reason for this appears to be the one-dimensional simulation; electrons along the magnetic field and ions almost perpendicular to it. Thus the $\underline{E}\times\underline{B}$-term in the coupling coefficient is not taken into account. As discussed above, it is not only an important term, it is the dominating one.

Multi-ion effects

Addition of more ion species into the plasma has two main effects on the physics described by the parametric dispersion equation (2); Firstly, differential streaming between the ion populations in the pump electric field can provide the coupling, and secondly, each new ion species increases the number of resonant wave modes available for decay processes.

In the LHPP's, the electron coupling dominates over the ion coupling, except if the density of the plasma is so low that the lower hybrid frequency is close to the ion cyclotron frequency, or if the electron currents cancel so completely that

$$|\mu_e|^2 < |\mu_{i1} - \mu_{i2}|^2 \qquad (10)$$

where i1 and i2 denote the different ion species. In the energy transfer mechanism we study in this work, these cases are not of importance.

From the point of view of our study, a more important multi-ion effect on LHPP's is the increase in the number of resonant modes that can participate in the processes. Under suitable conditions, this can make a resonant decay possible, e.g., between ion Bernstein modes of the different ion species. Such decay processes may have much smaller threshold values as nonresonant processes discussed in this work.

Discussion

We have used very simplified plasma models in this analysis of LHPP's. A more rigorous discussion would require that, e.g., the effects of the finite pump wave length, magnetic shear due to the field aligned currents, and density inhomogeneities should be carefully considered. Especially the density gradients may be important in determining the threshold electric fields. Porkolab [1977] has estimated such threshold values. If we apply his result for the channel LHW -> LHW + NRQM to our Model A, assume the wave length of the daughter waves to be 5 m and the gradient scale length 100 km, we find a threshold of about 4 mV/m. This analysis may not be rigorous enough but it demonstrates the importance of nonlocal effects if the plasma is not homogeneous.

In summarizing, we emphasize the main results of this work. Because of the difficulty in satisfying the wave vector matching conditions for resonant parametric decay channels of lower hybrid waves, parametric processes through nonresonant quasi-modes are expected to take place in the energy transfer process from precipitating keV electrons to ionospheric ions. The theoretically predicted threshold electric fields of the pump lower hybrid waves are estimated to be on the order of 1 mV/m for typical conditions at altitudes of 2000 - 3000 km on the auroral field lines. If the plasma consists of more than one ion species, it is expected that resonant decay is easier to occur and threshold values are probably lower.

There obviously still are problems to be solved before we really can understand the role of electron beam generated LHW's in the acceleration of ionospheric ions. This analysis of LHPP's suggests a solution to one of the problems, namely, the large phase velocities of electron beam generated LHW's as compared with thermal velocity of ionospheric ions.

Acknowledgments. A major part of this work was performed when I was visiting the Center for Space Research at MIT with the aid of financial support from AFGL through the contract F19628-83-C-0060 and from the Swedish Board of Space activities. It is a pleasure to thank the staff of the Center for hospitality and, especially, Tom Chang and Geoffrey Crew for many helpful discussions and valuable comments.

References

Banks, P., Collision frequencies and energy transfer; Electrons, Planet. Space Sci., 14, 1085, 1966.
Bernstein, I.B., Waves in a plasma in a magnetic field, Phys. Rev., 109, 10, 1958.
Chang, T., and B. Coppi, Lower hybrid acceleration and ion evolution in the suprauroral region, Geophys. Res. Lett., 8, 1253, 1981.
Gurnett, D.A., and L.A. Frank, A region of intense plasma wave turbulence on auroral field lines, J. Geophys. Res., 82, 1031, 1977.
Kaw, P.K., Parametric excitation of electrostatic waves in a magnetized plasma, Adv. Plasma Phys., 6, 179, 1976.

Koskinen, H.E.J., Lower hybrid parametric processes on auroral field lines in the topside ionosphere, J. Geophys. Res., 90, 8361, 1985.

Lotko, W., and J.E. Maggs, Damping of electrostatic noise by warm auroral electrons, Planet. Space Sci., 27, 1491, 1979.

Maggs, J.E., Electrostatic noise generated by the auroral electron beam, J. Geophys. Res., 83, 3173, 1978.

Maggs, J.E., and W. Lotko, Altitude dependent model of the auroral beam and beam-generated electrostatic noise, J. Geophys. Res., 86, 3439, 1981.

Ono, M., M. Porkolab, and R.P.H. Chang, Parametric decay into ion cyclotron waves and drift waves in multi-ion species plasma, Phys. Fluids, 23, 1656, 1980.

Porkolab, M., Parametric instabilities due to lower-hybrid radio frequency heating of tokamak plasmas, Phys. Fluids, 20, 2058, 1977.

Retterer, J.M., T. Chang, and J.R. Jasperse, Ion acceleration in the suprauroral region: A Monte Carlo model, Geophys. Res. Lett., 10, 583, 1983.

Retterer, J.M., T.S. Chang, and J.R. Jasperse, Ion acceleration by lower hybrid hybrid waves in the suprauroral region, This conference.

Rönnmark, K., WHAMP - Waves in homogeneous anisotropic multicomponent plasmas, KGI report no. 179, Kiruna Geophysical Institute, Kiruna, Sweden, 1982.

Rönnmark, K., Computation of the dielectric tensor of a Maxwellian plasma, Plasma Phys., 25, 699, 1983.

Temerin, M., C. Cattell, R. Lysak, H. Hudson, R.B. Torbert, F.S. Hudson, R.D. Sharp, and P.M. Kintner, The small-scale structure of electrostatic shocks, J. Geophys. Res., 86, 11278, 1981.

A NEW MECHANISM FOR EXCITATION OF WAVES IN A MAGNETOPLASMA
I. LINEAR THEORY

G. Ganguli and Y.C. Lee

Science Applications International Corporation, 8200 Greensboro Drive, McLean, VA 22102

P.J. Palmadesso

Naval Research Laboratory, Washington, DC 20375-5000

Abstract. We suggest a new mechanism for exciting waves in a magnetoplasma containing a transverse component of a nonuniform electric field. The energy density of the waves can become negative in the region where the transverse electric field is localized while it is positive outside this region. The coupling of the two adjoining regions through a nonlocal wavepacket allows a flow of energy from the negative energy region towards the positive energy region. This enables the mode to grow. We use ion cyclotron waves as an example to display the subtler features of the mechanism.

Introduction

Field aligned currents as a source of free energy for exciting the ion cyclotron instability [Drummond and Rosenbluth, 1962; and Kindel and Kennel, 1971] have been extensively discussed in the literature. Ion beams [Wiebel, 1970; Yamada et al., 1977] and electron neutral collisions [Chaturvedi, 1976] can also excite the ion cyclotron waves. A number of space observations of ion cyclotron waves can be satisfactorily explained by these mechanisms. However, it appears that some recent observations of ion cyclotron waves associated with magnetospheric shocks [Mozer et al., 1977; Temerin et al., 1981] and double layer experiments [Merlino et al., 1984; Alport et al., 1985] along with simultaneous observation of ion cyclotron and lower hybrid waves in a barium cloud release experiment [Koons and Pongratz, 1981; Pongratz, private communications] cannot be so easily resolved by the mechanisms known thus far. An interesting feature of these situations is the presence of a localized component of a static electric field transverse to the ambient magnetic field. None of the theoretical models mentioned above includes a transverse electric field. Further, a recent experiment [Nakamura et al., 1984] directly correlates the observed ion cyclotron waves with the two dimensional electric potential existing in their device.

Clearly the transverse component of the electric field plays an important role. This can easily be seen by examining the dispersion properties of the ion Bernstein modes with and without the electric field. The dispersion relation of the ion Bernstein modes without the electric field and for $k_z \sim 0$, where k_z is the wavevector along the external magnetic field, assumed in the z direction, is given by [Ganguli et al., 1985]

$$D(\omega, k) = 1 - \Gamma_0(b_1) - \sum_{n>0} \frac{2\omega^2 \Gamma_n(b_1)}{\omega^2 - n^2 \Omega^2} = 0. \qquad (1)$$

Here $\Gamma_n(b) = \exp(-b) I_n(b)$, $I_n(b)$ are the modified Bessel functions, $\Omega = eB/m_i c$ is the ion gyrofrequency, $b_1 = k^2 \rho_i^2 / 2$, $k^2 = k_x^2 + k_y^2$, $\rho_i = v_i / \Omega$ is the ion gyro radius, and v_i is the ion thermal velocity.

For simplicity we first consider an uniform electric field in the x direction. Consequently, there is a Doppler shift in the frequency in the laboratory frame, making it $\omega_1 = \omega - k_y V_E$, where $V_E = cE/B$ is the magnitude of the $\underline{E} \times \underline{B}$ drift. The dispersion relation (1) remains unaffected except for the replacement of ω by ω_1.

The electrostatic mode energy density, $U = (|E|^2/8\pi)(\partial \omega D/\partial \omega)$, for the ion Bernstein modes in the absence of the external electric field is [Ganguli et al., 1985],

$$U \propto \omega \left\{ \sum_{n>0} \frac{4\Gamma_n n^2 \Omega^2 \omega}{(\omega^2 - n^2 \Omega^2)^2} \right\} = \omega^2 \sigma(\omega). \qquad (2)$$

Since $\sigma(\omega)$ is a positive number, Bernstein modes are positive energy modes. When an uniform electric field is present the energy density is

298 IC WAVES DUE TO TRANSVERSE ELECTRIC FIELD

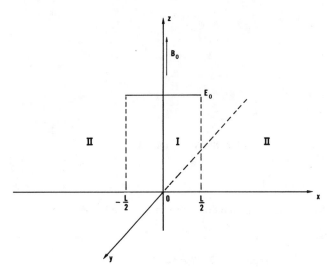

Fig. 1. A sketch of the electric field model.

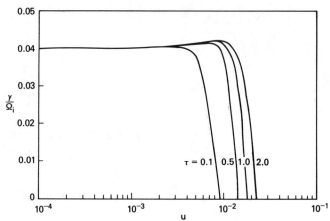

Fig. 3. A plot of γ against u for several τ. Here b = 0.61 and other parameters identical to Fig. 2.

changed to

$$U_1 \propto \omega\omega_1 \sigma(\omega_1). \qquad (3)$$

Thus for $0 < \omega < k_y V_E$, the mode energy density is negative. However, if the electric field is uniform the positive energy character can easily be recovered by a transformation to a frame moving with a velocity V_E. This is no longer possible if the electric field is nonuniform. For example, consider a very simple model of a nonuniform electric field (see Figure 1),

$$E(x) = \begin{cases} E, & -L/2 \leq x \leq L/2 \\ 0, & \text{otherwise} \end{cases}. \qquad (4)$$

We see two regions; with (region I) and without (region II) the electric field. Now the energy density of the Bernstein modes is negative in the region I but positive in regions II. Due to the inhomogeneity in the x direction the electrostatic potential will form a wavepacket which can span both the regions thereby coupling them; so that a flow of energy from the region of negative energy density towards the region of positive

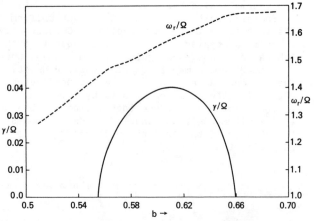

Fig. 2. Plot of the real and the imaginary parts of the eigenfrequency against b. Here $\tau = 0$, $\mu = 1837$, $\varepsilon = 0.3$, $u = 0.001$, $V_E = 2.9\, v_i$.

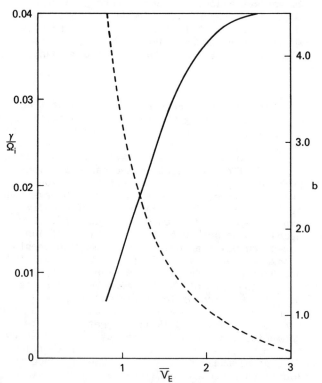

Fig. 4. A plot of γ maximized over b against $\overline{V}_E (= V_E/v_i)$. The dotted line gives the b where γ is maximum. Here $\varepsilon = 0.3$, $u = 0.0001$ and $\tau = 1$.

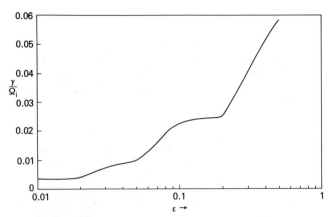

Fig. 5. A plot of γ against ϵ. The growth rate is maximized over b. Here $\tau = 1$, $V_E = 2.9\ v_i$, $u = 0.0001$ and $\mu = 1837$.

energy density will enable the mode to grow. A particle level physical description will be given in an accompanying paper [Palmadesso et al., 1985]. We now sketch the derivation of the nonlocal relation in the next section.

Theory

The presence of the nonuniform electric field in the x direction gives an average y drift which for $V_E \ll L\Omega$ can be approximated by V_E. Thus, we can use a Maxwellian for the initial distribution function with v_y shifted by V_E and obtain the general eigenvalue condition involving operators,

$$\left\{ k^2 + \sum_{n,\alpha} \frac{\Gamma_n(b_2)}{\lambda_\alpha^2} \left[1 + \left(\frac{\omega_1}{k_z v_\alpha}\right) Z\left(\frac{\omega_1 - n\Omega_\alpha}{k_z v_\alpha}\right) \right] \right\} \phi(x) = 0, \quad (5)$$

where $b_2 = \rho_i^2 (k_y^2 - \partial^2/\partial x^2)/2$, $Z(\zeta)$ is the plasma dispersion function and λ_α is the Debye length. Equation (5) is isomorphic to the dispersion relation of electrostatic waves in a homogeneous magnetoplasma but with two important differences: (i) $\omega_1 = \omega - k_y V_E$, is x dependent, and (ii) b_2 is an operator. An exact solution of (5) is beyond the scope of this paper. For $\rho_i \ll L$ we can expand $\Gamma_n(b_2)$ to $O(\partial^2/\partial x^2)$ and assume $(k_y \lambda_i)^2 \ll 1$ to reduce (5) to a second order differential equation,

$$\left\{ \frac{\partial^2}{\partial \xi^2} + \kappa_I^2(\omega_1, k_y) \right\} \phi(\xi) = 0 \quad (6)$$

where $\xi = x/\rho_i$, $\kappa_I^2 = -2Q(\omega_1, k)/(\partial Q/\partial b)$, $b = (k_y^2 \rho_i^2)/2$, $\tau = T_i/T_e$,

$$Q = 1 + \tau + \sum_n \Gamma_n(b) \left(\frac{\omega_1}{k_z v_i}\right) Z\left(\frac{\omega_1 - n\Omega}{k_z v_i}\right) + \tau \left(\frac{\omega_1}{k_z v_e}\right) Z\left(\frac{\omega_1}{k_z v_e}\right). \quad (7)$$

By solving (6) separately in the two regions and matching the logarithmic derivatives of the solutions at the boundary [Ganguli et al., 1985] we obtain the nonlocal dispersion relation

$$-\kappa_I \tan(\kappa_I/2\epsilon) = i\kappa_{II}, \quad (8)$$

where κ_{II} is identical to κ_I if ω_1 is replaced by ω and $\epsilon = \rho_i/L$. We now proceed to discuss the results in the following section.

Results

We plot the real (ω_r) and the imaginary (γ) parts of the complex eigen frequency normalized by Ω in Figure 2. Here $\tau = 0$, $\mu = m_i/m_e = 1837$, $\epsilon = 0.3$, $u = k_z/k_y = 0.001$ and $V_E = 2.9\ v_i$. The instability maximizes at $b = 0.61$ ($k_y \rho_i = 1.1$) and is highly localized in k_y making it very coherent. In contrast, the current driven ion cyclotron instability has a very flat spectrum.

Figure 3 is a plot of the growth rate against u for several values of τ. Here $b = 0.61$ and rest of the parameters are identical to the Figure 2. The growth rate peaks for $k_z \sim 0$. Even $u \sim 0.01$ is enough to provide sufficient electron Landau damping in order to suppress the instability. Further, for small enough u the growth rate is insensitive to the temperature ratio. This is in sharp contrast to the current driven ion cyclotron instability where increasing τ increases the critical drift.

In Figure 4 we plot γ maximized over b against V_E. The wave is unstable even for $V_E < v_i$. Here $\tau = 1$, $\epsilon = 0.3$ and $\mu = 1837$. The dotted line provides the value of b where the growth is maximum. Also, there are many branches of the

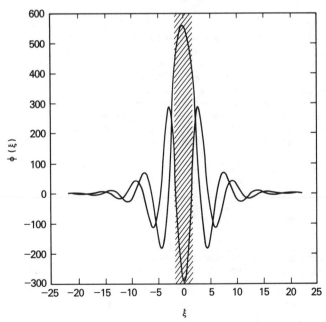

Fig. 6. The real and imaginary parts of a typical wave packet. Here $b = 0.61$, $\epsilon = 0.3$, $V_E = 2.9\ v_i$, where the electric field is localized.

unstable roots of the nonlocal dispersion relation (8). For example, Figure 4 represents a branch which gives $\omega/\Omega = 1.249 + .018i$ for $V_E = 1.16\ v_i$ and $b = 2.48$, while at the same time there exists another branch which gives $\omega/\Omega = 1.196 + .049i$ for V_E 1.2 v_i and $b = 1.98$. All possible branches of unstable roots of (8) have not yet been fully explored.

Figure 5 is a plot of γ maximized over b against ε. The value of ε is varied from 0.01 ($L = 100\ \rho_i$) to 0.3 ($L = 3.33\ \rho_i$). There is growth all over this interval with larger growth for strongly localized electric fields.

Figure 6 is a typical wavepacket $\phi(\xi)$ obtained by solving (6) numerically by employing a shooting code using the Numerov method. Here $b = 0.61$, $\tau = 1.$, $\mu = 1837$, $u = 0.0001$, $V_E = 2.9\ v_i$ and $\varepsilon = 0.3$. The region I where the electric field is localized is shaded. The corresponding eigenfrequency $\omega/\Omega = 1.5744 + .4002i$ is almost identical to that given by the analytical method (see Figure 2).

Conclusions

We have discussed a possible mechanism that can explain the short wavelength ($k\rho_i \geq 1$) high frequency ($\omega \geq \Omega$) oscillation arising in a magnetoplasma containing a static transverse electric field. The features of the waves are similar to the ones observed in the shock structures [Mozer et al., 1977; Temerin et al., 1981] i.e., the waves are very coherent with almost perpendicular propagation and the magnitude of the electric field is well above the minimum necessary for the onset of the instability. Similar observations in Q machine plasmas have also been reported [Alport et al., 1985]. The mechanism discussed here is general and can, in principle, be applied to a number of wave modes. We have also studied the growth of the lower hybrid waves by this method. In a recent experiment of barium cloud release, simultaneous observation of ion cyclotron and lower hybrid waves have been reported [Koons and Pongratz, 1981]. The mechanism described here can explain the simultaneous observation of both the waves at least in the linear stage. The nonlinear evolution of such a system is currently under investigation by means of particle simulation [Nishikawa et al., 1985]. Other possible applications have been discussed elsewhere [Ganguli et al., 1985a].

Acknowledgments. This work is supported by the Office of Naval Research and the National Aeronautics and Space Administration.

References

Alport, M.J., S.L. Cartier, and R.L. Merlino, Laboratory observation of ion cyclotron waves associated with a double layer in an inhomogeneous magnetic field, J. Geophys. Res., 91, 1599, 1986.

Chaturvedi, P.K., Collisional ion cyclotron waves in the auroral ionosphere, J. Geophys. Res., 81, 6169, 1976.

Drummond, W.E. and M.N. Rosenbluth, Anomalous diffusion arising from microinstabilities in a plasma, Phys. Fluids, 5, 1507, 1962.

Ganguli, G., Y.C. Lee, and P. Palmadesso, Electrostatic ion cyclotron instability due to a nonuniform electric field perpendicular to the external magnetic field, Phys. Fluids, 28, 761, 1985.

Ganguli, G., P. Palmadesso, and Y.C. Lee, A new mechanism for excitation of electrostatic ion cyclotron waves and associated perpendicular ion heating, Geophys. Res. Lett., 12, 643, 1985a.

Koons, H.C. and M.B. Pongratz, Electric fields and plasma waves resulting from a barium injection experiment, J. Geophys. Res., 86, 1437, 1981.

Merlino, R.L., S.L. Cartier, M. Alport, and G. Knorr, Observation of V-shaped double layers and ion cyclotron waves along diverging field lines, Proc. Second Symposium on Plasma Double Layers and Related Topics, p. 224, Innsbruck, Austria, 1984.

Mozer, F.S., C.W. Carlson, M.K. Hudson, R.B. Torbert, B. Parady, J. Yatteau and M.C. Kelly, Observation of paired electrostatic shocks in the polar magnetosphere, Phys. Rev. Lett., 38, 292, 1977.

Nakamura, M., R. Hatakeyama, and N. Sato, U-shaped double layers and associated ion cyclotron instability, Proc. Second Symposium on Plasma Double Layers and Related Topics, p. 171, Innsbruck, Austria, 1984.

Nishikawa, K.I., G. Ganguli, P. Palmadesso, and L.A. Frank, New mechanism for excitation of kinetic waves in a magnetoplasma: Particle simulation, Bull. Amer. Phys. Soc., 30, 1462, 1985.

Palmadesso, P., G. Ganguli, and Y.C. Lee, A new mechanism for excitation of waves in a magnetoplasma: Non linear theory, Proc. Chapman Conference on Ion Acceleration, Boston, 1985.

Temerin, M.C., C. Cattell, R. Lysak, M. Hudson, R.B. Torbert, F.S. Mozer, R.D. Sharp, and P.J. Kintner, Small scale structure of electrostatic shocks, J. Geophys. Res., 86, 11278, 1981.

A NEW MECHANISM FOR EXCITATION OF WAVES IN A MAGNETOPLASMA
II. WAVE-PARTICLE AND NONLINEAR ASPECTS

P. Palmadesso

Naval Research Laboratory, Washington, D.C. 20375

G. Ganguli and Y. C. Lee

Science Applications International Corporation
8200 Greensboro Dr., McLean, Va. 22102

Abstract. We present the results of a series of calculations of particle orbit dynamics in a growing electrostatic ion cyclotron wave driven by an external inhomogeneous transverse electric field. These calculations are intended to exhibit the essential physics underlying the instability mechanism, and to provide insights helpful in the study of energy exchange processes in the nonlinear regime, such as ion heating and processes relevant to the saturation of wave growth.

Introduction

The linear analysis methods employed in the paper by Ganguli et al. [1985], hereafter refered to as paper I, yield detailed information on growth rates and dispersive characteristics of electrostatic ion cyclotron (EIC) modes destabilized by an inhomogeneous transverse electric field in a magnetoplasma, but do not provide a clear and intuitively satisfying picture of the physical mechanisms underlying the instability. In this paper we shall attempt to construct such a picture by viewing the process at the level of a "typical" single particle interacting with a growing eigenmode. We shall solve the appropriate equations of motion and look at wave particle energy exchange processes, first in the linear and then in the nonlinear regime. Hopefully the understanding thus achieved will enable us to infer some of the properties of waves excited in this way, as they might be encountered in various applications. In particular, some properties of this instability as regards ion heating will be discussed. The calculations are intended to illustrate the most important mechanisms underlying the energy exchange as we currently understand them. We have aimed at simplicity rather than completeness and rigor.

We shall adopt the idealized ambient electric field model used in paper I, i.e.,

$$E(x) = \begin{cases} E & |x| \leq L/2 \quad \text{(region I)} \\ 0 & |x| \geq L/2 \quad \text{(region II)} \end{cases} \quad (1)$$

The wave envelope varies in the x direction, while the particle's drift motion is primarily in the y direction.

In a conventional microinstability the driving energy source is separated from the energy absorbing background particles by a displacement in velocity space; e.g., a beam-plasma interaction. In the present case the source and sink of energy are separated in configuration space. The most important effect of the inhomogeneity in the electric field is to allow the wave packet to conserve energy globally while the plasma loses or gains energy locally within each region. The energy exchange processes to be discussed here depend primarily on the magnitude of the particle's cross field drift motion within each region rather than on the local gradients. Thus we can capture the essence of this wave-particle interaction physics most simply by approximating the waveform as a plane wave within each region, with parameters matching those at the particle's guiding center, and then explaining the physical mechanisms which cause region I to act as a source of energy, i.e., particles in region I lose energy as the wave grows, while region II acts as a sink. Hopefully, it will then become clear that coupling of the two regions is a necessary condition for instability, and so the role played by the inhomogeneity will be clarified. Effects associated with the local gradients in wave amplitude, electric drift velocity, etc. can be significant in some cases, but a complete discussion of this issue is beyond the scope of the present work.

Model Equations

The equation of motion for a particle in an ambient electric field $E(x)$, an ambient magnetic field $B\hat{e}_z$, and an electrostatic plane wave of arbitrary type, with frequency ω, wavevector $\mathbf{k}$, and amplitude $E_k = k\phi_k$, is:

$$\frac{d\mathbf{u}}{d\tau} = -u_E \hat{e}_x + .5\varepsilon\kappa \sin(\kappa\cdot\zeta - \eta\tau) + \mathbf{u}\times\hat{e}_z \quad (2)$$

The equation has been made dimensionless by scaling with the initial value of the gyroradius for a thermal particle, ρ_0, and the gyrofrequency, $\Omega = qB/mc$, as follows: $u = v/\Omega\rho_0$, $\zeta = r/\rho_0$, $\tau = \Omega t$, $u_E = -(cE/B)/\Omega\rho_0$, $\eta = \omega/\Omega$, $\kappa = k\rho_0$, $\varepsilon = 2e\phi_k/m\Omega^2\rho_0^2$. For an EIC wave, the scaling is such that $\varepsilon \sim \delta n_i/n_i$, where δn_i is the ion density fluctuation associated with the wave.

It will be convenient to express the phase space coordinates as the sum of three components

$$\zeta = \zeta_0 + \zeta_1 + \zeta_2 \qquad u = u_0 + u_1 + u_2 \qquad (3)$$

which satisfy the following equations:

$$\frac{d\zeta_j}{d\tau} = u_j \qquad j = 0,1,2 \qquad (4a)$$

$$\frac{du_0}{d\tau} = -u_E \hat{e}_x + u_0 \times \hat{e}_z \qquad (4b)$$

$$\frac{du_1}{d\tau} = .5\kappa\varepsilon \sin(\kappa\cdot\zeta_0 - \eta\tau) + u_1 \times \hat{e}_z \qquad (4c)$$

$$\frac{du_2}{d\tau} = .5\varepsilon\kappa \left(\sin[\kappa\cdot(\zeta_1+\zeta_2)] \cos(\kappa\cdot\zeta_0-\eta\tau) - 2\sin^2[\kappa\cdot(\zeta_1+\zeta_2)/2] \sin(\kappa\cdot\zeta_0-\eta\tau) \right) + u_2 \times \hat{e}_z \qquad (4d)$$

These equations are exactly equivalent to eq. (2); ζ_2 and u_2 reduce to the second order terms in a perturbation solution of eq. (2) in the limit of small ε, but eq. (4d) has not been linearized and therefore ζ_2 and u_2 carry all the nonlinear effects for arbitrary ε.

One Dimensional Case

We would like to construct an analogue of the "wave energy" discussed in paper I, i.e., the change in the energy density of the plasma caused by the growth of the wave. We will discuss numerical solutions of (4a-d) for the EIC case, but it is useful to review some basic concepts and establish some terms of reference by first considering the simple case of a slowly growing Langmuir or ion acoustic wave propagating parallel to B, with no ambient electric field. In this case the wave energy consists of the space averaged electric field energy density $\langle |E_k|^2/8\pi \rangle$ plus the density weighted local space average of the shift in the particle kinetic energies due to the wave. In computing the contribution of a single particle to the latter, it can be shown that the density weighted space average may be replaced by an average in time over an oscillation period. The result is proportional to

$$\langle u^2/2 \rangle - \langle u_0^2/2 \rangle = \langle u_0 \cdot u_1 \rangle + \langle u_0 \cdot u_2 + u_1^2/2 \rangle + \langle u_1 \cdot u_2 \rangle + \langle u_2^2/2 \rangle \qquad (5)$$

where $\langle \rangle$ denotes the averaging process. The first order term averages to zero, and the last two terms on the right only contribute in the nonlinear regime. The remaining term contains the oscillation energy, always positive, and the $u_0 \cdot u_2$ term. It is useful to consider a relatively cold plasma drifting along the z axis with velocity u_D, in which case $u_0 \simeq u_D$. Solution of the motion equations to second order in $\varepsilon(\tau)$ then yields

$$\langle u_2 \rangle \simeq -1/[8\kappa^2(u_D - \eta/\kappa)^3] \hat{e}_z \qquad (6)$$

Eq.'s (5) and (6) indicate that in addition to the oscillation energy there is an energy shift associated with a secular change in the drift velocity in second order. This secular velocity shift is always antiparallel to the drift velocity as observed in the wave frame, $u_D - \eta/\kappa$, and represents the tendency of the particle velocity to approach the wave phase velocity as the wave amplitude grows. Ultimately $\langle u_2 \rangle + (u_D - \eta/\kappa) \to 0$ in the nonlinear regime as the particle becomes trapped. The secular part of the velocity u_2 may be understood as the result of conservation of an adiabatic invariant associated with the oscillation, but to aid the discussions in the next section, it is useful to interpret this as the response to a secular second order force, $F_2 = \langle d[u_2(\tau)]/d\tau \rangle$. The force F_2 arises because the wave field seen at the oscillating position of the particle does not quite average to zero when the wave amplitude is changing in time. The $u_0 \cdot u_2$ term makes a negative contribution to the wave energy in the lab frame when $u_D > \eta/\kappa$, and this term dominates when $u_D \gtrsim \eta/\kappa$.

The curves in figure 1 are based on numerical solutions of eq.'s (4), for such a case ($\eta = 1.5$, $\kappa = 1.28$, $u_D = 2$). The various contributions to the energy shift are normalized by dividing by ε^2. We expect the resulting quantities to remain constant in the linear regime, after an initial transient averaging time, so that entry into the nonlinear regime is obvious when it occurs. The dominant nonlinear effect in this case is particle trapping. The orbits are being time averaged over the fixed initial Doppler shifted wave period, but there is an amplitude dependent frequency shift in the nonlinear regime: this makes the orderly trapping behavior appear somewhat chaotic.

The correct expression for the wave energy can be constructed by completing the calculations outlined above and summing the oscillation and secular drift components of the particle energy shift, multiplying by the particle density, and adding in the field energy.

EIC Wave

Figure 2 shows time histories of some important terms in eq. (5), obtained by numerical integration of (4a-d), for a particle with a 30° pitch angle in a slowly growing EIC wave with parameters chosen to represent region I of the system discussed in paper I: $u_E = 2.9$, Re$\{\eta\} = 1.57$, $\kappa = 1.28$, $\kappa_y/\kappa = .86$, and $\kappa_z/\kappa = .0017$. In this case there are two incommensurable frequencies, ω_1 and Ω, associated with the motion so that the local time average is less effective at filtering out

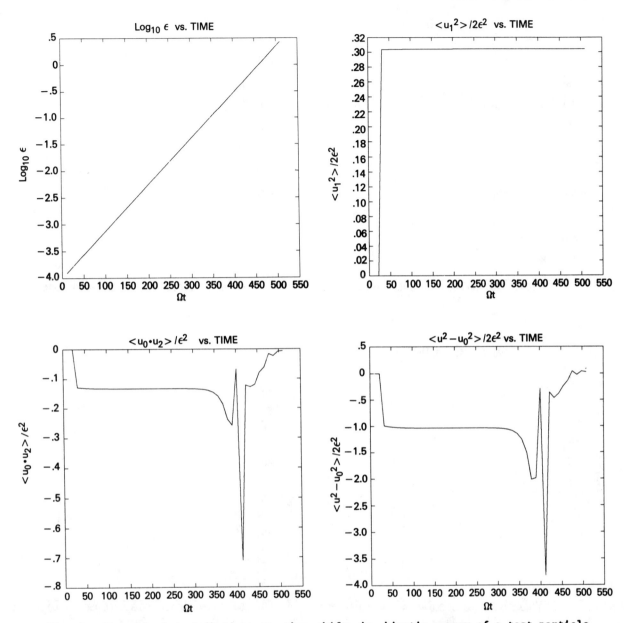

Fig. 1. Normalized contributions to the shift in kinetic energy of a test particle produced by the growth of an electrostatic wave propagating parallel to the magnetic field.

the oscillatory part of the motion, but the linear regime still stands out clearly.

In region I we have a transverse electric field which produces an electric drift velocity u_E in the y direction, and the parameters are such that $u_E > \eta/\kappa$. We expect the wave energy $U = [|E_k|^2/8\pi][\partial\omega D/\partial\omega]$, where D is the dielectric function, to be negative in this region, but note from fig. 2 that the kinetic energy shift is positive. As in the simple one dimensional case there is a secular force $F_2 \sim O(\nu\epsilon^2)$, where $\nu = \text{Im}\{\eta\} = \epsilon^{-1}d\epsilon/d\tau$, acting on the particle in second order, which opposes the drift as observed in the wave frame. In this case F_2 is essentially perpendicular to $B\hat{e}_z$. In the one dimensional case with $u_D > \eta/\kappa$, the particle response to F_2 took the form of an acceleration of order $\epsilon^2\nu$, which integrated to a secular velocity shift antiparallel to u_D and of order ϵ^2, which in turn resulted in a second order change in total kinetic energy. Here the force F_2 acts in a direction transverse to the ambient magnetic field, so that the response occurs in the form of a drift velocity $\sim O(\nu\epsilon^2)$. If $\nu \ll 1$, this drift velocity represents an insignificant shift in kinetic energy, but it

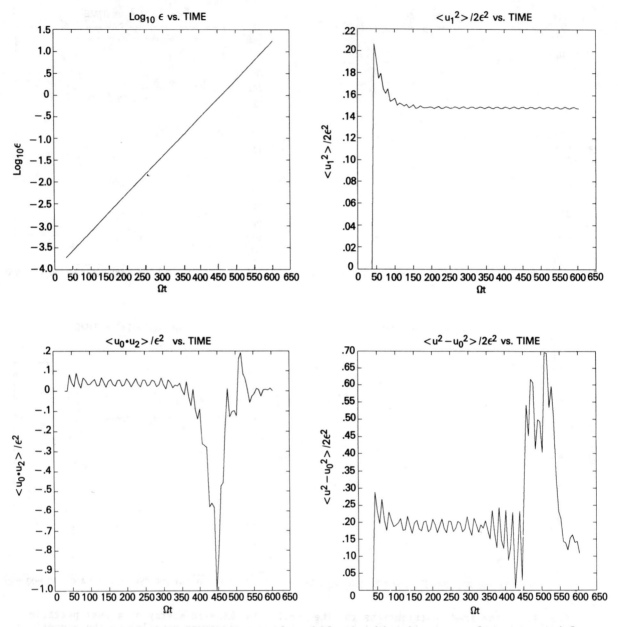

Fig. 2. Normalized contributions to the shift in kinetic energy of a test particle produced by the growth of an electrostatic ion cyclotron wave (region I parameters).

produces a spatial displacement of order ϵ^2. In particular, F_{2y} yields a shift in the x direction: since $\langle u_{2x} \rangle \sim O(\nu\epsilon^2)$, it follows that $\delta x = \rho_0 \int \langle u_{2x} \rangle d\tau \sim O(\epsilon^2)$. Rather than losing kinetic energy, the particle in this case is forced to drift across the equipotential surfaces associated with the ambient electric field, and the dominant effect is a loss of potential energy, $\delta\phi = -qE\delta x$, or $u_E \langle \zeta_{2x} \rangle$ in our dimensionless notation. The drift velocity $\langle u_{2x} \rangle$ carries a current J_{2x}, and so the ambient electric field is the ultimate energy source for wave growth via $\mathbf{J}_2 \cdot \mathbf{E}$. The region I wave-particle energy transfer physics is thus analogous to the operation of a magnetron.

In figure 3 these effects are exhibited via plots of relevant position and velocity shifts and the test particle's contribution to the wave energy, including the potential energy term. The "wave energy" is thus represented here by the quantity $\langle u^2/2 \rangle - \langle u_0^2/2 \rangle + u_E \langle \zeta_{2x} \rangle$. Again we have normalized by dividing by $\epsilon(\tau)^2$. Plots of these quantities with parameters appropriate to region II show similar behavior, except that δx is positive and does not produce a change in potential energy, since E is zero; thus the wave

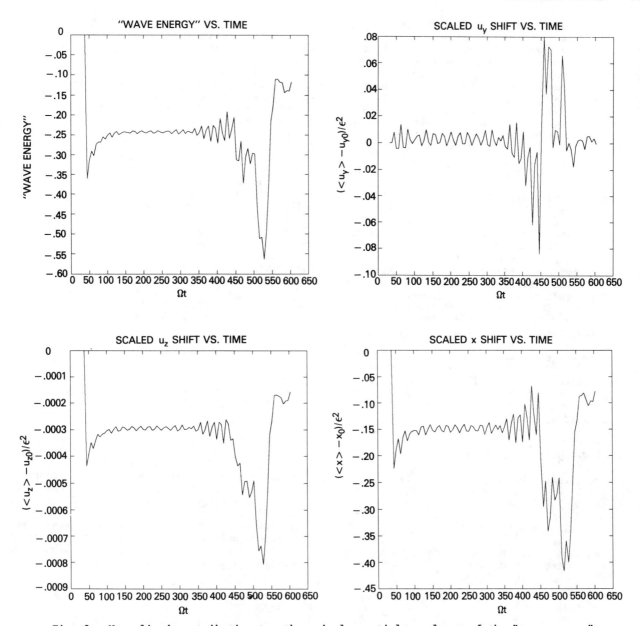

Fig. 3. Normalized contribution to the single particle analogue of the "wave energy" (upper left) for a test particle in region I in a growing EIC wave. The other curves show some relevant position and velocity shifts.

energy in region II is due to the kinetic energy shifts alone and is therefore positive. The test particle contributions to the wave energy corresponding to regions I and II are nearly equal in magnitude but opposite in sign, when parameters consistent with the linear analysis results in paper I are chosen.

Ion Heating

We now discuss the physics of resonant heating via the nonlocal wavepacket. Instead of the idealized electric field profile we consider a more realistic one, given by an arbitrary smooth function localized over a width L; $\mathbf{E}_0 = E_0(x)\,\hat{\mathbf{e}}_x$. The resonance condition is given by

$$|\omega_1(x) - k_z v_z \pm n\Omega_i| \lesssim \delta\omega \qquad (7)$$

where $\omega_1(x) = \omega - k_y V_E(x)$, and the resonance width in frequency space, $\delta\omega$ is a function of the wave amplitude.

By analogy with the results of the linear analysis for the idealized discontinuous electric field profile discussed in paper I, we expect that the Doppler shifted frequency $\omega_1(x)$ will vary from a value in excess of Ω_i in the positive energy part of the wavepacket to a value less than Ω_i in the negative energy region. Thus we expect there exist one or more points x_{Rj} within the wavepacket at which the condition $|\omega_1(x_{Rj})| = \Omega_i$ is satisfied. For an electrostatic wave such as an EIC wave or a lower hybrid wave near its resonance cone angle, the relationship between the mode amplitude and the resonance width $\delta\omega$ is [Palmadesso, 1972],

$$\delta\omega \sim \sqrt{ek|E_k|/m_i} \sim k\sqrt{e\phi_k/m_i} \sim kv_i\sqrt{e\phi_k/T_i} \quad (8)$$

where ϕ_k is the wave potential. Alternatively, since $e\phi_k/T_i \sim \delta n_i/n_i$ when $T_i \sim T_e$, we can write this in the form $\delta\omega \sim kv_i(\delta n_i/n_i)^{1/2}$. The observed values of $\delta n_i/n_i$ for EIC waves are frequently in the range 0.1 to 0.5.

It is characteristic of the instability mechanism being considered here that $k_z/k \ll 1$. Values of this ratio much less than 0.01 are typical [Ganguli et al., 1985]. Therefore, for a thermal ion, $\delta\omega \sim kv_i(\delta n_i/n_i)^{1/2} \gg k_z v_z$. It follows from Eq. (7) and the inequality $\delta\omega \sim \gg k_z v_z$ that within a finite sized interval Δx around x_{Rj} such that $|\omega_1 \pm \Omega_i| \lesssim \delta\omega$, all thermal ions are in cyclotron resonance with the wave.

To estimate the size of the resonant regions in x-space, we neglect the small Doppler shift $k_z v_z$ and expand V_E to first order in Δx (the locus of guiding centers of resonant particles) around x_{Rj}.

$$|\omega - k_y V_E(x_{Rj}) - k_y(\partial V_E/\partial x)\Delta x \pm \Omega_i| \lesssim \delta\omega \quad (9)$$

We replace $\partial V_E/\partial x$ by V_E/L in (9), recall that $\omega - V_E(x_{Rj}) \pm \Omega_i = 0$ and $\delta\omega \sim kv_i(\delta n_i/n_i)^{1/2}$, and solve for Δx to obtain

$$\Delta x \sim (k/k_y)(v_i/V_E)\sqrt{\delta n_i/n_i} \quad (10)$$

Thus, in the nonlinear regime resonant heating will take place within regions of significant size and will involve the bulk of the distribution rather than a few particles in the tail. This derivation presupposes that the electric field variation is approximately linear over distances of the order of a thermal gyroradius, so that $V_E(x_{Rj})$, which depends on the particle's tranverse thermal energy when finite Larmor radius effects are important, is approximately the same for all thermal particles. If this is not the case the situation is more complex, but it is clear that the physics of particle resonance is somewhat different here than in the usual local approximation case; i.e., the resonance in this case appears to be localized in configuration space rather than velocity space. The broad resonance provides heating of the ions perpendicular to the magnetic field. These waves are likely to be less sensitive to suppression by ion heating [see Palmadesso et al., 1974] than the Drummond and Rosenbluth [1962] waves.

Acknowledgements. This research was supported by the Office of Naval Research and the National Aeronautics and Space Administration.

References

Ganguli, G., Y. C. Lee, and P. Palmadesso, A new mechanism for excitation of waves in a magnetoplasma. I. Linear theory, Ion Acceleration in the Magnetosphere and Ionosphere (this monograph).

Drummond, W.E. and M.N. Rosenbluth, Anomalous diffusion arising from microinstabilities in a plasma, Phys. Fluids, 5, 1507, 1962.

Palmadesso, P.J., Resonance, particle trapping, and Landau damping in finite amplitude obliquely propagating waves, Phys. Fluids, 15, 2006, 1972.

Palmadesso, P.J., T.P. Coffey, S.L. Ossakow and K. Papadopoulos, Topside ionosphere ion heating due to electrostatic ion cyclotron turbulence, Geophys. Res. Lett., 1, 105, 1974.

HEATING OF LIGHT IONS IN THE PRESENCE OF A LARGE AMPLITUDE HEAVY ION CYCLOTRON WAVE

K.-I. Nishikawa[1] and H. Okuda[2]

[1]Department of Physics and Astronomy, The University of Iowa, Iowa City, Iowa 52242

[2]Plasma Physics Laboratory, Princeton University, Princeton, New Jersey 08544

Abstract. Heating of hydrogen ions has been investigated by plasma numerical simulations in the presence of an electrostatic oxygen cyclotron (EOC) wave which is observed on auroral field lines at low altitude (400-600km). Two types of instabilities have been found which can be driven by an EOC wave. One of them is an oscillating current-driven electrostatic hydrogen cyclotron instability whose frequency is near the hydrogen cyclotron frequency. The other is an oscillating lower hybrid two-stream instability whose frequency is near the lower hybrid, $\omega \approx \omega_{pH}$. Two-dimensional numerical simulations confirmed the presence of both instabilities resulting in strong heating of hydrogen ions. High energy tails of the hydrogen ions are observed in the perpendicular distribution.

Introduction

Energetic ions and electrons and electrostatic waves have been observed by the sounding rockets which passed through source regions of transversely accelerated ionospheric ions (TAI) at the several hundred km altitudes [Bering et al., 1975; Whalen et al., 1978; Yau et al, 1983]. Large fractional density fluctuations were observed and their spectral peaks in the 5- to 15-Hz range are consistent with O^+ ion cyclotron harmonics Doppler-shifted by the rocket velocity [Bering et al., 1975; Yau et al., 1983].

At such low altitude auroral zones, oxygen ions are the majority species, while hydrogen ions are the minority species [Yau et al., 1983]. Under such conditions, it has been shown by means of numerical simulations that the heating of oxygen ions is much stronger than that of hydrogen ions [Ashour-Abdalla and Okuda, 1984]. Heating of oxygen ions is associated with the current-driven electrostatic oxygen cyclotron (EOC) waves which grow to larger amplitude resulting in the strong heating of oxygen ions.

In this paper, we show that heating of hydrogen ions also can take place due to the excitation of oscillating current-driven electrostatic hydrogen cyclotron and oscillating lower hybrid two-stream instabilities. We investigate these two types of instabilities caused by the electron drifts in the presence of an oxygen cyclotron wave. The frequency of the oscillating electron current-driven electrostatic hydrogen cyclotron wave is near the hydrogen cyclotron frequency (Ω_H) and the wavelength is of the order of hydrogen gyro-radius ρ_H. For the oscillating lower hybrid two-stream instability, the wave frequency is near the lower hybrid $\omega_{lh} \approx \omega_{pH}$ and the wavelength is much shorter than the hydrogen gyro-radius ($\lambda \ll \rho_H$). We would like to emphasize that, contrary to our earlier works [Nishikawa et al., 1983, 1984], the decay instability is not possible here, since the frequencies of current-driven EHC and lower hybrid two-stream instabilities are much higher than that of EOC waves. The simulation model and results are presented in Section 2. In Section 3, we present concluding remarks and discussions.

Simulation Model and Results

We simulate the auroral plasma at the 900-1000 km altitudes region where the ion composition varies from 10% hydrogen to 25% hydrogen [Kindel and Kennel, 1971]. Under this condition, the hydrogen plasma frequency determines the magnitude of the lower hybrid frequency. We assume that only the electrostatic oxygen cyclotron waves are unstable due to their lower threshold current so that the EOC waves grow to large amplitude first [Ashour-Abdalla and Okuda, 1984]. At this stage, hydrogen ions do not play any role. We then study the second stage of the instability in which EHC and lower hybrid waves are excited by the large amplitude EOC waves in the present simulations. There, the first and second stages are clearly distinguishable in this case and the presence of oxygen ions in the simulations is not essential for the second stage. Furthermore, we assume that the oxygen ions (75-90%) are large enough to keep the amplitude of the oxygen cyclotron wave constant during the heating of hydrogen ions. Thus we consider a plasma consisting of electrons

308 HEATING OF LIGHT IONS

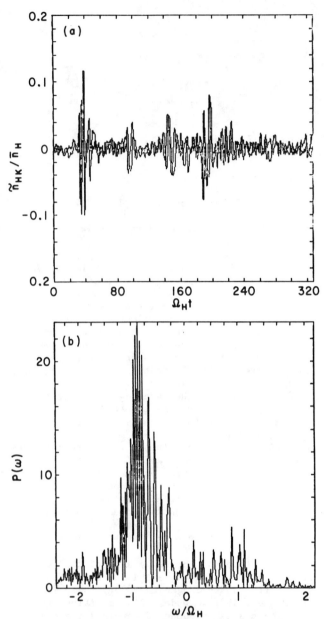

Fig. 1. Time evolution (a), and the power spectrum (b) of hydrogen ion density perturbation of the mode (0,1)th diagnosed during $0 < \Omega_H t < 328$ ($k_x = 0$, $k_y = 2\pi/L_y$). For the case $\omega_0 = 2\Omega_0$, $E_0^2/4\pi n_e T_e = 0.1$, $\tilde{B}_{oy}/B_0 = 0.1$, $m_H/m_e = 100$, and $\Omega_e/\omega_{pe} = 4$. This simulation is carried with the uniform recycling model for the electrons.

the frequency, and the wavevector of the applied EOC wave.

Under such conditions, two types of electrostatic instabilities may be excited [Nishikawa and Okuda, 1985b]. For both cases, it is the oscillating electron current along and across the magnetic field which are resposible for the instabilities.

Here we consider a two component plasma consisting of background electrons and hydrogen ions in the presence of a large amplitude pump EOC wave. We assume initially Maxwellian distributions for electrons and hydrogen ions with the same temperature. Electrons are treated by the guiding center approximation in the perpendicular direction while the parallel motion is treated exactly. Therefore, the electron cyclotron frequency should be larger than the electron plasma frequency. This code retains the full ion dynamics in 3-D velocity space. We use a system length $L_x = L_y = 64\Delta$ where Δ is the grid space which is equal to the electron Debye length, λ_e, $T_e = T_H$ and $n_e \lambda_e^2 = 9$. These parameters are kept the same for all the simulations reported in this article. As observed by the rocket [Yau et al., 1983], oxygen ions are much hotter than hydrogen ions and the observed electrostatic oxygen cyclotron wave has a long wavelength ($k \approx \rho_0^{-1}$). Therefore, a large amplitude oxygen cyclotron wave is applied externally in the form of $E_{ox,oy} \cos(\omega_0 t)$ where E_{ox}, E_{oy}, and ω_0 are the x and y components of amplitude, and $\omega_0 \approx \Omega_0$ is the frequency of the applied field. The amplitude of the pump field is increased linearly in time until $\omega_{pe} t = 200$.

First, in order to investigate an oscillating current-driven electro static cyclotron instability and hydrogen ions immersed in an external homogeneous magnetic field $\underset{\sim}{B}_0$ the presence of a large amplitude oxygen cyclotron wave (fixed amplitude). Taking the external magnetic field $\underset{\sim}{B}_0$ is in the y-z plane ($B_{oz} \gg B_{oy}$), we assume a presence of a large amplitude oxygen cyclotron wave in the form of $\underset{\sim}{E}_0 \cos(\omega_0 t - k_{ox} x - k_{oy} y)$ where $\underset{\sim}{E}_0 = (E_{ox}, E_{oy})$, ω_0, k_{ox} and k_{oy} are the wave amplitude,

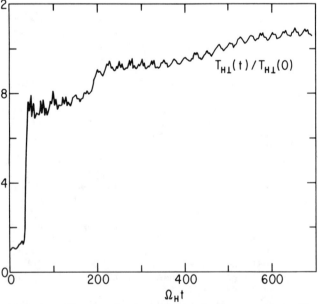

Fig. 2. Time evolution of the hydrogen perpendicular kinetic energy for the same case as Fig. 1.

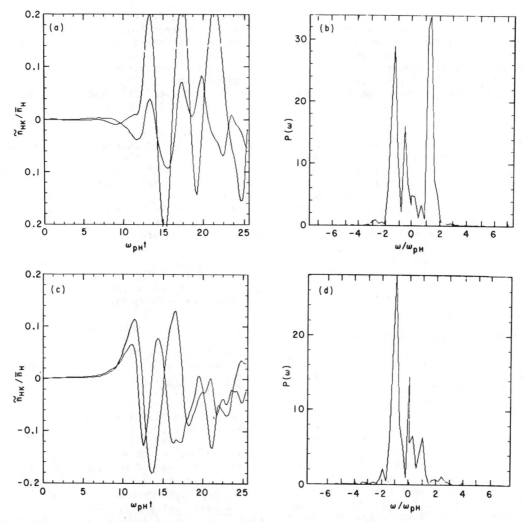

Fig. 3. Time evolutions and the power spectrum of hydrogen ion density perturbation of the mode (0,1)th (a) and (b) ($k_x = 0$, $k_y = 2\pi/L_y$), and the mode (1,1)th (c) and (d) ($k_x = 2\pi/L_x$, $k_y = 2\pi/L_y$) diagnosed during $0 < \Omega_H t < 25.6$, for the case $\omega_0 = 4\Omega_0$, $E_0^2/4\pi \bar{n}_e T_e = 0.025$, $B_{oy}/B_o = 0.01$, $m_H/m_e = 1600$ and $\Omega_e/\omega_{pe} = 2$.

ity, a simulation has been performed using the model described. The parameters of the simulation are $(E^2_{ox} + E^2_{oy})/4\pi n_e T_e = 0.1$, $\Omega_e/\omega_{pe} = 4$, $\omega_0 = 2\Omega_0$, $m_H/m_e = 100$, $m_0/m_H = 16$, $B_{oy}/B_o = 0.1$. In this case, the uniform recycling model is used in order to model the fresh electron flux along the auroral field line [Okuda and Ashour-Abdalla, 1983]. At each time step ($\omega_{pe}\Delta t = 4$), 0.6 percent of the total electrons are replaced by the initial cold electrons at random.

Heating of electrons and hydrogen ions have been observed. A drift is found in the electron parallel velocity distribution which is driven by the parallel component of the pump field. The oscillting electron drift along the magnetic field is given by $v^e_\parallel = (eE_{oy}/m_e\omega_0)(B_{oy}/B_o)\cos(\omega_0 t) = (E^2_{oy}/4\pi \bar{n}_e T_e)^{1/2}(\omega_{pe}/\Omega_e)(m_H/m_e)(\Omega_H/\omega_0)(B_{oy}/B_o)v_e \cos(\omega_0 t)$ (Nishikawa and Okuda, 1985b). Note that

$\omega_0 \approx \Omega_0$ is very small (especially for the realistic mass ratio $m_i/m_e = 1836$), so that even a modest amplitude of the pump field E_{oy} can give rise to a large electron drift which can excite EHC instability. The high energy tail is caused by the diffusions. While the parallel heating of hydrogen ions is weak, the strong perpendicular heating are observed.

The excitation of electrostatic hydrogen cyclotron waves is observed in the hydrogen ion density perturbation [Fig. 1 (a)]. The power spectrum shows the peaks around the hydrogen cyclotron frequency [Fig. 1 (b)]. These simulation results agree well with the theoretical predictions that the excited waves are EHC instability ($\omega \approx \Omega_H$, $k_\perp \approx \rho_H^{-1}$).

Time evolution of hydrogen perpendicular kinetic energy is shown in Fig. 2. We found that the large electron drift in the first cycle of the pump wave excited an EHC instability resulting in the perpendicular heating of hydrogen ions. After the initial strong heating of hydrogen ions, the heating rate becomes weak because of the reduced growth rate of the ECH waves. This is caused mainly by the generation of the temperatue anisotropy of hydrogen ions. Note, however, hydrogen perpendicular temperature increases monotonically in time without saturation until the end of the simulation. The final attaniable perpendicular temperature may be given by the marginal stability theory for a given electron drift speed which is about 4 v_e [Okuda and Ashour-Abdalla, 1983].

The lower hybrid two-stream instability can be driven by the $c\underline{E}_o \times \underline{B}_o/B^2$ perpendicular drift at $\omega \approx \omega_{pH}$ and $k_\perp \gg \rho_H^{-1}$. Since the gyro-radius of hydrogen ions is much larger than that of electrons, the electron $c\underline{E}_o \times \underline{B}_o/B_o^2$ drifts generates the cross-field current which is responsible for the lower hybrid two-stream instability [McBride et al., 1972].

We have performed a simulation for $(E^2_{ox} + E^2_{oy})/4\pi \bar{n}_e T_e = 0.05$, $\omega_o = 4\Omega_O$, $\Omega_e/\omega_{pe} = 2$, $m_H/m_e = 1600$, $m_O/m_H = 16$, $B_{oy}/B_o = 0.01$. In this case, the initial hydrogen ion gyro-radius is 20Δ and the electron $c\underline{E}_o \times \underline{B}_o/B_o^2$ drift is 4.5 v_H. The uniform recycling model for electrons is not used. Since the instability is a hydrodynamic, nonresonant type so that the exact shape of the electron parallel distribution is not important.

The excitation of lower hybrid waves are observed in the hybrid ion density perturbation. Two fundamental modes, (0,1)th and (1,1)th are observed in Fig. 3. The lower hybrid frequencies near ω_{pH} are found in both modes as shown in Fig. 3 (b) and (d). The similar power spectra are found in the electrostatic potential.

The electrons are heated in the parallel direction. While the parallel distribution for hydrogen ions remain almost unchanged, the perpendicular distributions are strongly heated. Furthermore, the high energy tail of the hydrogen ions is generated. The hydrogen ions are heated more strongly than by the parallel current-driven electrostatic cyclotron instability.

Conclusions

We have studied heating of hydrogen ions due to the two types of instability in the presence of an electrostatic oxygen cyclotron wave which is observed on auroral field lines at low altitudes. One of the instabilities is electrostatic hydrogen ion cyclotron instability driven by the oscillating electron parallel drift accelerated by the parallel component of the oxygen cyclotron wave. The other is the oscillating lower hybrid two-stream instability driven by the electron-ion relative drift across magnetic field.

We have performed several simulations with a two-dimensional code. The results of simulations indicate the excitation of electrostatic hydrogen cyclotron wave ($\omega \approx \Omega_H$, $k_\perp \approx \rho_H^{-1}$) driven by the electron parallel drift and lower hybrid two-stream instability ($\omega \approx \omega_{pH}$, $k_\perp \gg \rho_H^{-1}$). Both instabilities result in strong heating of hydrogen ions. It should be mentioned that the final ion temperature is determined by the balance between the heating due to the instabilities and the anomalous heat conduction associated with the ion transport. Generally speaking, anomalous transport is larger for light ions than for heavy ions so that it may be plausible that hydrogen ions remain at a lower temperature than oxygen ions at low altitude auroral zone.

Acknowledgments. This work is supported by the National Science Foundation Grant ATM83-11102 and the United States Department of Energy Contract No. DE-AC02-76-CHO-3073.

References

Ashour-Abdalla, M. and H. Okuda, Turbulent heating of heavy ions on auroral field lines, J. Geophys. Res., 89, 2235, 1984.

Bering, E. A., M. C. Kelley and F. S. Mozer, Observations of an intense field-aligned thermal ion flow and associated intense narrow band electric field oscillations, J. Geophys. Res., 80, 4612, 1975.

Kindel, J. M. and C. F. Kennel, Topside current instabilities, J. Geophys. Res., 76, 3055, 1971.

Nishikawa, K.-I., H. Okuda and A. Hasegawa, Heating of heavy ions on auroral field lines, Geophys. Res. Lett., 10, 553, 1983.

Nishikawa, K.-I., H. Okuda and A. Hasegawa, Heating of heavy ions on auroral field lines in the presence of a large amplitude hydrogen cyclotron wave, J. Geophys. Res., 90, 419, 1985a.

Nishikawa, K.-I. and H. Okuda, Heating of light ions in the presence of a large-amplitude heavy ion cyclotron wave, J. Geophys. Res., 90, 2921, 1985b.

Okuda, H. and M. Ashour-Abdalla, Acceleration of hydrogen ions and conic formation along auroral field lines, J. Geophys. Res., 88, 889, 1983.

Whalen, B. A., W. Bernstein and P. W. Daly, Low altitude acceleration of ionosopheric ions, Geophys. Res. Lett., 5, 55, 1978.

Yau, A. W., B. A. Whalen, A. G. McNamara, P. J. Kellogg and W. Bernstein, Particle and wave observations of low-altitude ionospheric ion acceleration events, J. Geophys. Res., 88, 341, 1983.

LINEAR EFFECTS OF VARYING ION COMPOSITION ON PERPENDICULARLY-DRIVEN FLUTE MODES: COMPARISON TO ROCKET OBSERVATION

David N. Walker

Plasma Physics Division, Naval Research Laboratory, Washington, DC 20375

Abstract. Electrostatic wave observations made during the injection of an Argon ion beam from the ARCS-I sounding rocket into the auroral ionosphere are compared to predictions of linear theory for a perpendicular ion beam-driven resonant instability. Assuming a two-ion component background plasma whose masses differ by at most a factor of two (e.g. O^+, O_2^+; O^+, NO^+), solutions are obtained as a function of varying ion composition both in the absence of a beam term (i.e., multicomponent ion Bernstein modes) and in the presence of an unmagnetized Ar^+ ion beam assuming small thermal spreading of the beam. Numerical solutions for resonantly driven oscillations using small beam densities and perpendicular beam velocities yield increasing growth rates for increasing wave number, k, for the first three modes studied. An analytical approximation to the growth rates following a procedure similar to one used by Roth et al. 1983, is consistent with this conclusion.

Introduction

A number of earlier laboratory investigations [Boehmer, 1975; Boehmer et al., 1976, Seiler et al., 1976] have studied excitation of electrostatic ion cyclotron waves in Q-machines by injection of an ion beam with a component perpendicular to the local magnetic field. More recently, active space plasma experiments and simulations [Kintner and Kelley, 1983, Lebreton et al., 1983, Pottelette, 1984] using perpendicular ion beams have concentrated on similar excitation of ion Bernstein modes in order to explain observations apparently related to the H^+ gyrofrequency. In particular it has been shown that the mode structure in a multicomponent plasma can be dominated by harmonics of the lightest ion [Hamelin and Beghin, 1976], sometimes at surprisingly small light ion concentrations. Since often the primary ion concentrations at altitudes of sounding rocket beam injection are O^+, O_2^+, NO^+, it is of interest to identify possible eigenmodes and growth rates associated with this excitation mechanism for varying concentrations of the separate ion species.

The ARCS I (Argon Controlled Release Study I) [Walker et al., 1980, Moore et al., 1982, Kaufmann et al., 1985] sounding rocket was the first in a series of three active auroral rocket experiments employing an Argon Ion beam neutral plasma generator. One purpose of the experiment was to study beam effects on the ambient plasma, particle distributions and the vehicle potential. The experiment differed from other active ion beam experiments (e.g., PORCUPINE) and from later ARCS experiments in that the ion gun was not separated from the experimental rocket payload. Instead the Argon ions were injected on the downleg of the trajectory from the spinning payload at an average energy of 25-ev into a $60°$ cone in different directions. The altitude at which the injections occurred are shown on the electron density profile of Figure (1). The payload was spin stabilized at a rate of 2.5 Hz and its axis was at an angle near $20°$ to the local magnetic field during the gun pulsing periods. The Argon neutral plasma generator was positioned at an angle of $45°$ from the forward payload central axis. From these descriptions it is clear that the nature of the injection process will produce a spread of velocities along a direction perpendicular to the ambient field.

The payload instrumentation consisted of energetic particle detectors covering the range 0-10 kev and mounted at various angles with respect to the payload axis; DC and AC electric and magnetic field receivers covering the range 0-10 MHz; an ion drift detector, and a pair of pulsed Langmuir probes [Szuszczewicz and Holmes, 1975, Holmes and Szuszczewicz, 1975]. In comparing predictions of the linear theory described above to observation, the power spectral frequencies of electrostatic oscillations observed by the Langmuir probes during plasma generator operation are analyzed. Clearly, the variety of possible low frequency electrostatic wave observations in a beam injection process as outlined, in addition to a lack of laboratory cross-correlation [Walker and Szuszczewicz, 1985] or, better, interferometric [Kintner et al., 1984] techniques makes an absolute mode identification through power spectra of $(\delta n/n)$ oscillations

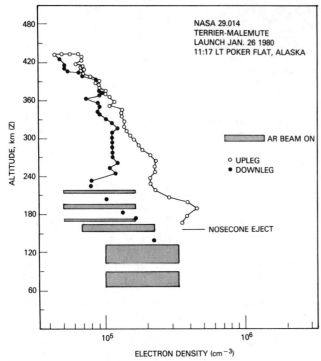

Fig. 1. Electron density vs. altitude showing altitudes of beam operation.

difficult at best. For these reasons the agreement between observations presented here and theoretical explanation is at best a necessary condition for a correct theory.

Theory

For an unmagnetized ion beam interaction with a two-component Maxwellian ion plasma, the collisionless flute mode ($k_\parallel = 0$) dispersion relation [Ichimaru, 1973, Seiler et al., 1976, Kintner and Kelley, 1983] derives from the plasma dielectric function,

$$\varepsilon = 1 - \sum_n \chi_{np} - \chi_B = 0 \qquad (1)$$

where χ_{np}, χ_B are plasma and beam susceptibilities. Expanding this expression and simplifying,

$$k^2 \lambda_{De}^2 - 2\alpha \left(\frac{T_e}{T_1}\right) \sum_n \frac{\Gamma_n(\lambda_1)}{\left(\frac{\omega}{n\Omega_1}\right)^2 - 1} - 2(1-\alpha)\left(\frac{T_e}{T_2}\right) \sum_n \frac{\Gamma_n(\lambda_2)}{\left(\frac{\omega}{n\Omega_2}\right)^2 - 1}$$
$$- \alpha_B \left(\frac{T_e}{T_B}\right) \frac{k^2 V_B^2}{(\omega - kV_s)^2} = 0 \qquad (2)$$

where we have used the small thermal spread approximation to the ion beam Z-function [Fried and Conte, 1961]

$$Z'\left(\frac{\omega - kV_s}{kV_B}\right) = \frac{k^2 V_B^2}{(\omega - kV_s)^2} \qquad (3)$$

and where we define (subscript 1 refers to lighter ion, 2 to heavier ion)

k = perpendicular wave number
ω = complex wave frequency
α = ratio of lighter ion (O^+) concentration to total ambient ion concentration
m_i = mass of ion i
Ω_i = gyrofrequency of ion i
$\lambda_2 = k^2 \rho_2^2$, $\lambda_1 = k^2 \rho_1^2$, ρ_i = gyroradius of ion i
ω_{pi} = plasma frequency of ion i
T_i = temperature of ion i
T_e = electron temperature
λ_{De} = electron Debye length
Ω_e = electron gyrofrequency
$\lambda_e = k^2 \rho_e^2$, ρ_e = electron gyroradius
α_B = ratio of beam ion (AR^+) concentration to total ambient ion concentration
ω_B = beam plasma frequency
m_B = mass of beam ion
T_B = beam ion temperature
V_s = perpendicular component of beam streaming velocity
  ($V_s = \bar{k} \cdot \bar{V}$, $\bar{V}$ = beam streaming velocity)
V_B = beam thermal velocity
$\Gamma_n(\lambda_i) = e^{-\lambda_i} I_n(\lambda_i)$ = modified Bessel function of order n.

We search for mode solutions such that $\omega \ll \omega_e$, so that χ_e, the electron susceptibility, is ignored in the dispersion relation where for strongly magnetized electrons ($\lambda_e \ll 1$) χ_e is a constant. In addition, we retain each of the magnetized ion sum terms for the thermal plasma but approximate the beam as demagnetized and, as noted, assume a small thermal spread. It should be noted that a surprising discovery of both later ARCS flights and the Porcupine experiment was that the beam ions were observed to expand in a perpendicular beam to large distances essentially as free particles [Kaufmann et al., 1985]. Also the relatively simple beam term in the linear disperion relation will assume a more complex form for more realistic beam distribution functions (e.g., peaked loss-cone distributions, distributions containing a parallel component, etc.). The further possibility of local density gradients can give rise to drift mode considerations [Walker and Szuszczewicz, 1975] which can affect Bernstein mode solutions and produce mode cross-coupling contributions [Simons, et al., 1980]. Finally, we assume the approximation of a collisionless plasma during the first two gun pulses considered ($\nu_{in}/\omega_i \sim 2 \times 10^{-2}$ at ~ 200 km [Risbeth and Garriot, 1969]). For later pulses (certainly the final two) there are serious reservations to using a collisionless model.

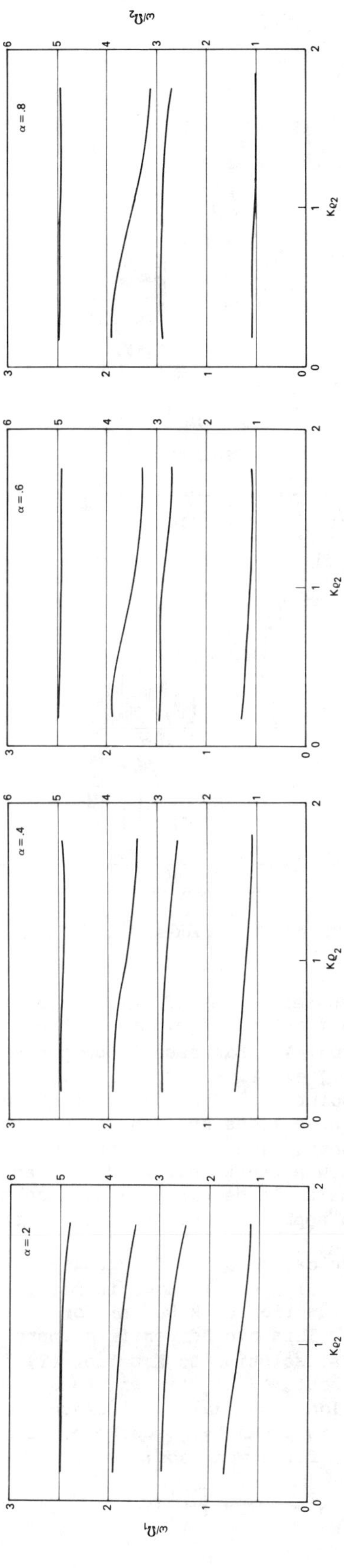

Beamless Results ($\alpha_B = 0$)

Approximate numerical solutions to the dispersion relation for the first four Bernstein eigenmodes ($\alpha_B = 0$) are presented in Figure 2 for $.2 \leq k\rho_2 \leq 1.8$ as a function of α, the relative density of the lighter ion (O^+). For ease in qualitative interpretation we have chosen $\mu=2$ even though for the case of NO^+ as the heavier ion this is only approximately true; for O_2^+ it is exactly correct. The left ordinate displays wave frequency units normalized to the O^+ gyrofrequency (Ω_1); the right hand ordinate units are multiples of the heavier ion gyrofrequency ($\Omega_2 = 2\Omega_1$). The abscissa is plotted in units of $k\rho_2$. Moving left to right through the figures one observes the effect on the eigenmodes of an increasing O^+ concentration: for low values of α each of the 4 Bernstein modes associated with the heavier ions provides a real solution at a given $k\rho_2$. Each solution is separated approximately from the next mode by Ω_2. As α increases one begins to observe a gradual coalescence, as it were, of the two innermost solutions for $k\rho_2 \geq 1.0$. The qualitative result pictured here is that as expected for large values of α solutions are separated by the lighter ion gyro-frequency (Ω_1). Generally then an observable effect of decreasing

Fig. 2. Ion Bernstein modes for a two-component ion plasma ($\Omega_1 = 2\Omega_2$) as a function of lighter ion concentration, α.

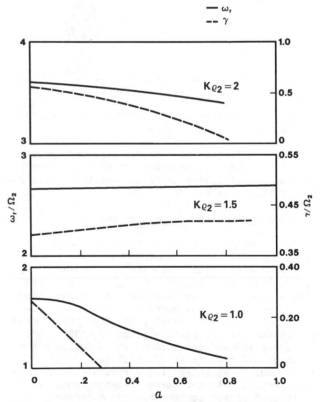

Fig. 3. Numerical solutions of the flute mode dispersion relation as a function of a lighter ion concentration, α, for the first three modes.

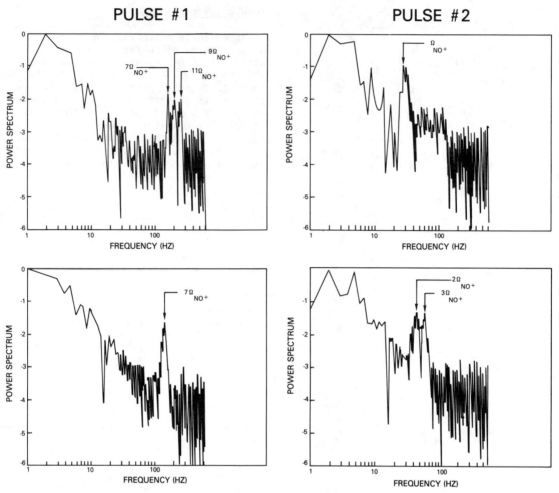

Fig. 4. Power spectra obtained during first two (2) gun pulse periods. ($\Omega_{NO^+} = \Omega_2$).

lighter ion density on electrostatic wave power spectra should be a narrowing of observable peak separation (if indeed waves are excited at all and linear theory is valid).

Multicomponent Plasma with Beam Excitation ($\alpha_B \neq 0$)

For resonant excitation of the beam eigenmodes it is necessary that the wave phase velocity be near the perpendicular component of ion beam speed, i.e., that Equation (2) have a small real contribution. Notice that this is distinct from the introduction of a perpendicular ion distribution of sufficient density to alter the thermal background characteristics [Kintner, 1980]. In fact, results and conclusions which follow are calculated for the case of small beam density so that, loosely speaking, the effect of the beam is a perturbation on the basic system eigenmodes.

For comparison to the beamless mode plots presented, we display the numerical solutions to Equation (2) in Figure (3). In this Figure frequencies and growth rates are plotted for the first 3 Bernstein modes of the multicomponent plasma as a function of plasma lighter ion (O^+) concentration, α. For ease in comparison we chose $T_s = 5\ T_i = 1$ ev, $T_B = 5\ T_i$, $\alpha_B = .05$ for numerical solution. The results for real ω for the first three modes are consistent with the beamless plots. The growth rates for the 1st and 3rd mode show a decrease toward zero as α increases as would be expected i.e. solutions are expected to approach $2\ \Omega_2$, $4\ \Omega_2$,...in the limit as $\alpha \rightarrow 1$.

The numerical solution for growth rates presented in Figure (3) shows increasing growth rates as a function of k (or n) for the modes considered. This can be verified approximately by an analytical solution to Equation (2) in which an expansion about $\omega = kV_s$ is performed. Performing this expansion, the analytical expression for the growth rate is given by, [see Roth, et al., 1983, Equation (6) for comparison]

$$\gamma^2 = -\delta\omega_r^2 + \left[\frac{-\omega_B^2\ \delta\omega r}{G_{12}}\right]^{1/2} \qquad (4)$$

where $G_{12} \propto 1/m$, hence for $\delta\omega_r < 0$, positive growth is possible and γ is generally observed to increase for increasing mode number at these frequencies.

Comparison to Observation

The electron density profile as a function of altitude is presented in Figure (1). The approximate range in altitude covered during plasma generator operation sequences is shown as cross-hatched bars for each of the six pulses. The power spectra shown in Figure (4) were taken during the first two pulses. The remaining pulsing periods show substantial power in the 200-300 Hz region and at least in two cases show peak separations for frequencies less than 100 Hz near Ω_{NO+}. We do not consider the later pulses at this time, however, since the spectra typically show a variety of higher frequency peaks and aliasing considerations related to harmonics of these signals are not yet clear. In addition, we assume a collisionless linear model, so that in highly collisional regimes, particularly below 120 km, any conclusions which emerge would be highly suspect. Returning to the power spectra of Figure (4) we observe strong low frequency peaks separated in the first pulse by approximately the 0^+ gyrofrequency. In pulse #2 the spectra show peaks with separation near $\Omega_{0+}/2$. If we interpret the peak separations as caused by transition through a region of varying α we would conclude that the rocket is passing from a lighter to heavier ion dominant region. Under active auroral conditions a transition from 0^+ to NO^+ dominance is not unreasonable at these altitudes [Rees and Walker, 1968; Swider and Narcisi, 1977; Kopp et al., 1985]. We therefore conclude that the low frequency ($\delta n/n$) power spectral peak separations are consistent with the linear theory as presented.

Conclusions

Solutions have been obtained to the linear dispersion relation for a reasonantly driven two-ion component plasma under the condition that $m_1 = m_2/2$. The solutions should be qualitatively valid for ionospheric parameter regimes which are characterized by 0^+, NO^+ dominance or 0^+, O_2^+ dominance. Conclusions which emerge from the beamless study are that eigenmodes tend to be separated by near harmonics of the light ion gyrofrequency (0^+) consistent with numerous past results. In addition, analytical approximations to the growth rates for wave frequencies near kV_s are consistent with numerical solutions obtained for low beam density. These solutions show generally increasing growth rates for the first ion Bernstein modes as a function of increasing wave number.

We have compared the ($\delta n/n$) power spectrum obtained by a Langmuir probe to theory and within the limits of difficulties outlined above find a favorable comparison to low frequency peak separation observed during the first two gun pulses.

Acknowledgements. The author wishes to thank P. Satyanarayana and M. Singh for helpful discussions and computer software. In addition, I wish to acknowledge the collaboration and support of my colleagues at the University of New Hampshire (Professor R. Arnoldy and Professor R. Kaufmann), the University of Minnesota (Professor L. J. Cahill) and Cornell University (Professor P. Kintner).

References

Boehmer, H., Excitation of ion cyclotron harmonic waves with an ion beam of high perpendicular energy, Phys. Fluids, 19, 1371, 1976.

Boehmer, H., J. P. Hauck and N. Rynn, Ion beam excitation of electrostatic ion-cyclotron waves, Phys. Fluids, 19, 450, 1976.

Fried, B. D. and S. D. Conte, The Plasma Dispersion Function, Academic Press, New York, 1961.

Hamelin, M. and C. Beghin, Electromagnetic and electrostatic waves in multicomponent plasma near the lower hybrid frequency, J. Plasma Phys., 15, 115, 1976.

Holmes, J. C. and E. P. Szuszczewicz, A versatile plasma probe, Rev. Sci. Instrum., 46, 592, 1975.

Ichimaru, S., Basic Principles of Plasma Physics: A Statistical Approach, W. A. Benjamin, 1973.

Kaufmann, R. L., R. L. Arnoldy, T. E. Moore, P. M. Kintner, L. J.Cahill, Jr., and D. N. Walker, Heavy ion beam - ionosphere interactions: electron acceleration, J. Geophys. Res., 90, 9595, 1985.

Kintner, P. M., J. LaBelle, M. C. Kelley, L. J. Cahill, Jr., T. Moore and R. Arnoldy, Interferometric phase velocity measurements, Geophys. Res. Lett., 11, 19, 1984.

Kintner, P. M. and M. C. Kelley, A perpendicular ion beam instability: solutions to the linear dispersion relation, J. Geophys. Res., 88, 357, 1983.

Kintner, P. M., On the distinction between electrostatic ion cyclotron waves and ion cyclotron harmonic waves, Geophys. Res. Lett., 7 (8), 585, 1980.

Kopp, E., L. Andre and L. G. Smith, Positive ion composition and derived particle heating in the lower auroral ionosphere, J. Atm. & Terr. Phys., 47, 301, 1985.

Lebreton, J. P., R. Pottelette, O. H. Bauer, J. M. Illiano, R. Truemann and D. Jones, Observation of waves induced by an artificial ion beam in the ionosphere, Active Experiments in Space, (ESA SP-195), 35-38, 1983.

Moore, T. E., R. L. Arnoldy, R. L. Kaufmann, L. J. Cahill, Jr., P. M. Kintner and D. N. Walker, Anomalous auroral electron distributions due to an artificial ion beam in the ionosphere, J. Geophys Res., 87, 7569, 1982.

Pottelette, R., J. M. Illiano, O. H. Bauer and R. Treumann, Observation of high frequency turbulence induced by an artificial ion beam in the ionosphere, J. Geophys. Res., 89, 2324, 1984.

Rees, M. H. and J. C. G. Walker, Ion and electron heating by auroral electric fields, Ann. Geophys., 24, 1-7, 1968.

Rishbeth, H. and O. K. Garriott, Introduction to Ionospheric Physics, Academic Press, 1969.

Roth, I., C. W. Carlson, M. K. Hudson and R. L. Lysak, Simulations of beam excited minor species gyroharmonics in the procupine experiment, J. Geophys. Res., 88, 357, 1983.

Seiler, S., M. Yamada, and H. Ikezi, Lower hybrid instability driven by a spiraling ion beam, Phys. Rev. Lett., 37, 700, 1976.

Simons, D. J., M. B. Pongratz and S. Peter Gary, Prompt striations in ionospheric barium clouds due to a velocity space instability, J. Geophys. Res., 85, 671, 1980.

Swider, W. and R. S. Narcisi, Auroral F-region: ion composition and nitric oxide, Planet. Space Sci., 25, 103, 1977.

Szuszczewicz, E. P. and J. C. Holmes, Surface contamination of active electrodes in plasmas: distortion of conventional Langmuir probe measurements, J. Applied Phys., 46, 5134, 1975.

Walker, D. N., J. C. Holmes and E. P. Szuszczewicz, Auroral Electrodynamics I: 1. Preliminary electron density profile and, 2. vehicle potential changes during and active experiment, NRL Memorandum Report, 4229, 1980.

Walker, D. N. and E. P. Szuszczewicz, Electrostatic wave observation during a space simulation beam-plasma discharge, J. Geophys. Res., 90, 1691, 1985.

THE DIRECT PRODUCTION OF ION CONICS BY PLASMA DOUBLE LAYERS

Joseph E. Borovsky

Los Alamos National Laboratory, Los Alamos, New Mexico

Glenn Joyce

Naval Research Laboratory, Washington, D. C.

Abstract. In this report, the electrostatic potential structures associated with multiple auroral arcs are briefly considered. The structure of these magnetized double layers are examined and the production of high-altitude, high-energy, gyrophase-bunched ion conics by ion acceleration upward through such magnetized double layers is discussed, as is the driving of electrostatic ion-cyclotron waves by these conics. The production of low-energy heavy-ion conics at low altitudes by moving arc structures is also briefly examined, as is the upward acceleration of ionospheric ions by very narrow ionospheric sheaths is discussed. Finally, the implications for future spacecraft measurements are discussed.

1. Overview

The moving electrostatic potential structures (sometimes unfortunately denoted as electrostatic shocks) associated with multiple auroral arcs have strong effects on the magnetosphere-ionosphere system. These narrow structures, with north-south extents of a few kilometers but with east-west extents of 1000's of km, reside within the large-sale inverted-V potential structure that fills the auroral-zone magnetosphere. One such auroral-arc potential structure is schematized in Figure 1. If the computer simulations of Borovsky and Joyce (1983) and Borovsky (1984) are correct, the structure is divided into two parts, a high-potential (1 - 15 kV) structure at high altitudes (> 1 R_e) and a low-potential (100's V) structure at low altitudes (100 km - 1 R_e).

The high-altitudes portions of such structures (1) produce upflowing energetic (1 - 15 keV) ion beams; (2) produce upflowing energetic (1 - 15 keV) gyrophase-bunched ion conics, with oxygen more conical than hydrogen; (3) produce downgoing sheet beams of energetic (1 - 15 keV) electrons that give rise to multiple auroral arcs; and (4) trap hot electrons that are backscattered off of the ionosphere. The gyrophase-bunched ion conics may drive large-amplitude electrostatic ion-cyclotron waves and the sheet beams of electrons may drive electrostatic electron-cyclotron and Langmuir waves.

The low-altitude portions of such structures can (1) draw oxygen and nitrogen ions upward from the ~ 100-km-altitude ionosphere and pre-accelerate them; (2) produce low-energy (~ 100 eV) heavy-ion and molecular conics as curls and rays move in an east-west manner along arcs; and (3) heat heavy ions in the ionosphere via the north-south drifts of the arcs.

2. Magnetized Plasma Double Layers

A double layer is a transition region between two plasmas with different electrostatic potentials. The temperatures, densities, and compositions of the two plasmas may also differ. An unmagnetized plasma double layer is depicted in the left-hand side of Figure 2. In the top panel the electrostatic potential is plotted in a cut normal to the double layer (y-direction). Obtained from a knowledge of the electrostatic potential (top panel), the number densities of electrons and ions are plotted in the middle panel. The number density of ions originating in the low-potential plasma (right side) drops to zero within the double layer; these ions not having enough energy to penetrate, they are reflected by the double layer. Likewise, the number density of electrons originating in the high-potential plasma (left side) drops to zero as they are reflected by the double layer. The number density of ions that originate in the high-potential plasma decreases as they are accelerated across the double layer and is low in the low-potential plasma where they form a fast, low-density beam. Likewise, the number density of electrons originating in the low-potential plasma drops across the double layer and is low in the high-potential plasma where they form a fast, low-density beam. Obtained from the electron and ion number densities (middle panel), the charge density is plotted in the bottom panel. A solution of Poisson's equation for this charge density self-

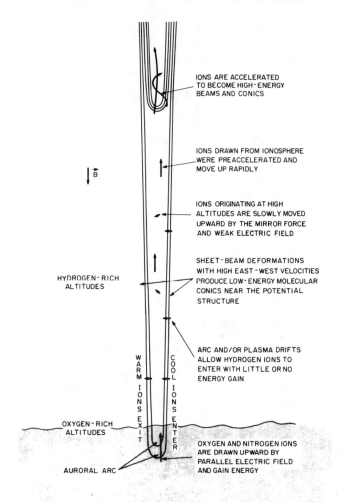

Fig. 1. A schematic of the electrostatic potential structure extending from altitudes of ~ 100 km to > 1 R_e associated with one component of a multiple auroral arc.

consistently gives the electrostatic potential that is depicted in the top panel. Thus, the time-stationary double-layer potential drop is supported by the charge layers created by inflowing particles. The structure of the double layer depends on the structure of these charge layers and the charge-layer structure depends only on the density of the exterior plasmas and the unit charge and kinetic energies (temperatures) of the particles therein. Thus, the dependence is on the Debye length $\lambda_{De} = (k_B T / 4\pi n_0 e^2)^{1/2}$.

To a remarkable degree, the structure of a strong plasma double layer is independent of the strength and orientation of a magnetic field (Borovsky, 1983). This is despite the fact that the magnetic field drastically affects the orbits of the plasma particles that make up the charge layers of the potential drop. A thorough study (Borovsky and Joyce, 1983b; Borovsky, 1983) yields the fact that double layers maintain the same Debye-length scaling with or without the presence of magnetic fields. Earlier reports were to the contrary (Swift, 1975, 1979; Mozer et al., 1977), indicating that magnetized plasma double layers should scale with ion gyroradii. Figure 3 displays one piece of evidence for Debye-length scaling. The points are the thicknesses of time-dependent oblique double layers obtained from a series of 26 computer simulations, plotted as a function of the plasma magnetic-field strength. As the magnetic-field strength varies by over 3 orders of magnitude, the double-layer thickness varies by only about 50%, and that variation is largely attributable to wave heating of the simulation plasmas. The curves are typical particle gyroradii in the simulations. All of these gyroradii vary from much less than the double-layer thickness to much greater than the thickness, over 3 orders of magnitude in each case. Meanwhile, the double-layer thickness remains at 20 - 30 λ_{De}, clearly demonstrating Debye-length scaling and not gyroradii scaling.

The magnetic-field independence of double-layer structure can be hueristically argued as follows. The structure of a double layer depends solely on the structure of the two charge layers created by the inflowing particles that are accelerated or reflected by the double layer. If the plasma is strongly magnetized, then the particles must drift at their thermal speed along the magnetic field to reach the double layer and form the charge layers. If the double-layer normal is directed an angle θ away from the direction of the magnetic field, then the number of particles entering a unit surface area of the double layer is reduced by the projection factor cosθ. Because of this, it might be anticipated that the charge density of the layers should be reduced by a factor of cosθ, but in fact this is compensated for because the time that a particle spends within the double layer is lengthened by a factor of 1/cosθ (the component of the electric field parallel to the magnetic field being reduced by cosθ and the pathlength though the double layer increased by 1/cosθ). Thus, the charge density within the oblique magnetized double layer is essentially the same as the charge density within an unmagnetized double layer. Non-heuristically, this can be seen by comparing the right-hand panels of Figure 2 to the left-hand panels. As is discussed in Borovsky (1983), polarization drifting and E × B drifting have only slight effects in the Poisson-Vlasov equations used to construct time-stationary magnetized-double-layer solutions. Gyrophase bunching of accelerated particles, observed in the computer simulations, has a larger but still non-dominant effect.

3. High-Altitude High-Energy Conics

The production of ion conics by stationary, oblique, magnetized double layers is a simple concept (Lennartsson, 1980; Borovsky, 1981, 1984; Borovsky and Joyce, 1983a; Greenspan, 1983, 1984). In a very strongly magnetized double layer, any ion entering from the high-potential plasma is

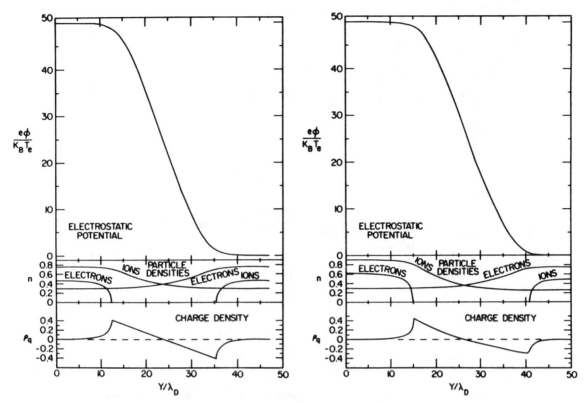

Fig. 2. Two time-stationary solutions to the Poisson-Vlasov equations for magnetized plasmas. In the left-hand panels the double layer is unmagnetized and in the right-hand panel the angle θ between the double-layer electric field and the magnetic field is 45°. The double-layer thickness is denoted as ΔL.

accelerated along the magnetic field like a bead on a wire, attaining a large velocity component parallel to the magnetic field. Such ions form a field-aligned beam emanating from the double layer. However, if the magnetic field is weak enough, an ion entering from the high-potential plasma will be accelerated by the double layer to gain a large velocity component parallel to the double-layer electric field. Such ions form a gyrophase-bunched conic emanating from the double layer. The transition from field-aligned ion beams to conical ions is simply formulated as (Borovsky, 1984)

$$\omega_{ci}\tau_{cr} > \pi \quad , \quad (1)$$

ω_{ci} being the ion gyrofrequency and τ_{cr} being the time required for an accelerated ion to cross the double layer. If Criterion (1) is satisfied, the ions will form conical distributions; if it is not, they will form field-aligned beams. Figure 4 depicts the pitch-angle distribution of double-layer accelerated ions from five variously magnetized simulations. It is seen here that Criterion (1) accurately predicts whether field-aligned beams (top 2 panels) or conics (panels 3 and 4) are produced.

The left-hand side of Criterion (1) being proportional to $m_i^{1/2}$, heavier ions are more likely to be conical than are light ions. From the double layers (electrostatic potential structures) in the high-altitude (> 1 R_e) auroral zone, oxygen conics will be produced more often than hydrogen conics (Borovsky, 1981, 1984; Borovsky and Joyce, 1983a). In crossing the double layer, both the beam ions and the conical ions will obtain kinetic energies equal to the double-layer potential drop, typically 1 - 15 kV.

Another mechanism commonly invoked to produce ion conics is perpendicular acceleration by electrostatic ion-cyclotron waves, an upflowing ion beam driving the waves and the waves turning the beam into a conic (as in Kintner et al., 1979). Contrary to this picture, in computer simulations of oblique double layers it was observed that large-amplitude, monochromatic, ion-cyclotron waves were driven by the free energy of energetic double-layer-produced gyrophase-bunched ion conics and that the effects of these waves were to reduce the pitch angles of the energetic ions, making the ion conics more like field-aligned beams (Borovsky, 1984).

4. Low-Altitude Low-Energy Conics

In computer simulations of magnetized two-dimensional double layers with partial electron

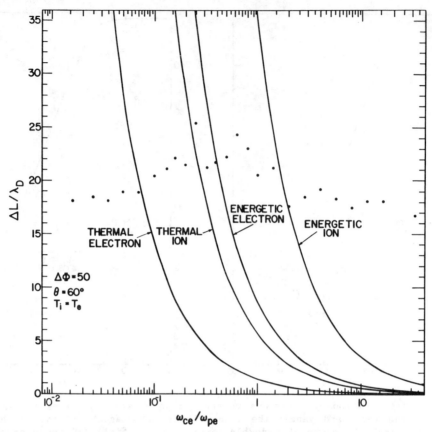

Fig. 3. The thicknesses of variously magnetized computationally simulated oblique double layers (points) are plotted as a function of the magnetic-field strength, a separate simulation having been run for each field value. Also plotted are estimates of typical particle gyroradii (curves). The angle θ between the double-layer electric field and the magnetic field is 60°.

backscattering off of the high-potential boundary, it was observed that a two-dimensional sheet of hot electrons became trapped between the double layer (auroral potential structure) and the backscattering boundary (ionosphere). Additionally, an electrostatic potential structure formed around the hot-electron sheet (Borovsky and Joyce, 1983a). If such a situation occurs because of ionospheric backscattering, then a multiple of east-west-aligned electrostatic structures such as the one in Figure 1 may form above multiple auroral arcs, the structures extending from ~ 100-km altitudes up to the electron-beam-producing potential structures at altitudes ≥ 1 R_e. The north-south width of these structures in the ionosphere should be somewhat larger than the width of an element of the multiple arc, which is only about 100-m wide (Maggs and Davis, 1968). In the ionosphere, the magnetic field is too strong for ion conics to form by ions falling through such potential structures, but if the arcs move relative to the ionospheric plasma then the potential structure will also move and the resulting time-varying electric fields can produce conical ion distributions. In particular, the motion of curls and rays (Hallinan and Davis, 1970) at velocities of 1 - 20 km/sec along the arc faces may produce rapid time variations that can produce heavy-ion conics. Using perpendicular electric fields similar to those observed by ionospheric satellites (Maynard et al., 1982), test-particle computer simulations (Borovsky, 1984) of the effects of a single curl moving along an arc have demonstrated the immediate production of heavy-ion and molecular conics with energies of up to 250 eV from cool topside-ionosphere plasma (Borovsky, 1984). Relatedly, simulations of the slow north-south drift of arcs found a slight heating of the heavier ionospheric ions in the wave of the arc.

Another effect of the potential structure surrounding the hot-electron sheet is the pre-acceleration of oxygen and nitrogen ions from the ionosphere. The acceleration is caused by the very narrow parallel-electric-field region occurring right in the region of electron backscatter (~ 100-km altitude). This portion of the electrostatic potential structure could be termed an ionospheric sheath. The sheath produces a rapid flow

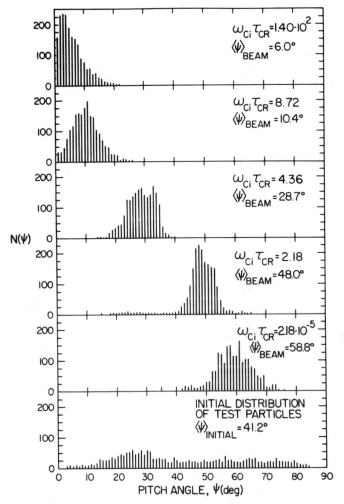

Fig. 4. The pitch angle distributions of the ions accelerated through 5 variously magnetized oblique double layers ($\theta = 60°$) are plotted in the top 5 panels and the pitch-angle distribution of the ions before they entered the double layers is plotted in the bottom panel.

of ions upward from the ionosphere toward the high-altitude portion of the electrostatic structure where they can be further accelerated to form the high-energy high-altitude beams and conics. Since the ionospheric sheath is driven by the energetic precipitating-electron beam of the arc, an increase in auroral activity should produce an increase in the amount of ionospheric ions entering the magnetosphere. Also, since oxygen and nitrogen ions enter the potential structure via the ionospheric sheath where they are pre-accelerated, whereas hydrogen ions enter at higher altitudes by moving across a perpendicular electric field (see Figure 1), after the ions pass through the high-altitude potential structure oxygen and nitrogen will have more kinetic energy that does hydrogen (Borovsky, 1984).

5. Implications for Future Satellite Instrumentation

Computer simulations and analytic theory indicate that auroral-zone double layers are structures with very small scale sizes. The thicknesses of these double layers are estimated to be ~ 1 km (Borovsky and Joyce 1983), thinner if the local current densities exceed 10^{-6} Amp/m^2. Present-day high-altitude satellites typically cover > 20 km during the time that they measure a pitch-angle distribution (Ghielmetti et al., 1978; Gorney et al., 1981; Burch et al., 1981). Double-layer-produced ion conics may be observed throughout such broad regions because proximate double layers will have similar magnetospheric and ionospheric boundary conditions and, hence, they can have similar properties and can produce similar conical distributions. It is predicted here that there is structure to the auroral-zone electron and ion distribution functions on spacial scales below those now measured.

Present-day electrical probes have time resolutions just at the limit needed for detecting these small-scale auroral-zone double layers (Mozer et al., 1977; Shawhan et al., 1981). However, the floating-double-probe technique may be erring when measuring strong electric fields in collisionless plasmas (Borovsky 1986). If the electron beams within the high-altitude auroral double layers have current densities exceeding ~ 10^{-4} Amp/m^2, then beam collection will overcome photoelectron emission and the probes will no longer float near the plasma potential. Very high-resolution particle data is needed to determine the current densities within the double layers themselves. In the low-altitude auroral zone, the earth's shadow will lead to a loss of photoelectron emission and the probes again will be dominated by electron-beam collection when they are within strong potential structures. In both cases, the electric-field value obtained by a double probe is much less than the actual value. Additionally, the obtained electric-field value differs for cylindrical and spherical probes.

If the picture discussed in this report is a true description of the multiple auroral arc, then it demands higher-resolution measurements in the auroral zone.

Acknowledgments. The authors wish to thank Dave Young for useful conversations. This work was supported by the NASA Solar-Terrestrial Theory Program and by the U. S. Department of Energy.

References

Borovsky, J. E., The Simulation of Plasma Double-Layer Structures in Two Dimensions, Ph. D. Thesis, University of Iowa, 1981.

Borovsky, J. E., "The Scaling of Oblique Plasma Double Layers", Phys. Fluids, 26, 3273, 1983.

Borovsky, J. E., "The Theory of Langmuir Probes in Strong Electrostatic Potential Structures", to appear in Phys. Fluids, 1986.

Borovsky, J. E., "The Production of Ion Conics by

Oblique Double Layers", *J. Geophys. Res.*, 89, 2251, 1984.

Borovsky, J. E., and G. Joyce, "Numerically Simulated Two-Dimensional Auroral Double Layers", *J. Geophys. Res.*, 88, 3116, 1983a.

Borovsky, J. E., and G. Joyce, " The Simulation of Plasma Double-Layer Structures in Two Dimensions", *J. Plasma Phys.*, 29, 45, 1983b.

Burch, J. L., J. D. Winningham, V. A. Blevins, N. Eaker, W. C. Gibson, and R. A. Hoffman, "High-Altitude Plasma Instrument for Dynamics Explorer-A", *Space Sci. Inst.*, 5, 455, 1981.

Ghielmetti, A. G., R. G. Johnson, R. D. Sharp, and E. C. Shelley, "The Latitudinal, Diurnal, and Altitudinal Distributions of Upward Flowing Energetic Ions of Ionospheric Origin", *Geophys. Res. Lett.*, 5, 59, 1978.

Gorney, D. J., A. Clarke, D. Croley, J. Fennell, J. Luhmann, and P. Mizera, "The Distribution of Ion Beams and Conics Below 8000 km", *J. Geophys. Res.*, 86, 83, 1981.

Greenspan, M. E., *Energization of Ions by Oblique Double Layers*, Ph. D. Thesis, University of California San Diego, 1983.

Greenspan, M. E., "Effects of Oblique Double Layers on Upgoing Ion Pitch Angle and Gyrophase", *J. Geophys. Res.*, 89, 2842, 1984.

Hallinan, T. J., and T. N. Davis, "Small-Scale Auroral Arc Distortions", *Planet. Space Sci.*, 18, 1735, 1970.

Kintner, P. M., M. C. Kelley, R. D. Sharp, A. G. Ghielmetti, M. Temerin, C. Cattell, P. F. Mizera, and J. F. Fennel, "Simultaneous Observations of Energetic (keV) Upstreaming and Electrostatic Hydrogen Cyclotron Waves", *J. Geophys. Res.*, 84, 7201, 1979.

Lennartsson, W., "On the Consequences of the Interaction Between the Auroral Plasma and the Geomagnetic Field", *Planet. Space Sci.*, 28, 135, 1980.

Maggs, J. E., and T. N. Davis, "Measurements of the Thicknesses of Auroral Structures", *Planet. Space Sci.*, 16, 205, 1968.

Maynard, N. C., J. P. Heppner, and A. Egeland, "Intense, Variable Electric Fields at Ionospheric Altitudes in the High Latitude Regions as Observed by DE-2", *Geophys. Res. Lett.*, 9, 981, 1982.

Mozer, F. S., C. W. Carlson, M. K. Hudson, R. B. Torbert, B. Parady, and J. Yatteau, "Observations of Paired Electrostatic Shocks in the Polar Magnetosphere", *Phys. Rev. Lett.*, 38, 292, 1977.

Shawhan, S. D., D. A. Gurnett, D. L. Odem, R. A. Helliwell, and C. G. Park, "The Plasma Wave and Quasi-Static Electric Field Instrument (PWI) for Dynamics Explorer-A", *Space Sci. Inst.*, 5, 535, 1981.

Swift, D. W., "On the Formation of Auroral Arcs and Acceleration of Auroral Electrons", *J. Geophys. Res.*, 80, 2096, 1975.

Swift, D. W., "An Equipotential Model for Auroral Arcs: The Theory of Two-Dimensional Laminar Electrostatic Shocks", *J. Geophys. Res.*, 84, 6427, 1979.

ON THE SPATIAL SCALE SIZE OF OBLIQUE DOUBLE LAYERS

W. Lennartsson

Lockheed Palo Alto Research Laboratory, Palo Alto, California 94304.

Abstract. Current literature on double layers in the space environment is largely based on different kinds of model calculations and is ambiguous on some of the key issues, notably on the scale size of oblique double layers. This paper briefly reexamines the latter issue under simplifying but fairly conventional assumptions, including: a. the electrons on the positive side of the double layer are characterized by a Debye length that is small in comparison to a typical gyro radius of the ions, b. the electric potential varies monotonically across the double layer, and c. the electric field is strictly time-invariant. The consequent limits on the transverse scale size are examined from two geometrical aspects, one considering the accessibility of particle trajectories in phase space, the other considering an adverse consequence of preserving the magnetic moment of the ions inside the double layer. These two aspects together may seem to imply that the scale size can be neither much smaller nor much larger than a typical gyro radius of ions approaching from the high potential side.

1. Introduction

The theory of double layers and the possible role these space charge structures play in the acceleration of auroral particles are still controversial topics in the literature on auroral phenomena. Double layers have been studied in laboratory plasmas for many years and are fairly well understood in that environment, at least in the case of unmagnetized plasmas (see review by Carlqvist [1979]). The knowledge gained in the laboratory is not readily transferred to the space environment, however, because the boundary conditions there are radically different and the plasma properties are much more complex.

Considerable efforts have been spent trying to predict the properties of double layers in space plasmas, but these efforts have had to rely in large parts on assumptions about the plasma environment, different in different studies, and have not yet produced a coherent picture [e.g., Block, 1972; Swift, 1975, 1976, 1979; Wagner et al., 1980; Witt and Lotko, 1983; Borovsky and Joyce, 1983; Borovsky, 1983; Singh et al., 1984; and references therein]. One area of some controversy has been the scale size of oblique double layers, that is oblique to the magnetic field lines. Several studies indicate that the thickness of such structures should in general be comparable to some characteristic ion gyro radius [e.g., Swift, 1979; Singh et al., 1984], but other studies may suggest values that are either much larger [e.g., Swift, 1975] or much smaller [e.g., Borovsky, 1983; Borovsky and Joyce, 1983].

This study is an attempt to isolate certain aspects of the scale size of oblique double layers from the obviously complicated problem of calculating, by numerical simulation or other means, a self-consistent model of such a structure. Two geometrical aspects are examined which together may suggest a scale size comparable to the gyro radii of ions approaching from the high potential side. The discussion is only indirect, however, pointing to certain difficulties with having either a very small thickness or a very large one, and does not establish the sufficient conditions for a stable self-consistent double layer. It does not necessarily apply to all possible combinations of particle distributions either, but it may still give helpful directions for interpreting more complex configurations.

2. Basic Assumptions

The discussion is limited to a simple geometry, including a nearly homogeneous and strictly time-independent magnetic field and, hence, a curl-free electric field:

$$\vec{B} \approx \text{constant} \qquad (1)$$

$$\text{curl } \vec{E} = 0 \qquad (2)$$

Also the electric field is assumed strictly time independent and, therefore, the sum of the kinetic energy $2^{-1}mv^2$ and the potential energy QV is a constant for each particle:

$$\frac{m}{2}v^2 + QV = \text{constant} \qquad (3)$$

The electric field is also assumed to be unidirectional:

$$\frac{d}{ds}V \leq 0 \qquad (4)$$

where s is a distance coordinate running from the high potential side to the low potential side of the double layer. This condition is automatically satisfied if the double layer consists of only two charge layers, one positive and one negative.

No explicit reference is made to the velocity distribution of the particles, but it may be helpful to keep certain "conventional" features in mind. Thus it may be assumed

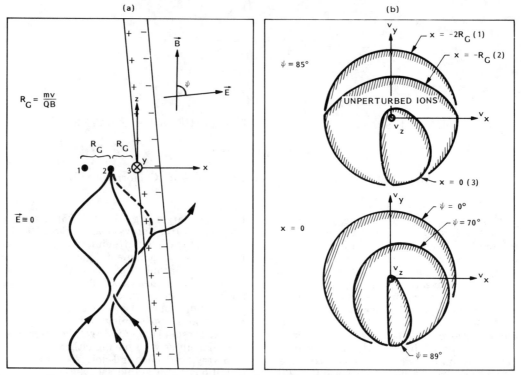

Fig. 1. a: Perturbation of ion trajectories by a thin oblique double layer (hypothetical), given ions with an arbitrary absolute velocity v and pitch angles $< 90°$. b: Projections onto the xy-plane of velocity regions ($|\vec{v}| = v$) containing unperturbed ions at varying x-coordinate but fixed electric field direction (top panel) or fixed x-coordinate but varying electric field direction (bottom panel).

that the initial velocity distribution (prior to acceleration) is at least gyrotropic,

$$f(\vec{v})_{\text{initial}} = f(v_\perp, v_\parallel) \quad (5)$$

where the perpendicular and parallel symbols refer to the magnetic field direction. It may also be assumed that a typical gyro radius of the ions is larger than that of the electrons of the same origin. More importantly, however, it is assumed that a characteristic Debye length λ_D can be ascribed to the electrons on the high potential side of the double layer which is much smaller than a typical ion gyro radius R_G (ion):

$$\lambda_D \ll R_G \text{ (ion)} \quad (6)$$

In Section 4, a comparison is made between an oblique double layer and one that has the electric field parallel to the magnetic field everywhere. This comparison presumes that the latter kind of double layer cannot have a scale size that is very much larger than λ_D.

3. Lower Limit on Scale Size

Imagine a planar double layer structure with a cross section as depicted in Figure 1a. The extension perpendicular to the plane of the figure, as well as above and below the figure, is large compared to the thickness. The electric field $\vec{E}$ is zero to the left and right and is perpendicular to the double layer on the inside (due to Eq. (2)). The angle between $\vec{E}$ and the magnetic field $\vec{B}$ (vertical) is ψ.

The oscillating lines illustrate representative trajectories of ions with a given (non-relativistic) initial energy $2^{-1}mv^2$, which would all converge at point 2 on the negative x-axis, if unperturbed. Any such trajectory passing through a point $x = x_0$, $y = 0$, and $z = 0$ at time $t = 0$ is defined by:

$$x = x_0 + R_G \sin \alpha (\sin \phi - \sin(\phi - \omega t))$$
$$y = R_G \sin \alpha (\cos(\phi - \omega t) - \cos \phi)$$
$$z = R_G (\cos \alpha) \omega t$$

where α is the pitch angle. ω is the gyro frequency QB/m, $R_G = \omega^{-1} v$, and ϕ is an arbitrary phase angle.

Whether or not a given trajectory can reach the point $(x_0, 0, 0)$ unperturbed by the double layer depends on the combination of α and ϕ. If the phase angle ϕ is defined to lie within the interval $-90° \leq \phi < 270°$, for example, the limiting combination is defined by $\phi - \omega t = 270°$ and $-z = x \tan \psi$. The limit corresponding to point 2 in Figure 1a, assuming $\psi = 85°$, is illustrated in the top panel of Figure 1b by the curve labelled $x = -R_G$. At point 1 ($x = -2R_G$) there is no such limit, and at point 3 ($x = 0$) the limit is narrower. The bottom panel of Figure

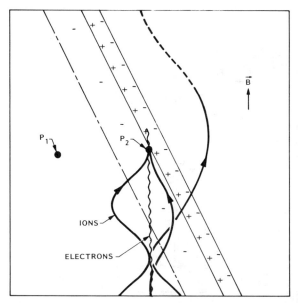

Fig. 2. A point P_1 sufficiently far from the double layer is accessible to any ion with a pitch angle $< 90°$, regardless of gyrophase angle. A point P_2 in the close vicinity can only be reached at certain combinations of pitch and gyrophase angles and, thus, has a reduced ion density. The electron density is the same at both points, however, implying a layer of negative charge contrary to the original assumptions.

1b shows a similar set of limits for different values of ψ, but all at point 3 $(x = 0)$.

Those combinations of α and ϕ that fall outside of the limits in Figure 1b can only be occupied by ions that have experienced the electric field at some earlier time, following trajectories further to the right than the unperturbed ions. These ions have regained the original energy (Eq. (3)) but have been accelerated upwards by the parallel component of $\vec{E}$ and thus have reduced pitch angles. Since initially $\alpha < 90°$ for all ions moving upwards, the angular space corresponding to perturbed ions in Figure 1b can only be partially filled and, hence, the total number of ions with a given energy is reduced at $x_0 > -2R_G$ (phase space density is preserved). The missing ions (missing from the initial distribution in (5)) are those that do not return to the left side of the double layer after the first encounter with the electric field. The number of such ions depends on the actual strength and distribution of $\vec{E}$, as well as the angle ψ.

The electrons approaching the double layer from below are repelled by the electric field and are not subject to the selective removal affecting the ions. Therefore, the electron density does not diminish in the vicinity of the double layer, as does the ion density, but forms a layer of negative charge there, as indicated in Figure 2. The same conclusion can be reached by merely referring to the smaller gyro radius of the electrons. This negative charge is contrary to the assumption of a thin double layer but can be removed, in principle, by making the double layer correspondingly thicker, of the order of a typical ion gyro radius (or larger).

This reasoning does not consider the particles originating above the double layer; however, it can be seen that only a carefully designed velocity distribution will allow particles from above (ions) to reduce the negative charge below the double layer in Figure 2.

4. Upper Limit on Scale Size

In a unidirectional electric field (Eq. (4)), the ion and electron densities n_i and n_e, as well as the charge density, can be expressed as functions of the electric potential V. Assuming only singly charged and positive ions, for simplicity, the charge density is proportional to $n_i(V) - n_e(V)$, and therefore

$$\frac{d^2}{ds^2}V \propto -(n_i(V) - n_e(V)), \quad (7)$$

provided the coordinate s is measured along the electric field vector. Particles entering the double layer from the high potential side initially experience a growing value of $n_i - n_e$ (if Eq. (4) is satisfied), and the faster this value grows as a function of V, the faster the potential V changes as a function of s, according to (7).

In the idealized case of a purely parallel electric field, the functional shape of the right-hand side of Eq. (7) may be expected to be the same as in the case of an unmagnetized double layer [cf., Block, 1972] and, by analogy, the potential V may be expected to change appreciably over a distance $\Delta s \sim \lambda_D$ [cf., Carlqvist, 1979]. The latter presumes of course that the "initial" particle distributions are somewhat analogous to distributions that have been studied in the context of unmagnetized double layers.

If the potential V is to change in a much smoother fashion for an oblique double layer, changing over a length scale Δs that is not only larger than λ_D but even much larger than R_G (ion), then the right-hand side of Eq. (7) must be made a much slower function of V. However, as will be argued below, the functional shape of $n_i(V) - n_e(V)$ cannot be substantially altered, given the same "initial" particle distributions, unless the magnetic moments of the ions are also changed. That in turn sets an upper limit on the distance Δs, over which the potential V changes appreciably from its value on the high potential side.

Figure 3a depicts the transformation of ion velocities during acceleration by a purely parallel electric field (upwards). The initial velocities at a potential V_0 are assumed to lie within a thin hemispherical shell in velocity space, although not necessarily filling this shell uniformly. At a potential $V < V_0$ the velocities lie within a spherical shell with larger radius v but smaller thickness dv (Eq. (3)). The velocities also lie within the intersection of this shell and a cylindrical surface defined by the maximum transverse velocities at V_0 (assumed equal to v_0 in the figure).

Figure 3b depicts the transformation of the same velocities in an oblique electric field under the condition that the ions preserve the magnetic moments (gyration energy proportional to B). This condition would be satisfied if the electric field did indeed have a spatial scale size large compared to ion gyro radii. That is, the transverse velocity $\vec{v}$ of each ion maintains the relation

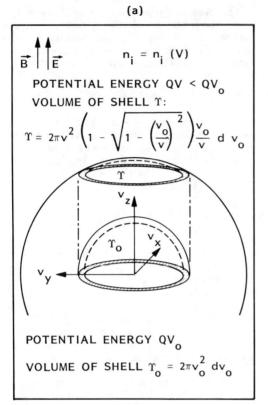
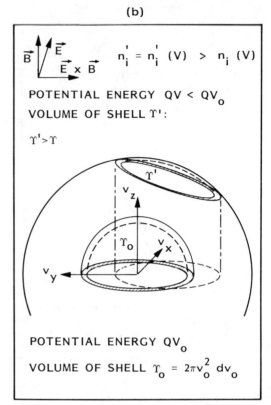

Fig. 3. a: Transformation of ion velocities in a parallel electric field, given a narrow range of (arbitrary) initial energy. b: Same for oblique electric field, assuming the magnetic moment is preserved.

$|\vec{v}_\perp - \vec{E} \times \vec{B} \, B^{-2}| =$ constant (cf., Eq. (1)). The velocities still lie within a spherical shell with the same radius v and thickness dv at a given potential V, and also within the same cylindrical surface. The cylinder, however, is displaced a distance equal to $|\vec{E} \times \vec{B} \, B^{-2}|$ from the center and, therefore, cuts out a larger surface than it does in a. Because the phase space density remains the same, the number density of these ions at potential V is larger in b than it is in a.

The higher density of upflowing ions at potential V in Figure 3b is not automatically counteracted by a corresponding reduction in the density of ions approaching from above. In particular, if the downflowing ions initially have an isotropic distribution of velocities these ions have an isotropic distribution also inside the electric field, as long as Equation (4) holds, and the density of these ions, therefore, depends only on the potential V, not on the direction of $\vec{E}$. It may be noted that the $\vec{E} \times \vec{B}$ drift of such ions does not affect the isotropy of the velocity distribution because it is counteracted by the density gradient.

Analogous considerations can be made for the electrons, but if the electrons have energies comparable to those of the ions of the same origin, the effect of the $\vec{E} \times \vec{B}$ drift on the electron velocity distribution is much smaller. In that case, at least, it follows that the difference $n_i(V) - n_e(V)$ within the positive layer of the double layer structure is no smaller in Figure 3b than it is in Figure 3a, but instead is slightly larger. Therefore, the assumption that the ions preserve the magnetic moments when entering the electric field from below is not self-consistent.

This somewhat circuitous reasoning points to the difficulty of obtaining a self-consistent distribution of charge if the magnetic moments of ions are preserved. However, these moments will not be altered in a substantial fashion unless the electric field has a characteristic scale length of the same order of magnitude as the initial gyro radii of the ions (or smaller), at least on the positive side of the double layer, so the same reasoning points to the difficulty of having an oblique double layer with a very smoothly distributed electric field [cf., Lennartsson, 1980]. The same kind of reasoning applied to the negative charge of the double layer would suggest that a typical gyro radius of electrons approaching from the low potential side is the determining scale length there, but the gyro radii of these electrons may not be large compared to a Debye length.

5. Concluding Remarks

It is worth noting that the two extreme geometries considered in Section 3 and 4, respectively, have opposing effects on the ion density. The sharply defined double layer in Section 3 leads to an artificial deficiency of ions on the positive side (Figure 2), whereas the smoothly defined double layer in Section 4 leads to an excess of ions. Some intermediate distribution of the electric field, having a scale length of the same order of magnitude as the gyro radii of the approaching ions, will therefore allow the ion density to better match the electron density.

It is also worth noting that the geometry considered in Section 3 requires that the ion gyro radii be small compared to the external boundaries of the plasma. It is, therefore, not a meaningful comparison to think of an unmagnetized double layer in terms of R_G (ion) $\to \infty$.

The full implication of these considerations can only be realized after a number of boundary conditions are specified, including the velocity distribution of the various particles approaching the double layer. It follows, for example, that the difficulty associated with a small scale size in Section 3 is alleviated if the upflowing ions all have small pitch angles to start with (cf., Figure 1b).

All interpretations, however, must take into account the assumptions in Section 2, especially Equation (3). Although this is a widely used assumption in plasma theory, it does not have the status of a natural law. Its validity is also related to the validity of Equation (4). Both assumptions are invalid if the electric field has temporally oscillating or fluctuating components and that may possibly account for the apparent discrepancy between the conclusion in Section 3 and the numerically simulated double layers of Borovsky [1983] and Borovsky and Joyce [1983].

Acknowledgments. This work was supported by the National Science Foundation under grant ATM-8317710 and the Lockheed Independent Research Program.

References

Block, L.P., Potential double layers in the ionsphere, *Cosm. Electrodyn., 3,* pp. 349-376, 1972.

Borovsky, J.E., The scaling of oblique plasma double layers, *Phys. Fluids, 26,* pp. 3273-3278, 1983.

Borovsky, J.E. and G. Joyce, Numerically simulated two-dimensional auroral double layers, *J. Geophys. Res., 88,* pp. 3116-3126, 1983.

Carlqvist, P., Some theoretical aspects of electrostatic double layers, in *Wave Instabilities in Space Plasmas,* edited by P.J. Palmadesso and K. Papadopoulos, D. Reidel Publ. Co., pp. 83-108, 1979.

Lennartsson, W., On the consequences of the interaction between the auroral plasma and the geomagnetic field, *Planet. Space Sci., 28,* pp. 135-147, 1980.

Singh, N., R. W. Schunk and H. Thiemann, Numerical simulations of double layers and auroral electric fields, *Adv. Space Res., 4,* pp. 481-490, 1984.

Swift, D.W., On the formation of auroral arcs and acceleration of auroral electrons, *J. Geophys. Res., 80,* pp. 2096-2108, 1975.

Swift, D.W., An equipotential model for auroral arcs, 2. Numerical solutions. *J. Geophys. Res., 81,* pp. 3935-3943, 1976.

Swift D.W., An equipotential model for auroral arcs: the theory of two-dimensional laminar electrostatic shocks, *J. Geophys. Res., 84,* pp. 6427-6434, 1979.

Wagner, J.S., T. Tajima, J.R. Kan, J.N. Leboeuf, S.-I. Akasofu, and J.M. Dawson, V-potential double layers and the formation of auroral arcs, *Phys. Rev. Lett., 45,* pp. 803-806, 1980.

Witt, E. and W. Lotko, Ion-acoustic solitary waves in a magnetized plasma with arbitrary electron equation of state, *Phys. Fluids, 26,* pp. 2176-2185, 1983.

DOUBLE LAYERS IN LINEARLY STABLE PLASMA

Robert H. Berman, David J. Tetreault and Thomas H. Dupree

Massachusetts Institute of Technology, Cambridge, MA 02139

Abstract. An unstable, highly intermittent state of turbulence has been observed to evolve from a quiescent, homogeneous simulation plasma that is linearly stable. This intermittent state consists of isolated phase space density holes that have double layer potential structures. The plasma is nonlinearly unstable because the holes grow in amplitude by accelerating to regions of higher average phase space density. The instability can be interpreted as a collection of colliding, growing holes. A series of simulations with a single isolated hole indicates that an isolated hole grows for any finite electron-ion drift velocity. The nonlinear hole instability growth rate is consistent with theoretical predictions. It is suggested that the instability can explain the observation of intermittent double layers on auroral field lines where the currents are small and the ion and electron temperatures are comparable.

Introduction

Double layers have been suggested as a particle acceleration mechanism for auroral plasma (Temerin, et al., 1982) and ion conics (Chang and Coppi, 1981). Evidence for this hypothesis comes from S3-3 satellite data which has shown that double layers exist on auroral magnetic field lines. The double layers appear intermittently; they are on the order of $32\lambda_D$ (λ_D is the Debye length) long and are separated from each other along the field lines by distances of some $10^3 \lambda_D$. One prominent explanation for the existence of these double layers involves the production of double layers from ion holes observed during the simulation of linear ion acoustic instability (Hudson et al., 1983). However, the measured values of T_e/T_i and v_D in the auroral region seem to be more consistent with stability than instability of ion acoustic waves. We report here on computer simulations that show the instability of phase space density holes (i.e., double layers) for plasmas that are stable to linear ion acoustic waves and suggest that hole instability is the cause of the observed double layers. The holes spontaneously form and become distributed intermittently in phase space. Isolated holes can grow for any finite current. Moreover, unstable ion holes have double layer potential structures - the acceleration and growth of an ion hole is due to the potential drop across the hole which results from the difference in the number of electrons reflected by the hole.

In a broader context, hole instability may be relevant to the release of particle fluxes in the earth's magnetotail. In the past, linear ion acoustic turbulence has been one of the mechanisms invoked to explain the anomalously large magnetic reconnection rates believed to be responsible for magnetospheric substorms (Papadopulos, 1979). While conditions in the plasma current sheet are apparently incompatible with linear ion acoustic instability, the conditions are consistent with hole instability.

The Simulations

The computer simulations were designed to investigate localized non-wave-like phase space density fluctuations in plasma. Such fluctuations - called "clumps" (Boutros-Ghali and Dupree, 1981; 1982) - have been shown to be an important component of the turbulence observed in recent computer simulations. For example, it was observed in one species simulations (Berman et al., 1983) that the decay of plasma fluctuations from a variety of initial conditions tends to produce clump fluctuations. Such clumps can be either depletions (holes; $\delta f \leq 0$) or enhancements ($\delta f \geq 0$) in the phase space density. Here, $\delta f = f - f_0$, where f is the phase space density and $f_0 = \langle f \rangle$ where $\langle \ldots \rangle$ denotes a spatial average. A depletion or hole in the phase space density tends to self-bind and, when in isolation, forms the trapped particle phase space eddy of a Bernstein-Greene-Kruskal (BGK) equilibrium (Dupree, 1982). Enhanced phase space material self-repels and fills the interstitial phase space regions between the holes. In related work (Berman et al., 1982), it was shown that the mean square fluctuation amplitude due to clumps can increase with time in a two-species plasma, i.e., clumps can be unstable. The condition for clump instability can be much less restrictive than that required for linear instability of waves. In particular, the simulations (Berman et al., 1982) show that the clump instability in a one dimensional, ion-electron plasma with electron drift can occur for electron drift velocities significantly below that required for the linear ion-acoustic instability. We show here that an isolated hole can grow for any finite drift velocity.

For our simulations we used a highly optimized, one-dimensional electrostatic code with $N_p = 102400$ particles per species and a spatially periodic system of length $L = 10\pi\lambda_D$. Therefore, the number of particles per Debye length was $n_0\lambda_D = 3259.5$ ($n_0 = N_p/L$ is the spatial particle density). Because of our limited computer time and our use of large $n_0\lambda_D$, we were forced to use a low value of ion to electron mass ratio m_i/m_e. Here, we report mainly on simulations where $m_i/m_e = 4$, though similar results from runs where $m_i/m_e = 100$ and $L = 512\lambda_D$ are also discussed (e.g., Barnes et al., 1985).

Our simulation experiments can be divided into two classes depending on the initial conditions used. In one class of runs (random starts), we simulated an initially quiescent plasma where the initial electrostatic energy was due to a random distribution of discrete particles. Spatially homogeneous ion and drifting electron Maxwellian velocity distribution functions were set up initially. In the second class (isolated hole runs), a single isolated phase space density hole was introduced initially into an ion Maxwellian distribution. Discrete particle fluctuations were suppressed initially by judicious rearrangement of the discrete particles (a so-called "quiet start").

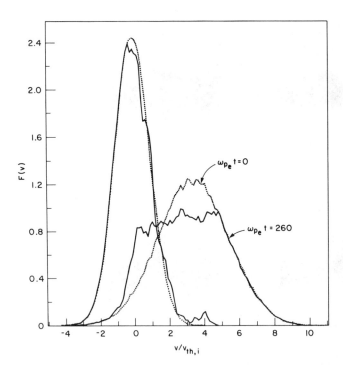

Fig. 1. The ion and drifting electron spatially averaged distribution functions at times $\omega_{p_e}t = 0$ (dotted line) and 260 (solid line) for the case of a linearly stable random start run with drift $v_D = 3.5 v_{th,i}$, $T_e = T_i$ and $m_i = 4 m_e$.

Random Start Runs and Development of Intermittency

For random start runs, with $T_e = T_i$ and $m_i/m_e = 4$, an instability occurs for electron drift velocities $v_D \geq 1.5 v_{th,i}$, i.e., significantly below the linear stability boundary of $v_D = 3.92 v_{th,i}$ for the simulation parameters used ($v_{th,i}$ is the ion thermal velocity). As the instability evolves, isolated phase space density holes emerge and – except for infrequent encounters with other holes – tend to persist and coexist with a smaller velocity scale background turbulence. The relative isolation or concentration of the holes in a small fraction of the available phase space – a feature we refer to as hole intermittency – appears to be a natural consequence of the turbulence. Once produced, the holes tend to grow in depth by accelerating to regions of higher average phase space density.

Results for a representative linearly stable run with $T_e = T_i$ and $v_D = 3.5 v_{th,i}$ are shown in Fig. 1. Figure 1 shows the electron and ion distribution functions at $\omega_{p_e}t = 0$ and at $\omega_{p_e}t = 260$. Considerable distortion of the ion distribution function and flattening of the electron distribution function is evident. A typical time sequence of the ion phase space density (not shown here for lack of space – see Berman et al., 1985) shows individual isolated holes with $e\phi/T_i \leq 1$, as well as "hole-hole" interactions resulting in tidal deformations and coalescing of holes. Runs above the linear stability boundary made with $T_e/T_i = 20$, $m_i/m_e = 100$, and $L = 32\lambda_D$ also show the development of intermittent ion holes (Berman et al., to be published). These results are similar to those obtained by others (e.g., Barnes et al., 1985; Sato and Okuda, 1981). In these linearly unstable runs, the unstable ion holes appear as a "large" double layer (have $e\phi/T_i \geq 1$) in contrast to the ion holes of the linear stable case (where $e\phi/T_i \leq 1$). This characterization of the ion hole potential in terms of T_i differs in the two cases because $T_e = T_i$ for the linearly stable runs while $T_e \gg T_i$ for the linearly unstable runs. This difference is superficial only, since both linearly stable and linearly unstable ion holes derive their free energy for growth from the electrons and therefore, both have $e\phi/T_e \approx 1$. As in the linearly stable runs (See next section below), the ion holes of the linearly unstable runs have double layer potential structures and grow and decelerate to regions of higher phase space density.

The growth rate of the mean square fluctuation level versus electron drift velocity v_D is shown in Fig. 2. It is clear that instability occurs for drift velocities significantly below the linearly stable boundary. The solid curve in Fig. 2 is a theoretical result obtained from Eq. (8).

Isolated Hole Starts And Hole Growth

In order to further study the phase space hole growth, we made a series of quiet start runs with an isolated ion hole initially present. These runs all have an initial ion hole whose depth is $\tilde{f}/f_{0i} = -0.6$ and whose size is $\Delta x = 6\lambda_D$ by $\Delta v = 0.2 v_{th,i}$. Holes with smaller $\tilde{f}$, Δx or Δv yielded similar results but were more difficult to study. Figure 3a shows the initial phase space density for a run with $v_D = 3.0 v_{th,i}$ and is typical of the initial conditions used in the isolated hole runs.

A typical case of linearly stable drift ($v_D = 3.0 v_{th,i}$, $m_i/m_e = 4$, $L = 10\pi\lambda_D$, and $T_e = T_i$) is shown in Figs. 3 - 4. Runs with $m_i/m_e = 100$ and $L = 512\lambda_D$ showed similar results (Berman et al., to be published). The ion hole placed initially at $(10\lambda_D, 1.5 v_{th,i})$ in Fig. 3A, evolves to $(22\lambda_D, 0.75 v_{th,i})$ at $\omega_{p_e}t = 120$ (see Fig. 3B). The deceleration process continues as the hole passes through $v = 0$ at $\omega_{p_e}t = 180$ (see Fig. 3G). From Figs. 3 and 4, we note that the hole depth increases with time for positive hole velocities but decreases for negative hole velocities. Referring to Fig. 1, it is apparent that the hole depth grows in regions where $f'_{0e}f'_{0i} \leq 0$ but decays where $f'_{0e}f'_{0i} \geq 0$ ($f'_0 = \partial f_0(v)/\partial v$). The cessation of hole acceleration after $\omega_{p_e}t = 280$ (see Fig. 4) can be traced to the plateauing of the electron distribution function as in Fig. 1. As in the random start runs discussed above, new ion holes emerge

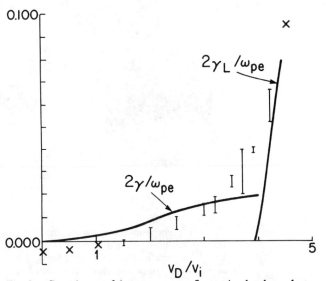

Fig. 2. Growth rate of the mean square fluctuation level vs. electron drift velocity v_D for the random start runs. γ_L is the growth rate of the fastest growing linearly unstable ion acoustic wave. γ is a theoretical curve discussed in connection with Eq. 13.

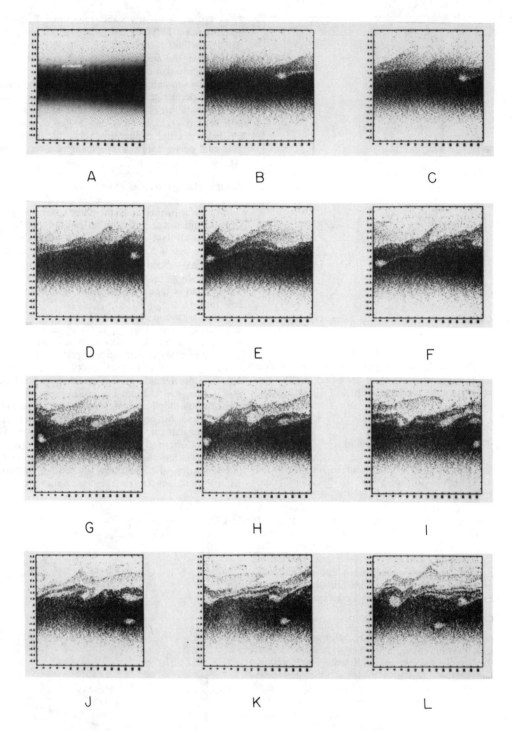

Fig. 3. A time sequence of the ion phase space for a linearly stable case with an initially prepared isolated ion hole (A) and $v_D = 3.0 v_{th,i}$. The sequence starts at (B), when $\omega_{p_e} t = 120$, with $10 \omega_{p_e}^{-1}$ between figures. (The figure for $\omega_{p_e} t = 150$ is excluded. Thus, 8G shows the ion phase at $\omega_{p_e} t = 180$.) The acceleration and growth of the ion hole is evident.

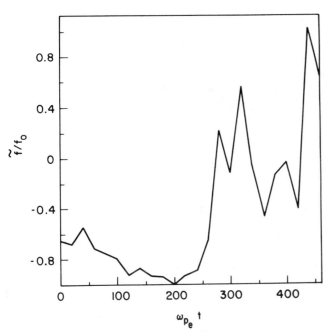

Fig. 4. The normalized ion hole depth $-\tilde{f}(t)/f_0[u(t)]$ for the isolated hole run of Fig. 3.

spontaneously out of the background plasma. One such hole is evident at $(13\lambda_D, 1.25v_{th,i})$ in Fig. 3F, while two others are observed at $(2\lambda_D, 1.25v_{th,i})$ and $(15\lambda_D, 1.5v_{th,i})$ in Fig. 3H. These latter holes tend to collide and coalesce into longer holes (see Figs. 3H - 3L).

A plot of the measured growth rate γ_H of the hole depth against v_D is shown in Fig. 5. Discrete particle collisions were negligible early in the run when the measurements were made. Moreover, since the growth rate depends on the hole depth (i.e., the growth rate is nonlinear), we made the measurements on holes of the same depth. The solid curve in Fig. 5 is a theoretical hole growth rate γ_h discussed next. Note that Fig. 5 implies that an isolated hole grows for any finite drift velocity. However, for the multi-hole, clump instability of Fig. 2, a marginal point at finite drift is evident. This difference is due to collisions between holes, i.e., the clump instability can only occur if the decay rate resulting from hole-hole collisions can be overcome by the growth of an individual hole.

Hole Model

The simulation results can be understood in terms of a collection of colliding, growing phase space density holes. Consider a two species electron-ion plasma. Roughly speaking, a hole can be characterized by its depth $\tilde{f}$, its spatial width Δx and its velocity trapping width $2\Delta v_t$. A single hole in isolation can be in equilibrium if its depth and therefore, its potential are sufficient to bind it. For an ion hole, this equilibrium condition is

$$\tfrac{1}{2} m_i \Delta v_{t,i}^2 = +e\phi_0 \tag{1}$$

where ϕ_0 is the minimum of the hole potential. For $e\phi_0/T_i \leq 1$, we can estimate $\tilde{f}$ by using Eq. (1) and Poisson's equation. The shielding length λ in the dielectric function $(1 + k^{-2}\lambda^2)$ is given by

$$\lambda^{-2} = -\sum_{e,i} \omega_p^2 \mathrm{PV} \int du \, \frac{f_0'(u)}{v-u} \tag{2}$$

(where PV means principal value). When $\Delta x \gg \lambda$, Poisson's equation is

$$-\lambda^{-2}\phi_0 \approx 4\pi n_i e \tilde{f}_i 2\Delta v_{t,i} \tag{3}$$

Combining Eqs. (3) and (1), we find that

$$\tilde{f}_i \approx -\frac{\Delta v_{t,i}}{4\lambda^2 \omega_{p_i}^2} \tag{4}$$

A more detailed calculation (Dupree, 1982) gives a similar value. The holes of Fig. 3 are consistent with Eq. (4).

Once produced, a BGK-like hole can become unstable (Dupree, 1983), i.e., get deeper, not fall apart. Consider an ion hole located at velocity u. Since the hole appears to be negatively charged, i.e., it is a region of plasma with a depletion of ions, electrons will be reflected by the ion hole's electric field. If there are more electrons moving faster than the hole than slower, i.e., $f'_{0e}(u) \geq 0$, then reflecting resonant electrons will impart momentum to the hole. Consequently, the ion hole moves (decelerates) to a region of velocity space with larger $f_{0i}(u)$, and therefore, the hole gets deeper, i.e., $f_0 - f = -\tilde{f}$ gets larger since f must remain constant. This situation is represented in Fig. 6.

One can compute the acceleration $\dot{u}$ and growth rate γ_h^i of the ion hole from momentum conservation (Dupree, 1983). The rate of change of the momentum of the ion hole is $M\dot{u}$ where $M \approx 2n_i m_i \tilde{f}_i \Delta x \Delta v_{t,i}$ is the ion hole mass. As the hole grows (at the rate γ_h), passing ions in a velocity layer $|v-u| \leq \Delta v$ gain momentum at the (approximate) rate

$$\gamma_h n_i m_i f'_{0i}(u) \Delta v_{t,i}^3 \Delta x. \tag{5}$$

The reflected electrons lose momentum at the rate

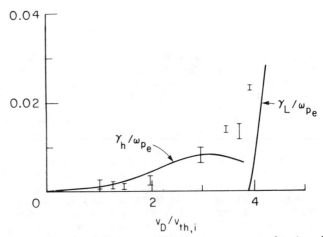

Fig. 5. The measured isolated hole growth rate γ_H as a function of electron drift v_D. The solid line is the theoretical growth rate γ_h obtained from Eq. (8).

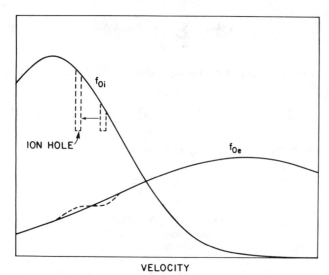

Fig. 6. A schematic picture of the deceleration and growth of an ion hole due to the reflection of electrons by the hole in a region where ion and electron velocity gradients are opposing.

$$\pi^{-1}\omega_{p_e}^2\phi_0^2 f'_{0e}(u). \quad (6)$$

The ion hole acceleration $\dot{u}$ can be obtained if we equate $M\dot{u}$ to the sum of Eqs. (5) and (6). The ion hole growth rate is defined through

$$\frac{\partial \tilde{f}_i}{\partial t} = \gamma_h^i \tilde{f}_i = -\dot{u} f'_{0i} \quad (7)$$

where the last equality results from the fact that the ion hole depth grows as it accelerates to velocities where the average phase space density f_{0i} is higher. Substituting the value for $\dot{u}$ into Eq. (7) and using Eq. (4) and the expression M, we can calculate the theoretical hole growth rate γ_h^i. Using $v_{th} = \lambda_D \omega_p$, we obtain the approximate expression

$$\gamma_h^i = -8 \frac{\Delta v_{t,i}}{\Delta x} \frac{v_{th,e}^2 f'_{0e}(u) v_{th,i}^2 f'_{0i}(u)}{(\lambda_D/\lambda)^4 + 8[v_{th,i}^2 f'_{0i}(u)]^2}. \quad (8)$$

A more rigorous calculation (Dupree, 1983), shown in Fig. 5, yields a result in accord with Eq. (8). Reasonable agreement with the simulations for small v_D is evident. As the plasma approaches linear instability, the hole growth rate formula (8) does not agree well with the simulation results (see Fig. 5) and a modified formula must be used. This regime is discussed elsewhere (Dupree I, to be published) where much improved agreement with the simulations is obtained.

If $f'_0(u) \neq 0$ for the non-hole species, e.g., electrons for an ion hole, then an electric field and a potential drop will occur across the hole. It is this field which accelerates the hole and leads to its growth. This potential structure is sometimes referred to as a double layer. The potential and electric field structure associated with an unstable ion hole are depicted in Fig. 7. We can calculate the magnitude of the potential drop $\delta\phi = \phi(x = +\infty) - \phi(x = -\infty)$ as follows. At $|x| = \infty$, $\partial^2\phi/\partial x^2 = 0$, so Poisson's equation may be written as

$$\lambda^{-2}\phi(x) = 4\pi n e \int dv\, \delta f \quad (9)$$

where δf is the perturbation in the distribution function due to reflected particles. Consider an ion hole. Electrons in a stream of velocity width $\Delta v_{t,i}$ are reflected from the hole. As a consequence, at $x = -\infty$,

$$\delta f(v) = \begin{cases} -2v f'_{0e}(u) & \text{for } -\Delta v_{t,e} \leq v - u \leq 0, \\ 0 & \text{otherwise,} \end{cases} \quad (10)$$

and at $x = +\infty$

$$\delta f(v) = \begin{cases} -2v f'_{0e}(u) & \text{for } \Delta v_{t,e} \geq v - u \geq 0, \\ 0 & \text{otherwise.} \end{cases} \quad (11)$$

Using these expressions for $\delta f(v)$ in Eq. (9), along with (1), we find

$$\frac{\delta\phi}{\phi_0} = -4\left(\frac{\lambda}{\lambda_{D,e}}\right)^2 v_{th,e}^2 f'_{0e}(u). \quad (12)$$

We have measured the potential drop across the hole in the simulations and found it to be generally consistent with Eq. (12).

We can model the effect of hole-hole collisions on the growth rate of an isolated ion hole by

$$\gamma = \gamma_h^i + \gamma_{hh} \quad (13)$$

where $\gamma_{hh} = -2pr(\Delta v/\Delta x)_i$ is the net hole-hole collision rate. The factor $r \leq 1$ models the tendency of holes to attract each other and recombine. Other simulations indicate $r \approx 1/3$ (Berman et al., 1983). The appearance of the packing fraction p in Eq. (13) takes into account the fact that the collision rate between holes is proportionally reduced as the hole packing fraction is reduced. A more detailed analysis from renormalized kinetic theory yields similar results (Tetreault, 1983). The kinetic theory predicts the multi-hole, clump instability growth rate shown as the solid curve in Fig. 2. Again, reasonable agreement with the simulations is obtained for small v_D.

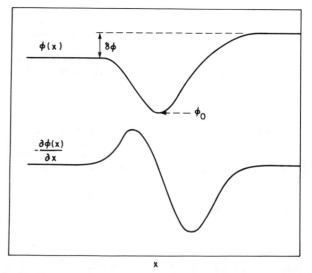

Fig. 7. The double layer potential $\phi(x)$ and electric field $-\partial\phi(x)/\partial x$ shown schematically for an unstable hole.

In the case of the linearly unstable runs, $e\phi_0/T_i \geq 1$ and the hole model deals with integral rather than local properties of the hole structure. One finds (Dupree II, to be published) that the ion hole still decelerates, but its growth rate is significantly reduced from that given by Eq. (8). For $e\phi_0/T_i \approx T_e/T_i$, the growth rate becomes

$$\gamma_h^i \approx \frac{v_i}{\Delta x} v_{th,e}^2 f'_{0e}(u) \qquad (14)$$

The hole acceleration and growth rate predicted by this $e\phi_0/T_i \geq 1$ model agress well with those observed in the linearly unstable runs by us and others (e.g., Barnes et al., 1983).

Acknowledgements. This work was supported by the U.S. Department of Energy, the National Science Foundation, and the U.S. Office of Naval Research. Certain calculations were performed with MACSYMA at M.I.T.

References

Barnes, C., Hudson, M.K., and Lotko, W., Weak Double Layers in Ion Acoustic Turbulence, *Phys. Fluids*, 28, 1055-1062, 1985.

Berman, R.H., Tetreault, D.J., Dupree, T.H., and Boutros-Ghali, T., Computer Simulation of Nonlinear Ion-Acoustic Instability, *Phys. Rev. Let.*, 48, 1249-1252, 1982.

Berman, R.H., Tetreault, D.J., and Dupree, T.H., Observation of Self-binding Turbulent Fluctuations in Simulation Plasma and their Relevance to Plasma Kinetic Theory, *Phys. Fluids*, 26, 2437-2459, 1983.

Berman, R.H., Tetreault, D.J., and Dupree, T.H., Simulation of Phase Space Hole Growth and the Development of Intermittent Plasma Turbulence, *Phys. Fluids*, 28, 155-176, 1985.

Berman, R.H., Tetreault, D.J., and Dupree, T.H., Simulation of Nonlinear Hole Instability in Linearly Stable and Unstable Plasma, to be published in *Phys. Fluids*.

Boutros-Ghali, T. and Dupree, T.H., Theory of Two-Point Correlation Function in a Vlasov Plasma *Phys. Fluids*, 25, 1839-1858, 1981.

Boutros-Ghali, T. and Dupree, T.H., Theory of Nonlinear Ion-Electron Instability, *Phys. Fluids*, 25, 874-883, 1982.

Chang, T. and Coppi, B., Lower Hybrid Acceleration and Ion Evolution in Superauroral Region, *Geophys. Res. Lett.*, 8, 1253-1256, 1981.

Dupree, T.H., Theory of Phase Space Density Holes, *Phys. Fluids*, 25, 277-289, 1982.

Dupree, T.H., Growth of Phase Space Density Holes, *Phys. Fluids*, 26, 2460-2481, 1983.

Dupree, T.H., I, Large Amplitude Ion Holes, to be published in *Phys. Fluids*.

Dupree, T.H., II, Growth Rate of Phase Space Density Holes near Linear Instability, to be published in *Phys. Fluids*.

Hudson, M.K., Lotko, W., Roth, I., and Witt, E., Solitary Waves and Double Layers on Auroral Field Lines, *J. Geophys. Res.*, 88, 916-926, 1983.

Papadopulos, K., Dynamics of the Magnetosphere, in *Dynamics of the Magnetosphere*, ed. S. Akasofu (D. Reidel: New York), 289.

Temerin, M., Cerny, K., Lotko, W., and Mozer, F., Observations of Double Layers and Solitary Waves in Auroral Plasma, *Phys. Rev. Lett.*, 48, 1175-1179, 1982.

Sato, T. and Okuda, H., Numerical Simulations on Ion-acoustic Double Layers, *J. Geophys. Res.*, 86, 3367-3368, 1981.

Tetreault, D.J., Growth Rate of the Clump Instability, *Phys. Fluids*, 26, 3247-3261, 1983.

EFFECTS OF WARM STREAMING ELECTRONS ON ELECTROSTATIC SHOCK SOLUTIONS

Earl Witt[†]

Air Force Geophysics Laboratory/PHG, Hanscom AFB, Massachusetts 01731

Mary Hudson

Department of Physics and Astronomy, Dartmouth College, Hanover, New Hampshire 03755

Abstract. In a previous paper we developed a quasineutral, one-dimensional fluid model of electrostatic shocks which, given a plasma model, allowed us to relate parameters describing the plasma to those defining shock solutions. Here we consider a plasma model consisting of warm downward streaming electrons, cool Boltzmann electrons, cold upward streaming ions and hot Boltzmann ions. We show that there is little difference between having a finite temperature in the electron beam and leaving the beam cold. This implies a greater utility for the cold electron beam model considered in an earlier paper, and adds weight to the idea that the smaller amplitude electrostatic shocks seen in conjunction with ion beams are nonlinear ion acoustic modes.

Introduction

Electrostatic shocks have been defined as latitudinally narrow (600m-60km) regions of intense (100-1000mV/m) electric field directed mainly perpendicular to the magnetic field [Mozer et al.,1977,1979,1980; Torbert and Mozer, 1978]. Such structures have been correlated with visual auroral arcs, field aligned current, and high energy ion beams or conics. In order to gain a better understanding of these structures, we have constructed a one-dimensional, quasineutral fluid model of electrostatic shocks that incorporates a constant magnetic field. This approach has implications for shock structure in general, and allows one to find 'shock' solutions from a general class of plasma models [Witt 1984; Witt and Lotko, 1983]. The approach involves treating one population of cool ions via the ion fluid equations and assuming a plasma model that describes the other particle species. The plasma model consists of expressions for the number density of each of the other species, as a function of the electrostatic potential, such as a Boltzmann density $n(\phi) = n_0 \exp(e\phi/T)$. In Witt and Hudson [submitted to JGR, 1986] shock solutions deriving from two different plasma models are found to have electric fields, scale lengths, and particle energization consistent with those observed in actual electrostatic shocks.

One of the two plasma models examined before consists of cold downward streaming electrons, hot Boltzmann ions, and cool Boltzmann electrons (the cold ions are described by the fluid equations). Here we examine the effects of finite electron beam temperature by modeling the beam electrons with a water-bag distribution. We consider the case in which the drift velocity of the electrons is greater than the thermal velocity. The model has no backscattered or reflected electrons; all beam electrons originate above the shock and propagate downwards.

We find that the addition of the electron temperature does not radically change the possible shock solutions. When double layers exist for both finite and zero beam temperature, all other plasma parameters being held fixed, the finite temperature solutions differ by at most a few percent from the zero temperature case. However, the addition of finite electron beam temperature causes solutions to vanish. By controlling which solutions are possible, the beam temperature can affect the wave quantities to a greater extent. For example, if we allow only the initial streaming electron energy to vary, and change T_{eb} of the electrons, the maximum possible shock potential decreases as much as 22% when the electron beam temperature is increased from 0.001 to 50 times the cool electron temperature (taken here to be around 10eV).

Below we briefly describe our shock model. We then describe the results of our study of shock solutions using the water-bag electron model.

Formalism

Much of this section comes from Witt [1984] (and references contained therein) and Witt and Hudson [submitted to JGR, 1986].

[†] Now at MRC, Post Office Drawer 719, Santa Barbara, California 93102

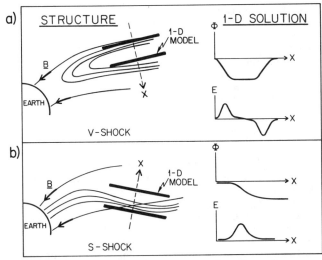

Fig. 1. Equipotential contours that have been proposed to explain some electrostatic shocks, and the associated one-dimensional model used here to describe these structures. Shown are V-shock contours (a), and S-shock contours (b).

Structures Being Modeled

Figure 1 indicates the types of structures we are modeling. A flattened solitary wave approximately models a V-shock, although it is impossible for the one-dimensional model to accurately give the electric field everywhere; if the model is correctly aligned with one half of the structure, it is not aligned with the other. However, since it turns out that the calculated shock parameters are insensitive to the shock orientation, the one-dimensional model is still relevant. For the S-shock pictured in Figure 1b, a better one-dimensional model, the double layer, should provide a good model of the local structure of the wave. The double layer is taken to model a narrow range in altitude of an electrostatic shock which itself spans several thousand kilometers in altitude. The parameters defining the plasma on the high potential side will change, depending on the altitude where we place the double layer. Here we are considering an altitude of about one earth radius.

Analysis

We assume a uniform magnetic field; a plasma model consisting of known expressions for the number densities, $n_s(\phi)$, for all species but the cold ions; quasineutrality; and that the waves are time stationary functions of a wave variable of the form $x - v_p t$, where v_p is the wave speed. Calculations are performed in the plasma frame in which the waves stream through initially stationary cool ions with wave speed v_p. Before assuming quasineutrality, the basic equations are the ion fluid equations, Poisson's equation, and the plasma model defined through the expressions for the number densities. These equations are given in normalized form below:

$$\partial_t N + \nabla \cdot N\mathbf{v} = 0 \quad (1)$$

$$(\partial_t + \mathbf{v}\cdot\nabla)\mathbf{v} = -\nabla\Psi + \mathbf{v}\times\mathbf{b} \quad (2)$$

$$(\Omega/\omega_i)^2 \nabla^2 \Psi = N_{eff}(\Psi) - N(\mathbf{x},t) \quad (3)$$

$$n_{eff}(\phi) = \sum_{s=i_{cool}} n_s(\phi)\cdot(-Q_s/e) \quad (4)$$

$$N_{eff}(\Psi) = n_{eff}(T_{eff}\Psi/e)/n_{eff}(0) \quad (5)$$

Here the velocity is normalized to the effective acoustic speed found in the absence of ion pressure, $c_s = (T_{eff}/m)^{1/2}$, which also defines the effective temperature T_{eff}, $\Psi = e\phi/T_{eff}$ is the normalized potential, the fluid ion density N is normalized to the unperturbed fluid ion density (N(0)=1), time is normalized to Ω^{-1}, where Ω is the ion cyclotron frequency, lengths are normalized to $\rho = c_s/\Omega$, ω_i is the ion plasma frequency, and **b** is a unit vector pointing along the magnetic field. n_{eff} is the unnormalized effective density expressed as a function of the unnormalized potential and consists of a weighted sum over the number densities of all species but the fluid ions. Q_s is the charge on species s. N_{eff} is the normalized version of (4) reexpressed as a function of the normalized potential.

Assuming that all variables are a function of the wave variable $\xi = (x - Mc_s t)/\rho$, where $M = v_p/c_s$, (1) and (2) may be put in the form of an oscillator equation:

$$(1/2)(\partial_\xi \Psi)^2 + F(\Psi) = 0 \quad (6)$$

where F is a function of Ψ, θ, M, $N(\Psi)$, $N'(\Psi)$ (where prime refers to differentiation by Ψ), and

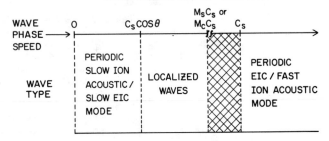

Fig. 2. Possible solutions of the fluid equations as a function of the wave speed, or Mach number = v_p/c_s. Periodic modes exist for $M < \cos\theta$ or $M > 1$. Localized modes, of interest for shock models, exist in the intermediate range. M_s is the Mach number for which a double layer exists. M_c is the Mach number for which a singularity develops in the electric field, and the equations break down.

Table 1. Calculated Shock Quantities

Ψ_s	normalized shock potential = $e\phi_s/T_{eff}$
E_{max}	maximum electric field in units of $T_{eff}/(e\rho)$
Δ_e/ρ	full spatial width at half maximum of the electric field divided by the acoustic Larmor radius $\rho = c_s/\Omega$
v_i/c_s	incoming cold ion speed, parallel to the magnetic field, divided by the acoustic speed $c_s = (T_{eff}/m)^{1/2}$

[1] Ω is the ion gyrofrequency, and m is the ion mass.

an integral of $N(\Psi)$. This equation cannot be solved until $N(\Psi)$ is found, and this is where quasineutrality is assumed to set the ion fluid density $N = N_{eff}$ (given by 5). Quasineutrality is a valid assumption provided the Laplacian in Poisson's equation is small; here we are aided by the fact that the factor in front of the Laplacian is around 1/49 [Mozer et. al,1980] at an altitude of around 1 R_e in the auroral zone.

One can show that the possible solutions to (6) depend on the wave speed, or Mach number, as is indicated in Figure 2. It is the localized modes which are of interest here. These are solutions with $E = 0$ and all other gradients vanishing at $\xi = \pm\infty$. Solitary waves exist for Mach numbers from $\cos\theta$ up to the minimum of 1, M_s, or M_c. Above $M=1$, no localized solutions exist. For $M=M_s$, the solitary wave degenerates into a double layer solution, and above M_s no localized solution exists; for $M \simeq M_s$, a flattened solitary wave is the result. For M approaching M_c, a singularity develops in the electric field of the corresponding solitary wave and this system of equations breaks down.

Shock Model

An electrostatic shock solution is defined to be an $M \simeq M_s$ (V-shock) or $M=\dot{M}_s$ (S-shock or double layer) solution in the wave frame. In this frame the cold ions stream into the structure with a speed $Mc_s/\cos\theta$ and are affected by the potential. The earth is assumed to be fixed in this frame. The consequences of this model are as follows. The cool ions must stream into the shock with a speed, parallel to the magnetic field, in excess of the local acoustic speed. The scale length of the structure varies with the acoustic Larmor radius. The shock potential depends only on the parallel incoming ion speed; if this quantity is fixed, shocks with two different angles of orientation would have the same potential jump (provided they could exist at the two angles). Finally, we have found numerically that the other shock parameters, such as the electric field and scale length, are insensitive to the angle of orientation of the shock. It must be noted that for quasineutrality to hold, θ must have a finite value which depends on the ratio $\Omega/\omega_i = \lambda_d/\rho$ (for small amplitude waves the criterion is $(\lambda_d/\rho)^2/\sin^2\theta \ll 1$; for $\Omega/\omega_i = 1/7$, θ_{min} is estimated to be around 10°).

To find a given shock solution one solves the oscillator equation numerically for a particular set of plasma parameters. Given cold fluid ions, one may solve directly for double layer solutions [Witt and Hudson, submitted to JGR, 1985]. Quantities that the model provides are given in Table 1. These include the normalized shock potential, the maximum electric field, the full spatial width at half maximum of the electric field, and the incoming ion streaming speed. We

Table 2. Water-Bag Plasma Model Parameters

water-bag electrons:	N_{eb}	n_{eb}/n_0, fraction of electrons of beam origin where $\phi=0$
	v_0	initial drift speed of beam electrons
	a	initial thermal speed of beam electrons
	T_{eb}	initial temperature of beam electrons = ma^2
	E_{de0}	initial drift energy = $mv_0^2/2$
Boltzmann electrons:	$(1-N_{eb})$	fraction of electrons which are Boltzmann at $\phi=0$
	T_c	cool electron temperature
fluid ions:	N_{ib}	n_{ib}/n_0, fraction of ions which are beam ions and cold at $\phi=0$
Boltzmann ions:	$(1-N_{ib})$	fraction of ions which are hot and Boltzmann at $\phi=0$
	T_{ih}	hot Boltzmann ion temperature

[1] In densities, lower case denotes unnormalized quantities.
[2] n_0 is the cold ion density where $\phi=0$.

Table 3. Model Parameters Used in Calculation

N_{eb}, N_{ib}	initial normalized number densities
T_c/T_{ih}, T_c/T_{eb}, E_{de0}/T_c	temperature and energy ratios
θ	angle electric field makes with magnetic field

also obtain via the plasma model the behavior of the various species throughout the shock structure.

Model Calculation

We investigate shock solutions deriving from the following plasma model. In the plasma frame, the plasma consists of hot Boltzmann ions, downward streaming water-bag electrons, cool Boltzmann electrons, and cold ions (described by the fluid equations). In going to the wave frame at the end of the calculation, we impart an upward velocity to all particles. This turns out to be negligible, in comparison with thermal speeds, for the electron populations. It can be significant for the ion populations, however, and indeed is a quantity of interest for the cold ions. This speed should be small compared to the hot ion thermal speed for us to treat the hot ions as Boltzmann. The effective density becomes:

$$n_{eff}(\phi) = \{(1-N_{eb})\exp(e\phi/T_c) + N_{eb}n_{wb}(\phi)/n_{wb}(0) - (1-N_{ib})\exp(-e\phi/T_{ih})\}/N_{ib} \quad (7)$$

where

$$n_{wb}(\phi) = (1/2)[\{(v_0/a + 1)^2 + 2e(\phi-\phi_s)/T_{eb}\}^{1/2} - \{(v_0/a - 1)^2 + 2e(\phi-\phi_s)/T_{eb}\}^{1/2}] \quad (8)$$

The symbols entering into these equations are defined in Table 2. The quantities actually used in calculations are given in Table 3.

Results

Once a solution is found, one needs to assume some values for the actual plasma parameters in order to obtain numbers for the solutions. The values assumed here are given in Table 4. The plasma density is assumed to be around 10cm^{-3},

Table 4. Assumed Plasma Parameters at 1 R_e Altitude

plasma density	10cm^{-3}
cool electron temperature	10eV
hot ion temperature	100eV - 1keV
ion beam temperature	cold (1-5eV)
ion cyclotron frequency	100 Hz
ion plasma frequency (beneath any shock)	700 Hz

the hot ion temperature from 100-1000eV, the temperature of the ion beam small enough to be considered cold (from 1-5eV), and the ion cyclotron frequency is 100Hz. The cold electron temperature is motivated by Temerin et al., [1982], while the range of ion and electron energies is motivated by Kaufman and Kintner [1982]. The streaming electron temperature is left as a free parameter. Previously, solutions have been found with electric fields from 100-1200mV/m, scale lengths from 1-3 km, and potential jumps of up to 2kV. We have found solutions deriving from the water-bag electron model that have similar values. The main focus here, however, is on the effects that finite temperature produces.

We find that the main effect of the finite electron beam temperature is to remove solutions

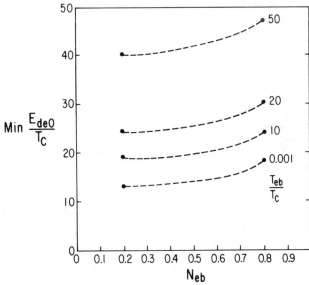

Fig. 3. Minimum initial electron drift energy versus electron beam density (at $\phi=0$) for several ratios of the beam temperature to the cold electron temperature, for various double layer solutions deriving from the water bag electron plasma model. The curves indicate how solutions disappear as the beam temperature is increased. For a given value of T_{eb}, solutions exist for values of E_{de0}/T_c greater than the indicated minimum, and no solutions exist outside the endpoints of a given curve. The larger shock potential drops occur for the smaller E_{de0}/T_c. For these curves, θ varies from $80° - 88°$, $N_{ib} = 0.995$, and $T_c/T_{ih} = 0.05$.

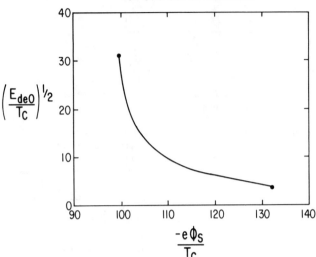

Fig. 4. $(E_{de0}/T_c)^{1/2}$ vs shock potential for a set of double layer solutions calculated using the water bag electron model. $\theta = 80°$, $N_{eb}=0.3$, $N_{ib}=0.999$, $T_c/T_{ih} = 0.05$, and $T_c/T_{eb} = 0.5$. The graph is effectively an I-V curve for this set of solutions.

that would be present in its absence. Figure 3 gives an illustration of this. We plot the minimum possible value of the initial electron streaming energy (below which no solutions exist) against the fraction of beam electrons N_{eb} for several ratios of the beam temperature to cool electron temperature. The hot ion temperature and density are fixed in these calculations. These particular solutions are of interest since the larger potential jumps occur for the smaller initial electron drift energies. As T_{eb} increases, solutions vanish. No solutions are found outside the density range between around $N_{eb} = 0.2$ and 0.8.

We have found that when a solution does exist for both zero and finite temperature, all other plasma parameters held fixed, that there is at most a few percent difference in any given shock quantity between the zero and finite temperature cases. However, the fact that some solutions disappear can affect the shock quantities more strongly. For example, consider the solutions in Figure 3 for $N_{eb} = 0.3$ as T_{eb}/T_c varies from .001 to 50. The shock potential at $E_{de0}/T_c=40$ is 22% less than that at the lower value of E_{de0}/T_c. This is about the maximum difference that was seen. Figure 4 shows a plot of shock potential vs $(E_{de0}/T_c)^{1/2}$ with the other plasma parameters held fixed, and indicates the range of variation of the shock potential, as a function of E_{de0}/T_c, for a different initial ion composition. This is effectively a plot of current, or initial electron drift speed vs the shock potential. As the square root of E_{de0}/T_c changes from 0 to 40, the shock potential changes by only about 30%. In the restricted range from 0 to 10 (corresponding to the spread in E_{de0}/T_c in Figure 2) the change is less than 22%. The conclusion is that adding finite electron beam temperature does not induce changes much greater than 20% from previous cold beam solutions.

Conclusions

The addition of finite temperature to downstreaming electrons in the plasma model consisting of downstreaming electrons, hot Boltzmann ions, cool Boltzmann electrons, and cold ions, does not appreciably change the resulting shock solutions. Therefore, provided that these fluid solutions are stable, this fluid shock model is a viable model for electrostatic shocks seen in conjunction with ion beams and with potential drops of up to 2kV. As a consequence of this model, cool ions stream into the shocks supersonically, the shock parameters are determined primarily by the incoming ion speed parallel to the magnetic field, and the width of these structures is of the order of 20-40 acoustic Larmor radii which, to be consistent with the quasineutrality assumption $(\Omega/\omega_i < 1/4)$, is greater than 40-160 Debye lengths.

Acknowledgements. This work was supported by NSF grant ATM-8445010, NASA grant NAGW-564, a grant from the University of California IGPP, and was partially sponsored by the Air Force Geophysics Laboratory, United States Air Force, under Contract F19628-83-C-0097. The United States Government is authorized to reproduce and distribute reprints for governmental purposes notwithstanding any copyright notation hereon.

References

Kaufman, Richard L., and Paul M. Kintner, Upgoing ion beams 1. Microscopic Analysis, J. Geophys. Res., 87, 10487, 1982.

Kletzing, C., C. Cattell, F. S. Mozer, S. I. Akasofu, and K. Makita, Evidence for electrostatic shocks as the source of discrete auroral arcs, J. Geophys. Res., 88, 4105, 1983.

Mozer, F. S., C. W. Carlson, M. K. Hudson, R. B. Torbert, B. Parady, J. Yatteau, and M. C. Kelley, Observations of paired electrostatic shocks in the polar magnetosphere, Phys. Rev. Lett., 38, 292, 1977.

Mozer, F. S., C. A. Cattell, M. K. Hudson, R. L. Lysak, M. Temerin, and R. B. Torbert, Satellite measurements and theories of low altitude auroral particle acceleration, Space Sci. Rev., 27, 155, 1980.

Temerin, M., K. Cerny, W. Lotko, and F. S. Mozer,

Observations of double layers and solitary waves in the auroral plasma, Phys. Rev. Lett., 48, 1175, 1982.

Torbert, R. B., and F. S. Mozer, Electrostatic shocks as the source of discrete auroral arcs Geophys. Res. Lett., 5, 135, 1978.

Witt, E., and W. Lotko, Ion-acoustic solitary waves in a magnetized plasma with arbitrary electron equation of state, Phys. Fluids, 26, 2176, 1983.

Witt, Earl Frederick, Localized Nonlinear Electrostatic Waves in the Auroral Plasma, Ph.D. Dissertation, University of California, Berkeley, 1984.

NUMERICAL SIMULATIONS OF AURORAL PLASMA PROCESSES: ION BEAMS AND CONICS

Nagendra Singh, H. Thiemann, and R. W. Schunk

Center for Atmospheric and Space Sciences, Utah State University, Logan, Utah 84322-3400

The characteristics of ion beams and conics, as seen in 2-D numerical simulations in which the plasmas are driven by current sheets of a finite thickness, are described. It is shown that the most energetic ions in the simulations have pitch angles near 90°, which implies a large perpendicular acceleration of the ions.

Introduction

In a companion paper [Singh et al., 1986], hereafter referred to as Paper 1, we describe the electric fields, both dc and ac, that were seen in our simulations. In this paper we discuss the associated parallel and perpendicular acceleration of ions. The simulation scheme and some definitions used in this paper are described in section 2 of Paper 1.

Upflowing Ions

In this section we discuss the energetics of the upflowing ions, which exist self-consistently with the quasi-steady electric fields and the waves discussed in Paper 1. The striking feature of these ions is that they are found to be the most energetic at pitch angles near 90°. This feature is consistent with the characteristics of auroral ion conics when they are observed in the region where they are energized [Kintner and Gorney, 1984].

We illustrate the perpendicular acceleration of the upflowing ions, as seen in our simulations, by presenting results from simulation S_1. Figure 1 shows phase-space plots of the upflowing ions. In these plots we have divided the ions into two groups; those lying inside and those lying outside the current sheet. Panels (a) and (b) show the plots for the ions outside the current sheet in $V_\perp$-Y and $V_\parallel$-Y planes, respectively, while panels (c) and (d) show similar plots for the ions inside the sheet. In these plots each dot represents an ion; the y coordinate shows the location (height) of the ions, while the $V_\perp$ and $V_\parallel$ coordinates show the perpendicular and parallel velocities, respectively.

Panel (d) of Figure 1 shows the presence of an upflowing ion beam inside the current sheet where the upward current flows. The energy of the beam increases with increasing height (y), as could be expected from the potential structure shown in Figure 5a of Paper 1. The presence of a beam suggests the existence of a positive slope in the parallel velocity distribution function of the ions. This is shown in Figure 2a. The distribution function is obtained by binning ions in parallel velocity bins of $\Delta \tilde{V}_\parallel = 0.1$. Thus, the results are shown in the forms of histogram. The solid-line histogram is for all the ions located at $\tilde{y} > 96$ inside the current sheet. On the other hand, the broken-line histogram is for the ions in the height range $65 < \tilde{y} < 96$ inside the sheet. A comparison of the two histograms shows that there is a well-defined beam over the later height range. Panel (d) of Figure 1 shows that at heights $\tilde{y} > 96$ there are some low-energy trapped ions, which tend to mask the beam feature, as shown by the solid line histogram in Figure 2a. We find that the maximum parallel energy (W_{bi}) of the upflowing ions inside the current sheet is comparable to the parallel potential drop $\Delta\phi_{int}$, which, in turn, is found to be comparable to the perpendicular potential drop $\Delta\phi_{\perp m}$ in Figure 5a of Paper 1. Such ion beam features are in agreement with satellite observations [Temerin et al., 1981a]. However, we find that $\Delta\phi_{\perp m} \sim \Delta\phi_{int}$ only in thin sheets ($l \lesssim \rho_i$), which correspond to thin auroral arcs [Torbert and Mozer, 1978].

We have studied the issue of the formation of ion beams in various simulation runs. We find that the relation $\Delta\phi_{\perp m} \sim \Delta\phi_{int} \sim W_{bi}$ is valid only in thin sheets. When the sheets are very wide ($l \gg \rho_i$) and the field-aligned current density remains subthermal, ion-beam features are hard to see. However, the upflowing ions are found to be heated considerably. Satellite observations indicate that in low altitude (< 2000 km) electrostatic shocks, ion beams are rarely seen [Temerin et al., 1981b]. At low altitudes, where the geomagnetic field is large and the ion thermal energy is low, it may not be possible to have narrow current sheets, with $l \lesssim \rho_i$. Thus, no significant parallel potential drop $\Delta\phi_{int}$ develops, and hence, no ion beam forms.

However, we find that if the field-aligned current becomes superthermal, $\Delta\phi_{int}$ develops even in wide sheets, but $\Delta\phi_{int}$ is still found to be appreciably smaller than $\Delta\phi_{\perp m}$. Thus, if ion beams form, their energies will be considerably smaller than the perpendicular potential drops estimated from the large perpendicular electric fields.

So far, our discussion has been on the parallel acceleration of the upflowing ions inside the current sheet. We now discuss the associated perpendicular acceleration of the ions. Figure 1c shows that the upflowing ions inside the current sheet have undergone a considerable perpendicular acceleration as well. The pitch angles of the ions in Figures 1c and 1d are shown in Figure 2b by plotting contours of the distribution function in the $\alpha_p - W$ plane, where α_p and W are the pitch angle and total energy of the ions. It is interesting to note that there are no ions at 0° pitch angle. The bulk of the ions occupy the pitch-angle range $10° \leq \alpha_p \leq 60°$. There are also some ions at pitch angles near 90°, which are found to be the most energetic. Thus, the upflowing ions within the sheet are not field-aligned. They have undergone a considerable perpendicular acceleration. However, their parallel velocity distribution function shows a positive slope in a certain height range.

The perpendicular acceleration of the ion beam in the region of the upward current is in agreement with satellite observations; perpendicular heating of ion beams in the auroral region have been reported by several authors [Temerin et al., 1981b; Collin et al., 1981]. The combined parallel and perpendicular acceleration of the ions in the current sheet appears to be very similar to the bimodal acceleration of ions seen from the DE-1 observations in the auroral plasma [Klumpar et al., 1984]. The bimodal acceleration simply implies the possiblity that the upflowing ions were accelerated in both the parallel and perpendicular directions; the parallel acceleration occurs via a potential drop, while wave-particle interactions cause the perpendicular acceleration. The

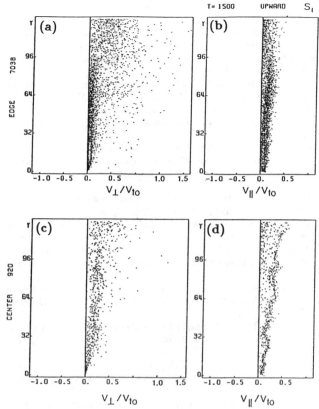

Fig. 1. Phase-space plots of the upflowing ions in simulation S_1. The top (a and b) and bottom (c and d) panels, respectively, show the ions lying outside and inside the current sheet at $\bar{t} = 1500$. Each dot represents an ion. The left and right hand side panels show the ions in the $V_\perp - Y$ and $V_\parallel - Y$ planes, respectively.

perpendicular acceleration of the ions leads us to a discussion of the ion conics, as seen in our simulation.

A comparison of panels (a) and (c) in Figure 1 shows that the ions lying outside the current sheet have undergone a much greater perpendicular acceleration than those lying inside the sheet. Furthermore, a comparison of the perpendicular and parallel velocities of ions outside the sheet shows that their perpendicular energies are considerably larger than their parallel energies. The pitch angles of these ions, as a function of their total energy, are shown in Figure 2c where the contours of constant distribution function are plotted in the $\alpha_p - W$ plane. The numbers on the contours show the natural logarithm of the magnitude of the distribution function on a relative scale. The most striking feature of Figure 2c is that the ions near $\alpha_p \simeq 90°$ are the most energetic.

A comparison of Figures 2b and 2c shows that the ions outside the sheet near $90°$-pitch angle are considerably more energetic than those inside the sheet. Figure 2d shows the perpendicular energy distribution function of the ions outside the sheet. The three linear asymptotes shown in the figure indicate that the composite ion distribution function can be approximated by three Maxwellian distributions, which represent a cold ion population with a temperature $T_i \sim 1.5\ T_0$, a population with an intermediate temperature $T_i \sim 10\ T_0$ and a very hot ion population with a temperature $T_i \simeq 51\ T_0$.

To summarize the above discussion on the perpendicular acceleration of the ions, we note that the most energetic upflowing ions are seen at pitch angles near $90°$. Although the above discussion is based on the results from simulation S_1 alone, the above conclusion is quite general and is supported by every simulation we conducted.

We note from Figure 2d that upflowing ions having perpendicular energies up to $120\ k_B T_0$ exist in the simulation plasma. When such ions leave the reservoir at $y = 0$, they are extremely cold ($T_i \simeq 0.002\ T_0$). Also, with a reservoir temperature of $T_i \sim T_0$ in many other simulations (Paper 1), we have seen ion acceleration of the same order of magnitude. We note that the magnitude of the ion acceleration seen in our simulations is in agreement with that seen in the simulations reported by Okuda and Ashour-Abdalla [1983]. As with our simulations (Figure 2d), they also reported a bulk heating along with the formation of an energetic tail. Now an important question arises regarding the mechanism responsible for the acceleration. It is well-known that a perpendicular acceleration of ions can be caused by dc electric fields, such as (a) localized perpendicular electric fields [Cole, 1976], and (b) the parallel and perpendicular electric fields associated with V-shaped double layers [Yang and Kan, 1983] and oblique double layers [Greenspan, 1984; Borovsky, 1984]. However, such acceleration mechanisms are limited to the net potential drop across the structure. In our simulations, the net potential drops ($\Delta\phi_{\perp m}$ and $\Delta\phi_{int}$) are only a few times ($k_B T_0/e$). Thus, acceleration by dc electric fields can account for only a fraction of the maximum acceleration (Figure 2d).

Other mechanisms which can cause the perpendicular acceleration involve wave-particle interactions. In this context, the ion heating by lower hybrid [Chang and Coppi, 1981; Singh and Schunk, 1984] and electrostatic ion-cyclotron waves [Lysak et al., 1980; Singh et al., 1983; Papadopoulos et al., 1980; Ashour-Abdalla and Okuda, 1984] have been extensively investigated. As noted earlier, such waves are excited in the simulation plasma. Because of the complex temporal and spatial evolution of the plasma, we find it difficult to determine the relative

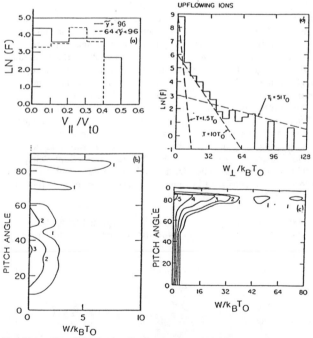

Fig. 2(a). Parallel velocity distribution function for the ions inside the current sheet (panel d, Fig. 1), (b) Energy distribution function (F) for the same ions; contours of constant F in the $\alpha_p - W$ plane are plotted, where α_p is the pitch angle and W is the total energy, (c) same as (b) but for ions lying outside the current sheet and (d) the perpendicular energy distribution for the ions in panel (a) of Figure 1.

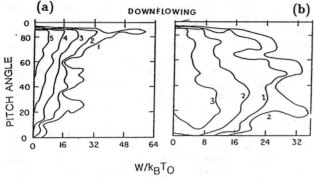

Fig. 3(a). Energy distribution function of the downflowing ions in the $\alpha_p - W$ plane for simulation S_1, (b) same as (a) but for simulation S_4.

importance of the various wave modes in the acceleration process. Our simulations have shown that the ion acceleration, a variety of wave modes, and parallel and perpendicular dc electric fields are integral parts of the complex plasma processes driven by a field-aligned current sheet.

Downflowing Ions

In the above discussion, we emphasized the energization of upflowing ions. We find that even the downflowing ions undergo a considerable energization. Figures 3a and 3b show the distribution function of the downflowing ions in simulations S_1 and S_4, respectively. The distributions are shown by plotting contours in the $\alpha_p - W$ plane, as was done for Figures 2b and 2c. These distributions display interesting contrasts. We recall that in S_1, the current sheet thickness $l \ll \rho_i$ and very weak downward electric fields develop in the regions outside the sheet. On the other hand, in S_4 $l \gg \rho_i$ and the downward fields are quite large (see Figures 5a and 5b in Paper 1). Figures 3a and 3b clearly manifest the difference in the magnitude of the downward electric fields in simulations S_1 and S_4. The most energetic ions in Figure 3a appear at pitch angles near 90°. This is not seen in Figure 3b. Instead, the ions tend to be more field-aligned and have a broad pitch-angle distribution.

We note that in our current simulations the magnetic field is uniform. Thus, the mirror force, which is an important feature of the geomagnetic field, is not included. If a mirror force were included in our simulations, it is possible that such ions would be turned upward and become upflowing conics. We hope to study the generation of upflowing ion conics including a magnetic mirror force in future simulations. This will allow us to self-consistently simulate the pressure cooker effect [Gorney et al., 1985].

Acknowledgement. This research was supported by NASA grant NAGW-77 and NSF grant ATM-8417880 to Utah State University.

References

Ashour-Abdalla, M. and H. Okuda, Turbulent heating of heavy ions on auroral field lines, *J. Geophys. Res., 89,* 2235, 1984.

Borovsky, J. E., The production of ion conics by oblique double layers, *J. Geophys Res., 89,* 2251, 1984.

Chang, T. and B. Coppi, Lower hybrid acceleration and ion evolution in the suprauroral region, *Geophys. Res. Lett., 8,* 1253, 1981.

Cole, K. D., Effects of crossed magnetic and (spatially dependent) electric fields on charged particle motion, *Planet. Space Sci., 24,* 515, 1976.

Collin, H. L., R. D. Sharp, E. G. Shelley, and R. G. Johnson, Some general characteristics of upflowing ion beams over the auroral zone and their relationship to auroral electrons, *J. Geophys. Res., 86,* 6820, 1981.

Greenspan, M. E., Effects of oblique double layers on upgoing ion pitch angle and gryrophase, *J. Geophys. Res., 89,* 2842, 1984.

Gorney, D. J., Y. T. Chiu, and D. R. Croley, Jr., Trapping of ion conics by downward parallel electric fields, *J. Geophys. Res., 90,* 4205, 1985.

Kintner, P. M., and D. J. Gorney, A search for plasma processes associated with perpendicular ion heating, *J. Geophys. Res., 89,* 937, 1984.

Klumpar, D. M., W. K. Peterson, and E. G. Shelley, Direct evidence for two stage (Bimodal) acceleration of ionospheric ions, *J. Geophys. Res., 89,* 10779, 1984.

Lysak, R. L., M. K. Hudson, and M. Temerin, Ion heating by strong electrostatic ion cyclotron turbulence, *J. Geophys. Res., 85,* 524, 1980.

Okuda, H. and M. Ashour-Abdalla, Acceleration of hydrogen ions and conic formation along auroral field lines, *J. Geophys. Res., 88,* 899, 1983.

Papadopoulos, K., J. D. Gaffey, Jr., and P. J. Palmadesso, Stochastic acceleration of large M/Q ions by hydrogen cyclotron waves in the magnetosphere, *Geophys. Res. Lett., 7,* 1014, 1980.

Singh, N., R. W. Schunk, and J. J. Sojka, Preferential perpendicular acceleration of heavy ionospheric ions by interactions with electrostatic hydrogen cyclotron waves, *J. Geophys. Res., 88,* 4055, 1983.

Singh, N. and R. W. Schunk, Energization of ions in the auroral plasma by broadband waves: Generation of ion conics, *J. Geophys. Res., 89,* 5538, 1984.

Singh, N., H. Thiemann, and R. W. Schunk, Numerical simulations of auroral plasma processes: Electric fields, This issue, 1986.

Temerin, M., N. H. Boehm, and F. S. Mozer, Paired electrostatic shocks, *Geophys. Res. Lett., 8,* 788, 1981a.

Temerin, M., C. Cattell, R. Lysak, M. Hudson, R. B. Torbert, F. S. Mozer, R. D. Sharp, and P. M. Kintner, Small scale structure of electrostatic shocks, *J. Geophys. Res., 86,* 11278, 1981b.

Torbert, R. B., and F. S. Mozer, Electrostatic shocks as the source of discrete auroral arcs, *Geophys. Res. Lett., 5,* 135, 1978.

Yang, W. H. and J. R. Kan, Generation of conics by auroral electric fields, *J. Geophys. Res., 88,* 465, 1983.

NUMERICAL SIMULATIONS OF AURORAL PLASMA PROCESSES: ELECTRIC FIELDS

Nagendra Singh, H. Thiemann, and R. W. Schunk

Center for Atmospheric and Space Sciences, Utah State University, Logan, Utah 84322-3400

The features of auroral potential structures, including the parallel ($E_\parallel$) and perpendicular ($E_\perp$) electric fields associated with electrostatic shocks and double layers, are described with the aid of 2-D numerical simulations in which the plasmas are driven by current sheets of a finite thickness (l). Large perpendicular electric fields, similar to those measured in electrostatic shocks, are primarily confined near the edges of the current sheets, where $E_\perp \gg E_\parallel$. Double layers with $E_\perp \sim E_\parallel$ that are spatially separated from the large $E_\perp$ form in the interior of wide current sheets, $l \gg \rho_i$, where ρ_i is the ion Larmor radius. Such double layers have upward electric fields in the region of the upward current. Downward electric fields develop outside the current sheet.

Introduction

Figure 1 shows a synopsis of the various auroral phenomena which have been observed from rockets and satellites. The phenomena relevant to the discussion of this paper are as follows. Electrostatic shocks have been identified as the occurrences of large localized dc perpendicular electric fields. Field strengths up to about 10^3 mV/m have been reported [Mozer et al., 1980]. The parallel electric fields associated with such shocks are comparatively very weak. Such electric fields are found to exist near the boundaries of the upward field-aligned currents.

To date, only weak double layers, with electric field strengths $E_\parallel < 15$ mV/m, have been observed in the upward field-aligned current region, where relatively weak ion beams also exist [Temerin et al., 1982]. Since the double layers are highly intermittent, both in time and in space, they are not likely to be observed with dc measurements. Only with wave measurements is it possible to detect such double layers, as was done with instruments on the S3-3 satellite [Temerin et al., 1982]. A possible reason for not observing double layers with amplitudes greater than 15 mV/m on the S3-3 satellite could be the saturation of the detectors.

It is well-known that upward electric fields exist, which accelerate the electrons downward, in the region of upward field-aligned currents. However, observations now indicate that downward electric fields, supporting potential drops of a few tens to about a few hundred volts, may exist in the region of the return currents [Gorney et al., 1985]. Such fields have been suggested to be responsible for creating the upward electron beams [Klumpar and Heikkila, 1982; Gorney et al., 1985].

The purpose of this paper is to present a brief summary of results from simulations relevant to the above observations on auroral electric fields. In a companion paper, the parallel and perpendicular acceleration of the ions is discussed. Various other phenomena, such as the linear and nonlinear behavior of high and low frequency waves, their effects on the potential structure, and the consequent electron acceleration will not be discussed here owing to the limitation on the size of the paper.

Simulation Scheme

A typical simulation geometry is shown in Figure 2. The simulations are carried out using a standard PIC technique. The simulation plasma occupies the two-dimensional space $-L_x/2 \leq x \leq L_x/2$ and $0 \leq y \leq L_y$. The electrons and ions, which are infinite rods extending in the z-direction in the two-dimensional model, are injected at the top ($y = L_y$) and bottom ($y = 0$) boundaries from Maxwellian plasma reservoirs with appropriate temperatures and drift velocities. The current sheet occupies the central region $-l/2 < x < l/2$ and extends along the magnetic field (Figure 2). The current carrying electrons are injected at the top boundary ($y = L_y$) with the appropriate flux and drift velocity V_{de} in order to maintain a desired current density. We use the electrostatic approximation in these simulations; the boundary conditions are $\phi(x, y = 0) = 0$, $E_y(x, y = L_y) = 0$, and $\phi(x = -L_x/2, y) = \phi(x = L_x/2, y)$, where ϕ is the electrostatic potential and $\mathbf{E} = -\nabla \phi$. Note the use of the floating (Neumann) boundary condition at $y = L_y$, which allows for the development of a self-consistent potential drop. The zero potential boundary condition at $y = 0$ simulates the conducting ionosphere.

The various particle populations used in these simulations are (1) electrons injected at the boundary $y = L_y$; temperature $T_{eu} = T_o$, (2) "hot" ions injected at the same boundary; $T_{iu} = T_H > T_o$, (3) electrons and ions injected at the lower boundary $y = 0$; temperatures $T_{il} = T_{el} = T_l \lesssim T_o$, and (4) current-carrying electrons injected with an average drift V_{de} at $y = L_y$ in the current sheet. In all of the simulations discussed here, the ions injected at $y = 0$ are given an average upward drift V_{di}.

We use the following definitions: λ_{do} is the Debye length based on the temperature T_o and on the initial density of $n_o = 4$ particles per cell, Ω_e is the electron-cyclotron frequency and $\omega_{po}^2 = n_o e^2/m\epsilon_o$, where ϵ_o is the permittivity of free space and m is the electron mass. The ion-electron mass ratio was chosen to be $M/m = 64$. We also use the following normalizations: distance $\tilde{y} = y/\lambda_{do}$, time $\tilde{t} = t\omega_{po}$, velocity $\tilde{V} = V/V_{to}$, potential $\tilde{\phi} = e\phi/k_B T_o$, electric field $\tilde{E} = E/E_o$, $E_o = k_B T_o/e\lambda_{do}$, and current density $\tilde{J} = J/(en_o V_{to})$, where $V_{to} = (k_B T_o/m_e)^{1/2}$. The numerical technique used here has been previously described in much greater detail by Singh et al. [1985].

We have carried out a variety of simulations by varying parameters, such as T_{el}, T_{il}, T_{iu}, J_o, ω_{po}/Ω_e, L_x, L_y and the current-sheet thickness [Singh et al., 1983, 1985]. However, in the present paper we have chosen to discuss results from simulations in which $L_x \times L_y = 64 \times 128 \lambda_{do}^2$, the largest size of the simulation plasma we could run on the computer. We discuss results from simulations S_1, S_2, S_3, and S_4, in which the current sheet thickness is 6, 12, 24, and $32\lambda_{do}$, respectively. In S_1 to S_3, $\tilde{J}_o \simeq 1.25$ and $\Omega_e/\omega_{po} = 2$, while in S_4 $\tilde{J} \simeq 0.6$ and $\Omega_e/\omega_{po} = 5$. The ion drift in the simulations was chosen to be $\tilde{V}_{di} = 0.14$. In simulations S_2 to S_4, $T_H = 5T_o$ and $T_{il} = T_{el} = T_o$, and in S_1, $T_H = 10T_o$ and $T_{il} = T_{el} = 0.002T_o$.

Quasi-steady Perpendicular Electric Fields

Before we present results on the field-aligned potential structures, we briefly discuss how perpendicular electric fields are generated near the edges of the current sheets. Kan et al. [1979] suggested that perpendicu-

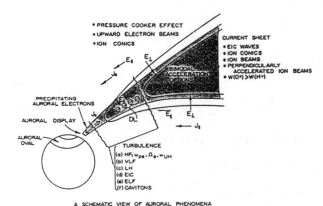

Fig. 1. A synopsis of auroral plasma processes. The shaded region indicates the current sheet in which a field-aligned upward current flows. The regions where parallel ($E_\parallel$) and perpendicular ($E_\perp$) electric fields, various wave modes, and particle acceleration occur, are shown. Downward currents flow on the flanks of the V-shaped potential structure.

lar electric fields, which are inherent to current sheets in a plasma, in combination with the conducting ionospheric boundary could produce V-shaped potential structures along auroral field lines. The perpendicular electric fields arise because of the difference in the Larmor radii of the electrons and ions. The auroral current sheets, in which the upward field-aligned currents flow, are populated by the current-carrying electrons and relatively hot ions, both species originate in the plasma sheet. Since the hot ions have much larger Larmor radii than the relatively cold electrons, which flow essentially along the field lines, the gyrating ions near the edges of the current sheet (within a Larmor radius) spend a considerably longer time outside the sheet, which produces a positive space charge there and leaves a negative space charge just inside the current sheet (Figure 3). Thus, dc perpendicular electric fields are generated near the edges. These fields are bipolar and point toward the sheet.

Figure 4 shows the electric fields from our simulations for several values of l and ρ_i, as indicated on the top of each panel. If the current sheet thickness $l \lesssim \rho_i$, the ion Larmor radius, an intense negative space charge develops within the sheet (Figure 3a). Thus, the large perpendicular electric fields permeate the entire width of the sheet so that $dE_\perp/dx \neq 0$, as seen in Figures 4a and 4b for which $l/\rho_i < 1$ and $\frac{4}{3}$, respectively. On the other hand, when $l \gg \rho_i$ the space charges are confined near the two edges (Figure 3b), leaving the interior of the current sheet quasi-neutral. Hence, the perpendicular electric fields are also confined near the edges (Figures 4c and 4d). Later, we show that this difference between the wide and narrow current sheets has important ramifications for the overall potential structure.

The strength of the electric fields shown in Figure 4 is roughly given by

$$E_{\perp m} \simeq E_o = k_B T_o / e \lambda_{do} = (n_o k_B T_o / \epsilon_o)^{1/2} \tag{1}$$

The $E_{\perp m}$ given above is a maximum in the sense that it decreases with decreasing height and goes to zero near the bottom boundary at $y = 0$.

Table 1 shows some estimates of $E_\perp$ when n_o and T_o are varied over a range that is appropriate for the auroral plasma at an altitude of about one earth radius (Re). The values of $E_\perp$ shown in the table (130 mV/m $<$ $E_{\perp m} <$ 1300 mV/m) are in fair agreement with satellite observations [Mozer et al., 1980]. We note that $E_\perp$ increases as the square root of the product of the density and temperature, and the electric field energy density is comparable to the electron thermal energy. At the locations of the large $E_\perp$, the parallel electric fields are found to be smaller by roughly an order of magnitude. This feature is consistent with measurements of electrostatic shocks in which $E_\perp \gg E_\parallel$. When we compare the maximum possible strengths of $E_\perp$ and $E_\parallel$ as seen in our simulations, irrespective of the locations of the maximum fields, it is found that $E_{\perp m} \gg E_{\parallel m}$ [Singh et al., 1983, 1985; Thiemann et al., 1984].

Figures 5a and 5b show the potential structures for a thin ($l < \rho_i$, S_1) and a wide ($l \gg \rho_i$, S_4) current sheet, respectively. In these figures the equipotential surfaces are shown in the x-y plane. Figures 5a and 5b show some interesting contrasts; in the case of the narrow sheet a relatively large negative potential valley develops near the top of the current sheet and, therefore, a correspondingly large parallel potential drop $\Delta \phi_{int}$ develops inside the sheet. Outside the sheet, the parallel potential drop $\Delta \phi_{ext}$ is small and of opposite polarity to that inside the current sheet. The subscripts "int" and "ext" refer to the regions interior and exterior to the current sheet. On the other hand, in the case of the wide sheet, exactly the opposite occurs; the potential drop inside the sheet ($\Delta \phi_{int}$) is weak and the strong parallel potential drop ($\Delta \phi_{ext}$) develops outside the sheet. In this case also, $\Delta \phi_{int}$ and $\Delta \phi_{ext}$ have opposite signs. Thus, we find that

$$\begin{aligned}\Delta\phi_{int} &\sim -\Delta\phi_{\perp m}, & |\Delta\phi_{ext}| &\ll \Delta\phi_{\perp m}, & l &\lesssim \rho_i \\ |\Delta\phi_{int}| &\ll \Delta\phi_{\perp m}, & \Delta\phi_{ext} &\sim \Delta\phi_{\perp m}, & l &\gg \rho_i\end{aligned} \tag{2}$$

where,

$$\Delta\phi_{\perp m} \simeq -\int_0^{\pm L_x/2} E_\perp dx \tag{3}$$

where the integration is carried out near the top of the simulation plasma and $\Delta\phi_{\perp m}$ is the potential drop associated with $E_\perp$ shown in

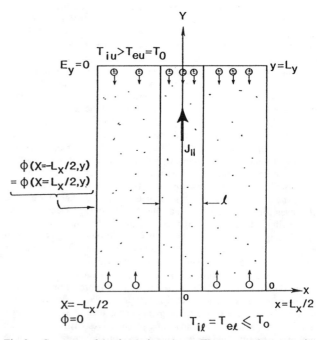

Fig. 2. Geometry of the simulation scheme. The current sheet occupies the central region $-l/2 \leq x \leq l/2$. The temperatures of the particle populations in the plasma reservoirs at $y = 0$ and $y = L_y$ are indicated. The hot ion population at $y = L_y$ has a temperature T_{iu}, which sometimes is referred to as T_H in the paper.

Double Layers

We find that both ϕ_{ext} and ϕ_{int} evolve into double layers. Normally, we find that ϕ_{int} is supported by a multiple double layer formation, which, for example, can be seen in Figure 5a. The parallel potential drop $\Delta\phi_{int}$ is taken up by two double layers; the one at the top, which can be inferred from the potential contours # 3, 4, and 5, has a strength of about $(k_B T_0/e)$, while the one below it (contours # 5 to 11) has a potential drop of $3k_B T_0/e$. We find that the top double layer occurs over a distance of about $20\lambda_{do}$. Thus, the magnitude of the electric field in this double layer is about

$$E_\parallel \sim 0.05\, k_B T_0/e\lambda_{do} \quad (4)$$

The bottom double layer is relatively strong in terms of the total potential drop, but is distributed over a longer distance and gives the same parallel electric field estimated above. Using the temperatures and densities in Table 1, we find that the parallel electric fields are expected to be an order of magnitude or more smaller than the large perpendicular

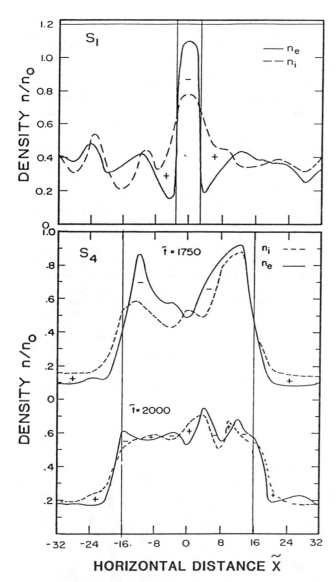

Fig. 3. Horizontal profiles of the electron and ion densities at $\tilde{y} = 100$ which show the creation of space charges. The top panel, which is from S_1 ($l < \rho_i$), shows the development of an intense negative space charge inside the sheet. The negative charge is balanced by positive space charges just outside the sheet. The profiles in the bottom panel are from S_4 and they show oscillatory features inside the sheet along with the space charges near the edges. Two sets of profiles are shown to illustrate that at times, as at $\tilde{t} = 1750$, a density cavity forms in the current sheet. At the time of the cavity formation, double layers are seen (see Figure 6).

Figure 4. Steady-state calculations [Kan et al., 1979] and some numerical simulations with very thin sheets ($l < \rho_i$) [Wagner et al., 1980; Singh et al., 1983] show that $\Delta\phi_{\perp m} < k_B T_H/e$, where T_H is the hot ion temperature at the upper boundary. When l is increased so that $l > \rho_i$, $\phi_{\perp m}$ exhibits a considerable temporal variation which is controlled by non-linear wave features [Singh et al., 1986]. The reason for the above contrasting features of the narrow and wide current sheets lies in the space-charge configurations associated with them, as shown in Figures 3a and 3b.

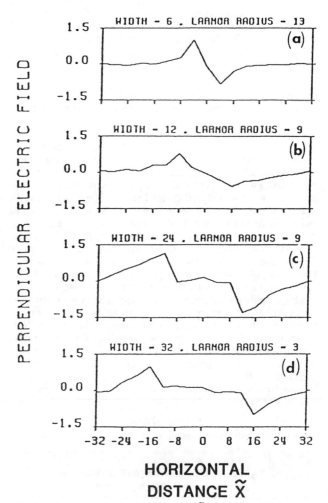

Fig. 4. dc perpendicular electric fields $\tilde{E}_\perp$ versus x at $\tilde{y} = 100$ in simulations S_1 to S_4. The current sheet width (l) and Larmor radius (ρ_i) in each case are indicated. $\tilde{E}_\perp$ peaks near the edges of the current sheet, where $E_\perp$ is of the order of unity. In wide sheets ($l > \rho_i$), $E_\perp$ in the central region is small and oscillatory, and near the edges $E_\perp \sim 1$.

346 ELECTRIC FIELDS

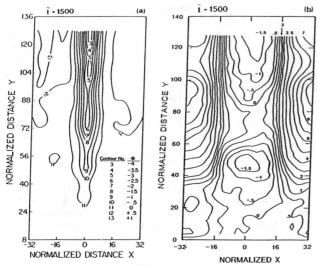

Fig. 5. Magnetic field-aligned potential structures. Equipotential surfaces are shown from (a) S_1 in which $\tilde{l}=6$, $\tilde{\rho}_i=13$, and (b) S_4 in which $\tilde{l}=32$, $\tilde{\rho}_i=3$ ($\tilde{l} \gg \tilde{\rho}_i$). Note that in panel (a) the normalized potentials of the contours are given in the legend, while in panel (b) they are marked on the contours.

electric fields ($E_{\perp m}$) occurring near the edges of the current sheet, which is consistent with electrostatic shocks in space. For the parameters in Table 1, the parallel electric fields in space are expected to lie in the approximate range of from 10 to 100 mV/m.

Because of the close proximity of double layers and electrostatic shocks associated with narrow current sheets, it is difficult to distinguish them. On the other hand, in wide sheets ($l \gg \rho_i$), it is possible to see double layers with comparable $E_\perp$ and $E_\parallel$ well-separated from the large perpendicular electric fields ($E_{\perp m}$) near the edges of the current sheet. An example of such a situation, taken from run S_4, is shown in Figures 6a and 6b. Figure 6a shows the distribution of the parallel potential along the axis ($x = 0$) of the current sheet. Two weak double layers with potential drops $\Delta\phi \simeq k_B T_o/e$ can be seen in the figure. The typical dimensions of these double layers along the magnetic field is about $L_\parallel \simeq 25\lambda_{do}$. Thus, the magnitude of the parallel electric field in the double layers is approximately the same as given by equation (4). The corresponding amplitudes of the perpendicular electric fields at $\tilde{y} = 40$ and 96 within the current sheet are shown in Figure 6b. We recall that in the present case the current sheet occupies the region $-16 < \tilde{x} < 16$. Note the occurrence of relatively large $E_\perp$ near the edges of the current sheet near the top ($\tilde{y} \simeq 96$), while relatively weak fields occur in the interior. Also note that the magnitude of $E_\perp$ near the edges decreases as $\tilde{y}$ decreases. The relatively large $E_\perp$ near the edges, when $y \gtrsim 96\lambda_{do}$, is given by equation (1). The typical magnitude of $E_\perp$ in the interior, where the parallel electric fields are given by (4), is

$$E_\perp \simeq 0.1 k_B T_o/e\lambda_{do} \quad (5)$$

which is comparable to the parallel electric field.

TABLE 1. Estimates of $E_\perp$ (mV/m) based on representative values of n_o and T_o.

T_o, n_o	1 cm^{-3}	5 cm^{-3}	10 cm^{-3}
1 eV	130	300	400
10 eV	400	900	1300
100 eV	1300		

The double layer observations reported by Temerin et al. [1982] indicate that $E_\perp < 15$ mV/m. Our simulations indicate that electric fields of this type could be generated when the density is low ($n_o < 5$ cm^{-3}) and the electron temperature $T_o \sim 1$ eV. Otherwise, the parallel electric fields in the simulated double layers exceed the observed values.

In connection with the weak double layers, Temerin et al. [1982] note that the polarization of the electric field is predominantly parallel, implying that $E_\perp \ll E_\parallel$. However, in the same paper they suggest that the parallel and perpendicular scale lengths of the double layers are comparable, implying that $E_\perp \sim E_\parallel$ if the parallel and perpendicular potential drops across the double layers are the same. Our simulations indicate that the latter condition ($E_\perp \simeq E_\parallel$) prevails.

The results shown in Figures 6a and 6b were from run S_4, in which the current inside the sheet was low, $\tilde{J}_o \leq 0.6$. In this case, double layer formation was infrequent. However, if the current density $\tilde{J}_o$ is increased, double layers form more frequently and their strength also increases [Singh et al., 1986].

The double layers shown in Figure 6a exist in conjunction with electrostatic ion-cyclotron waves in agreement with satellite observations [Temerin et al., 1982]. An example of low frequency fluctuations seen in association with the double layers in Figure 6a is shown in Figure 7; the top panel shows the time history of the ion density at the point (0, 64), while the bottom panel shows the frequency spectrum of the density fluctuations. The most dominant narrow-banded peak in the spectrum occurs at $\omega \simeq 1.3\Omega_i$. However, there are several other strong peaks at frequencies ranging from below Ω_i to about $4\Omega_i$. The fluctuations in the density are highly spiky. The spikes have periodicities that correspond to the frequencies of the two most dominant peaks in the spectrum; these peaks occur at $\omega_1 \simeq 0.38\Omega_i$ and $\omega_2 \simeq 1.3\Omega_i$. The time periods associated with these frequencies are $\tilde{\tau}_1 = 210$ and $\tilde{\tau}_2 = 60$. Such time periods of the spikes can be seen in Figure 7a.

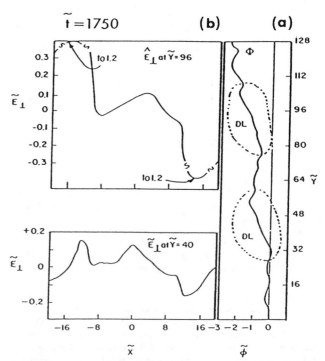

Fig. 6(a). Axial ($x = 0$) distributions of the potential for S_4. The double layers are encircled by dashed curves, (b) The perpendicular distribution of $E_\perp$ at $\tilde{y} = 40$ and 96; these locations are in the vicinity of the double layers shown in panel (a). The maximum value of $\tilde{E}_\perp$ at $\tilde{y} \simeq 96$ is about 1.2 at the edges of the current sheet.

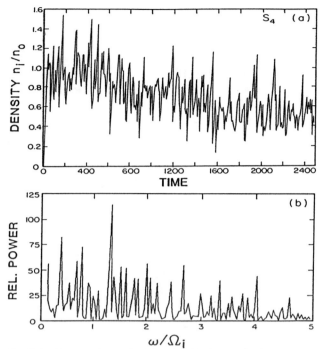

Fig. 7(a). Temporal evolution of the ion density at the point (0, 64), (b) Frequency spectrum of the fluctuations in the density.

Downward Parallel Electric Fields

The above discussion shows that the parallel electric field inside the current sheet is predominantly upward (positive). The potential structures in Figures 5a and 5b show that outside the sheet $E_\parallel$ is primarily downward (negative). We note that the large values of downward $E_\parallel$ outside narrow current sheets ($l \lesssim \rho_i$) are highly transient, dominated by the nonlinear features of electron-plasma oscillations. Normally, these fields are much weaker than those given by (4). However, when the sheets are wide ($l \gg \rho_i$), the large downward parallel fields are nearly a permanent feature, as can be deduced from the equipotential surfaces in Figure 5b. Downward electric fields in association with large perpendicular electric fields have been observed in low altitude electrostatic shocks [Mozer, 1980]. The downward electric fields accelerate the cold ionospheric electrons in the upward direction, forming upflowing electron beams [Klumpar and Heikkila, 1982; Gorney et al., 1985].

Summary

In addition to electrostatic shocks and double layers, we have seen a variety of other phenomena, which are mentioned here briefly. In a companion paper, we discuss ion beam and conic formation as seen in our simulations. Also we find that in the presence of parallel potential drops inside the current sheet, the downward accelerated electrons are neither monoenergetic nor perfectly field-aligned, which indicates a considerable scattering in both energy and pitch angle. This is expected owing to the fluctuations in the properties of the double layers supporting the parallel potential drop [Singh et al., 1986].

The turbulence seen in our simulations shows a hierarchy of time scales, corresponding to frequencies ranging from below the ion-gyrofrequency to the harmonic of the electron-plasma frequency. In real time, the turbulence appears as bursts of a high frequency wave modulated by waves of a multitude of time scales longer than that of the high frequency one. Sometimes, very low frequency ($\ll \omega_{pi}$ or Ω_i) oscillations appear in the potential structures. These oscillations are driven nonlinearly by the electron-plasma oscillations. A more complete description of the results from our simulations will be published elsewhere [Singh et al., 1986].

Acknowledgment. This research was supported by NASA grant NAGW-77 and NSF grant ATM-8417880 to Utah State University.

References

Gorney, D. J., Y. T. Chiu, and D. R. Croley, Jr., Trapping of ion conics by downward parallel electric fields, *J. Geophys. Res., 90*, 4205, 1985.

Kan, J. R., L. C. Lee, and S.-I. Akasofu, Two dimensional potential double layers and discrete auroras, *J. Geophys. Res., 84*, 4305, 1979.

Klumpar, D. M. and W. J. Heikkila, Electrons in the ionosphere source region: Evidence for runaway electrons as carriers of downward Birkeland currents, *Geophys. Res. Lett., 9*, 873, 1982.

Mozer, F. S., On the lowest-altitude S3-3 observations of electrostatic shocks and parallel electric fields, *Geophys. Res. Lett., 7*, 1097, 1980.

Mozer, F. S., C. A. Cattell, M. K. Hudson, R. L. Lysak, M. Temerin, and R. B. Torbert, Satellite measurements and theories of low altitude auroral particle acceleration, *Space Sci. Rev., 27*, 155, 1980.

Singh, N., H. Thiemann, and R. W. Schunk, Simulation of auroral current sheet equilibria and associated V-shaped potential structure, *Geophys. Res. Lett., 10*, 745, 1983.

Singh, N., H. Thiemann, and R. W. Schunk, Dynamical features and electric field strengths of double layers driven by currents, *J. Geophys. Res., 90*, 5173, 1985.

Singh, N., H. Thiemann, and R. W. Schunk, Simulations of auroral plasma processes, *Planet. Space Sci.*, to be submitted, 1986.

Temerin, M., K. Cerny, W. Lotko, and F. S. Mozer, Observations of double layers and solitary waves in the auroral plasma, *Phys. Rev. Lett., 48*, 1175, 1982.

Thiemann, H., N. Singh, and R. W. Schunk, Some features of auroral electric fields as seen in 2D numerical simulations, *Adv. Space Phys., 4*, 511, 1984.

Wagner, J. S., T. Tajima, J. R. Kan, J. N. Leboeuf, and J. M. Dawson, V-potential double layers and formation of auroral arcs, *Phys. Rev. Lett., 45*, 803, 1980.

Section VIII. MACROSCOPIC PROCESSES

MACROSCOPIC ION ACCELERATION ASSOCIATED WITH THE FORMATION OF THE RING CURRENT IN THE EARTH'S MAGNETOSPHERE

B. H. Mauk and C.-I. Meng

The Johns Hopkins University, Applied Physics Laboratory, Laurel, MD 20707

Abstract. As an illustration of the operation of macroscopic ion acceleration processes within the earth's magnetosphere, we review processes thought to be associated with the formation of the earth's ring-current populations. Arguing that the process of global, quasi-curl-free convection cannot explain particle characteristics observed in the middle (geosynchronous) to outer regions, we conclude that the transport and energization of the seed populations that give rise to the ring-current populations come about in two distinct stages involving distinct processes. Near and outside the geostationary region ($\gtrsim 6$ to $7\ R_e$), the energization and transport are always associated with highly impulsive and relatively localized processes driven by inductive electric fields. The subsequent adiabatic earthward transport is driven principally by enhanced, curl-free global convection fields.

1. Introduction

In general, macroscopic ion acceleration is synonymous with macroscopic ion transport, which means that particles are accelerated principally by being transported from one region of the magnetosphere to another. Two examples are adiabatic energization by means of convective transport within spatially or time varying magnetic configurations, and energization by means of transport that violates guiding center behavior in the vicinity of large-scale, electrified boundaries (such as shocks or neutral sheets). In principle, these processes are easy to understand as isolated entities. However, in applying these processes to specific problems, seldom can they be viewed in isolation. They must be viewed as components in the workings of complex systems, and the processes will likely not operate as simply as their underlying structures would imply.

As an illustration of the operation of macroscopic ion acceleration processes within the magnetosphere, we consider in this review the formation of the ring-current populations (tens of keV to hundreds of keV ions within the 2 to 5 R_e region of the earth's magnetosphere [Williams, 1983]). The simplest models of this formation involve the processes of adiabatic, convective transport and energization of populations that reside in the magnetospheric tail by means of enhanced, global, quasi-curl-free convection electric fields. While such processes certainly play an important role, we will argue that they are only part of the story. We will spend no time in this review accounting for the source of particles within the tail. The questions to be addressed are (a) how the transport from the outer to the inner regions takes place, and (b) whether the corresponding processes preserve the adiabatic invariants.

2. Global Convective Transport

It has become almost an accepted conclusion (see references below) that, during relatively quiet periods, plasmas that reside within the earth's magnetotail will be transported toward the earth in a quasi-time-stationary fashion because of the global, curl-free, cross-tail electric field imposed by the solar-wind/magnetopause dynamo. (With the concept of quasi-time-stationary convection, we assume that temporal fluctuations in the electric field average out to second-order effects on particle transport.) Figure 1, taken from Kavanagh et al. [1968], shows the low-energy plasma convection pattern expected to exist within the equatorial plane of the earth's magnetosphere from a uniform cross-tail plus a corotation electric field configuration. The solid line that forms the shape of a tear-drop marks the separatrix between open streamlines that connect to the tail regions and streamlines that circle around the earth forming closed orbits. The region of closed orbits is termed the "forbidden zone." As the plasmas outside the forbidden zone convect toward the earth, they are substantially energized. Figure 2, taken from Ejiri [1980], illustrates this effect. By conserving the first and second adiabatic invariants, large-pitch-angle particles, near 90°, are energized principally by betatron acceleration, while small-pitch-angle particles (near 0° or 180°) are energized principally by Fermi acceleration.

Quantitatively, the picture shown in Fig. 1 has been modified in recent years, principally because of the realization of the importance of the contribution of the polarization response of the trapped plasmas [e.g., McIlwain, 1972; Wolf et al., 1982]. However, the qualitative picture shown in Fig. 1 is still thought to be correct [Stern, 1977; Mauk and Meng, 1983b; Baumjohann et al., 1985]. We emphasize that, while the shapes of the convection orbits shown in Fig. 1 are thought to be qualitatively correct, the manner in which those convection orbits become populated with particles is still open to question.

Many authors have suggested [Kavanagh et al., 1968; Chen, 1970; Cowley and Ashour-Abdalla, 1976; Southwood and Kaye, 1979; Kivelson et al., 1980; Hultqvist et al., 1982; Horwitz 1984; and Fairfield and Viñas, 1984] that the plasma-sheet source population that resides within the magnetotail has direct access to the inner magnetospheric regions by means of the quasi-steady convection orbits shown in Fig. 1. The plasma-sheet populations, energized by the adiabatic transport, would thus populate the regions outside of the solid separatrix line shown in the figure. The most oft-quoted evidence for this idea is the work of Kivelson et al. [1980]. Kivelson et al. [1980] took advantage of the fact that the radial position of the separatrix line is energy (or, more precisely, magnetic moment) and species dependent because of the effect of the magnetic drifts. Because of these dependencies, Kivelson et al. [1980] noted that a geostationary satellite will observe energy cutoffs with values that depend on orbital position. Figure 3 shows those authors' test of this idea. The figure shows a grey-scale spectrogram of data taken by the geostationary ATS-5 satellite [DeForest and McIlwain, 1971]. The horizontal axis shows time (which translates into orbital position); the vertical axis shows energy for electrons (top grey-scale

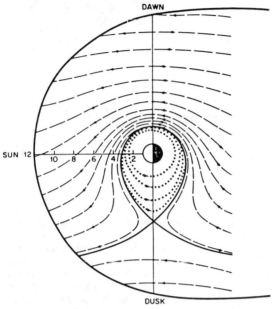

Figure 1: From Kavanagh [1968]. Magnetospheric, equatorial low energy particle convection pattern for a dipole magnetic configuration and an unshielded dawn-to-dusk plus corotation electric field configuration.

panel) and ions (bottom panel), where the ion energy scale is inverted (also the energy scale is nonstandard). The whiteness of this rather poorly reproduced figure is proportional to the count rate of curved-plate, electrostatic, energy/charge analyzers. The key feature of interest in Fig. 3 is the apparent encounter of new plasmas between ~ 1 and 6 UT, which corresponds to ~ 1900 to 2400 local time. The model calculations of Kivelson et al. [1980], with one fitting parameter, are shown in the figure as a solid black-on-white line. For the portion of the observed energy-time dispersion curve for which the model calculations have been performed, the model fits the data rather convincingly.

Quasi-time-stationary global convection, such as that discussed above, cannot itself populate the inner, ring-current regions during active time periods. It has long been recognized that time variations must invariably be associated with the processes that populate these regions [e.g., Fälthammer, 1965; Brewer et al., 1969]. It has, in fact, been common to adopt a diffusive approach in trying to incorporate such time variations into the transport models, at least for the more energetic particles [e.g., Birmingham, 1969; Cornwall, 1972; Lyons and Thorne, 1973; Spjeldvik, 1977]. Our goal, however, is to achieve a deterministic understanding of the particle transport. To the extent that time variations are responsible for the transport, we would like to know the form that the variations take and the manner in which the variations fit together with the time-stationary picture.

Given the apparent successes of the quasi-time-stationary convection picture during relatively quiet periods, it is tempting and natural to adopt a modified form of the process in order to account for the transport during active periods of time. According to the steady convection models, the degree of access that tail populations have to the inner regions depends on the size of the forbidden zone, which, in turn, depends on the magnitude of the cross-tail potential. Time dependence can be introduced in a deterministic way to the global convection models by allowing the cross-tail potential to vary with time resulting in simple global rescalings of the convection patterns. This point of view has been adopted by Walker and Kivelson [1975], Kaye and Kivelson [1979], Kivelson et al. [1980], and Chiu and Kishi [1984]; see also Grebowsky

[1971]. This time dependence in the global, curl-free convection patterns, coupled with the notions of direct convective access to the tail populations and adiabatic energization, allows one to construct a relatively complete model for the formation of the ring-current populations during active periods [see the discussion of Williams, 1983]. The successful model of Wolf et al. [1982], discussed in detail in a later section, while differing in important quantitative details, can be understood qualitatively on the basis of this overall framework. Additionally, to the extent that the authors speak of time fluctuations in the global electric field configuration, the diffusive models discussed in the preceding paragraph can also be understood on the basis of this framework.

The framework presented in the preceding paragraph depends strongly on the concept of quasi-time-stationary, curl-free convection, a process that is simply rescaled in a time dependent fashion during active periods. In the following section, we will present evidence that the magnetospheric transport processes are substantially more complex than this framework allows.

3. Problems with Quasi-Stationary, Curl-Free Convection

In recent years, the concept of quasi-time-stationary, curl-free convection within the earth's magnetosphere has been challenged. Simple global rescaling models to introduce time dependence are included in this challenge.

Erickson and Wolf [1980] and, more recently, Schindler and Birn [1982] have concluded that steady-state earthward convection of the plasma-sheet particles, as the plasmas approach the stronger fields close to the earth, is inconsistent with the pressure balance across the magnetotail. The argument is that, if pressure balance exists across the tail (with the external medium) at distant positions, and if the internal plasma-sheet plasmas convect in a classically adiabatic fashion (i.e., pressure $\propto$ (density)$^\gamma$) toward the earth, the pressure of the internal plasmas will rise so dramatically that a pressure balance cannot be maintained with the external medium. This conclusion depends on the magnetic configuration; Erickson and Wolf [1980] tested their hypothesis against many well-known standard magnetic field models. Schindler

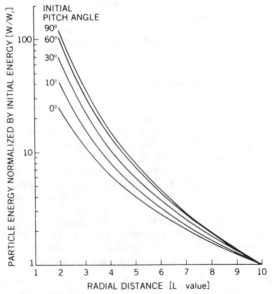

Figure 2: From Ejiri [1980]. Illustration of energization due to conservation of first two adiabatic invariants in a dipole magnetic field configuration and from a source of particles at 10 R_e.

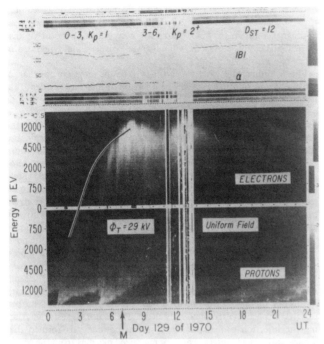

Figure 3: From Kivelson et al. [1980]. Geosynchronous satellite particle spectrogram plus a calculated dispersion curve (black-on-white line) based on time-stationary magnetospheric convection (see text).

and Birn [1982] have generalized the argument so that it can be divorced from specific field models, and have come to conclusions similar to those of Erickson and Wolf [1980]. Both sets of authors conclude that time-stationary convection does not occur, that some sort of time-dependent convection is responsible for earthward particle transport, and that such time-dependent convection is probably associated with substorm processes. Given the time-dependent solutions discussed by Schindler and Birn [1982], it is also clear that the introduction of time dependences in the form of a global rescaling of the convection patterns will not resolve the pressure balance problem found by these authors.

The conclusions of these authors are based on theoretical arguments alone. However, experimental support for their conclusions has been obtained by a more detailed examination of the geosynchronous data sets. The bottom panel of Fig. 4 shows a simulated geosynchronous particle spectrogram calculated based on unshielded, time-stationary convection. This panel shows electron and ion energies on the vertical axis and local time (or orbital position) on the horizontal axis. The shaded regions correspond to those energies that have access by means of steady convective transport to the tail regions. Based on the models of steady, earthward convection of the plasma-sheet tail populations, the shaded regions should be populated with particles.

The feature of interest to us in Fig. 4 (bottom panel) is the low-energy particle gap, which extends from ∼15 to ∼21 hr of local time for the particular cross-tail field value used. The gap corresponds to the passage of this circular orbit spacecraft into the forbidden zone diagrammed in Fig. 1. The size of the gap is energy dependent, and it was the energy dependence of the position of the portion of the gap near ∼21 hr local time that has been used to support the global, time-steady convection model, as is illustrated in Fig. 3. However, Fig. 4 shows a much more complete view of the expected steady convection signatures, and a reexamination of the geostationary signatures based on this more complete view is clearly warranted.

Examining first the portion of the gap near ∼21 hr local time, we note that, as one goes to higher and higher ion energies, one eventually comes to a point where the gap is closed off at ∼1 keV in the figure. The gap does not extend even close to the highest ion energies that are displayed for any reasonable values of the cross-tail potential that can be used. And yet, if one reexamines Fig. 3, the measured energy-time dispersion curve of interest extends to the highest energies sampled (50 keV). Had the calculated black-on-white line been extended to earlier times, it would have deviated sharply and qualitatively from the observed energy-time dispersion curve.

Examining next the portion of the gap near ∼15 hr in the bottom panel of Fig. 4, we expect to see, as the satellite travels around in its orbit, first the dropout of the higher energy electrons, followed by the dropout of low-energy particles, and followed still later with the dropout of the higher energy ions. Figure 5 shows what is actually being observed in the data. The feature of interest is the low-energy plasma dropout observed near 0430 UT on the left side of the figure. What one observes is the dropout of the higher energy ions followed by the dropout of low-energy particles, and followed still later with the dropout of higher energy electrons. The behavior shown in Fig. 5 is representative of very quiet to very active time periods [Mauk and Meng, 1983b]. Note that this behavior (i.e., the sense of the energy-time dispersion) is exactly the opposite of that which we expected, based on an examination of the bottom panel of Fig. 4.

We conclude that examinations of the complete ion-electron geostationary particle signatures do not support the models of time-stationary convection of particles from the tail regions. Mauk and Meng [1983b] have argued, additionally, that neither the introduction of simple time variability (in the form of variations in the cross-tail potential that rescale the convection patterns) nor the introduction of simple particle loss processes will alter this conclusion. The crucial point to recognize is that the observed dispersion patterns (and we will see others in the

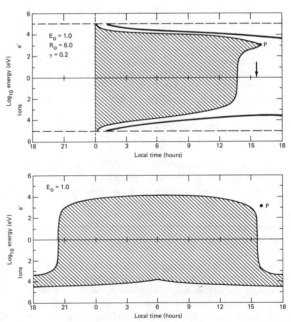

Figure 4: From Mauk and Meng [1983b]. (top) Convective evolution of injection boundary distributions as viewed in spectrogram form by a geosynchronous satellite. Note qualitative similarities with 0730 event on Figure 6. This is compared with the expectations based on time-stationary convection (bottom).

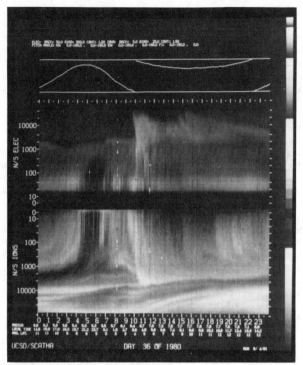

Figure 5: From Mauk and Meng [1983b]. Electron-ion spectogram sampled by the near geostationary SCATHA satellite during relatively quiet times. The feature of interest is the low energy plasma dropout near 0430 UT.

next section) are just as well ordered and just as repeatable as one might expect the patterns to be based on time-stationary convection. And yet, the patterns are inconsistent with the steady-convection picture. The observed patterns are not just smeared or jumbled versions of the expected steady-convection predictions. We will see that the basic transport and energization processes always appear to include intrinsically time-dependent mechanisms that are, in all likelihood, fundamentally distinct from the global, quasi-time stationary convection process.

4. Key Features of Dynamic Injections and the Injection Boundary Model

The dynamical (intrinsically time dependent) variability of particle features in the interior of the earth's magnetosphere, most particularly that associated with substorm processes, has, of course, been the subject of intensive study for many years [e.g., Akasofu, 1980]. In an effort to understand the processes responsible for earthward transport and energization of the near-earth magnetotail populations, there are several features of the so-called "substorm injections" that we consider to be key.

Figure 6, a particle spectrogram obtained using the ATS-6 geosynchronous satellite, illustrates several key features of substorm injections. We focus on the sudden brightening that occurs in both the electrons and the ions near 0730 UT ($\sim$0100 local time). A key feature of substorm injections, as noted by McIlwain [1974], is that the particle distribution, as characterized by a single spacecraft, can change "simultaneously" (compared to global convective time scales) over a very broad range of energies (including both electrons and ions). The injection is thereby "dispersionless." Not all injections appear dispersionless (as other features on Fig. 6 make clear), but it appears [Mauk and Meng, 1983a] that the dispersionless characteristic is observed only if the spacecraft is located in the proper region (most particularly near midnight). Subsequent quasi-stationary convection of the newly disturbed particle distributions will generate "dispersional" features in other regions of the magnetosphere.

Figure 7 illustrates what we consider to be the most crucial characteristic of injection signatures. The injection of interest is evident first near $\sim$0300 to 0400 UT ($\sim$2100 local time) in the ions and extends subsequently into the electrons near 0430 UT. These particles were injected relatively recently, but not at the spacecraft location. Thus the arrival times of the particles are dispersed with respect to energy, and the highest energy particles do not penetrate to the satellite location.

The key feature of interest on Fig. 7 is that the banded stucture of preexisting ion distributions (at ion energies $>$ 5 keV), which enters the picture at the extreme left, appears to pass through the newly injected ion dispersion pattern in a fashion that leaves the older structure essentially undisturbed. This feature was first noted by McIlwain [1974]. If we map the old and new populations back to earlier times, using quasi-stationary convective transport (thereby undoing the post-injection distortions), it is clear that some regions of space have been strongly disturbed by the injection process, whereas other regions have been left essentially undisturbed. The fact that, on the figure, the separation in energy between the old and new plasmas is extremely sharp means that the spatial separation between the disturbed and undisturbed regions of space is correspondingly sharp. The figure suggests that, during the injection process, a sharp spatial boundary is formed that separates the newly injected and/or energized plasmas from the preexisting, undisturbed plasmas. Given that old undisturbed plasmas are immediately adjacent in time to newly disturbed plasmas over a broad range of ion energies, it would be difficult (and we have tried) to conceive of the sharp spatial boundary as being differently located for different energies.

The feature that we have keynoted on Fig. 7, while not always observed because of the special conditions required, appears to be charac-

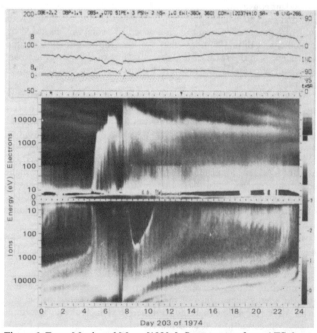

Figure 6: From Mauk and Meng [1983a]. Spectrogram from ATS-6 geosynchronous satellite showing key features of impulsive injection phenomena.

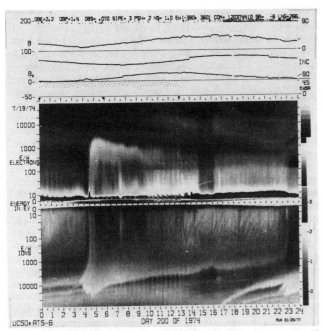

Figure 7: From Mauk and Meng [1983a]. Spectrogram from ATS-6 geosynchronous satellite showing more key features of impulsive injection phenomena (see text).

teristic of substorm injections [McIlwain, 1974; Mauk and Meng, 1983a]. An example is even observed on the much more disturbed display in Fig. 6. Note that between 1000 and 1400 UT, at ion energies between 8000 and 20,000 eV, there remains within the relatively dark portion of the display the faint remains of preexisting dispersive ion distributions that clearly constitute a distinct population from the newly injected population.

The characteristics of substorm injections described thus far lead McIlwain [1974] to suggest the so-called "injection boundary" model. The characteristics of the injection boundary have been subsequently explored by Mauk and McIlwain [1974], Konradi et al. [1975, 1976], Mauk and Meng [1983a,b], Strangeway and Johnson [1983], Quinn and Johnson [1985], Greenspan et al. [1985], and McIlwain [1985]. Figure 8 is one view of the injection boundary. The boundary shown is formed near the time of the initiation of a substorm expansion phase. The boundary forms very quickly as compared to global convective time scales. At the time of the boundary formation, all energies of the particle distribution have approximately the same common earthward edge (no tailward edge should be inferred from the figure). Finally, particle distributions earthward of the relatively sharp spatial boundary are left comparatively undisturbed.

The "double-spiral" shape of the injection boundary, first proposed by Konradi et al. [1975] and since updated by Mauk and Meng [1983a], and then McIlwain [1985], has been shown [Mauk and Meng, 1983a] to account for the many different types of particle energy dispersion signatures observed within the geostationary orbit. The top panel of Fig. 4 is one example of the many predicted dispersive signatures. This figure was generated by convolving the quasi-stationary convection of the injected plasmas with the motions of a geostationary satellite. The shaded regions show where new particles are expected to be observed (depending on the shape of the energy distribution of the injected plasmas). The unshaded regions map back to undisturbed spatial positions. At the time of the sudden injection, the satellite was located near midnight and was immersed in the newly injected plasmas. Figure 4 was not generated with any particular observation in mind, and yet, qualitatively, the dispersion edges look very similar to those of the signature that begins near 0730 UT on Fig. 6.

Two features in the top panel of Fig. 4 are of particular interest. The thick, solid, dispersive lines above and below the main shaded regions represent the first of many echoes; such ion echoes are observed on Fig. 6, if somewhat broadened [DeForest and McIlwain, 1971], and were previously observed at much higher energies by Lanzerotti et al. [1971]. The electrons do not appear to survive the trip around the earth nearly as intact as do the ions. Echoes are a natural consequence of the double-spiral configuration [see Fig. 9 of Mauk and Meng, 1983a]. Additionally, examining the region near 1300 to 1400 local time on Fig. 4 (top), where a plasma dropout occurs, the dispersion sense of plasma dropout observed so commonly in the data (recall the discussion in section 3 about Fig. 5) is now explained. This dispersion sense is a general feature of the convective evolutions of the injection boundary as viewed by a geostationary satellite. The comparison on Fig. 4 between the injection boundary signature (top) and the steady convection signature (bottom) emphasizes how very different the dropout dispersive signatures are for the two models.

The injection boundary model is not a physical model. No mechanisms are proposed for the formation of the boundary structure, although McIlwain [1974] did propose a local heating caused by field-aligned communications with the auroral ionosphere. It is a phenomenological model whose purpose is to place constraints on the range of physical mechanisms that can be considered as the injection process. Most

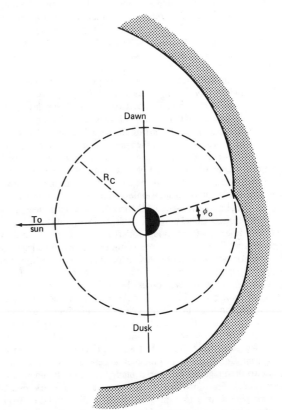

Figure 8: From Mauk and Meng [1983a,b]. Schematic of the injection boundary spatial configuration of the injection boundary model. Particle distributions tailward of the solid, double-spiral line are freshly injected and ready for convective reconfiguration. Distributions earthward are undisturbed.

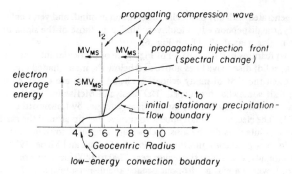

Figure 9: From Moore et al. [1981]. Schematic of injection front model of near geosynchronous particle injections (see text).

particularly, the energization and transport processes must be spatially localized to the extent that sharp spatial boundaries can be formed leaving undisturbed regions.

5. Injection Mechanisms

Another key observational characteristic of substorm injections was reported by Moore et al. [1981] who studied injections using two different radially displaced satellites (ATS-6 and SCATHA). It was noted that the appearance of dispersionless injection signatures at higher altitudes precedes their appearance at lower altitudes (at the same local time). Translating the time differences into radial propagation speeds, one obtains values between 15 and 100 km/s. The propagating, dispersionless particle feature was associated with a sharp propagating magnetic signature [first reported by Russell and McPherron, 1973] that results in a more compressed and a more dipolar (less tail-like) magnetic configuration as characterized at each of the satellite positions.

To explain their observations, Moore et al. [1981] proposed the "injection front" model shown in Fig. 9. It is proposed that injections correspond to some kind of compressional wave front that propagates earthward from disturbances occurring in the tail regions. The front is thought to be shock-like in that the speed of material entering the front from the upstream region is faster than that of the outflow, with the result that a spatially dispersed plasma boundary is substantially steepened and transported toward the earth. The steepening characteristic is used to explain the dispersionless characteristic of injections; the compressional characteristics of the front, along with the general transport into stronger global magnetic fields, will yield substantial energization. The injection-front model satisfies the injection-boundary model constraint discussed in the preceding section by hypothesizing that the injection boundary represents the position where the injection front stops propagating. There are problems with the model [see Moore et al., 1981], one of which we address below.

The driving force behind injections of the sort envisioned in this and the previous section must be transient electric fields; indeed transient electric fields perpendicular to the local magnetic field have been observed within the near geostationary magnetosphere by Shepard et al. [1980] and Aggson et al. [1983]. The transient electric fields are measured in the tail-side of the earth's magnetosphere, and, when encountered close to the magnetic equatorial plane, point predominently in the dawn-to-dusk directions. The strengths peak at 15 to 30 mV/m and last for time periods of 1 to 2 min. As shown in Fig. 10, taken from Aggson et al. [1983], the electric fields appear to be locally inductively driven. The bottom trace in the figure shows the dawn-to-dusk electric field component; its shape appears to be correlated with the time derivative of the locally measured radial magnetic field component shown on the top of the figure (the satellite was at a magnetic latitude of $\sim 20°$).

There are several features in Fig. 10 that are of particular interest. Calculations by Aggson et al. [1983] based on Fig. 10 indicate that variations in the locally measured magnetic field can account for the locally measured electric field transient if the generation region has a scale size of $\sim 2.5\ R_e$. Therefore, the transient electric fields are inductively generated in a rather confined region about the near geostationary ($\sim 7.5\ R_e$) measurement point and not directly by time variations occurring deeper within the tail. Assuming that substorms are initiated within the tail (say at 15 R_e; see the discussions), then those regions must generate some kind of electromagnetic structure that propagates toward the earth. This concept agrees well with the observations and model of Moore et al. [1981] discussed earlier.

However, if a propagating spatial structure exists, the shape of the electric field transient places strong constraints on its nature. Both Aggson et al. [1983] and Moore et al. [1981] note that the electric field signature of the propagating front model, in its simplest form at least, should be in the form of a step-function and not a delta-function. By suggesting that the observed electric field signature results from a sharp change in the magnetospheric current system, Aggson et al. [1983] appear to be suggesting that the transient is intrinsically time dependent within some spatial region and not the result of the passage of a spatial structure. However, since Shepard et al. [1980] have shown that the electric field transients are associated with particle injections, the injection boundary model constraints discussed in the preceding section would put constraints on the nature of this intrinsically time-dependent phenomenon. In particular, the region where the electric field acts must be very sharply bounded on the earthward side so as to leave a sharp separation between disturbed and undisturbed plasmas. It is difficult to understand how the electric field can be shielded so effectively across a narrow boundary ($\sim 0.1\ R_e$ [Moore et al., 1981]).

The above discussions make it clear that we are far from having a fully satisfactory model that explains all of the key aspects of substorm injections, and it still remains to be determined whether the injection

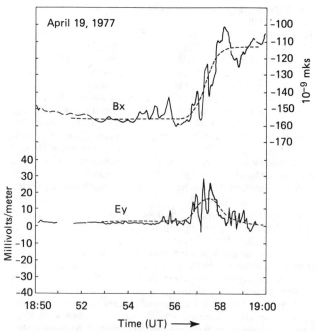

Figure 10: From Aggson et al. [1983]). Measurement of dawn-to-dusk impulsive electric fields (at $L = 7.5\ R_e$) apparently generated inductively by local variations in the magnetic configuration.

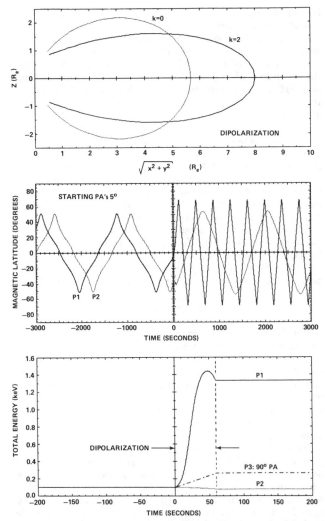

Figure 11: Generated for this paper, but inspired by the work of Quinn and Southwood [1982]. (top) Dipolarization of a flux tube quantified using the analytic field line mode of Quinn and McIlwain [1979] with free parameter k and the electric field measurements of Aggson et al. [1983]. (middle) Numerical calculations of trajectories of two particles (initial energies = 100 eV; initial pitch angles = 5°) with different bounce phases within the flux tube shown on the top panel. The dipolarization occurs between $t = 0$ at 60 s. (bottom) On a higher time resolution display, this panel shows the change in energy due to the dipolarization of the two particles (P1 and P2) shown on the middle panel, plus that of a 90° pitch angle particle (P3) (see the text for further information).

is in the form of a propagating spatial structure or in the form of a transient electric field applied to a "relatively fixed" region of space (albeit sharply shielded), or some combination of these two. (Additionally, we should not forget the local heating hypothesis of McIlwain [1974]).

Both concepts involve transient electric fields, however, and by taking a somewhat localized view we are in a position to discuss some of the consequences of such transients in terms of ion energization and transport. In order to explain what they call "bounce-phase-clustered" ion observations made using geostationary spacecraft, Quinn and South-

wood [1982] examined in general terms what they called the "convection surge." The convection surge is exactly what would be expected from the electric field signature shown in Fig. 10 if that signature acts on a single flux tube. It corresponds to a sudden displacement of a field line toward the earth over a very limited time span. The top panel of Fig. 11, showing field lines before and after the displacement, shows the concept envisioned by both Quinn and Southwood and by Aggson et al. [1983]. There are two key points that Quinn and Southwood [1982] make about the convective surge process. First, because the surge happens quickly it violates the second adiabatic invariants of the lower energy particles. Second, because of this violation, the particles that gain the largest multiplication in their energies will be those that have very small pitch angles and at the same time reside very close to the magnetic equator.

In order to confirm, illustrate, and quantify the points made by Quinn and Southwood [1982], we have for this discussion constructed a numerical computer model of the convection surge or dipolarization process. The field line shapes shown on the top panel of Fig. 11 were not just sketched but correspond to a simple, analytic field-line model presented by Quinn and McIlwain [1979]. The choice for a starting value of the free parameter $k = 2$ is reasonable, given field inclination angles measured in the 6 to 8 R_e region ($k = 0$ corresponds to a pure dipole). In order to model the subsequent dipolarization, we first calculate the time evolutions of the equatorial parameters $R_0(t)$ and $B_0(t)$, which are the equatorial field line radial position and the equatorial field strength at that position, respectively. The evolutions of these equatorial parameters are governed by a transient, equatorial, east-to-west electric field E_0 set equal to 20 mV/m for a time span T of 60 s. The starting equatorial parameters R_0, B_0, and β ($\equiv -(\partial B/\partial R)/B$) are 8 R_e, 50 nT, and $0.43/R_e$, respectively. The value E_0 is assumed to be independent of R_0.

Given that the equatorial convection velocity is $V_c = cE_0/B_0$, the variation of B_0 with time is easily determined from the frozen-in condition and from the fact that $\nabla_\perp \cdot \mathbf{V}_c \neq 0$, because of the nonzero value of the parameter β. Cylindrical symmetry is used for the calculation, and, with such symmetry, it turns out that the equatorial parameter β is invariant during the dipolarization process. Hence, our initial choice for β is just the dipolar value at ~6 R_e. Once the equatorial parameters are determined for each time step, the off-equatorial parameters, $R(\lambda,t)$ and $B(\lambda,t)$, are determined from the equatorial values, $R_0(t)$ and $B_0(t)$ by using the field line model and the derived expressions of Quinn and McIlwain [1979] (after correcting a typographical error in their presentation). It is assumed that the field-line model parameter "k" varies linearly between 2 and 0 during the time that the transient electric field is being applied. Additionally, the off-equatorial value of the transient electric field is computed self-consistently with the off-equatorial field-line motions. Out of the equator, the electric field will not in general equal 20 mV/m. In fact, as one moves away from the equator, the field switches from being east-to-west to being west-to-east, consistent with the measurements discussed in Aggson et al. [1983].

The second panel of Fig. 11 shows two sample particle trajectories calculated numerically using the model described above and the concepts of Quinn and Southwood [1982]. The vertical axis is magnetic latitude and the horizontal axis is time in seconds. The dipolarization occurs between 0 and 60 s near the middle of the figure. The two particles start out with nearly identical characteristics: energy equals 100 eV and equatorial pitch angle equals 5°. The difference between the starting characteristics lies only with the phasings of their bounce motions. This panel illustrates dramatically that the effect of the dipolarization on the particle depends critically on its position in its bounce motion during the application of the transient electric field. The bottom panel shows the effect even more dramatically. Here we show total energy versus an expanded time scale. The P1 particle, which on

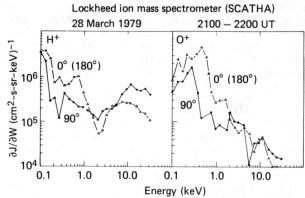

Figure 12: From Kaye et al. [1981]. H$^+$ (left) and O$^+$ (right) differential intensity versus energy spectra for pitch angles parallel (0°, 180°) and perpendicular (90°) to the local magnetic field vector. The measurements were made in the near geosynchronous regions by the SCATHA spacecraft.

the middle panel was close to the equator during the dipolarization, has increased its energy by well over an order of magnitude (to ~1.3 keV) whereas the P2 particle (the other particle in the middle panel) has actually decreased its energy. For comparison purposes, we have include in the bottom panel the energy gain of a 90° pitch-angle particle. The special particle with small pitch angle located close to the equator during the dipolarization has gained much more energy than the 90° pitch-angle particle, a substantial violation of the second adiabatic invariant for the configuration shown. Had the dipolarization occurred slow enough so that the second adiabatic invariants were preserved, the 5° particles would have increased their energies by factors of ~1.5, less than that gained by the 90° particle. Figure 11 confirms dramatically all of the points put forth by Quinn and Southwood [1982]. The effects become less dramatic at higher starting energies because the equatorial occupation times become smaller, but at several kiloelectronvolts starting energies they are still very substantial.

The convection surge process explains the bounce bunching, the field-aligned character, the typical energies involved, and the equatorial source properties of bounce-phase-clustered ions. Bounce-phase-clustered ions on the nightside are associated with magnetic signatures that are, in turn, associated with substorm injections [Quinn and McIlwain, 1979]. Also, the signature was found in 21 of 56 days of analysis [Quinn and Southwood, 1982]. Given the special observational conditions required for their observation [Quinn and McIlwain, 1979], it is not an unreasonable hypothesis that bounce-clusters are characteristic of substorm injections. If this hypothesis can be confirmed, it, along with the arguments given in sections 3 and 4, would help demonstrate that the convection surge process is a fundamental part of the transport and energization processes in the middle (near geosynchronous) regions.

Further indications of the possible key role of the convection surge process in geosynchronous transport may be discerned from an examination of the pitch-angle dependence of the energy spectra. Figure 12, taken from Kaye et al. [1981] shows the observational characteristics that are observed quite generally. Focusing first on the left panel, which displays H$^+$ characteristics, we see that at low energies, the field-aligned ions dominate over the field perpendicular ions, whereas at high energies, the field perpendicular ions dominate. In the past, the field-aligned characteristic at low energies has been taken as a signature of recent ionospheric extraction [e.g., Mauk and McIlwain, 1975; Lennartsson and Reasoner, 1978; Young, 1979; Horwitz, 1980; Fennel et al., 1981; Kaye et al., 1981]. We would like to point out that the characteristics shown on Fig. 12 (both at low and high energies) are just what

is expected using the convection surge process. At low energies, the second adiabatic invariant is violated and some low-energy field-aligned ions are energized very substantially, leading to a predominance of field-aligned ions if the initial distribution has a low enough temperature (< several hundred electronvolts [Mauk, 1985]). At higher energies, the second adiabatic invariant is preserved, leading to a predominance of energization in the perpendicular direction (see Fig. 2). We propose that the characteristics shown in Fig. 12 are signatures of the transport and energization mechanism. It cannot be assumed that the presence of field-aligned ions is a signature of recent ionospheric extraction. We note that, at the same energy, it is easier to violate the second adiabatic invariant of an O$^+$ ion than that of an H$^+$ ion. This fact could explain some differences between the left and right panels on Fig. 12.

So we see that several characteristics of the observed ion distributions point to the convection surge process (driven by powerful, transient, inductively generated electric fields) as being fundamentally associated with the transport and energization processes. We have, thus far, taken only a very localized view of the process, examining only the consequences to a single flux tube of particles. It still remains to be determined how the injection processes manifest themselves globally, whether in the form of a propagating front or in the form of a temporal transient over an extended region (albeit sharply bounded).

6. Ring Current Formation

Thus far, we have discussed the characteristics of particle transport and energization within the outer to middle (geosynchronous) regions of the magnetosphere. We have yet to generate the population (at 2 to 5 R_e) that gives rise to the predominant contributions to the ring current. We will discuss that generation in the context of the global magnetospheric storm modeling of Wolf et al. [1982]. We consider this work as being generally inclusive of simpler models and ideas that have been presented [e.g., Ejiri, 1980; Lyons and Williams, 1980; Chiu and Kishi, 1984].

The Wolf et al. [1982] study describes the results of a global computer simulation of the response of the magnetosphere to a sudden enhancement of and subsequent variations in the cross-polar-cap potential (related to the cross-tail potential) derived from an empirical relation between that potential and solar wind parameters. Based on the motion of particles within the magnetosphere, the currents generated by the curvature and gradient drift components of the motions (diamagnetic currents are not included), and the resulting variations in the ionospheric conductivities, the simulation computes the electric fields and the particle motions within the magnetosphere self-consistently. A very key assumption for us in the model is that there exists a continuous source of particles, ostensibly the plasma sheet population, that exists at a 10 R_e radial boundary.

Because of the enhanced global electric fields, the particles from the 10 R_e boundary are driven earthward very substantially. Based on the currents that result from the distribution of particles that are driven to the inner regions, the depression in the ground-based magnetic strength can be calculated and compared with actual observations. Figure 13 shows that there is a good agreement between the model (theory) and the data. A key conclusion of the simulation results is that inductive electric fields (calculated from the continuous, solar wind-driven scale changes in the chosen global magnetic field model) are small as compared to the curl-free electric fields and play almost no role in driving the plasma sheet populations earthward.

The results of Wolf et al. [1982] suggest that, in terms of the generation of the populations that contribute predominantly to the ring current, the impulsive, localized, inductively driven transport processes on which we have spent much time may, in fact, be irrelevant. A counter argument to that suggestion is that, in the simulation model, it is artificial to simply hypothesize the presence of a source population at 10 R_e. The evidence discussed in section 3 suggests that prestorm

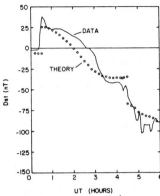

Figure 13: From Wolf et al. [1982]. Comparison of measured, ground-based magnetic Dst values (data) with the values computed using a global computer simulation (theory).

quasi-stationary convection does not necessarily provide for such a population. Additionally, it is well known that if quasi-stationary convection does provide for such a population, the spatial distribution will be quite different from being simply a sharp boundary at $\simeq 10\ R_e$ [Kivelson et al., 1980]. What is more, that spatial distribution will give rise to specific dispersive signatures within the inner and middle regions when the global, curl-free convective fields are enhanced during magnetic storm periods [again see Kivelson et al., 1980]. When the full ion-electron dispersive signatures within the geostationary environment are examined, it becomes clear that the expected dispersive signatures are not observed [sections 3 and 4, and Mauk and Meng, 1983a, b]. Finally, we note that the inductive electric field strengths calculated by Wolf et al. [1982] are about two orders of magnitude smaller than those observed by Shepard et al. [1980] and Aggson et al. [1983].

We suggest that the generation of the ring-current population is a two-stage process, and that Wolf et al. [1982] have done a good job of modeling the second stage. The first stage corresponds to the generation of the population that the Wolf et al. [1982] model takes for granted. From the evidence that we have presented in this review, it would appear that the generation (transport and energization) of particles within the outer and middle (geosynchronous) regions is always associated with impulsive, relatively localized, and inductively driven processes. The ring-current populations are then generated by the subsequent convective motions of this [using the words of McIlwain, 1974] newly reconstituted population, driven by the enhanced, global, curl-free electric fields modeled by the Wolf et al. [1982] simulations. We note that ground magnetometers will be most sensitive to the second stage of the ring-current generation, a fact that could explain the success of the simulations in reproducing ground measurements. A two-stage ring-current generation process would accommodate all of the observations that have been presented in this review.

7. Discussion

In this paper, we have examined the formation of the ring-current populations as an example of the operation of macroscopic ion acceleration processes within the earth's magnetosphere. Our point of view has been that macroscopic ion acceleration processes when viewed as isolated entities (e.g., convective transport, adiabatic energization) are relatively simple to understand, whereas, when those processes become components in the workings of complex systems, they do not operate as simply as their underlying structures would imply (e.g., impulsive, localized energization, and transport). Mirroring this statement, we have discussed how models for the formation of the ring current might be constructed by simple combinations of the concepts of adiabatic energization and the inward convection (of the ever-present tail populations) driven by enhanced, curl-free global electric fields. We have then shown that, when the full consequences of these concepts are examined [e.g., Erickson and Wolf, 1980; Schindler and Birn, 1983], and when observational evidences are examined fully, the formation of the ring-current populations is not as simple a process as these models would suggest. We then discussed the characteristics of the mechanisms that must be acting in concert with the simple mechanisms in order to explain the observational features of ion and electron transport within the middle (geosynchronous) to outer regions of the magnetosphere.

We have not discussed the controversial [e.g., Lui, 1980] model that injections are initiated by the formation of a reconnection X-line in the near-earth tail regions, $\simeq 15\ R_e$ [see Hones, 1984]. We argue only that earthward observations would place certain constraints on the process by which the X-region acts as a driver of the injection process. Because of geosynchronous injection observations and the observations presented by Moore et al. [1981] and Aggson et al. [1982], the X-region must generate some kind of self-sustaining electromagnetic structure that propagates toward the earth and performs the energization and transport. The inductive electric fields generated at the X-line probably do not penetrate directly (i.e., instantaneously) into the middle (geosynchronous) regions. It has been argued that the generation of particles with energies much higher than those considered in this report (> 1 MeV) may well require the formation of a near-earth, reconnection X-line [Pellinen and Heikkla, 1978; Baker et al., 1979; Sato et al., 1982]. If the X-line exists, it remains an important problem of how the energy is coupled into the near-earth energization process in a manner that explains the localized character of that energization (that character being the formation of sharp spatial boundaries between disturbed and undisturbed plasma and the localized generation of inductive electric fields).

As a final point, we would like to discuss the apparent contradictions inherent in the claim (as discussed in sections 2 and 3) that the convection patterns shown in Fig. 1 are qualitatively correct, and, at the same time, that the middle regions of the magnetosphere are not connected to the tail regions by means of the quasi-time-stationary convection streamlines. We believe, as is also stated by Kivelson et al. [1980], that during relatively quiet times, the separatrix boundary in Fig. 1 (if it exists as shown) lies well outside of the geosynchronous orbit region. During enhanced periods of activity, the separatrix region is driven to, or is formed at, the middle to inner regions, and in the absence of other processes coming into play, the ion and electron signatures would reflect the character of the quasi-stationary convection patterns. However, it appears that in one-to-one correlation with the activity that drives the separatrix boundary to the earthward regions is the initiation of the impulsive, localized, inductively driven processes that we have discussed so extensively. These impulsive processes dictate the character of the ion and electron signatures, and yet in between the occurrences of the impulses, the quasi-stationary convection patterns serve to determine the subsequent evolutions of the newly transported and energized plasmas.

Acknowledgments. We thank A. T. Y. Lui and D. G. Mitchell for helpful discussions. This work was supported by the Atmospheric Sciences Division, National Science Foundation grant ATM-8315041, by the Air Force Office of Scientific Research grant 84-0049, and by NASA contract to The Johns Hopkins University Applied Physics Laboratory and Department of the Navy under task I2U0S1P of contract N00024-85-C-5301.

References

Aggson, T. L., J. P. Heppner, and N. C. Maynard, Observations of large magnetospheric electric fields during the onset phase of a substorm, *J. Geophys. Res., 88*, 3981, 1983.

Akasofu, S.-I., editor, *Dynamics of the Magnetosphere*, D. Reidel, Hingham, Mass., 1980.

Baker, D. N., R. D. Belian, P. R. Higbie, and E. W. Hones, Jr., High-energy magnetospheric protons and their dependence on geomagnetic and interplanetary conditions, *J. Geophys. Res.*, 84, 7183, 1979.

Baumjohann, W., G. Haerendel, and F. Melzner, Magnetospheric convection between 0600 and 2100 LT: variations with Kp, *J. Geophys. Res.*, 90, 393, 1985.

Birmingham, T. J., Convection electric fields and diffusion of trapped magnetospheric radiation, *J. Geophys. Res.*, 74, 2169, 1969.

Brewer, H. R., M. Schulz, and A. Eviatar, Origin of drift-periodic echoes in outer-zone electron flux, *J. Geophys. Res.*, 74, 159, 1969.

Chen, A. J., Penetration of low-energy protons deep into the magnetosphere, *J. Geophys. Res.*, 75, 2458, 1970.

Chiu, Y. T., and A. M. Kishi, Kinetic model of auroral plasma formation by magnetospheric convection and injection, 1. Electrons, *J. Geophys. Res.*, 89, 5531, 1984.

Cornwall, J. M., Radial diffusion of ionized helium and protons: A probe of magnetospheric dynamics, *J. Geophys. Res.*, 77, 1756, 1972.

Cowley, S. W. H., and M. Ashour-Abdalla, Adiabatic plasma convection in a dipole field; proton forbidden zone effects for a simple electric field model, *Planet. Space Sci.*, 24, 821, 1976.

DeForest, S. E., and C. E. McIlwain, Plasma clouds in the magnetosphere, *J. Geophys. Res.*, 76, 3587, 1971.

Ejiri, M., R. A. Hoffman, and P. H. Smith, Energetic particle penetration into the inner magnetosphere, *J. Geophys. Res.*, 85, 653, 1980.

Erickson, G. M., and R. A. Wolf, Is steady convection possible in the earth's magnetotail? *Geophys. Res. Lett.*, 7, 897, 1980.

Fairfield, D. H., and A. F. Viñas, The inner edge of the plasma sheet and the diffuse aurora, *J. Geophys. Res.*, 89, 841, 1984.

Fälthammar, C.-G., Effects of time-dependent electric fields on geomagnetically trapped radiation, *J. Geophys. Res.*, 70, 2503, 1965.

Fennell, J. F., D. R. Croley, Jr., and S. M. Kaye, Low-energy ion pitch angle distributions in outer magnetosphere, ion zipper distributions, *J. Geophys. Res.*, 86, 3375, 1981.

Grebowsky, J. M., Model study of plasmapause motion, *J. Geophys. Res.*, 75, 4329, 1970.

Greenspan, M. E., D. J. Williams, B. H. Mauk, and C.-I. Meng, Ion and electron energy dispersion features detected by ISEE-1, *J. Geophys. Res.*, 90, 4079, 1985.

Hones, E. W., Jr., Plasmasheet behavior during substorms, in *Magnetic Reconnection in Space and Laboratory Plasmas*, AGU Geophysical Monograph 30, edited by E. W. Hones, Jr., p. 178, American Geophysical Union, Washington, D.C., 1984.

Horwitz, J. L., Conical distributions of low-energy ion fluxes at synchronous orbit, *J. Geophys. Res.*, 85, 2057, 1980.

Horwitz, J. L., Relationship of dusk-sector electric field to energy dispersion at the inner edge of the electron plasma sheet for nonequatorially mirroring electrons, *J. Geophys. Res.*, 89, 10865, 1984.

Hultqvist, B., H. Borg, L. A. Holmgren, H. Reme, A. Bahnsen, M. Jespersen, and G. Kremser, Quiet-time convection electric field properties derived from keV electron measurements at the inner edge of the plasma sheet by means of GEOS 2, *Planet. Space Sci.*, 30, 261, 1982.

Kavanagh, L. D., Jr., J. W. Freeman, Jr., and A. J. Chen, Plasma flow in the magnetosphere, *J. Geophys. Res.*, 73, 5511, 1968.

Kaye, S. M., and M. G. Kivelson, Time dependent convection electric fields and plasma injection, *J. Geophys. Res.*, 84, 4183, 1979.

Kaye, S. M., E. G. Shelley, R. D. Sharp, and R. G. Johnson, Ion composition of zipper events, *J. Geophys. Res.*, 86, 3383, 1981.

Kivelson, M. G., S. M. Kaye, and D. J. Southwood, The physics of plasma injection events, in *Dynamics of the Magnetosphere*, edited by S.-I. Akasofu, pp. 385-405, D. Reidel, Hingham, Mass., 1980.

Konradi, A., C. L. Semar, and T. A. Fritz, Substorm-injected protons and electrons and the injection boundary model, *J. Geophys. Res.*, 80, 543, 1975.

Konradi, A., C. L. Semar, and T. A. Fritz, Injection boundary dynamics during a geomagnetic storm, *J. Geophys. Res.*, 81, 3851, 1976.

Lanzerotti, L. J., C. G. Maclennan, and M. F. Robbins, Proton drift echoes in the magnetosphere, *J. Geophys. Res.*, 76, 259, 1971.

Lennartsson, W., and D. L. Reasoner, Low-energy plasma observations at synchronous orbit, *J. Geophys. Res.*, 83, 2145, 1978.

Lui, A. T. Y., Observations on plasmasheet dynamics during magnetospheric substorms, in *Dynamics of the Magnetosphere*, edited by S.-I. Akasofu, p. 563, D. Reidel, Hingham, Mass., 1980.

Lyons, L. R., and R. M. Thorne, Equilibrium structure of radiation belt electrons, *J. Geophys. Res.*, 78, 2142, 1973.

Lyons, L. R., and D. J. Williams, A source for the geomagnetic storm main phase ring current, *J. Geophys. Res.*, 85, 523, 1980.

Mauk, B. H., and C. E. McIlwain, Correlation of Kp with the substorm-injected plasma boundary, *J. Geophys. Res.*, 79, 3193, 1974.

Mauk, B. H., and C. E. McIlwain, UCSD auroral particles experiment, *IEEE Trans. Aerosp. Electron. Syst.*, AES-11, 1125, 1975.

Mauk, B. H., and C. E. McIlwain, Characterization of geostationary particle signatures based on the "injection boundary" model, *J. Geophys. Res.*, 88, 3055, 1983a.

Mauk, B. H., and C.-I. Meng, Dynamical injections as the source of near geostationary quiet time particle spatial boundaries, *J. Geophys. Res.*, 88, 10011, 1983b.

Mauk, B. H., Quantitative modeling of the convection surge process, American Geophysical Union Chapman Conference on Magnetotail Physics, The Johns Hopkins University Applied Physics Laboratory, Laurel, Md., 28-31 October 1985.

McIlwain, C. E., Plasma convection in the vicinity of the geosynchronous orbit, in *Earth's Magnetospheric Processes*, edited by B. M. McCormac, p. 268, D. Reidel, Hingham, Mass., 1972.

McIlwain, C. E., Substorm injection boundaries, in *Magnetospheric Physics*, edited by B. M. McCormac, p. 143, D. Reidel, Hingham, Mass., 1974.

McIlwain, C. E., Equatorial magnetospheric particles and auroral precipitations, submitted, *J. Geophys. Res.*, 1985.

Moore, T. E., R. L. Arnoldy, J. Feynman, and D. A. Hardy, Propagating substorm injection fronts, *J. Geophys. Res.*, 86, 6713, 1981.

Pellinen, R. J., and W. J. Heikkila, Energization of charged particles to high energies by an induced substorm electric field within the magnetotail, *J. Geophys. Res.*, 83, 1544, 1978.

Quinn, J. M., and R. G. Johnson, Observation of ionospheric source cone enhancements at the substorm injection boundary, *J. Geophys. Res.*, 90, 4211, 1985.

Quinn, J. M., and C. E. McIlwain, Bouncing ion clusters in the earth's magnetosphere, *J. Geophys. Res.*, 84, 7365, 1979.

Quinn, J. M., and D. J. Southwood, Observations of parallel ion energization in the equatorial region, *J. Geophys. Res.*, 87, 10536, 1982.

Russell, C. T., and R. L. McPherron, The magnetotail and substorms, *Space Sci. Rev.*, 15, 205, 1973.

Sato, T., H. Matsumoto, and K. Nagai, Particle acceleration in time-developing magnetic reconnection process, *J. Geophys. Res.*, 87, 6889, 1982.

Schindler, K., and J. Birn, Self-consistent theory of time-dependent convection in the earth's magnetotail, *J. Geophys. Res.*, 87, 2263, 1982.

Shepard, G. G., R. Bostrom, H. Derblom, C.-G. Fälthammar, R. Gendrin, K. Kaila, A. Korth, A. Pedersen, R. Pellinen, and G. Wrenn, Plasma and field signatures of poleward propagating auroral precipitation observed at the foot of the GEOS-2 field line, *J. Geophys. Res.*, 85, 4587, 1980.

Southwood, D. J., and S. M. Kaye, Drift boundary approximations in simple magnetospheric convection models, *J. Geophys. Res., 84*, 5773, 1979.

Spjeldvik, W. N., Equilibrium structure of equatorially mirroring radiation belt protons, *J. Geophys. Res., 82*, 2801, 1977.

Stern, D. P., Large-scale electric fields in the earth's magnetosphere, *Rev. Geophys. Space Phys., 15*, 156, 1977.

Strangeway, R. J., and R. G. Johnson, On the injection boundary model and dispersing ion signatures at near-geosynchronous altitudes, *Geophys. Res. Lett., 10*, 549, 1983.

Walker, R. J., and M. G. Kivelson, Energization of electrons at synchronous orbit by substorm-associated cross-magnetosphere electric fields, *J. Geophys. Res., 80*, 2074, 1975.

Williams, D. J., The earth's ring current: causes, generation and decay, *Space Sci. Rev., 34*, 223, 1983.

Wolf, R. A., M. Harel, R. W. Spiro, G.-H. Voigt, P. H. Reiff, and C.-K. Chen, Computer simulations of inner magnetospheric dynamics for the magnetic storm of July 29, 1977, *J. Geophys. Res., 87*, 5949, 1982.

Young, D. T., Ion composition measurements in magnetospheric modeling, in *Quantitative Modeling of Magnetospheric Processes*, edited by W. P. Olson, p. 340, Geophysical Monograph 21, American Geophysical Union, 1979.

ION ACCELERATION IN EXPANDING IONOSPHERIC PLASMAS

Nagendra Singh and R. W. Schunk

Center for Atmospheric and Space Sciences, Utah State University, Logan, Utah 84322-3400

Plasma expansion along the ambient magnetic field in regions of density gradients provides a mechanism for accelerating ions. A brief review of the basic phenomenon of plasma expansion is given. Estimates of the energies of the accelerated ions in an expanding ionospheric plasma along geomagnetic flux tubes are obtained by solving the time-dependent hydrodynamic equations. It is found that over certain altitude ranges each ion species can be the most energetic; the maximum energies (W) of the different ions are found to be limited to $W(H^+) \lesssim 10$ eV, $W(He^+) \lesssim 5$ eV, and $W(O^+) \lesssim 1.5$ eV.

Introduction

Satellite observations indicate that the ions in the magnetosphere of ionospheric origin are much more energetic than those in the ionosphere [Horwitz and Chappell, 1979; Singh et al., 1982; Sharp et al., 1977]. There are several mechanisms which have been proposed for the ion acceleration. Some of these involve wave-particle interactions along auroral field lines, as discussed in several papers in this monograph. In this paper we discuss another mechanism whereby ions are accelerated by a dc parallel electric field set up in the process of plasma expansion along magnetic field lines, but this mechanism is not limited to the auroral region. The process of ion acceleration in expanding plasmas has been known since the pioneering work of Gurevich et al. [1966]. Recently, Singh and Schunk [1982, 1983] suggested that such a mechanism could be responsible for producing the superthermal ions in the polar wind and in interhemispheric plasma flows after magnetic storms. In these studies, both single and multi-ion ionospheric plasma expansions were modelled by solving kinetic equations. Thus, the scale lengths of the expanding plasmas were severely limited. More recently, macroscopic single-ion plasma expansions along geomagnetic flux tubes were studied by solving the hydrodynamic equations for the plasma [Singh and Schunk, 1985]. In this paper, we present results from similar macroscopic calculations on expanding multi-ion ionospheric plasmas. We focus our attention on the ion-acceleration aspect of the expansion.

Basic Phenomenon of Plasma Expansion

Figure 1 shows a schematic of the initial plasma density configuration that has been frequently used in studies involving expanding plasmas. At $t = 0$, the half-space $x < 0$ is filled with a semi-infinite, electrically neutral, collisionless plasma. For $t > 0$, the plasma is allowed to expand into the vacuum and the subsequent temporal evolution is followed. The simplest treatment of the plasma expansion is obtained by assuming that the electrons always stay in an equilibrium with the developed electric fields; thus they obey the Boltzmann distribution,

$$N_e(x) = Z_i N_0 \exp\left[e\phi(x)/k_B T_e\right] \quad (1)$$

where N_e denotes the electron density, N_0 is the ion density in the unperturbed plasma, T_e is the electron temperature, k_B is the Boltzmann constant, Z_i is the ion charge number, and ϕ is the electrostatic potential. Assuming that the ions are cold, the ion continuity and momentum equations reduce to

$$\frac{\partial N_i}{\partial t} + \frac{\partial}{\partial x}(N_i V_i) = 0, \qquad \frac{\partial V_i}{\partial t} + V_i \frac{\partial V_i}{\partial x} = -\frac{Z_i e}{M_i}\frac{\partial \phi}{\partial x} \quad (2)$$

where V_i is the ion flow velocity, M_i is the ion mass, and N_i is its density.

If quasi-neutrality is assumed, the set of equations (1) and (2) allow self-similar solutions, which depend only on the ratio (x/t) of the independent variables x and t;

$$N_e = Z_i N_i = Z_i N_0 \exp\left[-(\xi+1)\right], \qquad V_i = C_s(\xi+1)$$

$$\phi = -(k_B T_e/e)(\xi+1), \qquad E = -\frac{d\phi}{dx} = \frac{k_B T_e}{e}\frac{1}{C_s t} H(\xi+1) \quad (3)$$

for $(\xi + 1) \geq 0$, where $\xi = (x/C_s t)$ is the self-similar variable and $C_s = (k_B T_e/M_i)^{1/2}$. H is the step function and E is the electric field. For $\xi + 1 < 0$, the plasma remains unperturbed.

Figure 1 shows the characteristics of the self-similar solution for the collisionless expansion of a single-ion plasma into a vacuum. As the expansion proceeds, a rarefaction wave propagates into the plasma at the ion-acoustic speed C_s. The polarization electrostatic field does not vary with position when $\xi \geq -1$, but its magnitude decreases inversely with time (see Equation (3)). Because of this electric field ion acceleration occurs, and depending on the electron temperature, energetic ions can be produced.

The basic assumptions in deriving the self-similar solutions given by the equations in (3) are as follows: (1) the ions are cold, (2) the electrons obey the Boltzmann distribution, and (3) charge neutrality prevails ($N_e = Z_i N_i$). The first assumption was examined by Gurevich et al. [1968], who obtained self-similar solutions by using the ion Vlasov equation in place of the continuity and momentum equations. The major modification due to the nonzero ion temperature is introduced near $\xi = -1$; the discontinuity at $x = -C_s t$ is smoothed out.

In the studies by Denavit [1979] and Mora and Pellat [1979], the assumption that the electrons obey the Boltzmann distribution was examined. These investigations showed that the self-similar solutions are valid as long as

$$\xi = \frac{x}{C_s t} \ll \left(\frac{M_i}{Z_i M_e}\right)^{1/2} \quad (4)$$

Physically, this condition implies that as long as the electron transit time across an expansion region with the electron thermal speed is much shorter than the expansion time with a speed C_s, the Boltzmann distribution is justified.

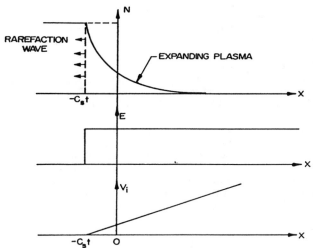

Fig. 1. Self-similar solutions for plasma expansion; density profile (top), electric field profile (middle), and velocity profile (bottom). Initially ($t = 0$), plasma occupied the region $x \leq 0$.

We now examine the third assumption about charge neutrality, which is valid when the scale length for the variation of the potential $L \gg \lambda_d$, where λ_d is the Debye length. The scale length L from the self-similar solution is easily found to be $L \simeq C_s t$. Since $\lambda_d \sim (T_e/N_e)^{1/2}$ and T_e remains nearly constant, using the expression for N_e in (3) and equating λ_d with $L \simeq C_s t$, the condition for validity of the self-similar solution is found to be

$$\xi \leq \xi_s = 2ln[\omega_{i0}t] - 1 \tag{5}$$

where ω_{i0} is the ion plasma frequency in the undisturbed plasma [Mora and Pellat, 1979]. We note that for values of ξ approximately in excess of ξ_s given by (5), the potential given by the self-similar solution (3) is not valid. Owing to the limitations imposed by equations (4) and (5), the self-similar solutions have a spatial-temporal restriction that can be expressed as $-C_s t \leq x \leq [2ln(\omega_{i0}t) - 1] C_s t$. Therefore, the self-similar solutions are valid over a finite range of x for $t \gtrsim \omega_{i0}^{-1}$. Physically, this is the time it takes the ions to respond to the rapid electron expansion and produce a quasi-neutral plasma flow over a limited extent.

In the region near $\xi = \xi_s$, where quasi-neutrality is violated, a positive-negative spacecharge separation occurs, causing the formation of a steep ion density front [Crow et al., 1975]. The speed of the front (V_F) can be estimated by substituting (5) into the relation $V_i = C_s(\xi + 1)$,

$$V_F \simeq 2C_s ln(\omega_{i0}t) \tag{6}$$

Recently, Singh and Schunk [1984] suggested that the moving ion front heading the expanding plasma is similar to the phenomenon of current-free double layers [Perkins and Sun, 1981]. We also note the limitation on ξ given by (4). Combining (4) and (6), we obtain the limiting value of the ion front velocity and the maximum acceleration of the ions;

$$V_i \ll \text{electron thermal velocity} \tag{7}$$

The above review of the phenomenon of plasma expansion is very brief. Further details can be found in Singh and Schunk [1982] and Samir et al. [1983].

Multi-ion Plasma Expansion Along an Open Field Line

We solve the time-dependent continuity and momentum equations for the ions along a magnetic flux tube;

$$\frac{\partial N_s}{\partial t} + \frac{1}{A}\frac{\partial}{\partial r}(AN_s V_s) = 0 \tag{8}$$

$$\frac{\partial V_s}{\partial t} + V_s \frac{\partial V_s}{\partial r} = -\frac{1}{M_s}\frac{1}{N_s}\frac{\partial P_s}{\partial r} + \frac{e}{M_s}E - g(r) \tag{9}$$

$$P_s = N_s k_B T_s, \qquad P_s N_s^{-\gamma} = \text{constant} \tag{10}$$

where t denotes time, r is the geocentric distance along a flux tube (see Figure 2), N_s and V_s are the number density and drift velocity of the ion species s, M_s is its mass, T_s is its temperature, $g(r)$ is gravity, P_s is pressure, k_B is the Boltzmann constant, and e is the magnitude of the electron charge. We assumed that the ion gas is isothermal ($\gamma = 1$). In (8), A is the cross sectional area of the magnetic flux tube. Near the poles of the Earth's magnetic field, $A \propto r^3$. The electric field E is calculated by assuming that the plasma is quasi-neutral and that the electrons obey the Boltzmann law. Accordingly, the electric field is given by

$$E = -\frac{d\phi}{dr} = -\frac{k_B T_e}{e}\frac{d}{dr} ln(\sum_s N_s/N_0) \tag{11}$$

where N_0 is the total plasma density at a reference location $r = r_0$ where the electrostatic potential ϕ is assumed to be zero, and the sum is over the ion species. We note that the assumption of quasi-neutrality is valid as long as we are interested in phenomena with spatial scale lengths $L \gg \lambda_d$, the plasma Debye length. In this paper we are concerned with such large-scale phenomena associated with transients in the outflow of plasma from the terrestrial ionosphere. The set of equations (8) to (11) are solved numerically using the flux-corrected-transport (FCT) technique [Boris and Book, 1976]. The details of the numerical procedure can be found in Singh and Schunk [1985].

We remark that there are a few recent works on large scale plasma flow along the geomagnetic flux tubes using the time-dependent fluid description of the plasma [Mitchell and Palmadesso, 1983; Khazanov et al., 1984; Gombosi et al., 1985]. However, the topics of emphasis in these papers are quite different from that in the present paper. Our main emphasis in the present paper is on ion acceleration in plasma expansions along the geomagnetic flux tubes.

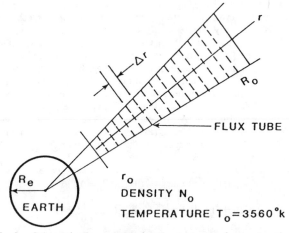

Fig. 2. Schematic diagram showing the geometry of our calcuations. An ionospheric boundary condition is imposed at $r = r_0$. Δr is the spatial grid size. In the present calculations $\Delta r = 37.5$ km and the corresponding time step is $\Delta t = 0.1$ second.

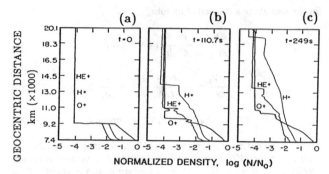

Fig. 3. Density profiles of O^+, He^+, and H^+ at three times. The profiles at $t = 0$ show the sudden depletion of the plasma for $r > 9188$ km. The profiles at later times show the expanding plasma.

The hydrodynamic model of the plasma provides a fairly good description of the laminar features of the expansion. Of course, some kinetic effects such as the excitation of certain wave modes and a proper treatment of discontinuities which may develop in the expansion are missing from this model. It is worth pointing out that in plasma expansion problems, where ions are expected to be highly supersonic, the pressure term in the ion momentum equation does not play a crucial role. Thus, we chose the simplest model for the ion temperature variation, namely, the ions remain isothermal. This assumption is justified for light ions such as He^+ and H^+, but may be questionable for heavy O^+ ions.

In our calculations, the base of the ionosphere is chosen to be $r = r_o = 7370$ km (see Figure 2) and the upper boundary is placed at $r = R_o = 20 \times 10^3$ km. Starting with arbitrary density profiles for O^+, H^+, and He^+ and with a prescribed relative concentration of these ions at the ionospheric base, the quasi-steady state density and velocity profiles are calculated for these ions. We qualify the profiles to be quasi-steady because the O^+-density and velocity profiles reach their steady states very slowly. The quasi-steady state density profiles at some instant of time, $t = 0$, are perturbed by depleting the plasma above the geocentric distance $r \simeq 9188$ km, while the velocity profiles are left unchanged; Figures 3a and 4a show the density and velocity profiles of the ions at $t = 0$. We note that in the calculations the relative concentration of the ions at the base ($r = r_o$) is taken to be $N_o(O^+)/N_o = 80\%$, $N_o(H^+)/N_o = 15\%$, and $N_o(He^+)/N_o = 5\%$.

The temporal evolution of the expansion of these ion species is shown in panels b and c of Figures 3 and 4, which show, respectively, the density and velocity profiles at times $t_1 = 110.7$ and $t_2 = 249$ seconds. We note from these figures that the expansion of each ion species is headed by a density front, which appears as a shock. Such shocks have been extensively studied in connection with the transient expansion of the solar wind. A further discussion of the shocks can be found in Hundhausen [1973] and in Singh and Schunk [1985]. Near the density fronts, velocity fronts also form. A considerable acceleration of all the ion-species occurs behind these fronts (Fig. 4b and 4c). The maximum flow velocities for the various ion species and the corresponding energies (W) are found to be as follows:

$$V_{max}(H^+) \simeq 43 \text{ km/s}, \quad W_{max}(H^+) \simeq 9.4 \text{ eV}$$

$$V_{max}(He^+) \simeq 16 \text{ km/s}, \quad W_{max}(He^+) \simeq 5.2 \text{ eV} \quad (12)$$

$$V_{max}(O^+) \simeq 4 \text{ km/s}, \quad W_{max}(O^+) \simeq 1.3 \text{ eV}$$

The spatial regions of these maximum accelerations move along with the moving density and velocity fronts.

We note that the energies given above are considerably larger than the 0.33 eV energy in the ionosphere. Thus, the relative maximum energization of the ions when compared with the ionospheric ion temperature T_i (~0.33 eV) is as follows: $W_{max}(H^+)/T_i \simeq 29$, $W_{max}(He^+)/T_i \simeq 16$, and $W_{max}(O^+)/T_i \simeq 4$. It is seen that the light H^+ ions undergo the maximum acceleration and the heaviest O^+ ions undergo the least acceleration. However, Figures 4b and 4c show an interesting feature regarding the relative energies of the ions in the expansion region; the maximum energization of the different ion species occurs at different altitudes. For example, we note from panel c of Figure 4 that H^+ has its maximum energy at $r \simeq 17400$ km, He^+ at $r \sim 13000$ km and O^+ at $r \simeq 10100$ km. The occurrence of the maximum ion energization at different altitudes, depending on the ion mass, is shown in Figures 5a and 5b by plotting the energies of the ions as a function of geocentric distance at the times $t = 110.7$ and 221.3 seconds. A comparison of Figures 5a and 5b shows that the regions of the maximum energization move upward.

The noteworthy feature of Figures 5a and 5b is that a heavy ion can be more energetic than a light ion over a certain altitude range. For example, in Figure 5a O^+ is the most energetic ion at $r = 9718$ km, He^+ is the most energetic at $r = 10797$ km and H^+ at $r = 12548$ km. Figure 5b shows that at a later time the heaviest O^+ ions cease to be the most energetic ion at any altitude, while this feature for He^+ ions persists. This difference between O^+ and He^+ ions arises because of the gravitational force. At early times, when the O^+ density gradients are large, acceleration by the polarization electric field dominates the gravitational acceleration. However, at later times as the O^+ density gradient decreases, the downward gravitational force becomes dominant and the acceleration of the O^+ ions by the polarization field ceases. On the other hand, in the case of light ions, such as He^+ and H^+, the gravitational force is found to have a negligible effect.

Recent DE-1 observations show interesting examples of field-aligned flows of He^+ and H^+ ion streams in the depleted dayside magnetosphere [Sojka et al., 1983]. These ion streams are the result of expanding ionospheric plasma. We find some interesting similarities between the satellite observations and the theoretical calculations discussed here. The observations suggest that the average parallel drift velocities of the ion streams were < 30 km/s for H^+ and ≤ 15 km/s for He^+. These drift velocities are close to those shown for the respective ions behind the He^+ expansion front in Figures 4b and 4c. It is important to note that above the He^+ expansion front, the He^+ density is expected to be negligibly small. Thus, if He^+ and H^+ are found to have densities of the same order of magnitude, as is the case in the observations reported by Sojka et al. [1983], it is quite possible that the observed streams were below the He^+ expansion front. Our calculations show that these streams could have

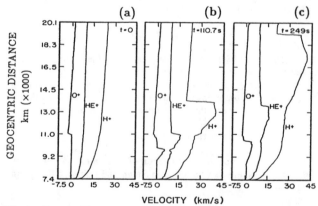

Fig. 4. Same as Figure 3, but the velocity profiles are shown. When the density profiles are perturbed at $t = 0$ by causing the depletion, the velocity profiles are left unchanged.

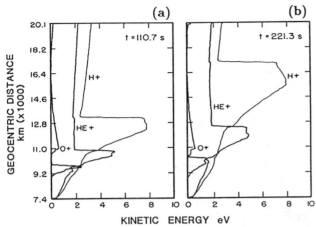

Fig. 5. Kinetic energy of the ions as a function of the geocentric distance are shown at two times to emphasize that each ion species has an altitude range over which it can be the most energetic. This range moves upward as the expansion proceeds.

been highly supersonic. In this connection, it is worthwhile to note that observationally it was not possible to unambiguously determine the partition of the ion energies into thermal and streaming components, and hence, uncertainty arose as to whether the streams were supersonic or subsonic [Sojka et al., 1983]. Just below the He^+ expansion front, the H^+ and He^+ Mach numbers can be as high as $M \simeq V_i/V_t(\alpha) \simeq 4$, where $V_t(\alpha)$ is the thermal velocity of the ions in the ionosphere. Our calculations predict that the streams should have been supersonic.

Acknowledgement. This research was supported by NASA grant NAGW-77 and National Science Foundation grant ATM-8417880 to Utah State University.

References

Boris, J. D., and D. L. Book, Solution of continuity equations by the method of flux-corrected transport, in *Methods in Computational Physics*, vol. 16, edited by B. Alder, S. Fernbach, and M. Rotenberg, p. 85, Academic, New York, 1976.

Crow, J. E., P. L. Auer, and J. E. Allen, The expansion of plasma into a vacuum, *J. Plasma Phys.*, 14, 65, 1975.

Denavit, J., Collisionless plasma expansion into a vacuum, *Phys. Fluids*, 22, 1384, 1979.

Gombosi, T. I., T. E. Cravens, and A. F. Nagy, A time-dependent theoretical model of the polar wind: Preliminary results, *Geophys. Res. Lett.*, 12, 167, 1985.

Gurevich, A. V., L. V. Pariiskaya and L. P. Pitaevskii, Self-similar solution of rarefied plasma, *Sov. Phys. JETP Eng. Transl.*, 22, 449, 1966.

Gurevich, A. V., L. V. Pariiskaya and L. P. Pitaevskii, Self-similar solutions of a low-density plasma, II, *Sov. Phys. JETP Eng. Transl.*, 27, 476. 1968.

Horwitz, J. L., and C. R. Chappell, Observations of warm plasma in the dayside plasma trough at geosynchronous orbit, *J. Geophys. Res.*, 84, 7075, 1979.

Hundhausen, A. J., Nonlinear model of high-speed solar wind streams, *J. Geophys. Res.*, 78, 1528, 1973.

Khazanov, G. V., M. A. Koen, U. V. Konikov, and I. M. Sidonov, Simulation of ionosphere-plasmasphere coupling taking into account ion inertia and temperature anisotropy, *Planet. Space Sci.*, 32, 325, 1984.

Mitchell, H. G., Jr., and P. J. Palmadesso, A dynamical model of the auroral field line plasma in the presence of field aligned current, *J. Geophys. Res.*, 88, 2131, 1983.

Mora, P., and P. Pellat, Self-similar expansion of a plasma into a vacuum, *Phys. Fluids*, 22, 2300, 1979.

Perkins, F. W., and Y. C. Sun, Double layer without current, *Phys. Rev. Lett.*, 46, 115, 1981.

Samir, V., K. H. Wright, Jr., and N. H. Stone, The expansion of a plasma into a vacuum: basic phenomenon and processes and applications to space plasma physics, *Rev. Geophys. and Space Phys.*, 21, 1631, 1983.

Sharp, R. D., R. G. Johnson, and E. G. Shelley, Observation of an ionospheric acceleration mechanism producing energetic (keV) ions primarily normal to the geomagnetic field direction, *J. Geophys. Res.*, 82, 3324, 1977.

Singh, N., W. J. Raitt, and F. Yasuhara, Low-energy ion distribution functions on a magnetically quiet day at geostationary altitude ($L = 7$), *J. Geophys. Res.*, 87, 681, 1982.

Singh, N. and R. W. Schunk, Numerical calculations relevant to the initial expansion of the polar wind, *J. Geophys. Res.*, 87, 9154, 1982.

Singh, N. and R. W. Schunk, Numerical simulation of counterstreaming plasmas and their relevance to interhemispheric flows, *J. Geophys. Res.*, 88, 7867, 1983.

Singh, N. and R. W. Schunk, The relationship between the electric fields associated with plasma expansion and double layers, Proceedings of the *Second Symposium on Plasma Double Layers and Related Topics*, 272, 1984.

Singh, N. and R. W. Schunk, Temporal behavior of density perturbations in the polar wind, *J. Geophys. Res.*, 90, 6487, 1985.

Sojka, J. J., R. W. Schunk, J. F. E. Johnson, J. H. Waite, and C. R. Chappell, Characteristics of thermal and suprathermal ions associated with the dayside plasma trough as measured by the Dynamics Explorer retarding ion mass spectrometer, *J. Geophys. Res.*, 88, 7895, 1983.

TIME-DEPENDENT NUMERICAL SIMULATION OF HOT ION OUTFLOW FROM THE POLAR IONOSPHERE

T.I. Gombosi*, T.E. Cravens, A.F. Nagy

Space Physics Research Laboratory, The University of Michigan, Ann Arbor, MI 48109

J.H. Waite, Jr.

Space Science Laboratory, NASA Marshall Space Flight Center, Huntsville, AL 35812

Abstract. Large outflows of low energy (<10eV) O^+ ions have been observed by instruments carried onboard the DE-1 satellite over the polar cap. Convection mapping calculations suggest an origin of these ionospheric outflows near the dayside polar cap boundary. In order to assess the role of high altitude ion or electron heating on hot plasma outflows from the ionosphere we carried out a representative set of calculations using a time-dependent polar wind code. In the model calculations the coupled time-dependent continuity, momentum and energy equations of a two ion (H^+ and O^+) quasi-neutral plasma were solved between 200 and 8,000 km. In both cases a significant O^+ upwelling was found together with a modest increase of the upward H^+ flux. In the ion heating case the O^+ temperature reached its highest value near the peak time of the heating process. During this period both ion species were hotter than the electron gas. Approximately 30 minutes after the short duration (~5 minutes) heating was initiated the ion temperatures decreased below 5,000 K, simultaneously with the disappearance of the O^+ upwelling. In the electron heating case the ions remained much colder than the electrons at all times and as a result of the heating a downward O^+ flux developed after about 30 minutes. It is our conclusion that ion rather than electron heating is probably the major source of the observed O^+ outflow from the polar ionosphere.

Introduction

The existence of high speed ionospheric plasma outflows along open geomagnetic field lines was originally suggested by Axford (1968) and Banks and Holzer (1968). A large number of papers dealing with various aspects of the polar wind have since been published (for a recent review see Raitt and Schunk (1983)). Over the last decade a variety of approaches have been adopted, including hydrodynamic, magnetohydrodynamic, kinetic, and semikinetic models. However, even though both the formulation of the governing equations and the methods of solution have been improved significantly since the initial work, until recently polar wind studies have been based on steady-state calculations. Mitchell and Palmadesso (1983) and Singh and Schunk (1985) have published time varying high speed calculations of the upper, collisionless polar wind region: the plasma parameters at the lower boundary (located well above the ionospheric peak) were free input parameters of the calculations. A time-dependent hydrodynamical model of plasma motion along closed plasmaspheric field lines was recently developed by Khazanov et al. (1984), who were able to employ large time-steps (> 15 min) in this region. Gombosi et al. (1985) have recently developed a new hydrodynamical numerical code to study the time-dependent behavior of ionospheric plasma along open magnetic field lines and applied it to transonic polar wind flows.

The first quantitative measurements of supersonic polar wind behavior were obtained by the retarding ion mass spectrometer (RIMS) carried onboard the Dynamics Explorer 1 (DE-1) satellite, which observed supersonic H^+ (Nagai et al., 1984) and O^+ ion flows (Waite et al., 1985). Large outflows (> 10^8 cm^{-2}s^{-1}) of low energy (<10 eV) O^+ ions have also been observed by the DE-1 particle detectors over the polar caps (Shelley et al., 1982; Waite et al., 1985). Convection mapping calculations by Waite et al. (1985) suggested an origin of these ionospheric outflows near the dayside polar cap boundary. The source region appears to produce large, transient outflows of not only O^+ ions, but also H^+, He^+, N^+, O^{++} and molecular ions as well. The observed ion outflows, which have been termed "upwelling ion events" by Lockwood et al. (1985), show signatures of strong ion heating, which probably takes place in the polar cusp. Recently Moore et al. (1985) have analyzed an individual upwelling ion event (see also Waite et al., this monograph) and reported that large O^+ upward fluxes were detected in the cusp region. The observations took place at ~1.8 R_e, where large fluxes (> 10^8 cm^{-2}s^{-1}) of upward moving slightly subsonic (M~0.9), hot O^+ ions (the estimated temperature was several tens of thousand K) were detected in the vicinity of the polar cusp.

An earlier steady-state high altitude (z>4500km) semikinetic calculation (Barakat and Schunk (1983)) has indicated that high electron temperatures (T_e~10,000K) can produce highly supersonic large outflows of energetic (~2eV) O^+ ions.

In this paper we apply a time-dependent hydrodynamic model (Gombosi et al., 1985) to investigate the effects of bulk plasma heating on ion flows in the polar ionosphere. The model does not allow for temperature anisotropies, field

---

*also at Central Research Institue for Physics, Hungarian Academy of Sciences, Budapest, Hungary

aligned currents, collisionless heat flow effects, wave excitations, and wave-particle interactions or large scale parallel electric fields. There is considerable evidence that ions are preferentially heated in the perpendicular direction (ultimately forming conics) and that this population carries a significant part of the heat flow. The present model assumes that preferentially heated particles undergo rapid pitch-angle scattering (for instance on magnetic field fluctuations) and become isotropized, so the global effect of particle heating can be approximated as a bulk heating mechanism. Such a hydrodynamic model can give us an insight for the global dynamical behavior of the polar ionosphere but of course cannot provide information on details such as the temperature anisotropies. This is true even if there is no rapid scattering and strong anisotropies may develop. We are aware of the limitations of our present model and the next step in our efforts to develop increasingly sophysticated models will be to include the effects of some of the important kinetic plasma behavior and field aligned currents. In this paper we describe the results of a parametric study of the transient plasma flows generated by short duration ion/electron heat source around an altitude of 3,000 km using our present hydrodynamic model. We find that ion rather than electron heating seems to be responsible for the upwelling ion events. The bulk heating was applied for 5 minutes, which approximately corresponds to the convection time through the cusp region.

Governing Equations

The numerical model simultaneously solves the time-dependent hydrodynamic continuity, momentum and energy equations for O^+ and H^+ ions along vertical magnetic field lines. It is assumed that the plasma is quasineutral, and that the ion and electron gases can be considered to be perfect fluids. The governing equations were described in our previous paper (Gombosi et al., 1985). Unfortunately several misprints appeared in that paper, therefore here we summarize the equations again.

The ion equations are the following:

$$\frac{\partial}{\partial t}(A\rho_i) + \frac{\partial}{\partial z}(A\rho_i u_i) = A S_i \qquad (1.a)$$

$$\frac{\partial}{\partial t}(A\rho_i u_i) + \frac{\partial}{\partial z}(A\rho_i u_i^2) + A\frac{\partial p_i}{\partial z} =$$

$$A\rho(\frac{e}{m_i}E_z - g) + A\frac{\delta M_i}{\delta t} + A u_i S_i \qquad (1.b)$$

$$\frac{\partial}{\partial t}(\frac{1}{2}A\rho_i u_i^2 + \frac{1}{\gamma_i-1}A p_i) + \frac{\partial}{\partial z}(\frac{1}{2}A\rho_i u_i^3 +$$

$$\frac{\gamma_i}{\gamma_i-1}A u_i p_i) = A\frac{\delta E_i}{\delta t} + A Q_i + \frac{\partial}{\partial z}(A \kappa_i \frac{\partial T_i}{\partial z}) +$$

$$A\rho_i u_i(\frac{e}{m_i}E_z - g) + A u_i \frac{\delta M_i}{\delta t} + \frac{1}{2}A u_i^2 S_i \qquad (1.c)$$

where ρ_i, u_i, T_i and p_i are the mass density, velocity, temperature and pressure of the i-th ion species, respectively. γ_i is the specific heat ratio, A is the cross-sectional area of the magnetic flux tube (A ~ 1/B, where B is the magnetic field), e/m_i is the charge to mass ratio, κ_i is the heat conductivity, E_z is the field-aligned electric field, and g is the local gravitational acceleration. S_i, $\partial M_i/\partial t$ and $\partial E_i/\partial t$ are the total mass production, the momentum exchange and the energy exchange rates due to collisions, respectively. Q_i is the external ion heating rate. The adopted collision terms are the same as those used by Raitt et al. (1975).

In the model local charge neutrality and the absence of field-aligned electric currents are assumed everywhere. The electron momentum equation is used to determine the field-aligned field, E_z. The electron energy equation is solved in the following form:

$$\rho_e \frac{\partial T_e}{\partial t} = \frac{(\gamma_e-1)m_e}{k A}\frac{\partial}{\partial z}(A \kappa_i \frac{\partial T_i}{\partial z}) - \rho_e u_i \frac{\partial T_e}{\partial z}$$

$$- T_e [S_e + \frac{A'}{A}(\gamma_e-1)\rho_e u_e \frac{\partial u_e}{\partial z}] \qquad (2)$$

$$+ (\gamma_e-1)(Q_e + \frac{\delta E_e}{\delta t})$$

where ρ_e, u_e, T_e and p_e are the electron mass density, velocity, temperature and pressure, respectively. A' is the spatial derivative of the area function, γ_e is the electron gas specific heat ratio, k is the Boltzmann constant, m_e is the electron mass, while S_e and Q_e are the total mass production and external heating rates, respectively. The collision term ($\delta E_e/\delta t$) and the electron heat conductivity (κ_e) were taken from Raitt et al. (1975).

The model of the neutral upper atmosphere included N_2, O_2, O and H (Raitt et al., 1975). O^+ ions are produced by photoionization, while H^+ ions were created by charge transfer. O^+ is chemically removed by reactions with N_2 and O_2, and H^+ by charge exchange with O.

The model ionosphere was bounded at the top and bottom by two infinite, external reservoirs. The ions in the stationary lower reservoir (located at 200 km) were assumed to be in chemical and thermal equilibrium with the neutral atmosphere, while the electron temperature was 1000 K. The upper reservoir (z=8,000 km) was taken to be a stationary, low pressure medium. The pressure in the upper reservoir was a time dependent free parameter of the calculations, helping us to control the plasma discharge from the ionosphere. The numerical code also allowed for time dependent, distributed ion and electron heat sources, which were input parameters for the calculations. In the calculations described in this paper a topside electron heat flux of 9.10^{-3} erg $cm^{-2}s^{-1}$ was also used in addition to the distributed heat source.

The coupled time-dependent partial differential equation system was solved with a combined Godunov scheme/Crank-Nicholson method. The details of this numerical method will be given in a subsequent paper.

Results and Discussion

In order to assess the effects of plasma heating on heavy ion outflows in the polar ionosphere, we considered two cases. The

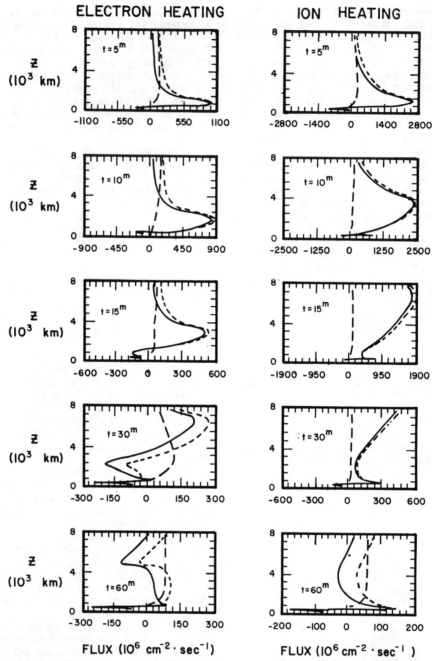

Fig 1. Temporal evolution of normalized particle fluxes (normalization altitude = 1000 km) obtained with electron and ion heating. O^+, H^+ ions and electrons are represented by solid, dashed, and dotted lines, respectively.

two cases differed only in the distributed plasma heating terms; all other processes were kept the same. The initial condition was a steady plasma outflow along open field lines. In both cases a Gaussian shaped heating profile (peak altitude = 3,000 km, half width = 250 km, total column heating rate = 0.1 ergs cm^{-2}s^{-1}) heated the ions and electrons, respectively. The heating was initiated at t=0 and increased to full strength over a time period of 5 minutes (the e-folding time was 150 sec) and then decreased with the same time constant. Five minutes is typical for the convection time of a flux tube through the polar cusp region into the polar cap itself. In the ion heating case the absorbed heat was at each altitude divided between H^+ and O^+ according to their local mass densities. In order to simulate open field lines the pressure in the topside reservoir was at all instances three times smaller than the topside ionospheric pressure.

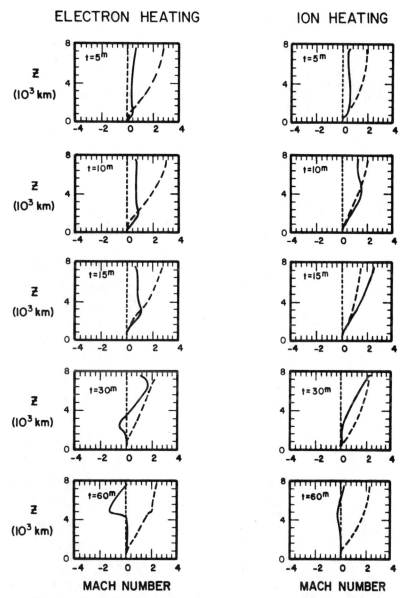

Fig. 2. Temporal evolution of hydrodynamic Mach numbers obtained with electron and ion heating. O^+, H^+ ions and electrons are represented by solid, dashed, and dotted lines, respectively.

Figures 1-3 show the temporal evolution of the particle flux (normalized to 1,000 km), Mach number (bulk velocity/thermal velocity), and temperature profiles. Time goes downward, starting at the top of the figures: altitude profiles are shown at t=5, 10, 15, 30, and 60 minutes, whith the electron and ion heating cases presented in the first and second columns, respectively.

At t=0 (not shown) the topside pressure drop (the "vacuum cleaner") was able to drive significant H^+ fluxes out of the ionosphere (the normalized flux was around 10^8 cm^{-2} s^{-1}). On the other hand the "vacuum cleaner" alone was not able to produce significant O^+ outflow. An additional energy source was needed to increase the ion scale heights at lower altitudes, and thus push large quantities of O^+ ions up to a region, where the effect of gravity becomes less important and the "vacuum cleaner" can drive large O^+ fluxes out of the ionosphere.

In the ion heating case the O^+ temperature quickly approached its peak value (~35,000K) and later slowly decreased to its initial level. The electron temperature remained practically unaffected, while the H^+ component reached its maximum temperature (~20,000K) about 15 minutes after the heating was initiated. At t=5min an O^+ upwelling started to form at an altitude of about 1,000km. The peak moved upward with a velocity of about 10 km/s, while simultaneously it became broader and broader. At ~5,000km the maximum O^+ upwelling flux was about 2×10^9 cm^{-2} s^{-1}. Below about 6,000 km the flux diminished after 30 minutes and the ionosphere slowly recovered from the disturbance.

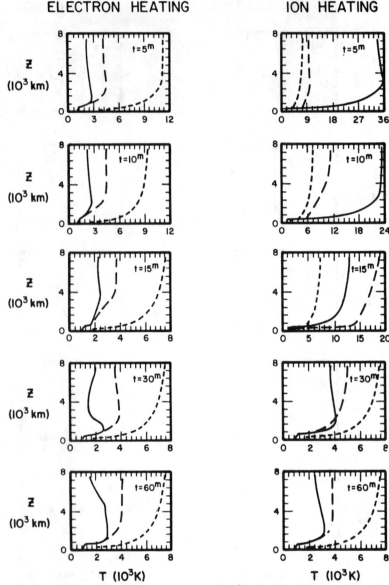

Fig. 3. Temporal evolution of particle temperatures obtained with electron and ion heating. O^+, H^+ ions and electrons are represented by solid, dashed, and dotted lines, respectively.

In the electron heating case the situation was different: hot electron and cold ion profiles were obtained. The maximum electron temperature was about 12,000K, while the ions were only indirectly heated with considerable time delay. This heating took place below 1,000 km, where the electron-ion collision frequencies are significant. After about 10 minutes an upward moving and broadening O^+ transient was formed. On the other hand, this upwelling moved about three times slower and had a factor of three smaller peak flux value than the one generated by ion heating. After the O^+ upwelling ended a downward flux was formed to refill the ionosphere.

The O^+ Mach numbers were close to unity during most of the upwelling event, and it remained below about 3 all the time for both cases. There were no significant differences between the Mach number magnitudes obtained in the ion and electron heating cases.

Our numerical models suggest that the ion heating case is able to reproduce the gross features of the upwelling events best (Moore et al., 1985, Waite et al., this monograph). Electron heating is also able to reproduce the observed magnitude of O^+ outflow flux. However, the calculation predicts that electron heating results in electron temperatures well exceeding the ion temperatures all the time; this prediction is inconsistent with observations showing highly elevated ion temperatures during upwelling events (cf. Lockwood et al, 1985).

Finally, it should be noted, that the temperature profiles exhibited significant spatial and temporal variations during each set of calculations: some profiles were fairly isothermal during extended periods, but there were non negligible deviations from constant temperature. Therefore numerical results which assume isothermal temperature profiles need to be treated with caution.

Summary

Coupled time-dependent continuinity, momentum and energy equations of a two ion (H^+ and O^+) quasi-neutral plasma were solved between 200 and 8,000 km for polar wind conditions. The model has several limitations (as discussed above) but it should still serve to give a feeling for the gross behavior of polar region plasma flows. Two specific cases (ion and electron heating) were considered: significant O^+ outflow was obtained in both cases. The calculated temperature and flow profiles (high T_{o^+}, low T_e and $M_{O^+} \sim 1$ in the ion heating case, low T_{o^+}, high T_e, and $M_{O^+} \sim 1$ in the electron heating case) suggest that ion rather than electron heating seems to be generating the upwelling ion events, for which large transient O^+ outflows, accompanied by elevated ion temperatures have been observed.

Acknowledgements. We wish to thank Drs. R.N. Schunk and W.J. Raitt for numerous helpful discussions. This work was supported by NASA grants NGR-23-005-015 and NAG5-472 and NSF grant ATM-8508753. The major fraction of the computations were carried out using the computing facilities of the National Center for Atmospheric Research, which is sponsored by the National Science Foundation.

References

Axford, W.I., The polar wind and the terrestrial helium budget, J. Geophys. Res., 73, 6855, 1968.

Banks, P.M. and T.E. Holzer, High-latitude plasma transport: The polar wind, J. Geophys. Res., 74, 6317, 1969.

Barakat, A.R. and Schunk, R.W., O^+ ions in the polar wind, J. Geophys. Res., 88, 7887, 1983.

Gombosi, T.I., Cravens, T.E. and Nagy, A.F., A time-dependent theoretical model of the polar wind: Preliminary results, Geophys. Res. Lett., 12, 167, 1985.

Khazanov, G.V., Koen, M.A., Konikov, U.V. and Sidonov, I.M., Simulation of ionosphere-plasmasphere coupling taking into account ion inertia and temperature anisotroy, Planet. Space Sci., 32, 585, 1984.

Lockwood, M., Waite, J.H., Moore, T.E., Johnsone, J.F.E., and Chappell, C.R., A new source of suprathermal O^+ ions near the dayside polar cap boundary, J. Geophys. Res., 90, 4099, 1985.

Mitchell, H.G. and Palmadesso, P.J., A dynamic model for the auroral field line plasma in the presence of field-aligned current, J. Geophys. Res., 88, 2131, 1983.

Moore, T.E., Lockwood, M., Chandler, M.O., Waite, J.H. Jr., Chappell, C.R., Persoon, A., and Sugiura, M., Upwelling O^+ ion source characteristics, submitted to J. Geophys. Res., 1985.

Nagai, T., J.H. Waite Jr., Green, J.L., Chappell, C.R., Olsen, R.C. and Comfort, R.H., First measurements of supersonic polar wind in the polar magnetosphere, Geophys. Res. Lett., 11, 669, 1984.

Raitt, W.J., Schunk, R.W. and Banks, P.M., A comparison of the temperature and density structure in the high and low speed thermal proton flows, Planet. Space Sci., 23, 1103, 1975.

Shelley, E.G., Peterson, W.K., Ghielmetti, A.G. and Geiss, J., The polar ionosphere as a source of energetic magnetospheric plasma, Geophys. Res. Lett., 9, 941, 1982.

Singh, N. and Schunk, R.W., Temporal behavior of density perturbations in the polar wind, J. Geophys. Res., 90, 6487, 1985.

Waite, J.H., Nagai, T., Johnson, J.F.E., Chappell, C.R., Burch, J.L., Killeen, T.L., Hays, P.B., Carignan, G.R., Peterson, W.K., and Shelley, E.G., Escape of suprathermal O^+ ions in the polar cap, J. Geophys. Res., 90, 1619, 1985.

Section IX. SUMMARY

IMPULSIVE ION ACCELERATION IN EARTH'S OUTER MAGNETOSPHERE

D. N. Baker and R. D. Belian

University of California, Los Alamos National Laboratory, Los Alamos, NM 87545

Abstract. Considerable observational evidence is found that ions are accelerated to high energies in the outer magnetosphere during geomagnetic disturbances. The acceleration often appears to be quite impulsive causing temporally brief (tens of seconds), very intense bursts of ions in the distant plasma sheet as well as in the near-tail region. These ion bursts extend in energy from tens of keV to over 1 MeV and are closely associated with substorm expansive phase onsets. Although the very energetic ions are not of dominant importance for magnetotail plasma dynamics, they serve as an important tracer population. Their absolute intensity and brief temporal appearance bespeaks a strong and rapid acceleration process in the near-tail, very probably involving large induced electric fields substantially greater than those associated with cross-tail potential drops. Subsequent to their impulsive acceleration, these ions are injected into the outer trapping regions forming ion "drift echo" events, as well as streaming tailward away from their acceleration site in the near-earth plasma sheet. Most auroral ion acceleration processes occur (or are greatly enhanced) during the time that these global magnetospheric events are occurring in the magnetotail. A qualitative model relating energetic ion populations to near-tail magnetic reconnection at substorm onset followed by global redistribution is quite successful in explaining the primary observational features. Recent measurements of the elemental composition and charge-states have proven valuable for showing the source (solar wind or ionosphere) of the original plasma population from which the ions were accelerated. New data relying on such methods hold great promise for illuminating the details of the substorm ion acceleration mechanism.

Introduction

The problem of particle acceleration in the earth's auroral zones has been attacked quite successfully over the past few years. Much of the progress in this area has been due to a strong interplay of observation and theory (e.g., Young, 1985; Klumpar et al., 1984; Chang and Coppi, 1981; Borovsky and Joyce, 1985; Ashour-Abdalla, 1985). The theoretical aspects of the auroral ion acceleration problem have been tractable since the acceleration appears to take place in a relatively localized region of space, the initial particle distributions can be reasonably well specified, the magnetic field and electric potential structures can be assumed with some confidence, and the accelerations occur only up to modest final energies.

It is important, however, to consider that auroral acceleration processes occur, for the most part, as only one component of the overall set of events that constitute the complex conditions associated with geomagnetic activity. During high levels of geomagnetic disturbances, large fluxes of ions are accelerated out of the terrestrial ionosphere and these may have a very significant subsequent effect on outer magnetospheric processes (e.g., Young, 1985; Baker et al., 1982a). Concomitantly with the outward acceleration of ions, there is a substantial acceleration of electrons toward the earth. These accelerated electrons produce intense auroral displays and are significant carriers of global magnetospheric currents. Together, the heating, transport, and loss of these auroral particles then constitutes one of the important energy dissipation mechanisms associated with geomagnetic activity.

When considering the larger context in which auroral ion acceleration occurs, it is useful to discuss an "elementary" episode of geomagnetic disturbance, viz., the magnetospheric substorm. In its simplest form, the substorm consists of a sequence of events such as shown in Fig. 1. In an idealized circumstance of an isolated substorm, the sequence of events will be initiated by the southward turning of the interplanetary magnetic field (IMF). This southward turning increases dayside magnetic reconnection substantially (McPherron, 1979) and over the course of the next hour or so large amounts of free energy (10^{21}-10^{23} ergs) are added to the earth's magnetotail in the form of increased magnetic flux. In entering this highly energized state, the magnetosphere becomes more and more unstable to the growth of large-scale instabilities which will tend to reconfigure the magnetosphere toward a much lower energy state (i.e., the initial ground state configuration). Present evidence (McPherron, 1979, and references therein) suggests that the massive reconfiguration and rapid energy reduction that occurs at substorm expansive phase onset is often due to the formation of a magnetic neutral line in the near-earth plasma sheet ($10 \lesssim r \lesssim 20\ R_E$) in the midnight sector. Such an interpretation is represented in the cross-sectional schematic illustration of Fig. 1.

The large azimuthal scale size of the substorm neutral line (and the strong electric field thought to exist parallel to it) implies a very large region of magnetic field annihilation, plasma jetting, and (potentially) rapid energy dissipation in the tail. It is in the context of these overall processes of the global interaction of the magnetosphere with the solar wind that auroral acceleration processes largely occur (see Baker et al., 1984, and references therein). In particular, we believe that the energy to drive the auroral acceleration and dissipation comes about because of the global processes that load large total plasma and magnetic field energies into the magnetotail and then rapidly convert this stored energy in the mid-tail at substorm onset.

Thus, perhaps in contrast to most of the papers in these *Proceedings*, our objective is to consider the larger context of overall geomagnetic activity as it pertains to particle acceleration processes. We wish to address the "driving engine" in the outer magnetosphere which, on the more limited scale, results in the acceleration of auroral particles to energies of order 1 to 10 keV. Specifically, we will address the sporadic and impulsive acceleration of ions in the earth's

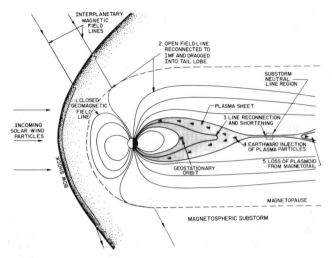

Fig. 1. A noon-midnight cross-sectional view of the earth's magnetosphere showing the possible sequence of events involved in a magnetospheric substorm. Enhanced dayside reconnection between the earth's field and the interplanetary magnetic field leads to "open" field lines which are dragged into the tail lobe on the earth's nightside. After a period (30 ~ 60 min) of such loading of magnetic flux into the tail lobes, a substorm neutral line forms in the near-earth plasma sheet causing large magnetospheric reconfiguration and earthward injection of hot plasma particles.

outer magnetosphere to very high energies ($\gtrsim 1$ MeV) during magnetospheric substorms. The observational characteristics of these populations will be reviewed, and an interpretation of such observations in the context of the model of Fig. 1 will be presented. Since a comprehensive theoretical understanding of the acceleration processes at play in the magnetotail does not presently exist, we will conclude this paper with some recommendations for future studies.

Review of Observations

In this section we will recount some of the principal observational results pertaining to ion acceleration in the outer magnetosphere. In particular we will review data obtained for ion burst events ranging in altitude from the geostationary orbit (6.6 R_E) to the deep magnetotail.

Ion Burst Time Scales

Spacecraft data in the earth's magnetotail (10 ~ 30 R_E) show that, in an average sense, substorm expansive phase onsets regularly produce hot, jetting plasmas in the plasma sheet (e.g., Bieber et al., 1982, and references therein). At the same time there frequently are also substantial increases in the fluxes of energetic particles in the plasma sheet (Scholer et al., 1985).

A particularly striking class of ion acceleration event is shown in Fig. 2 (Krimigis and Sarris, 1979; Sarris et al., 1976). The uppermost part of the diagram shows the occurrence of impulsive burst ion events in three energy ranges between 0.29 MeV and 1.85 MeV. As shown in the lower portion of Fig. 2, these ion bursts exhibit a strong (10^4:1) characteristic tailward streaming, directed along the local magnetic field line. These particular events were detected by IMP-7 instruments in the duskside magnetotail at a geocentric radial distance of ~35 R_E. Note that these ion bursts last only for a few 10's of seconds.

Simultaneous plasma data from the same spacecraft have been examined for many such events (Belian et al., 1981) and an example is shown in the middle panel of Fig. 3. The upper insets repeat the data for two of the impulsive burst events of Fig. 2. Two episodes of extreme plasma sheet thinning are indicated (Fig. 3) by the complete dropout of plasma sheet density on 26 March 1973 at ~1900 UT and again at ~1942 UT. The impulsive bursts of high energy ions were observed right at the time of the plasma dropouts, i.e., right at the edge of the thinning plasma sheet. In this instance, and in most other impulsive burst cases, the high energy ions occur right at the onset of the substorm expansive phase. This general feature is illustrated in the bottom panel of Fig. 3 which shows a magnetogram from the ground station at Kiruna, Sweden, which was near local midnight at the time of the magnetotail events. The ion bursts occurred near the beginning of an intense negative bay which is one of the classic signature of a substorm expansive phase onset.

Much nearer the earth, geostationary orbit spacecraft have shown that hot plasma and energetic particle populations become significantly more intense at substorm expansive phase onset. It has been supposed that such plasma increases at 6.6 R_E are due to the "injection" of particles into an azimuthally limited region of space near (and beyond) the geostationary orbit (McIlwain, 1974).

At high energies (E $\gtrsim$ 200 keV) geostationary orbit data show the common occurrence of ion "drift echo" events (Belian et al., 1978). An example of such an event is shown in Fig. 4 (Belian et al., 1981). In this figure we have plotted the proton differential flux versus universal time for a two-hour period of 14 April 1977. These data, from the geostationary orbit spacecraft 1976-059, correspond to two proton (ion) energy channels, viz, 0.4-0.5 MeV and 0.5-0.6 MeV. Spacecraft local time (LT) is shown at the top of the figure. The "drift echo" character, and hence the name of this phenomenon, comes from the periodic series of alternating peaks and valleys in the flux profile. Analysis on this and many similar events shows that a drift echo event corresponds to the injection of a single discrete bunch of energetic ions in the outer magnetosphere. These ions drift azimuthally around the earth many times (often 5 or more) maintaining a reasonable coherence. The recurrence periods of the "echoes" of the original ion burst depend on the gradient and curvature azimuthal drift times for each particle in the earth's confining magnetic field.

At the left-hand side of Fig. 4 we show the H-component magnetogram from Leirvogur for the period in question on 14 April. The sharp drop in the magnetogram at ~2235 UT indicates a strong substorm expansive phase. This time agrees to within a few minutes with the occurrence time for the drift echo event at 6.6 R_E. Hence, we have concluded that drift echo events are very much a substorm expansive phase phenomenon. Furthermore, as with impulsive bursts in the tail, the initial drift-echo burst is a temporally limited feature lasting several tens of seconds to a few minutes.

The similar energy, timing, duration, and energy spectra of impulsive burst ions in the deep tail and drift echo bursts at 6.6 R_E led Baker et al. (1979) to suggest a common origin for the two classes of event. A model (to be discussed below) was advanced to explain this common source.

Particle Anisotropies and Ion Source Locations

As shown in Fig. 2 above, measurements at ~30 R_E often show that impulsive burst ions are strongly streaming tailward. This free-streaming propagation along field lines has been taken as very convincing evidence of a source of energetic ions earthward of 30 R_E on field lines adjacent to the edge of the plasma sheet. A very consistent and plausible source for these particles would be the near-earth substorm neutral line (Sarris and Axford, 1979; Baker et al., 1979).

On the other hand, the data from geostationary orbit show rather

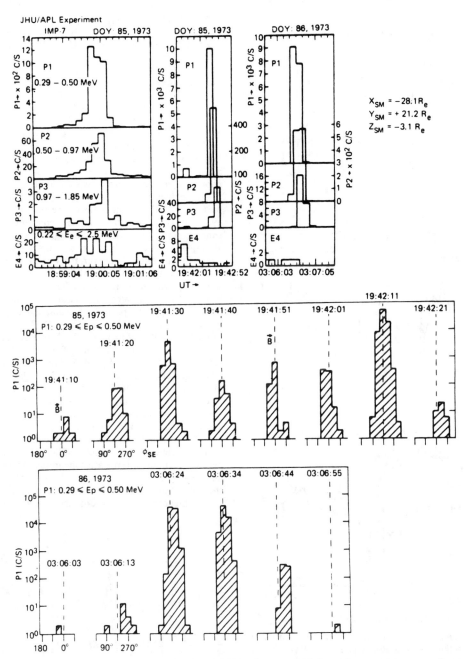

Fig. 2. IMP-7 energetic particle data for 26 and 27 March 1973. The top panel shows three proton energy channels and one electron channel at a basic time resolution of 10.24 sec. The middle and bottom panels show angular distributions of the impulsive ion bursts (see upper panel) in successive 10.24 sec time intervals. Large (~10^4:1) tailward streaming anisotropies of the 0.29 MeV ions are seen (from Krimigis and Sarris, 1979).

conclusively that injected energetic ion populations come from *outside* of 6.6 R_E (i.e., from locations at greater radial distances). The method employed to assess this issue is the flux gradient measurement. Because of their large gyroradii, ≳100 keV ions can provide good information about the strength and direction of gradients as may exist in injected clouds of hot plasma at 6.6 R_E at substorm onset.

The east-west gradient parameter is computed as

$$A_{EW} = (E - W)/(E + W)$$

where E is the flux measured in the sector wherein the sensor looks eastward and W is the ion flux with the sensor looking westward. Given the fact that the magnetic guiding center field is generally northward at geostationary orbit, $A_{EW} > 0$ implies a higher flux (or particle density) inside the spacecraft orbit, while $A_{EW} < 0$ implies a higher flux outside the spacecraft orbit.

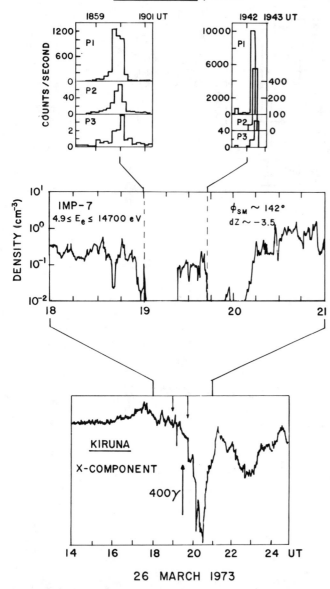

Fig. 3. (Top) Representative example of energetic ion impulsive bursts as recorded by the IMP 7 satellite (see Fig. 2). IMP 7 was in the plasma sheet near the dusk magnetopause at the time of these observations. The measured IMP 7 electron plasma density is shown in the middle panel while the nature of the concurrent auroral geomagnetic activity is shown by the Kiruna magnetogram in the lower panel (from Belian et al., 1981).

Baker et al. (1982b) used this technique to illustrate the general tendency of energetic ion injection pulses to arrive from outside of 6.6 R_E. These results are shown here in Fig. 5. The solid line (with scale on the left) shows the >145 keV proton flux measured by S/C 1977-007 on 29 July 1977. The dashed line (scale on right) shows the computed value of A_{EW}. At the leading temporal edges of the ion injections at 1200 UT and 1205 UT, the value of A_{EW} became strongly negative. This is a common characteristic for ion injection events and shows that injected ion bunches reach 6.6 R_E from a source region at substantially higher altitude than the geostationary orbit.

Substorm Recovery Events

Impulsive, temporally-limited injection events at substorm onset constitute essentially the only distinctive sort of ion flux increase at geostationary orbit. Of course, injection events do not always extend to very high energies and only under certain conditions are drift echo events seen (see Baker et al., 1979). Nonetheless, substorm expansive phase injection events seem to be the primary means of populating the outer magnetosphere ion trapping region.

On the other hand, thermal and suprathermal protons seem to be present virtually all the time in the plasma sheet (Scholer et al., 1985). It is found that the very energetic protons are greatly enhanced in intensity during geomagnetic activity and, as discussed above, impulsive bursts are closely associated with substorm expansive phase onset. However, the vast majority of ion "events" in the deeper magnetotail are not of the impulsive burst variety.

The majority of discrete high-energy ion events in the tail tend to be seen at substorm recovery and, roughly speaking, exhibit a rapid rise-slow decay time profile. An example of such an event is shown in Fig. 6 (Ipavich and Scholer, 1983). This figure shows proton fluxes from ~30 keV to over 300 keV in energy. It is generally found that the suprathermal protons (30 ~ 100 keV) closely follow the behavior of the thermal plasma sheet ions (Ipavich and Scholer, 1983; Scholer et al., 1985). Thus, as shown by the data of Fig. 6, the plasma sheet recovered (expanded over the ISEE spacecraft) at ~0320 UT on 26 March 1978 and remained present for the next several hours. Although the 30-36 keV protons diminished in absolute intensity very gradually over the period from 0330 to 0600 UT (probably due to a gradual plasma sheet cooling), there was no significant evidence in these ISEE-1 data of major changes in the plasma sheet bulk plasma after the initial recovery at 0320 UT. We note that ground magnetic records (e.g., at Great Whale River) showed a clear substorm recovery beginning at ~0300 UT.

The higher energy protons showed a substantially different

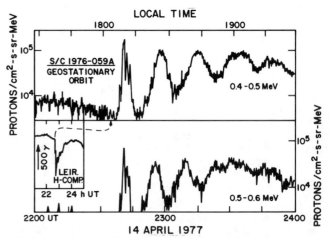

Fig. 4. An example of an energetic proton drift echo event as recorded by spacecraft 1976-059 at synchronous orbit. The upper panel shows 0.4-0.5 MeV proton fluxes while the lower panel shows 0.5-0.6 MeV fluxes. The inset at the left shows the auroral zone magnetogram from Leirvogur (from Belian et al., 1981).

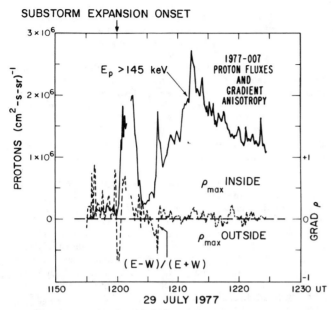

Fig. 5. A comparison of the >145 keV proton flux (solid line) and the associated east-west gradient anisotropy (dotted line) for a period on 29 July 1977. Strong gradient anisotropies occur (at ~1200 UT and ~1205 UT) as newly energized particles are injected near synchronous orbit (from Baker et al., 1982b).

such strong bursts of particles appear to originate in the near-earth plasma sheet region.

A successful substorm model must, of course, account for several distinct magnetospheric phenomena:
- Intensified polar ionospheric current systems;
- Global magnetospheric field reconfigurations;
- Strong plasma jetting in the near-earth plasma sheet;
- Large plasma injections into the vicinity of geostationary orbit; and,
- Overall typical energy dissipation rates of 10^{18}-10^{19} ergs/sec for periods of 30-60 min during the initial phases of substorm activity.

The model described in the Introduction and shown in Fig. 1 can account for many of these features primarily through the extraction and conversion of energy which was originally contained in the solar wind flow impinging on the Earth's magnetosphere. Many of the rapid dissipations and/or reconfigurations which are observed during substorms come about in the model of Fig. 1 due to the abrupt formation of the near-earth neutral line.

For the purposes of intense energetic particle acceleration, a

behavior than the lower energy plasma. Note in Fig. 6 that after the initial rapid rise in the >100 keV fluxes (lower three panels), all of these higher energy flux profiles exhibited a nearly monotonic decrease in intensity. Ipavich and Scholer (1983) showed that the decay time scale for the high energy ions is energy-dependent (as in Fig. 6) with the decay constant being shortest for the highest energies.

Thus, many (if not most) high-energy ion burst events in the magnetotail are due to entry of the observing satellite into the plasma sheet during substorm recovery periods. Baker et al. (1979) found that ~75% of the periods in which there were energetic proton events were characterized first by a plasma sheet thinning (substorm onset) followed by a plasma sheet expansion (substorm recovery). As the expanding plasma sheet envelopes the observing satellite, a rapid rise in energetic proton intensity is then observed. As pointed out by Scholer (1984), multiple burst ion events are often due to repeated entries and exits of the spacecraft into and out of the confined ion population in the plasma sheet. Certainly - under such circumstances - one need not and should not invoke a separate acceleration episode for each "burst" during the substorm recovery epoch. Nonetheless, any successful model of magnetotail dynamics must explain the presence in the plasma sheet of a greatly enhanced population of energetic particles during substorms.

Interpretation

Acceleration at Substorm Onset

As we have shown above, the observational evidence suggests quite strongly that acceleration of energetic particles during the substorm expansive phase is an intimate and integral part of the global substorm process. At least the highest energy and most intense ion fluxes tend to be closely associated with the substorm onset, and

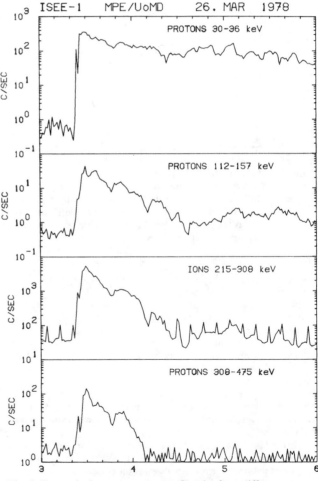

Fig. 6. Energetic ion count rate profiles in four different energy ranges as measured by ISEE-1 on 26 March 1978. An abrupt increase in all energy channels was seen at ~0320 UT when ISEE rapidly entered the plasma sheet (from Ipavich and Scholer, 1983).

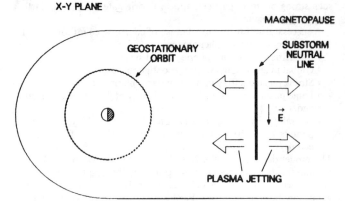

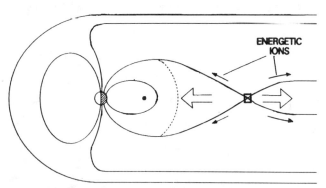

Fig. 7. Two cross-sectional views of the magnetosphere showing the substorm neutral line, its associated electric field pattern, and the jetting of hot plasma from the neutral line. Magnetic field lines are shown in the x-z plane in the lower panel with energetic ions streaming along field lines away from the neutral line both earthward and tailward. These features, associated with substorm onset, are taken as the primary type of impulsive ion acceleration in the model of Baker et al. (1979).

substorm model must be extended to include several features which have been outlined in this paper:
- Intense, brief bursts of $0.1 \lesssim E_p \lesssim 1.0$ MeV ions streaming tailward along boundary field lines in the plasma sheet at substorm onset;
- Injection of temporally and spatially limited bunches of ions in a similar energy range into the vicinity of geostationary orbit, again at substorm onset.

Realistically, a substorm model should account for these acceleration features in a fairly natural way rather than positing *ad hoc* acceleration sites throughout the magnetosphere independent of the dominant dynamical processes mentioned in the previous paragraph.

Baker et al. (1979) suggested a model which accounted in a reasonable way for most of the observations both in the plasma sheet and at geostationary orbit. An illustration of the substorm-onset features of this model is shown in Fig. 7. Both the observation of tailward-streaming impulsive bursts in the plasma sheet and drift echo events at geostationary orbit suggests a brief, intense acceleration process right at substorm onset. As illustrated by Fig. 7, we associate both of these particle populations with the near-earth substorm neutral line (see Fig. 1) which is also responsible for plasma sheet jetting, magnetic field reconfiguration, etc.

Earthward of the acceleration region the high-energy particles are introduced onto closed field lines. The observation of drift echo events (i.e., brief, narrow bursts of several hundred keV protons) implies that these particles are produced primarily in a brief period at the onset of the substorm. Those substorms producing such high energy particles typically occur during quite disturbed geomagnetic conditions. Thus the suggestion would be, as shown in Fig. 7, that under such circumstances, large induced electric fields would exist in the vicinity of the neutral line acceleration region. Total, integrated electric potentials along the electric field must be of order 1 MV, or more, in many such cases. The largest values of $\partial \mathbf{B}/\partial t$ must be quite transient since the drift echo events are typically quite short ($\lesssim 1$ min) in temporal extent.

Tailward of the acceleration region, the high-energy ions are introduced onto essentially open magnetotail field lines. The high-energy impulsive burst ions stream tailward from the neutral line region at the edge of the thinning plasma sheet as shown in Fig. 7. These bursts are of very short duration and, in fact, show "inverse" velocity dispersion (low energy particles appear before high energy particles). This dispersion effect (Sarris and Axford, 1979) represents a time-of-flight effect of different energy particles moving along field lines away from the fixed neutral line source region. This source is only active for a brief, limited time (Baker et al. 1979). As the plasma sheet thins more and more during substorm onset, the loci of higher and higher energy particles sweep over an observing spacecraft near the edge of the plasma sheet.

As noted, the tailward flowing impulsive burst protons escape essentially along open field lines down the tail, while drift echo protons subsequently drift in trapped orbits about the earth and disperse continually. As shown by the dotted line in the lower panel of Fig. 7, the magnetic field has changed from a highly taillike configuration (on the nightside) to a more dipole-like configuration during the substorm injection process. Given that the ions were accelerated deep in the tail, it may well be that the ion bursts maintain their coherent initial structure in drift echo events because of such a rapid field reconfiguration in the equatorial region. We would expect, as shown by the diagram in Fig. 7, that the earthward-injected energetic ions would travel along the original taillike field lines and would go deep toward the low-altitude foot of the field lines. Such nearly field-aligned particles spend much of their total bounce time near the mirror points. By the time the ions have mirrored and returned to the equatorial region, the field line could have collapsed substantially toward the earth. Hence, the ion bunch will have been brought effectively much closer to the earth but will not have undergone the strongly dispersive gradient and curvature drift that would be expected if the particles were brought closer to the earth as a bunch of largely equatorial ($\alpha \approx 90°$) particles.

This hypothesized effect for ions might also explain the relative absence of electron drift echo effects. Because of their much shorter bounce times at a given energy compared to protons ($\tau_p \sim 40\, \tau_e$), and because of their strong tendency to scatter in whistler-mode wave turbulence during the injection process, electrons will sense the equatorial collapse of the field line much more than the ions. Thus the electrons, even if they were bunched initially like the ions, will tend to be broadly dispersed by drift effects during the injection process.

A final outstanding issue about the high-energy acceleration process is why the energization seems so temporally limited at the highest energies. Certainly at lower energies (E < 100-200 keV) Baker et al. (1979) noted that a more "normal" type plasma heating and particle acceleration apparently continued as long as the neutral line remained in position in the near plasma sheet and as long as the substorm expansive phase was in progress.

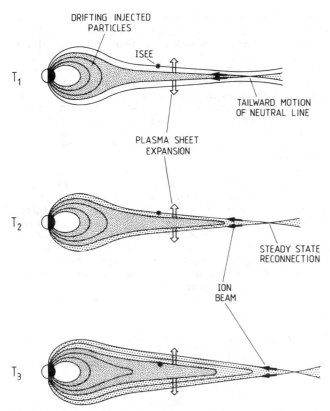

Fig. 8. A schematic illustration of the proposed sequence of events in the earth's magnetotail during a substorm recovery (Scholer et al., 1985). The model combines the tailward retreat of the substorm neutral line with the leakage of particles on closed field lines as originally proposed in the model of Baker et al. (1979).

Our present belief remains that the highest energy particle acceleration corresponds to an "explosive" phase of reconnection early in the substorm sequence. The rate of magnetic reconnection (e.g., Axford, 1984) is determined essentially by the Alfvén speed ($V_A = B/\sqrt{4\pi\rho}$) in the inflow region of the substorm neutral line. As long as plasma sheet field lines are being reconnected, then B is characteristic of the plasma sheet (5-10 nT) and the mass density ($\rho = nm_p$) is determined by the plasma sheet number densities ($n \sim 1$ cm^{-3}). As soon as reconnection has progressed to the stage shown in Fig. 7 where lobe-like field lines are reconnecting, then B at the inflow region would be much higher (20-40 nT) and plasma densities would be greatly diminished ($n \lesssim 10^{-2}$ cm^{-3}). Thus, we concur with the speculation of Axford (1984) that the high-energy substorm bursts probably correspond to the situation when the reconnection rate suddenly jumps from a rate determined by $V_A \sim 100$ km/s (deep in the plasma sheet) to a rate determined by $V_A = 1000\text{-}2000$ km/s at the plasma sheet-lobe interface. During this brief, transient transition interval, very large acceleration apparently takes place.

Particle Redistribution at Substorm Recovery

In essence the Baker et al. (1979) model assumes that as a result of the expansive phase acceleration and injection processes there exists a high density of quasi-trapped energetic particles in the outer radiation zones at the beginning of the substorm recovery. As the plasma sheet expands (thickens) during the substorm recovery phase, the plasma sheet eventually envelopes a relatively low-latitude observing spacecraft in the magnetotail. Then, as the plasma sheet increases in size, the model suggests that energetic ions from the outer magnetosphere ($6 \lesssim r \lesssim 10$ R$_E$) rapidly leak and/or diffuse outward to fill the newly expanded plasma sheet volume. Thus, rather than invoking a new, impulsive acceleration at substorm recovery, the model suggests that the plasma sheet is largely filled by redistributed particles which were produced at substorm onset. Belian et al. (1981) found that the thin, residual plasma sheet may be filled with particles continually leaking out of the trapping regions throughout the expansive phase.

Baker et al. (1979) and Belian et al. (1981) examined many substorm-recovery events and found substantial support for the above picture. They concluded that the energy spectra at 6.6 R$_E$ and in the recovering plasma sheet were very similar and absolute intensities at 6.6 R$_E$ were high enough to be consistent with the leakage hypothesis. Furthermore, this model also accounted quite well for the dawn-to-dusk anisotropy (Sarris et al., 1976) often seen in the recovering plasma sheet energetic ions. This anisotropy is consistent with a strong gradient of particle density equatorward and earthward of the magnetotail spacecraft. Of course, the model also explains the rapid rise-slow decay flux time profile seen during substorm recoveries (see Fig. 6); Baker et al. (1979) did some detailed modeling of this particular feature.

Scholer et al. (1985) suggested a specific extension of the above model primarily to account for the presence of energetic ($\lesssim 150$ keV) ion beams at the edge of the expanding plasma sheet. This amalgamated model is shown here as Fig. 8.

Scholer et al. pointed out that the tailward retreating neutral line (during substorm recovery) could be the source of the field-aligned beams which are injected toward the earth. But, as in the original Baker et al. model, the very high energy ions leak out of the outer radiation zones onto the newly reconnected (closed) field lines. As shown by the shading in Fig. 8, the very high energy particles are found on field lines nearer the center of the plasma sheet. The figure shows that as the neutral line moves tailward it creates more and more newly reconnected field lines and these tend to contain the lower energy, isotropized ions produced recently by the retreating neutral line.

Scholer et al. (1985) pointed out several concerns about the original Baker et al. model. These primarily hinge on the issues of the character of the spectrum (power law or exponential) of the recovering plasma sheet particles and whether the total population of quasi-trapped ions could supply the recovering plasma sheet. As noted above, Baker et al. (1979) did some modeling of the required trapped ion source strength and particle redistribution mechanism, but a complete numerical model was not employed. Thus, there remain several areas where more complete modeling is necessary and important.

Conclusions and Future Studies

We conclude from ongoing studies of energetic particles in the earth's outer magnetosphere that highly impulsive ion acceleration often occurs during substorms. As we have emphasized throughout this paper, this acceleration appears to be intimately related to the primary dynamical processes in the earth's magnetotail which cause global substorm effects. Although energetic ion acceleration in the magnetotail is not of dominant importance (either in energetics or dynamics), it *is* extremely important as a prototype of cosmic acceleration mechanisms. We conclude from the observations that:

- Acceleration occurs in a relatively localized region in the near-earth plasma sheet.
- The mechanism operates on a short (or even explosive) time-scale of 1 to a few minutes.

- The process accelerates some particles to very high energies ($\gtrsim 1$ MeV), i.e., well-above ordinary cross-tail potentials.

In the foregoing parts of this paper we have described a model which provides a qualitative description of the acceleration, injection, and redistribution of energetic ions during substorms.

Despite the general understanding that presently exists concerning energetic ion acceleration, and despite the two decades of observational work in this area, we still have much work to do. From an observational perspective, the present-day situation permits relatively large constellations of observing spacecraft at geostationary orbit and throughout the near-earth plasma sheet. These chance configurations of spacecraft should be exploited to pin down source locations and particle transport details. Furthermore, the modern instrumentation provides very substantial improvement in the available energy spectral resolution and in the 3-dimensional angular distribution measurement capabilities.

The role of energetic heavy ions in the magnetosphere has recently come under heightened consideration (e.g., Williams and Sugiura, 1985, and references therein). The advent of sensor systems capable of mass and charge identification are beginning to reveal the complex competition between the ionospheric and solar wind plasmas in controlling magnetospheric dynamics. With regard to the problem of high-energy ion acceleration, Blake et al. (1983) were able to use the earth's magnetic field as a large ion charge-state analyzer. By combining high-energy ion drift echo detection with the mass resolution capability of the SCATHA spacecraft, Blake et al. could identify the ultimate plasma source of a substorm-accelerated ion burst. They determined that in this case the very energetic ions produced at substorm onset were of solar wind, not ionospheric, origin. The new observational capabilities now available (as with the AMPTE spacecraft) hold the potential of answering most questions about the details of ion acceleration if the data sets can be fully exploited.

Despite the substantial emphasis given to the "neutral line" model of substorms in this review, we wish to caution readers that there certainly is not universal agreement concerning this model's appropriateness (e.g., Rostoker, 1984; Eastman et al., 1984). To a large extent the views minimizing the importance of the near-earth neutral line invoke a dominant role for the low-latitude boundary layer in driving all magnetotail dynamics, or else they rely upon acceleration of plasma beams by somewhat unspecified mechanisms in the distant magnetotail. Since these alternative viewpoints are only in rudimentary stages of development, they do not yet constitute fully testable models of magnetotail dynamics. In particular, to date these models have not really addressed the issue emphasized in this paper, viz., the intense acceleration of particles up to hundreds of keV at substorm onset.

One important theoretical model of particle acceleration that has been quite thoroughly developed is the "current sheet acceleration" model in the geomagnetic tail (Speiser, 1965; Lyons and Speiser, 1982). The theoretical work in this areas has used simple models of the tail current sheet geometry and then traces trajectories of single particles as they execute serpentine drift motions across the width of the tail. In the process of such drifts, particles (predominantly ions) can pick up large fractions of the cross-tail potential, i.e., several tens of keV total energy.

Lyons and Speiser (1982) have emphasized that this mechanism could be a major form of energy dissipation in the tail. They suggest that 5-10% of the incident solar wind energy flux impinging on the magnetosphere could go into this sort of particle energization. Such dissipation would be largely separate and apart from the substorm dissipation we have discussed above.

We certainly feel that the kind of current sheet acceleration discussed by Lyons and Speiser can, and probably does, occur in the distant magnetotail. It may very likely be an important source for the plasma ion beams emphasized by Eastman et al. (1984). Furthermore, this mechanism also may be quite important in the near tail as the plasma sheet thins prior to substorms (see Fig. 1). We are convinced, however, that this type of acceleration mechanism is not the dominant source of the very energetic ions that we have considered in this paper. It cannot readily account either for the energies observed or for the brief and bursty temporal character of the substorm-associated ion events. Thus, for these reasons we have stressed the need for large induced electric fields near the substorm neutral line.

To achieve full understanding, theoretical and numerical modeling efforts must be greatly increased. Given the rapidly increasing capabilities of present-day computers and the improving sophistication of numerical techniques, much more complete models of tail acceleration should be possible. Our present thinking is that a self-consistent, 3-D MHD model of the earth's magnetotail could be employed with the added feature of test particles to simulate the energetic ion population (J. Birn, private communication, 1984). Such modeling, using the detailed comparison of such results with new observations, holds genuine promise for revealing in a very precise way the workings of the powerful accelerator in the earth's magnetotail.

Acknowledgements. This work was done under the auspices of the U.S. Department of Energy. We thank Dr. D. T. Young for helpful discussions and numerous useful suggestions concerning this paper.

References

Ashour-Abdalla, M., Heavy ion dynamics: Sources and acceleration (this volume), 1985.

Axford, W.I., Magnetic field reconnection, in *Magnetic Reconnection in Space and Laboratory Plasmas* (Geophys. Monograph 30), p. 1, Amer. Geophys. Union, Washington, 1984.

Baker, D. N., R. D. Belian, P. R. Higbie, and E. W. Hones, Jr., High-energy magnetospheric protons and their dependence on geomagnetic and interplanetary conditions, *J. Geophys. Res.*, 84, 7138, 1979.

Baker, D. N., E. W. Hones, Jr., D. T. Young, and J. Birn, The possible role of ionospheric oxygen in the initiation and development of plasma sheet instabilities, *Geophys. Res. Letters*, 9, 1337, 1982a.

Baker, D. N., T. A. Fritz, B. Wilken, P. R. Higbie, S. M. Kaye, M. G. Kivelson, T. E. Moore, W. Studemann, A. J. Masley, P. H. Smith and A. L. Vampola, Observation and modeling of energetic particles at synchronous orbit on July 19, 1977, *J. Geophys. Res.*, 87, 5917, 1982b.

Baker, D. N., S. I. Akasofu, W. Baumjohann, J. W. Bieber, D. H. Fairfield, E. W. Hones, Jr., B. H. Mauk, R. L. McPherron, and T. E. Moore, "Substorms in the magnetosphere," Chapter 8 of *Solar Terrestrial Physics - Present and Future*, NASA Publ. 1120, Washington, D.C., 1984.

Belian, R. D., D. N. Baker, P. R. Higbie, and E. W. Hones, Jr., High-resolution energetic particle measurements at 6.6 R_E, 2, High-energy proton drift echoes, *J. Geophys. Res.*, 83, 4857, 1978.

Belian, R. D., D. N. Baker, E. W. Hones, Jr., P. R. Higbie, S. J. Bame, and J. R. Asbridge, Timing of energetic proton enhancements relative to magnetospheric substorm activity and its implication for substorm theories, *J. Geophys Res.*, 86, 1415, 1981.

Bieber, J. W., E. C. Stone, E. W. Hones, Jr., D. N. Baker, and S. J. Bame, Plasma behavior during energetic electron streaming events: Further evidence for substorm-associated magnetic reconnection, *Geophys. Res. Letters*, *9*, 664, 1982.

Blake, J. B., J. F. Fennell, D. N. Baker, R. D. Belian, and P. R. Higbie, Determination of the charge state and composition of energetic magnetospheric ions by observation of drift echoes, *Geophys. Res. Letters*, *10*, 1211, 1983.

Borovsky, J. E., and G. Joyce, The direct production of ion conics by double layers (this volume), 1985.

Chang, T., and B. Coppi, Lower hybrid acceleration and ion evolution in the suprauroral region, *Geophys. Res. Letters*, *8*, 1253, 1981.

Eastman, T. E., L. A. Frank, W. K. Peterson, and W. Lennartson, The plasma sheet boundary layer, *J. Geophys. Res.*, *89*, 1553, 1984.

Ipavich, F. M., and M. Scholer, Thermal and suprathermal protons and alpha particles in the plasma sheet, *J. Geophys. Res.*, *88*, 150, 1983.

Klumpar, D. M., W. K. Peterson, and E. G. Shelley, Direct evidence for two-stage (bimodal) acceleration of ionospheric ions, *J. Geophys. Res.*, *89*, 10779, 1984.

Krimigis, S. M., and E. T. Sarris, Energetic particle bursts in the earth's magnetotail, in *Dynamics of the Magnetosphere* (S.-I. Akasofu, Ed.), p. 599, D. Reidel, Hingham, Mass., 1979.

Lyons, L. R., and T. W. Speiser, Evidence for current sheet acceleration in the geomagnetic tail, *J. Geophys Res.*, *87*, 2276, 1982.

McIlwain, C. E., Substorm injection boundaries, in *Magnetospheric Physics*, ed. by B. M. McComac, D. Reidel Pub.. spCo., Dordrecht-Holland, P. 143, 1974.

McPherron, R. L., Magnetospheric Substorms, *Rev. Geophys. Space Phys.*, *17*, 657, 1979.

Rostoker, G., Definition of a substorm, hysical processes in a substorm, and sources of discomfort, *Magnetic Reconnection in Space and Laboratory Plasmas, Geophys. Monograph 30*, p.380, American Geophys. Union, Washington, D.C., 1984.

Sarris, E. T., and W. I. Axford, Energetic protons near the plasma sheet boundary, *Nature*, *77*, 460, 1979.

Sarris, E. T., S. M. Krimigis, and T. P. Armstrong, Observations of magnetospheric bursts of high-energy protons and electrons at ~35 R_E with IMP-7, *J. Geophys. Res.*, *81*, 2341, 1976.

Scholer, M., Energetic ions and electrons and their acceleration processes in the magnetotail, in *Magnetic Reconnection in Space and Laboratory Plasmas* (Geophys. Monograph 30), p. 216, Washington, D.C., 1984.

Scholer, M., N. Sckopke, F. M. Ipavich, and D. Hovestadt, Relation between energetic electrons, protons, and the thermal plasma sheet population: Plasma sheet recovery events, *J. Geophys. Res.*, *90*, 2735, 1985.

Speiser, T. W., Particle trjectories in model current sheets, 1, Analytical solutions, *J. Geophys. Res.*, *70*, 4219, 1965.

Speiser, T. W., Particle trajectories in model current sheets, 1, Analytical solutions, *J. Geophys. Res.*, *70*, 4219, 1965.

Williams, D. J., and M. Sugiura, The AMPTE Charge Composition Explorer and the 4-7 September 1984 geomagnetic storm, *Geophys. Res. Letters*, *12*, 305, 1985.

Young, D. T., Experimental evidence for ion acceleration processes in the magnetosphere (this volume), 1985.

EXPERIMENTAL IDENTIFICATION OF ELECTROSTATIC PLASMA
WAVES WITHIN ION CONIC ACCELERATION REGIONS

P. M. Kintner

School of Electrical Engineering, Cornell University, Ithaca, NY 14850

Abstract. The identification of electrostatic modes in the ionospheric and magnetospheric plasma is a difficult process. Some success has been achieved with electrostatic hydrogen cyclotron waves where Doppler broadening is insignificant and with zero-frequency turbulence where the spectrum is entirely Doppler shifted. However, it is not yet possible to identify specific modes in regions of transverse ion acceleration. If the modes are assumed to exist, some limits can be placed on their electric field amplitudes. An experimental technique to measure wavelength directly, thereby circumventing problems created by Doppler shifting, is reviewed.

I. Introduction

Perpendicular ion acceleration is a commonly measured phenomena in the high latitude ionosphere above 500 km altitude [Sharp et al., 1977; Shelley et al., 1976; Yau et al., 1983]. Nonetheless, the nature of the acceleration mechanism is not known. Three candidates have been suggested as the causitive mechanism: lower hybrid waves [Chang and Coppi, 1981], electrostatic ion cyclotron waves [Palmadesso et al., 1974; Ashour-Abdulla et al., 1982], and narrow potential jumps [Greenspan and Whipple, 1982; Greenspan, 1984; Borovsky, 1984]. Thus far no convincing experimental evidence exists to favor one mechanism over another. This is particularly true of attempts to identify specific normal modes or non-linear structures within the ionosphere.

In part, the identification problem arises from difficulty in placing a spacecraft within the acceleration region. While conical ion distributions may be measured far from their source, identification of the acceleration mechanism requires that sensors be located within the source. Furthermore, sounding rocket measurements indicate that the source is confined in altitude to a few hundred kilometers or less [Yau et al., 1983].

In part, the identification problem is produced by Doppler broadening of the wave field. That is, if the spacecraft velocity is the order of or larger than the wave phase velocity, then any discrete properties in the frequency domain will be destroyed. In principle, both the frequency and wavelength of a plasma wave should be measured to determine its mode of propagation, but in practice only the frequency is measured. In a few important cases, wavelength is also measured and we will discuss them in detail.

The remainder of this review will be divided into three sections. First, we will describe plasma wave receivers and then we will examine the experimental evidence for plasma waves associated with ion conics. In the last section, we discuss a technique for directly measuring wavelength and resolving any frequency shift ambiguities.

II. Interpretation of Electric Field Measurements

Virtually all measurements of fluctuating electric fields are performed using double probes. The probes consist either of wires, cylinders, or spheres separated by a few meters to over one hundred meters. The physical parameter measured is the potential difference between the two probes. The electric field is then inferred by dividing the potential difference by the antenna length. This simple calculation assumes that the wavelength is long compared to the probe separation and neglects the effect of any angle between the electric field vector and the antenna direction.

The full interpretation of the electric field measured within a propagating wave field is complex. The response of a double probe antenna to a propagating wave field is given by Fredricks and Coroniti [1976]. If the electric potential in the reference frame of the plasma is given by $\phi(\Omega,\underline{k})\exp\{i(\underline{k}\cdot\underline{x}-\Omega t)\}$ and the wave is assumed to be electrostatic so that $\underline{k}\times\underline{E} = 0$, then the measured electric field power spectrum is

$$S(\omega) = \int d^3k d\Omega \phi^2(\Omega,\underline{k})(\sin\underline{k}\cdot\underline{d})^2 \, \delta(\Omega-\omega-\underline{k}\cdot\underline{u})/d^2 \quad (1)$$

where $\underline{u}$ is the spacecraft velocity, $\underline{d}$ is the double probe separation vector, Ω is the plasma frame frequency, and ω is the spacecraft frame frequency.

In general this equation cannot be uniquely

Table 1
Typical Velocities in the Ionosphere

Experimental Technique	Velocity (km/s)
satellite	5-7
sounding rocket	.1-2
radar (relative to the drifting ionosphere)	1-2
Wave Phase Velocities	
Electrostatic ion cyclotron waves	
O^+ at .25 eV	2-10
H^+ at 1 eV	30-70
Lower Hybrid Waves at $k\rho=1$	
500 km altitude (f_{LHR}=3 kHz and O^+ at .25 eV)	90
5000 km altitude (f_{LHR}=300 Hz and H^+ at 1 eV)	30

inverted to find $\phi^2(\Omega,k)$ from $S(\omega)$. It does reduce in some limited cases. For example, if the wave phase velocity is independent of frequency and in a single direction, say parallel to $\underline{u}$, then ω can be related to a single wave vector ($\omega = \Omega-ku$) and

$$S(\omega) = \phi^2(\omega+ku,k)(\sin\underline{k}\cdot\underline{d})^2/d^2 \quad (2)$$

That is, the spectrum is simply Doppler shifted by ku. There are two other forms of equation (1) that interest us here. First, if the wave phase velocity is much larger than the spacecraft velocity ($\Omega/k \gg u$) then,

$$S(\omega) = \int d^3k \phi^2(\omega,k)(\sin \underline{k}\cdot\underline{d})^2/d^2 \quad (3)$$

In this case, one frequency in the plasma reference frame maps into the same frequency in the spacecraft reference frame, although there may be some ambiguity in the absolute amplitude since more than one wavevector may map into a single frequency in the spacecraft reference frame. The last case we will consider is for zero phase velocity ($\Omega=0$) and a two-dimensional distribution of wavevectors ($\phi(\underline{k}) = \phi(k_x,k_y)\delta(k_z)$). For these assumptions, equation (1) becomes,

$$S(\omega) = \int d^3k d\phi^2(\underline{k}) (\sin \underline{k}\cdot\underline{d})^2 \delta(-\omega-\underline{k}\cdot\underline{u})\delta(k_z)/d^2 \quad (4)$$

In this case, $S(\omega)=0$ when $\underline{k}\cdot\underline{d} = n\pi$ for $\omega+\underline{k}\cdot\underline{u}=0$ during the integral over k_x and k_y. If $\underline{k}$ and $\underline{d}$ and $\underline{u}$ are parallel, this simply implies that the antenna length is equal to a multiple of the wavelength when $S(\omega)=0$. A more detailed analysis of this limit is given by Temerin [1978].

Since the limiting forms of equation 1 are ordered by the ratio of the wave phase velocity to the spacecraft velocity we have listed in Table 1 some typical spacecraft or experiment velocities and wave phase velocities. The velocities for the lower hybrid mode are chosen for $k\rho=1$ at two different altitudes. The lower hybrid frequency is deduced from experimental measurements. We have chosen a minimum lower hybrid frequency which produces a minimum phase velocity.

Equation 1 is most difficult to interpret when the spacecraft velocity is the order of the wave phase velocity. From Table 1 we can see that this corresponds to satellite experiments trying to measure electrostatic oxygen cyclotron waves. In fact, a discrete spectra of electrostatic waves at multiples of the O^+ cyclotron frequency has never been measured from a satellite which indicate that, if the spectra exist at all, the spectra are Doppler broadened and not simply Doppler shifted. In the next sections we will discuss some observations which correspond to the limiting cases described above and then suggest an experimental solution to resolve the problems produced by Doppler broadening.

III. Measurements of Electric Field

Inspection of Table 1 suggests that satellite experiments should be able to measure electrostatic hydrogen cyclotron waves without any significant Doppler broadening. This has been done [Kintner et al., 1978] and an example of these waves is shown in Figure 1. In this case, the spectral power occurs at or just above the local H^+ gyrofrequency as predicted [Kindel and Kennel, 1971]. Furthermore, by comparing the width of a peak to its central frequency, we can estimate an upper bound on the Doppler broadening caused by a spacecraft moving at 7 km/s and a lower bound on the wave phase velocity. Since $\Delta f \simeq 20$ Hz and $f \simeq 100$ Hz, a lower bound on the wave phase velocity is $(f/\Delta f)$ (7 km/s) = 35 km/s which is consistent with Table 1. Further discussion of the electrostatic hydrogen cyclotron wave and

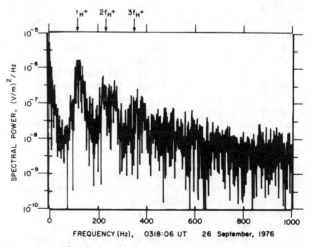

Fig. 1. Electric field power spectrum of electrostatic hydrogen cyclotron waves at 6500 km altitude in the auroral zone.

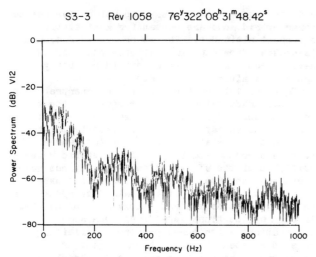

Fig. 2. Electric field power spectrum, of zero frequency turbulence. The nulls are produced when a multiple of the wavelength equals the antenna length.

sources of free energy can be found in Andre [1985a, 1985b], Bergmann [1984], and Kaufmann and Kintner [1982, 1984].

Equation 4 suggests that another type of wave should be identifiable from satellite measurements. That is, if fixed spatial electric field fluctuations ($\Omega=0$) exist in the ionosphere, they should be Doppler shifted into the spacecraft reference frame such that the power spectra exhibit nulls at $f = nu/d$, where n is an integer. An example of nulls occurring where the antenna length is a multiple of the wavelength is shown in Figure 2. For this experiment d is 37 m. The first null (n=1) at 200 Hz implies that the wave phase velocity (in the receiver reference frame) is 7.4 km/s which is the spacecraft velocity. By examining the width of the null, an upper bound on the wave phase velocity can be set at 700 m/s. Hence, the electric field receiver is simply responding to fixed spatial fluctuations.

Unfortunately, this creates a problem in the context of trying to identify electrostatic oxygen cyclotron (EOC) waves. First, the EOC mode should be found below 100 Hz where it appears that the zero frequency mode is dominant in Figure 2. Second, zero frequency turbulence is quite common in the high latitude ionosphere. Measurements of power below 100 Hz may or may not be related to EOC waves.

Next we will consider measurements of electric fields within regions of ion conic acceleration. At the time of this conference there exists only one publication reporting electric fields within the acceleration region [Kintner and Gorney, 1984]. They identified the acceleration region from ion phase space distributions peaked near 90° pitch angle.

A summary of the wave measurements is shown in Figure 3. These measurements were made near 2500 km in altitude at a magnetic local time of 9.0. The perpendicular ion conics were observed from 1017:30 U.T. to 1018:40 U.T. and the highest energy measured was about 1 keV. The power in three frequency bins is plotted. The frequency range of 40 Hz to 500 Hz covers the possibly Doppler broadened electrostatic ion cyclotron modes. The frequency range 800 Hz - 1400 Hz corresponds to the local lower hybrid resonance frequency and the frequency range 2000 Hz-5000 Hz corresponds to whistler mode waves above the LHR frequency. There was modest correlation between all three frequency ranges and the presence of perpendicular ion conics.

The total power in each frequency range can also be estimated. It was 4-9 mV/m (rms) for the 40 Hz-500 Hz range, .2-6 mV/m (rms) for the 800 Hz-1400 Hz range, and generally below 10 mV/m (rms) with a brief peak of 30 mV/m (rms) for the 2000 Hz-5000 Hz range.

Although knowing the power present in a specific frequency range is helpful for directing theoretical endeavors, we would really like to know the wave mode. For the low frequency range there was no hint of discrete structure produced by electrostatic ion cyclotron waves. The bursts of low frequency waves at roughly 10:19:30 U.T. and 10:20:00 U.T. appear to be zero frequency turbulence. In the higher frequency ranges there appeared to be a peak in spectral power at the lower hybrid frequency. However, the coupling efficiency of the waves in this frequency range to ions depends on the wavelength, which is unknown. Hence, it is not possible yet to determine which mode, if any, was responsible for accelerating the ions.

IV. Remedies to Determine Wave Dispersion Relations

Measurements in only the time domain produce much of the ambiguity in interpreting electric

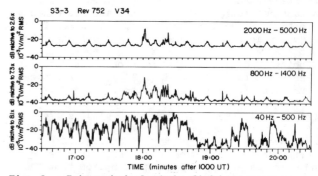

Fig. 3. Integrated electric field power over three frequency bands during a perpendicular ion acceleration event. The ion acceleration event begins at 10:17:30 U.T. and ends at 10:18:40 U.T. The three frequency ranges (40-500 Hz, 800-1400 Hz, and 2000-5000 Hz) correspond to electrostatic ion cyclotron modes, LHR modes, and modes above the LHR frequency respectively.

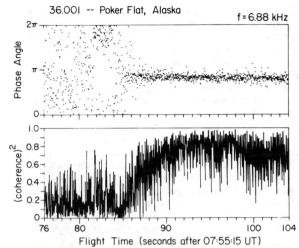

Fig. 4. Cross spectrum evaluated at 6.88 kHz during a period of auroral hiss. See text.

field measurements. If the wavevector associated with a specific frequency in the receiver reference frame is unique it is possible to determine the plasma wave dispersion relation [Kintner et al., 1984]. The experimental technique is to use two pairs of double probes to make the electric field measurement. Each pair is spatially separated and co-linear in an arrangement frequently called an interferometer. Then if a single wave crosses the two antenna pairs, there will be a phase difference between the measured electric fields. If the signals measured at probe pairs 1 and 2 are $S_1(\omega)$ and $S_2(\omega)$, the normalized cross spectrum may be calculated as

$$C_{12}(\omega) = \langle S_1 S_2^* \rangle / \sqrt{\langle S_1^2 \rangle \langle S_2^2 \rangle}$$

where $\langle \ \rangle$ implies an estimate of an ensemble average. The cross spectrum is a complex number consisting of two parts, an amplitude from 0 to 1 and a phase. If a single wave of frequency ω is present then $C_{12}(\omega) = \gamma \exp\{i \underline{k} \cdot \underline{d}\}$ where γ is the square of the coherence and $\underline{k} \cdot \underline{d}$ is the cross spectrum phase. If $\underline{k} \cdot \underline{d}$ is measured to be θ, then an upper bound for the wavelength can be found from $\gamma = (2\pi/\theta)d$. In practice the spacecraft is spinning so that $\underline{k} \cdot \underline{d}$ is changing in a sinusoidal fashion.

This technique has been successfully applied in one example thus far [Kintner et al., 1984]. In that experiment the wavelength of lower hybrid waves was measured in two separate cases. In the first example (Figure 4), the cross spectral amplitude and phase are plotted as a function of time. Separate time domain measurements show that VLF hiss began at 86 sec. At that time, the signal reached a high level of coherency and the phase organized itself. In this case we expected a phase of π radians and the offset from π can be explained with well understood systematic errors. More importantly the interferometer was spinning at .6 rps and, if $\underline{k} \cdot \underline{d}$ were non-zero, the spinning should have produced a finite modulation of the phase also at .6 rps. Hence, we can conclude that $\underline{k} \cdot \underline{d}$ was zero. This is consistent with VLF hiss propagating in a mode where $k\rho \ll 1$. In that case the wavelength should be much longer than the antenna separation (1.5 m). Futhermore, many wave vectors may be present as long as each wavelength is much longer than the antenna separation.

If the wavevector is the order of or somewhat larger than the antenna separation, $\underline{k} \cdot \underline{d}$ will have a finite value. If the cross spectrum is to have a unique interpretation at frequency ω, then only one $\underline{k}$ can produce the receiver response at ω. An example where only one $\underline{k}$ was present is shown in Figure 5. From this example, the wave energy was produced by an injected argon ion beam of energy 33 eV. The beam was turned off at 175.75 sec. Before the turn off the phase ($\underline{k} \cdot \underline{d}$) was modulated at the payload spin frequency and the modulation amplitude was $\pi/2$. This is consistent with a single wavevector producing the receiver response at 7.6 kHz. The inferred wavelength was 6 meters compared to the O^+ gyroradius of 4 meters.

V. Summary

The identification of plasma wave modes in the ionosphere and magnetosphere continues to be difficult. Electromagnetic modes can often be identified from their cut-offs and resonances [Gurnett et al., 1983]. However, Doppler shifting and broadening of electrostatic modes usually make the inferred frequencies ambiguous and destroy discrete features in the plasma frame spectra.

Even if the modes cannot be identified, it is possible to set limits on their amplitudes in specific situations. Kintner and Gorney [1984] have established limits for wave amplitudes in the presence of transverse ion acceleration. Their example is limited to only one event, the only one existing in the entire S3-3 data set. The limits are 4-9 mV/m (rms) for the possible electrostatic ion cyclotron modes, .2-6 mV/m (rms) for waves near the LHR frequency and less than 10 mV/m (rms) for waves propagating above the LHR frequency.

An effort to identify electrostatic modes, from their wavelengths is made by using spatially separated antennas (interferometers). Some success has been achieved by identifying long and short wavelength modes near the LHR resonance frequency.

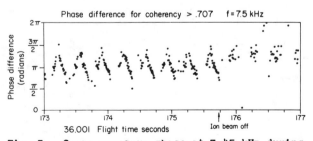

Fig. 5. Cross spectrum phase at 7.45 kHz during operation of an argon ion accelerator. See text.

Acknowledgements. I would like to thank F. Mozer for access to the S3-3 plasma wave data. The ARCS 2 ion beam experiment was a collaboration with L. J. Cahill, Jr., R. Arnoldy, and T. Moore. This research was supported by ONR Grant N00014-81-1-0018 and NASA Grant NAG5-601.

References

Andre, M., Ion waves generated by streaming particles, Ann. Geophysicae, 3, 73-80, 1985a.

Andre, M., Dispersion surfaces, J. Plasma Phys., 33, 1-19, 1985b.

Ashour-Abdalla, M., H. Okuda, and C. Z. Cheng, Acceleration of heavy ions on auroral field lines, Geophys. Res. Lett., 8, 795, 1982.

Bergmann, R., Electrostatic ion (hydrogen) cyclotron and ion acoustic wave instabilities in region of upward field-aligned current and upward ion beams, J. Geophys. Res., 89, 953-968, 1984.

Borovsky, J. E., The production of ion conics by oblique double layers, J. Geophys. Res., 89, 2251, 1984.

Chang, T. and B. Coppi, Lower hybrid acceleration and ion evolution in the superauroral region, Geophys. Res. Lett., 8, 1253, 1981.

Fredricks, R. W. and F. V. Coroniti, Ambiguities in the deduction of rest frame fluctuation spectrums from spectrums computed in moving frames, J. Geophys. Res., 81, 5591, 1976.

Greenspan, M. E. and E. C. Whipple, The effect of oblique double layers on particle magnetic moment and gyrophase, paper presented at Yosemite Conference on Origins of Plasmas and Electric Fields in the Magnetosphere, AGU, Yosemite, CA, 1982.

Greenspan, M. E., Effects of oblique double layers on upgoing ion pitch angle and gyrophase, J. Geophys. R, 89, 2847, 1984.

Gurnett, D. A., S. D. Shawhan, and R. R. Shaw, Auroral hiss, Z mode radiation, and auroral kilometric radiation in the polar magnetosphere: DE 1 observations, J. Geophys. Res., 88, 329, 1983.

Kaufmann, R. L., and P. M. Kintner, Upgoing ion beams, 1. Microscopic analysis, J. Geophys. Res., 87, 10487-10502, 1982.

Kaufmann, R. L., and P. M. Kintner, Upgoing ion beams, 2. Fluid analysis and magnetosphere-ionosphere coupling, J. Geophys. Res., 89, 2195-2210.

Kindel, J. M. and C. F. Kennel, Topside current instabilities, J. Geophys. Res., 76, 3055, 1971.

Kintner, P. M., M. C. Kelley, and F. S. Mozer, Electrostatic hydrogen cyclotron waves near one earth radius altitude in the polar magnetosphere, Geophys. Res. Lett., 5, 139, 1978.

Kintner, P. M., J. LaBelle, M. C. Kelley, L. J. Cahill, Jr., T. Moore and R. Arnoldy, Interferometric phase velocity measurements, Geophys. Res. Lett., 11, 19, 1984.

Kintner, P. M. and D. J. Gorney, A search for the plasma processes associated with perpendicular ion heating, J. Geophys. Res., 89, 937, 1984.

Palmadesso, P. J., T. P. Coffey, S. L. Ossakow, and K. Papadopoulos, Topside ionospheric ion heating due to electrostatic ion cyclotron turbulence, Geophys. Res. Lett., 1, 105, 1974.

Sharp, R. D., R. G. Johnson, and E. G. Shelley, Observation of an ionospheric acceleration mechanism producing energetic (keV) ions primarily normal to the geomagnetic field direction, J. Geophys. Res., 82, 3324, 1977.

Shelley, E. G., R. D. Sharp, and R. G. Johnson, Satellite observations of an ionosphere acceleration mechanism, Geophys. Res. Lett., 3, 654, 1976.

Temerin, M. A., The polarization, frequency, and wavelength of high-altitude turbulence, J. Geophys. Res., 83, 2609, 1978.

Yau, A. W., B. A. Whalen, A. G. McNamara, P. J. Kellogg, and W. Bernstein, Particle and wave observations of low-altitude ionospheric ion acceleration event, J. Geophys. Res., 88, 341, 1983.

A DIGEST AND COMPREHENSIVE BIBLIOGRAPHY ON TRANSVERSE AURORAL ION ACCELERATION

D. M. Klumpar

Lockheed Palo Alto Research Laboratories, Space Sciences
Laboratory, Palo Alto, California 94304

Abstract. An effort is made to unify the massive and somewhat extensive body of empirical evidence gleaned from a decade of study on the formation of conical distributions of heated ionospheric ions above the auroral ionosphere. Observations have been made at a wide range of locations and under a variety of geophysical conditions from both sounding rockets and satellites using various types of particle spectrometers operating over different energy ranges. Laboratory experiments have further verified the occurrence of transverse heating under conditions appropriate to the auroral ionosphere. While confirming the importance and the ubiquity of the transverse acceleration process, this wide diversity of empirical evidence has, at the same time, produced apparent inconsistencies and caused a certain degree of confusion regarding the fundamental acceleration mechanism and its relationship to other high latitude phenomena. The goal of this summary is to formulate an overview of our empirical understanding of the process or processes that produce conics, to attempt to resolve apparent contradictions, and to identify missing but important pieces of information that might be extracted by further analysis of existing data. Some discussion of our theoretical understanding in this field will also be presented with emphasis on points of disagreement between theory and observation. A lengthy bibliography of the primary literature on the subject is included.

Introduction

The objective of this paper is to bring together the extensive body of empirical facts and theoretical discussions on the formation of auroral ion conics and to arrive at a synopsis of our current understanding of the subject. As a formal review it is imperfect, incomplete, and intentionally does not contain the customary formal crediting of results and ideas to specific authors. In most instances the concepts expressed here have been distilled from many sources of information, both published and unpublished. Its intent is not to "review" the pertinent scientific literature, but rather, to present an overview of the current status of a sub-field of research as perceived by one individual through personal research, through personal contact with other researchers, and through the many scientific publications by which this field has advanced. Everyone who has conducted research in this field has contributed to the formulation of these impressions, often without this reporter's conscious recognition of their contribution. A thorough bibliography of research on transverse heating of ionospheric ions forms an integral part of this summary. The digest is intended not so much as a rigorous scientific treatise as it is a synopsis of the field culminating with the ideas expressed at this conference and documented by the papers contained in the *Proceedings of the Chapman Conference on Ion Acceleration in the Magnetosphere and Ionosphere.* More than one-third of the papers presented at the Chapman Conference on Ion Acceleration were concerned entirely or substantially with the generation of Transversely Accelerated Ions (TAI). Such a substantial fraction illustrates the interest in and breadth of research on the problem of transverse ion acceleration in the auroral topside ionosphere and the resultant conical ion distributions observed above the source region.

Before proceeding, it is appropriate to define the subject matter of this digest. Transversely accelerated ions (TAI) are ionospheric ions that appear to have been heated or accelerated over a relatively narrow altitude range, almost entirely in their gyro-component of motion in the earth's magnetic field. Due to the mirror force, as the ions gain kinetic energy, they are forced outward along the geomagnetic field. During the subsequent adiabatic motion of the ions, some of their transverse energy is converted into parallel energy and the ion distribution, initially confined to a small range of pitch angles centered on 90° begins to fold toward the field line, a process that continues as the ions move to greater altitudes. When observed by instruments above the source regions, the ion angular distribution is thus contained within the upgoing hemisphere and is peaked at some intermediate pitch angle, the value of which depends upon the altitude above the source region. The typical differential energy distribution of the ions monotonically decreases with increasing energy and the ions are tightly distributed along the surface of a cone in velocity space. For this reason, these measured distributions have also come to be known as conical ion distributions or just "conics."

At this point it is appropriate to establish some rules of terminology. The expression "Transversely Accelerated Ions" or 'TAI' will be used to denote the ions produced by transverse heating in a mirror geometry magnetic field and to discuss the mechanism or physical process involved in their production. The expression, "conical ion distributions" or the term "conic" will be applied to ion distributions measured at sufficient altitude above the source that their initial characteristics have been noticeably modified by transport. Chiefly, conics will have pitch angle distributions peaked along the surface of a cone whose apex half-angle is significantly less than 90°. As a general rule, then, conics are the measured distributions detected above a source of TAI after the ions have traveled a significant distance along the magnetic field.

The terms heating and acceleration will be used interchangeably in this digest despite the fact that there is a real physical distinction between the terms. Heating, albeit highly anisotropic heating, is probably a more accurate term to describe the current phenomenon, however, the final determination on this has yet to be made.

This report will summarize the principal, repeatable experimental facts that characterize TAI and the current status on the physical heating mechanism looking first at wave heating, the most popular category of proposed mechanisms, and then at alternatives to the wave heating scenario. The bulk of this report consists of a discussion of general impressions including questions, puzzles, inconsistencies, and problems. The report closes with an attempt to outline future research tasks, and a comprehensive bibliography.

Empirical Results

Morphology.

Transverse acceleration of ambient ions in a magnetized plasma in the presence of a field-aligned magnetic gradient results in the expulsion of a heated plasma along the magnetic field. The process is a ubiquitous one in the terrestrial auroral topside ionosphere and is also likely to occur in other astrophysical plasmas where similar electrodynamic conditions exist. In the terrestrial case, such ions are observed *in situ* by instrumentation on sounding rockets and satellites at altitudes from 400 km to more than 130,000 km. This altitude range represents the apparent source heights of various events observed and reported in recent years from sounding rockets and satellites capable of observing such distributions. The existence of such a wide diversity of source altitudes implies that the conditions that lead to the generation of TAI apparently can be created in nature through a wide variety of plasma environments. In contrast to the wide distribution in altitude, TAI appear to be rather confined in latitude to lie along a belt that, for all intents and purposes, coincides with the auroral oval. Thus, the conditions conducive to their generation appear to exist, in general, on flux tubes that are also associated with auroral electrodynamic processes.

Conics have been observed at all local times, but certain characteristic anisotropies in the local time distribution as a function of altitude and season give us additional clues on the generation of TAI. From the ISIS-2 satellite at an observing altitude of 1400 km, the occurrence of TAI was found to be limited to the winter season. During 1971 and 1972, the occurrence probability for observing TAI in the northern hemisphere dropped to near zero during the summer months. During the winter months at 1400 km, the presence of TAI were most probable in the nighttime; only a few observations of TAI occurred in the dayside cusp. Other measurements made at higher altitudes, up to 3500 km from ISIS-1 and even higher by S3-3, showed increased probabilities for the production of TAI in the cusp ionosphere. Taken together, these observations strongly suggest a control of the TAI generation mechanism effected by local plasma density. Where the density is highest (cusp, dayside, low altitudes), the TAI heating mechanism is suppressed; at lower ambient densities, the mechanism becomes more prolific.

Energy Distribution.

Another clue to the nature of the TAI acceleration mechanism is carried by the energy distribution of the ions. The differential flux is typically found to obey a power law in energy with an exponent near -1.8 over the energy range from a few eV to 15 keV. Such an energy distribution suggests a heating type mechanism rather than one that imparts a fixed energy to all ions.

Angular Distribution.

The shape of the angular distributions contains additional important clues: Since the conic distributions are confined along the surface of a single cone, we are led to conclude that the source region is narrowly confined in altitude. If the transverse acceleration were to be distributed in altitude, the measured distribution would contain ions that had been heated at different altitudes below the satellite. Those from the nearer or higher altitude part of the source would have experienced little adiabatic folding and hence would be detected near 90° pitch angle. At the same time, ions from the distant parts of the source below the satellite would experience significant adiabatic folding and arrive at pitch angles closer to the field line. In the extreme case, a source reaching from the ionosphere up to the satellite would produce a distribution that contained ions at all pitch angles between 90° and the edge of ionospheric source cone. Such broad distributions are not observed, so we conclude that the source is confined in altitude. No examples of double conics (or multiple conics) have ever appeared in the literature. In principle, such a distribution could exist and would be the signature of multiple sources on the flux tube below the satellite. That none have been reported indicates that on any given auroral flux tube there can be only one active region, *i.e.*, only one confined source of transversely accelerated ions at any one time. This suggests a localized instability along a field line for which the proper conditions only exist at a single point *i.e.*, over a single narrow slice of altitude.

The narrowness of the observed angular distributions also suggests that the source mechanism must accelerate the ions at precisely 90° to within a very small range of angles (perhaps ± 1°). If ions were to be generated at smaller pitch angles (say 80°), they would move downward until they reached their mirror point. When finally observed at higher altitude, they would be deduced to have come from a lower altitude than the actual source. At altitudes between the source and the mirror point downcoming conics would be observable. Likewise, ions generated at a symmetric pitch angle larger than 90° (*e.g.*, 100°) and observed at a higher altitude would also appear to come from lower altitude than the source. In fact, an acceleration mechanism that produces an ion distribution oblique to the field direction but symmetric in velocity space about $v_\parallel = 0$ would not be distinguishable from a pure transverse acceleration source when viewed from above. The fact that no downgoing conics have been reported argues against this later source distribution and reinforces the interpretation that the source mechanism that produces conics heats the ions in a purely transverse direction.

Association with Auroral Electrons.

The evidence that conics are found along the auroral oval naturally leads one to associate their presence with the precipitating auroral electrons (*i.e.*, inverted-Vs) that have so convincingly been associated with auroral arcs. Indeed, conics are often found among regions of auroral electron precipitation. There is evidence that TAI may be produced near the high and low latitude boundaries of inverted-Vs as well as below the potential drop associated with the inverted-V. On the other hand, there have been a number of instances reported in which conics were found widely separated from auroral electrons. The latitudinal scale size of many conic events is also inconsistent with the typical scale size of the intense electron precipitation regions that form auroral arcs. Conics are often observed for many hundreds of kilometers along a satellite orbit, while the structures associated with electron acceleration are seldom more than a hundred kilometers. One must conclude that the generation of TAI may occur under, but does not *require*, the presence of auroral electrons. Close scrutiny of some low altitude TAI events found in the presence of auroral electrons has even revealed a decrease in the electron flux at the same pitch angle associated with the presence of conical ions. Thus, any mechanism for production of TAI that requires the presence of significant fluxes of electrons of energies more than the order of 100 eV precipitating along the magnetic field must be carefully scrutinized. Recently, several authors have noted that low energy ($\leq$ 100 eV), highly field-aligned, upstreaming (or counterstreaming) electrons are often observed at the same time as ion conics. The ability to detect such low energy highly field-aligned electrons from a satellite requires specific circumstances that are not often present. Therefore, the limited number of such reported observations does not necessarily mean that such events are pathological or coincidental.

Association with Field-Aligned Currents.

There is a well documented association between the generation of TAI and the presence of field-aligned currents. Conics are observed on flux tubes characterized as being part of the Birkeland current system. Since field-aligned currents, both the primary current and the "return current," along auroral flux tubes are generally thought to be carried dominantly by electrons, we must finally conclude that the presence of conics is indeed associated with the presence of electrons. The primary distinction that is being made here in contrast to the previous discussion is that TAI do not require the presence of primary auroral electrons for their generation, but do seem to require a relative flow between electrons and the ambient plasma, creating a field-aligned current. Note, it may be sufficient that cold ionospheric electrons or the low energy upstreaming electrons discussed above carry this current.

A further observational fact relating TAI generation with field-aligned currents is an apparent relationship between the current density, ambient plasma density, and the presence of TAI. At 1400 km it is found that conics are observed when the field-aligned current density (either upward or downward) exceeds a threshold value that is itself proportional to ambient ion density. That is, the mechanism that generates transversely accelerated ions requires a relatively larger field-aligned current when the local plasma density is high. As one looks at successively more dilute plasmas, the field-aligned current density required to trigger the TAI mechanism becomes less. This result partially supports the previously discussed seasonal variation. In sunlight, the plasma density at a given altitude in the ionosphere is enhanced over its darkened state. Thus, at a particular altitude, generation of TAI will not proceed at the same value of field-aligned current density as it would have in a darkened ionosphere. In other words, for a given field-aligned current density, the active region for generation of TAI will occur at a higher altitude in the dayside relative to the nightside. This discussion raises a problem with respect to the limited altitudinal extent of the source which will be discussed in the following section.

Conics are observed in regions where the field-aligned current is upward as well as where it is downward. The mechanism does not seem to depend upon the relative sign of the field-aligned electron drift that carries the current. Conics are often observed at distinct boundaries separating regions of upward current from downward current, a fact that has not yet been adequately explained.

Composition.

Measurements of the composition of ion conics at energies above 200 eV have been performed and reported in the literature, but investigations aimed specifically at studying the mass composition of conics are few. Conical ion distributions are found involving all major terrestrial ion species (H^+, He^+, O^+), although not necessarily together. Conics have also been observed in He^+, sometimes in association with one or both of H^+ and O^+ with varying fractional concentrations, and sometimes alone. The presence of both H^+ and O^+ together in conics is relatively common. The relative intensities of the H^+ and O^+ can be quite variable from event to event and either species can be dominant. Clear conics have been observed in O^+ while simultaneous H^+ measurements show no evidence of transverse acceleration above the flux levels of ambient H^+. As mentioned, the published literature contains few studies where the composition of pure ion conics was specifically investigated for mass dependent effects. The very limited results are consistent with both components of mixed (H^+ and O^+) conics having been accelerated to the same temperature by a transverse acceleration process acting at the same altitude for both species. No mass dependent differences have been reported in pure conic events where both H^+ and O^+ were present. This is in strong contrast to the mass dependent differences associated with upflowing ion beams where significant differences between H^+ and O^+ have been reported in a number of studies. It is noteworthy that ion beams often contain mass dependent differences that have been attributed to transverse acceleration acting in addition to the parallel acceleration mainly responsible for these upstreaming ion beams, while in pure conics no mass dependence is reported. In the final analysis, it is not clear whether there is a particular transverse acceleration mechanism associated with ion beams that is mass dependent, or whether the mass dependencies observed in ion beams result from other processes occurring during the transport of the ions above the accelerator. There is a clear need for a definitive study of compositional characteristics of ion conics. It has not been established whether the composition of ion conic events has any relationship to source altitude, current density, electron flux, season, or any other geophysical parameter.

Solar Cycle Dependence.

It has recently been suggested that the transverse ion acceleration process may in some way be modulated in concert with the 11-year solar cycle. It is generally accepted that increased solar EUV heating increases the topside ionospheric scale height. There is concomitant strong evidence that the population of energetic ($\sim$ 1-17 keV) O^+ ions in the magnetosphere is also modulated with the 11-year solar cycle. It has been suggested that the TAI source mechanism might shift in altitude or be more or less effective at heating heavy (*i.e.*, O^+) ions as the solar EUV flux waxes and wanes, thus explaining (at least in part) the magnetospheric 11-year compositional variations. Again, hard evidence is difficult to come by and careful studies of a number of data sets involving a solar cycle, or a significant fraction thereof, will be required to understand this solar modulation.

Association with Plasma Waves.

Transversely accelerated ions are often found in loose association with plasma waves of various types, and much of the theoretical effort devoted to the TAI source mechanism has implicated wave–particle interaction in one way or another. Our empirical knowledge of wave–TAI associations is in an equally confusing state. Plasma waves are ubiquitous on auroral flux tubes. Given the large occurrence frequencies of conics in the same region of space, it should not then be too surprising to find associations between ion conics and plasma waves. The hard evidence for a one-to-one correspondence between transverse ion acceleration and any particular type of plasma wave responsible for the ion heating is lacking. A clear distinction between the type of waves associated with ion beams and the type associated with conics seems to be present in the S3-3 data. Electrostatic hydrogen cyclotron waves are highly correlated with upstreaming ions that have angular distributions peaked along the field line (beams). Often these ion beams show unexpectedly wide angular distributions indicating that perhaps the waves are interacting with the ion beams to effectively pitch angle scatter the ions. Pure ion conics on the other hand do not exhibit this close association with EIC waves. In the above sentence "pure" refers to conics not located on field lines where there is also evidence for parallel electric potentials below the satellite. There have been reported correlations between conics and the ion cyclotron harmonic mode, but the evidence in these instances is that the ion conic is the source of energy and that energy is being transferred from the ions to the wave. Other waves known as VLF saucers are also found to be associated with conics but the wave amplitudes are not sufficient to be the source of energy for the ions. One must recognize that there are a number of experimental difficulties that stack the deck against easily finding the waves that are responsible for generation of TAI (if, indeed, it is a wave driven process). Both the ions and the waves propagate. Typically one is not so lucky as to measure the ions in the source region, and thus the presence or absence of waves at the ion detection point may have no relationship to whether the waves are or are not present in the source. Furthermore, satellite velocities are comparable to the wave phase velocities, which brings up the complication that the responsible waves will be Doppler shifted, perhaps even outside of the range of detection. The definitive experiment, in which a number of TAI source regions are sampled and waves of a particular type having sufficient power to impart the observed energy to the ions are consistently detected, has yet to be done. In summary, the status of wave–conic association is that there is no conclusive empirical evidence that waves are responsible for the transverse acceleration of ionospheric ions in the high latitude ionosphere.

One exception to this dismal state of empirical data on wave–particle observations in perpendicular ion heating exists in the case of apparent transverse heating of He^+ ions in the equatorial plane. There, both experiment and theory seem to agree that electromagnetic ion cyclotron waves are present with sufficient amplitudes to transversely heat He^+ ions observed in this region of space.

The theoretical literature on the subject of perpendicular ion acceleration may be divided into two groupings for the purpose of the present discussion: A large group consists of those papers that advocate the TAI mechanism being due to heating of ambient ionospheric ions by plasma waves of one type or another with electrostatic ion cyclotron waves and lower hybrid waves being the two classes having been given the greatest attention. A second group consists of a substantially fewer number of authors who advocate transverse acceleration by small-scale size, quasi-static oblique electric field structures. Both groups contain successes and failures in explaining various aspects of the TAI heating processes, but to date no single theory has emerged that clearly explains the dominant features found in the observations.

Discussion

The objective in the following discussion is to qualitatively assess our current state of knowledge of transverse ion acceleration and the resultant conical ion distributions. It is the intent to expose inconsistencies where they exist, ask questions that seem to need asking, present puzzles, reveal unsolved problems, laud our accomplishments to date, and project into the future with a perhaps somewhat biased view of what should be done to ameliorate those questions, puzzles, and unsolved problems.

It has been just ten years since the first scientific presentation was made on those unanticipated but persistent stripes appearing in ISIS soft particle spectrograms and just eight years since the first publication of S3-3 observations confirming the existence of transverse acceleration of ionospheric ions. The field has experienced explosive growth since then: measurements have been expanded and significant interpretive theoretical effort has resulted in a number of potentially attractive theories for the generation of TAI.

Given the apparently wide diversity of plasma conditions under which transverse heating of ions can occur one must be open minded to the possibility that we are dealing with more than one physical mechanism. The low altitude TAI, those generated below the parallel acceleration region, may constitute a separate class from those seen as a persistent boundary or sheath layer at the edges of upward streaming ion beam structures. The conics found closely associated with electron inverted-Vs may have a separate origin from those observed in regions of upward current flow where no apparent auroral electrons exist. One difficulty with the hypothesis of multiple generation mechanisms is that experimentally there is no hard evidence that the observed conics have any intrinsic characteristic differences that might substantiate the existence of differing source mechanisms. It has been suggested, for example, that conics observed at the edges of inverted-Vs might owe their existence to oblique electric fields, while conics in high field–aligned current regions apart from inverted-Vs might have been generated by electrostatic ion cyclotron waves which are known to be unstable to field–aligned currents. If this is the case, might not one expect to find some differences in the angular or energy distributions of the accelerated ions? Yet no such distinctions have been reported.

Theoretical efforts to reproduce the ion distributions observed in conics have to a large extent concentrated on the source region generation mechanism without acknowledging that the measured conics have often traveled a substantial distance since leaving the source. During this transport, their distributions have evolved and been modified by a number of processes. Parallel acceleration or deceleration, additional transverse heating, cross field convection, velocity dispersion, two stream coupling and energy transfer, charge exchange, scattering, in addition to the gradient B force are all processes that will act to a greater or lesser extent to modify the ion distributions between the source and their detection at the satellite. Any theory that is to successfully reproduce the observed ion conic distributions must take these factors into account. Some progress has recently been made in this regard but it is clear that there is still room for significant improvement.

Apparent mass dependent differences in measured upflowing distributions have been reported and often misinterpreted by the casual reader as applying to ion conics. Theoretical work has often indicated that heavy ions may undergo preferential acceleration as a result of lower thresholds for instability than light ions. As discussed earlier, no such effects have been reported in pure conics dissociated with ion beams. It is not clear to this observer that the mass dependent effects seen in upflowing ion beams necessarily need to arise in the source. Is it not equally plausible that extraneous effects such as residence time, pressure cooker where a downward electric field constrains escape, transport dispersion, or energy transfer from light to heavy ions during transport might account for some or all of the species dependent difference? This question can be addressed by considering these additional effects in a comprehensive theoretical treatment as called out in the preceeding paragraph. Additionally, experimentalists need to look more carefully at measurements made in the source region where transport effects will not be present, or will at least be minimized.

Energetic (> 100 keV) molecular ions of terrestrial origin (NO^+, O_2^+, N_2^+) have recently been observed in the terrestrial magnetosphere following periods of high geomagnetic activity. The pathway by which such ions become heated to 10^5 times their mean thermal energy in the ionosphere and appear deep in the magnetosphere remains a mystery. Very recent observations of molecular ions at $1 - 3$ R_e in the polar cap at energies of tens of eV that suggest an energization process starting in the F–region might be an indication of at least part of that pathway. The possible role that transverse heating in the high latitude ionosphere might play in at least the initial stages of acceleration and ejection from their normal location in the E and F–region ionosphere needs both theoretical and empirical attention.

The question of how ionospheric ions, particularly the heavy component, reach up to the altitudes where in some cases they are observed to be heated has not been resolved. Current wisdom holds that H^+ is the dominant species above the $H^+ : O^+$ transition height which occurs no higher than about 2000 km for almost all situations. Yet O^+ conics are often reported coming from source altitudes well above the transition to H^+ where the expected ambient densities of O^+ should not be sufficient to produce the observed ion intensities. Similarly puzzling is the evidence that upward flowing ion beams containing significant O^+ fluxes appear to be accelerated upward in parallel potential regions reaching not much below 5000 km. It has not been determined what mechanism feeds these high altitude sources with heavy ions. It has been speculated that the transverse acceleration mechanism might be capable of feeding the high altitude source with sufficient ion fluxes. In the final analysis no hard data exist to support this conjecture. It is clear from sounding rocket observations and from the ISIS-2 data that the TAI source does indeed operate at sufficiently low altitudes where copious quantities of O^+ are available. Simultaneous and spatially separated observations within an upward flowing ion event and below the potential drop are needed to confirm whether or not low energy conical O^+ distributions are present directly below a source of upflowing O^+ beams. Perhaps analysis of data from the DE pair of auroral satellites will yet provide the evidence necessary to answer this important question.

There are important details of the acceleration process that are missing from the observations which could help to distinguish among the currently viable theories. Increased energy resolution at low energies should be able to distinguish between tail heating processes and core formation (bulk acceleration) when such measurements are made in the source region.

Summary

The transverse heating of ionospheric ions in the high latitude topside ionosphere is recognized as a fundamental process that supplies a significant flux of terrestrial plasma to the magnetosphere. During a decade of research, much has been learned about the occurrence and morphology of transversely accelerated ions, but more detailed investigations within the source regions are now required to determine the physical mechanism(s) responsible for the ion heating. Theoretical efforts have produced a number of potentially viable mechanisms but in many cases, the definitive measurements required to select among these theories are lacking. High spatial/temporal resolution measurements of plasma composition, densities, temperatures, and velocities in combination with full wave fields in the source regions must be performed. Simultaneous measurements at other locations on the flux tube above and below the active region will be required to formulate a complete picture that shows the intimate interrelationship between TAI and the other auroral processes known to exist on these flux tubes.

Bibliography

The following bibliography is intended to be an exhaustive reference to all primary refereed publications on the subject of transverse ion acceleration of ionospheric ions and their subsequent injection as ion conics into the magnetosphere. Secondary publications (e.g., reviews and republications as conference proceedings) are included to the extent possible. The bibliography also contains references on upflowing ionospheric ions (beams) at energies above the classical polar wind and at latitudes poleward of the low altitude plasmapause when the authors suggested or included transverse acceleration or ion heating. Ionospheric ions observed in the magnetosphere are included if the authors interpreted their measurements as resulting from direct injection of ionospheric ions. Papers dealing with waves, field-aligned currents, electric fields, electrons, etc. associated with ion conics and TAI are also included. Theoretical and experimental papers that suggest consequences resulting from the formation of conical ion distributions are included. Examples of this latter category included instabilities generated by the presence of conical ion distributions.

A few early theoretical papers that predate the discovery of transverse acceleration of ions in space plasmas but which, in retrospect, have contributed to our fundamental understanding of the associated plasma physics have also been incorporated in the listing.

Acknowledgments. The author acknowledges fruitful discussions with E.G. Shelley, W. Lennartsson, A.G. Ghielmetti, H.L. Collin, M. Temerin and C. Cattell in the preparation of this manuscript. This material is based upon work supported by the National Sciences Foundation, Division of Atmospheric Sciences, under grant number ATM-8516120.

Ashour-Abdalla, M. and H. Okuda, Transverse acceleration of ions on auroral field lines, in *Energetic Ion Composition in the Earth's Magnetosphere*, edited by R. G. Johnson, pp. 43-72, Terra Scientific Publishing Co., Tokyo, 1983.

Ashour-Abdalla, M. and H. Okuda, Turbulent heating of heavy ions on auroral field lines, *J. Geophys. Res., 89,* 2235, 1984.

Ashour-Abdalla, M., H. Okuda, and C.Z. Cheng, Acceleration of heavy ions on auroral field lines, *Geophys. Res. Lett., 8,* 795, 1981.

Barakat, A.R. and R.W. Schunk, O^+ ions in the polar wind, *J. Geophys. Res., 88,* 7887, 1983.

Barakat, A.R. and R.W. Schunk, O^+ charge exchange in the polar wind, *J. Geophys. Res., 89,* 9835, 1984.

Baugher, C.R., C.R. Chappell, J.L. Horwitz, E.G. Shelley, and D.T. Young, Initial thermal plasma observations from ISEE-1, *Geophys. Res. Lett., 7,* 657, 1980.

Bennett, E.L., M. Temerin, and F.S. Mozer, The distribution of auroral electro-static shocks below 8000-km altitude, *J. Geophys. Res., 88,* 7107, 1983.

Bergmann, R. and W. Lotko, Transition to unstable ion flow in parallel electric fields, preprint, 1985.

Borovsky, J.E., The production of ion conics by oblique double layers, *J. Geophys. Res., 89,* 2251, 1984.

Borovsky, J.E. and G. Joyce, Numerically simulated two-dimensional auroral double layers, *J. Geophys. Res., 88,* 3116, 1983.

Candidi, M., S. Orsini and V. Formisano, The properties of ionospheric O^+ ions as observed in the magnetotail boundary layer and northern plasma lobe, *J. Geophys. Res., 87,* 9097, 1982.

Cattell, C., The relationship of field-aligned currents to electrostatic ion cyclotron waves, *J. Geophys. Res., 86,* 3641, 1981.

Cattell, C., Association of field-aligned currents with small-scale auroral phenomena, in *Magnetospheric Currents*, edited by T.A. Potemra, *AGU Mono.* 28, p. 304, American Geophysical Union, Washington, D.C., 1984.

Cattell, C. and M. Hudson, Flute mode waves near ω_{LH} excited by ion rings in velocity space, *Geophys. Res. Lett., 9,* 1167, 1982.

Cattell, C., R. Lysak, R.B. Torbert, and F.S. Mozer, Observations of differences between regions of current flowing into and out of the ionosphere, *Geophys. Res. Lett., 6,* 621, 1979.

Chang, T. and B. Coppi, Lower hybrid acceleration and ion evolution in the suprauroral region, *Geophys. Res. Lett., 8,* 1253, 1981.

Chiu, Y.T., J.M. Cornwall, J.F. Fennell, D.J. Gorney, and P.F. Mizera, Auroral plasmas in the evening sector: Satellite observations and theoretical interpretations, *Space Sci. Rev., 35,* 211, 1983.

Cladis, J. B., Effect of magnetic field gradient on motion of ions resonating with ion cyclotron waves, *J. Geophys. Res., 78,* 8129, 1973.

Cladis, J.B. and W.E. Francis, The polar ionosphere as a source of the storm time ring current, *J. Geophys. Res., 90,* 3465, 1985.

Cladis, J.B. and R.D. Sharp, Scale of electric field along magnetic field in an inverted-V event, *J. Geophys. Res., 84,* 6564, 1979.

Cole, K. D., Effects of crossed magnetic and (spatially dependent) electric fields on charged particle motion, *Planet. Space Sci., 24,* 515, 1976.

Collin, H.L., R.D. Sharp, E.G. Shelley, and R.G. Johnson, Some general characteristics of upflowing ion beams over the auroral zone and their relationship to auroral electrons, *J. Geophys. Res., 86,* 6820, 1981.

Collin, H.L., R.D. Sharp, and E.G. Shelley, The occurrence and characteristics of electron beams over the polar regions, *J. Geophys. Res.*, 87, 7504, 1982.

Collin, H.L., R.D. Sharp, and E.G. Shelley, The magnitude and composition of the outflow of energetic ions from the ionosphere, *J. Geophys. Res.*, 89, 2185, 1984.

Collin, H.L. and R.G. Johnson, Some mass dependent features of energetic ion conics over the auroral regions, *J. Geophys. Res.*, 90, 9911, 1985.

Comfort, R.H. and J.L. Horwitz, Low energy ion pitch angle distributions observed on the dayside at geosynchronous altitudes, *J. Geophys. Res.*, 86, 1621, 1981.

Craven, P.D., R.C. Olsen, C.R. Chappell, and L. Kakani, Observations of molecular ions in the earth's magnetosphere, *J. Geophys. Res.*, 90, 7599, 1985.

Crew, G.B. and T.S. Chang, Asymptotic theory of ion conic distributions, *Phys. Fluids*, 28, 2382, 1985.

Curtis, S.A., Equatorial trapped plasmasphere ion distributions and transverse stochastic acceleration, *J. Geophys. Res.*, 90, 1765, 1985.

Curtis, S.A. and C.S. Wu, Gyroharmonic emissions induced by energetic ions in the equatorial plasmasphere, *J. Geophys. Res.*, 84, 2597, 1979.

Dakin, D.R., T. Tajima, G. Benford, and N. Rynn, Ion heating by electrostatic ion cyclotron instability: theory and experiment, *J. Plasma Physics*, 15, 175, 1976.

Dusenbery, P.B. and L.R. Lyons, Generation of ion-conic distribution by upgoing ionospheric electrons, *J. Geophys. Res.*, 86, 7627, 1981.

Dusenbery, P. B. and L. R. Lyons, The generation of ion-conics via quasi-linear diffusion, in *Physics of Auroral Arc Formation*, edited by S.-I. Akasofu and J. R. Kan, p. 456, AGU, Washington, D.C., 1981.

Fennell, J.F., P.F. Mizera, and D.R. Croley, Jr., Observations of ion and electron distributions during the July 29 and July 30, 1977 storm period, *Proc. of Magnetospheric Boundary Layers Conference*, Albach, 11-15 June 1979, p. 97, ESA SP-148, European Space Agency, Paris, 1979.

Fennell, J.F., D.R. Croley, Jr., and S.M. Kaye, Low-energy ion pitch angle distributions in the outer magnetosphere: ion zipper distributions, *J. Geophys. Res.*, 86, 3375, 1981.

Gendrin, R. and A. Roux, Energization of helium ions by proton-induced hydromagnetic waves, *J. Geophys. Res.*, 85, 4577, 1980.

Ghielmetti, A.G., R.G. Johnson, R.D. Sharp, and E.G. Shelley, The latitudinal, diurnal, and altitudinal distributions of upward flowing energetic ions of ionospheric origin, *Geophys. Res. Lett.*, 5, 59, 1978.

Ghielmetti, A.G., R.D. Sharp, E.G. Shelley, and R.G. Johnson, Downward flowing ions and evidence for injection of ionospheric ions into the plasma sheet, *J. Geophys. Res.*, 84, 5781, 1979.

Gombosi, T.I., T.E. Cravens, and A.F. Nagy, A time-dependent theoretical model of the polar wind: preliminary results, *Geophys. Res. Lett.*, 12, 167, 1985.

Gorney, D.J., An alternative interpretation of ion ring distributions observed by the S3-3 satellite, *Geophys. Res. Lett.*, 10, 417, 1983.

Gorney, D.J., A. Clarke, D. Croley, J. Fennell, J. Luhmann, and P. Mizera, The distribution of ion beams and conics below 8000 km, *J. Geophys. Res.*, 86, 83, 1981.

Gorney, D.J., S.R. Church, and P.F. Mizera, On ion harmonic structure in auroral zone waves: The effect of ion conic damping of auroral hiss, *J. Geophys. Res.*, 87, 10,479, 1982.

Gorney, D.J., Y.T. Chiu, and D.R. Croley, Jr., Trapping of ion conics by downward parallel electric fields, *J. Geophys. Res.*, 90, 4205, 1985.

Green, J.L. and J.H. Waite, Jr., On the origin of the polar ion streams, *Geophys. Res. Lett.*, 12, 149, 1985.

Greenspan, M.E., Effects of oblique double layers on upgoing ion pitch angle and gyrophase, *J. Geophys. Res.*, 89, 2842, 1984.

Gurgiolo, C. and J.L. Burch, DE-1 observations of the polar wind — a heated and an unheated component, *Geophys. Res. Lett.*, 9, 945, 1982.

Gurgiolo, C. and J.L. Burch, Composition of the polar wind — not just H^+ and He^+, *Geophys. Res. Lett.*, 12, 69, 1985.

Heelis, R.A., J.D. Winningham, M. Sugiura, and N.C. Maynard, Particle acceleration parallel and perpendicular to the magnetic field observed by DE-2, *J. Geophys. Res.*, 89, 3893, 1984.

Horita, R.E. and H.G. James, Source regions deduced from attenuation bands in VLF saucers, *J. Geophys. Res.*, 87, 9147, 1982.

Horwitz, J.L., Conical distributions of low-energy ion fluxes at synchronous orbit, *J. Geophys. Res.*, 85, 2057, 1980.

Horwitz, J., The ionosphere as a source for magnetospheric ions, *Rev. Geophys. Space Physics*, 20, 929, 1982.

Horwitz, J. L., Residence time heating effect in auroral conic generation, *Planet Space Sci.*, 32, 1115, 1984.

Horwitz, J.L., Features of ion trajectories in the polar magnetosphere, *Geophys. Res. Lett.*, 11, 1111, 1984.

Horwitz, J.L. and C.R. Chappell, Observations of warm plasma in the dayside plasma trough at geosynchronous orbit, *J. Geophys. Res.*, 84, 7075, 1979.

Horwitz, J.L., C.R. Baugher, C.R. Chappell, E.G. Shelley, and D.T. Young, Conical pitch angle distributions of very low-energy ion fluxes observed by ISEE-1, *J. Geophys. Res.*, 87, 2311, 1982.

Hudson, M.K., R.L. Lysak, and F.S. Mozer, Magnetic field-aligned potential drops due to electrostatic ion cyclotron turbulence, *Geophys. Res. Lett.*, 5, 143, 1978.

Hultqvist, B., On the origin of the hot ions in the disturbed dayside magnetosphere, *Planet. Space Sci.*, 31, 173, 1983.

Ionson, J.A., R.S.B. Ong, and E.G. Fontheim, Turbulent transport and heating in the auroral plasma of the topside ionosphere, *Planet Space Sci.*, 27, 203, 1979.

Kan, J.R., L.C. Lee, and S.-I. Akasofu, Two-dimensional potential double layers and discrete auroras, *J. Geophys. Res.*, 84, 4305, 1979.

Kaufmann, R.L., What auroral electron and ion beams tell us about magnetosphere-ionosphere coupling, *Space Sci. Rev.*, 37, 313, 1984.

Kaufmann, R.L. and P.M. Kintner, Upgoing ion beams: 1. Microscopic analysis, *J. Geophys. Res.*, 87, 10487, 1982.

Kaufmann, R.L. and P.M. Kintner, Upgoing ion beams:

2. Fluid analysis and magnetosphere-ionosphere coupling, *J. Geophys. Res.*, *89*, 2195, 1984.

Kaye, S.M., R.G. Johnson, R.D. Sharp, and E.G. Shelley, Observations of transient H^+ and O^+ bursts in the equatorial magnetosphere, *J. Geophys. Res.*, *86*, 1335, 1981.

Kaye, S.M., E.G. Shelley, R.D. Sharp, and R.G. Johnson, Ion composition of zipper events, *J. Geophys. Res.*, *86*, 3383, 1981.

Kindel, J.M. and C.F. Kennel, Topside current instabilities, *J. Geophys. Res.*, *76*, 3055, 1971.

Kintner, P.M., On the distinction between electrostatic ion cyclotron waves and ion cyclotron harmonic waves, *Geophys. Res. Lett.*, *7*, 585, 1980.

Kintner, P. M. and M. Temerin, Low frequency waves and irregularities in the low altitude boundary layer, *Proc. of Magnetospheric Boundary Layers Conference*, Alpbach, 11-15 June 1979, p. 209, ESA SP-148, European Space Agency, Paris, 1979.

Kintner, P.M. and D.J. Gorney, A search for the plasma processes associated with perpendicular ion heating, *J. Geophys. Res.*, *89*, 937, 1984.

Kintner, P.M., M.C. Kelley, F.S. Mozer, Electrostatic hydrogen cyclotron waves near one earth radius altitude in the polar magnetosphere, *Geophys. Res. Lett.*, *5*, 139, 1978.

Kintner, P.M., M.C. Kelley, R.D. Sharp, A.G. Ghielmetti, M. Temerin, C. Cattell, P.F. Mizera, and J.F. Fennell, Simultaneous observations of energetic (keV) upstreaming and electrostatic hydrogen cyclotron waves, *J. Geophys. Res.*, *84*, 7201, 1979.

Klumpar, D.M., Transversely accelerated ions: An ionospheric source of hot magnetospheric ions, *J. Geophys. Res.*, *84*, 4229, 1979.

Klumpar, D.M., Transversely accelerated ions in auroral arcs, in *Physics of Auroral Arc Formation*, edited by S.-I. Akasofu and J. R. Kan, p. 122, AGU, Washington, D.C., 1981.

Klumpar, D. M., Characteristics of the high-latitude ionospheric particle sources: transversely accelerated ions (TAI) at 1400 km, *Adv. Space. Res.*, *5*, 145, 1985.

Klumpar, D.M. and W.J. Heikkila, Electrons in the ionospheric source cone: evidence for runaway electrons as carriers of downward birkeland currents, *Geophys. Res. Lett.*, *9*, 873, 1982.

Klumpar, D.M., W.K. Peterson, and E.G. Shelley, Direct evidence for two-stage (bimodal) acceleration of ionospheric ions, *J. Geophys. Res.*, *89*, 10779, 1984.

Lennartsson, W., On the consequences of the interaction between the auroral plasma and the geomagnetic field, *Planet Space Sci.*, *28*, 135, 1980.

Lockwood, M., Thermal ion flows in the topside auroral ionosphere and the effects of low-altitude, transverse acceleration, *Planet Space Sci.*, *30*, 595, 1982.

Lockwood, M., Thermospheric control of the auroral source of O^+ ions for the magnetosphere, *J. Geophys. Res.*, *89*, 301, 1984.

Lockwood, M. and J.E. Titheridge, Ionospheric origin of magnetospheric O^+ ions, *Geophys. Res. Lett.*, *8*, 381, 1981.

Lockwood, M., J.H. Waite, Jr., T.E. Moore, J.F.E. Johnson, and C.R. Chappell, A new source of suprathermal O^+ ions near the dayside polar cap boundary, *J. Geophys. Res.*, *90*, 4099, 1985.

Lundin, R., B. Hultqvist, E. Dubinin, A. Zackarov, and N. Pissarenko, Observations of outflowing ions beams on auroral field lines at altitudes of many earth radii, *Planet. Space Sci.*, *30*, 715, 1982.

Lyons, L.R. and T.E. Moore, Effects of charge exchange on the distribution of ionospheric ions trapped in the radiation belts near synchronous orbit, *J. Geophys. Res.*, *86*, 5885, 1981.

Lysak, R. L., Electron and ion acceleration by strong electrostatic turbulence, in *Physics of Auroral Arc Formation*, edited by S.-I. Akasofu and J. R. Kan, p. 444, AGU, Washington, D.C., 1981.

Lysak, R.L., M.K. Hudson, and M. Temerin, Ion heating by strong electrostatic ion cyclotron turbulence, *J. Geophys. Res.*, *85*, 678, 1980.

Malingre, M. and R. Pottelette, Excitation of broadband electrostatic noise and of hydrogen cyclotron waves by a perpendicular ion beam in a multi-ion plasma, *Geophys. Res. Lett.*, *12*, 275, 1985.

Menietti, J.D. and J.L. Burch, "Electron conic" signatures observed in the nightside auroral zone and over the polar cap, *J. Geophys. Res.*, *90*, 5345, 1985.

Menietti, J.D., J.D. Winningham, J.L. Burch, W.K. Peterson, J.H. Waite, and D.R. Weimer, Enhanced ion outflows measured by the DE-1 high altitude plasma instrument in the dayside plasmasphere during the recovery phase, *J. Geophys. Res.*, *90*, 1653, 1985.

Mitchell, Jr., H.G. and P.J. Palmadesso, O^+ acceleration due to resistive momentum transfer in the auroral field line plasma, *J. Geophys. Res.*, *89*, 7573, 1984.

Miura, A., H. Okuda, and M. Ashour-Abdalla, Ion-beam-driven electrostatic ion cyclotron instabilities, *Geophys. Res. Lett.*, *10*, 353, 1983.

Mizera, P.F., J.F. Fennell, D.R. Croley, Jr., and D.J. Gorney, Charged particle distributions and electric field measurements from S3-3, *J. Geophys. Res.*, *86*, 7566, 1981.

Mizera, P.F., J.F. Fennell, D.R. Croley, Jr., A.L. Vampola, F.S. Mozer, R.B. Torbert, M. Temerin, R. Lysak, M. Hudson, C.A. Cattell, R.J. Johnson, R.D. Sharp, A. Ghielmetti, and P.M. Kintner, The aurora inferred from S3-3 particles and fields, *J. Geophys. Res.*, *86*, 2329, 1981.

Mizera, P.F., D.J. Gorney, and J.F. Fennell, Experimental verification of an s-shaped potential structure, *J. Geophys. Res.*, *87*, 1535, 1982.

Moore, T.E., Modulation of terrestrial ion escape flux composition (by low-altitude acceleration and charge exchange chemistry), *J. Geophys. Res.*, *85*, 2011, 1980.

Moore, T.E., Superthermal ionospheric outflows, *Rev. of Geophys. and Space Phys.*, *22*, 264, 1984.

Moore, T.E., C.R. Chappell, M. Lockwood, and J.H. Waite, Jr., Superthermal ion signatures of auroral acceleration processes, *J. Geophys. Res.*, *90*, 1611, 1985.

Mozer, F.S., C.A. Cattell, M. Temerin, R.B. Torbert, S. Von Glinski, M. Woldorff, and J. Wygant, The dc and ac electric field, plasma density, plasma temperature, and field-aligned current experiments on the S3-3 satellite, *J. Geophys. Res.*, *84*, 5875, 1979.

Mozer, F.S., C.A. Cattell, M.K. Hudson, R.L. Lysak, M. Temerin, and R.B. Torbert, Satellite measurements and theories of low altitude auroral particle acceleration, *Space Sci. Rev.*, *27*, 155, 1980.

Nagai, T., J.F.E. Johnson, and C.R. Chappell, Low-energy ($\leq$ 100 eV) ion pitch angle distributions in the magnetosphere by ISEE 1, *J. Geophys. Res.*, *88*, 6944, 1983.

Nagai, T., J.H. Waite, Jr., J.L. Green, and C.R. Chappell, First measurements of supersonic polar wind in the polar magnetosphere, *Geophys. Res. Lett.*, *11*, 669, 1984.

Nishikawa, K.-I. and H. Okuda, Heating of light ions in the presence of large-amplitude heavy ion cyclotron wave, *J. Geophys. Res.*, *90*, 2921, 1985.

Nishikawa, K.-I., H. Okuda, and A. Hasegawa, Heating of heavy ions on auroral field lines, *Geophys. Res. Lett.*, *10*, 553, 1983.

Nishikawa, K.-I., H. Okuda, and A. Hasegawa, Heating of heavy ions on auroral field lines in the presence of a large amplitude hydrogen cyclotron wave, *J. Geophys. Res.*, *90*, 419, 1985.

Okuda, H. and M. Ashour-Abdalla, Formation of a conical distribution and intense ion heating in the presence of hydrogen cyclotron waves, *Geophys. Res. Lett.*, *8*, 811, 1981.

Okuda, H. and M. Ashour-Abdalla, Acceleration of hydrogen ions and conic formation along auroral field lines, *J. Geophys. Res.*, *88*, 899, 1983.

Okuda, H. and K.-I. Nishkawa, Ion-beam-driven electrostatic hydrogen cyclotron waves on auroral field lines, *J. Geophys. Res.*, *89*, 1023, 1984.

Okuda, H., C.Z. Cheng, and W.W. Lee, Numerical simulations of electrostatic hydrogen cyclotron instabilities, *Phys. Fluids*, *24*, 1060, 1981.

Okuda, H., C.Z. Cheng, and W.W. Lee, Anomalous diffusion and ion heating in the presence of electrostatic hydrogen cyclotron instabilities, *Phys. Rev. Lett.*, *46*, 427, 1981.

Palmadesso, P.J., T.P. Coffey, S.J. Ossakow, and K. Papadopoulos, Topside ionosphere ion heating due to electrostatic ion cyclotron turbulence, *Geophys. Res. Lett.*, *1*, 105, 1974.

Papadopoulos, K., J.D. Gaffey, and P.J. Palmadesso, Stochastic acceleration of large m/q ions by hydrogen cyclotron waves in the magnetosphere, *Geophys. Res. Lett.*, *7*, 1014, 1980.

Pottelette, R., R. Treumann, O.H. Bauer, and J.P. Lebreton, Generation of electrostatic shocks and turbulence through the interaction of conics with the background plasma, *Geophys. Res. Lett.*, *12*, 57, 1985.

Pritchett, P.L., M. Ashour-Abdalla, and J.M. Dawson, Simulation of the current-driven ion cyclotron instability, *Geophys. Res. Lett.*, *8*, 611, 1981.

Quinn, J.M. and R.G. Johnson, Observation of ionospheric source cone enhancements at the substorm injection boundary, *J. Geophys. Res.*, *90*, 4211, 1985.

Redsun, M.S., M. Temerin, and F.S. Mozer, Classification of auroral electrostatic shocks by their ion and electron associations, *J. Geophys. Res.*, *90*, 9615, 1985.

Retterer, J.M., T. Chang, J.R. Jasperse, Ion acceleration in the suprauroral region: A Monte Carlo model, *Geophys. Res. Lett.*, *10*, 583, 1983.

Richardson, J.D., J.F. Fennell, and D.R. Croley, Jr., Observations of field-aligned ion and electron beams from SCATHA (P78-2), *J. Geophys. Res.*, *86*, 10105, 1981.

Roth, I. and M.K. Hudson, Particle simulations of electrostatic emissions near the lower hybrid frequency, *J. Geophys. Res.*, *88*, 483, 1983.

Roth, I. and M.K. Hudson, Lower hybrid heating of ionospheric ions due to ion ring distributions in the cusp, *J. Geophys. Res.*, *90*, 4191, 1985.

Roux, A., S. Perraut, J.L. Rauch, C. de Villedary, G. Kremser, A. Korth, and D. T. Young, Wave-particle interactions near Ω_{He^+} observed on board GEOS-1 and -2. 2. Generation of ion cyclotron waves and heating of He^+ ions, *J. Geophys. Res.*, *87*, 8174, 1982.

Shalimov, S.L., Auroral ion acceleration, Preprint #783, Space Research Institute, Academy of Sciences, USSR, 1983.

Sharp, R.D., Positive ion acceleration in the 1 R_e altitude range, in *Physics of Auroral Arc Formation*, edited by S.-I. Akasofu and J.R. Kan Geophys. Mono. Series, 25, p. 112, AGU, Washington, D.C., 1981.

Sharp, R.D., R.G. Johnson, and E.G. Shelley, Observation of an ionospheric acceleration mechanism producing energetic (keV) ions primarily normal to the geomagnetic field direction, *J. Geophys. Res.*, *82*, 3324, 1977.

Sharp, R.D., R.G. Johnson, and E.G. Shelley, Energetic particle measurements from within ionospheric structures responsible for auroral acceleration processes, *J. Geophys. Res.*, *84*, 480, 1979.

Sharp, R.D., E.G. Shelley, R.G. Johnson, and A.G. Ghielmetti, Counter-streaming electron beams at altitudes of $\sim$ 1 R_e over the auroral zone, *J. Geophys. Res.*, *85*, 92, 1980.

Sharp, R.D., D.L. Carr, W.K. Peterson, and E.G. Shelley, Ion streams in the magnetotail, *J. Geophys. Res.*, *86*, 4639, 1981.

Sharp, R.D., A.G. Ghielmetti, R.G. Johnson, and E.G. Shelley, Hot plasma composition results from the S3-3 spacecraft, in *Energetic Ion Composition in the Earth's Magnetosphere*, edited by R.G. Johnson, p. 167, Terra Scientific Publishing Co., Tokyo, 1983.

Sharp, R.D., W. Lennartsson, W.K. Peterson, and E. Ungstrup, The mass dependence of wave particle interactions as observed with the ISEE-1 energetic ion mass spectrometer, *Geophys. Res. Lett.*, *10*, 651, 1983.

Sharp, R.D., W. Lennartsson, and R.J. Strangeway, The ionospheric contribution to the plasma environment in near-earth space, *Radio Science*, *20*, 456, 1985.

Shelley, E. G., Ion composition in the dayside cusp: Injection of ionospheric ions into the high latitude boundary layer, in *Proceedings of Magnetospheric Boundary Layer Conference*, Alpach, 11-15 June, 1979, ESA SP-148, p. 187, European Space Agency, Paris, 1979.

Shelley, E.G., W.K. Peterson, A.G. Ghielmetti, and J. Geiss, The polar ionosphere as a source of energetic magnetospheric plasma, *Geophys. Res. Lett.*, *9*, 941, 1982.

Singh, N. and R.W. Schunk, Numerical calculations relevant to the initial expansion of the polar wind, *J. Geophys. Res.*, *87*, 9154, 1982.

Singh, N. and R.W. Schunk, Energization of ions in the auroral plasma by broadband waves: generation of ion conics, *J. Geophys. Res.*, *89*, 5538, 1984.

Singh, N. and R.W. Schunk, A possible mechanism for the observed streaming of H^+ and H^+ ions at nearly equal speeds in the distant magnetotail, *J. Geophys. Res.*, *90*, 6361, 1985.

Singh, N., R.W. Schunk, and J.J. Sojka, Energization of ionospheric ions by electrostatic hydrogen cyclotron waves, *Geophys. Res. Lett.*, *8*, 1249, 1981.

Singh, N., R.W. Schunk, and J.J. Sojka, Cyclotron resonance effects on stochastic acceleration of light ionospheric ions, *Geophys. Res. Lett.*, *9*, 1053, 1982.

Singh, N., R.W. Schunk, and J.J. Sojka, Preferential perpendicular acceleration of heavy ionospheric ions by interactions with electrostatic hydrogen cyclotron waves, *J. Geophys. Res.*, *88*, 4055, 1983.

Swift, D. W., An equipotential model for auroral arcs: the theory of two dimensional laminar electrostatic shocks, *J. Geophys. Res.*, *84*, 6427, 1979.

Temerin, M. and F.S. Mozer, Observations of the electric fields that accelerate auroral particles, *Proc. Indian Acad. Sci. (Earth Planet Sci.)*, *93*, 227, 1984.

Temerin, M., C. Cattell, R. Lysak, M. Hudson, R.B. Torbert, F.S. Mozer, R.D. Sharp, and P.M. Kintner, The small-scale structure of electrostatic shocks, *J. Geophys. Res.*, *86*, 11278, 1981.

Torr, M.R. and D.G. Torr, Energetic oxygen: A direct coupling mechanism between the magnetosphere and thermosphere, *Geophys. Res. Lett.*, *6*, 700, 1979.

Ungstrup, E., D.M. Klumpar, and W.J. Heikkila, Heating of ions to superthermal energies in the topside ionosphere by electrostatic ion cyclotron waves, *J. Geophys. Res.*, *84*, 4289, 1979.

Waite, Jr., J.H., T. Nagai, J.F.E. Johnson, C.R. Chappell, J.L. Burch, T.L. Killeen, P.B. Hays, G.R. Carignan, W.K. Peterson, and E.G. Shelley, Escape of suprathermal O^+ ions in the polar cap, *J. Geophys. Res.*, *90*, 1619, 1985.

Whalen, B. A., Low-altitude energetic ion composition observations, in *Energetic Ion Composition in the Earth's Magnetosphere*, edited by R. G. Johnson, 143-165, Terra Scientific Publishing Company, Tokyo, 1983.

Whalen, B.A., W. Bernstein, and P.W. Daly, Low altitude acceleration of ionospheric ions, *Geophys. Res. Lett.*, *5*, 55, 1978.

Yang, W.H. and J.R. Kan, Generation of conic ions by auroral electric fields, *J. Geophys. Res.*, *88*, 465, 1983.

Yau, A.W., B.A. Whalen, A.G. McNamara, P.J. Kellogg, and W. Bernstein, Particle and wave observations of low-altitude ionospheric ion acceleration events, *J. Geophys. Res.*, *88*, 341, 1983.

Yau, A.W., B.A. Whalen, W.K. Peterson, and E.G. Shelley, Distribution of upflowing ionospheric ions in the high-altitude polar cap and auroral ionosphere, *J. Geophys. Res.*, *89*, 5507, 1984.

Yau, A.W., P.H. Beckwith, W.K. Peterson, and E.G. Shelley, Long-term (solar cycle) and seasonal variations of upflowing ionospheric ion events at DE 1 altitudes, *J. Geophys. Res.*, *90*, 6395, 1985.

Yau, A.W., E.G. Shelley, W.K. Peterson, and L. Lenchyshyn, Energetic auroral and polar ion outflow at DE 1 altitudes: Magnitude, composition, magnetic activity dependence, and long-term variations, *J. Geophys. Res.*, *90*, 8417, 1985.

Yoshino, T., T. Ozaki, and H. Fukunishi, Occurrence distributions of VLF hiss and saucer emissions over the southern polar region, *J. Geophys. Res.*, *86*, 846, 1981.

Young, D.T., H. Balsiger, and J. Geiss, Correlations of magnetospheric ion composition with geomagnetic and solar activity, *J. Geophys. Res.*, *87*, 9077, 1982.

AUTHOR INDEX

Abe, Y. 191
Arnoldy, R. L. 201
Baker, D. N. 375
Belian, R. D. 375
Berman, R. H. 328
Boardsen, S. A. 43
Boehmer, H. E. 245
Borovsky, J. E. 317
Burch, J. L. 83, 98, 172
Cahill, L. J. 201, 206
Candidi, M. 164
Chan, C. 249
Chandler, M. O. 56, 61
Chang, T. 282, 286
Chapman, S. C. 179
Chappell, C. R. 50, 56
Cladis, J. B. 153
Coates, A. J. 136, 186
Collin, H. L. 67, 77, 83
Cornwall, J. M. 3
Cowley, S. W. H. 179
Cravens, T. E. 366
Crew, G. B. 286
DeCoster, R. J. 117
Dupree, T. H. 328
Eastman, T. E. 117
Erickson, K. N. 191
Erlandson, R. E. 201, 206
Frahm, R. A. 98
Frank, L. A. 117
Gallagher, D. L. 172
Ganguli, G. I. 297, 301
Ghielmetti, A. G. 67, 77
Goertz, C. K. 108
Gombosi, T. I. 366
Gurnett, D. A. 43, 108
Hershkowitz, N. 224
Horwitz, J. L. 56
Hudson, M. K. 271, 334
Hultqvist, B. 127
Jasperse, J. R. 282, 286
Johnstone, A. D. 136, 179, 186
Joyce, G. 317
Kaufmann, R. L. 92
Keath, E. P. 149
Kintner, P. M. 39, 201, 206, 384
Klumpar, D. M. 389
Koskinen, H. E. J. 291
Koslover, R. A. 245
Krimigis, S. M. 149
LaBelle, J. 201, 206
Lee, Y.-C. 297, 301

Lennartsson, W. 153, 323
Lockwood, M. 50, 56, 61
Ludlow, G. R. 92
Lui, A. T. Y. 149
Lundin, R. 127
Lysak, R. L. 261
Martin, R. F. 141
Mauk, B. H. 351
McEntire, R. W. 149
McWilliams, R. D. 245
Meng, C.-I. 351
Menietti, J. D. 172
Moore, T. E. 50, 56, 61, 201
Nagy, A. F. 366
Nishikawa, K.-I. 307
Okuda, H. 307
Orsini, S. 164
Palmadesso, P. J. 297, 301
Persoon, A. 61
Peterson, W. K. 43, 72
Pollock, C. 201
Reiff, P. H. 83, 98
Retterer, J. M. 282, 286
Rodgers, D. J. 136, 186
Roth, I. 271
Rynn, N. 235, 245
Samir, U. 254
Scales, W. A. 206
Schulz, M. 158
Schunk, R. W. 340, 343, 362
Sharp, R. D. 67, 77
Shelley, E. G. 43, 67, 72, 77, 83
Singh, N. 340, 343, 362
Smith, M. F. 136, 186
Southwood, D. J. 136
Stasiewicz, K. 127
Stenzel, R. L. 211
Stone, N. H. 254
Suguira, M. 61
Tetreault, D. J. 328
Thiemann, H. 340, 343
Waite, J. H. 50, 56, 61, 172, 366
Walker, D. N. 311
Weimer, D. R. 108
Whalen, B. A. 39
Williams, R. L. 172
Winckler, J. R. 191
Winningham, J. D. 83, 98
Witt, E. 334
Wright, K. H. 254
Yau, A. W. 39, 72
Young, D. T. 17